THE
HUMAN
ORGANISM

FIFTH EDITION

THE HUMAN ORGANISM

RUSSELL MYLES DeCOURSEY

Professor Emeritus of Zoology
Biological Sciences Group
University of Connecticut

J. LARRY RENFRO

Associate Professor of Biology
Biological Sciences Group
Physiology Section
University of Connecticut

McGRAW-HILL BOOK COMPANY

New York St. Louis San Francisco Auckland Bogotá Hamburg
Johannesburg London Madrid Mexico Montreal New Delhi
Panama Paris São Paulo Singapore Sydney Tokyo Toronto

THE HUMAN ORGANISM

1 2 3 4 5 6 7 8 9 0   DODO   8 9 8 7 6 5 4 3 2 1 0

This book was set in Laurel by York Graphic Services, Inc.
The editors were James E. Vastyan and James W. Bradley;
the designer was Jo Jones;
the production supervisor was Dennis J. Conroy.
New drawings were done by J & R Services, Inc.
The cover was designed by Rafael Hernandez.
R. R. Donnelley & Sons Company was printer and binder.

Library of Congress Cataloging in Publication Data

DeCoursey, Russell Myles, date
    The human organism.

    Includes index.
    1. Human physiology. 2. Anatomy, Human.
I. Renfro, J. Larry, joint author. II. Title.
QP34.5.D38 1980   612   79-19910
ISBN 0-07-016275-1

CONTENTS

UNIT FOUR   THE CIRCULATORY SYSTEM

PREFACE

The Human Organism, fifth edition, provides the student with a good background in human anatomy and physiology. The book has been largely rewritten to present new material and should be considered a suitable text for introductory students. Many new illustrations have been added, including flowcharts and diagrams. A greater number of illustrations are in color, including new color illustrations of the skull and other color plates.

The first three chapters have been rewritten and organized to provide a more unified introduction to fundamentals, including some elementary physics and chemistry. Terminology necessary to the understanding of a course in anatomy and physiology has been included also.

Chapters are grouped into units. For example, Unit 1, The Cell, includes three introductory chapters. Chapter 1 is entitled Basic Units of Structure, Chapter 2 takes up cellular structure and function, while Chapter 3 discusses acquiring, transforming, and using energy. Unit 2 contains chapters dealing with structural elements of the body, the tissues, skeletal structures and musculature. Unit 3 covers the controlling systems, with chapters on the nervous system and the endocrine glands.

An outline appears under Considerations for Study at the beginning of each chapter, and a summary is included at the end along with suggested reading. Suggested Reading affords an opportunity for the student to broaden his or her knowledge. For the most part selections have been made from readily available sources.

Muscle physiology, endocrinology, and lymph capillaries are among the subjects that have been added; the chapter on the kidneys has been rewritten and brought up to date; and new information on membrane physiology is included throughout the book. There is also a greater emphasis on injury to joints, since injuries of this sort have become very important in athletics. Common diseases and injuries are discussed under a new section called Clinical Aspects.

Developmental physiology has been introduced in appropriate chapters rather than being presented as a final chapter.

The Human Organism has been used extensively in community colleges, two-year colleges, schools of nursing, physical education, and physical therapy, and in four-year colleges and universities. This edition is also applicable to students in health-related programs such as the training of medical technicians or other paramedical programs.

The Human Organism is adaptable for one-semester, term system or a one-year course. A year course would permit more thorough study and more outside reading. A laboratory manual and an instructor's manual are available.

We appreciate the work of the illustrators and of the publishing staff of the McGraw-Hill Book Company. In particular, we should like to thank Barbara Tokay for her meticulous reading and her suggestions and our thanks also to our editor, Mr. James E. Vastyan, for his careful supervision.

RUSSELL MYLES DeCOURSEY
J. LARRY RENFRO

THE
HUMAN
ORGANISM

UNIT ONE

THE CELL

1

BASIC UNITS OF STRUCTURE

CONSIDERATIONS FOR STUDY

In this chapter we review several important principles of chemistry and physics and show how these principles may be applied to the study of anatomy and physiology. The following topics are reviewed:

1 Units of matter, including definitions of matter, energy, elements, atoms, ions, and molecules
2 Energy and its relationship to biological systems
3 Basic, physiologically important characteristics of water and organic compounds
4 Enzymes: how they work and some of their characteristics

We believe ourselves to be one of the most gifted organizations of matter and energy in the known universe. Among our gifts is an awareness that we are indeed significant and fully worthy of serious study. We are keenly aware of our environment and continually curious about ourselves, and perhaps more than any other living creature, we are capable of profiting by individual and collective experience.

The intricate structure and function of the human body is the product of continuously shifting interrelationships of atoms and molecules. No one knows today exactly how these relationships are organized, but this does not mean that understanding will not be forthcoming. The student's task is not to discover great truths but to obtain from available evidence the best possible answers. Even though incomplete, these answers become starting points from which more complete theories will be formulated. The study of human anatomy and physiology represents a synthesis of those aspects of chemistry and physics that may be applied to life processes. As such, the fields of anatomy and physiology encompass a tremendous body of knowledge. The attainment of that knowledge is a rewarding experience for many

3

students since every detail has immediate application and relevance to the individual as well as to humankind.

Because the structures and processes of anatomy and physiology are based upon chemical and physical principles, it is necessary to review some elementary concepts of the chemical and physical sciences before we begin our main study. Life fits into its environment because it is produced from the very atoms and molecules that compose the environment. To understand the relevance of basic chemistry and physics to living organisms, we have only to realize that the marvelously complicated human brain is made of only a dozen or so elemental building blocks.

UNITS OF MATTER

Everything we think of that is gaseous, liquid, or solid, or living or dead is matter. *Matter* is defined as that which has *mass* and *occupies space*. *Mass* is a measure of the *quantity* of matter and is independent of *weight,* which depends on gravitational pull. In the nuclear age, we are acutely aware that matter and energy are interconvertible—for example, in the reactions set off by exploding an atomic bomb. This interconversion is so slight in living matter, however, that we can ignore it and consider matter and energy as separate, fixed entities.

Energy is quantified by measuring its *effect upon matter.* In the simplest sense, energy is the *ability to do work,* and it exists in two basic states: *kinetic* and *potential. Kinetic energy* is energy in motion. It is the energy used to throw a ball into the air, for example. *Potential energy* is stored, or inactive, energy. It may be released as kinetic energy, just as kinetic energy can be captured and stored as potential energy. The energy we use to throw a ball, for example, is stored in chemical form in our muscles until we need it. All living matter depends upon the exchange of kinetic and potential energy with its environment to supply its energy requirements.

ELEMENTS The basic unit of matter is the *atom.* Atoms are composed of smaller particles called *protons, neutrons,* and *electrons.* These basic components are organized such that the protons and neutrons form a central, dense nucleus around which the electrons form a cloud. An *element* is matter consisting of only one type of atom, which is distinguished from other types by the number of protons, neutrons, and electrons present. For example, the *hydrogen* atom contains one proton in its nucleus and one electron orbiting the nucleus (Fig. 1.1). Hydrogen is the simplest atom and the only one that possesses no neutrons. Oxygen is a type of atom with eight protons and eight neutrons surrounded by eight electrons.

Elements are commonly referred to by the chemical symbols of their constituent atoms: O stands for oxygen; H for hydrogen; C for carbon; and N for nitrogen. Symbols are usually derived from the first two letters of the English or Latin name of the element; however, this is not always the case, and some symbols are more difficult to associate with the name of the element than are others. The following symbols are commonly used and will serve as examples: sodium, Na; potassium, K; copper, Cu; iron, Fe; and mercury, Hg. There are over 100 chemical elements included in the periodic table of chemical elements, which is arranged in order of increasing atomic numbers. The *atomic number* of an atom is equal to the number of protons the atom contains. Hydrogen, which has an atomic nucleus con-

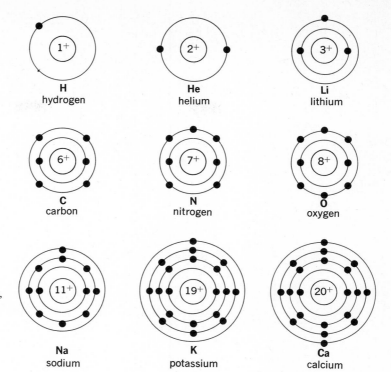

FIGURE 1·1
Diagrams of a few atoms selected from the periodic table, showing their structure. The number in the center of each diagram represents the positive nuclear charge, which is balanced by the negatively charged electrons.

taining one proton, is given an atomic number of 1, whereas uranium, with 92 protons, has an atomic number of 92, and the atomic number of lawrencium is 103.

ORGANIZATION OF THE ATOM The organization of particles in an atom is based on mutual attraction and repulsion. Protons carry a positive electric charge, while neutrons are uncharged (neutral), and electrons are negatively charged. Electric charge is a fundamental characteristic of protons and electrons and cannot be defined in terms of other characteristics of matter. Electric charge can be observed and quantified because particles with like charges repel each other and those with unlike charges attract each other. When they are not involved in reactions with one another, atoms possess equal numbers of protons and electrons so that they are electrically neutral.

The examples pictured in Fig. 1.1 show that the electrons are located in certain prescribed *orbitals*, or shells, indicated diagrammatically by circles. (Although a diagram of atomic orbitals commonly shows electrons in their shells or subshells revolving around the nucleus in planetary fashion, this is not entirely in keeping with modern theory. It is now thought that electrons move in certain volumes of space, and their movements exhibit wavelike properties.) Each shell holds a maximum number of electrons. The first shell nearest the atomic nucleus is limited to two electrons in the orbital, whereas the second shell has a maximum of eight electrons. Subshells, which contain electrons with slightly different energies, will not be discussed here, but the maximum number of electrons is known for each shell and subshell.

Protons have mass, but electrons have so little mass that it is considered negligible. Since the electrons moving in their orbitals are of such minute size, there is a relatively large amount of space within the boundaries of the atom. Different kinds of atoms have different numbers of electrons, and the electrons may be arranged in one to seven different orbitals.

When all the electron shells of an atom are completely filled, the atom is nonreactive. Let us consider the helium atom, an inert gas (Figs. 1.1 and 1.2). Helium's first shell contains the maximum number of electrons: two. The atom is, therefore, stable electrically, and it is almost completely inert (nonreactive) from a chemical standpoint.

The atom hydrogen, however, with one electron in the first shell may act as an *electron donor*, as in the reaction:

$$H_2 + Cl_2 \longrightarrow 2H^+ + 2Cl^-$$

Pairs of hydrogen atoms may also share electrons to form a hydrogen molecule, H_2. Many gaseous elements share electrons in this *diatomic* form. Oxygen, which has an incomplete second shell with six electrons, may complete the outer shell by sharing electrons with another oxygen atom to form a molecule of oxygen, O_2. In the formation of a molecule of water, H_2O, an oxygen atom shares its six outer electrons with electrons from two hydrogen atoms and, in this way, completes its outer shell of eight electrons. At the same time the hydrogen atoms acquire inner stable shells filled with two electrons (Fig. 1.3).

ATOMIC WEIGHT   Since atoms are unbelievably light, many millions of them would fail to register on even the most sensitive scales. It is not possible actually to weigh atoms, but chemical elements have been assigned an *atomic weight*, based upon their relative weight as compared with that of the carbon atom. The atomic weights of various atoms are given in the table of international atomic weights, which may be found in most chemistry textbooks. We list the atomic weights of some common chemical elements in Table 1.1.

Atomic mass is not synonymous with atomic weight. Whereas the atomic weight of an atom is assigned by its relation to the weight of the most commonly occurring carbon atom, carbon-12 (^{12}C), atomic mass is the sum of the total number of an atom's protons and neutrons. (Electron mass is so slight that it can be ignored.) Thus, you will note that the atomic mass of carbon is 12 (6 protons and 6 neutrons), but its atomic weight is 12.01115. The difference between mass and weight arises because carbon exists in

FIGURE 1·2
Diagrammatic representation of *a* the hydrogen atom and *b* the helium atom. These are not pictures of atoms but merely symbolic representations.

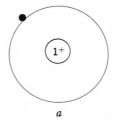

a

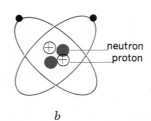

neutron
proton

b

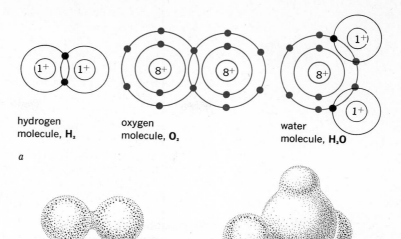

hydrogen
molecule, **H₂**

oxygen
molecule, **O₂**

water
molecule, **H₂O**

a

b

FIGURE 1·3

The formation of a molecule of water. *a* The covalent bonds between the oxygen atom and each hydrogen atom represent a pair of electrons shared between the atoms. *b* The two bulges in the oxygen atom below are exaggerated representations of the unshared pairs of electrons.

several forms, called *isotopes,* that vary in the number of neutrons they possess. Several isotopes of lithium and hydrogen are shown in Fig. 1.4. The difference in neutron number has no significant effect on the chemical behavior of the atom, but it accounts for its assigned atomic weight, since the atomic weight of a carbon atom is an *average* of all naturally occurring carbon atoms, including the isotopes.

TABLE 1·1

Some common chemical elements

Element	Symbol	Atomic number°	Atomic weight°	Common valences
Hydrogen	H	1	1.008	1
Carbon	C	6	12.01115	4
Nitrogen	N	7	14.0067	3
Oxygen	O	8	15.999	2
Sodium	Na	11	21.9898	1
Magnesium	Mg	12	24.305	2
Phosphorus	P	15	30.9738	3, 5
Sulfur	S	16	32.06	2
Chlorine	Cl	17	35.453	1
Potassium	K	19	39.10	1
Calcium	Ca	20	40.08	2
Iron	Fe	26	55.84	2, 3
Copper	Cu	29	63.54	1, 2
Iodine	I	53	126.9045	1

° The atomic number indicates the number of protons in the nucleus of the atom. The atomic weight is equal to the sum of the protons and neutrons and is an average of all naturally occurring isotopes of each element.

Molecular weight is equal to the combined atomic weights of all atoms composing the molecule. Water (H_2O) with two hydrogen atoms and one oxygen atom has a molecular weight of 18.

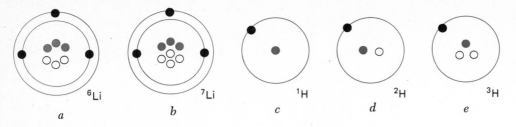

^{6}Li
a

^{7}Li
b

^{1}H
c

^{2}H
d

^{3}H
e

FIGURE 1·4

Lithium and hydrogen isotopes. *a* Lithium isotope, ^{6}L, atomic weight 6, three protons and three neutrons. *b* Lithium isotope, ^{7}L, atomic weight 7, three protons and four neutrons. *c* Hydrogen isotope, ^{1}H, atomic weight 1, one proton. *d* Hydrogen isotope, deuterium, ^{2}H, atomic weight 2, one proton and one neutron. *e* Hydrogen isotope, tritium, ^{3}H, atomic weight 3, one proton and two neutrons. Tritium-labeled compounds are used extensively in research.

ISOTOPES   Not all atoms of the same type have identical atomic weights, as we pointed out in the case of carbon. The number of isotopes, atoms of a certain element that vary in neutron number, varies for different elements. Oxygen has three stable isotopes: ^{16}O, ^{17}O, and ^{18}O. Carbon also has three isotopes: ^{12}C, ^{13}C, and ^{14}C. Of the carbon isotopes, ^{12}C and ^{13}C are stable, but ^{14}C, which occurs naturally as well, is *radioactive*. Carbon-14 undergoes a spontaneous emission of energy, a phenomenon called *radioactivity*. Hydrogen has isotopes ^{1}H, ^{2}H, and ^{3}H. The last, called tritium, occurs naturally and is unstable or radioactive. Calcium has six stable isotopes, iron has four, and so on.

As we noted previously, although there are physical differences in the isotopes of an element, the chemical properties of such isotopes are almost identical. This property of isotopes has been highly useful in the study of physiological processes. When radioactive isotopes are added to the nonradioactive ones ordinarily found in living matter, their progress within the organism can be monitored continuously by radiation detectors. The use of radioactive *tracers* allows the study of phenomena that would otherwise proceed undetected.

MOLECULES   Atoms combine with one another to form *molecules*. A *molecule* is a unit of matter that consists of a combination of two or more atoms. Atoms form molecules by creating *chemical bonds*. Bonds are formed either by the *sharing* or the *donation* of electrons by the involved atoms.

From the simple building blocks of atoms, a practically infinite array of molecules can be constructed by varying atomic composition and the sequence of the attachment of atoms. *Molecular weight* is simply the sum of the atomic weights of the atoms that compose each molecule.

CHEMICAL BONDING   *Ionic bonding*   When atoms gain or lose electrons, they become charged particles called *ions*, and the process is called *ionization*. Many *inorganic* (non-carbon-containing) *compounds* that dissolve in aqueous (water) solutions tend to ionize. In the transfer of electrons, if the atom gains an electron or electrons, it picks up a negative charge and is called an *anion;* if it loses an electron or electrons, it has a positive charge and is termed a *cation*.

An atom is normally neutral since the positive charges of the protons in the nucleus are balanced by the negative charges of the surrounding electrons. If the atom gains a negative charge by receiving an additional electron or electrons, an anion is formed. If an electron is lost, the balance of charge between protons and electrons changes to favor the positive charge of the protons. The particle then becomes a cation.

In an electric field such as that produced by a battery, anions move toward the *anode,* or positive pole, of the battery, whereas cations, being positively charged, move toward the *cathode,* or negative pole. Particles having a like charge will repel each other. The *force,* or work, required to move such like-charged atoms apart is a positive force, and a positive potential is indicated.

A negative potential is developed when the particles have opposite charges and attract each other. Such atoms are then held together by an *ionic bond* (Fig. 1.5). Sodium chloride, for example, is ionized in water into sodium ions (Na^+) and chloride ions (Cl^-). The plus and minus symbols indicate the sign of the ion's electric charge. Such an ion-containing solution can conduct an electric current and is therefore termed an *electrolytic solution.* Acids, bases, and salts are molecular compounds that dissociate into ions when dissolved in water; such compounds are termed *electrolytes.*

Covalent bonding Some molecules, such as those of the diatomic gases hydrogen (H_2) and oxygen (O_2), are composed of two identical atoms. Carbon dioxide (CO_2) is constructed from two different kinds of atoms, while the sugar glucose ($C_6H_{12}O_6$) has three different kinds of atoms. Some molecules, such as protein and fat molecules, have very complex structures. All these molecules share a common feature, however. When a molecule is formed by atoms that share electrons, the molecule is held together by *covalent* chemical bonds, the most common type of bonding in biological compounds (see Fig. 1.3). In this type of bonding, there is no actual transfer of electrons.

Certain kinds of atoms, usually in the nonmetal group, commonly share electrons and form covalent compounds. These atoms include hydrogen, oxygen, nitrogen, and carbon. Hydrogen, however, may either share its electron or transfer it. The number of electrons gained or lost by an atom is termed its *valence* or *electrovalence.*

Referring again to Fig. 1.1, we see that oxygen can accept two electrons in its outer shell; it therefore has a valence of 2. Nitrogen can accept three electrons and has a valence of 3; carbon can accept four electrons and ordinarily has a valence of 4 (see Table 1.1).

FIGURE 1·5

Two representations of the formation of sodium chloride by electron transfer. The sodium atom is shown with a single electron in its highest normal energy level. The chlorine atom is shown with seven electrons in its highest normal energy level (only the highest, or third, energy level is shown). If the chlorine atom gains one electron, it becomes a negatively charged chloride ion. If the sodium atom loses an electron, it becomes a positively charged ion.

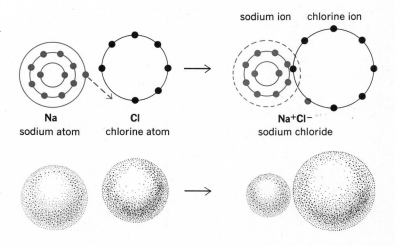

sodium ion   chlorine ion

Na
sodium atom

Cl
chlorine atom

Na⁺Cl⁻
sodium chloride

SOLUTIONS   Physiological solutions consist of a solid dissolved in a liquid. The liquid (*solvent*) may dissolve the solid substance without changing it chemically, as sugar dissolves in water. Such solutions are technically called mixtures. The solvent may, however, react chemically with a solute to form a true *solution*. The dissolved substance in a solution is the *solute*.

The concentration of solute in a solution may be expressed in several ways. Concentration is often expressed as a percentage, or parts per hundred. For example, the number of grams of the substance present in 100 milliliters (ml) [1/1000 liter (l) = 1 milliliter] may be measured. Concentration of a solute may also be expressed in *moles*. A *mole*, or *gram molecular weight*, is defined as the weight of a substance measured in grams that is equivalent to its molecular weight. If a substance has a molecular weight of 35, one mole of that substance would weigh 35 grams (g). (The number of molecules in a mole of any substance is 6.0225×10^{23}. This figure is known as *Avogadro's number*.)

A *molar* solution contains one mole of solute per liter. Note that in preparing a *molar* solution, one gram molecular weight of the solute is dissolved in a volume of solvent sufficient to make one liter of solution. In preparing a *molal* solution, the gram molecular weight of the solute is added to 1000 grams of solvent. This preparation may result in a solution volume greater than one liter. The millimole (0.001 mole) is often used as a basic unit of measurement in biological preparations where much smaller quantities are required.

WATER   Water is a compound quite unique in its ability to dissolve most substances. The characteristics of water that have made it the so-called medium of life result from the relationship of the hydrogen and oxygen atoms that constitute the molecule. Although the hydrogen and oxygen atoms share electrons, oxygen exerts a much stronger attraction for these electrons because of its much greater mass. As a result the shared electrons are attracted more toward the oxygen portion of the molecule, which becomes more negatively charged, thereby leaving the two hydrogen portions more positively charged. This uneven distribution of charge within the molecule creates a positive and negative pole. The water molecule is therefore called a *polar molecule*. The polarity of the water molecule causes a greater-than-expected attraction between water molecules in solution. Such *cohesiveness* of a water solution makes the boiling point, melting point, freezing characteristics, and solvation characteristics of water unusual when compared to other hydrides (hydrogen compounds) such as H_2S, NH_3, CH_3, and so forth.

The unique facility that water displays in dissolving substances can be seen when a substance such as sodium chloride (NaCl) is placed in it. The Na^+ and Cl^- ions are literally surrounded by water. The negatively charged portions of the water molecules tend to associate with the Na^+, and the positively charged portions with the Cl^-, as shown in Fig. 1.6. This characteristic of water maintains the sodium and chloride ions in solution.

Hydrogen bonding   The intermolecular attraction of water molecules is a good example of a third type of chemical bonding. As we have described above, the hydrogen nuclei (protons) of a water molecule become relatively more positively charged than the oxygen nucleus because of the stronger pull the

FIGURE 1·6

Water molecules are polar. Their positive portions tend to associate with chloride ions, while their negative portions tend to be attracted to the positively charged sodium ions. The polar nature of water is one reason it is such a good solvent.

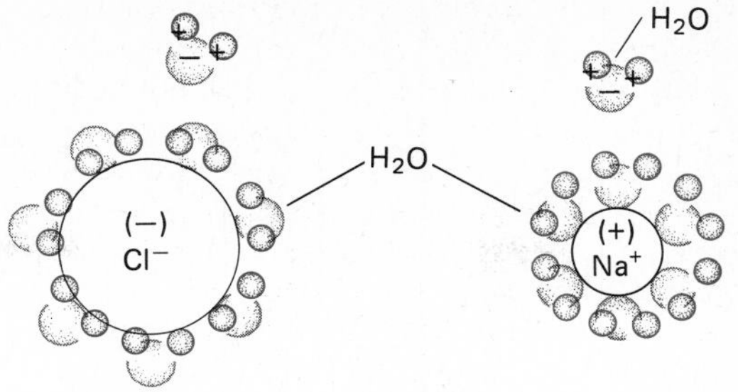

larger oxygen exerts upon the shared electrons. The positively charged hydrogens are then attracted to negatively charged oxygen ions in the region, forming very weak *hydrogen bonds* with such particles. Such hydrogen bonds account for the molecular interactions among water molecules and are the direct cause of water's comparatively high boiling point, high surface tension, and other unusual characteristics (Fig. 1.7). Hydrogen bonds are also found in a variety of other molecules and play important roles in determining the structure and reactivity of certain compounds.

HYDROGEN ION   The symbol pH refers to the hydrogen ion concentration of a solution. In pure water, the hydrogen ions (H^+) and hydroxyl ions (OH^-) are found in equal concentrations. Pure water, therefore, is neutral, or balanced. It is neither acidic ($[H^+] > [OH^-]$) nor basic ($[OH^-] > [H^+]$). Pure water has a pH of 7 at 25°C. *Basic,* or alkaline, solutions are those with a pH range of 7 to 14, and *acid* solutions are those with pH ranges between 7 and 0 (Fig. 1.8). *As the pH decreases, the hydrogen ion concentration rises, and the solution becomes more acidic.* For example, a solution of pH 3 is more acidic than a solution of pH 5.

Tissue fluids are usually close to pH 7. Saliva is usually slightly acid (pH 6.8); bile from the liver is alkaline (pH 7.6 to 8.6). Gastric fluid from the stomach is the most acid substance in the body. Samples of gastric juice as usually taken are diluted with water and contain mucin, enzymes, some organic matter, and electrolytes, but even this mixture has a pH range of about 1.6 to 2.6.

The hydrogen ion concentration of water under ordinary conditions is

FIGURE 1·7

The intermolecular attraction of water molecules accounts for its high boiling point and high surface tension.

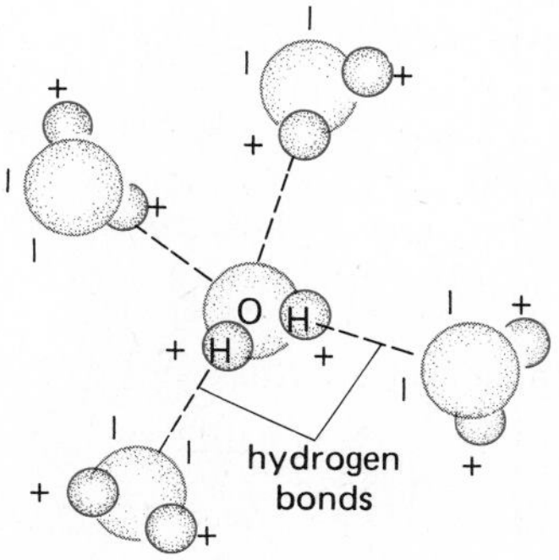

hydrogen bonds

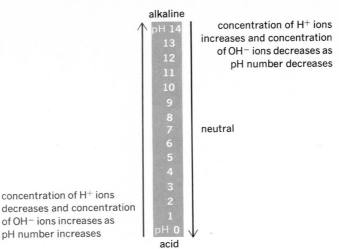

alkaline

pH 14
13
12
11
10
9
8
7
6
5
4
3
2
1
pH 0

acid

concentration of H$^+$ ions
increases and concentration
of OH$^-$ ions decreases as
pH number decreases

neutral

concentration of H$^+$ ions
decreases and concentration
of OH$^-$ ions increases as
pH number increases

FIGURE 1·8

The range of hydrogen ion concentration of solutions. A solution with a pH value of 7 is neutral.

close to 1/10,000,000, or 0.0000001, or 10^{-7} mole per liter. The pH of a solution is calculated from the formula

$$pH = \log \frac{1}{H^+ \text{ concentration}}$$

The *logarithm* of a number is the power to which 10 must be raised in order to produce that number. For example, the logarithm of 10 is 1. Since $10^2 = 100$, the logarithm of 100 is 2; the logarithm of 1000 (10^3) is 3. The pH system was devised to simplify the terminology used to express hydrogen ion concentrations. In this system, the number 10 is dropped and only the exponent is used. The minus sign is also dropped, and, instead, we take the *negative* logarithm of the hydrogen ion concentration. This quantity is represented by the symbol pH (see Table 1.2).

ACIDS, BASES, AND SALTS   *Acids*   An *acid* may be defined as a substance that releases hydrogen ions (H$^+$) when it dissociates upon being dissolved in water. A hydrogen atom, as we have seen, consists of a single proton and a single electron. When a hydrogen atom loses its electron, its proton, unable to exist alone in a water solution, tends to combine with some other ion or molecule. In a water solution, such protons really exist as *hydrated hydrogen ions*, that is, particles attached to water molecules. For most purposes, however, it is convenient to consider the proton as simply a hydrogen ion, symbolized by H$^+$.

There are strong and weak acids. Acids, such as hydrochloric acid (HCl) or sulfuric acid (H$_2$SO$_4$), that ionize almost completely and produce hydrogen ions in considerable quantities are *strong acids*. Sulfuric acid ionizes in two stages, since the bisulfate ion (HSO$_4^-$) formed in its first dissociation itself ionizes.

These reactions may be written out as follows:

(1) $\qquad\qquad$ HCl $\longrightarrow$ H$^+$ + Cl$^-$

(2) $\qquad\qquad$ H$_2$SO$_4$ $\longrightarrow$ H$^+$ + HSO$_4^-$

$\qquad\qquad\qquad\qquad$ HSO$_4^-$ $\longrightarrow$ H$^+$ + SO$_4^{2-}$

TABLE 1·2
Scale of pH, or hydrogen ion concentration, in moles per liter

Decimal	Exponent	pH
1.	10^0	0
0.1	10^{-1}	1
0.01	10^{-2}	2
0.001	10^{-3}	3
0.0001	10^{-4}	4
0.00001	10^{-5}	5
0.000001	10^{-6}	6
0.0000001	10^{-7}	7
0.00000001	10^{-8}	8
0.000000001	10^{-9}	9
0.0000000001	10^{-10}	10
0.00000000001	10^{-11}	11
0.000000000001	10^{-12}	12
0.0000000000001	10^{-13}	13
0.00000000000001	10^{-14}	14

Weak acids do not ionize as completely as strong acids, but all acids ionize to some extent. Such acids as acetic acid ($C_2H_4O_2$), citric acid ($C_6H_8O_7$), lactic acid ($C_3O_6H_3$), or carbonic acid (H_2CO_3) are considered to be weak acids. Acetic acid, usually written CH_3COOH, dissociates to form acetate and hydrogen ions as follows:

$$CH_3COOH \rightleftharpoons CH_3COO^- + H^+$$

Acetic acid is only moderately ionized, and relatively little of it dissociates into hydrogen ions and acetate ions at any one time. The ions formed during dissociation also readily recombine to produce CH_3COOH, thus diminishing ionization further. This *reversible* reaction is indicated by the two arrows in the above equation.

Bases A compound that dissociates in aqueous solution to yield hydroxyl ions (OH^-) is called a *base*. A strong base readily dissociates into ions. For example, sodium hydroxide dissociates into sodium and hydroxyl ions:

$$NaOH \longrightarrow Na^+ + OH^-$$

Strong bases combine readily with hydrogen ions, whereas weak bases combine slowly. Bases have a soapy feel and can be as destructive to living tissues as acids.

Salts When an acid and a base react chemically in solution, a *salt* is formed, as in the following reaction:

$$HCl + NaOH \longrightarrow NaCl + H_2O$$

The metallic sodium ion (Na^+) replaces the hydrogen ion (H^+) of the hydrochloric acid (HCl), and the salt, sodium chloride ($NaCl$), is formed.

The H^+ and OH^- ions combine to form water, H_2O. Other salts essential to body function are: potassium chloride (KCl), magnesium chloride ($MgCl_2$), calcium chloride ($CaCl_2$), calcium carbonate ($CaCO_3$), and calcium phosphate [$Ca_3(PO_4)_2$].

BUFFERS   Body fluids, such as the blood, spinal fluid, and tissue fluid, must maintain a rather constant *acid-base balance,* or pH level. To accomplish this goal, a buffering system is needed. A *buffer* is a substance that tends to maintain the hydrogen ion concentration of a solution at a certain level when either excess acid or alkali is added to the solution.

Within the body, the buffering action of certain constituents of the blood illustrates the operation of a buffering system. All acids ionize to a certain extent, and the blood contains relatively weak acids, such as carbonic and lactic acids. The blood also contains bases such as sodium bicarbonate. The weak acids and the bases tend to balance each other's ionizations, thus preventing blood pH from moving very far in either direction. The pH of the blood is slightly alkaline, and at normal body temperature, blood pH falls in the range of 7.35 to 7.45. Buffering systems will be discussed in more detail in Chap. 17.

BASIC ORGANIC CHEMISTRY OF LIVING MATTER

CARBON COMPOUNDS   The carbon atom is the key to the formation of the molecules characteristic of living matter. In the structural formula of methane, carbon has a valence, or bonding capacity, of 4. This means that carbon can form bonds with as many as four other atoms, such as atoms of hydrogen. Carbon atoms can also bond to other carbon atoms. Since such linkage of carbon atoms is principally associated with the products of living matter or with material that was alive at one time, these complex structures are called *organic compounds.* Carbon is a requisite for the construction of organic compounds. The basic categories of organic substances regarded as essential in the formation of living matter include the *carbohydrates* (sugars and starches), the *lipids* (fats), the *proteins* (amino acids), and the *nucleic acids.*

CARBOHYDRATES   Carbohydrates are composed of carbon, hydrogen, and oxygen atoms in a ratio of two hydrogens to one oxygen. The common names for these compounds are *sugars* and *starches.* Carbohydrates are both a quick source of energy and an energy reserve. The animal starch, glycogen, may be stored in muscle and liver, and these reserves may be mobilized as necessary. Sugars and starchy foods may also be converted to and stored as fats. There are three classes of carbohydrates: monosaccharides, disaccharides, and polysaccharides.

Monosaccharides   This carbohydrate class contains the simple sugars such as glucose, fructose, and galactose. These are six-carbon sugars, or *hexoses* (Fig. 1.9), and they are a good energy source for metabolic processes. Another group of monosaccharides is the *pentose* group, or five-carbon sugars, such as deoxyribose and ribose. These are important components of the genetic material deoxyribonucleic acid (DNA) and ribonucleic acid (RNA).

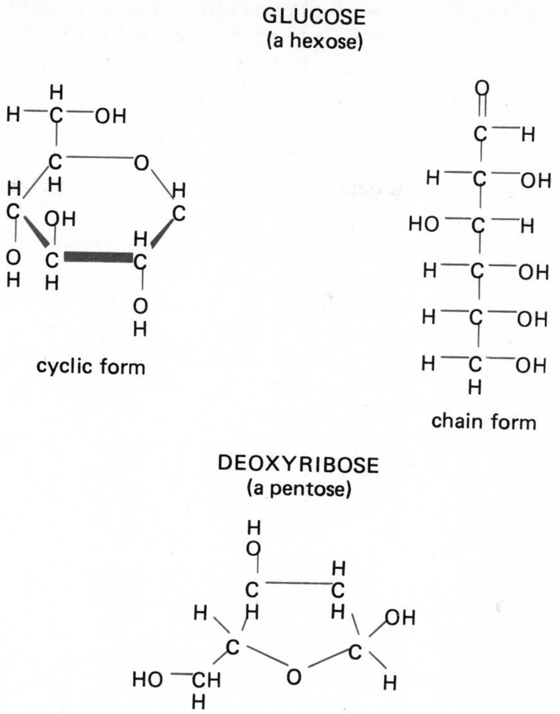

GLUCOSE
(a hexose)

cyclic form

chain form

DEOXYRIBOSE
(a pentose)

FIGURE 1·9

Structural formulas of two monosaccharides, glucose and deoxyribose.

Disaccharides These carbohydrates are composed of two monosaccharides that combine chemically, giving up one molecule of water in the process (Fig. 1.10). This reaction is shown in the following equation:

$$C_6H_{12}O_6 + C_6H_{12}O_6 \longrightarrow C_{12}H_{22}O_{11} + H_2O$$

Familiar disaccharides are sucrose (table sugar), lactose (milk sugar), and maltose (malt sugar). Malt sugar is an intermediate usually formed from glycogen by the action of the enzyme amylase.

Polysaccharides This class of carbohydrates contains compounds formed from groups of monosaccharides bound together, each molecule giving up one molecule of water as it combines with the others (Fig. 1.10). Since the number of molecules varies in different polysaccharides, their general formula is given as $(C_6H_{10}O_5)_n$, where n means "taken a number of times." Needless to say, polysaccharides are usually very large molecules. Examples are plant starch, glycogen, and cellulose. Polysaccharides are not usually sweet to the taste, and ordinarily they are not soluble in water. Starch is hydrolyzed by the enzyme amylase and breaks down to form dextrins and maltose. Cellulose is found in the walls of plant material and in woody tissues. It is not digestible by human beings since we have no digestive enzyme capable of breaking it down.

DISACCHARIDE

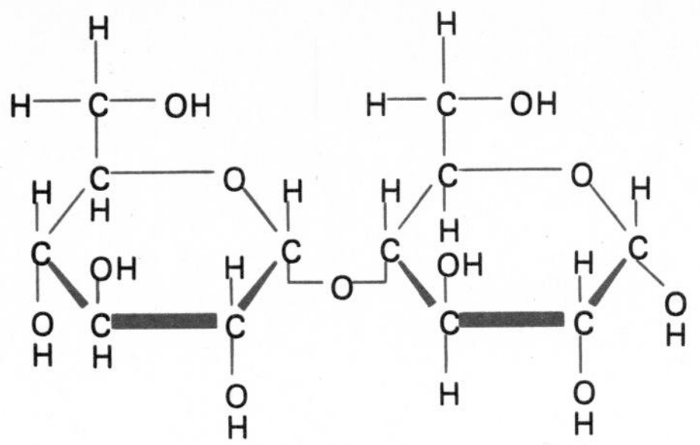

maltose

POLYSACCHARIDE

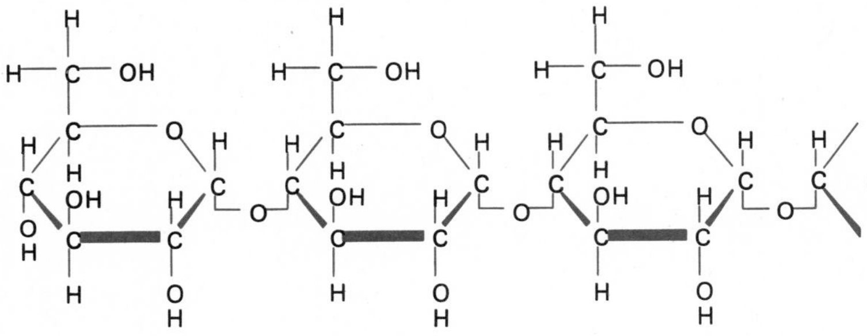

etc.

glycogen

FIGURE 1·10

Two glucose molecules may be linked by removal of one molecule of water to form the disaccharide maltose. If more than two monosaccharides are linked in this way, a polysaccharide is created.

LIPIDS   Three major groups of lipids are generally considered when dealing with living matter: triacylglycerols, phospholipids, and cholesterol. (Cholesterol is actually not a lipid but is handled by the body as if it were.) Triacylglycerols are perhaps the most numerous lipids in the body. These compounds consist of three fatty acid molecules attached to a molecule of glycerol (Fig. 1.11). Triacylglycerols form *nonpolar molecules* and thus are insoluble in water. The number of carbon atoms in each fatty acid determines the type of triacylglycerol. If the carbons in the fatty acid chain are linked by single bonds so that the maximum number of hydrogen atoms are bound to them, then the fat is said to be *saturated*. If the fatty acid chains do not contain the maximum number of hydrogens, the fat is said to be *unsaturated*.

An important property of fats, which will be considered further in Chap. 2, is that they do not mix with water. This property stems from the *nonpolarity* of the fat molecules. You will recall that covalent bonds are produced by sharing electrons, and the resulting molecule has no overall charge when the formation of the bond balances electrons and protons. However, as we saw with water, certain atoms of a molecule may attract the shared electrons to a greater extent than do the other atoms so that regions of positive and negative charge develop, producing a *polar* molecule. If all the atoms of the molecule attract the electrons equally, however,

TRIACYLGLYCEROL
(fat)

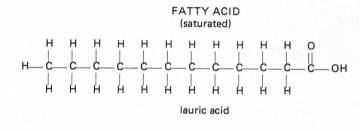

glycerol portion | fatty acid chains

FATTY ACID
(saturated)

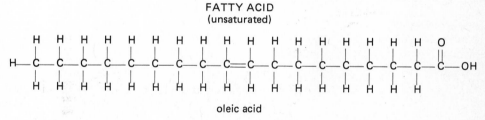

lauric acid

FATTY ACID
(unsaturated)

oleic acid

FIGURE 1·11

A triacylglycerol (fat) and a saturated and an unsaturated fatty acid.

then the molecule is termed *nonpolar*. In such molecules, no regions of positive and negative charge develop, and most lipids fall into this category. Because water is a polar solvent and lipids are usually nonpolar, there is no molecular interaction between them, and therefore, they separate when mixed. Substances that readily interact with water are said to be *hydrophilic* (from the Greek meaning "water loving"); substances which are repelled by water are termed *hydrophobic* (from the Greek meaning "water hating").

Phospholipids differ from triacylglycerols in that they are *amphipathic*. This term indicates that each molecule contains a water-soluble portion and a water-insoluble portion. This characteristic makes phospholipids especially important in functions of living organisms, as will be seen in Chap. 2.

PROTEINS Proteins form the greatest variety of molecules in living matter. These are usually large, complex compounds containing carbon, hydrogen, oxygen, and nitrogen. Many contain small amounts of sulfur and a trace of

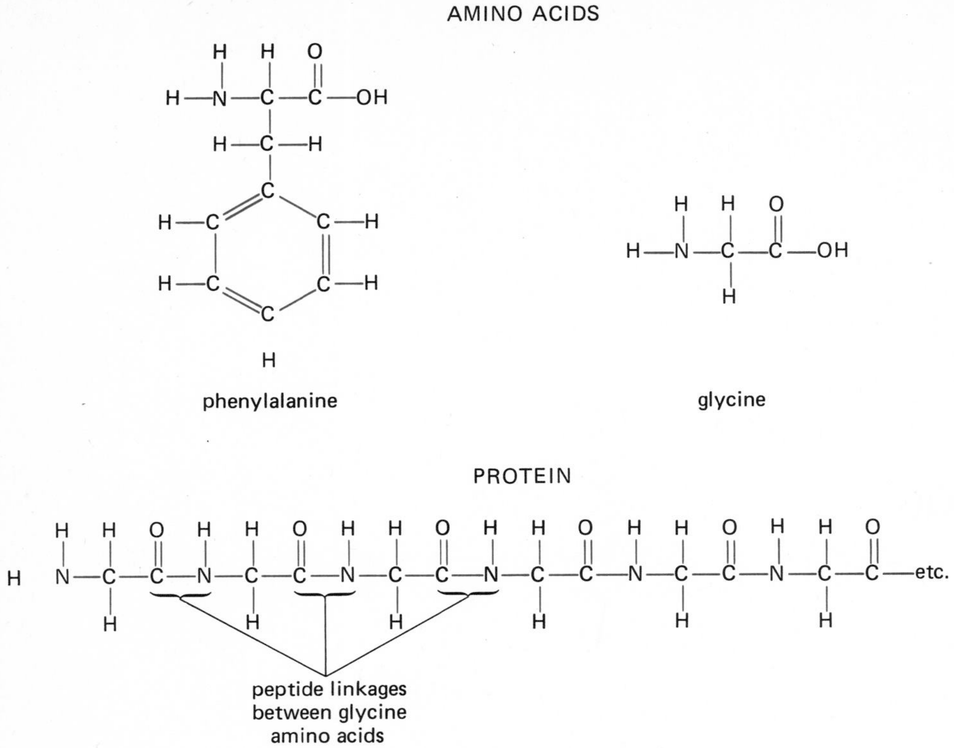

AMINO ACIDS

phenylalanine

glycine

PROTEIN

peptide linkages
between glycine
amino acids

FIGURE 1·12

Two amino acids, phenylalanine and glycine, and a protein.

such elements as phosphorus, zinc, iron, copper, and iodine. Proteins are built by linking amino acids through dehydration (loss of a water molecule) (Fig. 1.12). The linkage formed is termed a *peptide linkage*. When two amino acids are linked together, a *dipeptide* is produced; when three amino acids are bound together, a *tripeptide* is formed. The number of peptide linkages may be increased still more to create *polypeptides*. *Peptides*, then, are compounds composed of two or more amino acids containing one or more peptide groups.

Proteins, generally speaking, fall into two types: fibrous and globular. *Fibrous proteins* have elongate, pleated, or twisted structures and are relatively insoluble in water. Such proteins can extend and contract. A few examples of this type are the contractile fibers of muscle (actomyosin) and the proteins collagen, keratin, and fibrin.

Globular proteins are more soluble in water, and their structures are folded and shortened. Some have very elaborate structures; examples of complex globular proteins are hemoglobin, albumen, the myoglobin of muscles, the cytochromes concerned with electron transport, and most enzymes. Food proteins and amino acids will be discussed again in Chap. 19.

NUCLEIC ACIDS The nucleic acids are intimately involved in the transference and expression of genetic information. Nucleic acids are composed of long chains of *nucleotides*. Each nucleotide contains three subunits: a phosphate group (PO_4^-), a pentose sugar (ribose or deoxyribose), and a

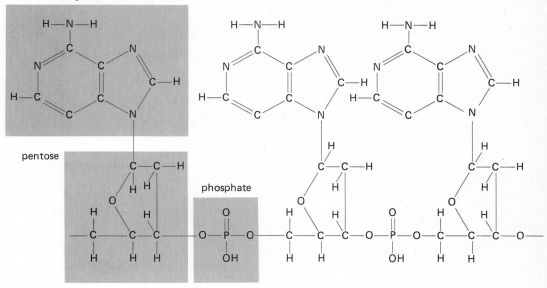

FIGURE 1·13

Deoxyribonucleic acid, a nucleic acid.

nitrogen base (pyrimidine or purine) (Fig. 1.13). Two types of nucleic acids are found in living matter, deoxyribonucleic acid (DNA) and ribonucleic acid (RNA). The general function of these molecules will be discussed further in Chap. 3.

ENERGY

As we stated earlier, energy is simply the ability to perform work. The biological meaning of "work" may sometimes be confusing. In physics, work is defined as a force acting through distance, but how is this type of "work" related to chemical biosynthesis, growth, and development? The relationship becomes clear if the basic concepts of potential energy and entropy are understood.

Potential energy can be illustrated in a variety of ways. A familiar example is the energy contained in a compressed spring. If the spring is allowed to uncoil, the energy is released as motion and can do work. Oil has potential energy also, which can be released by heating it. When oil burns, in the presence of oxygen, the oil is dehydrogenated (hydrogen atoms are removed) and energy is released in the form of heat. This energy was stored, as potential energy, in the chemical bonds between hydrogen and carbon in the oil molecules. Both the combustion of fuels and the metabolic process known as cellular respiration involve the breaking down of energy-containing compounds and the release of potential energy. The source of energy for most biological work is the potential energy stored in nutrients. Of course, the result of burning oil is fire, in which heat energy is released very rapidly. In living matter, nutrients are broken down slowly with a gradual release of energy. Rapid energy release is prevented by controlling substances called enzymes, which are discussed later in this chapter. A portion of the energy is lost as heat energy, as it is during combustion, but a large percentage is incorporated into still other chemical substances, to be released in stepwise fashion as the substance is degraded. As energy is redistributed to more usable forms by the enzymes found in living organisms, about 40 percent or more is trapped and thus available for biological work.

Entropy is a measure of randomness or disorganization; the greater the disorganization, the higher the entropy. In our example of burning oil, the burning of the fuel releases energy because the molecules of petroleum are dehydrogenated and broken down to simpler, less organized molecules (CO_2 and H_2O). The "randomness" of the atoms that made the oil molecules increases as molecular structures are broken down from complex to simpler forms. In order to decrease the entropy of the system, an input of energy and information is necessary.

Work is done by any process that *increases the potential energy and decreases the entropy of a system.* The metabolic processes that extract the energy from nutrients *increase* the entropy and *decrease* the potential energy of the nutrient materials. Synthesis of new compounds for growth and development, on the other hand, *increases* potential energy and *decreases* the entropy of those atoms in the newly synthesized molecules; thus, biological work has been done.

When molecules in fuel substances are built up (wood, coal, oil), chemical bonds are formed. The energy that goes into the formation of these bonds comes directly or indirectly from the sun. The same can be said of the energy incorporated into the molecules of food substances; such energy represents stored solar energy, which is captured in plants by the process of *photosynthesis.* This energy is measured in *calories.* One calorie is the amount of heat energy necessary to raise the temperature of one gram of water one degree centigrade.

ENZYMES AND COENZYMES

We have stressed that the extraction of energy from foodstuffs is controlled by enzymes, or organic *catalysts.* In fact almost every aspect of the work accomplished by physiological systems is controlled by such substances. To understand how enzymes accomplish this task, we must understand that chemical reactions within living matter proceed in the same manner as do all chemical reactions. A certain amount of kinetic energy, termed the *energy of activation,* is necessary to get a reaction started. Activation energy may take the form of heat. Increasing the heat of the reactants increases their kinetic activity (random motion) so that the particles collide more often. A dramatic example of the power of heat as activation energy is seen in the reaction of hydrogen with oxygen. If a room, held at room temperature, is filled with a mixture of these two gases, their rate of combination is extremely slow. If you strike a match in that room, however, the gases combine with an explosive release of energy. The match supplies the activation energy, a small push necessary to start the reaction. The reaction then proceeds spontaneously.

The rate at which a chemical reaction proceeds depends upon factors that increase or decrease the chances for collisions between molecules of the reactants. Human body temperature must be maintained within very narrow limits and therefore cannot be changed to supply activation energy for cellular reactions. Therefore, living matter carries on metabolic processes with the aid of *enzymes*. Catalysts, including enzymes, regulate the rate of chemical reactions in which they participate without being changed themselves in the reactions. Catalysts do not supply activation energy but *lower* the activation energy necessary to start a reaction.

Enzymes are generally named for the reactions they catalyze, and convention provides that the ending *-ase* be added. An enzyme that catalyzes an oxidation-reduction reaction is therefore termed an *oxidoreductase*, while an enzyme concerned with the digestion of proteins is a *protease*.

While hundreds of different enzymes are necessary to carry on the physiological reactions of living matter, the majority catalyze chemical reactions of two types: oxidation and hydrolysis. *Oxidation* reactions involve the removal of hydrogen atoms and are catalyzed by *oxidases* or *dehydrogenases*. *Hydrolysis,* on the other hand, involves the addition of one or more molecules of water to a compound, and such reactions are catalyzed by *hydrolases.* Hydrolytic reactions break chemical bonds to form simpler substances. Digestive enzymes, which break the larger food molecules into smaller molecules, are hydrolases. Hydrolases include carbohydrases, proteases, and lipases, as well as phosphatases and nucleases.

Enzymes are also involved in the *synthesis* (creation) of new materials. *Synthetases* are those enzymes that catalyze the formation of more complex molecules from simpler ones. They usually work by *dehydration,* or water removal.

Enzymes are, for the most part, quite specific in their action; that is, each enzyme works best in catalyzing a single reaction (or type of reaction). This greater "efficiency" in catalyzing one particular reaction is the basis of the concept of *enzyme specificity*. Thus, a certain enzyme may be essential for one step in a chemical reaction, but a different enzyme may be equally essential for any further action. The substance acted upon by the enzyme is called the *substrate*. Each enzyme catalyzes a particular chemical reaction with a particular substrate most rapidly, but it also catalyzes reactions with other substrates of the same general type, although, perhaps, somewhat more slowly.

All enzymes are probably proteins. Many have been prepared in the laboratory in highly purified form, and some of these have been crystallized, retaining their full activity. Since enzymes are protein in nature, they are inactivated by heat. Some enzymes require an associated metal or chemical compound, termed a *cofactor* or *coenzyme,* that must be present if the enzyme is to be active.

The processes that take place inside living organisms are often intricate and involved. The process of deriving and utilizing energy from food substances, for example, requires numerous enzymes and proceeds through an elaborate series of steps. Because of this, physiology is most easily studied in a step-by-step fashion, isolating each system and its complex functions. The student must apply this method of study with the understanding that each physiological system interacts with all others, producing characteristics of the whole organism that are not present in the parts. It is therefore important to integrate each newly learned fact into a general conception of the human organism.

SUMMARY

1 Matter exists in gaseous, liquid, and solid states and is composed of elements, or different types of atoms, each of which has a definite structure. Each atom consists of a nucleus and a particular number of electrons arranged in orbitals about the nucleus. The nucleus is composed of protons and neutrons. Protons are positively charged; electrons are negatively charged; neutrons have no charge.

2 Each element has a distinctive atomic mass or weight. Some elements have isotopes, which differ in atomic mass but have the same chemical properties. Radioactive isotopes have been very valuable in studying chemical, medical, and physiological processes.

3 If an atom gains an electron or electrons, it acquires a negative charge and is called an anion; if it loses an electron or electrons, it carries a positive charge and is called a cation. In an electric field, anions move toward the anode, or positive pole, while cations move toward the cathode, or negative pole. The process by which atoms gain or lose electrons and become charged particles is called ionization. Particles that have opposite charges attract each other and are held together by ionic bonds.

4 Atoms may share electrons and form covalent bonds. Such sharing of electrons is typical of biological reactions.

5 The symbol pH refers to hydrogen ion concentration of a solution. A pH of 7 is neutral, meaning that the concentrations of hydrogen ions (H^+) and hydroxyl ions (OH^-) are equally balanced. Solutions with a pH value above 7 are alkaline; solutions with a pH below 7 are acidic. A solution of pH 3 is more acidic than a solution of pH 4.

6 A buffer is a substance that, when added to a solution, tends to maintain the hydrogen ion concentration at a certain level when either excess acid or excess alkali is added to the solution.

7 Carbon compounds are of special importance in the study of anatomy and physiology since many of these organic substances are essential in the formation of living matter. The linkage of carbon atoms in rings and chains forms the backbone of complex molecules such as carbohydrates, fats, and proteins.

8 The source of energy for biological work is the potential energy of nutrients. Energy is released from nutrients in a stepwise fashion, in reactions controlled by enzymes. Entropy of physiological systems is decreased by synthetic activity in which large molecules are made from smaller ones. This process usually requires energy.

9 Enzymes are biological catalysts. They accelerate chemical reactions but are unchanged themselves by the reaction. Enzymes are proteins and often require metallic or chemical cofactors.

QUESTIONS

1 What is the value of radioactive isotopes in the study of physiological processes?

2 Which is more alkaline: a solution of pH 8 or one of pH 10? Which is more acid: a solution of pH 2 or one of pH 5? What is the pH of a neutral solution?

3 The normal pH of body fluids is about 7.4. What are the names of the substances that tend to minimize fluctuations in body fluid pH?

4 Why do we refer to HCl as a strong acid and H_2CO_3 as a weak acid? Can you give an example of a strong base?

5 What occurs during ionization? Distinguish between an anion and a cation. What is an electrolytic solution? Is a solution of glucose electrolytic?

6 Consider the structure of an atom. Most of the volume of an atom is composed of: (a) protons; (b) neutrons; (c) electrons; (d) empty space.

7 Distinguish between covalent, ionic, and hydrogen bonding. Give examples of each type.

8 Explain the difference between atomic weight and atomic mass. Why is the carbon atom especially significant in this question?

9 Refer to Table 1.1. What is the gram molecular weight of water (H_2O)? Sodium chloride (NaCl)? Calcium chloride ($CaCl_2$)? How many moles of water are there in 1 kg of water? How would you prepare a one molar solution of NaCl? A one molal solution?

10 Describe, in general, how enzymes facilitate biological reactions. To what class of organic compounds do enzymes belong?

SUGGESTED READING

Baker, J. J. W. and G. E. Allen: "Matter, Energy, and Life," 3d ed., Addison-Wesley Publishing Company, Inc., Reading, Mass., 1974.

Barrington, E. J. W.: "The Chemical Basis of Physiological Regulation," Scott, Foresman and Company, Glenview, Ill., 1968.

Cohen, A.: "Handbook of Cellular Chemistry," The C. V. Mosby Company, St. Louis, 1975.

Coult, D. A.: Molecules and Cells, chap. 2, in "Basic Chemistry for the Study of Living Matter," Houghton Mifflin Company, Boston, 1967.

Kleinsmith, L. J.: Molecular Mechanisms for the Regulation of Cell Function, *BioScience*, **22:**343–347 (1972).

(A number of good introductory textbooks of chemistry, biochemistry, and physics may be helpful to the student seeking additional information on the topics introduced in this chapter.)

2

THE CELL: CELLULAR STRUCTURE AND FUNCTION

In this chapter we begin the study of the cell. We have chosen to present the cellular organelles by discussing their structure and function together. The topics covered include:

1 Basic characteristics of cells: their composition and activities
2 Cellular organelles
3 Exchange processes between the cell and its environment, with emphasis on the role of cell membranes
4 Osmosis, diffusion, active transport, facilitated diffusion, and the relationship of such processes to cell membrane permeability
5 Function of the golgi apparatus, endoplasmic reticulum, and lysosomes
6 Homeostasis and the concepts of negative and positive feedback mechanisms

The characteristics of a unit of living matter depend not only upon the substances contained within it, but also upon the specific pattern of organization of its components. Biological systems create life from nonliving molecules by a continuous expenditure of energy. Energy, as we have stated before, is simply the ability to perform work, and in biological systems, energy expenditure must be balanced by energy input or the system will disintegrate. Therefore, each biological system must maintain a constant exchange of substances with its environment.

From molecules to organelles to cells to tissues to organs, there are many levels of organization within a complex organism such as the human being. The functional unit of all living organisms, however, is the *cell*, for it represents the lowest level of organization capable of performing all functions necessary for life. Proper physiological functioning of the whole organism depends upon the maintenance of the organizational integrity of this unit. Through the process of differentiation, cells in more complex organisms become *specialized;* each performs some vital function and, in turn, relies upon other cell types for support. Cellular activities are coordinated primarily by the organism's nervous and endocrine systems so that the needs of all cells in the body are met. Though many cell types exist among the trillions of cells that make up the human body, all cells share certain characteristics. In this chapter and in the following one we shall lay foundations for the discussion of complex physiological functions by describing a hypothetical, generalized cell and examining the structure, function, and interrelationships of its components.

BASIC STRUCTURAL CHARACTERISTICS OF CELLS

CELLULAR COMPOSITION The cell as an entity comprises a semifluid substance called *protoplasm* contained within a *plasma membrane.* The term "protoplasm" characterizes the entire living substance of the cell. The protoplasm of the cell is separated into two main divisions: the *nucleoplasm,* or protoplasm of the nucleus, and the *cytoplasm,* or protoplasm of the remainder of the cell. The basic constituents of protoplasm are water, electrolytes, proteins, lipids, and carbohydrates.

Water By far the largest single constituent of protoplasm is water, which is 70 to 85 percent of the cell's weight. Water is the major vehicle for dissolution, suspension, and transport of cellular materials, and it serves as the reaction medium for many cellular chemical processes. By virtue of its abundance, water determines cell volume, which directly influences chemical concentrations of various intracellular solutes, as well as the spatial relationships among and within the cellular organelles.

Electrolytes Most solutes dissolved within protoplasmic water are inorganic ions. The major electrolytes (particles that form charged ions) are potassium, magnesium, phosphate, sulfate, and bicarbonate, in addition to small amounts of sodium, chloride, and calcium in proportions shown in Table 2.1

TABLE 2·1
Representative concentrations of important electrolytes in extracellular and intracellular fluids

	Extracellular (in mmol)	Intracellular (in mmol)
Na	140	10
K	5	130
Cl	100	5
Mg	1.5	26
HCO_3	28	13
PO_4	2.5	40
SO_4	0.5	1.0

Electrolytes are critical for proper cellular functioning for several reasons.

1 They form the inorganic components of cellular reactions. Many important chemical reactions involving complex organic molecules cannot proceed properly in the absence of electrolytes.
2 Electrolytes function in such control mechanisms as the generation of electrochemical impulses in nerve and muscle.
3 By virtue of their large contribution to the total cellular solute, electrolytes also induce water to move in and out of the cell. This movement adjusts cellular volume (see the discussion of osmosis below).

Proteins On a weight basis, proteins constitute 10 to 20 percent of the cell's protoplasm. Two basic types of proteins are found in cells: *fibrillar* and *globular*. Fibrillar proteins usually compose the structural elements of cells; all intracellular organelles contain some structural protein. Globular proteins form the enzymes that control cellular metabolic functions. Finally, the *nucleoproteins*, proteins found in the cell nucleus, help regulate the functioning of the entire cell. Nucleoproteins may also be involved in control of cellular differentiation, which determines cell type.

Lipids The three major lipid components of the cell are triacylglycerols, phospholipids, and cholesterol. (As mentioned previously, cholesterol is actually not a lipid but is handled by the body as if it were.) Together, the lipids compose about 3 percent of the cell weight. Triacylglycerols are simply neutral fats that store cellular energy. Phospholipids and cholesterol are important structural components of cell membranes; they can restrict the movement of water and water-soluble molecules across cell membranes.

Carbohydrates About 1 percent of the cell's weight is carbohydrate. Most carbohydrate is used by the cell for nutrition. Glycogen (or animal starch) is the predominate carbohydrate in animal cells. The sugar glucose, which is the basic nutrient for practically all cells, is stored within cells as glycogen. Thousands of glucose molecules are linked, one to the other, to form the huge glycogen molecule.

PHYSIOLOGICAL CHARACTERISTICS OF CELLS

All types of cells exhibit certain basic physiological characteristics vital for the survival of each cell. These include irritability, motility, metabolic processes, growth, and reproduction. The degree to which a cell exhibits each of these basic physiological processes varies from one cell type to another and depends upon the cell's main function within the organism.

IRRITABILITY Cells are sensitive to certain alterations in their environment to which they respond, a characteristic termed *irritability*. Stimuli, which may be chemical, mechanical, thermal, and so forth, elicit certain changes on the cell membrane surface or within the cell that enable it to react in a manner characteristic of that type of cell. Some cells, such as those in sensory receptors, are highly specialized to respond to only one particular type of stimulation. Nerve cells, which exhibit a high degree of irritability, respond to stimulation by conducting a nerve impulse, a characteristic termed *conductivity*. Muscle cells respond to stimulation by contracting; such cells exhibit a high degree of *contractility*.

MOTILITY   The spontaneous motion of cells in their environment or the motion that occurs within cells is called *motility*. The protoplasm of all cells exhibits cyclic movement, or *cyclosis*. In some cells, such as the amoeba and certain white blood cells, the protoplasm "streams" to form *pseudopodia*, "false feet," enabling the entire cell to move. Ciliated cells propel either themselves or substances external to them by rhythmic movements of their whiplike *cilia*. The sperm cell, another very motile cell, covers considerable distances using swimming movements of the filament in its tail section. In the case of muscle cells, of course, the entire cell is specialized for motility; its main function is to contract.

METABOLISM   The term *metabolism* is applied to the sum of all life processes in general. Metabolism involves all chemical and physical activities of the cell or the organism. The metabolic process in which new materials are built up and energy is stored is called *anabolism*. Anabolic processes usually involve the expenditure of energy to synthesize large, complex molecules (in which energy may be stored), to contract muscle, and to transport materials across cell membranes. The metabolic process in which large molecules are broken down to smaller, simpler ones, usually accompanied by the release of energy, is termed *catabolism*. The cell stores the energy released during catabolism in the form of adenosine triphosphate (ATP), a compound that supplies energy directly for the metabolic processes that perform the work of the cell.

GROWTH AND MULTIPLICATION OF CELLS   When the synthesis of new cellular material exceeds the breakdown of material, the cell may increase in volume and mass, a process called *growth*. During anabolic processes, the cell takes in simple substances, such as amino acids, and assembles large complex molecules, such as proteins. After the cell grows to a certain size, it may undergo *mitotic division*, in which it splits into two equal cells. The resulting smaller cells eventually grow to full size. Tissues grow by the multiplication of cells, through mitotic division and the subsequent growth of these cells to normal size.

REPRODUCTION   The ability of cells to multiply by mitotic division in tissue growth should not be confused with the ability of an organism to *reproduce* new individuals. Sexual reproduction in the higher animals and in humans involves production of germ cells—spermatozoa and ova—that unite during the process of fertilization to form a new individual. The development of germ cells is discussed in Chap. 21.

RELATIONSHIPS
OF CELLULAR
STRUCTURE
TO FUNCTION

There are many kinds of cells, which may assume many shapes. In general, cells are microscopic masses of protoplasm containing a nucleus and more or less spherical in shape. But there are also the spindle-shaped cells of visceral muscle tissue, the thin flat cells of squamous epithelium, the cube-shaped cells of cuboidal epithelium, and many others (Fig. 2.1). Though most cells are microscopic in size, some egg cells (ova), which include large amounts of

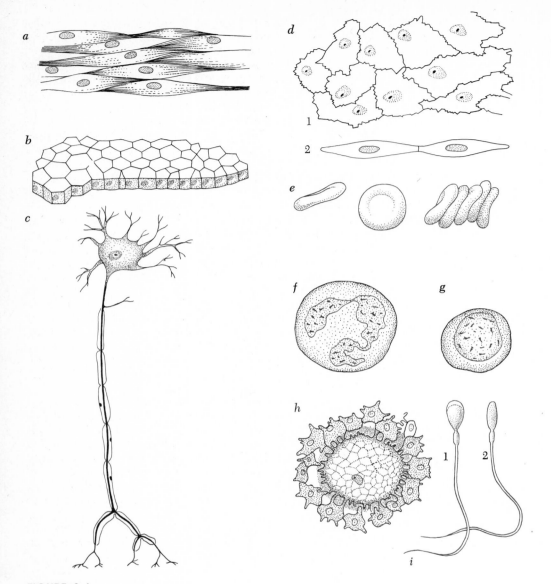

FIGURE 2·1

Representative cell types:
a smooth-muscle cells;
b cuboidal epithelium; *c* a
motor neuron; *d* simple squamous epithelium, *1* surface
view, *2* lateral view; *e* erythrocytes; *f* neutrophil; *g* lymphocyte; *h* human ovum; *i* human
spermatozoa, *1* surface view,
2 lateral view.

nutritive material, may be quite large. The human red blood cell is 0.008
millimeters (mm) in diameter, but voluntary muscle cells may be as long as
4 centimeters (cm). The axon process of a motor nerve cell may extend from
the lower part of the spinal cord to a muscle in the foot and be well over a
meter (m) in length. The diameter of the nerve cell body, however, is only
about 125 to 130 micrometers (μm) (0.125 to 0.130 mm) even for the larger
motor cell bodies of the spinal cord.

THE GENERALIZED CELL

As we have pointed out, cells vary greatly in shape and function, and in complex organisms, different cell types are specialized for varying functions. For this reason, different cellular organelles are found in varying numbers in different cell types. For purposes of introduction, however, it is often helpful to describe a (nonexistent) generalized cell, such as that depicted in Fig. 2.2, and the appearance of its organelles.

The *nucleus* is the distinct, differentiated area usually located near the center of the cell. It contains the chromosomes, along which the genes, the hereditary units that are passed from generation to generation, are located. The nucleus is the ultimate control center of the cell. In nondividing cells, the *nucleolus,* a nuclear organelle, is usually seen as an irregularly shaped, distinct area of the nucleus. The nucleus is surrounded by the *nuclear membrane,* a double membrane with large pores, which apparently controls movement of substances in and out of the nucleus.

Organelles of the cytoplasm, the cell's other major compartment, include the *endoplasmic reticulum,* a double-membrane structure that winds through the cytoplasm from the nuclear membrane to the plasma or cell membrane. The space separating the two membranes forming this structure is termed the *cisternae.* There are two types of endoplasmic reticulum: *rough* and *smooth.* This distinction is based upon the presence or absence of ribosomes. *Ribosomes* are tiny spherical granules located throughout the cytoplasm, and when they are attached to the endoplasmic reticulum, it is called *rough endoplasmic reticulum.* Ribosomes are the site of protein synthesis in the cell. *Smooth endoplasmic reticulum* has no ribosomes, thus presenting a smooth appearance.

The smooth regions appear to be continuous or intermittently continuous with the *golgi apparatus.* This organelle is composed of a series of shelflike membrane folds, which sometimes close upon themselves to form vesicles. Newly synthesized materials are packaged in the golgi apparatus and then transported throughout the cell's cytoplasm or out of the cell. The golgi apparatus is usually located near the nucleus. *Lysosomes* are vesicular bodies formed by the golgi complex that contain digestive enzymes. Those enzymes may destroy materials taken into the cell or even the cell itself.

The *centrioles* are two rodlike structures near the nucleus that are always arranged at right angles to each other. These organelles play a role in cell division. Spread throughout the cytoplasm are the small, sausage-shaped *mitochondria.* These are also double-membrane structures, whose internal folds are called *cristae.* Mitochondria are involved with energy production in the cell. The *plasma,* or *cell, membrane* forms the boundary of the cell and controls entry and exit of most materials.

All functions of the cellular organelles are interrelated, and they may be divided into three general categories: (1) exchange, (2) cellular respiration, and (3) protein synthesis and cellular reproduction. The exchange function involves the plasma membrane, endoplasmic reticulum, golgi apparatus, and the lysosomes, and in the remainder of this chapter, we will deal mainly with the structure and function of these organelles.

In Chap. 3, we will take up organelles that belong to the second and third major functional categories. The mitochondria are primarily concerned with extraction of chemical potential energy from nutrient materials, i.e., with the process of cellular respiration. The nucleus, nucleolus, and ribosomes are

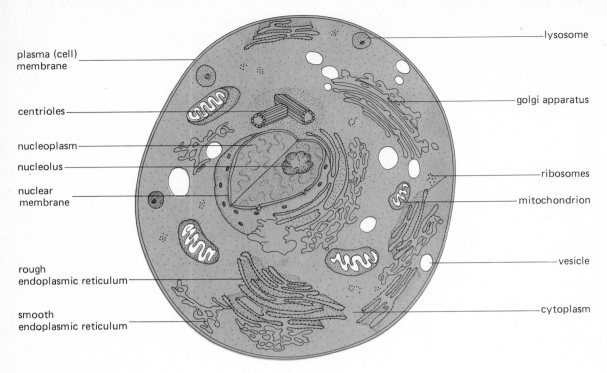

plasma (cell) membrane

centrioles

nucleoplasm

nucleolus

nuclear membrane

rough endoplasmic reticulum

smooth endoplasmic reticulum

lysosome

golgi apparatus

ribosomes

mitochondrion

vesicle

cytoplasm

FIGURE 2·2
Schematic representation of a generalized cell. The structures are illustrated as they are revealed by the electron microscope.

involved in protein synthesis and cell reproduction, and we will also mention the role of microfilaments and microtubules (not visible in Fig. 2.2) when we discuss these functions in Chap. 3.

THE PLASMA MEMBRANE

MEMBRANE STRUCTURE   All cells must acquire nutrients and oxygen and excrete waste products. Thus, cells are dependent upon a constant exchange of materials with their environment. This exchange occurs through and is regulated by the thin flexible membrane called the *plasma membrane,* or *cell membrane,* which surrounds the cytoplasm. This cellular organelle is not just a sac containing the cell's protoplasm; it is highly selective and regulates the ease with which substances permeate (or penetrate) it. The plasma membrane is a *selectively permeable membrane.* As seen in an electron micrograph, the plasma membrane is approximately 75 angstroms (Å) (Å = 0.0000001 mm) thick and appears as two dense layers separated by a less dense area (Fig. 2.3).

The cell membrane is primarily composed of phospholipids and proteins, with smaller amounts of cholesterol, lipoproteins (lipid plus protein), and glycoproteins (carbohydrate plus protein). The phospholipid molecules have a water-soluble "head" region and a lipid-soluble "tail" region (Fig. 2.4). The hydrophilic polar heads of the phospholipid molecules are attracted to and associate with water molecules in an aqueous environment, while the hydrophobic nonpolar tails are repelled by water, associating instead with each other. When mixed with water, phospholipid molecules form a fluid

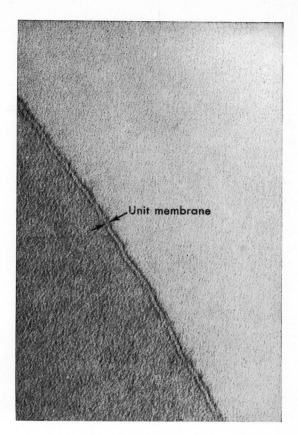

FIGURE 2·3

The plasma membrane. An electron micrograph showing the three-layered structure of an erythrocyte plasma membrane. ×280,000. (*Courtesy of J. D. Robertson, from Roy O. Greep, "Histology," McGraw-Hill Book Company, New York 1973.*)

membrane arranged in a bimolecular layer with water-soluble surfaces and a lipid-soluble center (Fig. 2.5). This *phospholipid bilayer* is the basic substance, or matrix, of all cell membranes.

The way in which the membrane's protein associates with the phospholipid bilayer is now fairly well understood. The membrane proteins extend either completely through the bilayer (ectoproteins) or only part way through it (endoproteins) (Fig. 2.6). All membrane proteins are exposed on either the inside, the outside, or both surfaces; that is, no proteins are completely enclosed by phospholipid. In addition, some proteins are free to move laterally through the surrounding phospholipid, while others are stabilized so that they always occupy a certain region of the cell surface. Membrane proteins are never symmetrically arranged. The portion of the protein protruding from the outer surface is always different from the portion protruding into the cell cytoplasm. Furthermore, the ectoproteins, which extend through both surfaces, cannot rotate within the membrane because the portions that extend through the phospholipid are hydrophilic, while the portion embedded in the phospholipid is hydrophobic. If a carbohydrate moiety is attached to a membrane protein (forming a glycoprotein), the carbohydrate always extends from the outer surface, never from the inner surface, of the membrane, as shown in Fig. 2.6.

PHOSPHOTIDYLSERINE

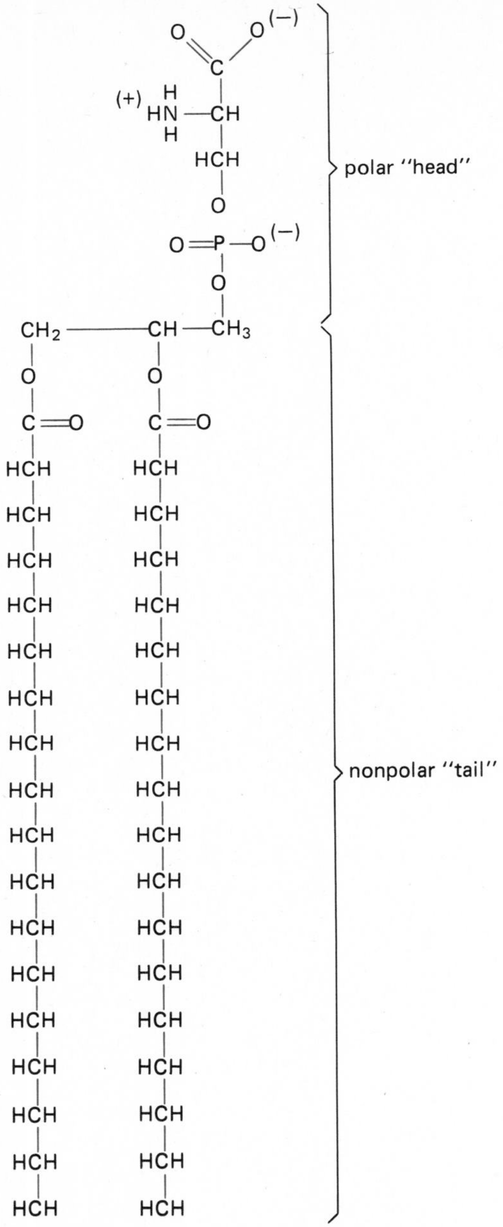

polar "head"

nonpolar "tail"

FIGURE 2·4
Phosphotidylserine, a phospholipid. Water will associate
with the polar "head" but not
the nonpolar "tail."

MEMBRANE PROTEIN FUNCTIONS Membrane proteins have various functions.
Some act as *transport proteins*, moving substances across the membrane;
some probably supply energy for transport processes; still others serve as
binding sites for chemical communicators such as hormones.

Living cell membranes behave as though they are porous, a characteristic

PHOSPHOLIPID BILAYER

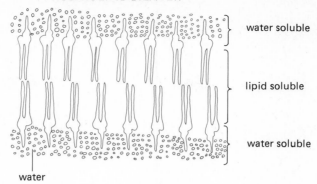

water soluble

lipid soluble

water soluble

water

FIGURE 2·5

A phospholipid bilayer. This type of structure forms the foundation for all cell membranes.

that makes transport of materials possible. Evidence that membranes contain pores comes from studies of the permeability of membranes to molecules of varying size and solubility in lipids (Fig. 2.7). Lipid-soluble (nonpolar) molecules, whether large or small, pass with relative ease through cell membranes because they can dissolve in the phospholipid portion. Ions and polar molecules, however, are not very soluble in lipids; but they readily penetrate most cell membranes if their diameter is no larger than about 8 Å. Polar and ionized molecules of only slightly larger size require assistance to pass through the cell membrane at all. Thus, cell membranes behave as though they contain pores that are about 8 Å in diameter. Pores of this type have never actually been observed in cell membranes, however. A possible explanation for the porous behavior of membranes is that the ectoproteins, which extend entirely through the phospholipid matrix, may provide a channel for small polar molecules and ions.

MOVEMENT THROUGH MEMBRANES

Having examined membrane structure, we will step back a bit and look at the ways in which substances move through cell membranes. We will also consider some physical factors governing molecular movement from one area to another.

FIGURE 2·6

Three-dimensional cross-sectional view of a plasma membrane. The proteins appear to float in a phospholipid "sea."

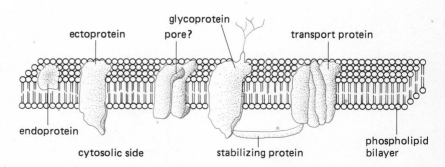

glycoprotein

ectoprotein   pore?   transport protein

endoprotein

cytosolic side   stabilizing protein   phospholipid bilayer

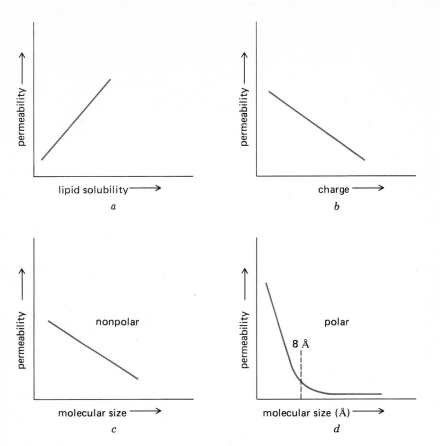

DIFFUSION   If a sugar cube is placed in a beaker of undisturbed water, the sugar molecules go into solution over time and spread slowly from the area of their greatest concentration toward the area of their lowest concentration until they are eventually distributed in uniform concentration throughout the solvent (Fig. 2.8). The tendency of the sugar molecules to move from an area of high concentration to an area of low concentration is expressed as the *force of diffusion. Diffusion*, whether of a solvent, a solute, or a colloid (suspended particle), is caused by the continuous, random, and independent motion of all particles. This motion depends upon the kinetic energy of the particles, which, in turn, is determined by the amount of heat energy they possess. Heat input will increase the kinetic energy of the particles and thus the rapidity of their motion. (This motion is what we measure when we take the temperature of a substance.) During their random movements, particles collide incessantly with other particles—like particles and particles of other substances—and the walls of their container. Net movement of a certain type of particle in solution depends upon its relative concentration in different parts of the solution.

In order for a substance to move by diffusion across a cell membrane, the membrane must be permeable to the substance, and a *concentration difference*, or *gradient*, for that substance must exist across the membrane for net

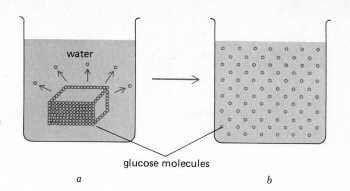

FIGURE 2·8

Diffusion. If a quantity of glucose is placed in a beaker of water *a*, the glucose molecules will disperse randomly until they are uniformly distributed *b*. This random thermal motion of molecules is expressed as the *force of diffusion*.

water

glucose molecules

a   *b*

movement to occur. As an example, cells receive a steady supply of oxygen from the blood by diffusion. The small, diatomic nonpolar oxygen molecules (O_2) easily penetrate cell membranes, and because the cells continuously use up molecular oxygen, the blood always contains a higher concentration of oxygen molecules than do cells. This situation creates a concentration gradient down which oxygen can diffuse.

Importance of diffusion   Diffusion is one of the most important fundamental forces that moves materials through cell membranes. The process works well for the distribution of substances over short distances; however, over distances of more than a fraction of a millimeter, the transport of substances by diffusion is impractical for living tissue. The laws of diffusion are such that every 10-fold increase in diffusion distance increases diffusion time 100-fold. Even though a molecule might diffuse 1 μm in 0.0001 seconds (s), it would require about 3 years (yr) for it to travel the length of an arm because of the great number of collisions it would encounter as it traveled that distance. This simple fact limits the size of most cells and makes an internal circulatory system necessary for all organisms larger than a certain size. If a cell exceeds a certain size, diffusion can no longer supply oxygen and other nutrients to the cell quickly enough to sustain normal cellular metabolism.

OSMOSIS AND BULK FLOW: WATER MOVEMENT THROUGH CELL MEMBRANES   As noted previously, cellular water content is critical for proper physiological functioning. It is essential, therefore, to understand some factors that influence the movement of water across the cell membrane. Two basic processes are involved. Both depend upon a difference in the kinetic activity (motion) of the water molecules in one region as opposed to another.

Bulk flow   The first process we shall describe is *bulk flow movement*. Water moves through a cell membrane along a *hydrostatic* (mechanical) *pressure gradient* imposed across the membrane. Imagine two cylinders filled with water and connected by a tube at their base. A piston is placed above the water in cylinder A, while cylinder B remains open (Fig. 2.9). When the piston is pressed down, creating a pressure wave at its bottom, energy is transmitted from the piston to each water molecule in cylinder A. When the *hydrostatic pressure* exerted on a given volume of water increases, the

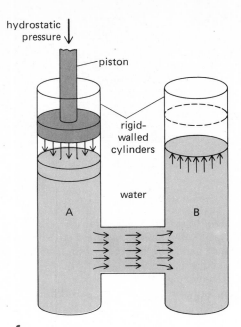

FIGURE 2·9

Bulk flow. Two rigid-walled containers are connected by a tube and initially contain equal amounts of water. Hydrostatic pressure applied to cylinder A will cause a bulk flow of water into cylinder B.

kinetic activity of each water molecule increases. In the setup depicted in Fig. 2.9, water then flows very rapidly into cylinder B. Its rate of flow depends upon the amount of pressure applied by the piston. A hydrostatic pressure differential thus causes a mass movement of water molecules, all in the same direction, from an area of high hydrostatic pressure to an area of lower hydrostatic pressure. This type of movement is referred to as *bulk flow*. For example, the blood flow through a blood vessel that is caused by the pressure wave generated with each contraction of the heart is bulk flow. Substances may also move through pores in a membrane by bulk flow if a hydrostatic pressure gradient exists.

Particle concentration: osmosis   The second factor that influences water movement is particle concentration. Water moves by the process of *osmosis* from an area of low particle (solute) concentration (high water activity) to an area of higher particle (solute) concentration (low water activity). The kinetic activity of water is greatest in pure water and is lowered by the addition of solute particles, which in effect dilute the pure water.

To illustrate this process we will modify the cylinders depicted in Fig. 2.9 so that they are connected by a membrane rather than a tube. The membrane in this setup is a thin, flexible, slightly elastic structure with pores that are large enough to permit water molecules (3 Å in diameter) to pass through but too small to permit glucose molecules (8.5 Å in diameter) to pass (Fig. 2.10). Now we will fill compartment A with pure water and compartment B with an equal volume of a glucose solution with a concentration of 100 mmol/l. We then seal the two cylinders at their tops. Water flows through the membrane in both directions, but the greatest flow, or *net flow*, is from the pure water in compartment A into the glucose solution in compartment B. The volume of compartment B increases, therefore causing

FIGURE 2·10

Osmosis and osmotic pressure. *a* Two rigid-walled cylinders are sealed and connected by a selectively permeable membrane that allows water, but not glucose, to pass. Cylinder A is filled with pure water and cylinder B with a 100 mmol glucose solution. Initial net flow of water by osmosis is from A to B, increasing the volume of B. *b* The increased volume of B causes distension of the membrane until its recoil exerts a hydrostatic pressure sufficient to stop net water flow. This hydrostatic pressure is the osmotic pressure of the 100 mmol glucose solution in B.

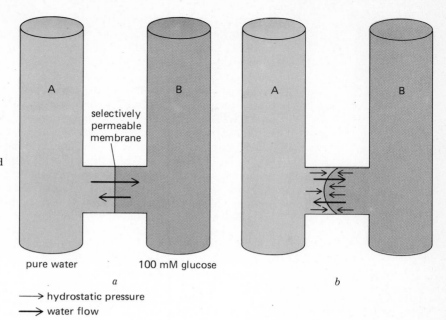

selectively permeable membrane

pure water   100 mM glucose

a   *b*

⟶ hydrostatic pressure
⟶ water flow

distension of the membrane. Net flow occurs in this direction because the presence of the solute particles—glucose molecules in this case—reduces the activity of the water molecules in compartment B.

Returning to our example, we find that the increasing volume in B exerts pressure on the membrane, causing it to distend. As volume in B increases, a point is reached at which the elastic membrane exerts an *equal* and *opposite pressure* against compartment B. This pressure is generated by the membrane's tendency to return to an unstretched condition (just as a spring recoils when stretched). The force exerted by the recoiling membrane is a hydrostatic pressure (just as though the membrane were a piston pushing on the volume in B). This pressure increases the kinetic activity of water in B so that it equals the kinetic activity of water in A, and the glucose solution is said to exert an osmotic pressure. The *osmotic pressure* of a solution is equal to the hydrostatic pressure that must be imposed upon it to prevent any net movement of water between the solution and pure water when these are separated by a membrane permeable only to water.

We know that water moves during osmosis according to the gradient of water activity; however, we have left unexamined the question of why particles dispersed in a solution lower the water activity. We can say with certainty that osmotic flow is far too rapid to be explained only by simple diffusion (random motion) of water molecules. Peter Scholander (see Suggested Reading at the end of this chapter) has presented evidence that osmosis is caused by *solute pressure* on a distensible membrane. According to his hypothesis, collision of the diffusing solute particles with the yielding boundary of the membrane slightly distends the membrane, thus decreasing hydrostatic pressure in the solution compared to that exerted by pure water. Returning to our illustration of bulk flow in Fig. 2.9, we can say the solute pressure effect "lifts" the piston rather than "pushes" it downward. The hydrostatic pressure gradient caused by solute pressure produces bulk flow of water, in addition to simple diffusion. Thus, this combination of flow processes explains the rapidity of osmotic flow of water through animal cell membranes.

Osmotic pressure, units Although the mechanism of osmosis is an interesting topic, our primary reason for discussing it is to emphasize that *osmosis occurs whenever two solutions having differing particle concentrations are separated by a membrane permeable to water but not to the solute particles.* When comparing two solutions, the one with the lower osmotic pressure is said to be *hypoosmotic,* while the one with the higher osmotic pressure is termed *hyperosmotic.* If the two solutions have the same osmotic pressure, they are *isosmotic.* The osmotic pressure of a solution is expressed in *osmoles,* abbreviated *osmol.* A one *osmolar* solution contains 6.0225×10^{23} particles (Avogadro's number) per liter of water. The osmolarity of human blood plasma is about 300 mosmol/l (0.3 osmol/l). Remember that some substances dissociate in water and produce more than one particle per molecule. For example, a 100 mmol/l solution of sodium chloride (NaCl) contains 100 mmol Na^+ and 100 mmol Cl^-; therefore, the osmotic pressure of this solution will approximate 200 mosmol/l.

TONICITY All body cells have the same internal osmotic pressure as the tissue fluid that bathes them. It must be emphasized that only those solutes that cannot penetrate cell membranes will exert an osmotic pressure. A living cell membrane is permeable to many substances other than water. If a living red blood cell is removed from the body and placed in a fluid bath, the behavior of the cell will depend not only upon the osmotic pressure of the bathing solution, but also upon the permeability of the cell membrane to the solute particles in the bath. For example, if a red blood cell is removed from blood, which has an osmotic pressure of 300 mosmol/l and placed in a sucrose solution with an osmotic pressure of 300 mosmol/l also, the cell will neither shrink nor swell because sucrose cannot penetrate the membrane. The osmotic flow of water is therefore effectively balanced (for a limited time). If the same red blood cell is transferred into a urea solution with an osmotic pressure of 300 mosmol/l, however, the cell will swell and burst because urea readily penetrates the cell membrane. Thus the urea bathing solution has a very low effective osmotic pressure. When a living cell is placed in a solution that causes no change in cellular volume, the solution is said, by definition, to be *isotonic* to the cell. In the above two examples, the sucrose solution was *both* isotonic and isosmotic to the cell, but although the urea solution was isosmotic to to the cell, it was *not* isotonic. When we consider osmotic relations in living tissue, therefore, we speak in terms of *tonicity.* If a solution is *hypotonic* to a cell, the cell will swell when placed in the solution. If the solution is *hypertonic* to a cell, the cell will shrink in the solution (Fig. 2.11).

Clinical aspects of solution tonicity The effect of solution tonicity on body tissue has important clinical consequences during intravenous (IV) feeding. The tonicity of fluids fed into the relatively small veins of the extremities must be very near that of plasma, or the resultant osmotic disturbance will produce pain at the IV site and eventual tissue deterioration in the area. For years,

FIGURE 2·11

Tonicity. *a* A red blood cell taken from blood plasma neither shrinks nor swells if placed in an isotonic solution. *b* A hypotonic solution causes swelling of the red blood cell. *c* A hypertonic solution causes the cell to shrink.

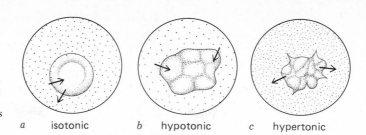

a   isotonic   *b*   hypotonic   *c*   hypertonic

the necessity of administering isotonic solutions prevented the delivery of adequate nourishment via IV feeding to patients who, for one reason or another, could not ingest or utilize food as it passed through their digestive tracts. Inclusion of sufficient concentrations of nutrients, especially fats and proteins, in the IV fluid made it hypertonic to plasma. This problem has been recently overcome by threading a catheter through the brachial vein in the arm into the superior vena cava, one of the major veins returning to the heart. Because of the large volume of blood in this vein, hypertonic nutrient solutions are rapidly diluted, preventing tissue trauma. This type of feeding, called *parenteral hyperalimentation,* has saved the lives of many patients who would formerly have died of malnutrition.

MEDIATED TRANSPORT   As we have seen, polar and ionized molecules may pass through the plasma membrane rather easily unless their size exceeds approximately 8 Å. Some nutrient materials, such as glucose and some amino acids essential to cell survival, are quite insoluble in lipids and are too large to diffuse through the membrane pores. These nutrients do obviously enter cells, however. Their entry is helped by the transport proteins in the cell membrane. As we mentioned earlier in this chapter, these are large proteins or glycoproteins that extend through the membrane's phospholipid bilayer. These transport proteins bind to nutrient (substrate) molecules and transport them through the membrane to the interior of the cell.

Transport protein molecules show *specificity;* that is, they interact specifically with only one type of substrate molecule. Mediated transport may be either *active* or *passive. Active transport* requires the continuous input of metabolic energy, while *passive transport,* also called facilitated diffusion, requires no energy input.

FACILITATED DIFFUSION   Glucose enters most cells by the process of *facilitated diffusion.* We do not yet understand how transport proteins function, but one proposed model of the mechanism by which facilitated diffusion proceeds may be described as follows. The transport protein probably undergoes a change in internal conformation (shape) when glucose binds to it (Fig. 2.12), and this change enables the protein to carry the glucose molecule through the membrane. Transport proteins continually capture and release glucose molecules. On the outside of the cell membrane, glucose concentration is high, and at any given moment, many transport protein molecules are engaged in capturing glucose molecules. On the inside of the membrane, glucose concentration is low, because glucose is used in metabolic processes, so at any moment, most transport proteins have given up their bound glucose. Facilitated diffusion works only when a concentration

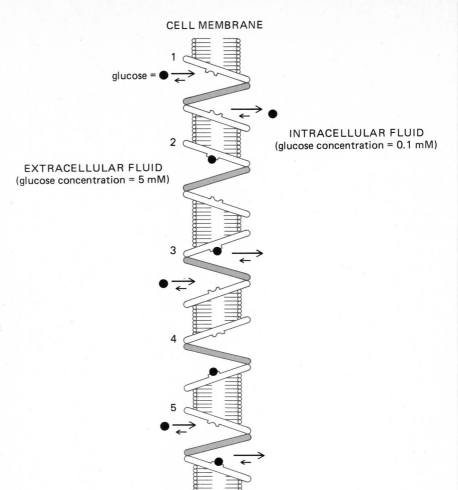

CELL MEMBRANE

glucose = ●

1

2

3

4

5

INTRACELLULAR FLUID
(glucose concentration = 0.1 mM)

EXTRACELLULAR FLUID
(glucose concentration = 5 mM)

FIGURE 2·12
Glucose is helped across the cell membrane by facilitated diffusion, the passive mediated transport process. In this hypothetical model, the transport protein binds glucose with equal affinity on either side of the membrane. Because more glucose is present outside, however, the net movement of glucose is into the cell. When the transport protein binds glucose, it undergoes a conformational change (step 2 to 3) and transports glucose to the opposite side of the membrane. Note that this process only proceeds when a glucose concentration gradient exists across the membrane.

gradient for the transported substance exists. This type of diffusion is distinguished from simple diffusion by the fact that the glucose concentration gradient is not the only factor to determine rate of diffusion. In facilitated diffusion the rate of entry of glucose into the cell increases only up to the point at which all transport proteins are in use (*saturated*) (Fig. 2.13). When no additional transport proteins are available for transport, the rate of glucose uptake levels off. In simple diffusion, diffusion rate continues to increase as long as the concentration gradient continues to increase.

ACTIVE TRANSPORT   In *active transport,* as opposed to facilitated diffusion, a transported substance may be moved *against* its concentration gradient. Although there may be a higher concentration of a substance inside the cell than outside, the cell can still accumulate more of the substance through active transport. For active transport to proceed, the transport protein must have a greater capacity to bind the substrate outside the cell membrane than

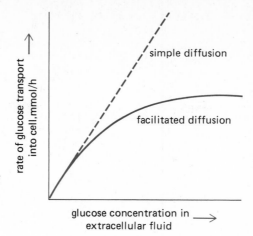

FIGURE 2·13

The rate of molecular transport in facilitated diffusion is dependent upon the amount of transport protein available. It is therefore a process with a saturation point, whereas simple diffusion is not.

inside. In other words, the *affinity* of the transport protein for the substrate must be asymmetrical.

Active transport requires an expenditure of cellular energy. Exactly how active transport takes place is still unknown, but several models that explain most of the known facts about the process have been proposed. One of these models is shown in Fig. 2.14. Active transport probably works by a mechanism similar to facilitated diffusion except that energy is required to

FIGURE 2·14

Hypothetical model of membrane active transport. A substance may be transported into a cell against its concentration gradient by the expenditure of energy. In this model, ATP hydrolysis produces a conformational change in the transport protein (steps 2 to 3 and 4 to 5).

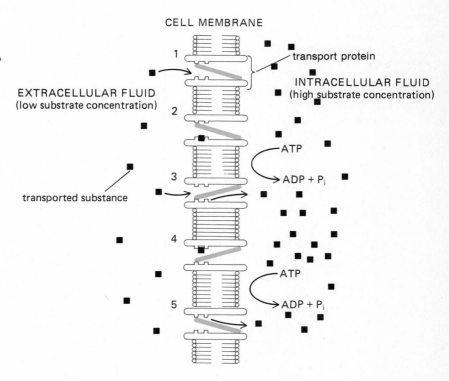

produce the internal conformational change. In the hypothetical example shown, the transport protein exposes a binding site for the substrate on the portion protruding outside the cell. After binding, the protein undergoes a conformational change that occludes the outer entry path for the substrate and opens an inner path, and the substrate is ejected into the cell's cytoplasm.

The sodium-potassium exchange pump   The *sodium-potassium exchange pump*, present in the membranes of all cells, is one of the most studied active transport mechanisms. The transport protein in this mechanism is *sodium-potassium–activated adenosine triphosphatase (Na, K–ATPase)*. This protein has now been isolated and purified in the laboratory. It consists of two identical subunits, each of which is composed of a small glycoprotein (molecular weight about 50,000) and a large polypeptide (molecular weight about 100,000).

The exact mechanisms by which Na, K–ATPase transports sodium and potassium across the membrane has not yet been determined. The molecule completely spans the membrane and does not rotate or migrate back and forth through the membrane. One hypothesis attempts to explain how the transport of sodium out of the cell is coupled to the transport of potassium into the cell. In this model, the two subunits are both transport molecules, and they are always 180° out of phase with each other (Fig. 2.15). The portion of each subunit located on the inner surface of the cell membrane has a high affinity for sodium, while the portion on the outer surface has a high affinity for potassium. When sodium and potassium bind to the high-affinity binding sites, a conformational change in each subunit, activated by ATP, ejects the bound ions on opposite sides of the membrane. The reciprocal conformational change may occur as shown in the model depicted in Fig. 2.15.

The sodium-potassium pump and electrical characteristics of the cell membrane   Many substances are actively transported into and out of cells, but transport of sodium and potassium is crucial to many cellular processes. The sodium-potassium exchange pump establishes large concentration gradients across the cell membrane for both sodium and potassium. These gradients are the basis for such cellular activities as nerve impulse conduction and initiation of muscle contraction. The sodium and potassium concentration gradients across the membrane establish an *electric potential difference* across the membrane. The sodium-potassium exchange pump must work continuously to maintain these concentration differences because sodium and potassium continuously diffuse down their respective concentration gradients. If such diffusion continues unchecked, intracellular fluid (high in potassium, low in sodium) will become similar in ion concentration to tissue fluid (high in sodium, low in potassium).

Sodium and potassium are both positively charged ions, and within any freely moving solution the number of positively charged ions must exactly equal the number of negatively charged ions. The cellular membrane is more permeable to potassium than to sodium, however, so potassium ions diffuse out faster than sodium ions can diffuse into the cell. Since the cell loses positive charges slightly faster than it gains them, the cell interior is *negatively charged* with respect to the cell exterior (Fig. 2.16). The cell

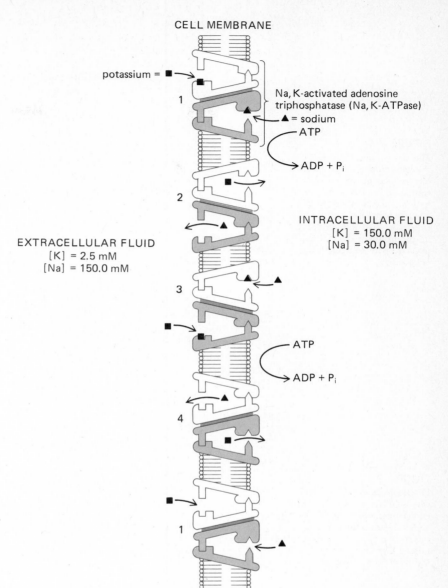

CELL MEMBRANE

potassium = ■

Na, K-activated adenosine
triphosphatase (Na, K-ATPase)
▲ = sodium
ATP

1

ADP + P$_i$

2

INTRACELLULAR FLUID
[K] = 150.0 mM
[Na] = 30.0 mM

EXTRACELLULAR FLUID
[K] = 2.5 mM
[Na] = 150.0 mM

3

ATP

ADP + P$_i$

4

1

FIGURE 2·15

Hypothetical model of the active sodium-potassium exchange pump of the cell membrane. Energy in the form of ATP is expended to produce a reciprocal conformational change in the two subunits of the transport protein (steps 1 to 2 and 3 to 4).

membrane thus establishes a separation of charge that produces an electric potential, or *voltage gradient,* of about 70 to 90 millivolts (mV). The functional significance of this voltage gradient for nerve impulse conduction and muscle contraction will be discussed in later chapters. The important point to be made here is that the diffusion of ions across the cell membrane is influenced not only by the chemical concentration gradient, but also by the electric potential difference. Thus, the passive distribution of an ion across the cell membrane depends on the *electrochemical gradient* for that ion. This phenomenon will be discussed again in Chap. 6.

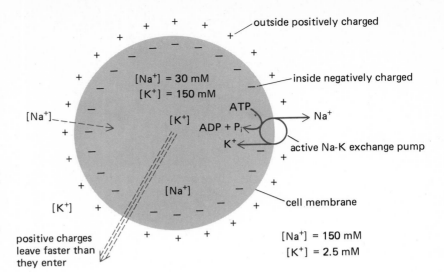

FIGURE 2·16

The sodium-potassium exchange pump maintains a high intracellular potassium concentration and a low intracellular sodium concentration. Because potassium diffuses out of the cell faster than sodium diffuses in, the inside of the cell is negative with respect to the outside.

outside positively charged

$[Na^+] = 30$ mM
$[K^+] = 150$ mM

inside negatively charged

$[Na^+]$

ATP
ADP + P_i $\longrightarrow$ Na$^+$

$[K^+]$

K$^+$

active Na-K exchange pump

$[Na^+]$

cell membrane

$[K^+]$

positive charges leave faster than they enter

$[Na^+] = 150$ mM
$[K^+] = 2.5$ mM

OTHER CELLULAR EXCHANGE MECHANISMS   The plasma membrane may invaginate (turn inward) and pinch off to form intracellular *vesicles* in a process called *endocytosis*. The reverse process, called *exocytosis*, in which an intracellular vesicle merges with the plasma membrane and empties its contents outside the cell, also occurs. Many cell types regularly take up substances dissolved or suspended in fluid by an endocytotic process termed *pinocytosis*, or "cell drinking." Cells may also take up solid, particulate matter by an endocytotic process called *phagocytosis* or "cell eating" (Fig. 2.17).

These processes may be thought of as another type of active transport. Substances contained in vesicles are not technically part of the intracellular

FIGURE 2·17

Phagocytosis. A white blood cell engulfs several bacteria by endocytosis. Once inside, the vesicle bearing the bacteria merges with lysosomes containing digestive enzymes, and the bacteria are destroyed.

PHAGOCYTOSIS

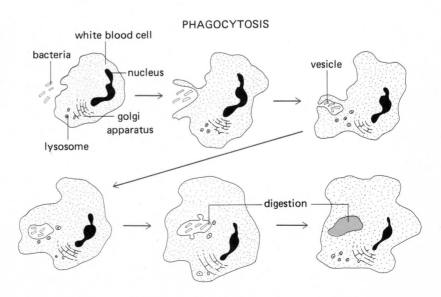

white blood cell
bacteria
nucleus
vesicle
golgi apparatus
lysosome
digestion

fluid because they are surrounded by fragments of the cell membrane. The exact process by which the vesicular contents are absorbed into the cytoplasm is unknown. In many instances, however, the vesicles merge with organelles called *lysosomes,* intracellular vesicles that contain digestive enzymes. When the contents of a vesicle are mixed with the digestive enzymes of a lysosome, much of the material is digested and absorbed into the cytoplasm.

OTHER CELL MEMBRANES

The cell membranes that are part of the cell's interior organelles are basically similar to the plasma membrane. All cell membranes are phospholipid bilayers. They differ from the plasma membrane mainly in the types of proteins they contain.

ENDOPLASMIC RETICULUM   One of the most important intracellular membranes that helps regulate exchange and distribution of materials within the cell is the *endoplasmic reticulum.* This structure forms a winding network of irregular, folded membranes spread throughout the cytoplasm (see Fig. 2.2). Between the doubled-membrane layers of this structure is the *cisternal space* that acts as a conduit for materials that are distributed within the cell. Proteins are synthesizd by the ribosomes on the rough endoplasmic reticulum and are then inserted through the membrane into the cisternal space. The fate of structural proteins—those destined to become part of the cell itself— is unclear at this time. Proteins that are secreted by the cell travel from the rough endoplasmic reticulum to the smooth endoplasmic reticulum (no ribosomes present). A lipid component may be added to such proteins here since the smooth endoplasmic reticulum is believed to be the site of cellular lipid synthesis.

GOLGI APPARATUS   The *golgi apparatus* (sometimes called the *golgi complex*) is a series of shelflike folds of double-membrane layers. The spaces between membranes are somewhat larger than those in the endoplasmic reticulum. The double membranes frequently "bud off," forming membrane vesicles. Various regions of the endoplasmic reticulum and the golgi apparatus are thought to be connected, or intermittently connected, by vesicles, which migrate between the two regions. Secretory products in the endoplasmic reticulum are passed to the golgi apparatus, where the materials may have a carbohydrate moiety added to them. These products are subsequently concentrated and packaged into secretory vesicles with their own bilayer membrane. Upon receipt of a proper stimulus, the secretory products are finally released from the cell by exocytosis in which vesicular and plasma membranes touch and the lipids of both membranes are enzymatically moved aside (Fig. 2.18). Secretory materials are thus discharged, and the vesicular membranes become part of the plasma membrane. Thus, the golgi apparatus is important in the regulation of the packaging and flow of materials through the cell membrane.

NUCLEAR MEMBRANE   The endoplasmic reticulum is also continuous with the outer side of the *nuclear membrane.* The inner layer of the nuclear membrane appears to be separate from the outer layer, but at intervals the two seem to fuse, forming thin, single layers. These thin areas are called pores,

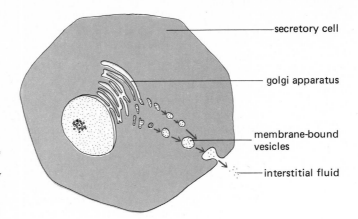

secretory cell

golgi apparatus

membrane-bound
vesicles

interstitial fluid

FIGURE 2·18

Exocytosis. Substances manufactured by the cell are packaged and distributed by the
golgi apparatus. Vesicles containing products "bud off" the
golgi membranes and migrate
to the cell surface, where they
merge with the plasma membrane and empty their contents into the interstitial fluid.

but they are apparently much larger than the pores of the plasma membrane
because larger messenger and transfer RNA molecules pass through them.

CELLULAR ENVIRONMENT AND HOMEOSTASIS

In order for the plasma membrane to exchange necessary materials with the
cellular environment, that environment must be stable. In the human body,
the cellular environment is the *interstitial,* or *intercellular, fluid* that bathes
the cells. Interstitial fluid is a *filtrate* of blood plasma, which is formed as the
blood is filtered across the walls of the smallest blood vessels, the capillaries.
Interstitial fluid is identical to blood plasma except that it contains very
little suspended protein.

The chemical reactions that occur inside cells are like any other chemical
reactions in that they proceed best under certain optimal conditions. For
example, the enzymes that catalyze the formation of ATP work best at
37°C, normal body temperature. Temperature variations of a few degrees
will decrease the efficiency of such enzymes. Other factors that influence
enzyme activity include pH, concentrations of oxygen and carbon dioxide,
substrate concentrations, osmotic pressure, hydrostatic pressure, and electrolyte concentrations. Alterations of these conditions, and others as well, in
the cellular environment may cause serious disruptions in cell function and
even death.

HOMEOSTATIC MECHANISMS   The processes involved in maintaining a stable
cellular environment are called *homeostatic mechanisms. Homeostasis* exists
when the physical and chemical characteristics of the body fluids are
maintained within the narrow limits compatible with proper cell function.
Over 100 years ago Claude Bernard, the noted French physiologist, first
pointed out the importance of the stability of the internal medium. Walter
B. Cannon later coined the term "homeostasis" to denote the state of
dynamic equilibrium in which the physical and chemical characteristics of
the interstitial fluid that bathes the cells are maintained. At the whole-body

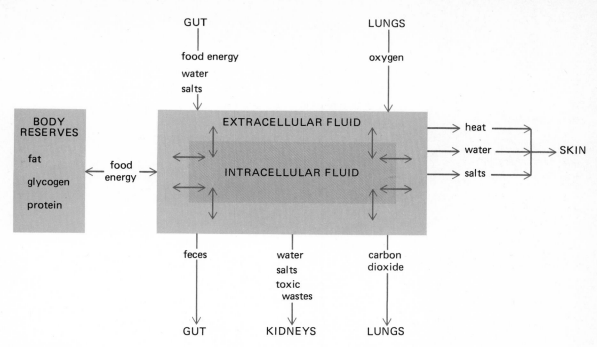

GUT   LUNGS

food energy   oxygen

water

salts

BODY RESERVES

fat

glycogen

protein

food energy

EXTRACELLULAR FLUID

INTRACELLULAR FLUID

heat

water   SKIN

salts

feces   water   carbon dioxide

salts

toxic

wastes

GUT   KIDNEYS   LUNGS

FIGURE 2·19
Maintenance of stable physical and chemical characteristics of the extracellular fluid is essential for proper cell function. The diagram shows the pathways of input and output necessary to balance extracellular fluid composition. Input and output may be intermittent, but because of body reserves and homeostatic mechanisms, extracellular fluid composition remains stable.

level, input and output are intermittent, but the immediate environment of the cells fluctuates very little because of the constant activity of myriad homeostatic mechanisms (Fig. 2.19). For example, food (nutrients, water, salts) is ingested only about three times a day. Thus the absorption of nutrients into the blood is intermittent. Because of this the chemical and nutrient levels in the blood might be expected to fluctuate widely. The composition of the blood, however, and thus that of the interstitial fluid, is quite stable because such organs as the liver and kidneys regulate blood composition continuously. As an illustration of this process, let us look at what happens if excess glucose is ingested. Glucose is rapidly taken up by the liver, as well as some other tissues, and converted to storage compounds, such as glycogen, which can be released gradually to sustain energy demands when the digestive tract is empty. Or, let us consider what happens if excess water or salts are ingested. In this case, the kidneys rapidly excrete excess water and salts in the urine. If the reverse happens—if these substances are in short supply—the kidneys respond by greatly restricting their excretion.

Negative feedback   The basic mechanism by which the body maintains the stability of its internal environment is called *negative feedback*. The principal components of a negative feedback system are *sensors, controllers,* and *effectors.* The sensors send information to the controllers, which interpret the information and direct the actions of the effectors.

To illustrate a negative feedback system in operation, we shall examine one factor involved in the regulation of arterial blood pressure (Fig. 2.20). Arterial blood pressure is, of course, one factor in the delivery of a continuous supply of nutrients to the tissues and in the continuous removal of

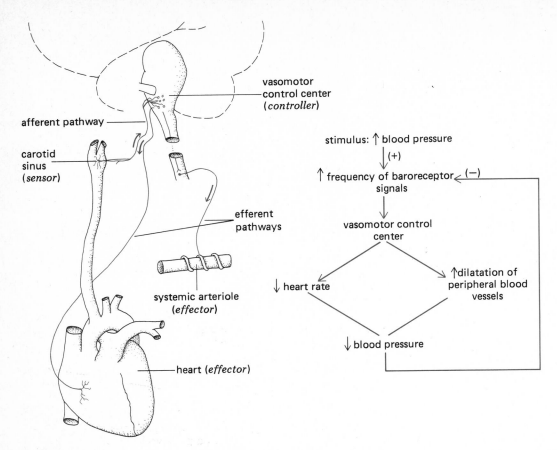

vasomotor control center (*controller*)

afferent pathway

carotid sinus (*sensor*)

efferent pathways

systemic arteriole (*effector*)

heart (*effector*)

stimulus: ↑ blood pressure
(+)
↑ frequency of baroreceptor (−)
signals

vasomotor control center

↓ heart rate

↑dilatation of peripheral blood vessels

↓ blood pressure

FIGURE 2·20

Homeostatic, negative feedback mechanism involved in arterial blood pressure regulation.

cellular wastes. In this example the sensors are a system of baroreceptors located in the walls of the major arteries. An increase in normal blood pressure causes the walls of these arteries to stretch. The baroreceptors detect the stretch as an increase in tension and respond by increasing the number of signals (impulses) they transmit, per unit of time, to the vasomotor control center, located in the brain's medulla. The vasomotor control center directs two main effectors: the *heart* and the smooth muscle in the walls of the smaller arteries called *arterioles*. Increased input from the baroreceptors causes the vasomotor control center to send signals that decrease heart rate and dilate the arterioles. The result is that the heart pumps blood more slowly into dilated vessels (a larger volume) and arterial blood pressure drops. The drop in blood pressure then decreases the tension in the large arteries, reducing stimulation of the baroreceptors. The effectors' response thus has a *negative* effect on the baroreceptors.

The basic principle of the negative feedback mechanism may thus be stated: *In response to stimulation from the sensors, the controller directs the effectors to remove or decrease the intensity of the stimulus that first affected the sensors.*

Another negative feedback mechanism is seen in the relationship between the hormone insulin and glucose (Fig. 2.21). After ingestion of

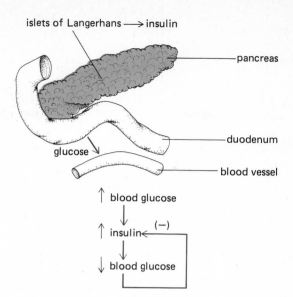

glucose, the sugar is absorbed from the digestive tract, and its concentration in the blood rises. Many tissues depend heavily upon insulin to stimulate the facilitated diffusion of glucose into the cells. Elevation of blood glucose concentration stimulates the release of insulin. As glucose is taken into the cells, the blood glucose level drops and insulin release also falls off.

Positive feedback   *Positive feedback* mechanisms are rare in normal physiological conditions and are most commonly seen in pathological or otherwise abnormal conditions. A good example of a positive feedback mechanism is seen during severe blood loss. When much blood is lost, the heart compensates for decreased blood volume by beating faster and with more force to maintain adequate oxygen delivery to the tissues, including the heart muscle itself. The harder the heart works, the more oxygen the heart muscle consumes. Because there is less blood to carry oxygen, however, less oxygen than normal can be supplied. As less oxygen is supplied to the tissues, low oxygen concentration stimulates the heart to work harder, increasing oxygen consumption and further weakening the heart, thus setting up a vicious cycle (Fig. 2.22). Positive feedback mechanisms are usually very rapid and explosive processes. In the situation described here, the heartbeat rapidly deteriorates to quick, feeble flutters, as oxygen supply fails and demand increases. A few positive feedback systems are seen under normal conditions. For example, the nerve action potential, which is an explosive change in the electrical characteristics of the nerve cell membrane, develops in a self-stimulating, or positive feedback, mechanism. The secretion of hormones during the menstrual cycle is controlled by another normal positive feedback mechanism. For a time during the menstrual cycle, the secretion of the hormones estrogen and follicle-stimulating hormone stimulate each other's secretion. These mechanisms will be discussed in more detail later. You should note for now that positive feedback mechanisms are usually deleterious in nature, and where they exist in normal physiological proc-

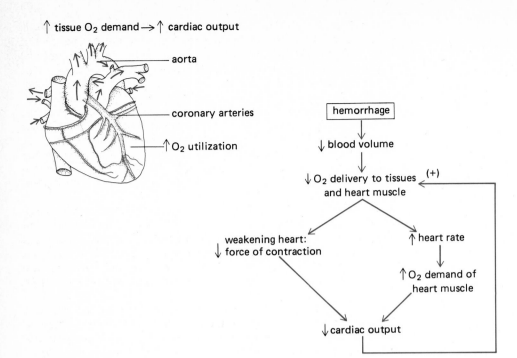

↑ tissue O_2 demand → ↑ cardiac output

aorta

coronary arteries

↑O_2 utilization

hemorrhage

↓ blood volume

↓ O_2 delivery to tissues (+)
and heart muscle

weakening heart:
↓ force of contraction

↑ heart rate

↑O_2 demand of
heart muscle

↓ cardiac output

FIGURE 2·22

Positive feedback mechanism
in severe blood loss. This "vicious cycle" will lead to total
heart failure.

esses, they are coupled with negative feedback mechanisms that control
their effects.

These examples of homeostatic mechanisms have been deliberately
oversimplified for purposes of discussion. Many negative feedback mechanisms are usually involved in the regulation of a characteristic such as blood
pressure or blood glucose concentration. We have selected these simplified
examples to illustrate that the state of homeostasis is not equivalent to
stagnancy. Small fluctuations in cellular environment are inevitable and,
indeed, necessary to stimulate sensory receptors. The body must maintain a
separation from the external environment, but it cannot isolate itself
entirely because it needs nutrients from external sources and it needs to rid
itself of waste products. A state of *dynamic equilibrium* exists, therefore, in
which physiological mechanisms promote a balanced movement of materials in and out of the tissues.

SUMMARY

1 The basic functional unit of all living organisms is the cell. Its major
constituents are water, electrolytes, proteins, lipids, and carbohydrates.
Some cells are highly differentiated and specialized in function, but most
types have certain organelles in common. The functions of cellular
organelles are interrelated and are basically involved with exchange of
nutrients and wastes, cellular respiration, protein synthesis, and cellular
reproduction.
2 All cells exhibit certain basic physiological characteristics necessary for
survival. These include irritability, motility, metabolic processes,
growth, and reproduction.

3 The exchange of materials between the cell and its surroundings is regulated primarily by the plasma membrane. Such membranes consist mainly of phospholipid bilayers and protein molecules. Ectoproteins extend all the way through the phospholipid bilayer, while endoproteins extend only part way through this layer. These proteins have many functions, including transport of substances through the cell membrane.

4 Lipid-soluble substances pass easily through cell membranes, but polar, or ionized, substances pass through with more difficulty. Cell membranes behave as though they contain pores about 8 Å in diameter.

5 Many substances diffuse through cell membranes. Diffusion is the tendency of particles (molecules) to move from an area of high concentration to an area of low concentration because of their continuous, random motion.

6 Osmosis is the movement of water from an area of low particle concentration (high water activity) through a membrane permeable only to water to an area of high particle concentration (low water activity). Osmotic pressure is equal to the hydrostatic pressure that must be imposed upon a solution to prevent any net movement of water into the solution when these liquids are separated by a membrane permeable only to water.

7 Transport proteins move certain substances across cell membranes. This mediated transport may be an active process, which requires input of metabolic energy, or a passive one, which does not require energy input. Facilitated diffusion is a passive process in which a transport protein moves a substance through a cell membrane only if a concentration gradient for that substance exists across the membrane. Active transport moves a substance across a cell membrane against an electrochemical gradient.

8 Substances may also be taken into cells by invaginations (vesicles) of the cell membrane that engulf substances. Pinocytosis and phagocytosis are two such processes. Cells may rid themselves of wastes or secretory products by a reverse process termed exocytosis in which a vesicle containing waste products merges with the plasma membrane, thus expelling its contents outside the cell.

9 The endoplasmic reticulum is a network of intracellular membranes that form folds throughout the cytoplasm. Rough endoplasmic reticulum has ribosomes on its cytoplasmic surfaces and is the site of protein synthesis.

10 Smooth endoplasmic reticulum and the golgi apparatus are involved with lipid synthesis and with manufacture of glycoprotein, respectively.

11 The golgi apparatus regulates the packaging and flow of materials into the cytoplasm or, by exocytosis, through the cell surface.

12 Proper cellular function requires stability of the intracellular environment. This stable state is referred to as homeostasis and is maintained principally by negative feedback mechanisms.

QUESTIONS

1 Why is the cell called the basic functional unit of the body?
2 Describe the structure of the cell membrane.
3 How can a solution be isosmotic and not isotonic?
4 Define osmosis.
5 What is osmotic pressure? How does it develop?

6 Discuss the rationale behind the hypothesis that plasma membranes contain 8-Å pores.

7 What is the energy source for diffusion?

8 Describe the hypothetical sodium-potassium exchange pump.

9 What is meant by the term "electrochemical gradient"?

10 What substances are responsible for the mediated transport of materials through cell membranes? Describe how two such transport mechanisms work.

SUGGESTED READING

Baker, J. J. W. and G. E. Allen: "Matter, Energy, and Life," 3d ed., Addison-Wesley Publishing Company, Inc., Reading, Mass, 1974.

Cohen, A.: "Handbook of Cellular Chemistry," The C. V. Mosby Company, St. Louis, 1975.

Glynn, I. M. and S. J. D. Karlish: The Sodium Pump, *Annu. Rev. Physiol.,* **37:**13–55 (1975).

Lehninger, A. L.: "Biochemistry," Worth Publishers, Inc., New York, 1975.

McLaughlin, S.: Lipid Bilayers as Models of Biological Membranes, *Bio-Science,* **26:**436–444 (1976).

Morris, J. G.: "A Biologist's Physical Chemistry," Addison-Wesley Publishing Company, Inc., Reading, Mass., 1968.

Satir, B.: The Final Steps in Secretion, *Sci. Am.,* **233:**28–38 (1975).

Scholander, P. F.: Osmotic Mechanism and Negative Pressure, *Science,* **156:**67–69 (1967).

Whittam, R. and K. P. Wheeler: Transport across Cell Membranes, *Annu. Rev. Physiol.,* **32:**21–60 (1970).

Rothman, J. E. and J. Lenard: Membrane Asymmetry, *Science,* **195:**743–753 (1977).

3

THE CELL: ACQUIRING, TRANSFORMING, AND USING ENERGY

In this chapter we discuss the means by which cells extract energy from foods and some of the ways that energy is used. The topics covered include:

1 ATP and the principles of oxidation-reduction
2 Cellular respiration, including a discussion of glycolysis, the tricarboxylic acid (TCA) cycle, and the respiratory chain
3 Structure and function of components of the cell nucleus, including replication of DNA and formation of messenger RNA (mRNA)
4 Protein synthesis and the activities of ribosomes
5 The mitotic process

The biological processes we have discussed in Chaps. 1 and 2 are those that provide suitable conditions for the occurrence of intracellular activities. Homeostatic regulation of the extracellular environment and the performance of various intracellular activities also require a continuous expenditure of energy. To do their work, cells must be able to extract chemical potential energy from nutrients and direct this energy toward the performance of specific functions. All homeostatic mechanisms and all energy gathering and expenditure are directed toward cell growth and reproduction. In this chapter we shall examine some cellular mechanisms involved in the extraction and utilization of energy for these purposes.

ACQUIRING ENERGY

As we stated earlier, energy is simply the ability to do work. Some types of work that cells and their organelles do have been discussed in Chap. 2. Energy is required for the transport of many types of substances into and out of cells. In addition, energy is required for the synthesis and breakdown of high-molecular-weight compounds such as enzymes and building materials for cell growth and maintenance. Every cell performs these activities. In addition, special cell types expend energy for more specific tasks, e.g., hormone secretion, contraction, synthesis of extracellular matrix material such as bone, and so forth.

Cellular metabolism is the sum of all cellular processes, and its foundation is the release and utilization of energy from foods. The three basic foodstuffs utilized by the body are carbohydrates, fats, and proteins. These substances are ingested in various forms and converted to simpler compounds in the digestive tract. Carbohydrates are usually ingested as polysaccharides and sugars; the digestive enzymes convert the complex starch and sugar molecules to simple sugars, mainly glucose. Fats are broken down to fatty acids and glycerol in the digestive tract, and proteins are degraded to amino acids. The products of digestion are absorbed through the gut wall into the blood and lymph, fluids which transport nutrients to the cells. Nutrients are taken into the cells primarily by facilitated diffusion and active transport, processes we discussed in Chap. 2.

The cell extracts chemical energy from all three classes of food substances, but we shall concentrate here on carbohydrate metabolism, pointing out its interrelationship with the metabolism of other types of nutrients. As we shall explain, energy stored in nutrient materials cannot be used directly for cellular work. Through step-by-step, enzymatically controlled processes, the cellular machinery incorporates some of the stored chemical potential of foodstuffs into a compound called *adenosine triphosphate* (*ATP*), which acts as a cellular energy store, or energy-carrier mechanism. When energy has been transformed to this form, it is available for almost all types of cellular work.

ADENOSINE TRIPHOSPHATE: ENERGY CURRENCY OF THE CELL The step-by-step dehydrogenation (removal of hydrogen atoms) of foodstuffs is enzymatically controlled so that much of the released energy is never really free, and only about half of it is lost from the cell as heat. Cells meet their energy requirements by converting the remaining energy in nutrient molecules into ATP. ATP is then withdrawn and used by the cell as needed for metabolic processes.

Adenosine, the base of the ATP molecules, is composed of a *purine* called *adenine* (a nitrogenous base) and ribose, a five-carbon sugar (Fig. 3.1). If one phosphoric acid group is added to adenosine, one energy-containing phosphate bond is formed and the structure is called *adenosine monophosphate* (AMP), a relatively low-energy compound. But adenosine monophosphate may also form one or two additional bonds with phosphoric acid molecules, thereby increasing its bound energy tremendously. This process of adding phosphoric acid molecules is called *phosphorylation*. With one additional phosphate bond, AMP becomes *adenosine diphosphate* (ADP). Adding a third phosphate bond to ADP produces *adenosine triphosphate* (ATP).

The electrons of the ATP molecule are organized in such a way that a large amount of energy—8000 calories (cal)—is released when the third phosphate bond is broken. This same amount of energy is added to ADP when the additional phosphate bond is formed. ATP is such a valuable energy currency because of the availability of enzymes that transfer phos-

adenine   +   ribose   ⟶ adenosine

a purine   pentose sugar

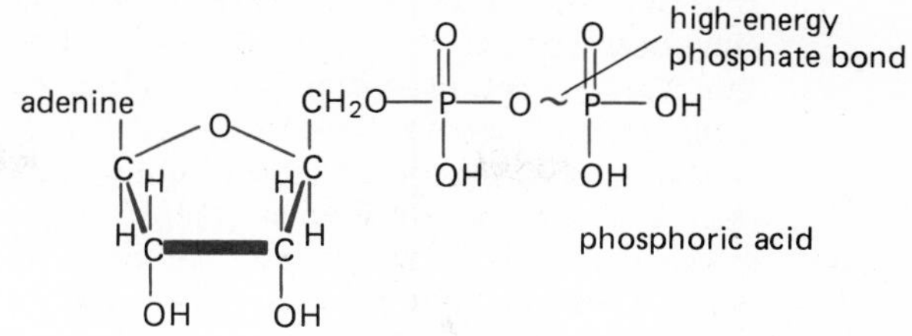

adenosine diphosphate (ADP)

FIGURE 3·1

Adenosine triphosphate is composed of adenine, ribose, and phosphate. Attachment of the two terminal high-energy phosphates increases the bond energy of the molecule tremendously.

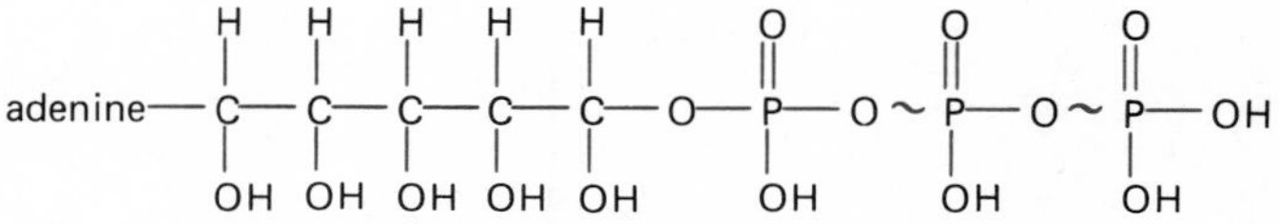

adenosine triphosphate (ATP)

phate groups, including their energy capacity, to other substances. We will see later that energy from ATP plays a key role in muscle contraction, indirectly provides for transmission of the nerve impulse, and takes part in hundreds of cellular processes. ATP energy also goes into the resynthesis of many compounds, including ATP reserves. Thus, ATP is the pivotal energy-exchange mechanism in the cell. Once energy is in the form of ATP, it is immediately available for use in a one-step reaction, or it may be held in reserve. ATP is a common energy source that may be used gradually for efficient cellular metabolism.

Synthesis of ATP   The energy supplied to the cell's ATP energy-carrying mechanism must ultimately come from foods. The oxidation of the breakdown products of carbohydrates, fats, and proteins provides the needed energy for ATP synthesis. Phosphate groups are derivatives of phosphoric acid, H_3PO_4, but in equations representing energy transfer, the phosphate group, PO_4^{3-}, is commonly symbolized P, and the high-energy phosphate bond is written $\sim P$.

Oxidation-reduction   The energy in fuel foods may be transferred to ATP in several ways, but the most effective way is by *oxidation*. Oxidation reactions account for 95 percent of all energy transferred from foodstuffs to the *adenylic acid* (ATP) *system*. Biological oxidation in cells is usually summarized by the equation:

$$C_6H_{12}O_6 + 6O_2 \longrightarrow 6CO_2 + 6H_2O + \text{energy}$$

glucose   oxygen   carbon   water

dioxide

In this reaction, glucose is oxidized and loses electrons. As glucose is broken down, energy is released and trapped in the terminal phosphate bond of ATP. Although it is true that glucose is a fuel, that oxygen is utilized, and that the end products of this reaction are carbon dioxide, water, and energy, this equation is not only oversimplified, but actually inaccurate. It implies that oxygen combines directly with glucose, a situation that never occurs in living cells. When oxygen is added in a single step to a substrate like glucose in a test tube, a large amount of energy is released, mostly in the form of heat. In cells, the oxidation of glucose involves many steps. In each of these steps, hydrogen atoms are removed from the substrate as carbon-to-carbon bonds are broken. A large part of the energy released at these steps is then captured in the energy-rich phosphate bonds of ATP.

The chemical process of oxidation is defined as the loss of electrons, and the process of reduction is defined as the addition of electrons. In iron, for example,

$$\text{Fe}^{2+} \underset{\text{reduction}}{\overset{\text{oxidation}}{\rightleftarrows}} \text{Fe}^{3+} + \text{electron}$$

In living cells, the steps in oxidation (and reduction) not only involve transfer of electrons from one molecule (the electron donor) to another (the electron acceptor), but the simultaneous transfer of the hydrogen ions (protons) formed as the electron donor surrenders an electron. When we recall that the hydrogen atom contains one proton and one electron, we can see why oxidation may also involve the removal of hydrogen (and its electron) and reduction may be obtained by the addition of hydrogen (and its electron). Whether oxidation is accomplished by the release of hydrogen from a compound or by the addition of oxygen, or both, the process always involves the loss of electrons and the release of energy. Oxidations, therefore, are described as *exergonic;* that is, energy is given off as the molecule is *degraded,* or broken down. A familiar example of biological oxidation is the oxidation of succinic acid to fumaric acid.

COOH   COOH
|
CH$_2$ CH
| $\xrightarrow{\text{oxidation}}$ ‖
CH$_2$ CH   + 2H$^+$ + 2 electrons
|
COOH   COOH

    succinic acid   fumaric acid

In this example succinic acid donates two hydrogen atoms and is oxidized (dehydrogenated) by their loss.

Reduction is *endergonic;* that is, it requires the addition of energy. This process involves either the loss of oxygen or the addition of hydrogen, or both. Reduction includes the addition of electrons. Whenever an oxidation reaction occurs, there is a corresponding reduction reaction. Since electrons are never set free but must always be accepted by another atom, whenever an atom or a group of atoms is oxidized, another atom or group of atoms is reduced. It is characteristic of metabolic reactions that two hydrogen ions and two electrons are removed from the substrate molecule (succinic acid in the above reaction) in oxidation. The general form for oxidation of a substrate (S=2H) is:

$$S{=}2H \xrightarrow{\text{oxidation}} S + 2H^+ + 2 \text{ electrons}$$

This simplified equation more accurately portrays the known facts of the step-by-step oxidative process in living tissues than the equation involving glucose and oxygen we cited earlier.

The overall process in which most cellular ATP is formed may be summarized as follows. Enzyme systems present in the cell must:

1 Remove pairs of hydrogen atoms from the substrate
2 Pass electron pairs from substrate to oxygen
3 Pass the pair of H^+ ions to the oxygen molecule that has accepted the electron pair
4 Pass the bond energy of the hydrogen pairs to a high-energy phosphate bond

The final result of this process is a decrease in the number of hydrogen atoms present in the substrate, a decrease of bond energy in the substrate, an increase in energy stored in phosphate bonds, and the formation of water and carbon dioxide.

CELLULAR RESPIRATION Having examined some of the basic principles of energy transfer by cells, we shall now take up the more specific chemical reactions involved in *cellular respiration*. *Respiration* is defined as the oxidation of organic materials through a series of steps in which molecular oxygen serves as the final electron acceptor. This, then, is an *aerobic* process; that is, it requires oxygen. Several *metabolic pathways*, or sequences of enzyme reactions, are found in living cells. All such pathways are interconnected, but we shall approach the subject by first following a molecule of glucose through three major metabolic processes: glycogenesis, glycolysis, and the tricarboxylic acid cycle.

Glycogenesis The process of cellular respiration may begin with *glycogenesis*, a process in which glucose is converted to glycogen (Fig. 3.2). Once inside the cell, a glucose molecule is immediately phosphorylated. ATP is hydrolyzed to ADP, and an ionized phosphate is attached to the glucose molecule. Thus, energy is expended, and glucose 6-phosphate is formed. The attach-

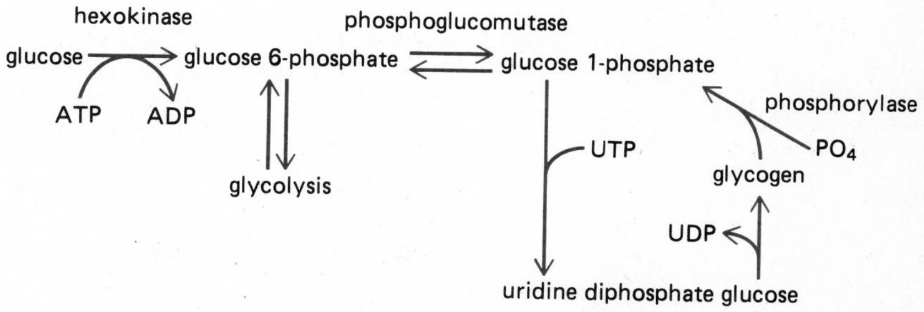

FIGURE 3·2

Glycogenesis, the pathway for the conversion of glucose to glycogen. Glycogenolysis converts glycogen to glucose 1-phosphate.

ment of the ionized phosphate group to the glucose molecule serves two functions. First, the cell membrane is practically impermeable to the phosphorylated glucose, so the sugar is thus trapped inside the cell; and, second, phosphorylation is the initial reaction for the entry of glucose into several metabolic pathways.

Under certain conditions glucose 6-phosphate is not immediately broken down but is stored as glycogen. It should be noted that most excess glucose is converted to and stored as fat; however, in muscle and liver tissue, glycogen is an important storage form of glucose. Glycogenesis involves a rearrangement of glucose 6-phosphate to glucose 1-phosphate. Uridine triphosphate (UTP), another high-energy compound, is used together with glucose 1-phosphate to form uridine diphosphate glucose, which is subsequently converted to glycogen. The reverse process, re-formation of glucose 1-phosphate from glycogen, is called *glycogenolysis*. Glycogen formation and breakdown are closely controlled by several hormones as part of the regulation of blood glucose concentration.

If glucose 6-phosphate is not converted to glycogen, it may be metabolized by either of two basic pathways:

1 *Glycolysis*, which leads to the tricarboxylic acid cycle
2 The *phosphogluconate oxidative pathway*, which is of minor importance in most cells

Glycolysis *Glycolysis* is the process in which glucose is broken down to lactic acid. Glycolysis requires no oxygen, so it is called an *anaerobic process*. It produces relatively little ATP. In glycolysis, a six-carbon glucose molecule is broken down to two 3-carbon lactic acid molecules. The details of the process are shown in Fig. 3.3. Glycolysis involves several important points. First, activation energy is required. Phosphorylation of the intermediate products formed briefly during glucose breakdown requires the expenditure of two ATPs per glucose molecule. Second, almost all glycolysis reactions are enzyme-catalyzed, and most are reversible. (All are directly or indirectly reversible in the liver.) Third, four ATPs per glucose molecule are produced during glycolysis by *substrate phosphorylation;* no molecular oxygen is required for this. In substrate phosphorylation, energy is directly transferred from one of the intermediate breakdown products of glucose to ADP, forming ATP.

Although glycolysis is not the major energy-supplying pathway, circumstances do arise in which the molecular oxygen present is insufficient for oxidative phosphorylation, the major reaction, to occur. A notable example is the situation that develops in cells of very active muscle tissue. During heavy exercise, muscle cells rapidly expend their ATP supply, as well as their energy reserves, which are stored as creatine phosphate. Oxygen cannot be supplied fast enough to maintain ATP production by oxidative phosphorylation, that is, by respiration. Muscle tissue continues heavy activity in spite of this energy deficiency by relying on glycolysis for ATP production.

Glycolysis is efficient in this situation for two reasons. First, under anaerobic conditions, glycolytic intermediates do not build up because lactic acid, the end product, leaves the cells and diffuses into the blood.

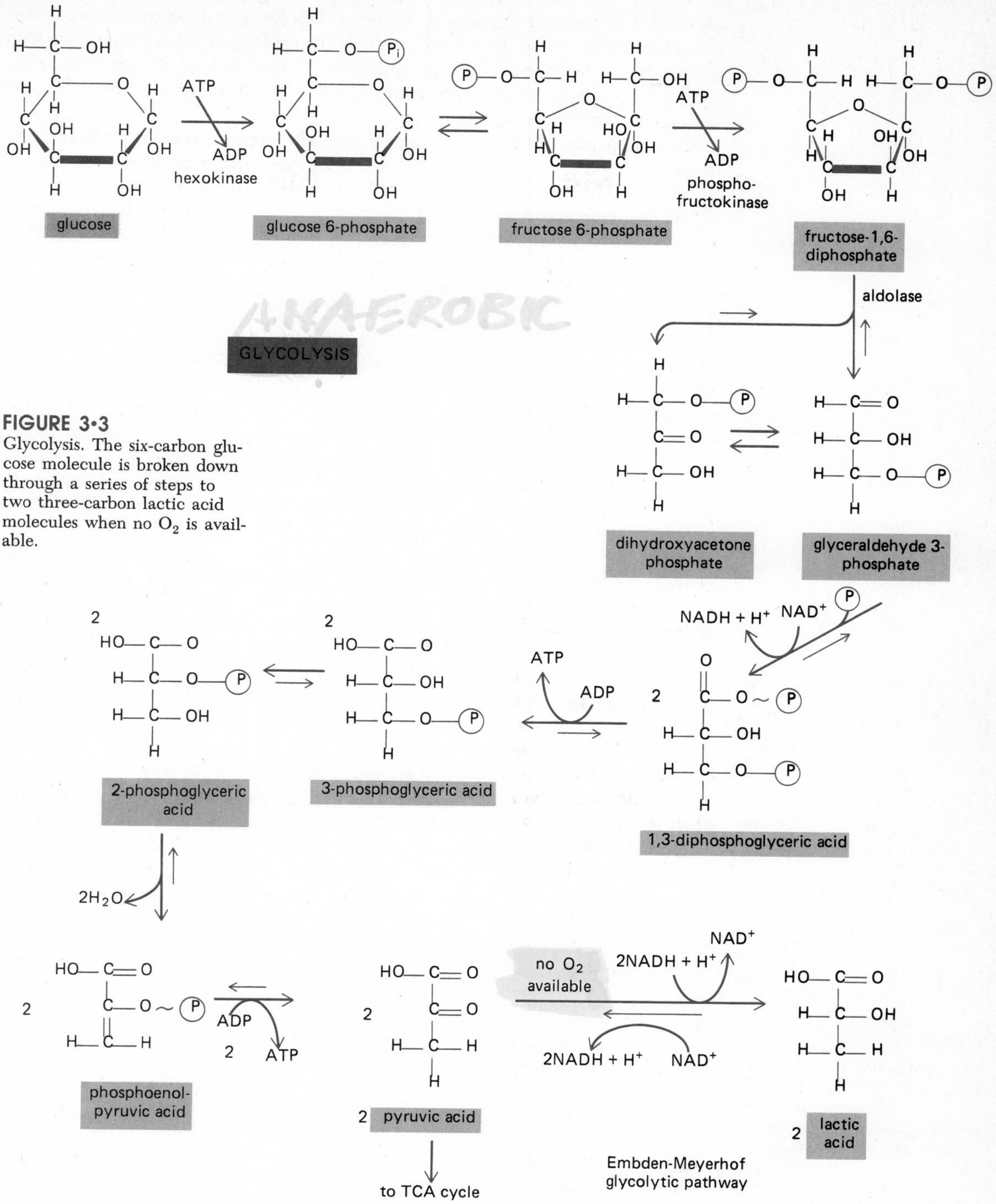

ANAEROBIC

GLYCOLYSIS

FIGURE 3·3

Glycolysis. The six-carbon glucose molecule is broken down through a series of steps to two three-carbon lactic acid molecules when no O_2 is available.

Embden-Meyerhof glycolytic pathway

During lactic acid formation the crucial coenzyme, *nicotinamide adenine dinucleotide* (NAD$^+$), which is reduced by hydrogen atoms freed during glycolysis, is returned to the oxidized state and thus does not limit the reaction. If all the NAD$^+$ were reduced, glycolysis would stop. Under anaerobic conditions, net ATP production by glycolysis is only two ATP per glucose molecule. The hydrogens used to form lactic acid from pyruvic acid during glycolysis are not lost to the system but can be transferred back to the NAD$^+$ when oxygen is once more available. This reaction is summed up in the following equation:

$$2 \text{ pyruvic acid} \xrightleftharpoons[\text{2NADH} + \text{H}^+ \; \text{2NAD}^+]{\text{2NADH} + \text{H}^+ \; \text{2NAD}^+} 2 \text{ lactic acid}$$

Second, although the yield of ATP per glucose molecule is small, ATP can be supplied in considerable quantity because the glycolytic process is very fast in comparison to oxidative pathways. Muscle tissue is specially adapted to operate anaerobically for short periods of time; most organs, such as the brain, however, can survive only a few seconds if deprived of oxygen.

When oxygen is present, lactic acid is not formed because pyruvic acid is immediately degraded in the mitochondria through the process of oxidative phosphorylation.

Mitochondria and cellular respiration   In glycolysis only a small portion of the potential energy of the glucose molecule is released. The cell obtains more than 90 percent of the energy supplied by glucose through the reactions of the *tricarboxylic acid (TCA) cycle*. This is an *aerobic* (requiring oxygen) process terminating in the release of CO_2 and H_2O. The reactions take place only in the *mitochondria*, organelles located in the cytoplasm. Each cell contains between several hundred and several thousand of these minute bodies.

MITOCHONDRIA STRUCTURE   Mitochondria vary in size and shape but are usually elongate, sausage-shaped structures approximately 15,000 Å long and 5000 Å wide. They are visible, when properly stained, under the high power of a light microscope, but their detailed structure is seen only with an electron microscope (Fig. 3.4). Mitochondria have a double-membrane structure composed of an outer membrane that encloses an inner membrane. Each membrane is composed of the typical phospholipid bilayer. An outer compartment between the outer and inner membranes is thought to contain a gel-like liquid. The inner membrane is folded so that shelflike projections

FIGURE 3·4

The mitochondrion, magnification about x400,000 (with an electron microscope). Inner surface particles can be seen on these rat-liver mitochondria.

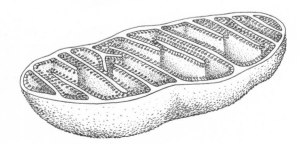

extend toward the interior of the structure. These folds form sacs, called *cristae*, which are fluid-filled (Fig. 3.5). The space enclosed by the inner membrane contains a more dense material called the *matrix*. The outer membrane in most cases appears to be smooth, but the inner surface of the inner membrane is studded with particles 8 to 10 nanometers (nm, one-billionth of a meter) long mounted on stalks. The stalks are 5 nm long and 3 nm wide and are attached to other proteins within the phospholipid bilayer. These particles have been shown to have a molecular weight of 350,000 to 400,000, and they are called ATPases for reasons that will be explained below. Located within the inner membrane are the components of the *respiratory chain*.

MITOCHONDRIA: SEMIAUTONOMOUS ORGANELLES   Good evidence exists that mitochondria are semiautonomous organelles that reproduce themselves and have their own genetic material and enzymatic resources. Mitochondrial DNA (the genetic blueprint material) replicates itself and directs synthesis of several proteins. Such protein synthesis is performed by ribosomes within the mitochondrion itself. Not more than 10 proteins, which contribute 5 to 10 percent of the total mitochondrial protein, are made by mitochondria. Thus, the mitochondria must rely on the cytoplasm for synthesis of the bulk of its proteins. All mitochondrial protein products are found on the mitochondria's inner membrane. The enzymes of the inner membrane are

FIGURE 3·5

a Electron micrograph of a mitochondrion; arrows indicate cristae. x64,000. (*Courtesy of James A. Freeman.*) *b* Stalked subunits projecting from cristae. x192,000. Inset: subunits at higher magnification. Spacing between units is 100 Å. C, christae; ims, inner membrane subunits. [*Courtesy of Donald F. Parsons, Science, 140:985–987 (1963). Copyright, 1963, by the American Association for the Advancement of Science.*]

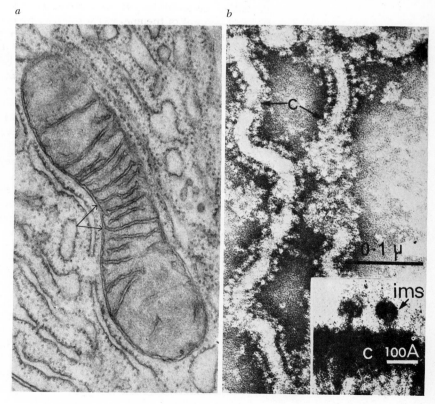

a

b

derived from both mitochondrial and cytoplasmic protein synthesis, and some subunits of critical respiratory enzymes are supplied by the mitochondrion itself.

The tricarboxylic acid cycle (TCA cycle) The enzymes involved in the TCA cycle are located in the inner compartment of the matrix of the mitochondria shown in Fig. 3.5. The inner membrane, which forms the inner compartment, is selectively permeable to a very great degree, and most substances pass in and out of the inner compartment by substrate-specific, mediated transport. The three-carbon pyruvic acid molecules, the end product of aerobic glycolysis, are transported to the interior of the matrix. There they are oxidized and decarboxylated (carbon dioxide removed) to form acetyl coenzyme A, carbon dioxide, and nicotinamide adenine dinucleotide ($NADH + H^+$), the reduced form of the hydrogen carrier. Fatty acids (from fats) and amino acids (from proteins) may also be converted to acetyl coenzyme A and enter the energy cycle at this point. The vitamin thiamine acts as a coenzyme in pyruvic acid metabolism, and the vitamin pantothenic acid forms part of the coenzyme A molecule. Through the loss of one carbon, the pyruvic acid becomes acetic acid, a two-carbon molecule, which is the acetyl portion of acetyl coenzyme A. This two-carbon molecule is next combined with oxaloacetic acid, a four-carbon molecule, to form the six-carbon citric acid molecule, the starting point for the TCA cycle (Fig. 3.6).

In the first reaction of the TCA cycle, the citric acid molecular structure is changed slightly by the loss of water to form *cis*-aconitate and then by the addition of water to form isocitrate. As the TCA cycle progresses, isocitrate is oxidized to oxalosuccinic acid. The enzyme involved in this reaction is isocitrate dehydrogenase, and the coenzyme is NAD^+. At this point, NAD^+ is reduced to $NADH + H^+$. The subsequent decarboxylation of oxalosuccinic acid produces ketoglutaric acid, and CO_2 is given off.

A series of complicated reactions follow. Ketoglutaric acid undergoes oxidative decarboxylation, and CO_2 is released again in the process. Coenzyme A is again involved in this reaction, and the first step produces succinyl coenzyme A. NAD^+ is reduced to $NADH + H^+$ with a subsequent release of energy. Coenzyme A is removed from succinyl CoA by enzymatic action to form succinic acid, and the energy produced is used to form ATP by substrate phosphorylation.

Succinic acid then undergoes dehydrogenation to form fumaric acid. Flavin adenine dinucleotide (FAD), another hydrogen carrier molecule, is reduced in this reaction to form $FADH_2$, with a subsequent production of energy.

Fumaric acid reacts with H_2O, catalyzed by the enzyme fumarase, to form malic acid. Malic acid is then oxidized, and oxaloacetic acid re-forms. This is not the same oxaloacetic acid molecule that started off the TCA cycle, but with its replacement, the cycle is ready to begin again.

In the reactions that take pyruvic acid into the TCA cycle and through it to the regeneration of oxaloacetic acid, events may be summarized as follows: Three decarboxylations form three CO_2 molecules, which are equivalent to the original three-carbon pyruvic acid molecule. Pairs of hydrogens have been removed from intermediates at five points, reducing four NAD^+ to four $NADH + H^+$ and one FAD to $FADH_2$. In the formation

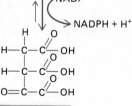
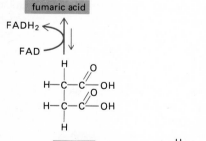

FIGURE 3·6
The TCA cycle takes place in the inner mitochondrial compartment. After the intermediate step of conversion of pyruvic acid to acetyl CoA, the TCA cycle begins with the combination of acetyl CoA with oxaloacetic acid.

of succinic acid, one ATP is produced by substrate phosphorylation. Thus, at the end of the TCA cycle, one glucose molecule (six-carbon) has been completely oxidized to two pyruvic acid molecules (three-carbon) to six CO_2 molecules, with a net formation of three ATP. The student should note that the TCA cycle spins *twice* for each glucose molecule completely metabolized.

Nicotinamide adenine dinucleotide As foodstuffs are dehydrogenated, most released hydrogen atoms are accepted by the coenzyme nicotinamide adenine dinucleotide (NAD^+) or, in some cases, a phosphorylated version ($NADP^+$). Nicotinamide adenine dinucleotide has a positive charge in the oxidized state. It is a hydrogen carrier that accepts two electrons and one proton to become NADH, its reduced form, plus H^+. Electrons from the hydrogen atoms taken up by NAD^+ are passed from this coenzyme along a series of catalysts called the *respiratory chain*, which is located on the inner membrane of the mitochondrion.

The respiratory chain The inner membrane of the mitochondrion is a marvelously complex phospholipid bilayer with at least 60 different types of ecto- and endoproteins. Each of these proteins has a specific function, but, directly or indirectly, they work to form ATP from ADP and inorganic phosphate. The hydrogen atoms obtained from the TCA cycle that takes place in the inner matrix of the mitochondria energize this ATP synthesis.

The precise process by which ATP is synthesized has not been determined; however, the mechanism described below is in accord with almost all experimental data accrued to date. The overall process can be visualized quite simply. The electrons from the hydrogen atoms spun off from the TCA cycle are passed from one inner membrane electron carrier to another (Fig. 3.7). These electrons supply the energy for the active transport of hydrogen ions (protons) from the inner mitochondrial compartment to the outer compartment, thus building a rather steep electrochemical gradient for H^+ across the inner membrane. The protons tend to diffuse back into the inner matrix, and they are allowed to pass through the stalked ATPases within the mitochondrial membrane. The ATPases use the energy of this backdiffusion to phosphorylate ADP, thus forming ATP.

The reaction cycle in the respiratory chain properly begins with reduced NADH + H^+ (Fig. 3.8). This hydrogen carrier delivers its two electrons to flavine mononucleotide (FMN), which, thus reduced, transports two hydrogen protons from the inner mitochondrial compartment to the outer. Reduced FMN is then oxidized as it passes its two electrons, one at a time, to the iron-sulfur proteins (so named because they contain these substances). These pass one electron to each of two ubiquinone molecules (a chemical relative of vitamin K). The reduced ubiquinone molecules each take up an H^+ from the inner compartment. Each ubiquinone molecule then gains an electron from cytochrome b, which recycles electrons between itself and ubiquinone, and again each ubiquinone takes up a proton from the inner compartment. Two reduced ubiquinone molecules, each with two hydrogen atoms, are produced in these reactions. Reduced ubiquinone is highly mobile, and it migrates to the outer surface of the inner membrane where each molecule transfers one electron to cytochrome c_1 and a proton to the outer compartment. Thus, two more H^+ (a total of four so far) are trans-

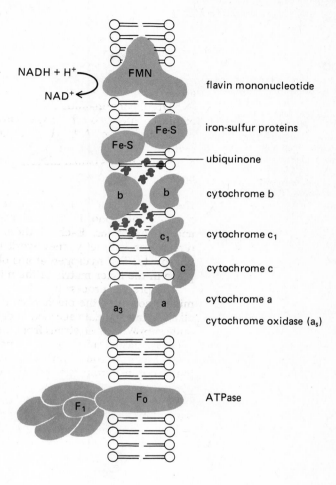

INNER MITOCHONDRIAL
MEMBRANE

INNER COMPARTMENT · · · · · OUTER COMPARTMENT

$NADH + H^+$

NAD^+

FMN — flavin mononucleotide

Fe-S — iron-sulfur proteins

Fe-S

— ubiquinone

b · b — cytochrome b

c_1 — cytochrome c_1

c — cytochrome c

a_3 · a — cytochrome a

cytochrome oxidase (a_s)

F_1 F_0 — ATPase

FIGURE 3·7
Electron transport by the respiratory chain begins with $NADH + H^+$. The energy gained by the transfer of electrons through this sequence is used to pump H^+ from the inner to the outer compartment. The F_1 and F_0-ATPase is the site of ATP formation. The depicted organization of the respiratory chain components is hypothetical.

ported to the outer compartment. The ubiquinone molecules next return their borrowed electrons to cytochrome b and transport another proton to the outer compartment. Now a total of six H^+ have been transported to the outer compartment. Cytochrome b recycles its electrons. Cytochrome c_1 passes its pair of electrons to cytochrome c, which passes them to cytochrome a, which passes them to cytochrome a_3. Cytochrome a_3 passes the electron pair to molecular oxygen in the inner compartment. This chain of reactions produces two hydroxyl ions as follows:

$$\tfrac{1}{2}O_2 + H_2O + 2 \text{ electrons} \longrightarrow 2OH^-$$

Thus, two OH^- are formed for each electron pair transferred. These OH^- combine with H^+ in the inner compartment to form water.

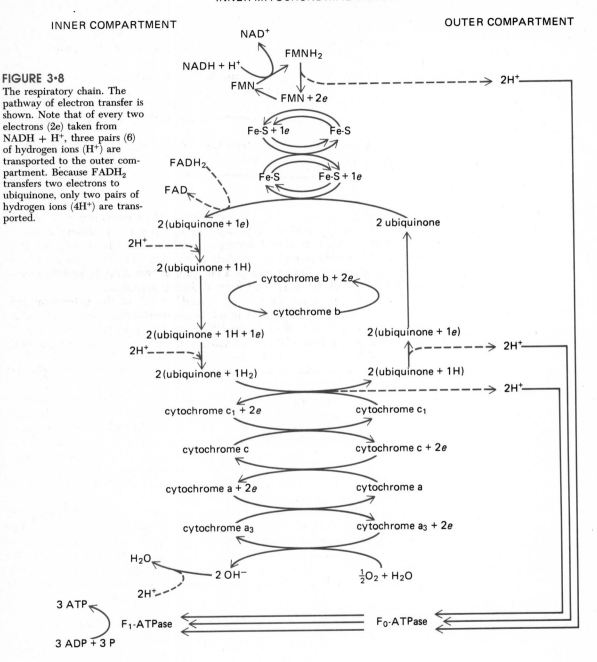

FIGURE 3·8

The respiratory chain. The pathway of electron transfer is shown. Note that of every two electrons (2e) taken from $NADH + H^+$, three pairs (6) of hydrogen ions (H^+) are transported to the outer compartment. Because $FADH_2$ transfers two electrons to ubiquinone, only two pairs of hydrogen ions ($4H^+$) are transported.

The sequence of reactions described here indicates that for every two electrons transferred from NADH + H$^+$ through the respiratory chain, six hydrogen ions are transported from the inner compartment to the outer compartment of the mitochondrion.

The hydrogen ion gradient across the inner membrane supplies the energy for ATP formation. The exact mechanism for ATP manufacture is still unknown. It is known, however, that the process occurs on the large-stalked ATPases that project into the inner compartment. For every two H$^+$ that diffuse down their gradient into the inner compartment, one ATP molecule is formed. The ATP formed is then transported from the mitochondrion to the cytoplasm in exchange for ADP by a mediated transport process.

Cellular respiration, therefore, is largely a function of the mitochondria. As hydrogen electrons are passed along the series of mitochondrial catalysts, ATP is formed. This process requires oxygen and is referred to as *oxidative phosphorylation*.

Efficiency of glucose metabolism The breakdown of glucose thus proceeds in stages. We can approximate the amount of energy captured from glucose as ATP at each stage of the breakdown process. In aerobic glycolysis, four ATP are formed by substrate phosphorylation, but two ATP are expended in the reaction; glycolysis thus produces a net gain of two ATP. In addition, two molecules of NADH + H$^+$ are formed during glycolysis.

The hydrogen atoms carried by NADH + H$^+$ in the cytoplasm are transported into the inner mitochondrial compartment, where they enter the respiratory chain at ubiquinone. Ubiquinone transports only two pairs of H$^+$ and thus only two ATP are formed per electron pair. Therefore, a total of four ATP are formed from the extramitochondrial NADH + H$^+$. For aerobic glycolysis there is a net production of about six ATP per glucose molecule.

Taking two pyruvate molecules from glycolysis to acetyl coenzyme A and then on through the TCA cycle produces two ATP by substrate phosphorylation, as well as eight NADH + H$^+$ and two FADH$_2$. From the eight NADH + H$^+$, 24 ATP are produced (a ratio of three ATP to two electrons). However, the two FADH$_2$ enter the respiratory chain at ubiquinone, so only two ATP are produced for two electrons, providing a maximum of four ATP. The complete oxidation of each glucose molecule thus yields about 36 ATP molecules (Table 3.1).

The complete oxidation of one mole of glucose releases 686,000 cal of energy. Since 8000 cal is stored when ADP is transformed to ATP, the oxidation of one mole of glucose, which produces 36 ATP, results in the storage of 288,000 cal (36 × 8000 cal). The efficiency of energy transfer is around 40 percent (288,000/686,000 × 100 percent). The remaining 60 percent of the energy obtained in the oxidation of glucose is released as heat and is therefore not available for cellular metabolism.

ACQUIRING ENERGY FROM FATS AND PROTEINS Although up until this point we have concerned ourselves with energy production from glucose, fats (or triacylglycerols) and proteins can also be used for energy production. In fact, during the postabsorptive state in which no food is present in the digestive tract, the major source of energy for most body tissues is stored fat.

TABLE 3·1

Production of ATP from glucose during aerobic respiration

Substrate phosphorylation			
Glycolysis	(4 ATP − 2 ATP)	=	2 ATP
TCA cycle		=	2 ATP
Respiratory chain			
Glycolysis	$(2\ NADH + H^+)°$	=	4 ATP
TCA cycle	$(8\ NADH + H^+)$	=	24 ATP
	$(2\ FADH_2)°$	=	4 ATP
Total			36 ATP

° Enters at ubiquinone.

(An important exception to this is the brain.) Triacylglycerols contain more than twice as much potential energy per gram as carbohydrates, and fats do not contain water. Thus, such compounds are a compact, energy-rich food source for the body. Body protein may also be broken down for energy. About 100 g of protein is normally oxidized per day; the minimum dietary requirement for protein is, therefore, about this amount.

Energy from fats Fats are first broken down to fatty acids and glycerol. This reaction is controlled by lipases found in the cytoplasm of most cells. Glycerol (glyceraldehyde 3-phosphate) is one of the metabolic intermediates in glycolysis and thus enters the metabolic pathway at that point. Fatty acids must first undergo transformation by a process called *beta-oxidation* to acetyl coenzyme A, and in this form they subsequently enter the TCA cycle (Fig. 3.9). Their further oxidation is accomplished in the mitochondria.

Energy from proteins Proteins are first broken down to their constituent amino acids. The amino acids are subsequently *deaminated;* that is, the amine (NH_2) part of the molecule is removed. Deamination produces ammonia (NH_3), a highly toxic substance that is detoxified by conversion to urea in the liver. Deamination also produces keto acids, which can enter the glycolytic and TCA pathways at various points (see Fig. 3.9). Since the human body cannot store protein as it does carbohydrate and fat, the small amount of protein used each day means that some structural or enzymatically functional component of the tissues has been broken down. This protein must be replaced by intake of dietary protein. Protein is degraded to supply energy for the brain. The brain normally utilizes blood glucose for its energy source. When the digestive tract is empty, glycogen stores may be rapidly converted to glucose to maintain blood glucose levels. Fats cannot be utilized to produce glucose, although glucose can be converted to acetyl coenzyme A and subsequently to fat. Proteins, however, supply certain amino acids that can undergo *gluconeogeneis;* that is, they can be used to form glucose. Thus, when carbohydrate stores dwindle, proteins must be broken down to supply energy for the brain.

USING ENERGY

The cell expends much of its energy for synthesis of proteinaceous enzymes and structural material, for growth, and for reproduction. We have discussed

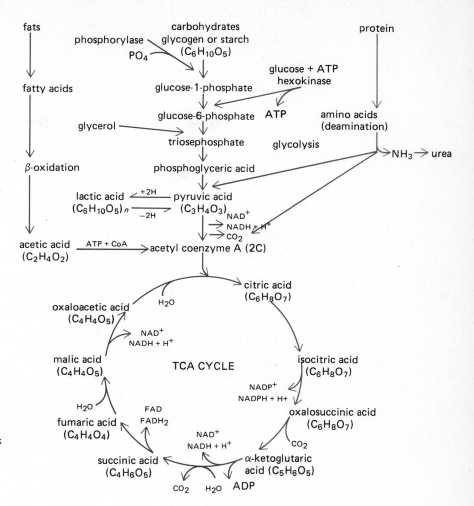

FIGURE 3·9

Outline of glycolysis and the tricarboxylic acid cycle. One molecule of glucose produces two molecules of pyruvic acid; therefore two turns of the cycle are necessary to accomplish the breakdown of one molecule of glucose.

the means by which cells acquire raw materials and energy; we shall now examine the processes within the cell that direct the utilization of these resources. These activities are largely determined by the organelles of the nucleus.

STRUCTURE OF THE NUCLEUS   The nucleus is important because it contains the blueprint information that directs protein synthesis and mitotic activity. The *nucleoplasm* is the fluid protoplasm, or nuclear sap, surrounded by the nuclear membrane. The nucleoplasm contains a small spherical body, the *nucleolus* (little nucleus). The nucleus may contain two or more nucleoli. In fixed and stained preparations the nucleus usually appears as a dark body in the pale cytoplasm and the nucleolus as a still darker body within the nucleoplasm. The nucleolus is composed of dense granules and fibrils of ribonuclear protein and has no surrounding membrane. There are large amounts of ribonucleic acid in the nucleolus, and there is good evidence that ribosomal ribonucleic acid (rRNA) is synthesized there. This same evidence also indicates that the nucleolus plays a part in the biogenesis of cytoplasmic ribosomes. The nucleus contains the chromosomes, which are in an extended state when the cell is not dividing. During cell division, the chromosomes are seen as thickened, compact bodies.

Chromosomes   In the human body, every cell nucleus contains 46 chromosomes with the exception of egg and sperm cells, which contain 23 chromosomes, the *haploid* number. Chromosomes are composed primarily of proteins and deoxyribonucleic acid (DNA), the genetic material. The protein is of two basic types: *histones* and *nonhistone component* proteins (NHC). Both types play a role in regulating the transcription of genetic information. Some RNA is always present in the chromosomes, although the amount varies from cell to cell. Several enzymes, such as DNA and RNA polymerases, are also associated with the chromosomes. Chromosomes contain *genes,* the essential hereditary factors responsible for the development and regulation of the specific metabolic processes performed by each cell. Each gene is composed of a sequence of *nitrogenous bases,* which form part of the DNA molecule.

The genes determine the types of proteins that will be synthesized by the cell. Enzymes are protein in nature. Genes therefore control enzyme synthesis, and through this synthesis, genes direct hundreds and perhaps thousands of specific chemical reactions that play roles in all phases of cellular metabolism.

The DNA molecule   The DNA molecule is the key to life processes. It is a complex molecule able to replicate itself. This nucleic acid is constructed as a double, rather than a single, helix (a spiral, somewhat like the thread of a screw) chain of nucleotides. Each strand of the double helix is the molecular complement of the other. The DNA molecule resembles a ladder twisted into a helix. The sides of the ladder are composed of deoxyribose sugar groups, which alternate with phosphate groups. The two sides of the ladder show reverse polarity with reference to the sugar and phosphate units. The rungs of the ladder, which extend between two sugar groups, contain nitrogenous bases held together by hydrogen bonds. The nitrogenous bases of DNA are *adenine, thymine, guanine,* and *cytosine.* The bases are arranged so that adenine is paired with thymine, and guanine is paired with cytosine (Fig. 3.10). In 1953, while working together at the Cavendish Laboratory in Cambridge, England, Watson and Crick developed the currently accepted model of the basic structure of the DNA molecule. Models, such as that in Fig. 3.11, show only a small part of the DNA chain of nucleotide units, which may be well over 10,000 units in length.

The ratio of adenine to thymine or of guanine to cytosine is constant in any given organism. The sequence of these bases provides the *genetic code* for each organism. For example, the code specifies which amino acids are to be linked into polypeptide chains in the formation of certain proteins.

Adenine and guanine are purines, whereas cytosine, thymine, and uracil are pyrimidines. Adenine is probably more familiar as adenosine, a combination of adenine with the pentose sugar ribose. Adenosine is a nucleo*side.* If a phosphate group is added to the sugar part of adenosine, the RNA nucleo*tide* adenosine monophosphate (AMP), also called adenylic acid, is formed. Pyrimidines and purines contain both carbon and nitrogen in their

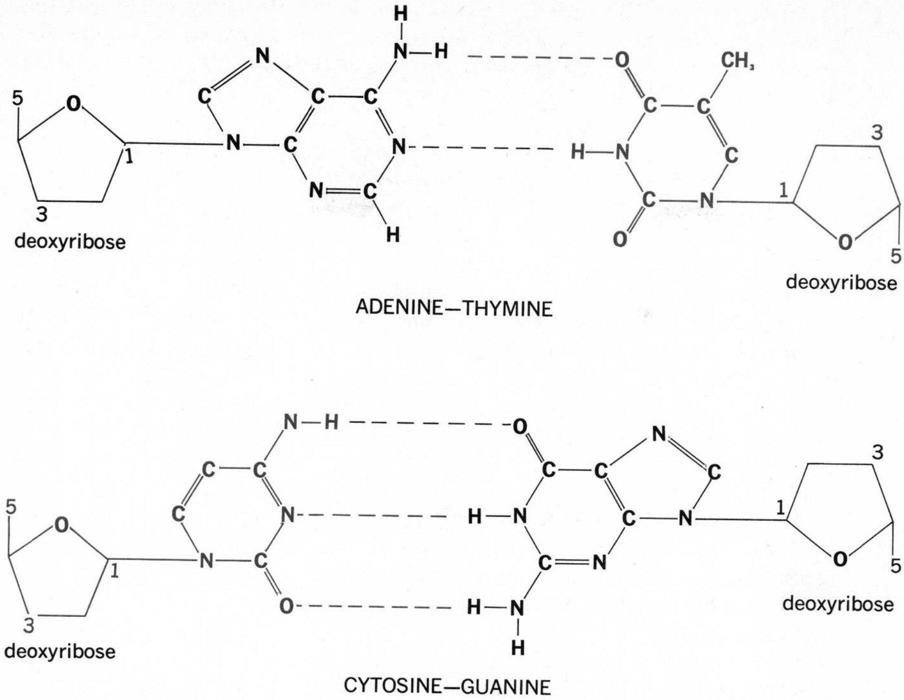

FIGURE 3·10
Structural formulas for adenine-thymine and cytosine-guanine.

ring structure. Pyrimidine structure is represented by a six-sided ring with four carbon atoms and two nitrogen atoms. Purine structure retains the pyrimidine ring but includes a side ring with one carbon atom and two nitrogen atoms (see Fig. 3.10). These bases combine with phosphorylated pentose sugars (Fig. 3.12) to form more complex molecules called *nucleotides*. Polymerization of nucleotides produces nucleic acids such as DNA and RNA.

Replication of DNA (duplication of the genetic material) occurs just preceding cell division in all cells capable of dividing. The genetic code, then, is duplicated at each cell division, and the original DNA molecule acts as a template for the new molecule. The 46 chromosomes in the human cell comprise 23 pairs. For a gene located at a certain point on one chromosome, there exists another gene resembling it at the same point on the other chromosome of the pair. During fertilization the male parent contributes 23 chromosomes in the sperm cell and the female parent contributes 23 chromosomes in the ovum. These chromosomes pair up, and in every mitotic cell division thereafter each chromosome divides (see Fig. 3.17).

Functions of DNA and RNA If the DNA molecule is to bear the hereditary characteristics, as contained in the gene, it must perform at least two functions:

1 It must be able to replicate itself, maintaining the integrity of the genetic code.
2 It must provide a code that directs the production of proteins.

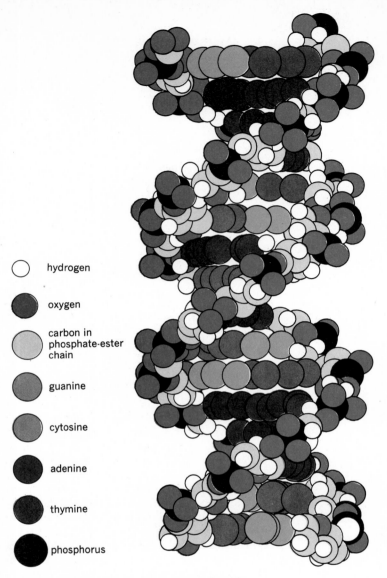

○ hydrogen

● oxygen

◐ carbon in phosphate-ester chain

◐ guanine

◐ cytosine

● adenine

● thymine

● phosphorus

FIGURE 3·11
Molecular model of DNA.
[*From L. D. Hamilton, A Bulletin of Cancer Progress, CA,* **5**:*159* (1955).]

a nitrogen base + a pentose sugar → a nucleoside
pyrimidines or purines + pentose + phosphate → nucleotides

hundreds or thousands of nucleotides $\xrightarrow{\text{polymerization}}$ nucleic acids

DNA does replicate itself precisely, although replication does not begin at one end of the threadlike chromosome and proceed to the other end. Instead, segments of each chromosome replicate according to a definite program. Many important facets of the DNA replication process in the cells of higher animals remain to be determined, but it is known that DNA replication is a complicated process that requires the coordinated activity of

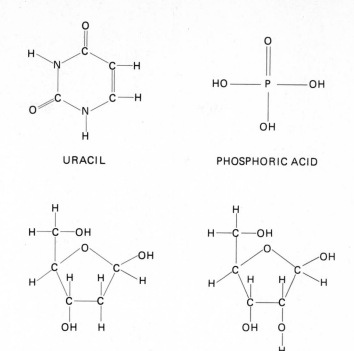

FIGURE 3·12

Structural formulas of uracil, phosphoric acid, deoxyribose, and ribose.

URACIL

PHOSPHORIC ACID

DEOXYRIBOSE

RIBOSE

many enzymes and cofactors. Replication apparently begins at many *initiation points* on the DNA double helix; these points are recognized by specific *initiation proteins*. Other proteins, called *unwinding proteins*, bind the DNA at these points and separate and uncoil the two strands.

Phosphorylated mononucleotides—the high-energy compounds, ATP (adenosine triphosphate), GTP (guanosine triphosphate), CTP (cytosine triphosphate), and TTP (thymine triphosphate)—align themselves on the DNA template strands according to the pattern of base pairing described above. At the initiation points an enzyme called *RNA polymerase*, so called because it promotes the linkage of one nucleotide to the next, is directed by the DNA to generate a short *priming* strand of RNA. DNA polymerization begins behind the short RNA primer. The energy for polymerization is supplied by the high-energy phosphate bonds.

DNA is made in short strands of 100 to 150 bases. The RNA primer is excised, and the enzyme *DNA ligase* joins the short DNA fragments. Gradual uncoiling and rotation of the chromosome results in the replication of a DNA molecule exactly like the original (Fig. 3.13). Note that adenine and thymine form two hydrogen bonds between them, whereas guanine and cytosine are held together by three hydrogen bonds.

In addition to replicating itself, the DNA molecule also acts as a template for the formation of ribonucleic acid, thus controlling the synthesis of proteins. Just as the enzyme DNA polymerase is essential to catalyze the polymerization of DNA, so the enzyme *RNA polymerase* is necessary to catalyze the synthesis of RNA from its DNA template. The four essential

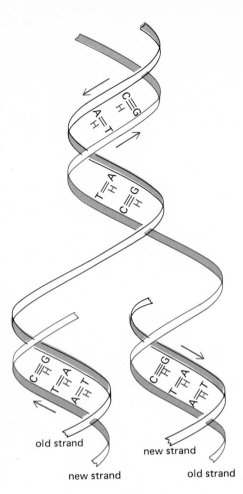

old strand

new strand

new strand

old strand

mononucleotides in RNA are ATP, UTP, CTP, and GTP. RNA is a single strand; it is formed as a complementary copy of only one of the two DNA strands. Probably only a part of a DNA strand is coded at any one time. The DNA molecule is remarkably stable because of constant repair work by special enzymes, but RNA breaks down in a short time, and new RNA is synthesized. The RNA molecule resembles a hairpin loop, with the single strand twisted on itself, leaving some regions unpaired. The base pairing is believed to be incomplete, probably because complementary bases are lacking on an opposing portion of the strand. The single strand consists of repeating units of the pentose sugar ribose and phosphate. One of the four RNA bases is attached to each sugar unit. Three of the bases are identical with those found in DNA. The fourth base is *uracil*, which replaces the thymine of the DNA molecule. It is closely related chemically to thymine, which is illustrated in Fig. 3.12.

Approximately 75 percent of the cytoplasmic RNA in animal cells is contained in the ribosomes; this type is called *ribosomal RNA*. Some RNA is found also in the nucleus. Nuclear RNA constitutes about 15 percent of the

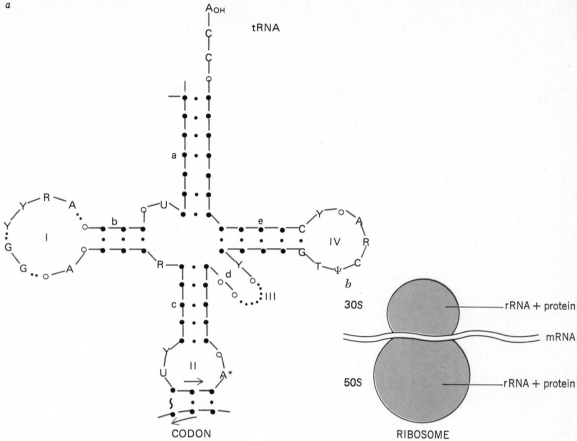

tRNA

A_OH

30S
rRNA + protein

mRNA

50S
rRNA + protein

RIBOSOME

FIGURE 3·14

a Standard orientation of transfer RNA in the "cloverleaf" form, as proposed by B. F. C. Clark, W. E. Cohn, and T. H. Jukes. I, II, III, and IV are "loops" or unpaired base regions; a, b, c, d, and e represent helical (base-paired) regions or "arms." The upper arrow indicates position of the anticodon. Loop III is not always present. [*Courtesy of T. J. Jukes and Lila Gatlin, Recent Studies Concerning the Coding Mechanism, Prog. Nucleic Acid Res. Mol. Biol. 11:303–350 (1971).*] *b* A ribosome. 30S and 50S are sedimentation constants for eukaryotic cells with 80S ribosomes.

total cellular RNA in animal cells. It is found principally in the chromosomes and in the nucleolus. There is also a small amount in the nuclear sap. The amount of nuclear RNA varies in different kinds of cells, so the amount of RNA is not directly related to the amount of cellular DNA, which always remains constant.

Three kinds of RNA are commonly recognized; these are ribosomal RNA (rRNA), messenger RNA (mRNA), and transfer RNA (tRNA) (Fig. 3.14).

RNA: Ribosomes and ribosomal RNA In Chap. 2 we noted that the endoplasmic reticulum was studded with ribosomes on its cytoplasmic surfaces. Ribosomes are spheroid bodies containing RNA and protein. As we have indicated, the synthesis of ribosomal RNA occurs in the nucleolus. The structural RNA of the ribosome apparently takes no part in the genetic coding involved in protein synthesis. Ribosomes are composed of two subunits, which are usually described in Svedberg units. A Svedberg unit is based on the sedimentation rate of particles dispersed in the ultracentrifuge. The sedimentation constant of bacterial ribosomes is 70S. For the two subunits, including their associated protein, the sedimentation constants are 30S and

50S. The 50S subunit is the largest, with a molecular weight of 1.6 million. It contains one each of 34 different proteins and two molecules of RNA. The 30S subunit contains up to 21 proteins and one RNA molecule. Its molecular weight is about 1 million.

Protein synthesis takes place on groups of ribosomes called *polyribosomes* or *polysomes*. Electron microscopy reveals that a fine filament connects ribosomes; this filament has been shown to be mRNA. There seems to be some evidence of an opening between the subunits of ribosomes. This has been interpreted as a "tracking groove" through which the mRNA passes. Such a space may also permit tRNA to reach mRNA at this point. The 50S and 30S ribosomal subunits come together for the synthesis of proteins, but at other times they exist as individual subunits. The main function of the ribosome is to coordinate the translation of the coded information contained in mRNA.

Messenger RNA Messenger RNA, the second type of RNA, is nuclear RNA that travels from the nucleus into the cytoplasm, carrying information necessary for the synthesis of proteins. Presumably it moves out through nuclear pores. When it reaches the cytoplasm, mRNA attaches itself to ribosomes. Each mRNA, which may contain hundreds of nucleotides, is formed on a portion of a DNA strand and represents the transcription of the genetic information contained in that portion of DNA. The sequence of bases in the mRNA molecule specifies the arrangement of amino acids in the protein whose synthesis will be directed by that mRNA. Each set of three bases, a unit called a *codon*, specifies a particular amino acid or serves some other coding function.

The mRNA in some bacteria has a very rapid turnover, in a matter of minutes. This may not be true for the cells of higher animals, but the attachment of mRNA to ribosomes and the subsequent formation of polyribosomes is quite rapid.

Transfer RNA Transfer RNA is the third form of RNA. It is a low-molecular-weight molecule, containing around 80 nucleotides. The small tRNA molecule is composed of a chain of nucleotides arranged in a cloverleaf pattern with some base pairings between the usual RNA four bases (Table 3.2). In addition, tRNA has several unusual bases such as inosinic acid (which can substitute for guanine in base pairing), pseudouracil, methylguanine, and others.

Certain structural characteristics are common to all tRNA molecules. One end of the nucleotide chain always ends with the bases CCA. This is the 3′ end. The base guanine is always found at the other free end, the 5′ end. The loop opposite the free ends contains the *anticodon*, which consists of three unpaired bases complementary to the three bases that specify a particular amino acid on a segment of mRNA. Referring to Fig. 3.14, we see that the solid circles represent bases in the helical regions of the tRNA molecules. They are usually unpaired. R and Y represent purine-pyrimidine nucleosides; T represents ribothymidine; psi, pseudouridine; °, a modified base. The base sequence that specifies all amino acids is known. The first sequence to be unraveled was that of alanine, in research performed by Holley and his associates. The anticodon for alanine was found to be IGC, which matches up with the base sequence CCG on mRNA.

TABLE 3·2
Comparison of bases and sugars in DNA and RNA

Bases	DNA	RNA
	CG	CG
	AT	AU

Sugars	Deoxyribose	Ribose
	CG	CG
	GC	GC
	AT	AU
	TA	UA

NOTE: A, adenine; C, cytosine; G, guanine; T, thymine; U, uracil.

The function of tRNA is to carry activated amino acids, which are attached to their 3' end, to a complementarily coded area on mRNA. The amino acids are first activated by ATP to form a higher-energy compound. The activating enzyme is amino acyl synthetase, and the energy released by the breakdown of ATP enables the amino acid and tRNA to form a chemical bond. The synthetase enzymes are specific for each amino acid. The synthetases recognize both the conformation of a certain amino acid and the correct tRNA (Fig. 3.15).

RNA AND PROTEIN SYNTHESIS   Protein synthesis begins when mRNA from the nucleus forms a complex with the ribosome 30S subunit, several substances called initiation factors, and a tRNA that carries N-formylmethionine, an initiating amino acid (f-Met-tRNA). This initiating complex combines with the ribosome 50S subunit to form the 70S ribosome. After the tRNA attaches to complementary bases on a strand of mRNA, the amino acids carried there by the tRNAs form peptide bonds with other amino acids, and a polypeptide chain of amino acids begins to grow.

Having deposited their amino acids in the correct sequence, the tRNA molecules move away, but the polypeptide chain is always attached to a tRNA until the complete sequence of mRNA has been read. The ribosome moves along the mRNA while tRNA molecules assemble amino acids according to the genetic code read from the mRNA molecule. The formation of polypeptide chains is remarkably rapid, requiring only a few seconds, or perhaps a minute for longer chains. Releasing factors break the protein

FIGURE 3·15

The functions of messenger and transfer RNA. The genetic code of DNA is transcribed to mRNA, which then combines with protein and moves into the cytoplasm, becoming associated with polyribosomes. Amino acids activated by specific synthetases and ATP are carried to coded areas on mRNA by transfer RNA. Anticodons carried by transfer RNA recognize their complementary codons on mRNA, and the amino acids carried by the tRNAs form a chain of polypeptides. It is assumed that tRNA units are released after amino acids are located on mRNA and polypeptide chains are formed.

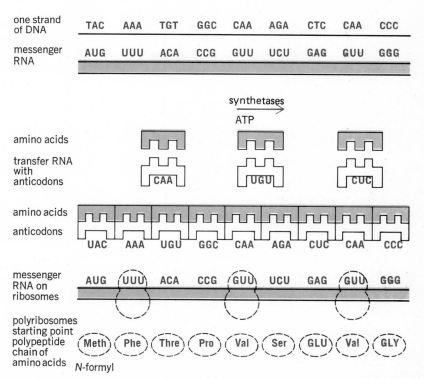

chain from the terminal tRNA when synthesis is completed. The tRNA, mRNA, and 30S and 50S subunits then all separate.

The "letters" of the genetic code The arrangement of base pairs in the DNA molecule determines the arrangement of amino acids in proteins whose synthesis is guided by that DNA strand. We have seen how DNA acts as a template for the transcription of triplet complementary codons on an mRNA strand. Since only four bases must designate, or "code for," the 20 amino acids involved in protein synthesis, more than one base must be used to code for an amino acid. If the code used only two bases, only 16 different combinations would be possible. The minimum number of bases in the code, therefore, must be *three*. Using three bases, 64 combinations are possible, far more than the number needed, but it has been found that in some cases as many as four different sets of bases code for the same amino acid. A code of this type is said to be "degenerate."

In 1961 Nirenberg and Matthei prepared the first synthetic "messenger" RNA, which contained only the base uracil. When this synthetic polymer of uridylic acid (poly U) was added to an active cell-free mixture derived from *Escherichia coli*, a colon bacillus, it was found that only the amino acid phenylalanine was produced. It seemed evident that the RNA code for phenylalanine was UUU. It was subsequently discovered, however, that UUC also codes for phenylalanine. In the intervening years, various experimental methods have been used to work out the codons for all 20 amino acids (Table 3.3).

A few triplet-base combinations do not code for amino acids; these were originally termed "nonsense" codons. It now appears that UAA, UGA, and UAG are *terminator codons* that indicate the end of synthesis for certain polypeptide chains. Codons that initiate synthesis have been more difficult to determine, but AUG has been identified as the *starting codon* in protein synthesis in bacteria. It is likely that these codons will apply to cells of other organisms as well.

TABLE 3·3
The amino acid code

UUU Phenylalanine	UCU Serine	UAU Tyrosine	UGU Cysteine
UUC Phenylalanine	UCC Serine	UAC Tyrosine	UGC Cysteine
UUA Leucine	UCA Serine	UAA Chain Termn.	UGA Chain Termn.
UUG Leucine	UCG Serine	UAG Chain Termn.	UGG Tryptophan
CUU Leucine	CCU Proline	CAU Histidine	CGU Arginine
CUC Leucine	CCC Proline	CAC Histidine	CGC Arginine
CUA Leucine	CCA Proline	CAA Glutamine	CGA Arginine
CUG Leucine	CCG Proline	CAG Glutamine	CGG Arginine
AUU Isoleucine	ACU Threonine	AAU Asparagine	AGU Serine
AUC Isoleucine	ACC Threonine	AAC Asparagine	AGC Serine
AUA Isoleucine	ACA Threonine	AAA Lysine	AGA Arginine
AUG Methionine	ACG Threonine	AAG Lysine	AGG Arginine
GUU Valine	GCU Alanine	GAU Aspartic acid	GGU Glycine
GUC Valine	GCC Alanine	GAC Aspartic acid	GGC Glycine
GUA Valine	GCA Alanine	GAA Glutamic acid	GGA Glycine
GUG Valine	GCG Alanine	GAG Glutamic acid	GGG Glycine

Cellular differentiation Since every cell in the body of a given individual contains the same genetic material, how is it possible for so many types of cells to develop? Or, to rephrase the question, how does cellular differentiation take place? Experiments in which the chromatin material of the nucleus has been reconstituted in the test tube have shown that factors in the nonhistone component proteins (NHC) of the chromosomes determine the availability of DNA as the template for mRNA transcription during the cell development. Several experimenters have shown that tissue-specific mRNAs are dependent on the NHC proteins associated with the chromosomes. For example, liver cells synthesize globin proteins, and thus necessarily must have the ability to produce mRNA for globin. Brain cells do not normally produce globin proteins. If the DNA material belonging to brain cells is mixed with the NHC proteins of liver, however, brain chromosomes can then synthesize globin. On the other hand, if liver chromosomes are mixed with brain NHC proteins, the ability to synthesize globin mRNA is lost. From evidence such as this it can by hypothesized that variation in the expression of the genetic code in different cell types is due to the masking, or covering up, by NHC protein of all codes but those necessary in a certain cell type.

DNA operator-repressor system Which NHC proteins are involved with the masking of genes and how such proteins work have not been determined. A system in the virus bacteriophage lambda may be used to describe how such a masking system might work in higher organisms. Remember that gene expression depends upon the transcription of the DNA code onto an mRNA molecule. This transcription is accomplished by the enzyme RNA polymerase. Certain regions of a gene are called *operators* because they signal the RNA polymerase where to begin transcription of the DNA code for a particular protein. *Terminator regions* of a gene signal the end of the particular genetic sequence.

Other genes, termed *repressors,* code for the synthesis of an mRNA involved in the manufacture of a *repressor* protein, presumably one of the NHC proteins (Fig. 3.16*a*). When repressor genes are active, the repressor proteins bind to the operator region of a particular sequence of DNA and prevent RNA polymerase from synthesizing RNA from that portion of the chromosome. Thus, that entire gene is "silenced." Transcription of that genetic sequence resumes only when the repressor is inactivated.

Inducers are substances that induce the cell to synthesize certain types of enzymes. For example, hormones sometimes elicit changes in cellular metabolism by inducing the synthesis of particular enzymes. Inducers are thought to act by *inactivating* the repressor proteins (Fig. 3.16*b*). *Suppressors* can cause production of certain types of enzymes to stop. Suppressors are thought to act by reactivating the inactivated repressor protein (Fig. 3.16*c*).

THE CELL CYCLE AND MITOTIC CELL DIVISION Another major use of cellular energy is seen in cell growth and division. During the growth and development of an individual, all cells go through a cycle. The cycle is broken into four main phases, G_1, S, G_2, and M. Immediately after a cell is produced by the parent cell, it goes through *phase G_1,* a period of growth corresponding to the time between the formation of the cell and the beginning of DNA

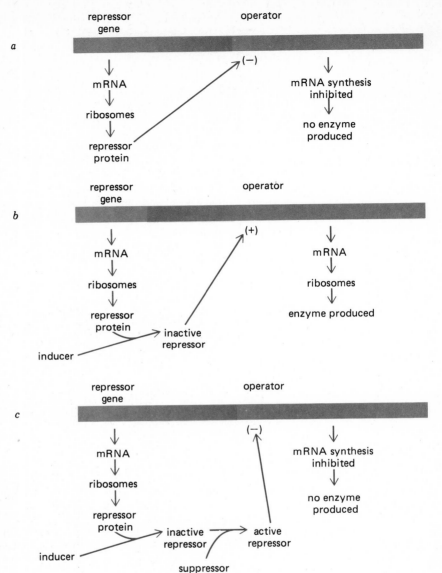

FIGURE 3·16

DNA operator-repressor system. *a* Repressor genes cause the production of repressor proteins that inhibit the operator region of a gene, thus preventing synthesis of that enzyme. *b* Inducers and *c* suppressors affect this process in different ways.

replication. This is usually the longest phase in the cell cycle. When DNA replication begins, the cell is said to be in the synthesis or S *phase*. Even though DNA replication is complete at the end of the S phase, the cell is still not ready to divide. The period between the S phase and the beginning of cell division is the G_2 *phase*, or second growth phase. This is another period of growth for the cell during which further preparations are made for the beginning of the *M phase* or mitosis, the period of cell division. The amount of time spent in each phase varies in different cell types. Most of the variability, however, is in the G_1 phase, which may take hours, days, or weeks. The S phase in most mammalian cells lasts 6 to 8 h and the G_2 phase lasts 3 to 5 h.

Mitosis The process of mitosis involves a series of changes in the cell's nucleus that result in the accurate division of the cell's chromosome complement. Actually the process involves a continuous series of changes, but for convenience, it is usually divided into four phases: prophase, metaphase, anaphase, and telophase. The stages of a typical mitotic cell division are shown in Figs. 3.17 and 3.18.

Interphase includes phases G_1, S, and G_2 of the cell cycle, those phases during which the cell is not dividing. Cells that show no outward signs of division are frequently called resting cells, although this term is not very appropriate since it implies that the cell's metabolism is inactive. Actually, the cell is very active during this period. DNA is replicated and tremendous cell growth occurs during interphase. Rapidly growing tissues contain many cells in various stages of mitotic division, but in older tissues most cells may be in a more or less permanent interphase because they never enter the S phase. Such cells no longer participate in cell cycles.

It has been estimated that in the adult human most somatic cells pass through 50 mitotic divisions during the growth of the individual. Some cells (for example, neurons) do not continue to divide once the system is mature. Enzymes control the cell cycle, and the cellular production of any enzyme is, of course, controlled by the genes that code for that enzyme. Although not yet proven in higher organisms, apparently specific genes control the steps in the cell cycle. It is not known what initiates the S phase, but it is known that the cell receives a signal of some sort before DNA replication begins. Initiation of the M phase, that is, condensation of the chromosomes in preparation for cell division, is also dependent upon some sort of signal, the nature of which is as yet undetermined.

Stages of mitosis PROPHASE The appearance of the cell changes as it enters the early stages of mitotic division. Considerable activity is seen in the cytoplasm around the *centrioles,* where lines of flow radiate outward. These lines form the *aster.* The centrioles are visible under the high power of a light microscope as a dense area in the cytoplasm, commonly referred to as the *centrosome* (central body). The electron microscope, however, has revealed the rather unusual structure of the centriole. Each centriole is paired, and the two parts usually lie at right angles to each other; that is, in electron micrographs one centriole usually appears in cross section and the other in longitudinal section or position (Fig. 3.19). Each consists of nine groups of cylindrical bodies arranged in a circle, each group commonly containing two or three cylindrical fibers. Centrioles are approximately 0.5 μm long and 0.15 μm in diameter. They are self-replicating bodies with no limiting membrane. They appear to play an important part in the formation of the aster or spindle during mitotic division.

In the late interphase or early prophase, the pairs of centrioles start to move away from each other, and filaments of the spindle form between them. Each chromosome, still in the extended state, has formed two strands. By late prophase, when the centrioles have migrated to opposite poles, a fully developed spindle lies between them. The spindle appears to be a fibrous structure, protein in nature, and the filaments are often described as

FIGURE 3·17

Stages of mitosis in animal cells. [*Courtesy of Daniel Mazia, How Cells Divide, Sci. Am.,* **205**:*101–120 (1961). Copyright, 1961, by Scientific American, Inc. (All rights reserved.)*]

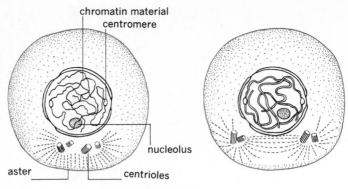

chromatin material
centromere

nucleolus

aster   centrioles

INTERPHASE

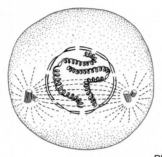 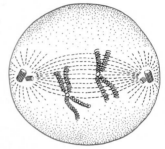

PROPHASE

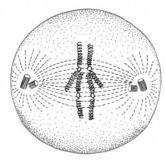 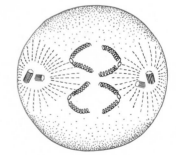

METAPHASE   ANAPHASE

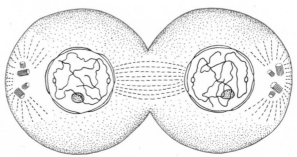

TELOPHASE

82 THE CELL: ACQUIRING, TRANSFORMING, AND USING ENERGY

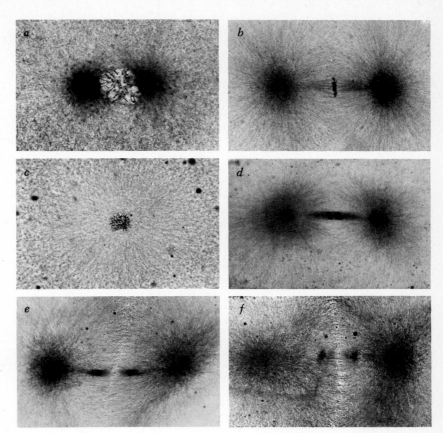

FIGURE 3·18

Photomicrographs of animal mitosis. *a* Early prophase; *b* metaphase; *c* metaphase, polar view; *d* early anaphase; *e* late anaphase; *f* telophase. Note plane of cleavage and formation of daughter nuclei. (*Courtesy of General Biological Supply House, Inc., Chicago.*)

microtubules. Microtubules are cellular organelles composed of 13 protofilaments, which are made up of proteins called *tubulins* (Fig. 3.20). Tubulins are assembled and disassembled by the cell in an as yet poorly understood way. In a mitotically dividing cell, microtubules seem to radiate from the centrioles and attach to the chromosomes. Microtubules are believed to play a role, as yet not understood, in chromosome movement. Drugs such as colchicine, which inhibit the formation of microtubules, also inhibit spindle formation and mitosis.

FIGURE 3·19

Centrioles. These structures are self-replicating and are believed to be responsible for formation of the spindle during mitosis.

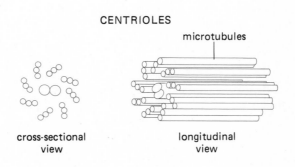

CENTRIOLES

microtubules

cross-sectional view

longitudinal view

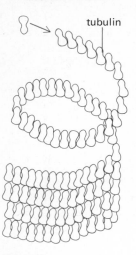

tubulin

FIGURE 3·20
Microtubules. These organelles probably form the cytoskeleton (cell support) and are in part responsible for cell shape. They may also serve as points of attachment for contractile microfilaments. They are composed of beadlike chains of the protein tubulin; they are arranged spirally and are constantly broken down and re-formed.

During prophase the chromosome strands coil tightly, and the chromatin material condenses around them so that the coils are no longer visible. The chromosomes divide longitudinally into chromatids, which lie side by side at the earliest stage of visibility. A minute pale area in the chromatid, called the *centromere* or *kinetochore,* indicates the point of attachment to the spindle fiber. This area remains unstained by the dyes ordinarily used for staining chromosomes. The two centromeres, one in each chromatid, lie side by side and face opposite poles of the spindle. The nuclear membrane and nucleolus ordinarily disappear at about this time, and mitosis progresses to the metaphase.

METAPHASE This is usually a brief but important state that leads up to the separation of the chromatids. The chromosomes arrange themselves so that their centromeres are lined up in the equatorial plane of the spindle. The centromeres divide in the late metaphase, but the actual separation of the chromatids is considered to be part of the next phase, anaphase.

ANAPHASE The chromatids start to move away from each other toward the opposite centrioles at the beginning of anaphase. They separate at the centromeres first, the ends often presenting a V-shaped appearance as though the chromosomes were being pulled by some force at the middle. As the two chromatids move away from each other, they may properly be called chromosomes since each will become an individual chromosome in the daughter cells produced in the last stage of cell division. The late anaphase is indicated when the centromeres have migrated far apart toward their respective spindle poles.

TELOPHASE The two groups of chromosomes never quite reach the poles of the spindle because at the completion of their migration the poles and spindle tend to disappear as the two new nuclei are organized. The nuclear membranes and the nucleoli again become visible, and the chromosomes become invisible as the chromosomal filaments uncoil in the newly formed nuclei. As stated earlier, a new centriole is produced by each centriole, and the two new cells assume the characteristics of cells in interphase.

Cytokinesis A constriction forms in the cytoplasm between the two developing nuclei, and the constriction develops until the cell is completely divided. The constriction forms exactly where the equator of the spindle was located at metaphase. This division is termed *cytokinesis,* and it is accomplished by the action of many tiny filaments called *microfilaments.* These are composed of contractile elements similar to actin and myosin found in abundance in muscle cells. The two daughter cells grow rapidly until they have attained their typical size, after which they may, in turn, undergo mitotic division.

Timing of mitosis The time necessary for the accomplishment of a mitotic division varies greatly in different tissues. An average time may be about 3 h, although some tissues have been observed to complete mitosis in 30 min and others may require several hours. An optimum temperature facilitates the process. Prophase usually consumes the longest period of time, whereas the metaphase is one of the shortest phases.

The biological significance of the elaborate process of mitosis is the provision for equal quantitative and qualitative division of the DNA chromatin material. The result is that each daughter cell has identical hereditary characteristics.

CLINICAL ASPECTS

THE CELL CYCLE AND CANCER   Normally the cell cycle is closely controlled by the body. After an initial period of growth and multiplication, some cells (such as nerve and skeletal muscle cells) reach maturity and stop cycling. Other cells, such as those forming red blood cells, cycle continuously throughout the organism's life. *Cancer* is a pathological condition in which the cell cycle in a particular tissue is not normally controlled. The phases of the cell cycle are perfectly normal, but the cycle's initiation goes out of control. The most critical event in cycle initiation is the start of the S phase; upon replication of DNA, the cell is usually committed to division. *Carcinogens,* agents that cause cancer, induce noncycling cells to initiate their S phases and thus reenter the cell cycle. It is perfectly normal for some cells to go from a noncycling to a cycling condition. For example, cells that form antibodies to combat a particular disease organism start to cycle when that antibody is needed. But, again, this process is carefully controlled.

The influence that all body cells exercise over the individual cell cycles can be demonstrated outside the body in isolated cell cultures. Cell cultures live and grow normally in a culture dish as long as they are artificially maintained in a medium that simulates the normal internal environment of the body. Cell cultures undergo *contact inhibition,* a phenomenon in which their growth and multiplication (cycling) stops when they have formed a single layer of adjacent cells that covers the surface of the culture medium. If the culture is infected with a cancer-causing virus, contact inhibition disappears. Under this condition, the cell cycle is reinitiated, and the cells multiply uncontrollably, forming many layers and crowding each other off the culture. It is not yet known how the virus transforms noncycling cells into cycling cells. One major difference in the transformed cells has been observed, however. Contact-inhibited cells form junctions with each other through which ions and small molecules may pass from cell to cell. Virus-transformed cells lose the ability to form such junctions. This change is apparently related to certain differences in the characteristics of the cell membrane.

In addition to viruses, x-rays, other ionizing radiation, and certain chemicals can induce cancerous growth. Cancer cells invade surrounding tissues by a process called *metastasis.* Cancer cells may spread throughout the body by way of blood or lymph vessels. At present there is no certain cure for many types of cancer, but early surgery, chemotherapy, and radiation therapy are often effective in producing temporary or long-term remissions.

RECOMBINANT DNA AND MANIPULATION OF GENES   Since 1973 investigators have been working on ways to artificially combine portions of DNA molecules from one organism with those of another. The procedure for preparation and propagation of *recombinant DNA* has become very simple now that several key enzymes that control DNA activity have been identified. Four basic elements of the procedure are as follows. The first step involves utilization of enzymes that break and rejoin DNA molecules from different sources. DNA molecules are usually broken with the enzyme *restriction endonuclease*. When DNA fragments from two species are mixed in the presence of *DNA ligase*, various recombinations of the DNA molecules occur. The second step involves the use of a suitable gene carrier that replicates itself as well as a foreign DNA segment. Viruses are often used as such gene carriers. In the third step, recombinant DNA is introduced into functional cells, using viruses as a vehicle. In the fourth step, a clone of recipient cells that has acquired the recombinant DNA is selected.

Recombinant technology has circumvented the barriers in nature that prevent interchange of genetic information between unrelated species. The technique holds both great promise and great risk. Those investigators most knowledgeable in the field, however, believe that with proper controls, the promise is far greater than the risk.

Recombinant DNA studies have already provided much information about bacterial resistance to antibiotics. Such studies may also lead to the construction of bacterial strains that can manufacture important biological products such as antibodies, vaccines, hormones, and vitamins, as well as many other substances. Bacteria that can produce human insulin have recently been developed. Recombinant DNA methodology also provides a means for the production of possibly virulent new pathogens. For this reason, strict governmental guidelines have been established for investigations in this area.

SUMMARY

1 The major energy carrier of the cell is ATP. A small amount of ATP is formed in glycolysis by substrate phosphorylation, which requires no molecular oxygen. Most ATP is formed from the transfer of hydrogens through the respiratory chain in the mitochondria. Most hydrogens are produced in the TCA cycle, which is an aerobic process terminating in the release of CO_2 and water. The TCA-cycle reactions also take place in the mitochondria.

2 Mitochondria are minute organelles consisting of an outer membrane surrounding an inner, folded membrane. The inner membranes are studded with stalked spheres called ATPases. This membrane also contains many enzymes of the electron transport system involved in the respiratory chain. The major function of the mitochondria is to produce ATP, the high-energy fuel, from ADP. Mitochondria are self-replicating and can direct the synthesis of several critical mitochondrial proteins.

3 The nucleus is a small body, usually spherical in shape, located near the center of the cell. The nuclear protoplasm is called the nucleoplasm, and it is surrounded by the nuclear membrane. The nucleus contains at least one nucleolus, or little nucleus. Chromosomes are located in the nucleus.

4 Chromosomes contain genes, which control hereditary characteristics. There are 46 chromosomes, or 23 pairs, in the human cell. The body of chromosomes is composed chemically of nucleoprotein. The principal nucleic acid is DNA, although RNA is present in lesser amounts. By directing enzymatic action, genes determine protein synthesis.

5 The DNA molecule, constructed in the form of a helix, is a double-stranded chain of nucleotides. It may be compared to a ladder, in which the sides are composed of deoxyribose sugar groups that alternate with

phosphate groups. The rungs extend between sugar groups and contain nitrogenous bases held together by hydrogen bonds. There are four such bases in DNA: adenine, thymine, guanine, and cytosine. In the DNA molecule adenine is always paired with thymine, and guanine with cytosine. The sequence of these bases provides the genetic code for the synthesis of proteins. Adenine and guanine are purines; cytosine and thymine are pyrimidines.

6 Ribonucleic acid (RNA) is closely related chemically to DNA. It is single-stranded but twisted on itself to form a double helix. The single strand contains repeating units of ribose sugar and phosphate. There are four bases in RNA, as in DNA; three of the bases are the same as those found in DNA, but the fourth base is uracil, a pyrimidine that replaces the thymine of DNA.

7 There are three kinds of RNA: ribosomal, messenger, and transfer RNA. About 75 percent of the RNA found in the cytoplasm is ribosomal. Another 15 percent is located in the nucleus and is associated with the chromosomes and the nucleolus.

8 DNA acts as a template for the synthesis of mRNA. Messenger RNA leaves the nucleus and enters the cytoplasm, where it becomes attached to ribosomes. The chemical bases of mRNA are complementary to those in one strand of DNA, with the exception that uracil replaces thymine. The sequence of bases in mRNA determines the arrangement of amino acids in the polypeptide chains of the proteins synthesized.

9 Transfer RNA aligns activated amino acids in sequence according to a complementary coded area on mRNA. This process is directed by ribosomes. Transfer RNA is a single-stranded molecule with the nucleotides arranged in a cloverleaf pattern. The free end of the nucleotide chain at which amino acids are attached ends with the bases CCA. The other free end contains guanine. The loop opposite the free ends carries the three unpaired bases of the anticodon. These bases are complementary to three bases on mRNA designating a particular amino acid.

10 The "letters" of the genetic code are formed by using three bases in differing combinations. In this way amino acids may be arranged according to a coding of bases on mRNA.

11 Protein synthesis takes place on groups of ribosomes rather than on individual ones. Groups of ribosomes are called polyribosomes or polysomes.

12 The cell cycle includes the G_1, S, G_2, and M phases. The G_1 phase includes the initial growth that occurs after cell division. S is the phase during which DNA replication occurs. G_2 is a growth period following the S phase, which prepares the cell for mitosis, or the M phase.

13 Cells divide by the process of mitosis, which includes the following stages: prophase, metaphase, anaphase, and telophase. A cell that is not dividing is in interphase, a period that includes G_1, S, and G_2.

QUESTIONS

1 Why is ATP considered the energy "currency" of the cell? What property of ATP allows it to act as an energy carrier?

2 In what way are the mitochondria involved in energy metabolism? Where, within the mitochondria, are the TCA-cycle enzymes and the respiratory-chain enzymes located?

3 Describe the structure of ribosomes. What are their functions?

4 How are DNA and RNA interrelated?

5 What are the two main functions of DNA? What factors control gene expression?

6 Describe the complete process of protein synthesis beginning with the genetic code on the DNA molecule.

7 Describe the stages of the cell cycle and mitosis. How does control of the cell cycle in cancer cells differ from that of normal cells?

8 Discuss one possible mechanism for the formation of ATP by the inner membrane of the mitochondria.

9 What is the ultimate source of the energy in foodstuffs?

10 On average, how many moles of ATP are produced from one mole of glucose during anaerobic metabolism? During aerobic metabolism?

SUGGESTED READING

Bronk, J. R.: Membrane Adenosine Triphosphatases and Transport Processes, *Biochem. Soc. Spec. Publ. No. 4.*, London (1974).

Byan, J.: Microtubules, *BioScience,* **24:**701–714 (1974).

Campbell, A. M.: How Viruses Insert Their DNA into the DNA of the Host Cell, *Sci. Am.,* **235:**102–114 (1976).

Cohen, S. N.: The Manipulation of Genes, *Sci. Am.,* **233:**24–34 (1975).

Crick, F. H. C.: The Genetic Code III, *Sci. Am.,* **215:**55–62 (1966).

De Busk, A. Gib: "Molecular Genetics," The Macmillan Company, New York, 1968.

Green, D. E., and G. A. Blondin: Molecular Mechanism of Mitochondrial Energy Coupling, *BioScience,* **28:**18–24 (1978).

Hardman, J. G., G. A. Robison, and E. W. Sutherland: Cyclic Nucleotides, *Annu. Rev. Physiol.,* **33:**311–336 (1971).

Hinkle, P., and R. E. McCarty: How Cells Make ATP, *Sci. Am.,* **238:**104–123 (1978).

Holley, R. W.: The Nucleotide Sequence of a Nucleic Acid, *Sci. Am.,* **214:**30–39 (1966).

Kornberg, A.: The Synthesis of DNA, *Sci. Am.,* **219:**64–78 (1968).

McElroy, W. D.: "Cell Physiology and Biochemistry," 3d ed., Prentice-Hall, Inc., Englewood Cliffs, N.J., 1971.

Maniatis, T., and M. Ptashe: A DNA Operator-Repressor Skystem, *Sci. Am.,* **234:**64–80 (1976).

Mazia, D.: The Cell Cycle, *Sci. Am.,* **230:**54–68 (1974).

Miller, O. L., Jr.: The Visualization of Genes in Action, *Sci. Am.,* **228:**34–42 (1973).

Moore, P. B.: Neutron-Scattering Studies of the Ribosomes, *Sci. Am.,* **235:**44–65 (1976).

Nomura, M.: Ribosomes, *Sci. Am.,* **221:**28–35 (1969).

Satir, B.: The Final Steps in Secretion, *Sci. Am.,* **233:**28–37 (1975).

Spooner, B. S.: Microfilaments, Microtubules, and Extracellular Materials in Morphogenesis, *BioScience,* **25:**440–451 (1975).

Stein, G., and J. Stein: Chromosomal Proteins: Their Role in Regulation of Gene Expression, *BioScience,* **26:**488–498 (1976).

Tzagoloff, A.: Genetic and Translational Capabilities of the Mitochondrion, *BioScience,* **27:**18–25 (1977).

UNIT TWO
STRUCTURAL ELEMENTS

In the next four chapters we shall take up those systems that contribute to the structural organization of the body.

In the first unit of this text, we took a brief look at cells—their structures, functions, and metabolism. We noted that the cell was the basic unit of structure and organization in all living organisms.

In Unit Two we shall discuss tissues, groups of similar cells found in close association with one another, and the structural support systems of the body, the skeleton and the muscular system.

There are four general types of tissues: epithelial tissue, connective tissue, muscle tissue, and nerve tissue. Each type of tissue is made up of specialized cells and cellular products; each type performs specialized functions.

We shall discuss epithelial and connective tissues in Chap. 4, and we shall show how these two tissue types combine to form the skin, or integument, the major protective covering of the body. In Chap. 5, we shall take up the skeleton, the body's major internal support system, which is composed of bone, cartilage, joints, ligaments, and tendons, all special types of connective tissue. In the final two chapters of this unit, Chaps. 6 and 7, we shall discuss the anatomy and physiology of the muscular system, which together with the skeletal system supports the body and makes movement possible.

We shall leave a discussion of nervous tissue for Unit Three, in which we shall discuss the nervous system.

4

EPITHELIAL AND CONNECTIVE TISSUES: THE SKIN

CONSIDERATIONS FOR STUDY

In this chapter, we cover the following topics:

1 How the structure of the body is organized from tissues, organs, and organ systems
2 The types of epithelial tissue: where they are found and what functions they perform
3 The types of connective tissues: how their structure differs from that of epithelial tissue and their functions
4 The integument: how epithelial and connective tissue unite in this important organ system, what tissues comprise it, and how it protects the internal environment of the body and warms it and cools it as well.

STRUCTURAL ORGANIZATION OF THE BODY

The basic unit of body structure is the *cell*, whose structure and functions we discussed in Chaps. 2 and 3. There are many kinds of cells, as we have mentioned. When similar cells are grouped together to form a structural and functional unit, they are termed *tissues*. Even though you may think that a complicated organism must be built up from many different kinds of tissues, there are really only four general types: *epithelial, connective, muscular,* and *nervous* tissues.

TISSUES *Epithelial tissues* are groups of cells, usually delicate, that form the lining of such internal structures of the body as the mouth, pharynx, vagina, and anus. Epithelial tissues also form part of the outer cover of the body, the skin. This type of epithelial tissue is much rougher than the inner-lining layers.

Connective tissues are supporting tissues. Such tissues are found in the walls of organs and throughout the body. The skeletal system is formed almost entirely of different types of connective tissue such as bone, cartilage, ligaments, and tendons.

Muscle tissues enable the body and its internal organs to move. Skeletal muscles, together with the skeleton, support the body and allow voluntary movement. Muscle tissue of the heart enables it to pulsate, and visceral or smooth muscle tissues allow the walls of the stomach, intestine, and the uterus to move in involuntary contractions.

Nervous tissue is characterized by its ability to translate stimulation into electrochemical nervous impulses. The unit of structure of this tissue is the *nerve cell,* or *neuron.*

ORGANS   As we move up in complexity of organization, we come to *organs,* which are structures made up of various kinds of tissues, all of which work together to perform specialized functions. The heart, the stomach, the liver, the kidneys, and the lungs are examples of organs. If we group organs of similar function together, we form *organ systems.* When we discuss body functions—the subject matter of physiology—we usually divide them by organ systems.

Organ systems   Traditional classification schemes list nine organ systems: skeletal, muscular, nervous, circulatory, respiratory, digestive, excretory, endocrine, and reproductive. The *skin* and its related structures (hair, nails, and glands) form the integument and may be considered as a separate system. We shall discuss this system later in the chapter. The lymphatic system, which is closely associated with the circulatory system, may also be studied as a separate system. Before we begin our in-depth study of tissues, however, we shall first present a brief overview of the organ systems and their function.

The *skeletal system* is a structural system composed of bones, cartilages, tendons, ligaments, and joints. This system supports and protects the body.

The *muscular system* comprises the voluntary and involuntary muscles. These organs provide movement, support, and coordination. They also contribute to the development and preservation of body heat.

The *nervous system* originates and transmits nerve impulses. This system includes the brain, the spinal cord, and the nerves. Special parts of the nervous system are responsible for sight, hearing, taste, smell, and the cutaneous senses. The autonomic nervous system is responsible for the nervous regulation of the body's internal environment.

The *circulatory system,* also called the *blood vascular system,* is composed of the blood, the lymph, the heart and the blood vessels. This system transports oxygen to cells and tissues, removes carbon dioxide from cells, transports nutrient materials to cells, removes waste products from cells, and helps regulate acid-base balance and many other body functions.

The *respiratory system* is responsible for respiration and breathing, or ventilation. Structures of this system include the lungs and their subcomponents, the larynx, the vocal folds, the trachea, and the bronchial tubes.

The *digestive system* comprises the alimentary tract, which extends from the mouth to the anus. Its structures include the salivary glands, the stomach, the liver, the pancreas, and the intestine. This system digests and absorbs nutrients necessary for energy and growth and eliminates solid food waste products.

The *urinary*, or *excretory system*, eliminates liquid wastes. The system is composed of the kidneys, the ureters, the urinary bladder, and the urethra.

The *endocrine system* comprises glands of internal secretion—also called ductless glands—such as the thyroid, the hypophysis (pituitary), the adrenals, and the gonads. This system is one of the body's two major internal control systems. It regulates physiological functions such as growth, metabolism, and reproduction through secretion of hormones.

The *male reproductive system* is composed of the testes, which produce spermatozoa, the prostate gland, and several duct systems. The *female reproductive system* is composed of the ovaries, the fallopian tubes, and the uterus. The ovaries produce ova. The fetus develops in the uterus.

EPITHELIAL TISSUES

Epithelial tissues
Simple squamous
Stratified
Cuboidal
Simple columnar
Pseudostratified

In Unit I, we studied the structure and functions of individual cells; we turn now to the assemblies of cells, termed tissues. Muscular and nervous tissues will be discussed in subsequent chapters, but in this chapter we shall take up epithelial and connective tissues and the integument, an organ composed of these tissues. We shall first consider the epithelial tissues. Their descriptions are summarized in Table 4.1.

Epithelial tissues cover the surface of the body, forming the skin, and form the delicate linings of body cavities that open directly or indirectly to the surface. Since the tissues that line the thoracic and abdominal cavities have a different embryological origin, they are referred to as *mesothelial tissues*, or *mesothelium*. Likewise, delicate tissues that line the heart and blood vessels are called *endothelial tissues*, or *endothelium*. The word "epithelium" may be used, in an elementary sense, to refer to all such tissues.

The cells of epithelial tissues are placed close to one another, with very little intercellular substance between them. The tissues formed by epithelial cells are usually delicate, especially when they line a cavity, and are often only one cell layer thick.

All epithelial tissues have a *free surface* and a *basal surface*. The basal surface lies on tissue formed from the condensation of the ground substance produced by the underlying connective tissue. This *basement membrane* supports the epithelium and attaches it to the underlying tissue.

Epithelial tissues are classified by their shape into three types: *squamous*, *cuboidal*, and *columnar*. If the cells are arranged in a single layer, the tissue is called *simple epithelium*. If there is more than one layer of cells, the tissue is termed *stratified epithelium* (Fig. 4.1).

SIMPLE SQUAMOUS EPITHELIUM *Simple squamous epithelium* consists of flat cells arranged in a single layer, somewhat like tiling laid in a mosaic pattern (see Fig. 4.1*a*). When the cell is viewed from above, the multiple nuclei usually appear oval in shape, but when the cell is seen from the side, the nuclei appear as a raised area on the free surface of the very thin cell. This type of epithelium forms the thin serous linings of the alveoli of the lungs. It is found also in the lens of the eye and in the lining of the membranous labyrinth of the inner ear. The mesothelium, which lines the body cavities, and the delicate endothelial lining of the heart and blood vessels may be classified as simple squamous epithelium.

TABLE 4·1

Epithelial and connective tissues

Type	Where found	Description
Epithelial		
Simple squamous	Thin serous linings of alveoli of lungs	Flat oval cells in a single layer
	The mesothelium which lines body cavities	
	The endothelial lining of the heart and blood vessels	
Stratified squamous	Lining of cheek	Several layers of cells; the surface layer consists of flat cells, the deeper layers tend toward cuboidal
	The external protective layer of the skin	
Cuboidal	Lining of glandular ducts and tubules	Somewhat cube-shaped
Simple columnar	Lining of organs of digestive tract (inner surface layer)	A column of cells, each cell taller than wide; goblet cells present, especially in intestine
Connective		
Loose fibrous, or areolar	Around organs; between organs and other structures	Fibers and fibroblasts in a semifluid base or matrix
Dense fibrous	Ligaments, tendons, aponeuroses, and fasciae	Mostly collagenous white fibers with rows of cells between bundles of fibers
Reticular elastic	Walls of blood vessels, respiratory structures such as lungs and bronchioles, vocal folds	Collagenous fibers but with a greater amount of elastic fibers
Adipose or fat	Subepidermal, on membranes and around organs	Signet-ring cells, the nucleus lies in a thin ring of cytoplasm
Cartilage		
Hyaline	Covers the ends of bones at joints. Larynx, nose, rings of trachea, and bronchi (the forerunner of bone)	Intercellular substance composed of white collagenous fibers. The cells lie in lacunae. (All types of cartilage covered by perichondrium)
Yellow elastic	External ear	Yellow elastic fibers in a network of white fibers. Cells in lacunae
	Auditory tube	
White fibrocartilage	Intervertebral disks. Symphysis between pubic bones	Bundles of wavy white fibers. Cells in lacunae
Bone	Skeletal structures	A hard connective tissue covered with periosteum. Bone cells arranged in a haversian system

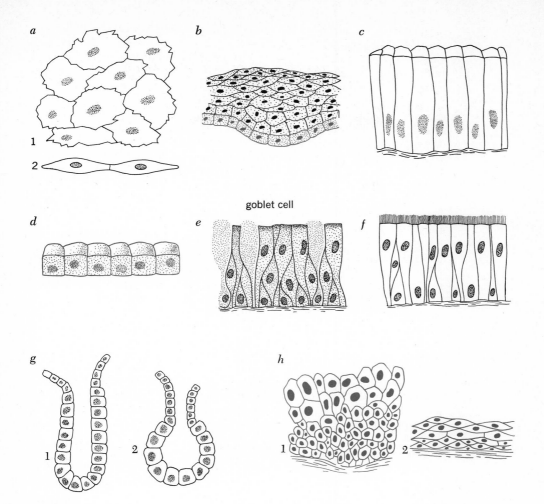

goblet cell

FIGURE 4·1

Diagrammatic representation of various types of epithelial tissues: *a* simple squamous epithelium, *1* surface view, *2* lateral section; *b* stratified squamous epithelium in longitudinal section; *c* simple columnar; *d* cuboidal; *e* columnar epithelium with goblet cells; *f* ciliated pseudostratified columnar epithelium; *g* cuboidal epithelium in *1* simple tubular and *2* alveolar glands; *h* transitional epithelium, as in the urinary bladder, *1* relaxed, *2* extended.

STRATIFIED SQUAMOUS EPITHELIUM   *Stratified squamous epithelium* is composed of several cell layers (Fig 4.1*b*). The deeper cell layers tend to be cuboidal, but the layers close to the free surface flatten out. If we scrape the lining of the cheek with a toothpick and place the scrapings under a microscope, we will see the thin, flat squamous cells of the surface layer. The cells of the deeper layers multiply by mitotic divisions and move up to the surface so that the surface layer is constantly replenished as it wears away. The basement membrane of this type of tissue does not form a straight line, and the basal layer of epithelial cells is not arranged in a regular pattern. Underlying tissues project up into the basal epithelial layer in numerous rounded elevations called *papillae*. The basement membrane therefore appears as an irregular line in histological sections. Stratified squamous epithelium forms the external protective layer of the skin; it also lines the mouth, pharynx, esophagus, vagina, and anal canal.

A variation of stratified epithelium is found in the lining of the urinary bladder, the ureters, and the basal parts of the urethra. This is the so-called *transitional epithelium,* which can relax or extend. In the relaxed condition, transitional epithelial cells tend to be spherical in shape, but when the bladder or tube is distended, the cells of the upper layers become flattened and elongated, resembling stratified squamous epithelium. No distinct basement membrane is present in this kind of tissue (Fig. 4.1*h*).

CUBOIDAL EPITHELIUM   *Cuboidal epithelium,* a type of simple epithelium, is composed of cube-shaped cells; it lines the ducts of many glands (Fig. 4.1*d*). When we look down on a row of cuboidal cells, their surfaces appear square, representing one side of a cube. The nuclei are spherical and located near the center of each cell. Cuboidal epithelium is found in the lining of the smaller ducts and tubules of the salivary glands, the pancreas, the liver, and the kidneys. Although cuboidal epithelium is not considered a distinct type by many authorities, it is a useful designation for the characteristic cells lining the ducts of glands.

SIMPLE COLUMNAR EPITHELIUM   When the cells in a single row of epithelial tissue are laterally compressed so that they are taller than they are wide, they form *simple columnar epithelium* (Fig. 4.1*d*). The oval-shaped nucleus of such cells is located toward the base of the cell. This type of tissue is commonly found throughout the digestive tract, notably in the lining of the stomach and the intestine, in the gallbladder, and in the larger ducts of digestive glands. These cells secrete digestive fluids and absorb food materials.

Goblet cells   These enlarged flask-shaped cells are interspersed among typical columnar cells. They may be considered unicellular glands that produce a mucoid secretion. They are well represented in the epithelial lining of the intestine, becoming abundant in the large intestine (Fig. 4.2).

PSEUDOSTRATIFIED EPITHELIUM   *Pseudostratified epithelium* is a modification of simple columnar epithelium (Fig. 4.1*f*). As a result of compression, the cells of this type of epithelial tissue are of different heights, and their nuclei, therefore, are in irregular layers. The majority of the cells are long and slender and extend from the basement membrane to the free surface, but a considerable number are confined to an area close to the basement membrane. The distribution of the cell nuclei at different levels is characteristic of this type of tissue. This form of epithelium lines the nose, the trachea, and the larger bronchial tubes of the lungs. It also lines portions of the male and female reproductive structures.

EPITHELIAL CELL SURFACES   *Ciliated epithelium*   Columnar epithelium is often ciliated. Hairlike processes called *cilia* project from the free surface of cells and wave continuously, producing currents in the extracellular fluid. Ciliated epithelium lines the trachea and bronchial tubes. Movements of the cilia of these cells produce a current that moves upward and assists in clearing the lungs of foreign matter. Ciliated cells are also found in the fallopian tubes, where they help move the ovum downward toward the uterus. The epididymis and the vas deferens of the male reproductive system also are lined with ciliated epithelium.

Electron micrographs reveal the minute structure of cilia (Fig. 4.3). In

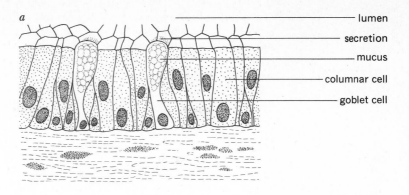

a

— lumen
— secretion
— mucus
— columnar cell
— goblet cell

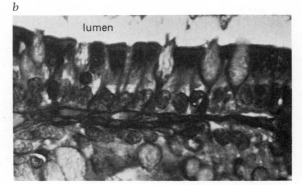

b

lumen

FIGURE 4·2

a Columnar epithelium of the intestine, including goblet cells. *b* Photomicrograph of columnar epithelium. (*Courtesy of General Biological Supply House, Inc., Chicago.*)

FIGURE 4·3

Electron micrograph showing cilia on the surface of dog tracheal epithelium. x11,500. Note the structure of cilia in cross sections. (*From James A. Freeman, "Cellular Fine Structure," McGraw-Hill Book Company, New York, 1964.*)

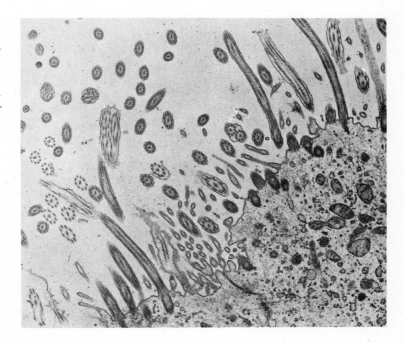

cross section they are composed of a central pair of *filaments* surrounded by nine groups of filaments. Each group usually contains two filaments, but sometimes there are three. Centrioles have a similar structure, as we have seen, but lack the central pair of filaments. If they are viewed in longitudinal sections a *basal body,* or *basal corpuscle,* can be seen at the base of each cilium. The basal body resembles a centriole. Numerous mitochondria are located below the basal body.

Microvilli   The minute projections on the free-surface membrane in epithelial cells, shown by electron microscopy, are called *microvilli* (Fig. 4.4). They are present especially in cells concerned with absorption, such as the columnar epithelial cells that line the intestine and in the "brush border" of cells in the proximal convoluted tubules of the kidneys. Wherever present, microvilli greatly increase the absorption surface of the cell.

EPITHELIAL CELL JUNCTIONS   *Desmosomes (macula adherens)*   The junctions of certain epithelial cells, especially those in the squamous epithelium of the skin and in muscles of the heart, form firm attachments from one cell to another. Electron micrographs indicate a disk of dense material on either side of opposing cell membranes at these places (Fig. 4.5). The dense areas are called *desmosomes.* Fine *tonofibrils* enter the desmosome from the cytoplasm but do not cross the intercellular space of the desmosome.

MULTICELLULAR GLANDS   Multicellular glands, which are composed of epithelial tissue, often become exceedingly complex in their structure and

FIGURE 4·4

Electron micrograph of microvilli in the absorptive epithelial cells of rat ileum. x37,000. *(From James A. Freeman, "Cellular Fine Structure," McGraw-Hill Book Company, New York, 1964.)*

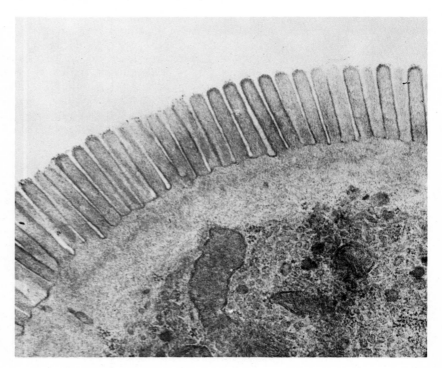

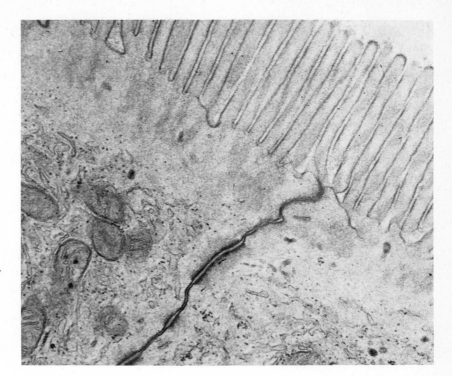

FIGURE 4·5
Electron micrograph showing desmosomes between squamous epithelial cells. The desmosome consists of fine fibrils which converge into a dense thickening of each plasma membrane. x39,000. (*From James A. Freeman, "Cellular Fine Structure," McGraw-Hill Book Company, New York, 1964.*)

function. Their classification and functions are described in Table 4.2. A large number of epithelial cells are involved in gland structure; these specialize to become secretory cells. Two structural types of glands produced by simple infolding of the epithelium are usually recognized: simple tubular and simple alveolar glands (Fig. 4.1g). The *simple tubular gland* is an inpocketing, shaped like a test tube. The *simple alveolar gland* is slightly more complex; it is enlarged at the base and shaped more like a flask. Glandular structure becomes more complex when these simple multicellular types branch, and often the tubular and alveolar types are combined in a single gland.

Sweat glands *Sweat glands* are examples of simple multicellular glands. Sweat glands are found over almost all the skin surfaces of the body but are most abundant on the forehead, in the axillae of the arms, the palms of the hands, and the soles of the feet. There are two kinds of sweat glands, *merocrine* (eccrine) and *apocrine* glands (Fig. 4.6). Merocrine glands are unbranched tubular glands with a long duct opening at a pore in the skin. The coiled tubular portion is the secretory part, while the long narrow duct conveys the secretion to the surface. The merocrine glands pour water (sweat) onto the skin surface, which cools the skin through evaporation. They also moisten the palms of the hands and the soles of the feet since these are friction surfaces and a little moisture provides a better grip. In addition to water, salt and some other electrolytes are lost in perspiration. Merocrine glands are larger and generally more active in men than in women.

TABLE 4·2
Types of multicellular glands

Type	Functional description	Examples
Endocrine	Internally secreting and ductless	Endocrine pancreatic glands, adrenal glands
Exocrine		
Merocrine	Secretion does not include secretory cells	Salivary glands, exocrine pancreatic glands, eccrine sweat glands
Apocrine	Secretions accumulate near free ends of glands and contain parts of cells, but basal portions remain intact and functional	Mammary glands and large, specialized sweat glands
Holocrine	Secretions accumulate within the cell itself. The cell and the secretion are discharged together. The secretory cells are replaced by new cells	Sebaceous glands

FIGURE 4·6

Section through the skin, illustrating a merocrine (eccrine) gland, an apocrine gland, and a hair follicle. Detail of the subepithelium is shown only in part. The melanocytes are pigment cells.

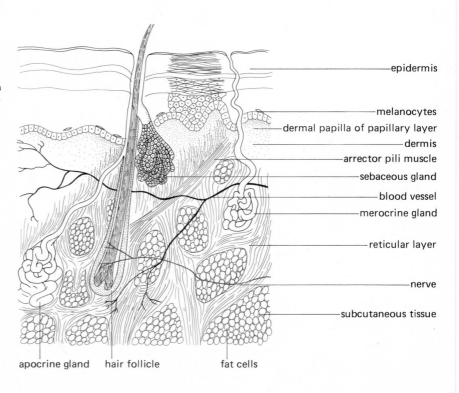

epidermis

melanocytes

dermal papilla of papillary layer

dermis

arrector pili muscle

sebaceous gland

blood vessel

merocrine gland

reticular layer

nerve

subcutaneous tissue

apocrine gland   hair follicle   fat cells

Apocrine glands of the skin are large, specialized sweat glands. They probably evolved originally from scent glands, and they are still partly responsible for the odorous substances in sweat. The action of bacteria produces most of the objectionable odor of perspiration. Such glands are located predominantly in the axillae and anogenital areas. Unlike the merocrine glands, they are not directly concerned with temperature regulation, but they respond to emotional states, stress, and sexual excitement with increased secretion. The glands enlarge at puberty, and in the female they are active during menstruation. Apocrine glands are usually larger than merocrine glands, but they produce only a small amount of fluid. Profuse sweating in the armpits tends to spread the apocrine fluid.

Sebaceous glands   Sebaceous glands, another type of simple multicellular gland, are ordinarily associated with hair follicles. Their ducts open into the upper part of the hair follicle. They supply an oily or waxy secretion called *sebum* to the hair and skin. Sebum protects the skin against becoming too dry and also reduces absorption through the skin. Such glands are called *holocrine glands.*

Other glands   Glands called *ceruminous glands* are located in the skin of the external canal of the ear. These glands secrete earwax. They are modified sweat glands. The mammary glands, another type of apocrine gland, will be considered in Chap. 21, in which the reproductive system is discussed.

Endocrine glands, or glands of internal secretion, do not have ducts and ordinarily deliver their secretions directly into the bloodstream, which distributes the secretions throughout the body. They will be considered in later chapters.

EPITHELIAL MEMBRANES   Epithelial membranes are of two general types: *mucous membranes* and *serous membranes*. Such membranes are composed of various types of epithelial cells deposited upon a layer of connective tissue.

Mucous membranes   *Mucous membranes* line the structures of organ systems that open to the exterior, such as the respiratory, digestive, and urogenital organ systems. A mucous membrane does not necessarily secrete mucus; for example, the lining of the urinary bladder is not mucus-secreting. However, the surface epithelium of most organs of these systems is kept moist by the mucus secretion of glands found in this type of epithelial tissue.

Serous membranes   *Serous membranes* are composed of two layers: the *parietal layer,* which lines a cavity, and the *visceral layer,* which covers an organ. Serous membranes line body cavities that do not open to the exterior and cover organs that lie within these cavities. An exception to this distribution is found in the peritoneal cavity in the female; this is not a closed cavity, since the *uterine tubes* open into it, but it is lined with a serous membrane. Serous membranes produce a thin fluid that lubricates and moistens the areas and the organs covered.

The serous membrane lining the abdominal cavity is called the *peritoneum.* Its visceral layer covers most organs of the abdomen. It is composed of simple squamous epithelium on a base of connective tissue. The epithelial layer is called a *mesothelium* because it is derived from mesoderm. The peritoneum forms the *mesentery,* a thin, double-layered membrane that supports the intestine, and the *omentum,* an apronlike fold which extends downward from the stomach and lies over the intestine.

The *pleura* is the serous membrane that lines the thoracic cavity and covers the lungs. The pleural cavity is a potential space between the parietal and visceral muscle layers. It contains a minute amount of fluid, which enables the two moist and lubricated surfaces to move against each other freely and smoothly.

The *pericardium* is the serous membrane that covers the heart.

A type of serous membrane also lines the interior of the heart, the blood vessels, and lymphatic vessels. This membrane, composed of simple squamous epithelium, is called an *endothelium.* The endothelium forms a very thin smooth layer that lines the chambers of the heart and the larger blood vessels. In the capillaries, the endothelium functions as a selectively permeable membrane between the blood and the tissues.

CONNECTIVE TISSUES

Connective tissues are found throughout the body. They help to form the framework of organs, and they pervade tissue spaces, filling in and connecting organs and various other structures.

Connective tissues are characterized by a small number of cells and an abundant amount of intercellular substance. Connective tissues contrast sharply with epithelial tissues in this respect. The intercellular substance, which varies considerably in different kinds of connective tissues, makes it possible to classify these tissues as areolar, adipose, fibrous, cartilage, or bone.

EMBRYONIC DEVELOPMENT Typical embryonic connective tissue cells, with the exception of cartilage and bone cells, are large, star-shaped cells with numerous processes. They are called *fibroblasts,* and they produce the connective tissue fibers found in adult connective tissues. Fibroblasts are thought to arise from an early embryonic tissue known as *mesenchyme,* in which irregular cells and their protoplasmic processes form a network enclosing a fluid material, or matrix. The number of cells in the young embryonic material is greater than in differentiated, adult connective tissues, and there is relatively less intercellular substance.

As connective tissues develop, the cells lose their stellate appearance and are often separated by more abundant intercellular material. This material varies from a fluid, or semifluid, mucoid substance to a firm ground substance, such as that in cartilage or the rigid matrix of bone.

Fibrous elements
  Collagenous fibers
  Elastic fibers

FIBROUS ELEMENTS OF CONNECTIVE TISSUE Characteristic connective tissue fibers lie within the intercellular material or matrix. They are of two general types: *collagenous* and *elastic fibers.* Collagenous, or white, fibers are found in wavy bundles (Fig. 4.7). They contain *collagen,* an albuminoid protein. The individual fibrils do not branch and are typically found in relatively large bundles. They show very little elasticity and are therefore used abundantly in strong structures such as ligaments and tendons.

Collagen, the principal constituent of connective tissue, is the major structural protein of the body. It is synthesized primarily by fibroblasts, and

FIGURE 4·7

Collagenous fibers in rat mesentery. [*From Roy O. Greep (ed.), "Histology," McGraw-Hill Book Company, New York, 1973.*]

collagenous fibers are found abundantly in the connective tissues of the skin and in ligaments, tendons, cartilage, and bone. The electron microscope shows that collagen fibrils are cross-striated (Fig. 4.8). Each fibril is composed of long-chain polypeptide molecules of tropocollagen. It is thought that tropocollagen molecules, the precursors of collagen, are synthesized by the fibroblasts and extruded from the cell, where they form collagen fibrils by polymerization. The tropocollagen molecules overlap slightly in linear array; this gives them a striated appearance. If newly formed collagen molecules are dissolved in a cold salt solution and then the solution is warmed, the molecules will re-form into typical striated collagen.

Elastic fibers occur singly and branch to unite with similar fibers. They are larger, individually, than collagenous fibrils, are slightly yellow when grouped in fresh material, and take a straight, not undulating, course through the tissue. They contain the protein, *elastin,* which is responsible for their elasticity. Since they are highly elastic and since they differ in appearance from collagenous fibers, they are commonly referred to as *yellow elastic fibers* (Fig. 4.9).

Cellular elements
Mast cells
Macrophages
Plasma cells

CELLULAR ELEMENTS IN CONNECTIVE TISSUE *Mast cells* are fairly large cells found in fibrous connective tissues. Such cells are characterized by dense cytoplasmic granules and resemble granular white blood cells to a certain extent. The cytoplasmic granules are somewhat smaller in mast cells than in blood cells, however, and since they are soluble in water, they cannot be stained with ordinary aqueous staining techniques.

Mast cells are fairly well represented in subcutaneous areolar connective tissues. They contain a blood anticoagulant called *heparin.* As a result of injury, they also release *histamine,* a substance that stimulates vasodilatation in small arterioles and capillaries and also increases the permeability of capillaries to protein (Fig. 4.10a).

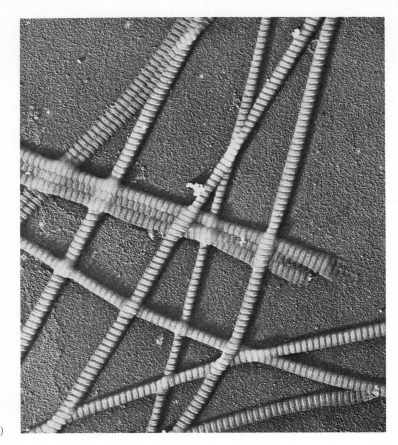

FIGURE 4·8
Electron micrograph of
collagen. Note cross striations.
(Courtesy of Dr. Jerome Gross.)

FIGURE 4·9
Network of elastic fibers in rat
mesentery. [*From Roy O.
Greep (ed.), "Histology,"
McGraw-Hill Book Company,
New York, 1973.*]

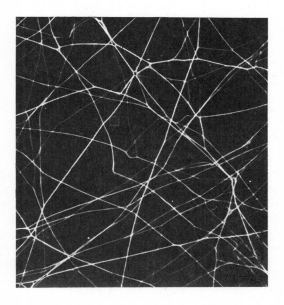

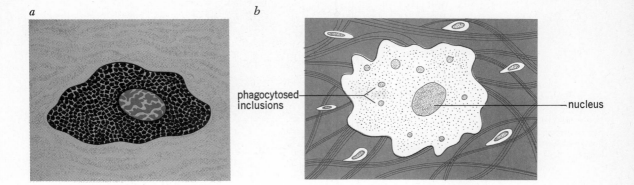

a   b

phagocytosed inclusions

nucleus

FIGURE 4·10

a Mast cell. *b* Macrophage with cellular inclusions by phagocytosis.

Macrophages are active phagocytic cells seen in loose fibrous connective tissues. They resemble monocytes, a type of white blood cell, and are closely related to them. Macrophages are not found in the blood, however; they are wandering cells seen in tissues, or fixed cells lining passageways in such structures as the lymph nodes and the spleen. In healthy connective tissues, macrophages have cytoplasmic processes that extend outward, giving the cell a somewhat stellate appearance; but when they become active phagocytic cells, as they are in an area of inflammation, for example, they lose their cytoplasmic processes and appear as oval-shaped cells. The scanning electron microscope shows these cells to have a rough, membranous surface studded with microvilli.

Macrophages in certain organs are often given special names. The *Kupffer cells* of the liver and the "dust cells" of the alveoli of the lungs are examples. In lymph nodes, spleen, and bone marrow, macrophages are commonly called *reticuloendothelial cells*. All macrophages seem to be closely related regardless of location, and all ingest large amounts of foreign material (Fig. 4.10*b*).

Plasma cells also are present in loose fibrous connective tissue. Such cells are associated with lymphocytes and may be derived from them. Plasma cells have an abundant supply of rough endoplasmic reticulum and produce antibodies. These cells will be considered further in our discussion of lymphocytes and immunity in Chap. 14.

Connective tissues
  Loose fibrous, or areolar
  Dense fibrous
  Reticular
  Elastic
  Adipose
  Cartilage
  Synovial membranes
  Bone

LOOSE FIBROUS, OR AREOLAR, CONNECTIVE TISSUE Loose fibrous, or areolar, tissue is the most abundant of the connective tissues and the most widely distributed. It is composed of rather primitive cells in a semifluid base, or *matrix*. These cells are largely fibroblasts or fibrocytes, but *mast cells, macrophages,* and *plasms cells* also may be present. Within the matrix are wavy bundles of collagenous and elastic fibers, which form an interlacing network (Fig. 4.11). The spaces, or areolae, between the fiber bundles vary with the type of tissue. *Loose fibrous,* or *areolar, tissue* consists of relatively few fibers with wide spaces between the fibers. The skin of animals is attached to the body by strands of loose areolar tissue. A variation of loose areolar tissue, *dense areolar tissue,* is found in the deeper layer of the skin itself, forming the dermis. Areolar tissue is found throughout the body, surrounding organs, supporting them, and attaching them to various structures. It invests blood vessels and nerves and is found between organs and

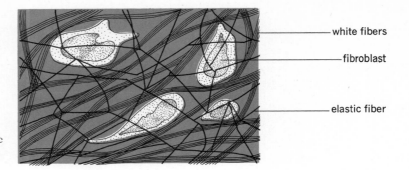

white fibers

fibroblast

elastic fiber

FIGURE 4·11

Loose fibrous, or areolar, connective tissue, including elastic fibers, white fibers, and fibroblasts.

other structures. Areolar tissue is continuous throughout the body and may therefore serve as a pathway for the spreading of infections. It is also involved in the regulation of the water balance of tissues, since it can absorb and hold a very considerable amount of water. Under certain abnormal conditions, water retention may result in swelling (edema) of the ankles or other parts of the body.

DENSE FIBROUS CONNECTIVE TISSUE   Although connective tissues generally contain fibers, the fibrous portion of such structures as ligaments and tendons is so great that they are commonly classified as *dense fibrous tissues*. In tendons, which are the best examples of this tissue type, large wavy bundles of white collagenous fibers are seen in longitudinal section, and between the bundles of fibers are rows of tendon cells (Fig. 4.12). Cytoplasmic layers extend outward from cells and tend to encircle bundles of fibers. In cross section, slender processes of tendon cells branch out and anastomose around bundles of fibers.

Fibrous connective tissue is a very strong, tough tissue laid down in glistening white cords. It is pliant, yet has very little elasticity. The blood supply is not abundant. If this tissue is injured, in a sprain, for example, it may take a long time for the injury to heal. Collagen takes part in the healing of wounds, and more precursors of this substance are produced by fibroblasts after an injury. Some researchers consider *collagenous fibers* themselves to be a special type of very strong connective tissue, as is found in tendons and ligaments.

Tendons attach muscles to bones. They can be observed in the wrist or ankle and also along the back of the hand or in the foot, where they appear as long cords that extend out to the digits.

Ligaments help to hold bones together at the joints. Ligaments may take the form of flat bands or capsules. Though most ligaments are composed largely of white fibers, some ligaments contain elastic fibers in abundance.

Aponeuroses are thin tendinous sheets of fibrous tissue that ordinarily attach to flat muscles.

Fasciae cover muscles and help to hold them in place. They are thin sheets of white fibrous connective tissue.

RETICULAR TISSUE   *Reticular tissue* contains delicate white fibers that form a network with primitive connective tissue cells. The reticular cells have numerous protoplasmic processes, and their nuclei are flat, oval disks. In

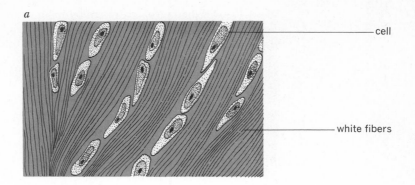

a

cell

white fibers

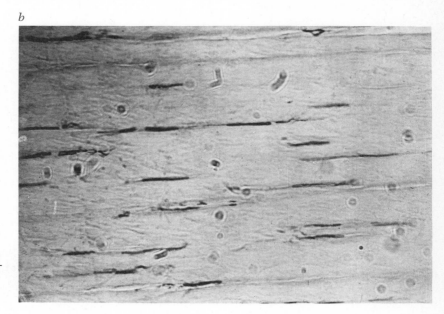

b

FIGURE 4·12
a Tendinous tissue. *b* Photomicrograph of fibrous connective tissue. (*Courtesy of Carolina Biological Supply Company.*)

lymphoid or adenoid reticular tissue, lymphocytes are found in the fibrous network. Reticular tissue is found in lymph nodes, in the spleen, and in bone marrow.

ELASTIC CONNECTIVE TISSUE Sometimes considered as a distinct type, *elastic connective tissue* is a fibrous tissue in which elastic fibers are more abundant than white collagenous fibers. Cells are located between the fibers in the ground substance. The predominance of yellow fibers gives the tissue a slightly yellow color. Elastic tissue has the properties of elasticity and extensibility. It is found in the walls of blood vessels, in respiratory structures such as the lungs and bronchioles, in the vocal folds, and between the cartilages of the larnyx. It is present in a few elastic tendons.

ADIPOSE TISSUE *Adipose*, or fat, *cells* are found in association with loose connective tissue. When such tissue contains large numbers of adipose cells, it is termed *adipose tissue.* Subepidermal areolar tissue is commonly filled with fat cells. Within a framework of fibers are single cells or groups of cells; each cell is filled with fat. The cells become so distended with fat that the cytoplasm consists of only a thin ring encircling each cell. The nucleus lies in the ring of cytoplasm, which resembles the setting in a signet ring (Fig. 4.13).

Although some adipose tissue is found in the embryo, where it appears to function as a fat-storage organ, adipose cells of adult areolar tissues of the skin and mesenteries are considered to be fibroblasts modified by the accumulation of oil droplets within their cytoplasm. As more fat is deposited, the cell increases in size until groups of large fat cells make up the greater part of adipose tissue. The matrix, containing fibers, becomes a slender framework between masses of fat-laden cells. If the accumulated fat is used up in body metabolism, these cells resume the fibroblast appearance.

FIGURE 4·13

a Diagrammatic representation of adipose tissue. *b* Photomicrograph of adipose tissue. (*Courtesy of Carolina Biological Supply Company.*)

a

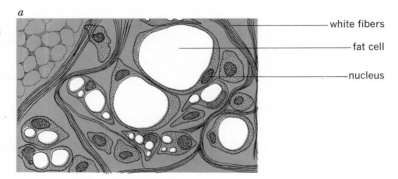

white fibers

fat cell

nucleus

b

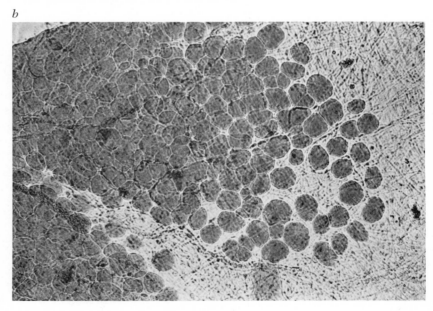

Fat is not deposited indiscriminately over the body. It is found primarily as a subcutaneous layer or as a deposition on membranes such as the mesenteric or greater omentum (see Chap. 17). The kidneys are usually embedded in fat, which helps to hold them in place. Fat is also found around the intestine and in the furrows of the heart. The eyeball lies embedded in a padding of fat. There is a layer of fat around the joints of the skeletal system, and fat is present in the marrow of long bones. It fills in between muscles and helps to support the blood vessels and the nerves.

The subcutaneous layer of fat acts as an insulating layer which prevents excessive heat loss from internal organs. It may be assumed that those individuals with a thick subcutaneous fat layer, other factors being favorable, are better insulated against cold. Women commonly have a somewhat thicker layer of subcutaneous adipose tissue than men. Subcutaneous fat tends to be deposited in certain areas, for example, in the ventral abdominal area.

Adipose tissue is an important source of reserve energy, since stored fat can be oxidized and used as food via the metabolic pathways described in Chap. 3. When we discuss the utilization of foods in Chap. 20, we shall find that a great deal more energy can be obtained from fats per unit quantity than from any other kind of food.

Excessive amounts of adipose tissue may become something of a liability, especially in older persons, when the circulatory system is forced to carry blood to nourish all the extra pounds of fat. The action of the heart may also be hampered somewhat by a deposition of fat on its surface, since in seriously overweight persons, fat is deposited around internal organs as well as in the subcutaneous layers.

CARTILAGE   *Cartilage* is a dense, white connective tissue consisting of modified connective tissue cells surrounded by a massive matrix of intercellular material, which contains both white and yellow fibers. The great bulk of the intercellular substance is composed of a dense mass of white fibers, however, but individual fibers are not visible under the light microscope. In embryonic development, cartilage arises from mesenchyme.

Within the matrix are cartilage cells, or *chondrocytes*, which lie singly or in small groups in cavities called *lacunae*. Chondrocytes secrete a protein-polysaccharide substance that gels and gives cartilage its tough gristlelike consistency. The lacunae lie in a *capsule* of mucopolysaccharide substance that is somewhat more dense than the surrounding matrix. Cartilage growth comes first from the proliferation of chondrocytes in an inner mass of cells and second from the differentiation of *chondroblasts*, which form chondrocytes in the inner part of the covering membrane.

Cartilage is ordinarily found in relatively thin sheets covered by a membrane called the *perichondrium*. Since cartilage itself is essentially nonvascular, the perichondrium, which is well supplied with blood vessels, acts as a nourishing membrane. Lymphatics and nerves are also confined to the perichondrium. Nutrient fluids diffuse slowly through the dense intercellular substance to nourish the cells.

There are three forms of cartilage in the body: hyaline, yellow elastic, and white fibrocartilage. These three types are distinguished by variations in the structure of their intercellular substances.

Hyaline cartilage The most abundant and most widely distributed type is *hyaline cartilage.* Its structure corresponds to the general description of cartilage given above (Fig. 4.14). The relatively few cells are located in lacunae, and the intercellular substance is composed of white collagenous fibers. Hyaline cartilage has a glistening white or pearl-like appearance, but in thin sections it takes on a slightly bluish tinge. This type of cartilage is the forerunner of bone in the embryo. Hyaline cartilage is found in the nose, in the larynx, and in the rings of the trachea and bronchial tubes. It is the type of cartilage attached to the ribs, and it covers the ends of bones at joints. In the latter case, cartilage promotes the smooth action of joints and helps to absorb some of the shock of walking, running, and jumping.

Yellow elastic cartilage Since it develops from hyaline cartilage, *yellow elastic cartilage* is somewhat similar to hyaline cartilage in appearance, but great numbers of yellow elastic fibers are found within a network of white collagenous fibers (Fig. 4.15). The cells lie in lacunae, and a perichondrium

FIGURE 4·14
a Hyaline cartilage. *b* Photomicrograph of hyaline cartilage. (*Courtesy of Carolina Biological Supply Company.*)

a

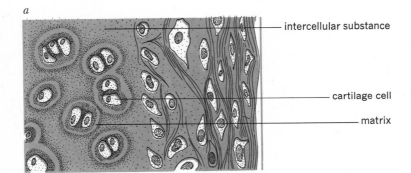

— intercellular substance

— cartilage cell

— matrix

b

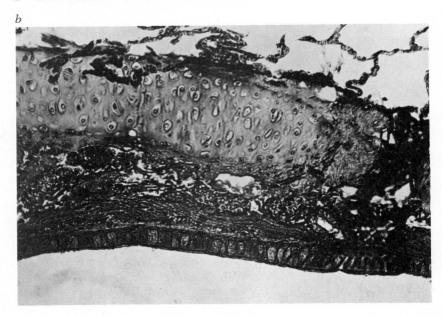

a

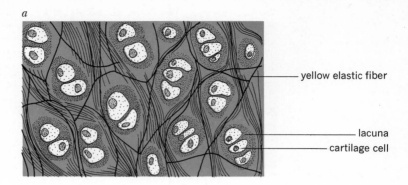

yellow elastic fiber

lacuna
cartilage cell

b

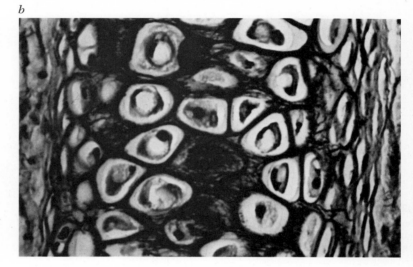

FIGURE 4·15

a Elastic cartilage. *b* Photomicrograph of elastic cartilage. (*Courtesy of Carolina Biological Supply Company.*)

is always present. The elastic fibers impart a certain degree of elasticity to this tissue. Elastic cartilage is found in the external ear, in the epiglottis, in the eustachian tube, and in some of the small laryngeal cartilages.

White fibrocartilage   *White fibrocartilage* resembles fibrous connective tissue in general appearance. It is composed of bundles of wavy, white collagenous fibers, the bundles more or less parallel to one another. Cartilage cells lie in lacunae in the matrix, surrounded by bundles of fibers (Fig. 4.16).

Fibrocartilage is found notably between the vertebrae in the intervertebral disks and in the cartilage between the two pubic bones. It is placed around the outer borders of movable joints, where it tends to make the joint cavities deeper. This tissue is often not well defined, grading off into areolar or fibrous connective tissue. When fibrocartilage is associated with hyaline cartilage in a joint or in the intervertebral disks, hyaline cartilage always makes contact with the bone and gradually grades into fibrocartilage. The plates of fibrocartilage between the vertebrae form a padding that absorbs much of the shock transmitted through the vertebral column when a person walks or jumps.

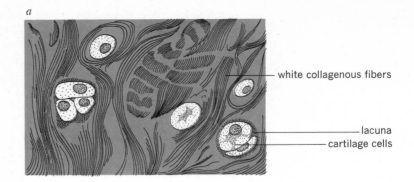

white collagenous fibers

lacuna
cartilage cells

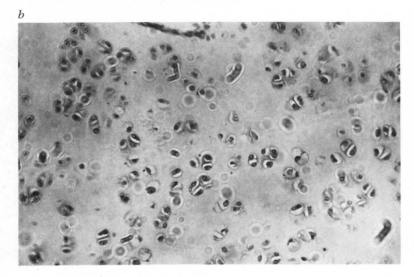

FIGURE 4·16

a Fibrocartilage. *b* Photomicrograph of fibrocartilage. (*Courtesy of Carolina Biological Supply Company.*)

SYNOVIAL MEMBRANES   These membranes are composed of connective tissues. Unlike mucous membranes, *synovial membranes* have no epithelial layer. They line the articular capsules of joints and secrete synovial fluid into the articular cavity. Synovial fluid is a thin, clear fluid which lubricates the joints. *Bursae* and *tendon sheaths,* sacs inserted to cushion structures that move against each other (for example, tendons moving over bones), are also formed by connective tissue membranes lined by synovial membranes. These structures also contain synovial fluid.

BONE   *Bone,* or *osseous tissue,* is the hardest of the connective tissues. Its hardness is due largely to the deposition of inorganic salts in an organic matrix called *ossein.* The two most abundant salts are calcium phosphate and calcium carbonate.

The crystalline mineral found in bone is hydroxyapatite. Apatite is calcium phosphate fluoride, $CaFCa_4(PO_4)_3$, a mineral commonly found in rocks. In hydroxyapatite, the fluorine atoms (F) are replaced by hydroxyl groups (OH). The crystals of mineral matter are embedded in the organic matrix composed largely of crystalline collagen. Hyaluronic acid is also found in the matrix.

Ossein includes fibers embedded in a matrix containing collagen, bone cells, blood vessels, and cartilaginous substances, as well as the deposits of inorganic salts.

We shall continue our discussion of bone in Chap. 5, when we take up the skeletal framework of the body.

THE INTEGUMENT

Integument
  Epidermis
  Dermis
  Skin Derivatives

The *integument,* or *skin,* with its derivatives is often regarded as an organ, but from an anatomical standpoint, it fits nicely into the study of epithelial and connective tissues. Its functions are varied and complex, as the skin includes such derivatives as hair, nails, glands, and sensory receptors.

The skin forms a remarkable protective layer over the outer surface of the body. In embryonic development, it folds inward and becomes continuous with the mucous linings of the mouth, nose, urethra, vagina, and anus at mucocutaneous junctions. The skin has two principal layers: (1) an outer *epidermis,* formed by stratified squamous epithelium, and (2) an inner *dermis,* or *corium,* which is a fibrous connective tissue layer. The *hypodermis,* a loose connective tissue layer beneath the dermis, is often filled with subcutaneous adipose tissue, but it is not a part of the skin.

EMBRYONIC DEVELOPMENT OF THE INTEGUMENT   The integument of the embryo is, at first, a single layer of cuboidal epithelium, derived from ectoderm, over a layer of mesenchyme, an embryonic tissue of mesodermal origin. The epithelial layer soon becomes stratified and develops into several specialized layers common to the epidermis (Figure 4.17). The deeper layer

FIGURE 4·17
Stages in the development of the skin in the embryo.

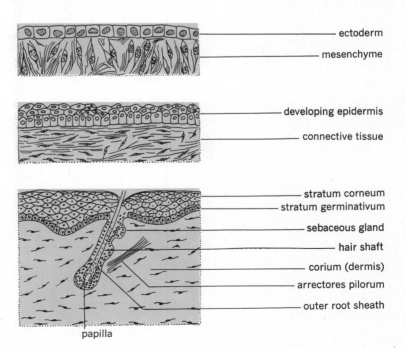

— ectoderm
— mesenchyme

— developing epidermis
— connective tissue

— stratum corneum
— stratum germinativum
— sebaceous gland
— hair shaft
— corium (dermis)
— arrectores pilorum
— outer root sheath

papilla

of mesenchyme gives rise to the dermis. Connective tissue fibers attach the dermis to the superficial and deep fascia covering skeletal muscles, tendons, and bones.

Development of nails and hair The nails and hair are derivatives of the skin. The nails arise from thickenings of the epidermis at the ends of the fingers and toes. The nails develop during the third month of embryonic development and grow slowly, reaching the finger tips during the ninth month. The toenails are a little later in attaining completion.

The hair follicle originates as a downgrowth of epidermal cells into the dermis below. The invading cells become club-shaped and invaginated at the lower end, where mesodermal tissue pushes in to form the follicle papilla. The hair shaft begins to form at the base of the downgrowth and grows outward to penetrate the skin surface. The hair follicle develops at an angle to the surface of the skin, and along its lower side are two outgrowths of cells. The upper one is the primordium of a sebaceous gland, which produces an oily secretion at the base of the hair. The lower outgrowth is called the *epithelial bed* and is associated with the rapidly growing hair root (see Fig. 4.6). Also attached on the lower side of the hair follicles are the arrectores pilorum muscles, which pull on the bases of the hairs and cause small elevations in the skin (gooseflesh) in reflex reactions to cold or fear.

The first body-hair covering of the fetus is a soft, downy coat called the *lanugo.* It appears in about the seventh month of gestation and is shed before birth. The body hair that replaces it is somewhat coarser but remains delicate and inconspicuous in children and in the mature female to a certain extent. The changes associated with puberty produce coarser body hair in both sexes and the growth of pubic and axillary hair and also the beard in the male.

Development of sweat glands The sweat glands arise from an invagination of epidermal cells into the dermis. The cylindrical downgrowth is solid at first, but later a cavity develops, which then forms a long hollow tube. The base of this tube coils extensively in the mature gland. Glands of the axilla, eyelids, and external auditory canals are regarded as specialized sweat glands. The sweat glands begin to form in the fourth month and are well established in the seventh month of embryonic development.

Young children have a delicate skin with sweat glands poorly developed, with soft hair, and with sebaceous glands scarcely functioning. The young skin has abundant blood vessels and heals readily from small cuts and bruises. There is constant change in the skin throughout life, but during adolescence and puberty, the skin becomes much more active.

THE EPIDERMIS The outer layer of the epidermis is composed of dead cells. Such dead cells produce a surface layer that protects the body's internal environment against air, water, most foreign substances, and microorganisms. This layer also wards off excessive ultraviolet radiation. Cholesterol in the dead cells enhances their waterproof character. Although the epidermis protects against most substances, certain gases, some oils and fats, and many insecticides and fungicides may penetrate through this layer and injure the body.

On most body surfaces, the epidermis is a thin flexible layer, but on the palms of the hands and on the soles of the feet, it becomes much thicker and rougher. The ridged or rough epidermal surface on these parts of the body provides more gripping power. Each individual has distinctive ridges, whorls, and arches on the fingertips and soles of the feet which remain constant throughout life. Fingerprinting is thus a reliable means of identification. Footprints are commonly used to identify newborns.

THE DERMIS   The *dermis,* or *corium,* is composed of fibrous connective tissue. It extends from the layer of dermal papillae just underneath the epidermis to a reticular layer that is continuous with the subcutaneous hypodermis. The papillary layer contains more elastic than collagenous fibers. The reticular layer is composed of a network of collagenous fibers. The dermis contains blood vessels, nerves, sweat glands, hair follicles, and cutaneous sensory receptors (see Fig. 4.6).

SKIN DERIVATIVES   *Hair*   Humans are essentially naked animals although our bodies are covered with very fine, colorless hairs. *Hair* grows from follicles deeply embedded in the skin. Its rate of growth has been determined to be 0.3 mm/day.

If hair is studied under the light microscope, straight hair is seen to be round in cross section, whereas wavy or curly hair is alternately round and oval. Kinky hair is flat, somewhat like a ribbon.

The number of hair follicles does not differ greatly between men and women, but body hair on women is usually soft, short, and essentially colorless, and therefore escapes notice. At puberty, hair follicles enlarge, especially in the male, and body hair becomes more prominent. At this time a boy's beard begins to grow.

SHAVING AND BALDING   Hair itself is not alive, so shaving hair will not cause it to become coarse, as some people believe. Baldness is not connected to cutting or to not cutting hair. Physiological changes in the body, probably combined with hereditary tendencies, cause the typical male-pattern baldness. In this type of balding, hair follicles fail to grow hair at a normal rate, although some fine, thin hair may persist. After the follicles degenerate, new hair cannot grow. Hair follicles can be transplanted, and these transplanted follicles will grow hair, but this is a slow and painful process.

Nails   The tough, horny plates that grow out of a bed of tissue at the distal ends of fingers and toes are *nails.* The *nail bed,* or root, contains a layer of epithelial cells and produces the body of the nail. The white, half moon-shaped growth area at the nail base is called the *lunule.* The nails are semitransparent and appear pink because they reflect the capillary bed underneath except at the lunule. The longitudinal ridges that appear in the nail add strength to it. Nails may become brittle with age, and toenails especially may become hard and thick.

Sensory receptors   Sense organs that supply us with information about the surrounding environment are located in the skin. We receive pain impulses from naked nerve endings, but sensations of touch (tactile), pressure, cold, and warmth arise from specialized sensory receptors. Obviously it is quite important to our welfare that we are able to interpret these sensations. We shall consider these receptors in detail when we discuss the special senses in Chap. 11.

SKIN FUNCTIONS The skin helps to maintain body fluid content and composition, to maintain blood pressure, and to regulate body temperature. Sensory receptors in the skin react to tactile stimuli, pressure, and temperature. The skin repairs its own minor wounds, an important function since without an intact skin, humans cannot survive long.

Temperature regulation One of the most important functions of the skin lies in its role as a regulator of body temperature. The body ordinarily maintains a temperature of about 37°C, or 98.6°F. In order to maintain a constant temperature, the body must regulate the production of heat as well as heat loss. We shall learn later that heat control centers are located in the hypothalamus, a structure located in the lower part of the brain. Most body heat is produced by the oxidation of food, but oxidation processes in all tissues, especially in the skeletal muscles, contribute to the overall amount.

The regulation of heat loss is one of the most important functions of the skin. A small amount of heat is lost in exhaling, a small amount by the elimination of urine and fecal material, but the greatest amount (about 85 percent) is lost from the body surface.

The skin contains a vast network of arterioles (small arteries), capillaries (the smallest blood vessels), and nerves. As the environmental temperature rises, arterioles near the surface dilate, permitting more blood to reach the cooler skin surface. Merocrine sweat glands then pour water (sweat) onto the skin's surface, where it evaporates and cools the skin.

The mechanism by which rising temperature causes skin blood vessels to dilate has not been fully determined, and vasodilator nerve fibers to these blood vessels have not been found. It has been observed, however, that when sweat glands are stimulated, vasodilatation also occurs. There is evidence that a polypeptide called *bradykinin*, a substance known to be a potent vasodilator, is released in the blood plasma.

The skin also conserves heat by contracting blood vessels near the body surface, thus reducing heat loss. Epinephrine, a hormone secreted by the adrenal glands, mimics the action of the sympathetic nervous system and controls blood flow to the capillary network. Sympathetic stimulation of the smooth muscles in the skin causes vasoconstriction and a consequent rise in blood pressure. At the same time, perspiration is largely depressed. At room temperature invisible sweating (insensible sweating), which is readily evaporated but tends to keep the skin soft and moist, takes place.

SKIN COLOR The color of the skin and hair depends largely upon the amount of *melanin* present. Melanin is a dark-colored pigment normally found in minute intracellular organelles called *melanosomes*. These organelles are found in a type of cell termed a *melanocyte* (Fig. 4.6). In embryonic development, melanoblasts (immature forms) migrate from neural crest cells which lie above the neural tube (see Fig. 4.18). As they mature into melanocytes, they locate in the basal layer of the skin, with their filaments reaching into the outer layers of the epidermis. Melanin is formed within melanocytes by the action of a polypeptide enzyme *tyrosinase* which catalyzes the hydroxylation of the amino acid *tyrosine* to dehydroxyphenylalanine, commonly known as *dopa*.

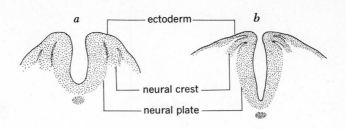

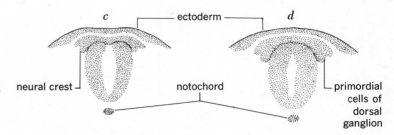

FIGURE 4·18

Developement of the neural tube and neural crest (transverse section): *a* neural groove open; *b-d* gradual closure to form the neural tube. (*Redrawn from Bradley M. Patten, "Human Embryology," 3d ed., McGraw-Hill Book Company, New York, 1968.*)

Albinism Because they are unable to produce melanin, human albinos have pale straw-colored hair, pink irises, and fair, reddish skin. Albinism is inherited as a homozygous recessive character. Individuals with normal pigmentation will be either homozygous *AA* or heterozygous *Aa* for this character. The recessive homozygous condition *aa* produces the albino. Albinos are unable to form melanin because the tyrosine enzyme system is lacking. They have a normal number of melanocytes, but the recessive gene apparently does not provide for the formation of the enzyme tyrosinase, which is essential for the production of melanin. Albinos are very sensitive to light because of the lack of pigment in the irises, which exposes the eyes to too much light. The skin is sensitive to ultraviolet light rays since there is a loss of protective pigment.

Normal Parent *Aa* x *Aa*
Children *AA Aa Aa aa* albino

CLINICAL ASPECTS

SKIN DISORDERS   Injuries to and diseases of the skin are numerous, and many are of a serious nature. Only the more common will be considered here.

Acne This skin problem affects a high percentage of young people at puberty. Its primary cause appears to lie in the increasing androgenic activity of the gonads, or sex glands, at this age. Because of this hormonal change, the skin tends to be oily, and the skin glands overactive. Pimples and blackheads appear, mostly on the face, when the ducts of sebaceous glands become obstructed and their secretions collect. Blackheads receive their coloration from exposure of the glandular substance to air, not from an accumulation of dirt.

Washing regularly with mild soap and warm water is usually recommended as treatment. Some vitamin A preparations may prove to be efficient. Severe cases should be treated by a physician, who may prescribe antibiotics.

Burns Burns may cause serious injury to the skin and be life-threatening if severe enough. Burns may follow direct exposure to fire, to a hot surface, to scalding liquids, to injurious chemicals such as strong acids or bases, or to ionizing radiation. *Local effects* of burns include the changes that occur at the burn site, such as redness, blisters, and localized pain. *Systemic effects* involve the whole body. Fluid loss is a primary concern in severe burns. An increase in fluids input tends to offset fluid and electrolyte loss. Steps must also be taken to prevent infection. Body metabolism generally increases in moderate and severe cases, and the patient's increased nutritional requirements must be met.

Burns are classified by degree of skin involvement. *First-degree burns* involve only the epidermal layer of the skin and cause only a painful, localized inflamed area. A severe sunburn may cause a first-degree burn.

Second-degree burns include both the epidermis and the dermis, but regeneration of the epidermis is possible so long as the dermis is not completely destroyed. Second-degree burns usually produce large blisters.

Third-degree burns are the most severe type. Both epidermis and dermis may be largely destroyed. The body surface may appear blackened or charred. Regeneration of skin is slow since it must progress inward from the surrounding edges of uninjured tissue. Skin grafts may help the healing process by supplying a foundation for skin regeneration, but they may produce considerable scar tissue. Skin grafts also help to prevent evaporation and drying out of the underlying tissue, so they are often used in treating third-degree burns.

Sunburn Following undue exposure of unprotected skin, ultraviolet rays from sunlight may cause serious burns. Sunburn is not noticed until several hours after exposure when the skin begins to turn red. People with fair skins are more susceptible to sunburn than are those with dark complexions who have a greater amount of skin pigmentation, which protects against ultraviolet radiation. Tanning increases the amount of pigmentation and thickens the skin. There is some risk in acquiring a suntan, especially in older people, since skin cancer may develop following long and constant exposure to sunlight and its ultraviolet radiation. To avoid sunburn you should allow only about 15 min for the first exposure, or start the first exposure late in the day, after 4 P.M. Suntan lotions and sunscreens protect the skin from ultraviolet rays and permit longer exposure. Prolonged exposure to sunlight over the years is inadvisable since it tends to damage the skin, as well as increase the risk of skin cancer.

Frostbite and freezing of the skin Overexposure of body parts to extreme cold may cause frostbite and freezing. The affected areas first become numb and turn red. If real freezing has occurred, the skin turns white or gray. In the early stages of exposure to cold, peripheral vasoconstriction occurs, and the body tends to conserve heat, but when exposure continues over a period of time, the skin capillaries dilate. This action would ordinarily warm the skin and prevent damage, but if the body is exposed to very cold temperature and surface blood vessels dilate, a great amount of body heat will be lost. People exposed to severe cold have reported feeling warm before nearly freezing.

In treating frostbite and freezing, the affected parts are gently warmed with warm water or moderate heat. The frozen part is *never* rubbed with snow. If frostbite is severe, a doctor should be consulted as soon as possible.

Bacterial infections of the skin   The bacteria involved in skin sores and eruptions are commonly of the staphylococci or streptococci types. *Pimples* may become infected by bacteria, especially if they are scratched and thus opened to infection. *Boils* and *carbuncles* are raised areas of local skin inflammation and bacterial infection. They contain pus, which comes to a yellowish head. The head ruptures eventually, releasing the pus. A carbuncle is usually formed from a group of boils; it is more severe than a single boil and is based more deeply in the underlying tissue. Both conditions are quite painful and may require treatment by a physician. Boils and carbuncles are usually caused by *Staphylococcus aureus*.

Fungal infections   A fungus is responsible for the condition known as *athlete's foot* (*tinea pedis*). Athlete's foot causes intense itching and is sometimes painful. The fungus thrives under the moist conditions produced by sweating of the feet accompanied by poor ventilation in the footwear. A fungal infection of the scalp known as *ringworm* (*tinea capitis*) occurs primarily in children. In ringworm infections the hair falls out in patches and the skin scales.

Eczema   This condition is not considered to be a skin disease in itself, but is rather a symptom of disease. It is described as a reddening of the skin accompanied by swelling and itching.

Psoriasis   This condition is a fairly common skin disease of unknown cause. Usually red, elevated lesions covered with silvery scales are present. The patient may pass through periods of remission, but there is no known cure for this condition. Special ultraviolet treatments have recently shown some promise in selected cases.

Warts   The yellow or brown elevated areas known as warts may appear rather suddenly on the surface of the skin, and they may also disappear spontaneously. They are caused by a herpes-type virus, as are fever blisters, or cold sores, of the mouth and lips. They are not harmful unless they develop in an area that may be easily irritated by rubbing against clothing or another portion of the body. If an inflamed area develops around a wart, the wart should be removed.

Moles   These are usually benign growths, which are distributed over the body. They arise as flat, light brown growths, but after adolescence they tend to enlarge, turn dark brown, and become elevated. In older people, moles that enlarge and become darker, especially if surrounded by any inflammation or irritation, may become cancerous and should be removed. This form of cancer is called *malignant melanoma*.

Birthmarks   Birthmarks, or congenital discolorations of the skin, are vascular anomalies that occur infrequently. They vary in size and color, but are usually pink, red, or bluish. One type is permanent and does not disappear as the child grows to adulthood. Another type, termed a *strawberry birthmark,* often regresses or disappears with age.

Insect infestations An infestation of lice is called *pediculosis.* Pediculosis of the scalp refers to an infestation involving the *head louse.* This condition is more apt to occur in children and causes itching of the scalp. The *body louse* lives predominantly in clothing but travels to the body to feed. Eggs are laid, especially in the seams of clothing. The body louse can transmit typhus fever. The third type of human louse is the pubic or crab louse that attaches to hairs of the pubic region. Infestations of lice do not usually occur when there is the opportunity for frequent bathing and changes of clothing. Head lice infestation may be an exception since children often pick up lice from contacts with infected children at school.

The harvest *mite,* or "chigger," may cause eruptions of the skin when the larval stage is picked up. Eggs are laid on vegetation, and a person walking through this vegetation may become a host for the larvae which crawl onto the host, burrowing into the skin with their heads. Their feeding causes severe itching and local reddening of the skin.

Insect bites and stings The skin is also the target for the bites of insects and for stings of bees and wasps, but these usually cause only temporary pain, reddening, and swelling. In people who are allergic to them, insect bites and stings may cause serious anaphylactic reactions.

Contact dermatitis This is a reaction to material that comes in contact with the skin. Many plants produce secretions that irritate the skin; the best known are probably *poison ivy* and *poison oak.* Contact with these plants may cause severe inflammation and itching. The stinging sensation of nettles causes an almost immediate local reaction but does not last long. Dermatitis caused by use of cosmetics is a common problem. The chemicals used in the preparation of such cosmetics as nail lacquer, hair dyes, lipstick, cleansers, detergents, and various plastics may cause local skin irritations.

Scarring The skin is stretched over the body, and in deep wounds the edges tend to pull apart. It may be necessary then to take stitches in the edges or to use metal clips to hold the skin over the wound. Deep injuries may fill in with fibrous scar tissue. In pregnancy or in obesity, the skin, especially over the abdomen, may become so greatly extended that long red scars called stretch marks may develop. Even though the skin may return to normal, the scars remain.

Skin cancer An eruption on the skin that bleeds and does not heal should be viewed with suspicion. Fortunately, most skin cancers can be treated successfully. There are two common types: *basal cell* and *squamous cell* cancers. The basal cell type develops slowly and does not metastasize. A crust forms over a small lesion, and if the covering material is rubbed off or shed, the undersurface bleeds. Squamous cell cancers are usually larger, deeper, and more serious than the basal type. If untreated, they increase in size and may metastasize. They are often found around the mouth and on the lips, face, ears, and neck. Exposure to sunlight may aid their growth. They should be treated by a physician without delay.

SUMMARY

1 Tissues are groups of cells that are similar in origin, structure, and function. Epithelial tissues are thin and usually delicate and contain very little intercellular substance. Types of epithelial tissues include simple and stratified squamous, cuboidal, and columnar. Columnar epithelium may be ciliated, as it is in the lining of the trachea and bronchial tubes; it may contain goblet cells, as it does in the lining of the intestine.

2 Epithelial cell surfaces may have cilia, which consist of nine groups of filaments surrounding a central pair. Mitochondria are numerous at the base of each cilium. Epithelial cells may have minute projections from their free surface. These are microvilli, which increase the absorptive surface of the cell.

3 Junctions between epithelial cells have been demonstrated. These disks of dense material are desmosomes.

4 Epithelial tissues form various types of multicellular glands. There are two kinds of sweat glands, merocrine (eccrine) and apocrine. Merocrine glands are concerned with temperature regulation, and their ducts emerge in the pores of the skin. Apocrine glands are stimulated to secrete under certain conditions such as emotional states, stress, and sexual excitement. The sebaceous glands of hair follicles are holocrine glands.

5 There are two types of epithelial membranes. Mucous membranes are epithelial membranes that line the respiratory, digestive, and urogenital tracts. Various types of glands in these membranes keep the tract surface moist. Mucus is commonly secreted by these membranes. Serous membranes line body cavities that do not open to the exterior, such as the abdominal and the thoracic cavities. The pleural cavity is a potential cavity between the parietal and visceral layers of the pleura. The pericardium is a double layer covering the heart. Serous membranes also line the interior of the heart, the blood vessels, and the lymphatic vessels. Serous fluid is thin and watery.

6 Synovial membranes are composed of connective tissues and, unlike mucous membranes, have no epithelial layer. They line the articular capsules of joints and secrete synovial fluid. Bursae and tendon sheaths are also lined by synovial membranes and contain synovial fluid.

7 Connective tissues are tough and fibrous and usually contain an abundance of intercellular substance. Loose fibrous, dense fibrous, reticular, elastic, adipose, cartilage, and bone are types of connective tissues. Ligaments and tendons are examples of dense fibrous connective tissue. There are three kinds of cartilage: hyaline cartilage, yellow elastic cartilage, and white fibrocartilage.

8 There are two fibrous elements in connective tissue: collagenous and elastic fibers. Collagen has been called the major structural protein of the body. Collagenous fibers are abundant in the skin and in ligaments, tendons, cartilage, and bone. When highly magnified, collagen has a distinctly striated appearance. Elastic fibers contain the protein elastin, which has elastic properties.

9 Cellular elements of connective tissue include several cell types. Mast cells are fairly large cells found in fibrous connective tissues. The cytoplasm is granular, and mast cells resemble somewhat the granular white cells of the blood. They produce heparin, a blood anticoagulant, and also release histamine, a vasodilator. Macrophages are phagocytic cells found in loose connective tissues. They may be wandering cells or

fixed cells lining passageways in such structures as lymph nodes and spleen. In the liver they are called Kupffer cells. They constitute the reticuloendothelial cells of lymph nodes, spleen, and bone marrow. They resemble the monocytes of the blood and are closely related to them, but macrophages are not found in the blood.

10 Bony tissue is the hardest connective tissue. Its hardness is due largely to the deposition of the inorganic salts calcium phosphate and calcium carbonate. Bones also contain a large amount of collagen.

11 The skin, or integument, is composed of two principal layers, the epidermis and the dermis. The epidermis is composed of stratified squamous epithelium, whereas the dermis is a fibrous connective tissue layer. The hypodermis lies beneath the dermis, but is not a part of the skin.

12 Epidermal cells produce the fibrous protein keratin, which is the principal constituent of the surface layer. Nails and hair are built from this protein.

13 The color of the skin depends largely on the amount of pigment melanin present. The pigment is produced by cells called melanocytes.

14 The dermis, or corium, contains blood vessels, nerves, sweat glands, hair follicles, and the cutaneous sensory receptors.

QUESTIONS

1 Describe the outstanding structural characteristics of epithelial tissues. How do they differ from connective tissues?
2 Mast cells have what function?
3 Where would you find squamous, cuboidal, or columnar epithelium?
4 Where is ciliated epithelium found?
5 Compare the minute structure of cilia with that of centrioles.
6 Discuss the differences in structure and function between merocrine and apocrine glands.
7 Where are sebaceous glands found and what is their function?
8 Describe the differences between mucous, serous, and synovial membranes.
9 Where is collagen found and what are its functions?
10 How do macrophages function?
11 Discuss some of the functions of adipose tissue.
12 What are the individual uses of tendons, ligaments, aponeuroses, and fasciae?
13 Discuss three types of cartilage, indicating differences in structure and function. Locate various types of cartilage in the body.
14 Discuss the structure and function of the skin.
15 How does the skin act as a regulator of body temperature?
16 Where are melanocytes located?
17 A first-degree burn involves what layer of the skin?
18 List some diseases of the skin.

SUGGESTED READING

Several textbooks on histology are listed at the back of the book.

Fraser, R. D. B.: Keratins, *Sci. Am.*, **221:**86–96 (1969).

Greep, R. O. (ed.): "Histology," 3d ed., McGraw-Hill Book Company, New York, 1973.

Harrison, R. J. and William Montagna: "Man," Appleton Century Crofts, New York, 1969.

Montagna, William: The Skin, *Sci. Am.,* **212:**56–66 (1965).

Warfel, A. H. and S. S. Elberg: Macrophage Membranes Viewed through a Scanning Electron Microscope, *Science,* **170:**446–447 (1970).

Windle, W. F.: "Textbook of Histology," McGraw-Hill Book Company, New York, 1976.

5

OSTEOLOGY: THE STUDY OF BONES AND SKELETON

CONSIDERATIONS FOR STUDY

In this chapter, we shall continue our study of body tissues and support systems, as we take up the following subjects:

1 The development and structure of bone tissue: how different types develop during early life until the structure of finished bone is achieved
2 The anatomy of the skeleton: the composition of the axial and appendicular skeletons
3 Articulations: an introduction to joints and how they work
4 Clinical considerations: a survey of the most common diseases, syndromes, and injuries of the skeletal system

The skeleton is the bony framework of the body. It is the principal support for other parts of the body and protects the vital organs. Bones are a form of connective tissue, and they have undergone many adaptive changes to perfect them for their present use. The bones in the arm of a human, for example, are homologous to the bones in the wing of a bird or those in the wing of a bat. Bones are not to be regarded, therefore, as mere pieces of framework, but as living, active organs that have a long phylogenetic history of development and are subject to considerable change during an individual's period of growth and even in later life.

It is most important that the skeletal system be regarded as a living organ system and not as a collection of dried bones. The skeleton functions in body metabolism by storing calcium and phosphorus. Calcium, as we shall see later, is essential for the proper functioning of the nervous system, the contraction of muscles, and the clotting of the blood. The skeleton is also the

124

site of red blood cell production; such cells are formed in the red marrow of bones.

Hormones from the thyroid and parathyroid glands, together with vitamins A, C, and D, regulate the calcium level of the blood and largely control the deposition of calcium in bones. Under certain conditions, calcium may be withdrawn from bones in order to maintain the calcium level of the blood, or additional calcium may be absorbed from the intestine under the influence of vitamin D. Some authorities consider vitamin D to be more of a hormone than a vitamin, since the body can synthesize it if the skin is irradiated with ultraviolet light; vitamin D is also distributed by way of the bloodstream, as are hormones. Like hormones, it is effective in very small amounts. For most humans, however, the major source of vitamin D is in food, and it is therefore still considered to be a vitamin.

The regulation of calcium distributed throughout the body offers a fine example of homeostasis, the concept we introduced in Chap. 2. Under the influence of parathyroid hormone and vitamin D, calcium is stored in bones. As needed by the body, calcium is withdrawn from bones and carried by the blood to supply the needs of muscles, nerves, and many other tissues. This is a hormone feedback mechanism (see Chap. 2).

Phosphate is essential both in carbohydrate metabolism and in the energy pathways in which adenosine triphosphate (ATP) is produced by *oxidative phosphorylation* (see Chap. 3). Phosphorus also plays a role in the formation of various organic compounds such as phosphocreatine, phospholipids, and phosphoproteins.

BONE STRUCTURE AND DEVELOPMENT

BONE STRUCTURE   Bone is not the solid homogeneous material that it may appear to be upon superficial examination (Fig. 5.1). Only the walls of the bone, which form the *diaphysis*, or shaft, are composed of hard bone; the ends, or *epiphyses*, are filled with a porous, spongy network called *cancellous* bone. Within the shaft is a long *medullary canal* containing *bone marrow*.

There are two types of bone marrow. The type in the medullary canal is *yellow marrow*. It contains connective tissue, blood vessels, immature blood cells, and a great many fat cells, which give it a yellow color. *Red marrow* is found in the ends of long bones in the cancellous tissue. It is more abundant in the larger flat bones and in the ribs. The red marrow contains individual fat cells, but these are distributed among a great number of developing blood cells. After birth the red blood cells and granular white blood cells are produced in the red marrow.

A study of a very thin transverse section taken from the shaft of a bone shows that this osseous tissue is composed of *concentric lamellae*, or rings. The centers of the rings are called *haversian*, or *central*, *canals*. These contain blood vessels and small amounts of connective tissue. Between the concentric layers of bone are cavities called *lacunae*, in which bone cells lie. Radiating out in all directions from the lacunae are tiny canals, termed *canaliculi*, which branch and anastomose until they form a network throughout the bone (Fig. 5.2). Minute protoplasmic filaments grow outward from *osteocytes*, or bone cells, into the canaliculi. Thus bone itself becomes a living tissue, containing cells and connecting strands of cytoplasm even in the hardest and most compact portions of its structure. Haversian

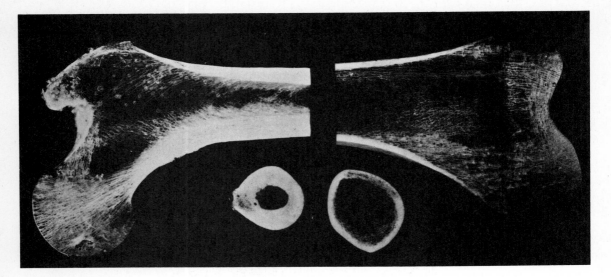

FIGURE 5·1

Section of proximal portion of the human femur showing internal structure. (*Courtesy of Otto Kampmeier, Ph.D., M.D., and the Upjohn Company.*)

canals are not isolated; they are joined by communicating canals. The concentric arrangement of layers of bone and bone cells around a canal is called a *haversian system,* or an *osteon* (Fig. 5.3). Areas between haversian systems are filled in with *interstitial lamellae,* which do not have a concentric arrangement.

The *periosteum* is a tough fibrous membrane that covers bone except at the *articulating,* or joining, surfaces. The periosteum has two poorly defined layers. The outer layer is composed of connective tissue filled with blood vessels, lymphatic vessels, and nerves. The inner layer contains elastic fibers, cells, and some blood vessels, although it is not as vascular as the outer layer. The cells of the inner layer are thought to develop into bone cells. In the bones of a young person, many osteoblasts are found on the inner layer of the periosteum next to the surface of growing bone, but in an adult such cells are inactive unless an injury occurs. If a bone is fractured, this osteogenic layer is activated, and it supplies new bone to replace the old, thus repairing the break.

FIGURE 5·2

Microscopic structure of bone, showing lamellae and central canal. Enlargement *1* shows an osteocyte, while enlargement *2* shows lacunae and canaliculi.

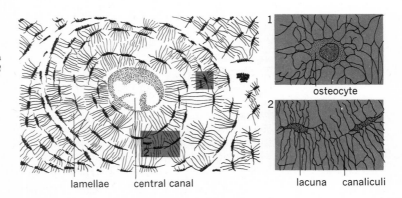

lamellae   central canal

1
osteocyte

2
lacuna   canaliculi

FIGURE 5·3
Thin cross section of compact bone, illustrating a haversian system, or osteon. (*Courtesy of General Biological Supply House, Inc., Chicago.*)

Blood vessels and nerves from the periosteum penetrate into the hard substance of bone through canals (*Volkmann's canals*). These minute passageways pass through the lamellae as secondary vascular canals. Some connect with blood vessels in haversian canals, and others follow an irregular path through solid bone. Nutrient arteries supply the bone marrow; such arteries commonly penetrate the shaft of a long bone near the middle and divide, sending branches into the marrow toward either end of the bone. Veins are also found in the marrow cavity, many of them emerging through the cancellous tissue at the ends of long bones.

BONE CELLS *Osteoclasts* are the largest of the bone cells. Such cells commonly have three or four nuclei and cytoplasmic extensions that reach out into the surrounding bone. Osteoclasts break down bony structure. For example, osteoclasts play a role in the enlargement of the marrow cavity of long bones during the period of growth. The enzymatic action of acid phosphatase probably enables osteoclasts to act on bone, breaking it down and releasing calcium. A state of balance exists between the use of calcium by osteocytes in building bone and the withdrawal of calcium by osteoclasts during the destruction of bony tissue.

Osteoclasts are strongly influenced by the lowering of blood calcium. When blood calcium concentration decreases, a hormone from the parathyroid gland acts to increase the number and the activity of the osteoclasts, thus causing calcium and phosphate to be released. In order to guarantee homeostasis of blood calcium concentration, a second hormone called *calcitonin* is secreted by the thyroid gland. This hormone causes a rapid lowering of blood calcium concentration by stimulating the deposition of calcium in bone and by restricting the activity of osteoclasts in the breakdown of bone. Under the influence of calcitonin, the number of osteoclasts is reduced, and there is some indication that the osteoclasts are converted into osteoblasts.

Osteoblasts, a second type of bone cell, function in the development of bone. When they are found in lacunae they are called *osteocytes.*

OSTEOGENESIS: BONE DEVELOPMENT   *Membrane and cartilage bone*   Most of the bones of the cranium and face develop between membranes in the embryo and are called *membrane bones.* These bones are not preformed in cartilage. The remainder of the embryonic skeleton, however, is first formed of cartilage, which is later replaced by bone. Bones formed in this way are commonly called *cartilage bones.* There are, therefore, two methods of bone formation in the embryo: *intramembranous,* for example, the formation of skull bones, and *intracartilaginous,* for example, the replacement of cartilage by bone in the remainder of the skeleton.

DEVELOPMENT OF INTRAMEMBRANOUS BONE   The bones of the cranium develop from centers of ossification in connective tissue. Fibroblasts, located in fibrous connective tissue, change to osteoblasts, the *bone-forming cells.* The intercellular substance of the connective tissue becomes gelatinous in consistency. Spicules of bone appear, and calcium salts are deposited in the intercellular matrix. Finally, long, thin plates of bone surrounded by osteoblasts appear in the connective tissue. Some osteoblasts locate individually in lacunae; these are then called osteocytes. They maintain connections with each other and with the nourishing membrane through the bone canaliculi.

The connective tissue around the developing bone also forms the periosteum, the nourishing membrane that can produce bone. The inner layer of this fibrous covering is lined with osteoblasts and is known as the *osteogenic layer.* More compact bone is deposited immediately under the periosteum, whereas the intermediate portion of the bone is composed of bone spicules in loose, spongy arrangement. The skull bones are not completely ossified at birth. Membranous areas called *fontanels* are found between the skull bones already formed. The thickest portion of the bone is at the center of *ossification,* or bone growth; the bones are paper thin at the edges.

Development of cartilaginous bone   The skeleton of the developing embryo, with the exception of most of the skull bones, is formed of hyaline cartilage. After the second month of development, cartilage is gradually replaced by bone. The perichondrium becomes the periosteum and forms a cylinder of bone at the middle of the shaft in the long bones. The cartilage within this ring of bone begins to degenerate. The cartilage cells appear swollen, and calcification occurs in this area. The cartilage cells die as they are released by the degeneration of cartilage.

Endochondral bone then begins to replace the degenerating cartilage at the center of ossification. The shaft is gradually converted into bone as

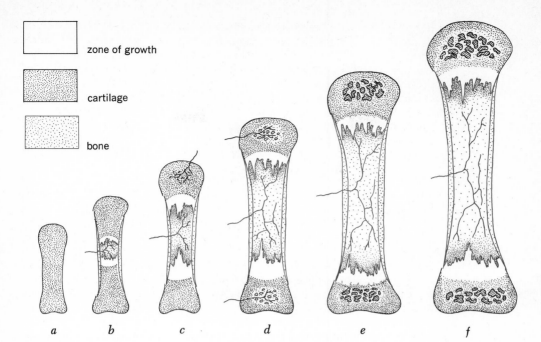

zone of growth

cartilage

bone

a   *b*   *c*   *d*   *e*   *f*

FIGURE 5·4

a to *f* Stages in the ossification of a long bone.

ossification proceeds away from the center toward the ends of the bone (Fig. 5.4).

There are evidently two kinds of bone formation in cartilage bones: endochondral and perichondral. In perichondral bone formation, bone grows around cartilage. Within the bone, however, bone always arises from connective tissue membranes, even in the case of cartilage bone.

While ossification progresses from the center of the shaft, the cartilage ends continue to grow and the bone increases in length. There is necessarily considerable reorganization inside the bone. Primary marrow cavities are formed by the inward growth of perichondral tissue into degenerating cartilage. Blood vessels accompany the buds of periosteal tissue. Cells of perichondral tissue form the primary bone marrow, which fills the marrow cavities. Finally, the primary marrow spaces are united to form the single medullary cavity of long bones. A thin connective tissue membrane called the *endosteum* lines the marrow cavity. It is thinner than the periosteum and is more evident in the larger yellow-marrow cavities.

The marrow cavity in the bone of an adult is considerably larger than its original cartilage antecedent in the embryo. If a metal band is placed around a bone of a young animal, the metal band will be located in the bone marrow of the adult. This indicates that the marrow cavity is enlarged as the bone grows and that all the original cartilage and the bone that was formed around it has been resorbed during the growth of the bone. Bones also increase in diameter as successive layers of bone are produced by the osteogenic layer of the periosteum.

As bone development proceeds, there may be several centers of ossification. The ossification of the shaft progresses toward the epiphyses, or ends (Fig. 5.5). The epiphyses are mostly cartilaginous at birth; they begin to

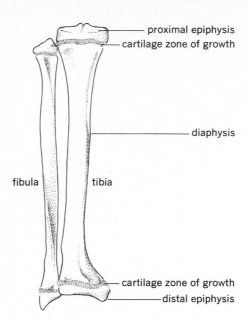

proximal epiphysis
cartilage zone of growth

diaphysis

fibula   tibia

cartilage zone of growth
distal epiphysis

FIGURE 5·5

The tibia and fibula of a child about five years old.

ossify at various times after birth. Ossification can proceed from more than one center in the epiphyses of the larger bones. Bone-forming tissue invades the cartilage, and spicules of bone appear, producing a network of spongy bone.

Both the diaphysis and the epiphyses change to bone before growth is completed, but a cartilage plate, the *epiphyseal plate*, separates them during the entire period of growth. Proliferation of cartilage cells and the formation of new matrix proceed on the epiphyseal side of the plate. The cartilage toward the diaphysis is gradually replaced by bone. These zones of growth are maintained in the long bones of the female until between the seventeenth and nineteenth years, at which time they have been entirely replaced by bone, and no further growth in the length of bones or in the height of the individual occurs. Most young women commonly attain essentially all their growth in height by the time they are fifteen to sixteen years old. In the male, the zones of growth are replaced by bone between the nineteenth and twenty-first years. For practical purposes the period of growth in height in the male is largely completed at about sixteen to eighteen years of age, although complete fusion of all epiphyses may not occur until twenty-five years of age (Fig. 5.6).

Living bones may react and change their shape somewhat in response to undue demands on them for support. On the other hand, if they are not used they may become greatly weakened. Bones require a certain amount of stress in order to maintain their strength.

TERMS OF REFERENCE

Before we begin our study of the skeleton, we must first learn a few terms of reference that describe position and direction of the body and outline general anatomical divisions. In animals that walk on four legs, the head is said to be *anterior*, while the tail is *posterior*, or *caudal*. The back of the

FIGURE 5·6

Differences between the fetal and the adult skeleton. The figures for the fetal skeleton indicate the stage of development in the fetal months; epiphyses begin to appear about this time. The figures for the adult skeleton indicate the wide range of years during which the zones of growth disappear and the epiphyses are completely fused with the diaphysis. [*After Graca and Noback in Barry J. Anson (ed.), "Morris' Human Anatomy," 12th ed., McGraw-Hill Book Company, New York, 1966.*]

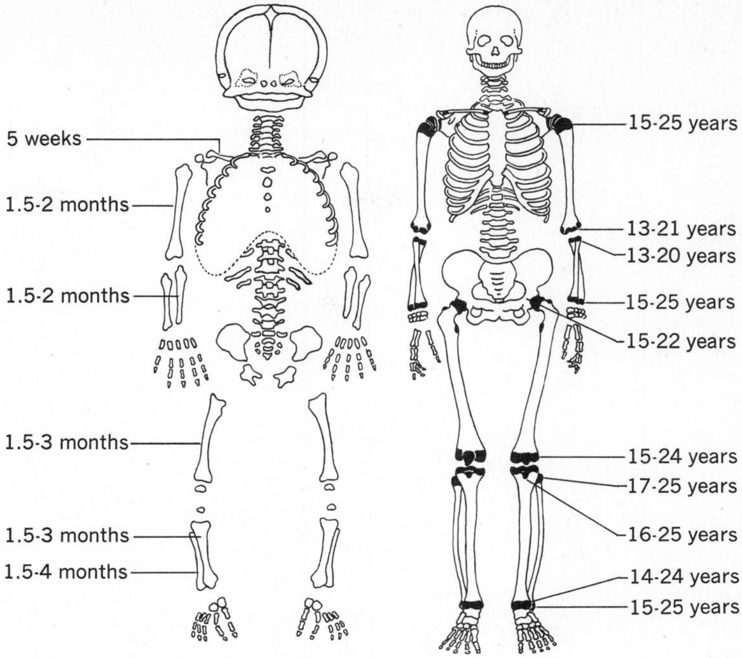

animal is termed *dorsal*, and the underside is called *ventral*. The human being, however, walks in an upright position, and terms applied to the quadruped do not always apply neatly to the biped.

The human figure, in anatomy, is studied in the *anatomical position:* erect and facing the viewer (*frontal view*) or with the back to the viewer (*posterior view*). The hands are held at the side with the thumbs pointing away from the body. In this position, the head is located at the *superior*, or *cranial*, end of the body and the feet are said to be at the *inferior* end. *Cephalic* may be used also in referring to the position of the head. The term *anterior*, or *ventral*, is used to refer to the frontal aspect of the body, whereas *posterior*, or *dorsal*, designates the back. Important positional and directional terms are summarized in Table 5.1.

Proximal and *distal* are two other directional terms in common use. *Proximal* refers to the area closest to the body or closest to the midline within the body; *distal* designates the portion farthest away.

Other locational terms include *medial, lateral, superficial, deep, parietal, visceral,* and *peripheral*. Structures in, or close to, the midline are labeled *medial*, while those at the side are termed *lateral*. *Superficial* structures are those close to the body surface; *deep* structures are those farther away from the surface. *Parietal* usually refers to a body wall or to a membrane lining a wall; for example, the outer membrane, lining the thoracic or abdominal cavities is a parietal membrane. *Visceral* is a term used to indicate an internal organ; for example, the visceral peritoneum is a membrane that covers the viscera, or internal organs, of the abdominal cavity. *Peripheral* refers to a position toward an external surface.

Directional terms relating to the hand and foot are sometimes difficult to

TABLE 5·1

Terms of reference

Terms	Definition	Example
Superior	Toward the head	The head is superior to the thorax.
Inferior	Toward the feet	The feet are inferior to the knees.
Anterior, or ventral	Front	The nose is anterior to the ear.
Posterior, or dorsal	Back	The vertebral column is posterior to the sternum (breastbone).
Medial	The midline	The linea alba lies along the mesal, or midsagittal, plane.
Lateral	Toward the side	The arms are lateral to the trunk.
Proximal	Nearer the midline or nearer the trunk	The shoulder is proximal to the elbow.
Distal	Farther from the midline or trunk	The fingers are distal to the wrist.
Parietal	Refers to a body wall or to the membrane lining the wall	The thoracic wall is lined by a membrane, the parietal pleura.
Visceral	Refers to an internal organ	The visceral pleura covers the lungs.
Peripheral	Toward the external surface	The skin contains peripheral sensory receptors.
Superficial	Structures closer to the surface, such as superficial muscles	The external oblique muscle is superficial to the internal abdominal oblique muscle.
Deep	Structures more internal	The interossei are deep muscles of the hand.

understand. The palm of the hand is the *volar,* or *palmar,* surface, while the back of the hand is *posterior,* or *dorsal.* The sole of the foot is termed the *plantar* surface, whereas the opposite, or upper, side is called *dorsal.* To interpret these terms, it helps to think of the location of the corresponding structure on a four-legged animal.

PLANES OF REFERENCE   It is often useful to consider *planes of reference,* as well as locational terms, when speaking of regions of the body. Organs and various body structures are easy to locate when the plane in which they are viewed is specified. For example, we may speak of *sagittal sections,* meaning that the body or organ has been cut into right and left portions. A *midsagittal plane* divides a structure into equal right and left halves. *Parasagittal* refers to a plane made on either side, parallel to the median plane.

Coronal, or *frontal,* sections divide the body into anterior, or ventral, portions and posterior, or dorsal, parts. A coronal section of the head, for example, divides the head into an anterior portion, which includes the face, and a posterior portion, which includes the back, or occipital, region. These sections are illustrated in Fig. 5.7.

A *horizontal,* or *transverse plane,* divides the body into superior and inferior portions. Such a plane lies at right angles to the longitudinal planes (sagittal and coronal). Horizontal sections of organs and tissues are commonly called *cross sections.* If tissue sections are cut along the longitudinal axis, they are termed *longitudinal sections.* Oblique tissue sections also may be made.

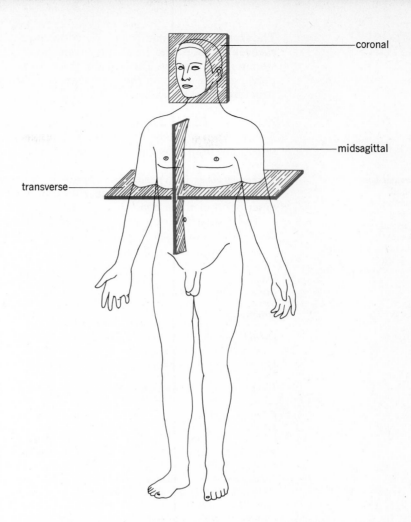

coronal

midsagittal

transverse

FIGURE 5·7
Anatomical planes of
reference.

BODY CAVITIES   The cavities of the body are divided into *dorsal* and *ventral*
cavities. Each of these cavities is, in turn, divided into two parts. It should be
understood that these cavities are filled completely with organs and body
fluids. The dorsal cavity is subdivided into a *cranial cavity,* which contains
the brain, and into a *spinal cavity,* which contains the spinal cord. The
ventral cavity is subdivided into the *thoracic* and *abdominopelvic* cavities.
The thoracic cavity contains the lungs, the heart with its attached blood
vessels, and several other structures, such as the thymus, esophagus, trachea,
nerves, and lymphatic vessels.

The abdominal cavity is separated from the thoracic cavity by the
thoracic diaphragm. This large cavity contains the stomach, intestine, liver,
pancreas, kidneys, and spleen. There is no distinct separation between the
abdominal and pelvic cavities, and they are commonly considered together
as the abdominopelvic cavities. Within the pelvic cavity we find the urinary
bladder, the internal reproductive organs, and the terminal portion of the
large intestine.

The skeleton is commonly considered to consist of two divisions. The *axial skeleton* includes the head, the hyoid bone, the vertebral column, the sternum, and the ribs. The bones of the arms and legs plus the shoulder and pelvic girdles make up the *appendicular skeleton.* Terms related to bone structure are summarized in Table 5.2.

THE SKULL   The skull bones form the skeletal structure of the head and face. For purposes of study, bones included in the cranium may be considered separately from those that form the skeletal structure of the face.

The cranial bones
Frontal
Occipital
Parietal
Temporal
Sphenoid
Ethmoid

The cranial bones   The parietal and temporal bones are paired, while the other cranial bones are single, making a total of eight bones in the cranium. A description of these bones is summarized in Table 5.3.

The cranial bones are joined by irregular lines of fibrous tissue that unite the edges of the bones, forming a *suture:* such joints are known as *immova-*

TABLE 5·2
Bone structure terms

Structure	Example
Condyle, a rounded eminence of bone	The occipital condyle
Crest, a prominent ridge	The crest of the tibia
Epicondyle, an eminence on or above a condyle	The epicondyles of the femur
Facet, a small, smooth articulating surface	The articulating surface of a vertebra
Fissure, a groove or cleft	The fissures of the sphenoid bone
Foramen, an opening (foramina, plural)	The foramen magnum in the occipital bone
Fossa, a depression	The coronoid fossa of the humerus
Fovea, a small pit	As in the head of the femur
Head, a rounded articulating part of a bone. There is often a narrowed neck portion adjacent to it	The proximal articulation of the humerus or femur
Lacuna, a small or minute cavity	The space occupied by a cartilage or bone cell.
Line, a low, linear ridge	The linea aspera of the femur
Notch or *incisura,* a deep indentation, notably along the border of a bone	The greater sciatic notch of the ilium
Process, a prominent, roughened outgrowth	The coracoid process of the scapula
Sinus, a cavity or air space within certain skull bones	The frontal sinus
Spine, a bony projection	The spine of the scapula
Spinous process, often a sharp, pointed projection	The spinous process of a thoracic vertebra
Sulcus, a furrow or groove in a bone	The radial sulcus of the humerus
Trochanter, a large bony process	The greater and lesser trochanters of the femur
Tubercle, a small, rounded prominence	The tubercles of the humerus
Tuberosity, a large prominence, usually roughened	The deltoid tuberosity of the humerus

TABLE 5·3

Cranial bones

Name	Description
Single bones	
Frontal	Forms the anterior part of the cranium or forehead; also the upper part of the orbits or eye sockets. Contains the frontal sinuses.
Occipital	Forms the posterior part of the cranium and part of the base of the skull. Contains a large opening, the foramen magnum.
Sphenoid	Constitutes a part of the base of the skull; also a part of the cranial wall. The body of the bone is wedged in just anterior to the occipital, and the greater wings form a part of the side wall anterior to the temporal bone. Contains the sella turcica.
Ethmoid	Located anterior to the body of the sphenoid. Contains the ethmoidal sinuses. The crista galli and cribriform plates are a part of this bone.
Paired bones	
Parietal	Lie between the frontal and occipital bones on either side and form part of the sides and top of the cranium.
Temporal	Located below the parietal bones on either side. Contains the inner ear.
Auditory ossicles **6 paired bones** Malleus (2) Incus (2) Stapes (2)	Tiny inner ear bones located in the temporal bones. They function in the transmission of sound.

ble joints. The principal sutures are the *coronal,* between the frontal and parietal bones; the *lambdoidal,* between the occipital and parietal bones; and the *sagittal,* which extends along a midsagittal line from the base of the nose to the posterior fontanel. The skull bones, shown in Fig. 5.8, have not completed their ossification at birth. In the skull of a newborn infant, the sagittal suture divides the frontal bone into two parts (Fig. 5.9). The space between the frontal bones and the parietal bones is the *anterior fontanel.* This is the largest of several spaces that occur at the angles of union of the skull bones. Membranous tissue is found between the cranial bones at the fontanels and the sutures. The anterior fontanel normally closes at about eighteen months of age. The posterior fontanel is located at the posterior end of the sagittal suture between the occipital and parietal bones. It is much smaller than the anterior fontanel and closes a few months after birth.

The *frontal bone* forms the anterior part of the cranium, that is, the forehead. It also forms the upper part of the *orbits,* or eye sockets, and a small anterior part of the nasal cavity. Above the orbits are the supraorbital ridges, and beneath these ridges lie two cavities termed the *frontal sinuses* (Fig. 5.10).

The *occipital bone* forms the back of the cranium and a considerable part of the base of the skull. The large opening at the base of the skull through which the spinal cord passes is the *foramen magnum* of the occipital bone. The two *occipital condyles* located on either side of the foramen present a smooth *articular,* or joining, surface where the skull rests on the first cervical

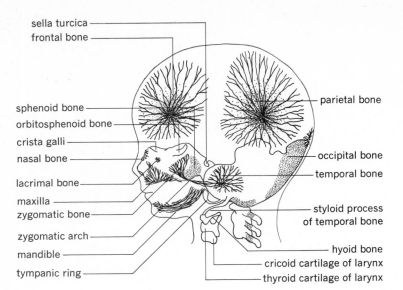

FIGURE 5·8

Skull of human embryo of about 12 weeks. (*After Hertwig and Patten, in Bradley M. Patten, "Human Embryology," 3d ed., McGraw-Hill Book Company, New York, 1968.*)

Labels in figure:
sella turcica
frontal bone
sphenoid bone
orbitosphenoid bone
crista galli
nasal bone
lacrimal bone
maxilla
zygomatic bone
zygomatic arch
mandible
tympanic ring
parietal bone
occipital bone
temporal bone
styloid process of temporal bone
hyoid bone
cricoid cartilage of larynx
thyroid cartilage of larynx

vertebra. The external occipital protuberance can be felt along the median line posteriorly.

Between the frontal and occipital bones are the right and left *parietal bones*. These form a considerable part of the sides and top of the cranium.

The *temporal bones,* located below the parietal bones on either side, form a part of the sides and base of the skull (Fig. 5.11, Plate 1). The thick *petrous portion* contains the hearing apparatus. The *styloid process* projects downward from the undersurface. Above and anterior to the petrous portion, the bone becomes relatively thin and platelike. This is the *squamous part* of the temporal bone. From its lower lateral surface, the *zygomatic process* extends outward to meet the temporal process of the zygomatic bone, with which it forms the zygomatic arch. Below the base of the zygomatic process is the *mandibular fossa,* a depression into which the condyle of the *mandible,* or lower jaw, articulates.

FIGURE 5·9

Skull of a newborn infant, showing development of bones and fontanels.

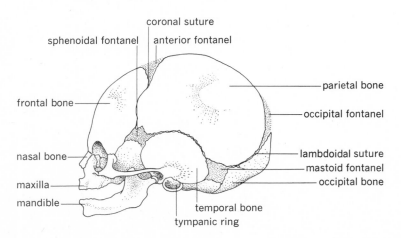

Labels in figure:
sphenoidal fontanel
coronal suture
anterior fontanel
frontal bone
nasal bone
maxilla
mandible
temporal bone
tympanic ring
parietal bone
occipital fontanel
lambdoidal suture
mastoid fontanel
occipital bone

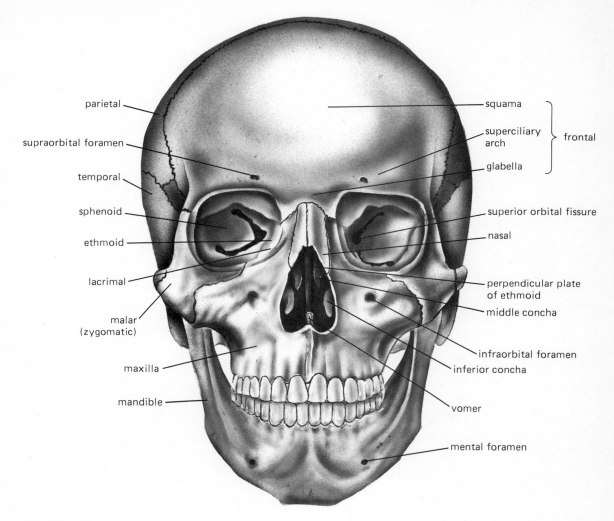

parietal

supraorbital foramen

temporal

sphenoid

ethmoid

lacrimal

malar
(zygomatic)

maxilla

mandible

squama

superciliary
arch

glabella

frontal

superior orbital fissure

nasal

perpendicular plate
of ethmoid

middle concha

infraorbital foramen

inferior concha

vomer

mental foramen

FIGURE 5·10
Anterior view of the skull.

The *mastoid portion* of the temporal bone lies posterior to the external auditory canal. The *mastoid process* projects downward from this portion. This part of the bone contains the *mastoid air cells,* air spaces which are connected with each other and with the cavity of the middle ear (Fig. 5.12). Infections of the middle ear can enter these air spaces, causing the condition known as *mastoiditis.* Since there is only a thin plate of bone between the mastoid air cells and the membrane covering the brain, an uncontrolled mastoid infection may spread to this membrane and cause a much more serious inflammation. When this happens, a mastoid operation, which consists of cutting into the mastoid process to permit external drainage of the infection, may be necessary.

Three small bones that form part of the hearing mechanism are located within the middle ear. These bones are the *malleus, incus,* and *stapes.* They will be considered later when we discuss the ear and audition.

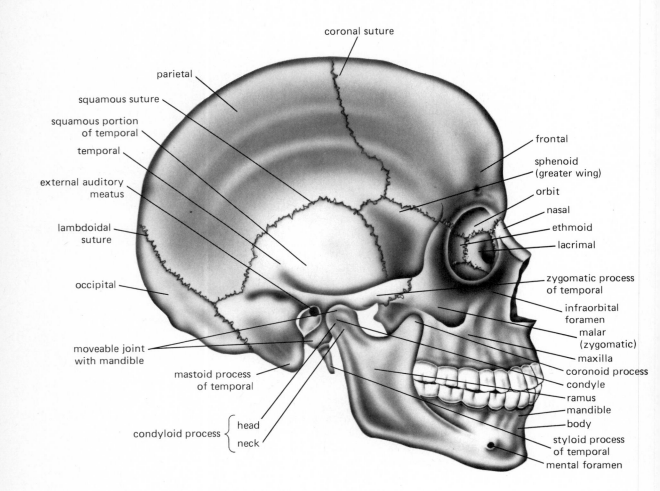

coronal suture

parietal

squamous suture

squamous portion
of temporal

temporal

external auditory
meatus

lambdoidal
suture

occipital

moveable joint
with mandible

mastoid process
of temporal

condyloid process { head
                    neck

frontal

sphenoid
(greater wing)

orbit

nasal

ethmoid

lacrimal

zygomatic process
of temporal

infraorbital
foramen

malar
(zygomatic)

maxilla

coronoid process

condyle

ramus

mandible

body

styloid process
of temporal

mental foramen

FIGURE 5·11

Lateral view of the skull: the
mandible has been disarticu-
lated and slightly turned. Portions of the lateral sides of the
temporal and lacrimal bones
can be seen in the skull; complete lateral views are shown
separately.

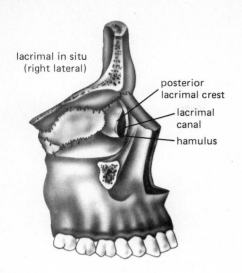

lacrimal in situ
(right lateral)

posterior
lacrimal crest

lacrimal
canal

hamulus

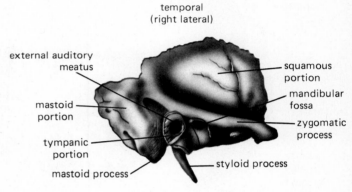

temporal
(right lateral)

external auditory
meatus

squamous
portion

mandibular
fossa

mastoid
portion

zygomatic
process

tympanic
portion

mastoid process

styloid process

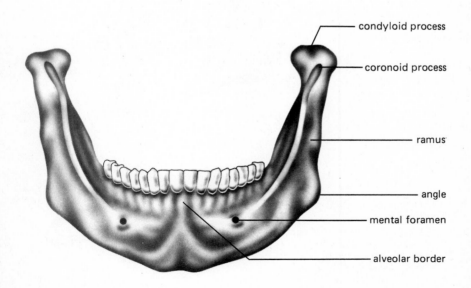

condyloid process

coronoid process

ramus

angle

mental foramen

alveolar border

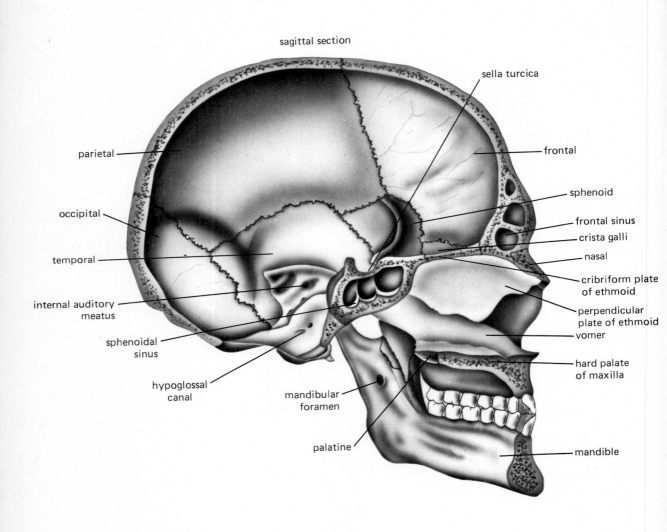

sagittal section

sella turcica

parietal

frontal

occipital

sphenoid

frontal sinus

crista galli

temporal

nasal

cribriform plate
of ethmoid

internal auditory
meatus

perpendicular
plate of ethmoid

vomer

sphenoidal
sinus

hard palate
of maxilla

hypoglossal
canal

mandibular
foramen

palatine

mandible

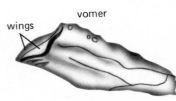

vomer

wings

FIGURE 5·12

Starting with a sagittal section,
the interior bones have been
progressively removed to expose the left wall of the skull.

parasagittal section

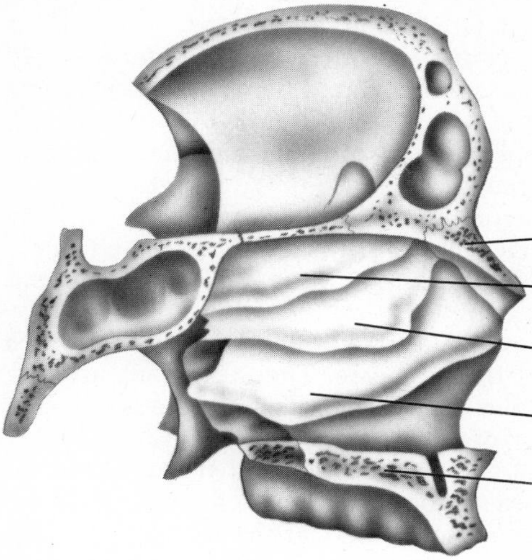

nasal

superior
concha

middle
concha

inferior
concha

hard palate
of maxilla

medial view of right wall

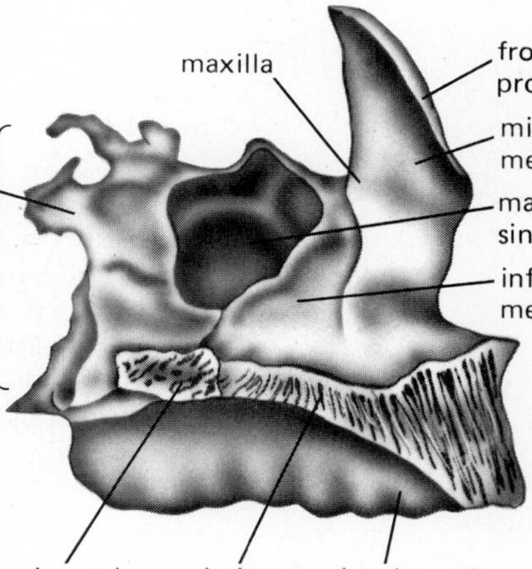

maxilla

frontal
process

middle
meatus

maxillary
sinus

inferior
meatus

palatine

perpendicular
plate

horizontal
plate

palatine
process

alveolar
process

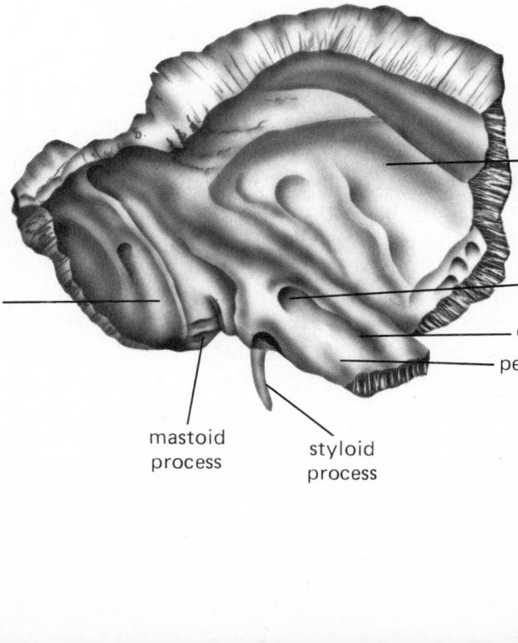

squamous
portion

right medial
view of
temporal

mastoid
portion

internal
auditory meatus

carotid canal

petrous portion

mastoid
process

styloid
process

The *sphenoid bone* is somewhat more difficult to locate. It is wedged in between other cranial bones to form a part of the anterior cranial structure. The name *sphenoid* is taken from a Greek word that means "wedge." The bone consists of a *median body portion* and three sets of lateral paired processes: the *greater wings,* the *lesser wings,* and the *pterygoid processes* (Fig. 5.13).

The body of the sphenoid lies between the ethmoid and occipital bones. It contains two large air spaces, the *sphenoidal sinuses,* that are separated by a thin layer of bone. The sinuses drain into the nasal cavity. A depression, or fossa, called the *sella turcica,* meaning "Turk's saddle," is located on the upper body of the sphenoid. The *hypophysis,* or pituitary body, rests in this fossa.

The greater wings of the sphenoid apread out from the body and form a part of the anterolateral floor of the braincase. They then turn upward, where they can be seen from the outside, and form a part of the side wall of the skull just anterior to the temporal bone on either side (Fig. 5.14).

The lesser wings are somewhat anterior and on a higher plane than the greater wings. They are best seen as forming a part of the floor of the braincase, but they also form the posterior part of the roof of the orbit.

The pterygoid processes project downward from the body of the sphenoid, where they are located in the posterior part of the side walls of the nasal cavity.

The *ethmoid bone* is a cranial bone of lightweight construction, located anterior to the body of the sphenoid bone. Its lateral masses contain labyrinths of air spaces called the *ethmoidal sinuses.* Its perpendicular plate forms the upper part of the nasal septum. The *cribriform plates* are horizontal portions of the ethmoid bone that lie in the floor of the braincase anteriorly. They also form the roof of the nasal cavity. These plates are perforated by tiny openings that permit the passage of the fibers of the olfactory nerve from the nasal mucosa to the olfactory bulbs of the brain. Between the cribriform plates, a crest of bone rises to furnish attachment for the *falx cerebri,* a portion of the membrane covering the brain. The crest of bone is called the *crista galli* (cock's comb). The ethmoid bone forms a part of the orbit of the eyes. From the sides of the labyrinth portion, two pairs of thin, scroll-shaped bones project into the nasal cavity. These are the superior and middle nasal conchae.

The facial bones
Mandible
Maxillae
Palatine bones
Vomer
Zygomatic bones
Lacrimal bones
Nasal bones
Inferior nasal conchae

The facial bones The bones of the face are all paired except for the mandible and the vomer. Their descriptions are summarized in Table 5.4.

The *mandible* is the lower jawbone. It has a U-shaped horizontal part and two vertical parts, known as the *rami.* Each ramus is divided into an anterior *coronoid process* and a posterior *condyloid process.* The temporal muscle is attached to the coronoid process. The condyle of the condyloid process articulates on the temporal bone. The teeth of the lower jaw are embedded in cavities in the upper part of the alveolar border. There is an anterior opening on either side of the mandible, called the *mental foramen,* that permits the passage of the inferior mental nerve.

At birth the right and left sides of the mandible are united anteriorly by fibrous tissue. The two sides start to grow together during the first year, and the union is usually complete by the end of the second year. At birth the rami are short and slope away from the body instead of standing at a sharp

TABLE 5·4
The facial bones

Name	Description
Single bones	
Mandible	The lower jaw. Holds the lower set of teeth. There are two processes, an anterior coronoid and a posterior condyloid.
Vomer	Forms the lower and posterior part of the nasal septum.
Hyoid	Suspended from styloid process of temporal bone. The tongue and several tongue muscles attach to it (not a true skull bone).
Paired bones	
Maxilla	The upper jaw. The alveolar process supports the upper set of teeth. Contains the maxillary sinus.
Palatine	Forms the posterior part of the hard palate, a part of the floor and lateral wall of the nasal cavity, and a portion of the orbital floor.
Zygomatic	The cheekbone. Forms the anterior part of the zygomatic arch and a part of the lateral wall of the orbit.
Lacrimal	Located at the inner angle of the orbit. A thin, slender bone. A groove provides for the passage of the tear duct.
Nasal	Forms the bony part of the nasal bridge.
Inferior nasal concha	A thin, scroll-like bone projecting into the nasal cavity below the superior and middle conchae of the ethmoid.

angle as they do in the adult. The body of the bone lengthens as the child grows in order to accommodate the molar teeth. The mandible also deepens as the upper alveolar border, which supports the teeth, is built up. In old age the alveolar border is absorbed after the teeth are lost, and the chin then becomes more prominent. Some old persons are therefore able to raise the chin until it comes close to the nose.

The *maxillae* are paired bones that unite to form the upper jaw. There are four processes arising from each bone: the *palatine, alveolar, zygomatic,* and *frontal processes*. The maxillae form the greater part of the floor of the orbits, the side walls and the floor of the nasal passage, and the anterior part of the roof of the mouth (Fig. 5.15). The palatine processes form a part of the floor of the nasal passage and the anterior part of the hard palate. The condition known as *cleft palate* occurs when the two palatine processes fail to unite before birth. This condition varies in severity. In the simple form the uvula is divided. In a much more complicated form, the cleft may extend through both the soft and hard palates and involve the lips. A cleft in the lip, sometimes referred to as *harelip,* may appear on either side or both, but never appears in the medial position. The median nasal process is occasionally absent.

The alveolar process supports the teeth of the upper jaw.

The zygomatic process articulates with the zygomatic bones.

The frontal process arises almost vertically from the body of the maxilla and forms the greater part of the skeleton of the outer nose.

Two large cavities located on either side of the nasal passage are called the *maxillary sinuses.* They are located in the body of each bone, and they open into the nasal cavity.

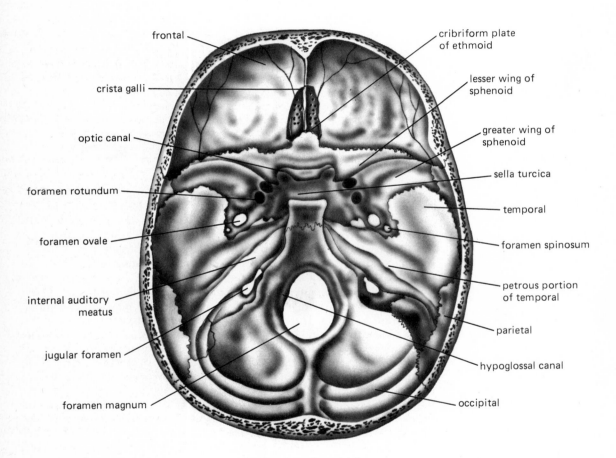

frontal

cribriform plate
of ethmoid

crista galli

lesser wing of
sphenoid

optic canal

greater wing of
sphenoid

sella turcica

foramen rotundum

temporal

foramen ovale

foramen spinosum

petrous portion
of temporal

internal auditory
meatus

parietal

jugular foramen

hypoglossal canal

foramen magnum

occipital

superior view of cranial floor

FIGURE 5·13
Portions of the superior views
of the ethmoid and sphenoid
bones can be seen in the floor
of the cranial cavity.

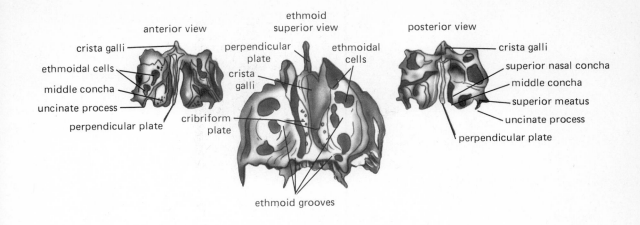

anterior view

crista galli
ethmoidal cells
middle concha
uncinate process
perpendicular plate

ethmoid superior view

perpendicular plate
crista galli
cribriform plate
ethmoidal cells

ethmoid grooves

posterior view

crista galli
superior nasal concha
middle concha
superior meatus
uncinate process
perpendicular plate

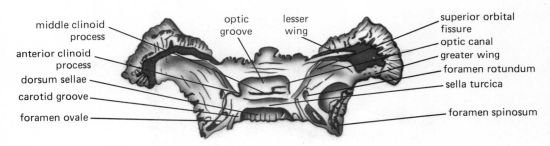

sphenoid superior view

middle clinoid process
anterior clinoid process
dorsum sellae
carotid groove
foramen ovale

optic groove
lesser wing

superior orbital fissure
optic canal
greater wing
foramen rotundum
sella turcica
foramen spinosum

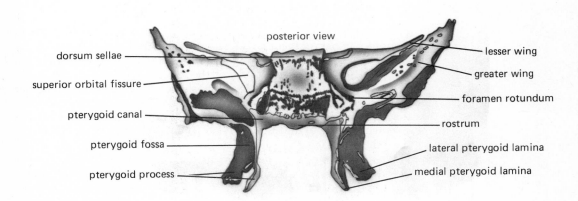

posterior view

dorsum sellae
superior orbital fissure
pterygoid canal
pterygoid fossa
pterygoid process

lesser wing
greater wing
foramen rotundum
rostrum
lateral pterygoid lamina
medial pterygoid lamina

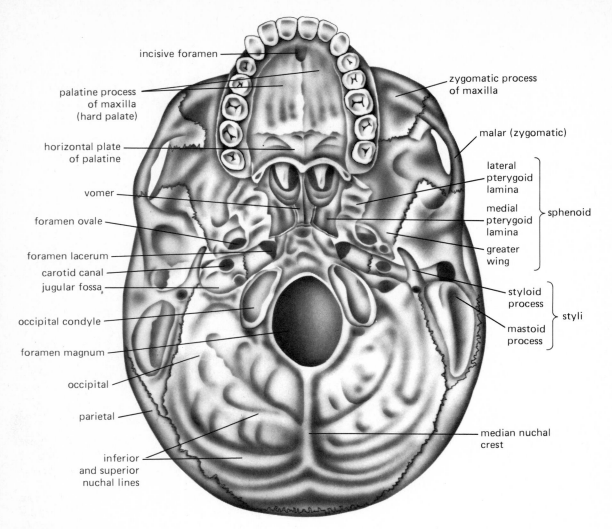

incisive foramen

palatine process
of maxilla
(hard palate)

horizontal plate
of palatine

vomer

foramen ovale

foramen lacerum

carotid canal

jugular fossa

occipital condyle

foramen magnum

occipital

parietal

inferior
and superior
nuchal lines

zygomatic process
of maxilla

malar (zygomatic)

lateral
pterygoid
lamina

medial
pterygoid
lamina   sphenoid

greater
wing

styloid
process
                  styli
mastoid
process

median nuchal
crest

FIGURE 5·14
Inferior view of the skull.

The *palatine bones* form the posterior part of the hard palate; vertical
portions form a part of the lateral walls of the nasal cavity, posterior to the
maxilla. The vertical portions extend upward as far as the orbits, where the
orbital processes form a small part of the floor of the orbits.

The *vomer* (plowshare bone) is a thin, single bone that forms the lower
part of the nasal septum. Its lower anterior border is grooved to receive the
cartilage septum of the nose; its upper anterior border fuses with the
perpendicular plate of the ethmoid bone. The vomer often turns to one side
or the other, making the nasal passages of unequal size (see Fig. 5.12).

The *zygomatic,* or *malar, bones* are known commonly as the *cheekbones.*
They form a prominence below the orbits and also form a part of the outer
border and floor of the orbits. The temporal process of each bone projects

146 OSTEOLOGY: THE STUDY OF BONES AND SKELETON

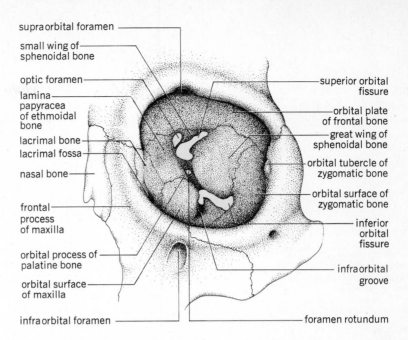

supraorbital foramen

small wing of sphenoidal bone

optic foramen

lamina papyracea of ethmoidal bone

lacrimal bone

lacrimal fossa

nasal bone

frontal process of maxilla

orbital process of palatine bone

orbital surface of maxilla

infraorbital foramen

superior orbital fissure

orbital plate of frontal bone

great wing of sphenoidal bone

orbital tubercle of zygomatic bone

orbital surface of zygomatic bone

inferior orbital fissure

infraorbital groove

foramen rotundum

FIGURE 5·15
Left orbital cavity.

posteriorly to meet the zygomatic process of the temporal bone. These two processes form the zygomatic arch on either side of the skull.

The *lacrimal bones* are located at the inner angle of the orbits directly posterior to the frontal processes of the maxillae. These are the smallest bones of the face. Each bone is grooved opposite a similar groove in the frontal process of the ethmoid bone, thus forming the opening through which the tear duct passes into the nasal cavity.

The *nasal bones* are two small, elongate bones that lie side by side to form the upper part of the bridge of the nose. The nasal cartilages form the lower part of the structural framework and are largely responsible for its shape.

The *inferior nasal conchae* (turbinated bones) are thin, shelflike plates projecting into the nasal cavity on either side below the superior and middle conchae of the ethmoid bone. They are often described as resembling a scroll of parchment, since they turn on themselves. In the living person, the conchae are covered with mucous membrane. All the nasal conchae warm and moisten the air that passes over their warm, moist membranes. The air is also partially cleansed, since some of the dust particles adhere to the mucous membrane.

The paranasal sinuses of the frontal, ethmoid, sphenoid, and maxillary bones also open into the nasal cavity or into the nasopharynx by means of small openings. The nasal mucous membrane that lines the nose extends through these openings and lines the cavities of these bones. The sinuses are ordinarily air cavities. When infections of the upper respiratory tract, such as colds, cause the mucous membrane to secrete abnormally, the paranasal sinuses may be filled with secretions.

The *hyoid bone* is shaped like the letter U (Fig. 5.16). It is located in the neck just above the larynx (Adam's apple) and is the attachment site for the

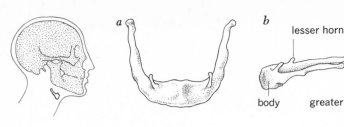

FIGURE 5·16

The hyoid bone: *a* frontal or anterior view; *b* lateral view.

tongue and several of the tongue muscles. The greater horns of the hyoid extend posteriorly from the body of the bone and are suspended by ligaments from the styloid processes of the temporal bones. The hyoid bone arises in the embryo from the cartilages of the second and third visceral arches. It is, therefore, not strictly a skull bone but a bone derived from the embryonic visceral skeleton, or splanchnocranium.

The vertebral column
7 cervical vertebrae
12 thoracic vertebrae
5 lumbar vertebrae
5 sacral vertebrae
4 coccygeal vertebrae

THE BONES OF THE TRUNK   *The vertebral column*   The *vertebral column,* or backbone, is composed of individual bones called vertebrae (Fig. 5.17). The column is divided into 7 cervical, 12 thoracic, 5 lumbar, 5 sacral, and 4 coccygeal vertebrae in the human embryo. Their descriptions are summarized in Table 5.5. Before the sacral vertebrae fuse to form the sacrum, and before the coccygeal vertebrae fuse, the total number of vertebrae is 33. In the adult, however, the sacrum is fused into a single bone and the coccyx is usually more or less fused, thus reducing the number of vertebrae to 26.

TABLE 5·5

Bones of the vertebral column and thorax

	No. of bones	Description
Vertebral column		
Cervical vertebrae	7 (the atlas, axis, and 5 others)	The atlas is the first cervical (neck) vertebra. It turns on the second cervical vertebra, or axis. The first two are highly specialized to permit turning the head
Thoracic vertebrae	12	Have facets for the attachment and support of the 12 pairs of ribs
Lumbar vertebrae	5	The largest vertebrae, located between the thorax and pelvis, is in the small of the back
Sacrum	5	Fused vertebrae
Coccyx	4	Fused, small, poorly developed
Thorax		
Ribs	12 pairs	True ribs, seven pairs attached to sternum by costal cartilages; the eighth, ninth, and tenth pairs attached upward by costal cartilages; the eleventh and twelfth ribs are not attached anteriorly and are floating ribs
Costal cartilages	10 pairs	Attach to first 10 pairs of ribs
Sternum	1	The breastbone, composed of three parts: the manubrium, body, and xiphoid, or ensiform, process

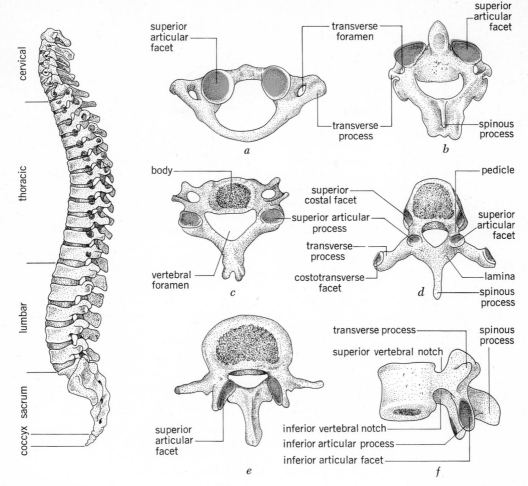

cervical

thoracic

lumbar

coccyx sacrum

superior
articular
facet

transverse
foramen

superior
articular
facet

transverse
process

spinous
process

a

b

body

superior
costal facet

superior articular
process

vertebral
foramen

transverse
process

costotransverse
facet

pedicle

superior
articular
facet

lamina

spinous
process

c

d

transverse process

spinous
process

superior vertebral notch

superior
articular
facet

inferior vertebral notch

inferior articular process

inferior articular facet

e

f

FIGURE 5·17

The vertebral column, illustrating individual vertebrae: *a*
atlas; *b* axis; *c* fourth cervical;
d sixth thoracic; *e* third lumbar, superior view; *f* lateral
view.

The vertebral column supports the trunk and protects the spinal cord.
The individual vertebrae vary in structure, but in general they consist of a
body, or *centrum,* which supports the weight of the trunk, and an *arch,*
which, with the body, forms the *vertebral foramen,* the opening through
which the spinal cord passes.

Each vertebra has a spinous process that projects dorsally. Ligaments and
muscles of the back attach at this process. Two transverse processes project
laterally and dorsally from a typical vertebra; muscles attach also at these
processes.

Articular processes, which form joints, join the vertebrae into a series or
column. The articulating surfaces are covered with cartilage, which promotes smooth motion of the vertebrae.

There are typically seven *cervical vertebrae* in the neck region of mammals. The first cervical vertebra is the *atlas* (Fig. 5.18). It is specialized to
support the skull and has large smooth surfaces that articulate with the

FIGURE 5·18
The atlas, anterior view.

ATLAS

occipital condyles to support the skull. The body of the atlas is not well developed, and it appears as an almost circular bone with two transverse processes. A posterior tubercle takes the place of a dorsal spinous process.

The *axis* is the second of the cervical vertebrae (Fig. 5.19). The *odontoid process,* a strong bony process, arises from its upper surface. The atlas pivots on this process when the head is turned. This process is strikingly specialized to permit freedom of head movement. The dorsal spine of the axis is thick and strong compared with the spines of other cervical vertebrae.

The seventh cervical vertebra has a long, narrow dorsal spine, which is not forked. In the cervical vertebrae the transverse processes on either side form a foramen, which transmits the vertebral artery.

The 12 thoracic vertebrae are larger and stronger than the cervical vertebrae. The dorsal spinous processes are long and project downward. The transverse processes are large and help to support the ribs. These transverse processes, except for those of the eleventh and twelfth thoracic vertebrae, have facets that articulate with the tubercles of the ribs. Demifacets on the body of each vertebra articulate with a rib head on either side.

The five *lumbar vertebrae* are large, and their dorsal processes are shaped somewhat like an ax. The transverse processes are thin. The fifth lumbar vertebra is not typical, since it is modified for articulation with the sacrum.

The *sacrum* is formed by the fusion of the five sacral vertebrae. It is somewhat triangular in shape, with the apex downward, and is held between the hipbones by very strong ligaments. The inner surface is smooth and concave. The outer surface is very irregular and convex. The sacral canal is the passageway for the posterior end of the spinal cord, where the roots of the sacral and coccygeal nerves lie.

The *coccyx* consists of four rudimentary vertebrae. The first coccygeal vertebra usually remains free until past middle age, when all four vertebrae fuse. The coccyx in humans may be regarded as a vestigial tail.

FIGURE 5·19
The axis.

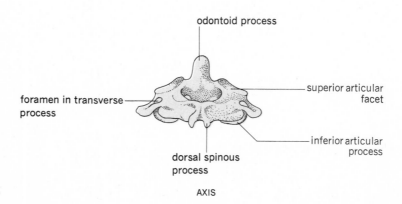

odontoid process

foramen in transverse process

superior articular facet

inferior articular process

dorsal spinous process

AXIS

The *vertebral column* is the dorsal support of the trunk, but it is not a rigid column. Individual vertebrae are separated by cartilage disks that absorb the shocks of walking and jumping. The vertebrae are bound together by ligaments. Muscles attached to the dorsal spines and to transverse processes act as levers and permit limited movement. The column of vertebrae can be bent forward or backward or to either side, and considerable rotation is possible.

We are often admonished to keep our spines straight, but such advice must be taken while keeping the natural lateral curvatures of the spine in mind. The column bends in natural dorsoventral curvatures. There are two primary curvatures: *thoracic* and *sacral*. Both are concave anteriorly and are present at birth. The cervical curvature is a secondary one and develops as the child begins to hold its head up and sit upright. It is convex anteriorly. The lumbar curvature is also a secondary, or compensating, curvature and is convex anteriorly. It forms the hollow of the back just above the sacrum in the lumbar region.

Good posture requires that the body be held erect so that none of the vertebral curvatures is exaggerated. When a person stands with head erect, chest up, and lower part of the abdomen held in, the ear, shoulder, hip, knee, and ankle should be in vertical alignment (Fig. 5.20).

In poor posture, the head tips forward, the shoulders and chest sag downward, and the abdomen protrudes. Under these conditions, the tho-

FIGURE 5·20

The position of the skeleton in lateral view: *a* good posture; *b* poor posture.

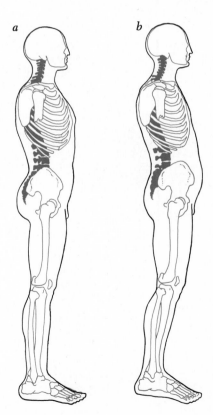

a   *b*

racic curvature may be accentuated and cause a round-shouldered or even a hunchbacked condition known as *kyphosis*.

In some individuals the lumbar curvature may be greater than normal, a condition called *lordosis*. This curvature is normally more pronounced in the female than in the male.

Scoliosis is a lateral curvature. The vertebral column often shows a slight curvature toward the right, but this curvature may become exaggerated by poor working conditions, such as writing at a desk that is too low. Scoliosis may produce a condition in which one shoulder is lower than the other or in which the hipbones are not level.

The thorax
Thoracic vertebrae
Ribs
Costal cartilages
Sternum

The thorax The *thorax* is formed by the thoracic vertebrae, the ribs, the costal cartilages, and the sternum (Fig. 5.21). Their descriptions are summarized in Table 5.5. It protects the heart, the thoracic blood vessels, and the lungs and helps support the bones of the shoulder girdle. The diaphragm forms the floor of the thoracic cavity.

There are 12 pairs of *ribs* in both male and female. The true ribs articulate with the thoracic vertebrae and are attached to the sternum by the costal cartilages. The first seven pairs are *true ribs*. The next three pairs—the eighth, ninth, and tenth—are *vertebrochondral ribs*. They attach indirectly to the sternum anteriorly by cartilages that join each other. The eleventh and twelfth pairs of ribs are tipped with cartilage, but they do not attach anteriorly. They are *vertebral*, or *floating ribs*.

A typical rib is a long, slender, curved bone. It articulates posteriorly with two vertebrae (Fig. 5.22). The first, tenth, eleventh, and twelfth ribs each articulate with a single vertebra. A tubercle just beyond the neck region of the rib articulates with the transverse process of the vertebra. The eleventh and twelfth ribs have no tubercle for articulation with a transverse process. At the anterior end, the rib is attached by a *costal cartilage*. The eleventh and twelfth ribs again are exceptions.

FIGURE 5·21
The thoracic cage.

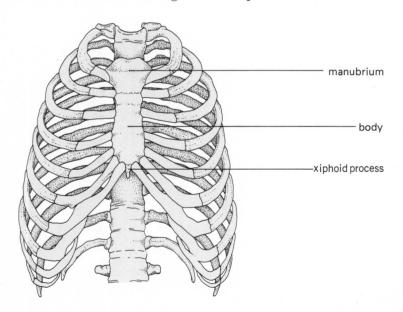

manubrium

body

xiphoid process

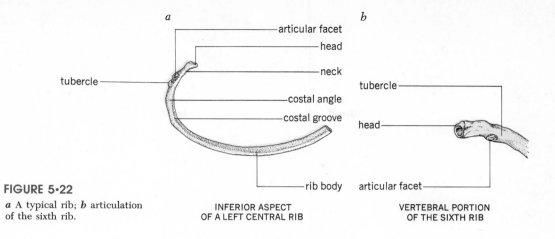

FIGURE 5·22

a A typical rib; *b* articulation of the sixth rib.

Labels on figure a (INFERIOR ASPECT OF A LEFT CENTRAL RIB): tubercle, articular facet, head, neck, costal angle, costal groove, rib body

Labels on figure b (VERTEBRAL PORTION OF THE SIXTH RIB): tubercle, head, articular facet

The *sternum* is the breastbone. It is a thin, flat bone located in the anterior part of the thorax along a median line. It is composed of three parts: the manubrium, the body, and the xiphoid, or ensiform, process. The *manubrium* lies at the top of the thorax; the clavicle (or collarbone) and the cartilages of the first pair of ribs attach to it. The cartilages of the second pair of ribs attach partly to the manubrium and partly to the body. The *body* of the sternum is the longest part. The costal cartilages of the second to the seventh pairs of ribs attach to it. The *xiphoid process* projects downward from the body of the sternum. It varies considerably in shape, and in youth and until middle age is composed of cartilage. During middle age it gradually ossifies, and in old age it becomes entirely bone and tends to fuse with the body.

THE APPENDICULAR SKELETON

The *appendicular skeleton* comprises the remaining bones (Fig. 5.23).

The appendicular skeleton
  Shoulder girdle, arms, hands
  Clavicles
  Scapulae
  Humerus
  Ulna
  Radius
  Carpal bones
  Metacarpal bones
  Phalanges

SHOULDER GIRDLE, ARMS, AND HANDS   The vertebrate skeleton of an animal that walks on all four legs may be described as a bridge of vertebrae supported by four pillars, the legs. The fact that man walks erect and that the arms have therefore undergone structural modification has not altered the basic skeletal plan. The arms articulate with the shoulder girdle; the legs support the pelvic girdle (Plates 1 and 2).

The shoulder girdle is composed of two clavicles and two scapulae (Fig. 5.24). It is not a complete girdle, since the sternum separates the clavicles anteriorly and muscles attach the scapulae to the trunk posteriorly. Descriptions of these bones are summarized in Table 5.6.

The *clavicles* are the collarbones. They can be felt anteriorly just above the thorax on either side. They are slender, curved brace bones that form the anterior part of the shoulder girdle. They articulate anteriorly with the manubrium and posteriorly with the acromion processes of the scapulae.

The *scapulae* present broad, flat surfaces for the attachment of shoulder muscles (Fig. 5.25). The inner surface of each bone is somewhat concave. The outer, or dorsal, surface is divided unequally by a ridge called the *spine*. The spine terminates as the *acromion process*. The clavicle articulates with

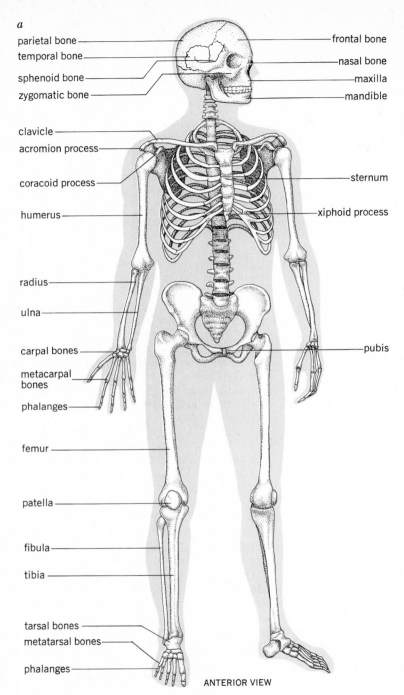

a

parietal bone —————

temporal bone —————

sphenoid bone —————

zygomatic bone —————

frontal bone

nasal bone

maxilla

mandible

clavicle —————

acromion process—————

coracoid process —————

humerus —————

sternum

xiphoid process

radius —————

ulna —————

carpal bones —————

metacarpal bones —————

phalanges —————

pubis

femur —————

patella —————

fibula —————

tibia —————

tarsal bones —————

metatarsal bones —————

phalanges —————

ANTERIOR VIEW

FIGURE 5·23

The skeleton: *a* anterior view;
b posterior view.

b

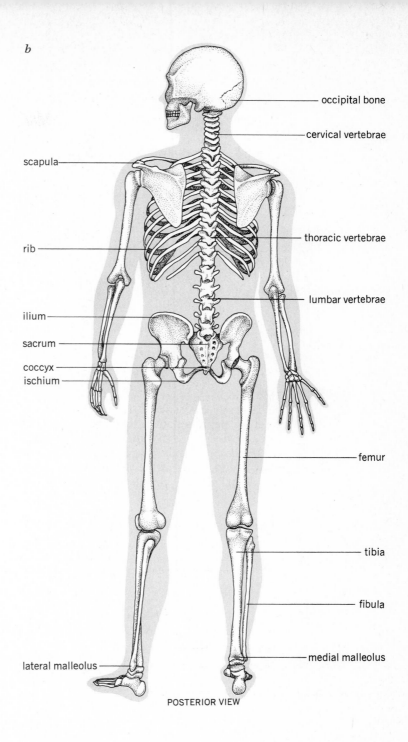

occipital bone

cervical vertebrae

scapula

thoracic vertebrae

rib

lumbar vertebrae

ilium

sacrum

coccyx

ischium

femur

tibia

fibula

medial malleolus

lateral malleolus

POSTERIOR VIEW

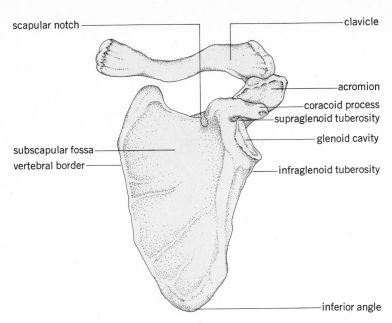

scapular notch

clavicle

acromion

coracoid process

supraglenoid tuberosity

glenoid cavity

subscapular fossa

vertebral border

infraglenoid tuberosity

inferior angle

FIGURE 5·24

Left shoulder girdle (anterior view).

THE LEFT SCAPULA, ANTERIOR VIEW

this process. Another process, the *coracoid*, projects upward and forward underneath the clavicle. Below these two processes there is a smooth, circular, depressed area: the *glenoid fossa*. The head of the humerus (the bone of the arm above the elbow) articulates in this cavity.

The *humerus* is the long bone of the upper arm. Its rounded head fits into the glenoid cavity of the scapula and forms a *ball-and-socket joint* at the shoulder (Fig. 5.26). Such joints permit rotating movements. Just beyond the

TABLE 5·6
Bones of the shoulder girdle, arms, and hands

Location	Name	Total no. of bones	Description
Shoulder	Clavicle (1)	2	The collarbone
	Scapula (1)		The shoulder blade, a large, thin, flat bone
Arm	Humerus (1)	2	The long bone of the arm
Forearm	Radius (1)	2	The bone that rotates; at its base it is on the thumb side
	Ulna (1)		Forms hinge joint at elbow; at its base it is on the little-finger side
Wrist		8	Two transverse rows of small bones in the wrist
Hand	Metacarpal (5)	19	Long bones of the palm
	Phalanges (14)		Bones of the fingers; two in the thumb, three in each finger

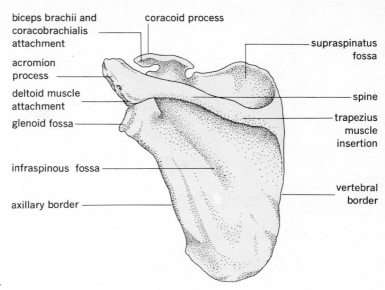

FIGURE 5·25
Left scapula (posterior view).

biceps brachii and coracobrachialis attachment

coracoid process

acromion process

deltoid muscle attachment

glenoid fossa

infraspinous fossa

axillary border

supraspinatus fossa

spine

trapezius muscle insertion

vertebral border

constricted neck region are two bony prominences; the *greater tuberosity* and the *lesser tuberosity* (Fig. 5.27). The greater tuberosity is a lateral projection and can be felt as the point of the shoulder. The lesser tuberosity projects anteriorly when the arm is resting at the side.

The area for a short distance below the tuberosities has been called the *surgical neck* of the bone because many fractures occur there.

The tendon for the long head of the biceps muscle lies in the groove between the tubercles, termed the *intertubercular groove*.

The shaft of the humerus is long and rather slender. The *deltoid tuberosity* is located about midway on the lateral margin. The deltoid muscle is inserted there.

The distal extremity of the humerus widens and flattens. There are two smooth articular surfaces. The outer *capitulum* articulates with the radius,

FIGURE 5·26

A synovial joint. Section through the shoulder joint, an example of a freely movable ball-and-socket type.

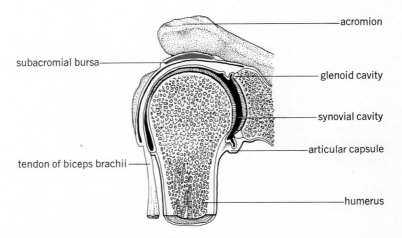

subacromial bursa

tendon of biceps brachii

acromion

glenoid cavity

synovial cavity

articular capsule

humerus

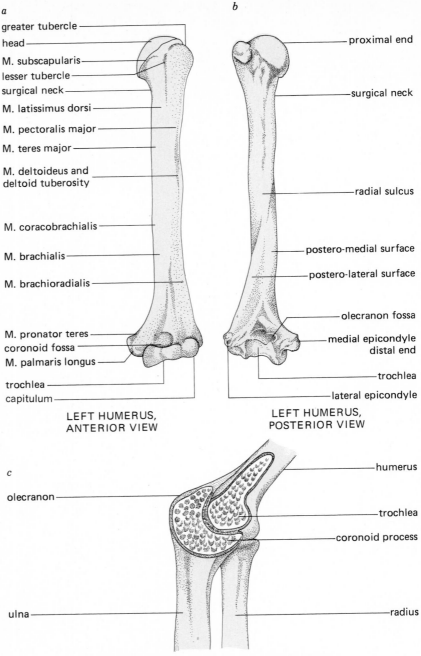

a

greater tubercle
head
M. subscapularis
lesser tubercle
surgical neck
M. latissimus dorsi
M. pectoralis major
M. teres major
M. deltoideus and
deltoid tuberosity

M. coracobrachialis

M. brachialis

M. brachioradialis

M. pronator teres
coronoid fossa
M. palmaris longus

trochlea
capitulum

LEFT HUMERUS,
ANTERIOR VIEW

b

proximal end

surgical neck

radial sulcus

postero-medial surface
postero-lateral surface

olecranon fossa
medial epicondyle
distal end

trochlea

lateral epicondyle

LEFT HUMERUS,
POSTERIOR VIEW

c

olecranon

ulna

humerus

trochlea

coronoid process

radius

SECTION THROUGH THE LEFT ELBOW JOINT

FIGURE 5·27

a Left humerus (anterior view) indicating muscle attachments; *b* left humerus (posterior view); *c* section through left elbow joint.

and the inner *trochlea* articulates with the ulna. The radius and the ulna are the bones of the forearm (Fig. 5.28). Above the capitulum on the anterior side is a shallow fossa, which receives the margin of the head of the radius when the arm is fully flexed.

On the posterior side, the *olecranon fossa* receives the olecranon process of the ulna when the arm is fully extended. The olecranon fossa is separated from the coronoid fossa by a thin, bony partition.

The *ulna* is the longer bone of the forearm. It can be felt along the back of the forearm from the elbow to the wrist. The proximal end of the bone presents two processes, the *olecranon* and the *coronoid;* between them is the *trochlear notch.* The olecranon is the point of the elbow, and although it is a strong process, it is often broken. When the arm is extended, the anterior tip of the process is received by the olecranon fossa of the humerus. The trochlear notch articulates against the trochlea of the humerus and forms a hinge joint. A *hinge joint* permits movement in only one plane. It may be of interest that the elbow is normally bent outward at an angle of about 10° in men and about 15° in women.

On the inner side of the coronoid process of the ulna, there is a smooth articular surface that articulates with the head of the radius. This is the radial notch. The distal extremity consists of the head of the ulna and a small

FIGURE 5·28
Left radius and ulna (anterior view).

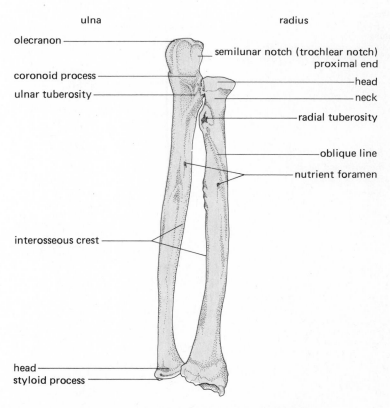

ANTERIOR VIEW OF THE BONES OF THE LEFT FOREARM

lateral process called the *styloid process*. It can be felt just above the wrist on the little-finger side.

The *radius* is a cylindrical bone resembling the spoke of a wheel. It lies *laterad* (toward the side) to the ulna on the same side of the forearm as the thumb. The head is a smooth, hollowed-out disk, which articulates with the outer capitulum of the humerus. As the hand is rotated, the radius pivots on the capitulum. This is a good example of a *pivot joint*. Below the constricted neck of the bone on the medial side, the tendon of the biceps brachii muscle attaches to the radial tuberosity. The lower extremity is flattened and presents two articular surfaces. One articulates with wrist bones, and the other, the *ulnar notch*, permits rotating movement on the ulna.

There are eight small bones in the wrist known as carpal bones (Fig. 5.29). (For a study of the individual wrist bones the reader is referred to a reference text on human anatomy.) The bones of the *carpus*, or wrist, are bound together by ligaments, yet their arrangement in rows permits some movement.

The *metacarpal bones* make up the skeletal framework of the hand. There are five bones in each hand. At their distal ends they articulate with the proximal row of *phalanges*, or finger bones. There are 14 phalanges, three for each finger and two for the thumb, or *pollex*.

The pelvic girdle, legs, and feet
Hipbones
 Ilium
 Ischium
 Pubis
Pelvis
Femur
Patella
Tibia
Fibula

THE PELVIC GIRDLE, LEGS, AND FEET   Descriptions of these bones are summarized in Table 5.7.

The *hipbones* (ossa coxae) are two large, flat bones that form the pelvic girdle. They are the broadest bones in the skeleton. They form a *symphysis* (a slightly movable joint) anteriorly, and the sacrum completes the girdle posteriorly (Fig. 5.30). The posterior appendages are attached to the pelvic girdle.

Each hipbone develops in the fetus as three separate bones: the ilium, the ischium, and the pubis. The *ilium* is the uppermost and largest of the three

FIGURE 5·29

Bones of the wrist and hand (palmar surface of hand).

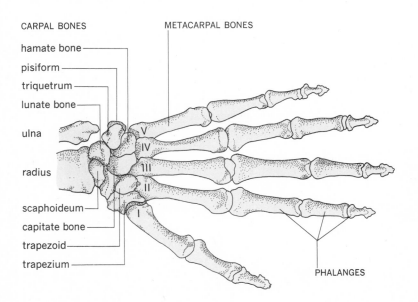

CARPAL BONES

METACARPAL BONES

hamate bone

pisiform

triquetrum

lunate bone

ulna

radius

scaphoideum

capitate bone

trapezoid

trapezium

PHALANGES

TABLE 5·7
Bones of the pelvic girdle, legs, and feet

Location	Name	Total no. of bones	Description
Hipbones	Ossa coxae (1)	1	The ilium, ischium, and pubis fuse to form the hipbone. The skeletal pelvis includes the sacrum and coccyx
Thigh	Femur (1)	1	The longest bone in the body
Knee	Patella (1)	1	The kneecap
Leg	Tibia (1)	2	Anteromedial bone of leg, the shin bone
	Fibula (1)		Lateral leg bone
Ankle	Calcaneus (1)	7	Ankle and heel bones
	Talus (1)		
	Cuboid (1)		
	Navicular (1)		
	Cuneiform (3)		
Foot	Metatarsal (5)	19	Bones that help to form the arches; form the framework of the foot
	Phalanges (14)		Toe bones; two in great toe, three in each of the others

parts. Its crest can be felt along the upper border of the hipbone. The anterior superior spine, from which anatomical and surgical measurements are made, can be located by following the crest forward to the most prominent anterior projection. Before the three parts of the hipbone fuse, the ilium is separated from the ischium and pubis by a suture that passes through the *acetabulum,* the cavity that receives the head of the femur (Fig. 5.31).

The *ischium* is the lower posterior portion of the hipbone. The *ischial tuberosities* are large, bony prominences that support the body in a sitting

FIGURE 5·30

Pelvic girdle: male pelvis (anterior view).

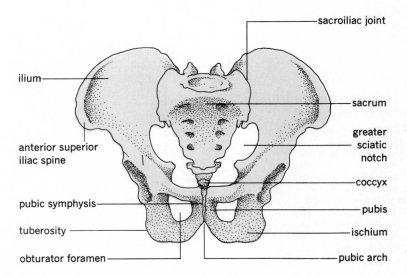

THE APPENDICULAR SKELETON   **161**

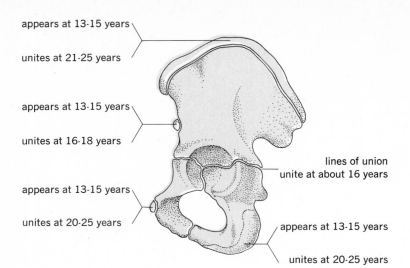

appears at 13-15 years

unites at 21-25 years

appears at 13-15 years

unites at 16-18 years

lines of union
unite at about 16 years

appears at 13-15 years

unites at 20-25 years

appears at 13-15 years

unites at 20-25 years

FIGURE 5·31

The hipbone during growth (lateral view). [*Redrawn from J. Parsons Schaeffer (ed.), "Morris' Human Anatomy," 12th ed., McGraw-Hill Book Company, New York, 1966.*]

position. The *obturator foramen*, formed by the ischium and the pubis on either side, is the largest foramen in the skeleton.

The lower anterior portion of each hipbone is the *pubis*. The pubic bones meet anteriorly along a median line and form a slightly movable joint, the *symphysis pubis*. A disk of fibrocartilage is placed between the two bones, and they are held together by ligaments. The symphysis is a useful landmark for pelvic measurements. Below the symphysis the pubic bones form the pubic arch.

The *pelvis* is considered to consist of two parts: an upper, or greater, pelvis, bounded laterally by the broad upper portions of the ilia, and the true pelvis, which lies below the brim. The true pelvis is bounded by the lower parts of the hipbones and posteriorly by the sacrum and coccyx.

The female pelvis differs from that of the male in many respects. The differences are due largely to the adaptation of the true pelvis of the female to childbearing. The female pelvis is wider, and the cavity is larger and shorter. The sacrum is shorter and broader, and the coccyx is more freely movable. The ischial tuberosities, as well as the acetabula, are farther apart. The pubic arch is much wider in the female pelvis than in the male (Fig. 5.32).

FIGURE 5·32

Differences between the pelves of the human female and male. Note that the femur is nearly vertical in the male but slopes inward from the wider female pelvis. The pubic arch is wider in the female pelvis than in the male, and the opening of the true pelvis of the female is larger and more oval-shaped than that of the male.

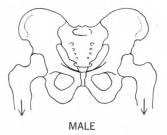

MALE

FEMALE

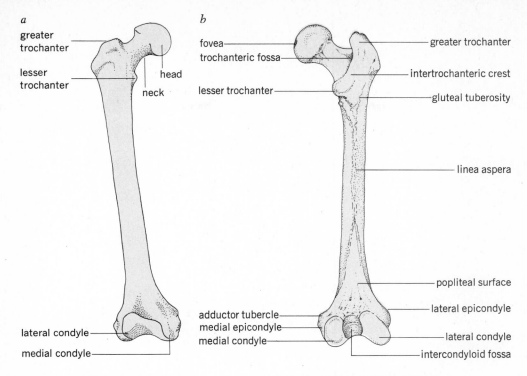

a
greater trochanter
lesser trochanter
head
neck

b
fovea
trochanteric fossa
lesser trochanter
greater trochanter
intertrochanteric crest
gluteal tuberosity
linea aspera
popliteal surface
lateral epicondyle
lateral condyle
intercondyloid fossa
adductor tubercle
medial epicondyle
medial condyle

lateral condyle
medial condyle

RIGHT FEMUR, ANTERIOR VIEW

THE RIGHT FEMUR, POSTERIOR VIEW

FIGURE 5·33
The femur (thigh bone):
a right femur (anterior view);
b posterior view.

The *sacroiliac joint* is a symphysis at the back of the pelvis. Tall, slender persons are especially subject to strain at this joint between the sacrum and the ilium, which may cause severe backache. The surface area above the sacrum is marked off by the posterior superior iliac dimple in the skin on either side.

The *femur,* or thigh bone, is the bone of the leg above the knee. It is a strong, heavy bone and the longest in the body. At the upper end of the bone there is a rounded head, which forms a ball-and-socket joint with the acetabulum in the hipbone. Below the long, constricted neck region there are two processes, the greater and lesser trochanters, that afford attachment for some of the muscles of the thigh and gluteal regions. The intertrochanteric crest lies between the two trochanters (Fig. 5.33).

Along the posterior side of the shaft there is a ridge, the linea aspera, to which muscles are attached. The lower end of the femur widens, and at the extremity there are two bony prominences, the *lateral* and *medial condyles.* The smooth articular surfaces form a *hinge joint* with the tibia. The patella also articulates with the lower extremity of the femur on the anterior patellar surface. The femur is inclined inward because of the width of the pelvis. The degree of inclination is greater in women than in men.

The *patella,* or kneecap, is a flat sesamoid bone embedded in the tendon of the quadriceps femoris muscle (Fig. 5.34). It is the largest sesamoid bone in the body, measuring about 2 in in both length and width, and is sur-

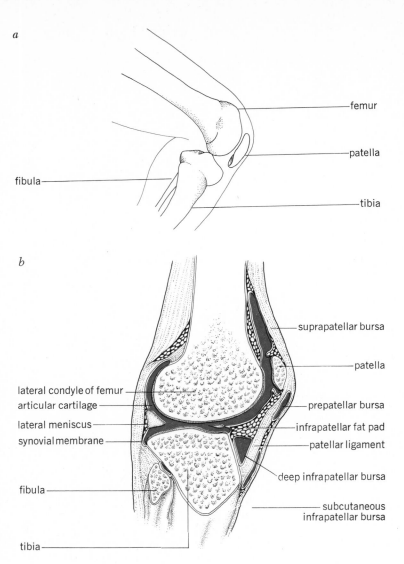

a

femur

patella

fibula

tibia

b

suprapatellar bursa

patella

lateral condyle of femur

articular cartilage

lateral meniscus

synovial membrane

prepatellar bursa

infrapatellar fat pad

patellar ligament

deep infrapatellar bursa

subcutaneous infrapatellar bursa

fibula

tibia

FIGURE 5·34

The knee joint: *a* lateral view; *b* section through knee joint.

rounded by bursae. The patella and the bursae serve to protect the knee joint. When one bends the knee, the patella becomes fixed against the intercondyloid notch of the femur. The ligamentum patellae is a strong ligament that extends from the patella to the tibia. When one kneels, the weight of the body rests partly on this stretched ligament and partly on the tubercle of the tibia rather than on the patella. Persons who work on their knees are subject to inflammations of the bursae (bursitis). The condition commonly referred to as "water on the knee" can result from injury to the bursae.

The *tibia*, or shin bone, is the larger bone of the leg below the knee (Fig. 5.35). The upper end of the bone widens and forms the *lateral* and *medial* condyles. Between the condyles there is a sharp projection, the *inter-*

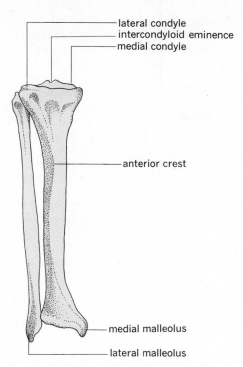

lateral condyle
intercondyloid eminence
medial condyle

anterior crest

medial malleolus

lateral malleolus

RIGHT FIBULA — RIGHT TIBIA
(ANTERIOR VIEW)

FIGURE 5·35

Right tibia and fibula (anterior view).

condyloid eminence. The articular surfaces that form the knee joint with the condyles of the femur are slightly concave. The shaft of the bone is roughly triangular, with the anterior apex sharp and well defined. The sharp anterior border of the shaft is known as the *shin.* It can be felt as a ridge along the anterior part of the leg. The lower end of the tibia articulates with the *talus,* an ankle bone. The lower end is not so wide as the upper extremity. There is a long process on the medial side known as the *medial malleolus.* It can be felt as the inner prominence of the ankle. On the opposite side there is a depressed articular surface where the lower extremity of the fibula articulates.

The *fibula* is a long, slender bone lateral to the tibia. Its head articulates with the tibia, but the articulation is below the knee joint. The lower end projects downward below the tibia and forms the *lateral malleolus,* the outer ankle bone. The talus is confined between the lateral malleolus of the fibula and the medial malleolus of the tibia (Fig. 5.35).

The foot bones
  Tarsal bones
  Metatarsal bones
  Phalanges

The foot bones Seven bones form the tarsus. The tarsal bones are the *calcaneus* (or heel bone), the *talus* (astragalus), *navicular bone, cuboid bone,* and the first, second, and third *cuneiform bones.* The tarsal bones are larger and more specialized than the carpal bones. The calcaneus, the largest of the tarsal bones, helps to support the weight of the body and affords attachment for the large muscles of the calf of the leg. The talus, which lies above the calcaneus and articulates with the tibia and fibula, is at the top of the longitudinal arch of the foot. The navicular bone is immediately anterior to the talus on the inner side of the foot. The cuboid bone is just anterior to the calcaneus on the lateral side. The three cuneiform bones form a row in front of the navicular, with the first cuneiform bone on the inner side (Fig. 5.36).

The five bones that make up the *metatarsus* are homologous with the matacarpal bones of the hand. The metatarsal bones articulate at their proximal ends with the tarsal bones to form the longitudinal arch of the foot. The metatarsal bones also form the longitudinal arch of the foot. The metatarsal bones also form the transverse arch. The metatarsal bones articulate distally with the phalanges.

The phalanges are the bones of the toes. They are homologous with the bones of the fingers. There are two phalanges in the great toe, and three in each of the other toes.

The foot has been modified to support the body in an erect position. The bones are so arranged that they form two *arches:* a *longitudinal* and a *transverse* arch. The arches are not rigid but yield as weight rests upon them, springing back into place as the weight is lifted. A well-functioning arch, therefore, gives a certain spring to the step.

FIGURE 5·36
a Homology of hand and foot; *b* arches of the foot (lateral view).

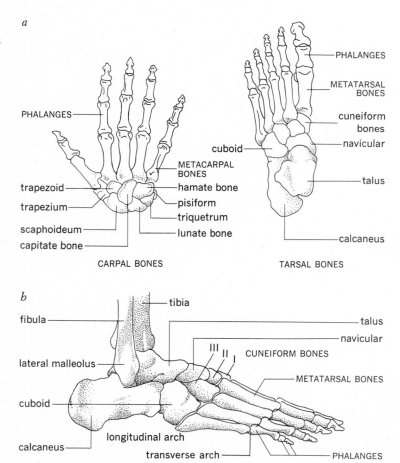

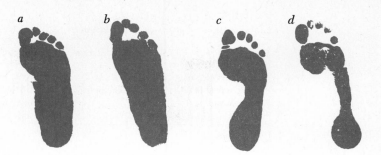

FIGURE 5·37

Prints of the foot: *a* flatfoot; *b* pronated flatfoot; *c* normal, moderate arch; *d* high arch. (*Courtesy of Paul W. Lapidus, M.D.*)

The bones of the foot that form the arches are bound together by ligaments. The arches are also supported by the muscles and tendons of the foot. If the muscles and tendons become weakened from malnutrition or occupational conditions that require long hours of standing without much exercise, the ligaments may allow the arches to sag and produce the condition known as *flatfoot.* The tarsal bones are often forced toward the inner side of the foot in a pronated position (Fig. 5.37).

Muscular exercises, such as standing on the toes, can do much to strengthen the musculature of the foot and can aid in restoring strength to the arch. Artificial arch supports should be used only if there seems to be no chance of restoring the normal condition of the arch by strengthening the muscles.

Weakening of the transverse arch often causes pain at the base of the toes. Wearing high-heeled shoes forces most of the weight of the body onto the ends of the metatarsal bones. They may enlarge and push sideways out of line. The inner line of the shoe should be straight, not curved outward at the toe. The painful enlargement at the joint between the first metatarsal bone and the great toe is called a *bunion.*

ARTICULATIONS

Various kinds of joints or articulations have been mentioned in considering the skeletal system (see Table 5.8). Joints are commonly classified according to the amount or kind of movement associated with the articulation. The study of articulation is called *arthrology.* The better-known term *arthritis* refers to inflammation of the joints.

Joints may be described as immovable, slightly movable, and freely movable. They may be classified also, in the same order, as fibrous, cartilaginous, and synovial joints.

Immovable joints
  Cranial sutures

IMMOVABLE JOINTS There are several kinds of *immovable joints.* The sutures between cranial bones are an example. These bones are fitted together by an interlocking arrangement and are held together by fibrous connective tissue. The sutures often become more ossified in the later years.

Slightly movable joints
  Intervertebral disks
  Pelvic articulations
    (pregnancy)

SLIGHTLY MOVABLE JOINTS These are articulations that have a pad of fibrocartilage between two bones. The intervertebral disks, examples of *slightly movable joints,* have a core of more pliable tissue, which permits some flexibility in the vertebral column. The symphysis between the pubic bones is also of this type, although there is practically no movement. The

TABLE 5·8

Classification of joints or articulations

Type	Kind of articulation	Example
Immovable	Fibrous	Sutures between cranial bones
Slightly movable	Fibrocartilaginous	Joints between the vertebrae
Freely movable	Synovial	Hinge joints, ball-and-socket joints, pivot joints

sacroiliac articulation is a similar example. At the termination of pregnancy these pelvic articulations are slightly flexible occasioned by structural changes in the fibrocartilage.

Freely movable joints
  Ball-and-socket joints
    Shoulder joint
    Hip joint
  Hinge joints
    Elbow
    Knee
  Pivot joints

FREELY MOVABLE JOINTS   Articulations of this type fall into several categories (see Table 5.9). The *ball-and-socket* type are the most freely movable, but *hinge joints* permit free movement in a single plane. *Pivot joints* permit rotary movements, such as the atlas turning on the axis or the proximal end of the radius pivoting around the ulna. The articulating ends of bones are covered with hyaline cartilage. The joint is enclosed in a fibrous articular capsule, which is supported by the capsular ligament. The capsule is lined by a synovial membrane, which secretes synovial fluid to lubricate the articulation. The small space between the articular cartilages and enclosed by the synovial membrane is called the *articular cavity*. It contains synovial fluid. The articular cavity often is continuous with the cavities of the bursae.

SYNOVIAL BURSAE

Bursae are sacs of fibrous connective tissue lined with a synovial membrane. They are commonly located at joints to prevent the friction of one surface moving upon another. They may be subcutaneous; examples are the bursa over the patella or the olecranal bursa. They are found also between tendons or between muscles and bones, wherever friction is likely to occur. The synovial membrane secretes a viscid lubricating fluid. Mechanical injuries to bursae or infections can cause the fluid to accumulate. Inflammation of the bursa is called *bursitis*. Common locations of bursitis are subdeltoid (above the shoulder joint), at the elbow, around the calcaneus tendon, and around the knee.

TABLE 5·9

Types of freely movable joints

Type	Description	Example
Ball-and-socket joint	The ball-shaped head of one bone fitting into a socket in another bone	The head of the humerus fits into the scapula to form the shoulder joint
Hinge joint (ginglymus)	A rounded convex surface fitting into a concave surface; movement in one plane only	Elbow joint between the humerus and the ulna
Pivot joint	A shallow depression at the end of one bone rotating on a rounded surface	The radius pivots on the capitulum of the humerus

CLINICAL ASPECTS

DISEASES AND INJURIES OF THE BONES AND JOINTS *Sprain* When the ligaments and tendons of a joint have been stretched or torn away from their attachment to a bone, the condition is called a *sprain*. Sprains heal slowly and may require the attention of a physician for several weeks.

Tennis elbow "Tennis elbow," lateral epicondylitis, is an injury to a small bony prominence at the distal end of the humerus where some of the extensor muscles of the forearm arise. The strain on these muscles may develop from any rotating motion and not necessarily from playing tennis. Pain is felt on the outer side of the elbow and down the lateral side of the forearm. A similar but less well known condition on the medial side of the forearm is called "golfer's elbow."

Heat often relieves the pain, and the arm is sometimes placed in a sling to permit healing.

Fracture A *fracture* is a break in a bone. The bones of elderly persons are sometimes very brittle and break easily. The bones of children, which are not yet fully formed, contain a great deal of organic matter, so that when a child fractures a bone, the bone may not break in two but some of the fibers may be broken. Such a fracture is called a *greenstick fracture* because it resembles the way a green stick breaks. If the broken ends of a bone protrude through the skin, the fracture is classified as a *compound fracture*. If the skin is not broken, the fracture is called a *simple fracture* (Fig. 5.38).

Fractures must be treated by a physician. To prevent greater injury, it is often advisable not to move a fracture patient until the physician arrives. If the patient must be moved, splints and simple bandaging should be applied to support and immobilize the injured parts. In the case of a compound fracture, great care should be exercised to prevent infection (Fig. 5.39).

A physician should set the fracture, preferably with the aid of x-rays so that the ends of the bone can be placed together properly. A callus of new bone forms around the broken ends and holds them together. The periosteum plays an important part in the regeneration of new bony tissue.

Dislocation *Dislocations* occur when bones are forced away from their proper position at a joint. A dislocation usually involves strained or torn ligaments and considerable inflammation. Reduction of the dislocated bones should be performed by a physician, except perhaps in the case of dislocated fingers, which can be put back into place by a firm pull (Plates 37 and 38).

Ruptured disk *Ruptured*, or *slipped*, *disk* is a common back injury. The intervertebral disks of modified cartilage provide excellent padding between the vertebrae, but there is some degeneration and loss of elasticity of the disks with age. This condition may cause mild lower back pain. Sharp pain indicates that the soft center of the disk has been pushed outward. The disk usually pushes dorsally and exerts pressure on the roots of spinal nerves, causing pain. Surgical removal of the protruding part may be necessary.

Whiplash *Whiplash* usually occurs as a result of an automobile accident in which the head is suddenly and violently forced forward or backward. Muscles, ligaments, or tendons are injured at about the level of the fifth cervical vertebra. The term "whiplash" has been somewhat abused in

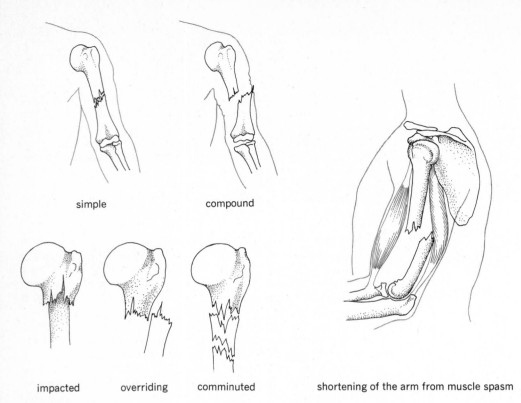

simple

compound

impacted

overriding

comminuted

shortening of the arm from muscle spasm

FIGURE 5·38

General classification of fractures. (*After William W. Stiles, "Individual and Community Health," McGraw-Hill Book Company, New York, 1953.*)

liability suits, since a sprain or tendon injury is difficult to diagnose accurately. The neck muscles of a whiplash sufferer may be supported by a soft sponge-rubber collar until the damage is repaired.

Cartilage injury   Cartilages and ligaments of the knee are often subject to injury. The knee is the site of numerous injuries because it is a weight-bearing joint that is poorly constructed for its function and thus very vulnerable. Knee cartilages may be torn or displaced by abrupt sideways movement. The knee joint is held together by ligaments, and since it is a hinge joint, it does not permit rotation. Injuries may be produced by a strong force applied laterally or from the front so as to overextend the joint. Such stresses may cause serious damage to ligaments and cartilage. Pieces of cartilage may block knee action, making it impossible to fully extend the leg at the knee. Torn ligaments usually require surgical repair.

Rickets   Vitamin D deficiency causes *rickets,* a condition that produces a disturbance in the normal calcification of bones. When vitamin D is absent, the long bones of the body do not grow straight. Common manifestations of rickets are bowlegs, knock-knees, or a very narrow chest with deformities of the sternum and costal cartilages.

The diet of children should include foods rich in vitamin D, such as egg yolk, butter, and whole milk. It may be advisable to supplement the diet with fish-liver oils or vitamin D concentrates. Fresh vegetables should also

a

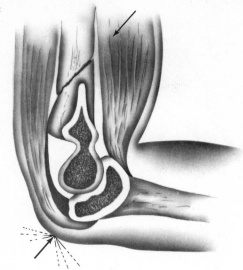

Forces.

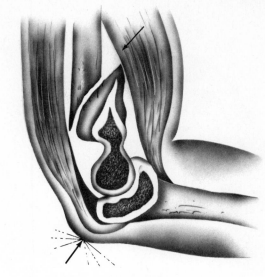

Resulting displacement.

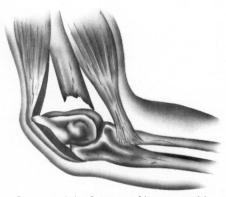

Supracondylar fracture of humerus with
backward displacement of distal fragment.

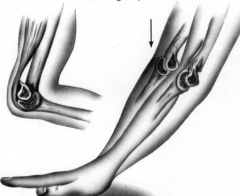

Forces resulting in this type of fracture.

PRODUCTION OF SUPRACONDYLAR FRACTURE OF HUMERUS

b

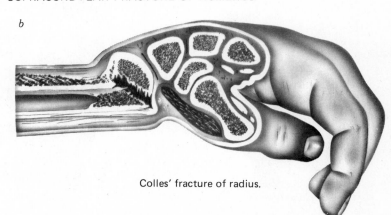

Colles' fracture of radius.

FIGURE 5·39
Specific types and causes of
various fractures and dislocations.

CLINICAL ASPECTS **171**

c

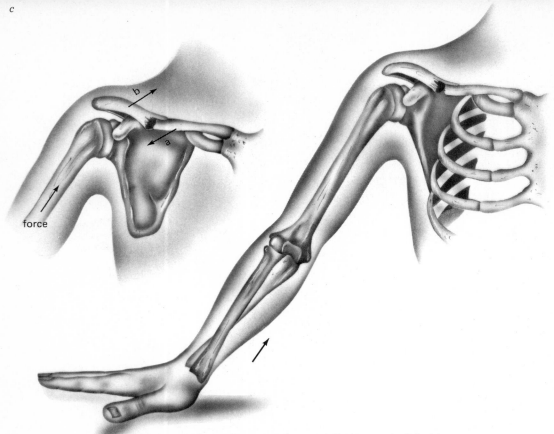

b

a

force

Mechanism producing fracture and upward displacement of clavicle.

d

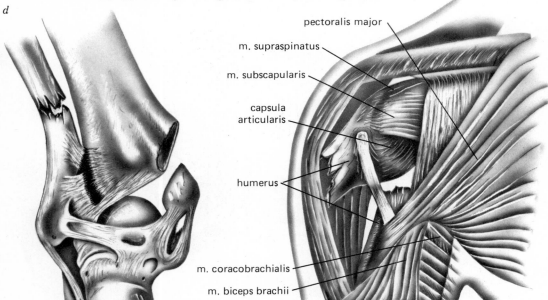

pectoralis major

m. supraspinatus

m. subscapularis

capsula
articularis

humerus

m. coracobrachialis

m. biceps brachii

Pott's fracture of ankle.

Fracture of surgical neck of humerus.

_e_

Displacement of fractured radius by contraction of pronator teres.

Displacement of fractured radius by contraction of pronator quadratus.

Fracture of both bones of forearm.

FRACTURES

_f_

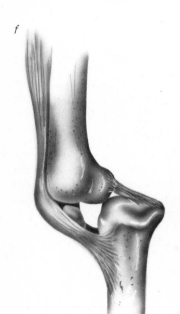

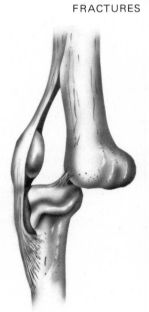

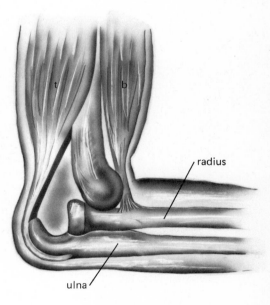

radius

ulna

Posterior dislocation of leg.

Anterior dislocation of leg.

Posterior dislocation of elbow.

DISLOCATIONS

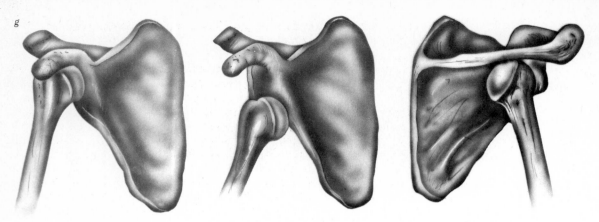

Dislocation of humerus. (Left) Subcoracoid. (Center) Subglenoid. (Right) Subspinous.

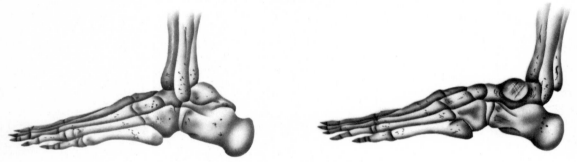

Dislocation of foot. (Left) Backward. (Right) Forward.

be included in the diet to provide calcium and phosphorus, which are necessary for the proper hardening of bone.

Finally, dietary vitamin D can be supplemented by exposing the skin to the direct rays of the sun. When the surface of the body is irradiated with ultraviolet light, it manufactures its own vitamin D. For further discussion, see Chap. 20.

Osteoporosis This condition results from the body's failure to maintain the integrity of the bone. This condition usually occurs in the elderly, and more often in women than in men. In osteoporosis the bones are more porous than normal and may become seriously weakened. Older people may shrink somewhat in height, largely because of changes in shape of the vertebrae caused by osteoporosis. Osteoporosis seems to be caused by some reduction in the output of hormones affecting bony structure and by vitamin D and calcium deficiencies. Areas most commonly affected are the spine, the pelvis, and the thigh bones.

Osteomyelitis *Osteomyelitis* is a disease of the bone marrow. It is usually caused by a staphylococcus infection and more rarely by other coccal bacteria. The disease is much more common in children than in adults. The primary infection may be at some distant location, such as in an ulcerated tooth. The bacteria are carried by the blood to the affected marrow. It usually appears in one of the long bones, near a joint such as the knee joint.

The inflammation is very painful. The administration of antibiotics will usually control the infection.

Arthritis   *Arthritis* is a term applied to inflammation of the joints. There are many different kinds of joint inflammations. The most common types are osteoarthritis, rheumatoid arthritis, and gouty arthritis. These diseases cause a great deal of disability and pain, especially in older people.

Osteoarthritis is a degenerative disease affecting cartilage, primarily that of the weight-bearing joints. It usually affects people over age forty. Since there are no pain receptors in cartilage, the condition is not painful in the early stages as the cartilage wears away and becomes thinner. When the articular cartilage covering the ends of bones begins to break up, however, spurs of bone move into the joint, and movement of the joint produces pain. The joints most commonly affected are the knees, the hips, and those of the spine.

Rheumatoid arthritis is a crippling disease of the hands, elbows, wrists, and feet. This type of arthritis affects more women than men and usually has its onset between twenty-five and thirty-five years of age. Hips and knees may become involved at a later age. This is a generalized, degenerative disease of unknown cause. The disease attacks the synovial membranes, producing inflammation and thickening. Synovial fluid accumulates, and the resulting pressure produces swelling of the joint and causes pain. As the disease progresses, the cartilage of the joint is destroyed, and fibrous tissue invades the areas. Eventually, bony tissue forms and immobilizes the joint. At present, there is no cure for rheumatoid arthritis, but various drugs are used to relieve pain and make the individual more comfortable.

Gouty arthritis is a form of arthritis caused by excess production of uric acid. During the normal synthesis of the nucleic acid purine, uric acid is given off as a waste product and excreted by the kidneys. If an excess of uric acid is produced, however, or if the kidneys fail to excrete all of it, some uric acid will be carried in the blood. Crystals of sodium urate may then be deposited in the joint cartilages. The joint at the base of the great toe seems to be a common location for such deposits. The intense pain of gout is caused by the infiltration of urate crystals into joint cartilage, resulting in inflammation and swelling. Several drugs may be used to treat the gouty condition. A drug called colchicine, derived from the fall crocus *Colchicum*, has been used for many years.

Bone cancer   Several cancers of bone and connective tissue are common. *Sarcoma* is the name given to malignant tumors derived from connective tissue cells. Osteogenic sarcoma is a cancer of developing bone cells; this type of cancer is most common in young people. Considerable destruction of bone, accompanied by swelling and pain, occurs.

Chondrosarcomas are slow-growing malignancies of cartilage, usually located in the ribs, the pelvis, and in the proximal ends of the femur and humerus. Such cancers are more likely to affect older people.

Multiple myeloma is the most common type of bone marrow malignancy. Tumors produce lesions in the red marrow, destroying its blood cell–producing capabilities and causing anemia. In advanced cases, the skeleton is greatly weakened by osteoporosis, and spontaneous fractures may occur. The disease is far more common in males than in females and occurs most often in those past middle age.

Aplastic anemias Bone marrow transplantation may hold a bright promise for the future in the control of certain types of aplastic anemia (anemias in which no blood cells are produced). Such anemias are caused by defects and degenerative changes in the bone marrow. In order for bone marrow transplants to work, however, the immune response system of the recipient must be completely suppressed so that the body will accept the foreign tissue. As in other organ transplants, the body rejects the new tissue cells, treating them as an invasion of foreign protein. Considerable progress has been made in the field of immunology, however, and eventually aplastic anemia and certain types of cancer may be brought under control by such treatments.

SUMMARY

1 The skeleton is the bony framework of the body. Bone must be thought of as a living, developing organ with many functions other than support, however.

2 The bones of the cranium that do not preform as cartilage are membrane bones. The bones of the appendicular skeleton preform as cartilage and are called cartilage bones. The haversian system of bone structure consists of concentric layers of bone and bone cells around the haversian canal. Bone cells lie in cavities called lacunae, from which minute canals (canaliculi) radiate outward. The periosteum, a tough fibrous membrane, is the protective membrane that covers bone except at the articulating surfaces. The inner layer of the periosteum is osteogenic.

3 The marrow cavity is lined by a thin connective tissue membrane called the endosteum. Large, multinucleated cells, called osteoclasts, are found in areas in which bone is being resorbed. They are concerned with the breaking down of bony structure and therefore aid in enlarging the marrow cavity. The osteogenic layer of the periosteum provides for the increase in diameter in bones by producing successive layers on the outside.

4 Membrane bones develop directly from centers of ossification in connective tissue. Cartilage bones develop from hyaline cartilage, which is gradually replaced by bone. Cartilage zones of growth remain between the diaphysis and the epiphyses in growing bones. Zones of growth in the long bones ordinarily change to bone between the fifteenth and twenty-first years, after which there is no further growth in height.

5 Terms of reference and planes of reference are described to provide a guide to direction and positions in the body.

6 The axial skeleton comprises the head, the hyoid bone, the vertebral column, the sternum, and the ribs.

7 The cranial bones are the frontal, parietal, occipital, temporal, sphenoid, and ethmoid bones. The occipital bone contains the foramen magnum and the occipital condyles. The anterior fontanel is a space located medially between the frontal and parietal bones. The posterior fontanel is located at the posterior end of the sagittal suture between the occipital and parietal bones. The temporal bones contain the middle and internal ear structures, and each bone has a mastoid process. Outstanding features of the sphenoid bones are the sella turcica and the greater and lesser wings. The crista galli, cribriform plate, perpendicular plate, and superior and middle nasal conchae are ethmoidal structures.

8 The facial bones are the mandible, maxilla, palatine bone, vomer,

INDEX OF ANATOMICAL PLATES

PLATE 1

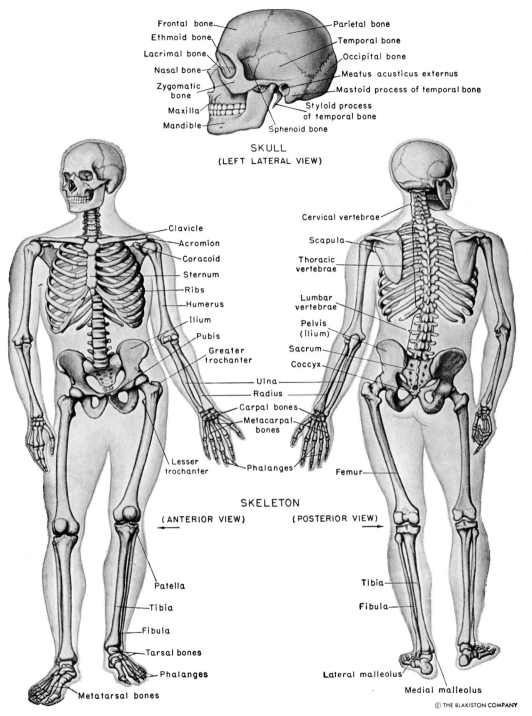

Frontal bone
Ethmoid bone
Lacrimal bone
Nasal bone
Zygomatic bone
Maxilla
Mandible

Parietal bone
Temporal bone
Occipital bone
Meatus acusticus externus
Mastoid process of temporal bone
Styloid process of temporal bone
Sphenoid bone

SKULL
(LEFT LATERAL VIEW)

Clavicle
Acromion
Coracoid
Sternum
Ribs
Humerus
Ilium
Pubis
Greater trochanter
Ulna
Radius
Carpal bones
Metacarpal bones
Lesser trochanter
Phalanges

Cervical vertebrae
Scapula
Thoracic vertebrae
Lumbar vertebrae
Pelvis (ilium)
Sacrum
Coccyx
Femur

SKELETON

(ANTERIOR VIEW)   (POSTERIOR VIEW)

Patella
Tibia
Fibula
Tarsal bones
Phalanges
Metatarsal bones

Tibia
Fibula
Lateral malleolus
Medial malleolus

© THE BLAKISTON COMPANY

SKELETON

PLATE 2

SHOULDER JOINT

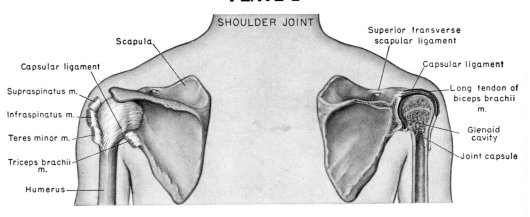

Scapula

Capsular ligament

Supraspinatus m.

Infraspinatus m.

Teres minor m.

Triceps brachii m.

Humerus

Superior transverse scapular ligament

Capsular ligament

Long tendon of biceps brachii m.

Glenoid cavity

Joint capsule

HIP JOINT

(ANTERIOR VIEW)

(POSTERIOR VIEW)

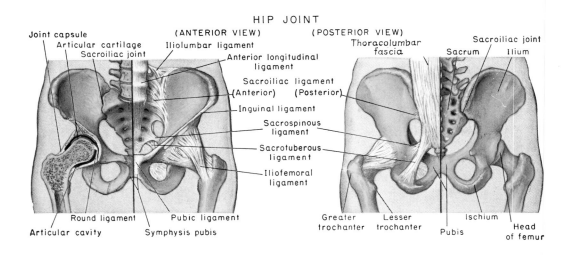

Joint capsule

Articular cartilage

Sacroiliac joint

Iliolumbar ligament

Anterior longitudinal ligament

Sacroiliac ligament
(Anterior) (Posterior)

Inguinal ligament

Sacrospinous ligament

Sacrotuberous ligament

Iliofemoral ligament

Round ligament

Pubic ligament

Articular cavity

Symphysis pubis

Thoracolumbar fascia

Sacrum

Sacroiliac joint

Ilium

Greater trochanter

Lesser trochanter

Ischium

Pubis

Head of femur

KNEE JOINT AND LIGAMENTS

(ANTERIOR VIEW)

(LATERAL VIEW)

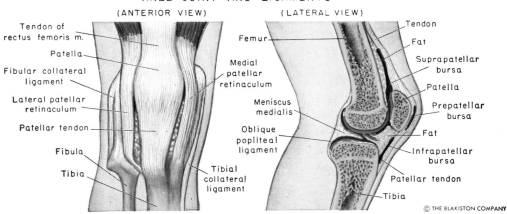

Tendon of rectus femoris m.

Patella

Fibular collateral ligament

Lateral patellar retinaculum

Patellar tendon

Fibula

Tibia

Femur

Medial patellar retinaculum

Meniscus medialis

Oblique popliteal ligament

Tibial collateral ligament

Tendon

Fat

Suprapatellar bursa

Patella

Prepatellar bursa

Fat

Infrapatellar bursa

Patellar tendon

Tibia

© THE BLAKISTON COMPANY

JOINTS

PLATE 3

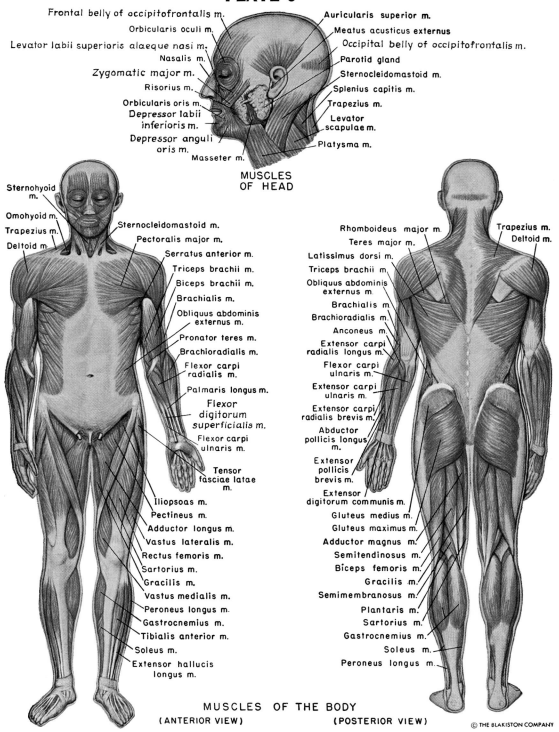

MUSCLES OF HEAD

Frontal belly of occipitofrontalis m.
Orbicularis oculi m.
Levator labii superioris alaeque nasi m.
Nasalis m.
Zygomatic major m.
Risorius m.
Orbicularis oris m.
Depressor labii inferioris m.
Depressor anguli oris m.
Masseter m.

Auricularis superior m.
Meatus acusticus externus
Occipital belly of occipitofrontalis m.
Parotid gland
Sternocleidomastoid m.
Splenius capitis m.
Trapezius m.
Levator scapulae m.
Platysma m.

MUSCLES OF THE BODY
(ANTERIOR VIEW)

Sternohyoid m.
Omohyoid m.
Trapezius m.
Deltoid m.
Sternocleidomastoid m.
Pectoralis major m.
Serratus anterior m.
Triceps brachii m.
Biceps brachii m.
Brachialis m.
Obliquus abdominis externus m.
Pronator teres m.
Brachioradialis m.
Flexor carpi radialis m.
Palmaris longus m.
Flexor digitorum superficialis m.
Flexor carpi ulnaris m.
Tensor fasciae latae m.
Iliopsoas m.
Pectineus m.
Adductor longus m.
Vastus lateralis m.
Rectus femoris m.
Sartorius m.
Gracilis m.
Vastus medialis m.
Peroneus longus m.
Gastrocnemius m.
Tibialis anterior m.
Soleus m.
Extensor hallucis longus m.

(POSTERIOR VIEW)

Rhomboideus major m.
Teres major m.
Latissimus dorsi m.
Triceps brachii m.
Obliquus abdominis externus m.
Brachialis m.
Brachioradialis m.
Anconeus m.
Extensor carpi radialis longus m.
Flexor carpi ulnaris m.
Extensor carpi ulnaris m.
Extensor carpi radialis brevis m.
Abductor pollicis longus m.
Extensor pollicis brevis m.
Extensor digitorum communis m.
Gluteus medius m.
Gluteus maximus m.
Adductor magnus m.
Semitendinosus m.
Biceps femoris m.
Gracilis m.
Semimembranosus m.
Plantaris m.
Sartorius m.
Gastrocnemius m.
Soleus m.
Peroneus longus m.

Trapezius m.
Deltoid m.

MUSCLES

PLATE 4

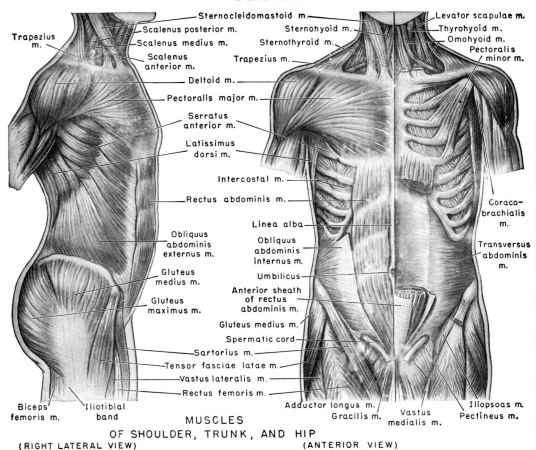

Trapezius m.

Sternocleidomastoid m.
Scalenus posterior m.
Scalenus medius m.
Scalenus anterior m.
Deltoid m.
Pectoralis major m.
Serratus anterior m.
Latissimus dorsi m.
Intercostal m.
Rectus abdominis m.
Linea alba
Obliquus abdominis externus m.
Obliquus abdominis internus m.
Gluteus medius m.
Umbilicus
Gluteus maximus m.
Anterior sheath of rectus abdominis m.
Gluteus medius m.
Spermatic cord
Sartorius m.
Tensor fasciae latae m.
Vastus lateralis m.
Rectus femoris m.

Sternocleidomastoid m.
Sternohyoid m.
Sternothyroid m.
Trapezius m.

Levator scapulae m.
Thyrohyoid m.
Omohyoid m.
Pectoralis minor m.

Coraco-brachialis m.

Transversus abdominis m.

Biceps femoris m.
Iliotibial band

Adductor longus m.
Gracilis m.
Vastus medialis m.

Iliopsoas m.
Pectineus m.

MUSCLES
OF SHOULDER, TRUNK, AND HIP
(RIGHT LATERAL VIEW)   (ANTERIOR VIEW)

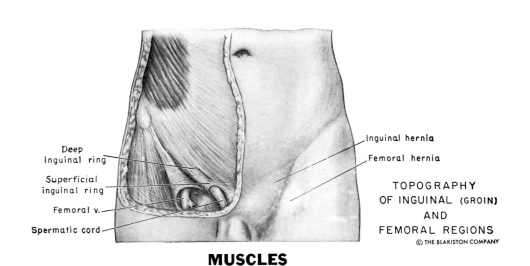

Deep inguinal ring
Superficial inguinal ring
Femoral v.
Spermatic cord

Inguinal hernia
Femoral hernia

TOPOGRAPHY OF INGUINAL (GROIN) AND FEMORAL REGIONS

MUSCLES

PLATE 5

SCHEME OF CIRCULATION

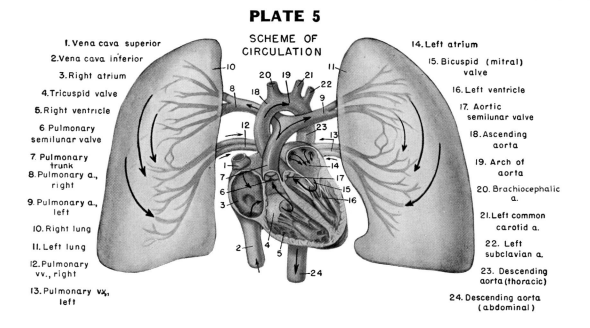

1. Vena cava superior
2. Vena cava inferior
3. Right atrium
4. Tricuspid valve
5. Right ventricle
6. Pulmonary semilunar valve
7. Pulmonary trunk
8. Pulmonary a., right
9. Pulmonary a., left
10. Right lung
11. Left lung
12. Pulmonary vv., right
13. Pulmonary vv., left

14. Left atrium
15. Bicuspid (mitral) valve
16. Left ventricle
17. Aortic semilunar valve
18. Ascending aorta
19. Arch of aorta
20. Brachiocephalic a.
21. Left common carotid a.
22. Left subclavian a.
23. Descending aorta (thoracic)
24. Descending aorta (abdominal)

LEFT VENTRICLE OF HEART, OPENED

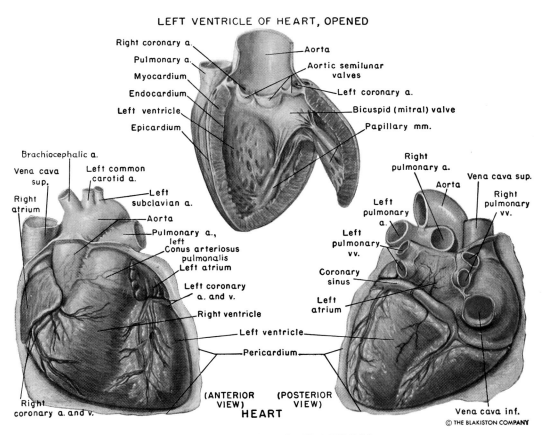

Right coronary a.
Pulmonary a.
Myocardium
Endocardium
Left ventricle
Epicardium

Aorta
Aortic semilunar valves
Left coronary a.
Bicuspid (mitral) valve
Papillary mm.

Brachiocephalic a.
Vena cava sup.
Right atrium
Left common carotid a.
Left subclavian a.
Aorta
Pulmonary a., left
Conus arteriosus pulmonalis
Left atrium
Left coronary a. and v.
Right ventricle
Right coronary a. and v.

Right pulmonary a.
Aorta
Vena cava sup.
Right pulmonary vv.
Left pulmonary a.
Left pulmonary vv.
Coronary sinus
Left atrium
Left ventricle
Pericardium
Vena cava inf.

(ANTERIOR VIEW)   (POSTERIOR VIEW)

HEART

BLOOD CIRCULATION

PLATE 6

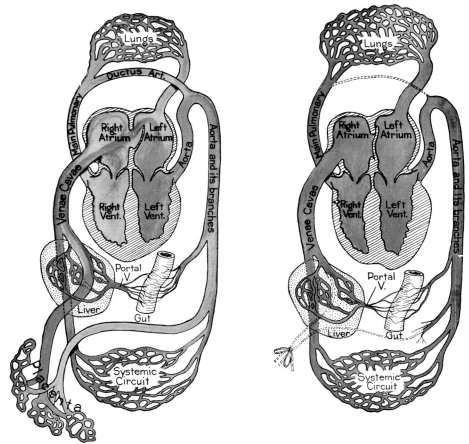

(Left) Prenatal (fetal) circulation. (Right) Postnatal circulation.

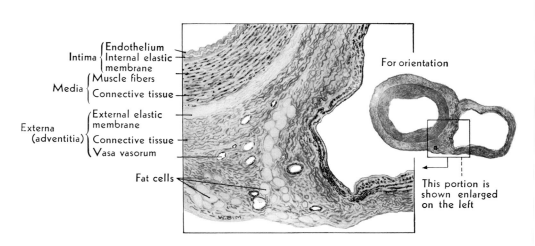

Section through human ulnar artery and vein, showing (left) wall of artery and (right) wall of vein.

PRE- AND POSTNATAL BLOOD CIRCULATORY SYSTEMS

PLATE 7

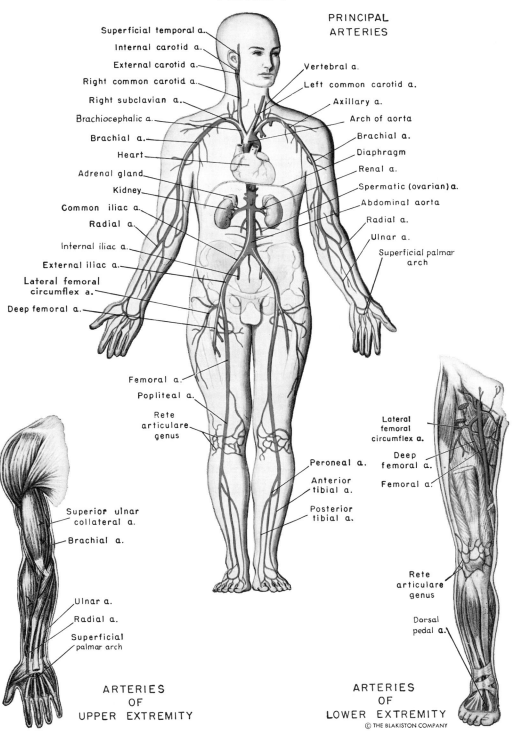

PRINCIPAL ARTERIES

Superficial temporal a.
Internal carotid a.
External carotid a.
Right common carotid a.
Right subclavian a.
Brachiocephalic a.
Brachial a.
Heart
Adrenal gland
Kidney
Common iliac a.
Radial a.
Internal iliac a.
External iliac a.
Lateral femoral circumflex a.
Deep femoral a.

Vertebral a.
Left common carotid a.
Axillary a.
Arch of aorta
Brachial a.
Diaphragm
Renal a.
Spermatic (ovarian) a.
Abdominal aorta
Radial a.
Ulnar a.
Superficial palmar arch

Femoral a.
Popliteal a.
Rete articulare genus

Peroneal a.
Anterior tibial a.
Posterior tibial a.

Lateral femoral circumflex a.
Deep femoral a.
Femoral a.

Rete articulare genus

Dorsal pedal a.

Superior ulnar collateral a.
Brachial a.

Ulnar a.
Radial a.
Superficial palmar arch

ARTERIES
OF
UPPER EXTREMITY

ARTERIES
OF
LOWER EXTREMITY

© THE BLAKISTON COMPANY

ARTERIES

PLATE 8

SUPERFICIAL ARTERIES OF HEAD AND NECK

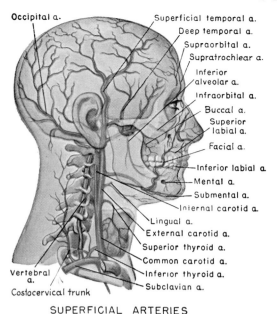

Occipital a.
Superficial temporal a.
Deep temporal a.
Supraorbital a.
Supratrochlear a.
Inferior alveolar a.
Infraorbital a.
Buccal a.
Superior labial a.
Facial a.
Inferior labial a.
Mental a.
Submental a.
Internal carotid a.
Lingual a.
External carotid a.
Superior thyroid a.
Common carotid a.
Inferior thyroid a.
Subclavian a.
Vertebral a.
Costocervical trunk

CIRCULATION OF ABDOMINAL VISCERA

Liver
Hepatic a.
Cystic a.
Celiac trunk
Left gastric a.
Lienal a.
Spleen
Gall bladder
Gastro-duodenal a.
Right gastric a.
Stomach
Duodenum
Right gastro-epiploic a.
Left gastro-epiploic a.
Great omentum
Intestines

CIRCULATION OF DIGESTIVE TRACT

Ileocolic a.
Right colic a.
Middle colic a.
Superior mesenteric a.
Ascending colon
Omentum
Transverse colon
Transverse mesocolon
Left colic a.
Abdominal aorta
Inferior mesenteric a.
Descending colon
Superior rectal a.
Sigmoid a.
Sigmoid colon
Bladder
Rectum

DEEP CIRCULATION OF ABDOMEN

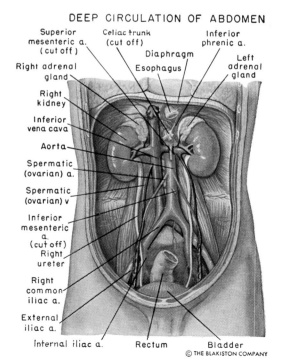

Superior mesenteric a. (cut off)
Celiac trunk (cut off)
Inferior phrenic a.
Diaphragm
Right adrenal gland
Esophagus
Left adrenal gland
Right kidney
Inferior vena cava
Aorta
Spermatic (ovarian) a.
Spermatic (ovarian) v.
Inferior mesenteric a. (cut off)
Right ureter
Right common iliac a.
External iliac a.
Internal iliac a.
Rectum
Bladder

ARTERIES

PLATE 9

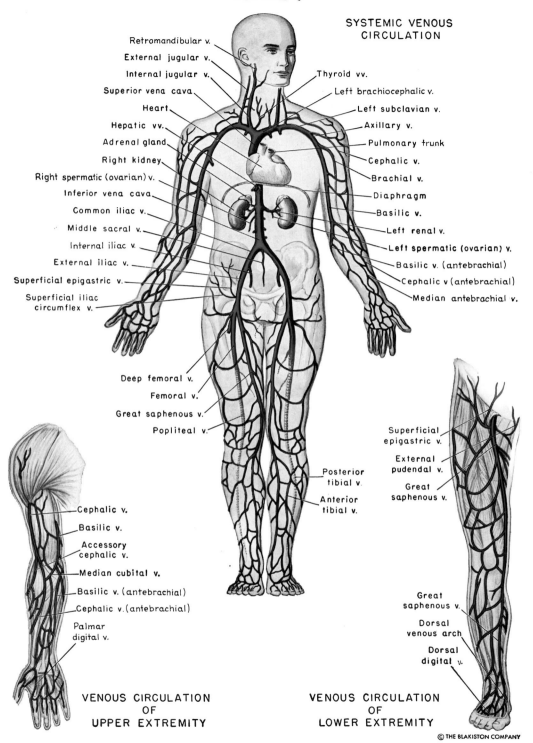

SYSTEMIC VENOUS
CIRCULATION

Retromandibular v.
External jugular v.
Internal jugular v.
Superior vena cava
Heart
Hepatic vv.
Adrenal gland
Right kidney
Right spermatic (ovarian) v.
Inferior vena cava
Common iliac v.
Middle sacral v.
Internal iliac v.
External iliac v.
Superficial epigastric v.
Superficial iliac
circumflex v.

Thyroid vv.
Left brachiocephalic v.
Left subclavian v.
Axillary v.
Pulmonary trunk
Cephalic v.
Brachial v.
Diaphragm
Basilic v.
Left renal v.
Left spermatic (ovarian) v.
Basilic v. (antebrachial)
Cephalic v (antebrachial)
Median antebrachial v.

Deep femoral v.
Femoral v.
Great saphenous v.
Popliteal v.

Posterior
tibial v.
Anterior
tibial v.

Superficial
epigastric v.
External
pudendal v.
Great
saphenous v.

Great
saphenous v.
Dorsal
venous arch
Dorsal
digital v.

Cephalic v.
Basilic v.
Accessory
cephalic v.
Median cubital v.
Basilic v. (antebrachial)
Cephalic v. (antebrachial)
Palmar
digital v.

VENOUS CIRCULATION
OF
UPPER EXTREMITY

VENOUS CIRCULATION
OF
LOWER EXTREMITY

© THE BLAKISTON COMPANY

VEINS

PLATE 10

PRINCIPAL VEINS OF HEAD AND NECK

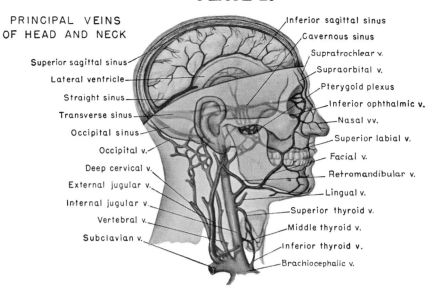

Inferior sagittal sinus
Cavernous sinus
Supratrochlear v.
Supraorbital v.
Pterygoid plexus
Inferior ophthalmic **v.**
Nasal vv.
Superior labial v.
Facial v.
Retromandibular v.
Lingual v.
Superior thyroid v.
Middle thyroid v.
Inferior thyroid **v.**
Brachiocephalic v.

Superior sagittal sinus
Lateral ventricle
Straight sinus
Transverse sinus
Occipital sinus
Occipital v.
Deep cervical v.
External jugular v.
Internal jugular v.
Vertebral v.
Subclavian v.

PORTAL VEIN AND PRINCIPAL TRIBUTARIES

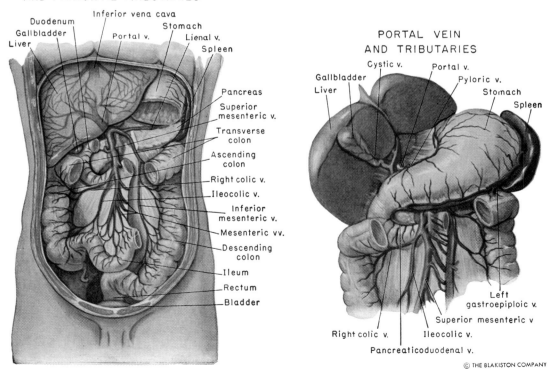

Duodenum
Gallbladder
Liver
Inferior vena cava
Portal v.
Stomach
Lienal v.
Spleen

Pancreas
Superior mesenteric v.
Transverse colon
Ascending colon
Right colic v.
Ileocolic v.
Inferior mesenteric v.
Mesenteric vv.
Descending colon
Ileum
Rectum
Bladder

PORTAL VEIN AND TRIBUTARIES

Cystic v.
Portal v.
Gallbladder
Liver
Pyloric v.
Stomach
Spleen

Left gastroepiploic v.
Superior mesenteric v
Right colic v.
Ileocolic v.
Pancreaticoduodenal v.

VEINS

PLATE 11

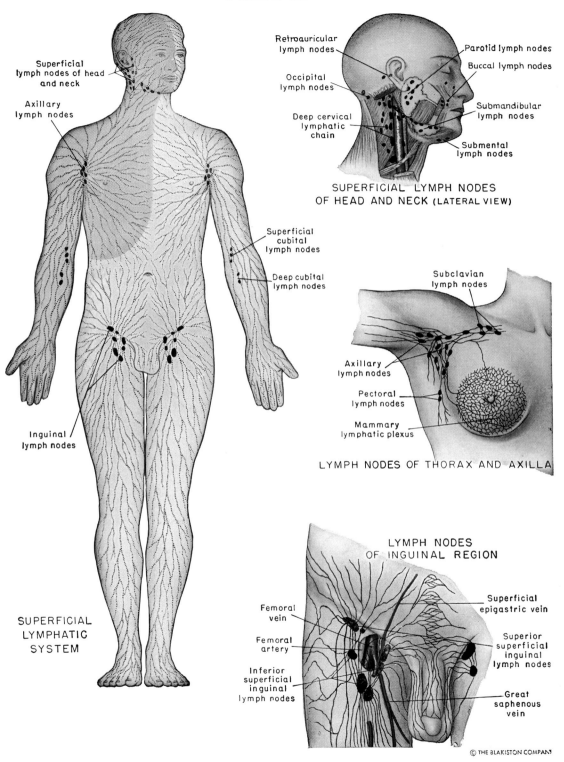

Superficial
lymph nodes of head
and neck

Axillary
lymph nodes

Superficial
cubital
lymph nodes

Deep cubital
lymph nodes

Inguinal
lymph nodes

SUPERFICIAL
LYMPHATIC
SYSTEM

Retroauricular
lymph nodes

Occipital
lymph nodes

Deep cervical
lymphatic
chain

Parotid lymph nodes

Buccal lymph nodes

Submandibular
lymph nodes

Submental
lymph nodes

SUPERFICIAL LYMPH NODES
OF HEAD AND NECK (LATERAL VIEW)

Subclavian
lymph nodes

Axillary
lymph nodes

Pectoral
lymph nodes

Mammary
lymphatic plexus

LYMPH NODES OF THORAX AND AXILLA

LYMPH NODES
OF INGUINAL REGION

Femoral
vein

Femoral
artery

Inferior
superficial
inguinal
lymph nodes

Superficial
epigastric vein

Superior
superficial
inguinal
lymph nodes

Great
saphenous
vein

© THE BLAKISTON COMPANY

LYMPHATIC SYSTEM

PLATE 12

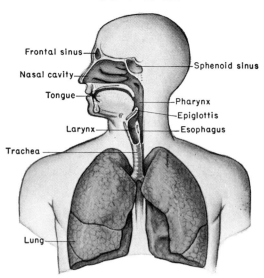

RESPIRATORY TRACT
(SCHEMATIC)

Frontal sinus
Nasal cavity
Tongue
Larynx
Trachea
Lung
Sphenoid sinus
Pharynx
Epiglottis
Esophagus

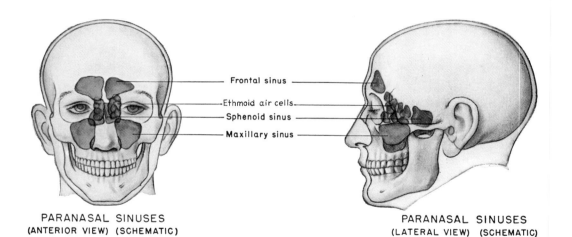

Frontal sinus
Ethmoid air cells
Sphenoid sinus
Maxillary sinus

PARANASAL SINUSES
(ANTERIOR VIEW) (SCHEMATIC)

PARANASAL SINUSES
(LATERAL VIEW) (SCHEMATIC)

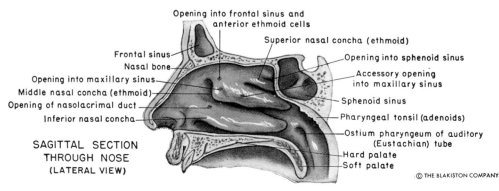

Opening into frontal sinus and
anterior ethmoid cells
Superior nasal concha (ethmoid)
Frontal sinus
Nasal bone
Opening into maxillary sinus
Middle nasal concha (ethmoid)
Opening of nasolacrimal duct
Inferior nasal concha
Opening into sphenoid sinus
Accessory opening
into maxillary sinus
Sphenoid sinus
Pharyngeal tonsil (adenoids)
Ostium pharyngeum of auditory
(Eustachian) tube
Hard palate
Soft palate

SAGITTAL SECTION
THROUGH NOSE
(LATERAL VIEW)

© THE BLAKISTON COMPANY

RESPIRATORY TRACT

PLATE 13

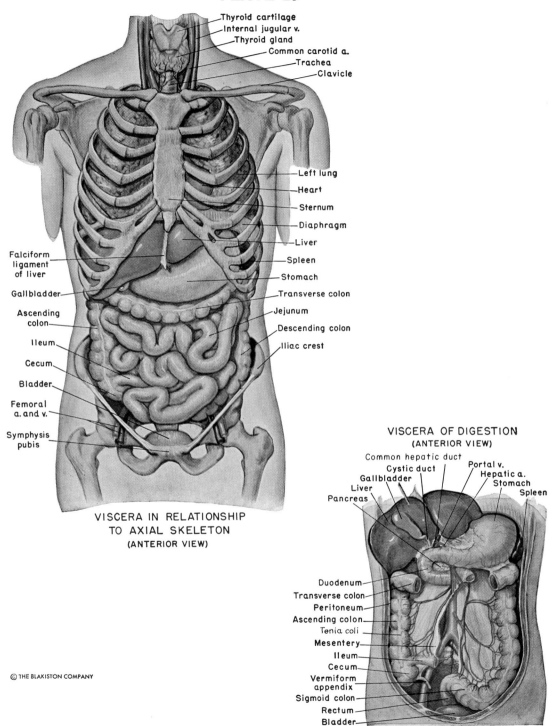

Thyroid cartilage
Internal jugular v.
Thyroid gland
Common carotid a.
Trachea
Clavicle

Left lung
Heart
Sternum
Diaphragm
Liver
Spleen
Stomach
Transverse colon
Jejunum
Descending colon
Iliac crest

Falciform
ligament
of liver
Gallbladder
Ascending
colon
Ileum
Cecum
Bladder
Femoral
a. and v.
Symphysis
pubis

VISCERA IN RELATIONSHIP
TO AXIAL SKELETON
(ANTERIOR VIEW)

VISCERA OF DIGESTION
(ANTERIOR VIEW)

Common hepatic duct
Cystic duct
Gallbladder
Liver
Pancreas
Portal v.
Hepatic a.
Stomach
Spleen

Duodenum
Transverse colon
Peritoneum
Ascending colon
Tenia coli
Mesentery
Ileum
Cecum
Vermiform
appendix
Sigmoid colon
Rectum
Bladder

VISCERA

PLATE 14

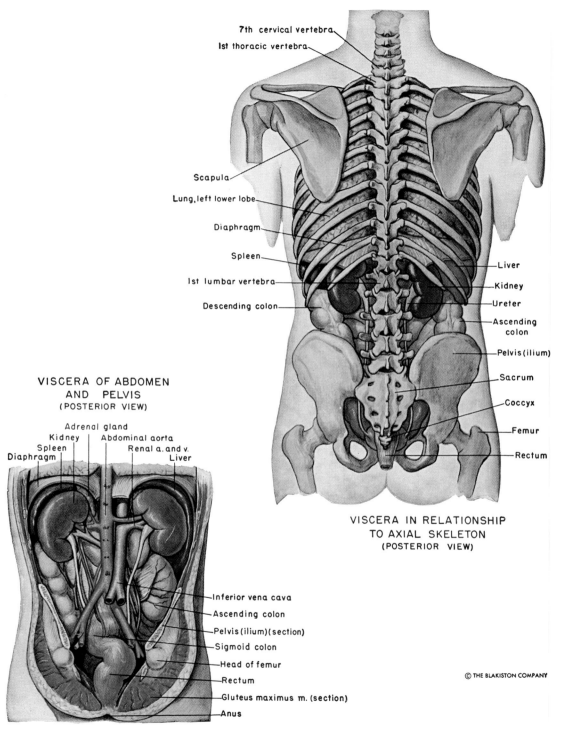

7th cervical vertebra

1st thoracic vertebra

Scapula

Lung, left lower lobe

Diaphragm

Spleen

1st lumbar vertebra

Descending colon

Liver

Kidney

Ureter

Ascending colon

Pelvis (ilium)

Sacrum

Coccyx

Femur

Rectum

VISCERA IN RELATIONSHIP TO AXIAL SKELETON
(POSTERIOR VIEW)

VISCERA OF ABDOMEN AND PELVIS
(POSTERIOR VIEW)

Adrenal gland

Kidney   Abdominal aorta

Spleen   Renal a. and v.

Diaphragm   Liver

Inferior vena cava

Ascending colon

Pelvis (ilium)(section)

Sigmoid colon

Head of femur

Rectum

Gluteus maximus m. (section)

Anus

VISCERA

PLATE 15

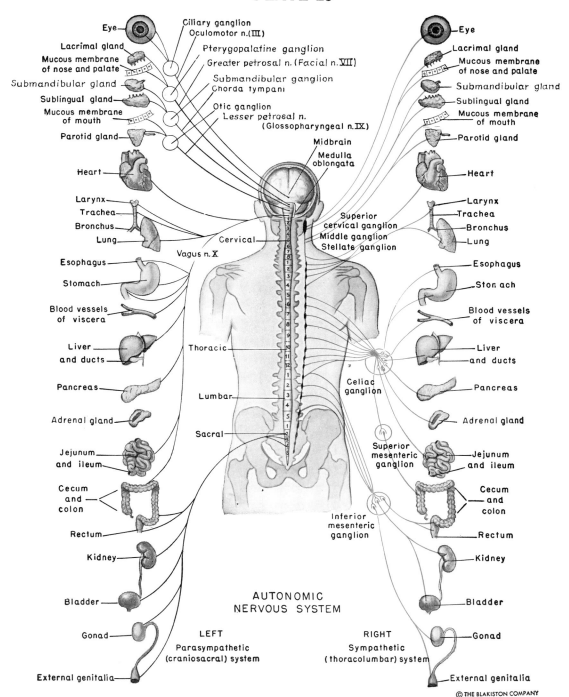

Ciliary ganglion
Oculomotor n.(III)
Pterygopalatine ganglion
Greater petrosal n. (Facial n. VII)
Submandibular ganglion
Chorda tympani
Otic ganglion
Lesser petrosal n.
(Glossopharyngeal n. IX)
Midbrain
Medulla oblongata

Eye
Lacrimal gland
Mucous membrane of nose and palate
Submandibular gland
Sublingual gland
Mucous membrane of mouth
Parotid gland
Heart
Larynx
Trachea
Bronchus
Lung
Cervical
Vagus n. X
Esophagus
Stomach
Blood vessels of viscera
Liver and ducts
Pancreas
Adrenal gland
Jejunum and ileum
Cecum and colon
Rectum
Kidney
Bladder
Gonad
External genitalia
Thoracic
Lumbar
Sacral

Superior cervical ganglion
Middle ganglion
Stellate ganglion
Celiac ganglion
Superior mesenteric ganglion
Inferior mesenteric ganglion

Eye
Lacrimal gland
Mucous membrane of nose and palate
Submandibular gland
Sublingual gland
Mucous membrane of mouth
Parotid gland
Heart
Larynx
Trachea
Bronchus
Lung
Esophagus
Stomach
Blood vessels of viscera
Liver and ducts
Pancreas
Adrenal gland
Jejunum and ileum
Cecum and colon
Rectum
Kidney
Bladder
Gonad
External genitalia

AUTONOMIC
NERVOUS SYSTEM

LEFT
Parasympathetic
(craniosacral) system

RIGHT
Sympathetic
(thoracolumbar) system

© THE BLAKISTON COMPANY

AUTONOMIC NERVOUS SYSTEM

PLATE 16

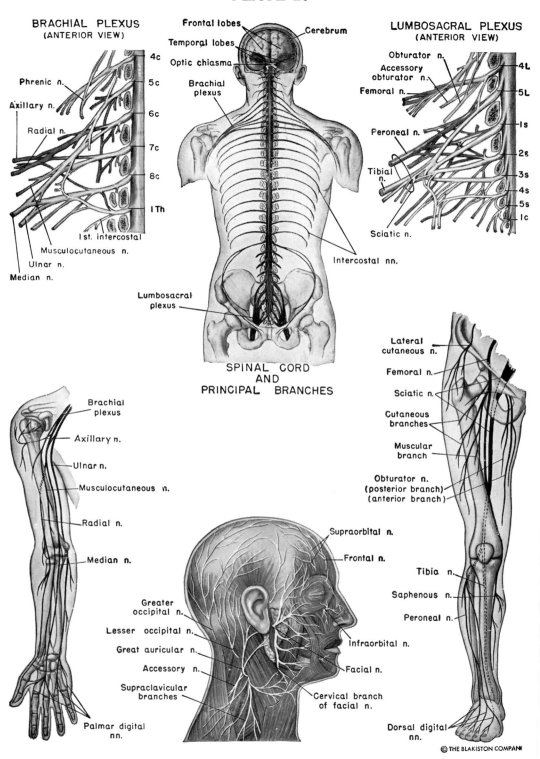

BRACHIAL PLEXUS
(ANTERIOR VIEW)

Phrenic n.
Axillary n.
Radial n.

4c
5c
6c
7c
8c
I Th

1st. intercostal
Musculocutaneous n.
Ulnar n.
Median n.

Frontal lobes
Temporal lobes
Optic chiasma
Brachial plexus
Cerebrum

Lumbosacral plexus

SPINAL CORD
AND
PRINCIPAL BRANCHES

Intercostal nn.

LUMBOSACRAL PLEXUS
(ANTERIOR VIEW)

Obturator n.
Accessory obturator n.
Femoral n.

Peroneal n.

Tibial n.

Sciatic n.

4L
5L
Is
2s
3s
4s
5s
Ic

Brachial plexus
Axillary n.
Ulnar n.
Musculocutaneous n.
Radial n.
Median n.

Palmar digital nn.

Greater occipital n.
Lesser occipital n.
Great auricular n.
Accessory n.
Supraclavicular branches

Supraorbital n.
Frontal n.

Infraorbital n.
Facial n.
Cervical branch of facial n.

Lateral cutaneous n.
Femoral n.
Sciatic n.
Cutaneous branches
Muscular branch
Obturator n.
(posterior branch)
(anterior branch)

Tibia n.
Saphenous n.
Peroneal n.

Dorsal digital nn.

© THE BLAKISTON COMPANY

NERVOUS SYSTEM

PLATE 17   BASE OF BRAIN

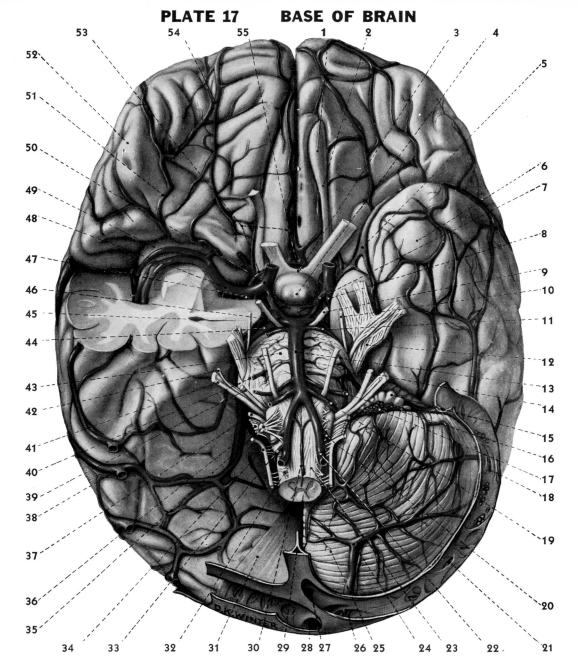

Accessory n. (spinal accessory) (XI), **34**. Vestibulocochlear n. (VIII), **41**. Anterior cerebral a.,**53**. Anterior communicating a., **55**. Anterior inferior cerebellar a., **19**. Anterior spinal a., **30**. Basilar a.,**13**. Brainstem, **23**. Cerebellar v., **24**. Cerebellar v.—opening of into sinus, **26**. Cerebellum, **25**. Choroid a., **8**. Choroid plexus of 4th ventricle, **17**. Confluens sinuum—opening of straight sinus into, **27**. Facial n. (VII), **39**. Flocculus,**15**. Frontal lobe,**52**. Ganglionic branches,**47**. Glossopharyngeal n. (IX),**37**. Hypoglossal n. (XII), **35**. Inferior frontal v.,**54**. Occipital diploic v. to transverse sinus,**38**. Insula (island of Reil),**49**. Intermediary n. (Wrisberg),**42**. Internal auditory a. (a.auditiva interna),**16**. Internal carotid a.,**50**. Lateral sinus—junction of with sigmoid sinus,**18**. Lateral sinus—opening of a superficial vein into, **21**. Lateral ventricle—inferior cornu,**45**. Middle cerebral a.,**48**. Middle cerebral a.—cortical branch of,**51**. Occipital sinus,**29**. Oculomotor n.(III),**9**. Olfactory bulb, **1**. Olfactory tract,**2**. Optic chiasma, **3**. Hypophysis (pituitary), **4**. Pons,**43**. Pontine branch of basilar a.,**14**. Posterior cerebral a.,**44**. Posterior cerebral v., **46**. Posterior communicating a., **7**. Posterior inferior cerebellar a., **33**. Posterior spinal a., **20**. Roots of 1st spinal n., **32**. Trigeminal ganglion,**12**. Sigmoid sinus—junction of with transverse sinus,**18**. Sinus—opening of cerebellar v. into, **26**. Straight sinus—opening of into confluens sinuum, **27**. Superior cerebellar a., **11**. Superior sagittal sinus—opening of into confluens sinuum, **28**. Deep middle cerebral v.,**5**. Temporal pole,**6**. Tentorium cerebelli,**31**. Trigeminal n.(V)—motor root,**40**. Trochlear n. (IV),**10**. Vagus n. (X),**36**. Vertebral a.,**22**.

PLATE 18

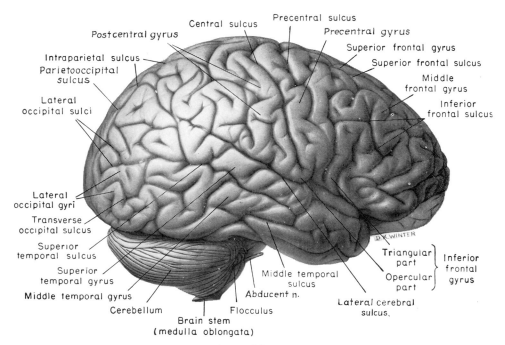

Central sulcus
Precentral sulcus
Precentral gyrus
Postcentral gyrus
Superior frontal gyrus
Intraparietal sulcus
Superior frontal sulcus
Parietooccipital sulcus
Middle frontal gyrus
Lateral occipital sulci
Inferior frontal sulcus
Lateral occipital gyri
Transverse occipital sulcus
Superior temporal sulcus
Triangular part
Inferior frontal gyrus
Superior temporal gyrus
Middle temporal sulcus
Opercular part
Middle temporal gyrus
Abducent n.
Lateral cerebral sulcus.
Cerebellum
Flocculus
Brain stem (medulla oblongata)

D.K.WINTER

Lateral View

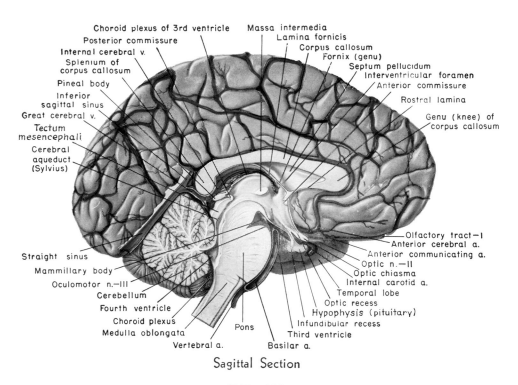

Choroid plexus of 3rd ventricle
Massa intermedia
Posterior commissure
Lamina fornicis
Internal cerebral v.
Corpus callosum
Splenium of corpus callosum
Fornix (genu)
Septum pellucidum
Pineal body
Interventricular foramen
Inferior sagittal sinus
Anterior commissure
Great cerebral v.
Rostral lamina
Tectum mesencephali
Genu (knee) of corpus callosum
Cerebral aqueduct (Sylvius)
Olfactory tract—I
Anterior cerebral a.
Straight sinus
Anterior communicating a.
Mammillary body
Optic n.—II
Oculomotor n.—III
Optic chiasma
Cerebellum
Internal carotid a.
Fourth ventricle
Temporal lobe
Choroid plexus
Optic recess
Medulla oblongata
Hypophysis (pituitary)
Pons
Infundibular recess
Vertebral a.
Third ventricle
Basilar a.

Sagittal Section

BRAIN

PLATE 19

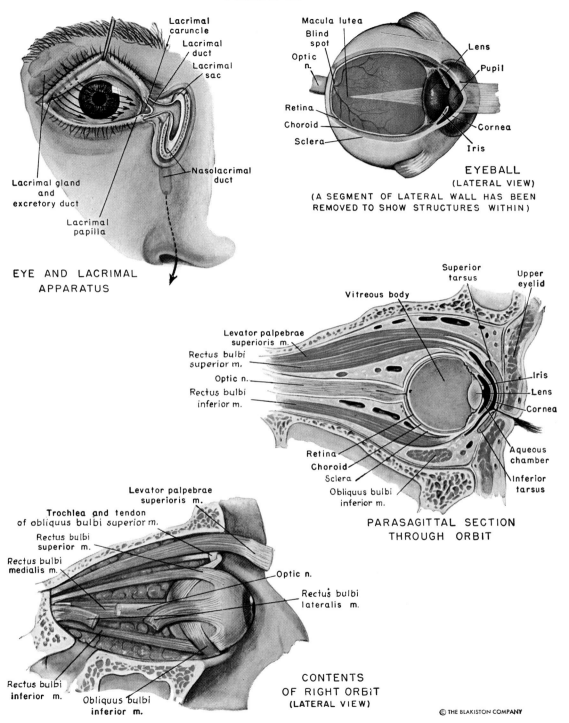

Lacrimal caruncle
Lacrimal duct
Lacrimal sac
Nasolacrimal duct
Lacrimal gland and excretory duct
Lacrimal papilla

EYE AND LACRIMAL APPARATUS

Macula lutea
Blind spot
Optic n.
Lens
Pupil
Retina
Choroid
Sclera
Cornea
Iris

EYEBALL
(LATERAL VIEW)
(A SEGMENT OF LATERAL WALL HAS BEEN REMOVED TO SHOW STRUCTURES WITHIN)

Superior tarsus
Upper eyelid
Vitreous body
Levator palpebrae superioris m.
Rectus bulbi superior m.
Optic n.
Rectus bulbi inferior m.
Iris
Lens
Cornea
Retina
Choroid
Sclera
Obliquus bulbi inferior m.
Aqueous chamber
Inferior tarsus

PARASAGITTAL SECTION THROUGH ORBIT

Levator palpebrae superioris m.
Trochlea and tendon of obliquus bulbi superior m.
Rectus bulbi superior m.
Rectus bulbi medialis m.
Optic n.
Rectus bulbi lateralis m.
Rectus bulbi inferior m.
Obliquus bulbi inferior m.

CONTENTS OF RIGHT ORBIT
(LATERAL VIEW)

EYE

PLATE 20

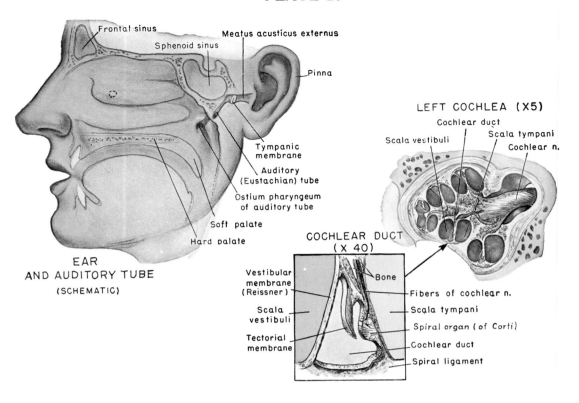

Frontal sinus
Sphenoid sinus
Meatus acusticus externus
Pinna
Tympanic membrane
Auditory (Eustachian) tube
Ostium pharyngeum of auditory tube
Soft palate
Hard palate

EAR
AND AUDITORY TUBE
(SCHEMATIC)

LEFT COCHLEA (X5)
Cochlear duct
Scala vestibuli
Scala tympani
Cochlear n.

COCHLEAR DUCT
(X 40)
Vestibular membrane (Reissner)
Bone
Scala vestibuli
Fibers of cochlear n.
Scala tympani
Tectorial membrane
Spiral organ (of Corti)
Cochlear duct
Spiral ligament

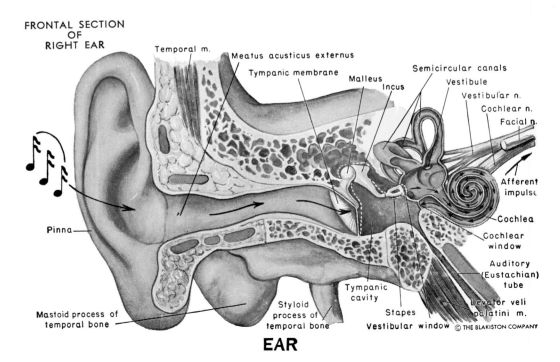

FRONTAL SECTION
OF
RIGHT EAR

Temporal m.
Meatus acusticus externus
Tympanic membrane
Malleus
Incus
Semicircular canals
Vestibule
Vestibular n.
Cochlear n.
Facial n.
Afferent impulse
Cochlea
Cochlear window
Auditory (Eustachian) tube
Levator veli palatini m.
Tympanic cavity
Stapes
Vestibular window
Pinna
Mastoid process of temporal bone
Styloid process of temporal bone

© THE BLAKISTON COMPANY

EAR

PLATE 21

DECIDUOUS DENTITION

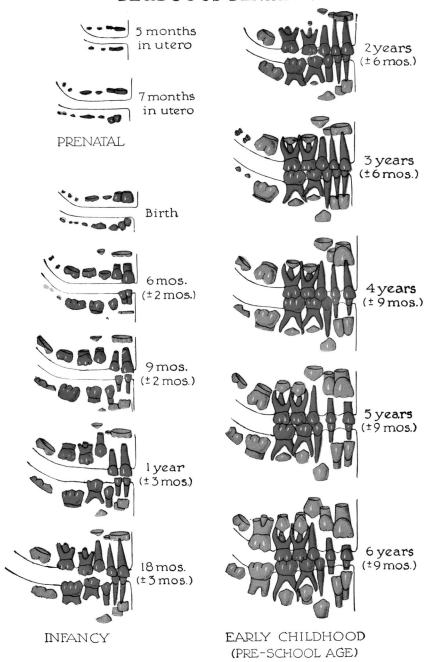

5 months in utero

7 months in utero

PRENATAL

Birth

6 mos. (±2 mos.)

9 mos. (±2 mos.)

1 year (±3 mos.)

18 mos. (±3 mos.)

INFANCY

2 years (±6 mos.)

3 years (±6 mos.)

4 years (±9 mos.)

5 years (±9 mos.)

6 years (±9 mos.)

EARLY CHILDHOOD (PRE-SCHOOL AGE)

DEVELOPMENT OF HUMAN DENTITION

(Courtesy, Schour and Massler, American Dental Association.)

PLATE 22

MIXED DENTITION

PERMANENT DENTITION

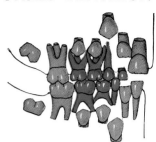

7 years
(± 9 mos.)

11 years
(± 9 mos.)

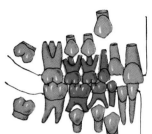

8 years
(± 9 mos.)

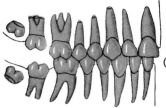

12 years
(± 6 mos.)

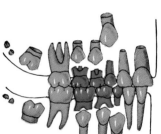

9 years
(± 9 mos.)

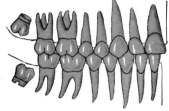

15 years
(± 6 mos.)

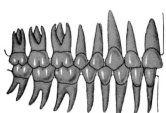

21 years

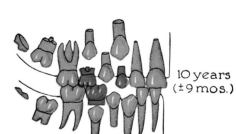

10 years
(± 9 mos.)

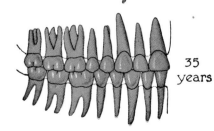

35 years

LATE CHILDHOOD
(SCHOOL AGE)

ADOLESCENCE
and ADULTHOOD

DEVELOPMENT OF HUMAN DENTITION

(Courtesy, Schour and Massler, American Dental Association.)

PLATE 23

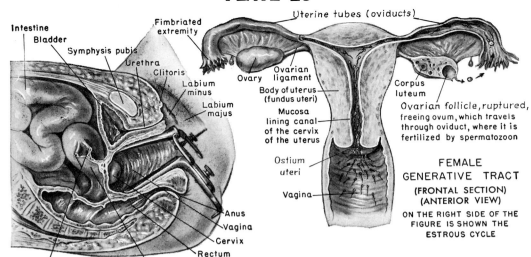

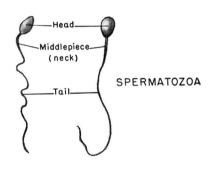

**EXTERNAL AND INTERNAL
FEMALE GENITALIA**

Labels (External and Internal Female Genitalia)

Intestine
Bladder
Symphysis pubis
Urethra
Clitoris
Labium minus
Labium majus
Anus
Vagina
Cervix
Rectum
Cut ends of uterine tube and round ligament of the uterus
Uterus
Fimbriated extremity

Female Generative Tract

Uterine tubes (oviducts)
Ovary
Ovarian ligament
Body of uterus (fundus uteri)
Mucosa lining canal of the cervix of the uterus
Ostium uteri
Vagina
Corpus luteum
Ovarian follicle, ruptured, freeing ovum, which travels through oviduct, where it is fertilized by spermatozoon

**FEMALE
GENERATIVE TRACT
(FRONTAL SECTION)
(ANTERIOR VIEW)
ON THE RIGHT SIDE OF THE
FIGURE IS SHOWN THE
ESTROUS CYCLE**

SPERMATOZOA

Head
Middlepiece (neck)
Tail

SPERMATOZOA

EARLY PREGNANCY

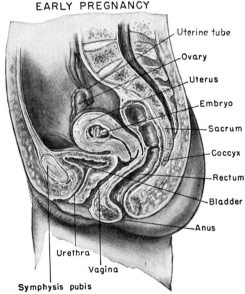

Uterine tube
Ovary
Uterus
Embryo
Sacrum
Coccyx
Rectum
Bladder
Anus
Urethra
Vagina
Symphysis pubis

PREGNANT UTERUS AT TERM

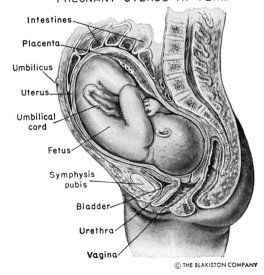

Intestines
Placenta
Umbilicus
Uterus
Umbilical cord
Fetus
Symphysis pubis
Bladder
Urethra
Vagina

GENITALIA—FEMALE

PLATE 24

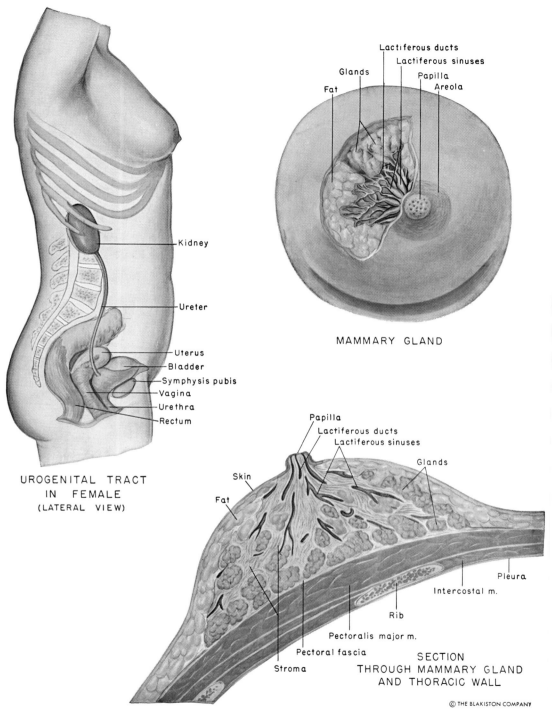

Lactiferous ducts
Lactiferous sinuses
Glands
Papilla
Areola
Fat

MAMMARY GLAND

Kidney

Ureter

Uterus
Bladder
Symphysis pubis
Vagina
Urethra
Rectum

UROGENITAL TRACT
IN FEMALE
(LATERAL VIEW)

Papilla
Lactiferous ducts
Lactiferous sinuses
Glands
Skin
Fat

Pleura
Intercostal m.
Rib
Pectoralis major m.
Pectoral fascia
Stroma

SECTION
THROUGH MAMMARY GLAND
AND THORACIC WALL

UROGENITAL TRACT—FEMALE; MAMMARY GLAND

PLATE 25

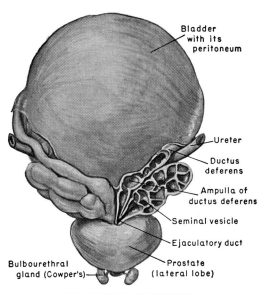

Bladder
with its
peritoneum

Ureter

Ductus
deferens

Ampulla of
ductus deferens

Seminal vesicle

Ejaculatory duct

Bulbourethral
gland (Cowper's)

Prostate
(lateral lobe)

MALE BLADDER, PROSTATE,
AND SEMINAL VESICLES
(POSTERIOR VIEW)

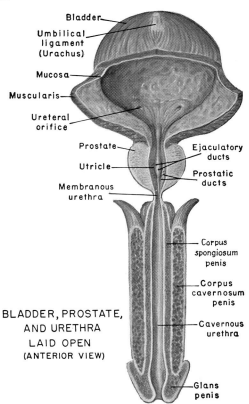

Bladder

Umbilical
ligament
(Urachus)

Mucosa

Muscularis

Ureteral
orifice

Prostate

Utricle

Membranous
urethra

Ejaculatory
ducts

Prostatic
ducts

Corpus
spongiosum
penis

Corpus
cavernosum
penis

Cavernous
urethra

Glans
penis

BLADDER, PROSTATE,
AND URETHRA
LAID OPEN
(ANTERIOR VIEW)

VISCERA OF MALE PELVIS
(LATERAL VIEW)

Ductus deferens

Spermatic vv.

Bladder

Ureter

Seminal
vesicles

Rectum

Prostate

Anus

Epididymis

Testis

Scrotum

Tunica
vaginalis

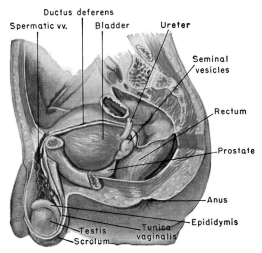

MALE PELVIS
(SAGITTAL SECTION)

Seminal vesicle

Ejaculatory duct

Bladder

Symphysis
pubis

Urethra

Penis

Testis

Prostate

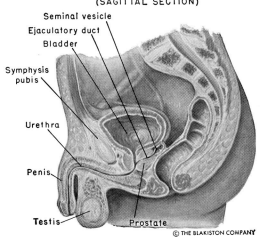

UROGENITAL TRACT—MALE

PLATE 26

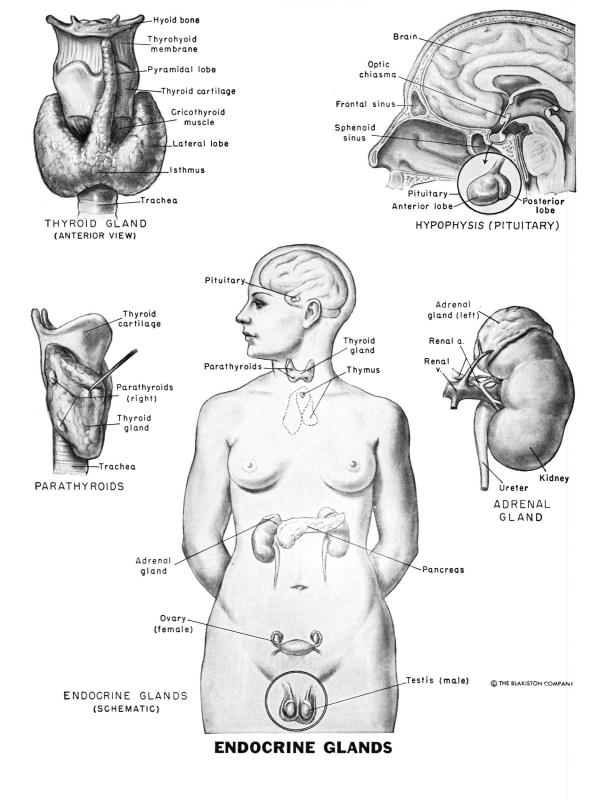

THYROID GLAND
(ANTERIOR VIEW)

Hyoid bone
Thyrohyoid membrane
Pyramidal lobe
Thyroid cartilage
Cricothyroid muscle
Lateral lobe
Isthmus
Trachea

HYPOPHYSIS (PITUITARY)

Brain
Optic chiasma
Frontal sinus
Sphenoid sinus
Pituitary
Anterior lobe
Posterior lobe

PARATHYROIDS

Thyroid cartilage
Parathyroids (right)
Thyroid gland
Trachea

ADRENAL GLAND

Adrenal gland (left)
Renal a.
Renal v.
Ureter
Kidney

Pituitary
Parathyroids
Thyroid gland
Thymus
Adrenal gland
Pancreas
Ovary (female)
Testis (male)

© THE BLAKISTON COMPANY

ENDOCRINE GLANDS
(SCHEMATIC)

ENDOCRINE GLANDS

zygomatic bone, lacrimal bone, nasal bone, inferior nasal concha, and hyoid. These are all paired bones except the mandible and the vomer. The mandible is the lower jaw; the maxillae form the upper jaw. The palatine bones form the posterior part of the bony roof of the mouth. The vomer is the lower portion of the median nasal septum. The zygomatic bones are the cheekbones. The lacrimal bones are fragile, grooved bones located medially, just inside the bony orbit. The nasal bones form the bridge of the nose, and the inferior nasal conchae are scroll-like bones located below the ethmoidal nasal conchae in the nasal passageway. The bones containing the paranasal sinuses are the frontal, ethmoid, and sphenoid bones and the maxillae. The hyoid bone is located in the neck just above the larynx.

9 In the human embryo, the vertebral column is composed of 33 vertebrae: 7 cervical, 12 thoracic, 5 lumbar, 5 sacral, and 4 coccygeal. The first cervical vertebra is the atlas, and the second is the axis. The sacral vertebrae fuse to form the sacrum in the adult, and the coccyx consists of four rudimentary vertebrae, more or less fused.

10 The appendicular skeleton comprises the remainder of the bones of the body.

11 The thoracic cage is formed by 12 thoracic vertebrae, 12 pairs of ribs, the costal cartilages, and the sternum.

12 The shoulder girdle consists of two clavicles and two scapulae. The clavicles are the collarbones. The head of the humerus articulates with the scapula.

13 The bones of the arm are the humerus, radius, and ulna. There are eight carpal bones in the wrist. Metacarpal bones form the framework of the hand. Finger bones are phalanges.

14 The hipbones form the pelvic girdle. Each hipbone develops as three separate bones: the ilium, ischium, and pubis.

15 The thigh bone is the femur. The patella, or kneecap, protects the knee joint anteriorly. The tibia and fibula are bones of the leg.

16 There are seven tarsal bones. Of these, the heel bone is the calcaneus, and the bone that articulates the ankle is the talus. The five metatarsal bones of the foot are homologous with the metacarpal bones of the hand. The phalanges are the toe bones. The tarsal and metatarsal bones form the longitudinal arch of the foot. The metatarsal bones also form the transverse arch.

17 Bursae are sacs of fibrous connective tissue lined with synovial membrane. They contain synovial fluid and are placed between moving parts to prevent friction.

18 Articulations are classified as immovable, slightly movable, and freely movable. The sutures between cranial bones are immovable joints. The intervertebral disks permit some movement between the vertebrae and are slightly movable joints. Ball-and-socket, hinge, and pivot joints are classified as freely movable.

19 Some diseases of the bones and joints are osteoporosis, osteomyelitis, arthritis, and cancer.

QUESTIONS

1 Describe the skull of a newborn infant.
2 Discuss the relationship of calcium and phosphorus metabolism to bone formation and resorption.

3 Locate the fontanels.

4 Which bones contain paranasal sinuses?

5 Name and locate the natural curvatures of the vertebral column.

6 What bone forms the bony prominence above the shoulder joint? The bony prominence at the elbow?

7 When the hand is turned from palm side up to palm side down, which bone rotates at the elbow?

8 What are some of the differences between male and female pelves?

9 What is the function of the bursae?

10 What bone forms the large bony prominence on the inner side of the ankle?

11 Compare the structure of the hand with that of the foot. How do they differ?

12 Discuss the structure and function of the two arches of the foot.

13 Discuss the structure of bone, including the embryonic development of different kinds of bone. How do bones grow?

14 Where do the red blood cells originate in the adult?

15 Describe the differences between three kinds of arthritis.

SUGGESTED READING

Congdon, C. C.: Bone Marrow Transplantation, *Science,* **171:**1116–1124 (1971).

Old, Lloyd J.: Cancer Immunology, *Sci. Am.,* **236:**62–79 (1977).

Several textbooks on histology and anatomy are listed at the end of the text. A few are given here.

Edwards, L. F., and G. R. L. Gaughran: "Concise Anatomy," 3d ed., McGraw-Hill Book Company, New York, 1971.

Gray, H.: "Anatomy of the Human Body," 29th ed., Lea & Febiger, Philadelphia, 1973.

Greep, R. O. (ed.): "Histology," 3d ed., McGraw-Hill Book Company, New York, 1973.

Schlossberg, L., and G. D. Zuidema: "The Johns Hopkins Atlas of Human Functional Anatomy," The Johns Hopkins Press, Baltimore, 1977.

6

MYOLOGY: THE PHYSIOLOGY OF MUSCLES

CONSIDERATIONS FOR STUDY

Muscle tissue is specialized for movement, and the ability of muscle tissue to produce movement stems from its property of contractility. In this chapter we shall discuss the anatomy, physiology, and biochemistry of muscle contraction, including:

1 The structure of skeletal muscle fibers and the sliding-filament theory of contraction
2 The control of muscle contraction through excitation-contraction coupling
3 The sources of energy for muscle contraction
4 The types of muscle contraction and the definition of the motor unit
5 The basic properties of the three types of muscle tissue: skeletal, smooth, and cardiac
6 The physiological basis of some common clinical disorders

MUSCLE TISSUE

Muscle tissue is the third type of body tissue we shall study. This tissue forms the muscles of the body organs that are responsible for movement. *Skeletal muscles* work against the inert skeletal framework to produce such movements as walking, running, and jumping. Supporting this activity is the special kind of muscle tissue found in the heart—*cardiac muscle*. The constant contraction and relaxation of cardiac tissue causes the heart to beat, thus supplying blood to all parts of the body. Still a third kind of muscular tissue makes up the *viscera*, the internal organs of the body such as the stomach, the intestine, and the uterus. *Visceral* or *smooth muscle* is

TABLE 6·1
Muscle types and characteristics

Type	Contraction	Muscle cell	Striations	Nuclei
Skeletal	Voluntary	Long cylindrical	Strongly striated	Multiple nuclei around periphery of fiber
Cardiac	Involuntary	Fibers short and branching	Finely striated	Elongate-oval between intercalated disks
Smooth	Involuntary	Spindle-shaped	Nonstriated	Single nuclei, usually in middle of cell

responsible for such internal activity as churning the food in the stomach and passing it through the intestine. Muscle tissue accounts for 40 to 50 percent of the body weight, the percentage being slightly greater in men than in women. The characteristics of the three types of muscle are summarized in Table 6.1.

Muscle tissue is composed of specialized cells that develop from embryonic cells called *myoblasts*. The properties characteristic of muscle tissue are essentially the same as those of the specialized cells that make up this tissue. One characteristic of muscle tissue is *excitability*, the capability to react to a stimulus; another characteristic is *contractility*, the capability to become shorter and thicker, or to contract. *Work* is accomplished by the contraction of muscles. Muscle tissue also exhibits *extensibility*, the ability to stretch out. Skeletal muscles are usually arranged in antagonistic pairs so that when one muscle contracts, the opposing muscle extends. The actions of the muscles that flex the arm or the leg illustrate this principle (Fig. 6.1).

FIGURE 6·1
The biceps brachii and the triceps brachii as examples of antagonistic muscles.

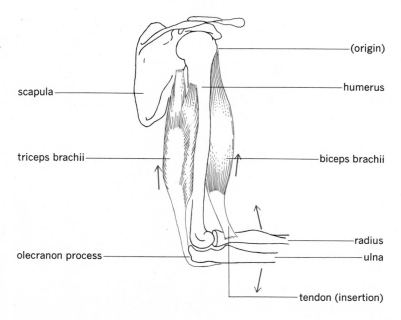

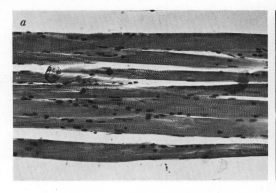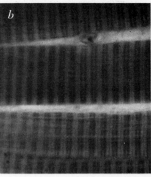

FIGURE 6·2

Skeletal muscle tissue. Note striations and nuclei. *a* Striations shown moderately magnified. (*Courtesy of General Biological Supply House, Inc., Chicago.*) *b* The resolution is much greater, and various zones and bands can be identified. Compare with Fig. 6.4*d*. (*Courtesy of Dr. W. M. Copenhaver.*)

Cardiac muscle tissue exhibits extensibility when the chambers of the heart are distended as they fill with blood. Visceral muscle shows great extensibility; for example, the stomach may be greatly distended by food and the urinary bladder may distend to contain large amounts of accumulated urine. Another characteristic is the capability of muscle tissue to return to its normal length after it contracts or extends.

STRUCTURE OF SKELETAL MUSCLE TISSUE

SKELETAL MUSCLE STRUCTURE   The muscles that are designed primarily to move the skeletal framework are called *skeletal muscles,* and they represent a considerable portion of the body weight. A piece of beefsteak is largely composed of skeletal muscle. These muscles are ordinarily attached to bones by tendons. Since the gross action of sketetal muscles may be directed by higher centers of the brain, the movement they participate in is voluntary, and they are often called the *voluntary muscles.*

Structure of a skeletal muscle fiber   Muscle tissue has a very complex structure. Under a microscope, a bit of skeletal muscle tissue appears to be composed of many fibers that extend along the length of muscle. Each fiber is finely *striated* with alternating light and dark bands (Figs. 6.2 and 6.3). The fibers are the basic structural units of muscle tissue. They are cylinders of small diameter, 0.01 to 0.1 mm, about the same size as a human hair. They are 1 to 40 or 50 mm in length. Some fibers are inserted in the muscle tendon. Others, however, are located only in the body of the muscle itself.

Each muscle fiber is enclosed in a delicate but strong sheath called the *sarcolemma,* which is essentially the cell membrane (Fig. 6.4). This sheath encloses the semifluid protoplasm of the fiber, the *sarcoplasm.* There are many nuclei in the sarcoplasm of a single fiber, and these lie just underneath the sarcolemma. Numerous mitochondria that produce ATP are present in these fibers.

Structure of muscle fibrils   Muscle fibers are composed of *myofibrils,* or muscle fibrils. These are threadlike structures, cross-striated like the fiber in which they lie. Fibrils have no surface membrane, since they are composed of protein molecules.

The electron microscope reveals that a myofibril is composed of still smaller filaments. These filaments, shown in Fig. 6.4, are of two kinds, thick

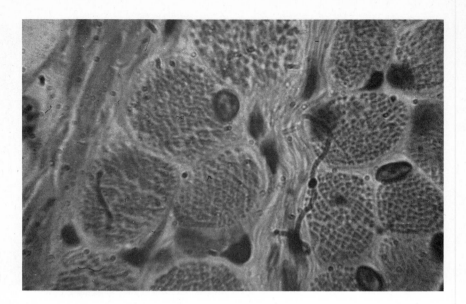

and thin. *Thick filaments* are about 0.01 μm in diameter and 1.5 μm in length. *Thin filaments* are about 0.005 μm in diameter and about 2 μm long. In cross section each thick filament is surrounded by six thin filaments in a hexagonal arrangement (Fig. 6.5). The thick and thin filaments are composed of different proteins.

Filament proteins *Myosin* is the protein of the thick filaments. The myosin molecule is large and elongate with a long straight portion that ends in paired globular heads. The structure resembles somewhat a minute golf club. The molecules are placed so that they face in opposite directions within the thick filament; that is, the heads point in one direction in half of the filament and in the other direction in the other half, as shown in Fig. 6.5. The globular heads form the cross bridges of the fibril seen in Fig. 6.6, but there are no cross bridges in the middle of the myosin filament. The heads project in groups of three around the thick filament. For simplicity, the heads are commonly illustrated as a single figure.

Myosin is an enzyme that plays a part in the dephosphorylation of adenosine triphosphate (ATP), the process in which chemical energy is transformed into mechanical energy. (See Chap. 3 for description.) Myosin is, therefore, an ATPase. This enzyme activity takes place in the globular heads of the myosin molecule, which also contain an actin-binding site.

The thin filament is composed of the protein *actin*. The actin filament consists essentially of two strands of small globular molecules wound around each other in a helical arrangement. The actin filament is generally only 1 μm long and is composed of 300 or more actin molecules. Although actin is the principal protein of the thin filaments, two additional proteins are present in small amounts. These are *tropomyosin* and a *troponin complex*.

The long thin filament that winds around the actin filament and extends over seven actin molecules is composed of tropomyosin (Fig. 6.7). Tropo-

FIGURE 6·4

Diagrammatic representation of skeletal muscle greatly enlarged: *a* longitudinal section; *b* cross section; *c* diagram illustrating a sarcomere; *d* electron micrograph of striated muscle tissue from rabbit (original magnification ×24,000). The broad, dense areas are A bands. The less dense area passing through the middle of the A band is the H zone. The wide, light-colored areas are the I bands. The black band passing through the I band is the Z line. (*Courtesy of H. E. Huxley.*)

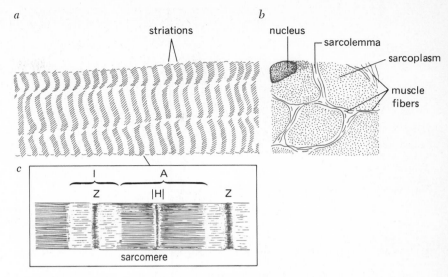

myofibrils

myosin molecules are long and slender, and are attached head to tail. They form the two tropomyosin filaments located in grooves on the actin filament.

The troponin complex is composed of globular proteins spaced at intervals along the thin filament. The troponin complex is closely associated with tropomyosin and is located near one end of the molecule. Troponin complexes have been shown to be located about every 400 Å along the 1-μm (10,000-Å) length of the thin filament.

Organization of muscle fibrils Thick and thin filaments overlap for part of their length, producing the dark, dense, *anisotropic*, or A, band shown in Fig. 6.4.

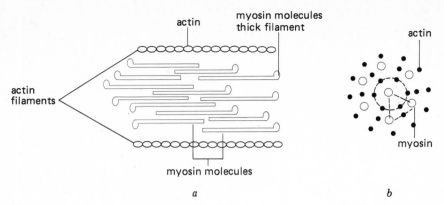

FIGURE 6·5

a Longitudinal section of muscle fiber. *b* A cross section through a muscle fibril shows that each myosin filament is surrounded by three myosins.

A lighter colored, less dense area, called the *isotropic,* or I, band, is composed solely of thin filaments, in relaxed muscle, whereas the narrow H zone contains only thick filaments. A narrow zone of dense material, known as the Z line, lies in the middle of each I zone and divides the fiber into segments called *sarcomeres.* In other words, the segment limited by a Z membrane at either end is a sarcomere.

The sliding-filament theory According to the *sliding-filament theory* developed by A. F. Huxley, H. E. Huxley, and J. Hanson, contraction of the sarcomere is caused by the two sets of filaments sliding over each other. It has been observed that the A band, composed of thick filaments, retains a constant length during either contraction or extension of the sarcomere. This indicates that the thick filaments do not contract or lengthen. Since the thin filaments are attached to the Z line, sarcomere shortening contracts the muscle fiber by sliding the thin filaments into the H zone of the A band (Fig. 6.8). It appears that the thin filaments divide at the Z line, making a Y-shaped, staggered appearance at their ends. Each branch of the Y makes a connection with another thin filament in the next sarcomere.

FIGURE 6·6

Proposed action of cross bridges: *a* in resting muscle; *b* during contraction the cross bridges are thought to reach out from myosin filaments and make contact with actin filaments; *c* the head of the myosin molecule slants with the movement of the filaments. (*After Hugh Huxley.*)

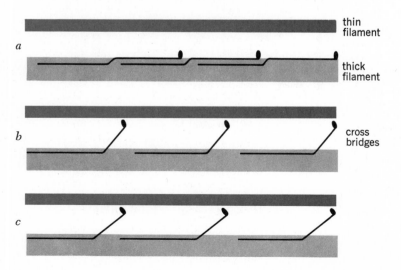

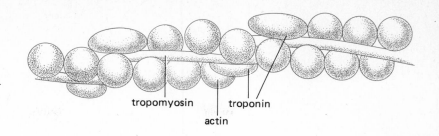

FIGURE 6·7

A coiled chain of actin molecules and a tropomyosin filament with its associated troponin.

tropomyosin   troponin

actin

As we have seen, during contraction the Z lines come closer together. The various bands and zones have a different appearance during contraction than during the resting state. During a rest period, the I band and H zone are somewhat lighter in color, since there is no overlap of thick and thin filaments in these areas when the myofibril is stretched. But during contraction the I-band area contracts greatly and the H zone also narrows and almost disappears as thin filaments invade this region. In a maximum contraction the ends of thin filaments come together and may even overlap each other for a short distance.

MUSCLE FIBER INTERNAL MEMBRANE SYSTEMS   The *sarcoplasmic reticulum* is a part of the internal membrane system of skeletal muscle fibers. It consists of a network of channels and vesicles within the muscle fiber that is similar in many ways to the endoplasmic reticulum of other cells. The network extends along the muscle fibrils in the sarcoplasm of the muscle cell. This network can be seen in cardiac tissue, but it reaches its highest development in skeletal muscle. It does not seem to be present in smooth muscle. In

FIGURE 6·8

a Diagram of muscle zones and filaments: resting and contracted muscle; schematic representation of myosin molecules indicating that the heads point in one direction in one half of a thick filament and in the opposite direction in the other half. The heads of these molecules represent the cross bridges projecting out from thick filaments to thin filaments. *b* In contraction the two sets of filaments slide over each other and the Z lines move closer together. A sarcomere is the area between two Z lines. The Z lines pass through I bands. H zones are located in A bands.

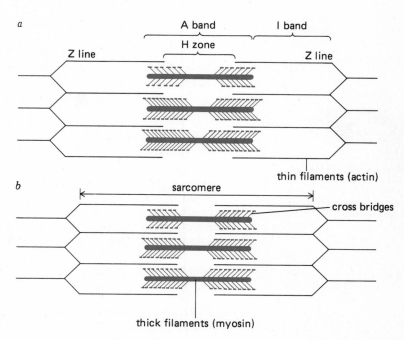

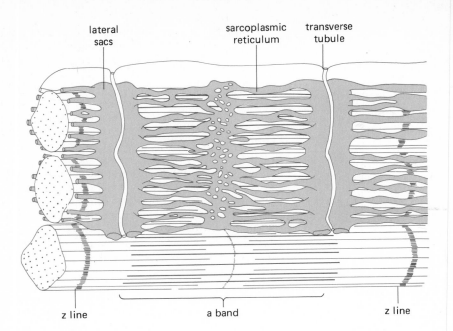

lateral sacs   sarcoplasmic reticulum   transverse tubule

z line   a band   z line

FIGURE 6·9
Sarcoplasmic reticulum and transverse tubule system. Note the lateral sacs in close association with the transverse tubules.

addition to the tubules that run parallel to the muscle fibrils, a system of *transverse tubules* exists. These make up the second part of the internal membrane system (Fig. 6.9). The transverse tubular system is a direct infolding of the surface membrane. In human skeletal muscle the transverse channels enter the fiber at the junction of the A and I bands. The transverse tubules are not necessarily at the same position in all muscle types. The *lateral sacs* of the sarcoplasmic reticulum associated with the transverse system form *triads*, which consist of sacs or vesicles on either side of the transverse tubule near the surface membrane. Lateral sacs are also called *lateral cisternae*, or outer vesicles, by some physiologists. The sarcoplasmic reticulum and the channels of the transverse system are not continuous; there is no indication that they open into each other. Nevertheless, they evidently cooperate in the initiation of skeletal muscle contraction.

MUSCLE CONTRACTION

Small electrical gradients generated by cell membranes provide a signaling system that provides communication between the cells of the nervous system and between nerve cells and effector cells such as those in muscle and gland tissue. The processing of information within the nervous system and the exercise of control by the nervous system over effectors are accomplished in large part by alteration of the electrical properties of cell membranes. In order to discuss the physiology of skeletal muscle contraction, therefore, we must first examine the origin and effect of the electric signals which control and initiate muscle contraction.

MOLECULAR EVENTS   *Diffusion potentials*   In Chap. 2 we noted that an electric potential difference exists across the plasma membranes of all cells. This electric potential difference is primarily a potassium diffusion potential. *Diffusion potentials* build up whenever different types of ions that can penetrate a membrane are initially distributed across the membrane in different concentrations (Fig. 6.10). For example, let us take an *artificial* situation in which a 100 mM KCL solution is separated from a 100 mM NaCl solution by a membrane that is permeable to K$^+$ but not Na$^+$ or Cl$^-$ and see how a diffusion potential may develop. In this example, a chemical gradient exists for K$^+$ to diffuse through the membrane from the side with 100 mM KCl concentration into the side with 0 mM KCl concentration. When K$^+$ ions move across the membrane, they carry electric charge from one side of the membrane to the other, so that when K$^+$ moves from 100 mM to 0 mM, the amount of positive charges relative to the amount of negative charges on the 0 mM side increases. As more positive charges move, the net positive charge on the 0 mM side becomes greater. Potassium continues to diffuse down its concentration gradient until the amount of net positive charge on the 0 mM side is sufficient to repel K$^+$ at the same rate as it is diffusing into that side. When this point is reached and the electrical driving forces are strong enough to balance the chemical driving forces, the system is said to be at *electrochemical equilibrium*. The electric potential difference in this example will persist indefinitely if there is no external energy input. Because the potential difference is caused by ion diffusion, it is a *diffusion potential*.

If we now measure the chemical concentrations of KCl in the 100 mM and 0 mM sides, we will find that the initial concentrations are unchanged. Only a few K$^+$ ions have moved, and we have no chemical analysis sensitive enough to detect the concentration changes. As we pointed out in Chap. 2,

FIGURE 6·10

The semipermeable membrane separates two salt solutions but is permeable only to K$^+$. The ion diffuses down its concentration gradient creating a type of electric potential, called a diffusion potential across the membrane.

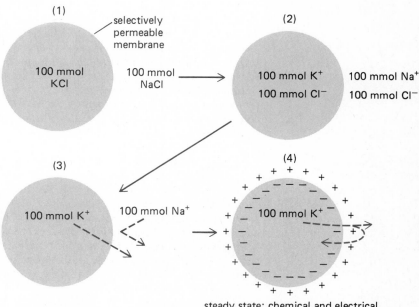

(1)

selectively permeable membrane

100 mmol KCl

100 mmol NaCl

(2)

100 mmol K$^+$
100 mmol Cl$^-$

100 mmol Na$^+$
100 mmol Cl$^-$

(3)

100 mmol K$^+$

100 mmol Na$^+$

(4)

100 mmol K$^+$

steady state: chemical and electrical gradients cause equal and opposite effect on K$^+$ movement.

the amount of positive and negative charges in a solution must be equal; that is, electroneutrality must be preserved. In the artificial system just described, however, the selectively permeable membrane caused a microscopic violation of electroneutrality by separating a few positive and negative charges. Because positive and negative charges tend to attract each other, such separation creates a *potential* for the charges to come together. The potential energy held in the attraction of such potential charges may be transformed into kinetic energy and thus do *work*. Charge separation is the fundamental means by which electric potentials are developed across cell membranes.

Diffusion potentials in muscle tissue   The plasma membranes of living cells are permeable to many ions, not just potassium. Careful measurements have demonstrated, however, that the *resting* muscle sarcolemma is 50 to 100 times more permeable to potassium than to sodium: indeed, almost all plasma membranes share this characteristic. This high permeability to potassium coupled with the sodium-potassium exchange pump (discussed in Chap. 2) maintains a high intracellular potassium concentration and a low intracellular sodium concentration. These diffusion potentials explain the electric potential that exists across all cell membranes. The sodium-potassium pump creates and maintains a concentration gradient for potassium such that the potassium ions tend to diffuse from their high concentration inside the cell to their low concentration outside the cell faster than sodium ions can diffuse from their high concentration outside the cell to their low concentration inside the cell. Thus, the resting membrane potential of the muscle cell is normally near the potassium diffusion potential; that is, an electrical gradient of -70 to -90 mV is present across the cell membrane. Because more positive charge is leaving the cell than is entering, the inside of the cell is negative with respect to the outside.

The action potential: alterations of membrane potential   The permeability characteristics of resting muscle and neuron plasma membranes are those described above. Upon proper stimulation of the membrane, however, membrane permeability of these excitable cells changes rapidly so that the membranes are more permeable to sodium (Fig. 6.11). This change allows sodium ions to rush into the cell faster than potassium ions are leaving and rapidly changes the electric potential difference across the cell membrane from the inside negative resting condition toward a positive potential.

In the "resting" state the cell membrane is *polarized* so that the inside is negative. Upon proper stimulation the membrane *depolarizes* and the inside becomes less negative with respect to the outside. If the membrane potential is sufficiently depolarized, it reaches a value known as the *threshold membrane potential*. At this point, a positive feedback process develops in which sodium permeability increases, causing more depolarization of the membrane, which further increases sodium permeability, causing greater depolarization, and so forth. Thus, an explosive change in membrane potential occurs. When the membrane potential reaches a value of about $+20$ mV, with the inside of the membrane being positive, the sodium permeability suddenly returns to the resting level. This is accompanied by a brief sharp increase in permeability of the membrane to potassium that produces *repolarization*. During repolarization the membrane potential is returned to

FIGURE 6·11

The relative size of the arrows in the upper part of the graph indicates the magnitude of Na⁺ and K⁺ fluxes across the cell membranes during an action potential. *a* In the resting membrane K⁺ movement out of the cell predominates, and the inside of the cell is about −70 mV. *b* Upon stimulation of the cell membrane a great increase in Na⁺ movement into the cell takes place, causing depolarization of the membrane to the threshold potential (−50 mV) followed by a further rapid depolarization to about +20 mV. *c* Rapid repolarization occurs when Na⁺ permeability returns to normal and K⁺ permeability temporarily increases. *d* After repolarization the K⁺ permeability returns to normal and the cell membrane remains at the resting potential of −70 mV.

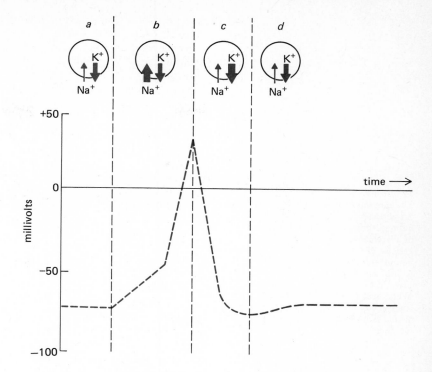

the resting condition of −70 to −90 mV, inside negative. These changes in membrane potential are called the *action potential.*

The action potential occurs first on the membrane at the point at which it is stimulated. This "wave" of depolarization spreads, however, from the point of origin over the entire cell surface. It travels because the locally induced action potential causes an *electrotonic* effect on adjacent areas of membrane (a flow of electric current between the two areas), producing a sudden increase in sodium permeability in these regions. In other words, the action potential is said to be *propagated* along the cell surface. In neurons the propagated action potential is called a *nerve impulse* and transmits information within the nervous system and between neurons and effector cells. In muscle cells the action potential initiates the cellular events that lead to muscle fiber contraction.

Excitation-contraction coupling In skeletal muscle, contraction is initiated when an action potential induces excitation at a *neuromuscular junction* (Fig. 6.12). At the neuromuscular junction a branch of a nerve cell process called an *axon* is closely associated with a skeletal muscle fiber. The axon forms an *axonal end terminal* that faces, but does not touch, a specialized region of the sarcolemma called the *motor end plate.* These two structures form a *synaptic connection*, i.e., an area specialized for communication. When an action potential travels over the axonal branch and arrives at the axonal end terminal, a chain of events occurs that eventually produces a muscle contraction. The nerve cell action potential produces an increased permeability to calcium ions in the axonal end terminal that somehow causes

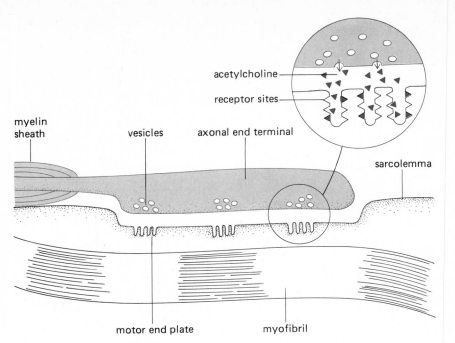

acetylcholine

receptor sites

myelin sheath

vesicles

axonal end terminal

sarcolemma

FIGURE 6·12
The neuromuscular junction. An action potential in the axonal end terminal causes release of acetylcholine present inside vesicles of the terminal. The neurotransmitter diffuses to receptors on the motor end plate of the sarcolemma. When acetylcholine binds these receptors, an action potential is produced in the muscle fiber.

motor end plate

myofibril

agitation of small vesicles located in the terminal. These vesicles contain a neurotransmitter substance called *acetylcholine*. When these vesicles are stimulated, some merge with the plasma membrane and, by exocytosis, empty their contents into the *synaptic cleft,* or space, between the nerve terminal and muscle cell. The acetylcholine then diffuses across the synaptic cleft and interacts with the motor end plate of the sarcolemma. The motor end plate contains receptor molecules that bind momentarily with acetylcholine. The interaction causes a brief increase in sodium ion permeability of the sarcolemma and initiates an action potential, called the *end plate potential,* that subsequently spreads over the entire surface of the muscle fiber. The acetylcholine is quickly destroyed by acetylcholinesterase, an enzyme always present in the neuromuscular junction.

The action potential not only spreads over the muscle fiber surface, but it also travels to the interior of the fiber via the transverse tubule system. Although the t tubules are not connected to the lateral sacs of the sarcoplasmic reticulum, an action potential in the t tubules causes stored calcium ions to move rapidly out of the lateral sacs of the sarcoplasmic reticulum into the cytoplasm of the muscle fiber.

Triggering contraction with calcium ions Calcium ions play a very important part in triggering muscle contraction. Calcium ions are found in low concentrations in the cytoplasm of the resting muscle fiber, but after an action potential stimulates the fiber, calcium is released by the sarcoplasmic reticulum, and its concentration rises rapidly. Calcium ions appear to bind to the troponin complex of the fiber's thin filaments. When calcium binds with the troponin complex, some sort of change occurs in tropomyosin, the long thin filament found wrapped around the actin filament, and this change is then transmitted to actin. The transformed actin combines with myosin at the binding sites on the myosin cross bridges. This event starts contraction.

Our muscles are not always contracted, however. What prevents the actin and myosin from constantly interacting and causing contraction? When the concentration of calcium is low, the troponin complex and tropomyosin act as inhibitors, preventing actin from combining with myosin. But when calcium ions are available in quantity, they bind to the troponin complex and somehow change it, which allows tropomyosin to interact with actin. The combination of actin and myosin activates the myosin cross bridges, and the thick and thin filaments slide over one another, causing the fiber to contract.

Recent studies suggest that tropomyosin may physically block the attachment of the head of the myosin molecule to actin when the calcium level is low. When the calcium level rises, however, the helical strand of tropomyosin molecules moves deeper into the groove in the actin molecule, thus permitting the myosin heads, or cross bridges, to bind to sites on the actin filament.

As we have indicated earlier, the globular head of the myosin molecule has an active site that reacts with a similar site on the actin molecule. The myosin molecule also contains an active site that enables the molecule to function as an ATPase. The myosin ATPase, activated by actin, splits ATP to ADP plus P, thus releasing energy for movement of the cross bridges. Magnesium ions greatly influence the activity of this enzyme.

There are many cross bridges in each muscle fiber, but the movement of a single cross bridge produces only a minute sliding movement of thick and thin filaments. During contraction only about half of the cross bridges of a fiber are bound to thin filaments at any one time; the other cross bridges are in intermediate stages. In order to produce noticeable contraction, each cross bridge must be in continual action. It must break its bond to an actin filament and then form a new bond at a new site in a repeating cycle. ATP may function in producing this cycle. ATP binds to myosin, breaking the bond between actin and myosin and freeing the cross bridges for another cycle.

Relaxation If muscles are to cause movement, they must relax as well as contract. The flooding of calcium ions through the sarcoplasmic reticulum initiates contraction. But how does the muscle relax, or what prevents muscle from remaining in a contracted state? The release of calcium ions from the lateral sacs of the transverse tubules requires only a few milliseconds, whereas forcing calcium ions back into the lateral sacs may require several *hundred* milliseconds. Therefore, the cross bridges continue to form and break bonds with actin filaments until the calcium concentration reaches a low-enough level to permit the troponin-tropomyosin system to inhibit the interaction of actin and myosin. In other words, relaxation occurs only when enough calcium ions are returned to the lateral sacs to block the systems that cause contraction. The process of "pumping" calcium ions back into the lateral sacs for storage requires energy, which is supplied again by ATP.

Muscles become temporarily hard and rigid a short time after death, a condition called *rigor mortis*. Rigor mortis is caused by the inability of dead muscle cells to produce ATP. Myosin cross bridges bind to actin, but without the energy derived from the breakdown of ATP, these bonds cannot be broken. When muscles have been contracting strongly or are greatly fatigued just before death, rigor mortis sets in more promptly.

The events in skeletal muscle contraction are summarized in Table 6.2.

SOURCES OF ENERGY FOR MUSCLE CONTRACTION The immediate source of energy for the contraction of muscle comes from the breakdown of ATP as represented by the equation

$$ATP + H_2O \xrightarrow{\text{ATPase}} ADP + P_i + \text{energy}$$

The amount of ATP stored in muscle is not large, and gathering a supply from anaerobic and aerobic metabolism would probably take too long to support rapid movement. There is, however, another high-energy phosphate compound that releases energy, *creatine phosphate*, abbreviated CP. Much of the energy released by CP is used in the resynthesis of ATP, thus making up for ATP loss during exercise.

$$ADP + CP \xrightleftharpoons{\text{creatine kinase}} ATP + \text{creatine}$$

TABLE 6·2
Skeletal muscle contraction and relaxation

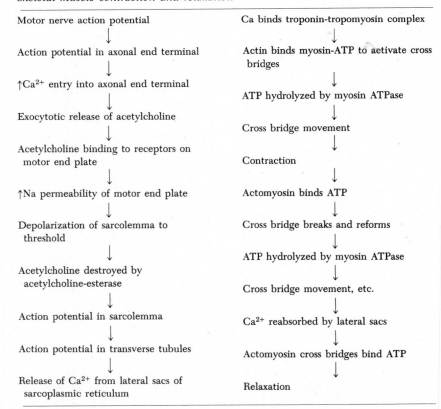

Motor nerve action potential
↓
Action potential in axonal end terminal
↓
↑Ca^{2+} entry into axonal end terminal
↓
Exocytotic release of acetylcholine
↓
Acetylcholine binding to receptors on motor end plate
↓
↑Na permeability of motor end plate
↓
Depolarization of sarcolemma to threshold
↓
Acetylcholine destroyed by acetylcholine-esterase
↓
Action potential in sarcolemma
↓
Action potential in transverse tubules
↓
Release of Ca^{2+} from lateral sacs of sarcoplasmic reticulum

Ca binds troponin-tropomyosin complex
↓
Actin binds myosin-ATP to activate cross bridges
↓
ATP hydrolyzed by myosin ATPase
↓
Cross bridge movement
↓
Contraction
↓
Actomyosin binds ATP
↓
Cross bridge breaks and reforms
↓
ATP hydrolyzed by myosin ATPase
↓
Cross bridge movement, etc.
↓
Ca^{2+} reabsorbed by lateral sacs
↓
Actomyosin cross bridges bind ATP
↓
Relaxation

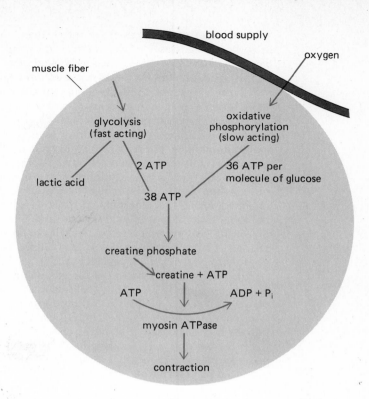

blood supply

oxygen

muscle fiber

glycolysis
(fast acting)

oxidative
phosphorylation
(slow acting)

lactic acid

2 ATP

36 ATP per
molecule of glucose

38 ATP

creatine phosphate

creatine + ATP

ATP

ADP + P_i

myosin ATPase

contraction

FIGURE 6·13
Phosphorylation in muscle.

The enzyme involved in this reaction is *creatine kinase,* also called creatine phosphoryltransferase. This reaction is reversible, and during rest periods or periods of reduced activity CP may be rebuilt from ATP and creatine (Fig. 6.13).

Glycogen, another source of energy, is usually abundant in muscle tissue. In the process of glycogenolysis and glycolysis, glycogen is broken down to pyruvic acid in an anaerobic reaction. The steps involved in these pathways have been detailed in Chap. 3. The energy released during these reactions is relatively low.

When pyruvic acid is formed, however, it enters the numerous mitochondria found in muscle cells and is oxidized in the tricarboxylic acid cycle. In this metabolic pathway, ATP is formed as hydrogen and electrons are passed along a series of catalysts and cytochromes in the respiratory chain. CO_2 is also given off during the tricarboxylic acid cycle.

During muscle contraction, when ATP breaks down to ADP, ADP enters the mitochondria, where it acts as a phosphate acceptor and is resynthesized to ATP.

Vigorous exercise requires energy, and this energy must be produced by large amounts of oxygen and ATP. The body depends upon the circulatory system to deliver adequate amounts of oxygen to muscles. Oxidative phosphorylation produces large amounts of ATP (36 ATP molecules from one glucose molecule) but at a relatively slow rate. Glycolysis, on the other hand, produces only a small amount (two ATP molecules from one glucose molecule), but these are produced at a much faster rate. In vigorous exercise, ATP may be used up faster than it can be produced by the tricarboxylic acid cycle, and muscle fatigue may set in. Glycolysis takes much less time than the tricarboxylic acid cycle does; it thus continues to supply ATP, but the nutrients required for glycolysis are soon exhausted.

Glycogen is stored in the liver, and muscle glycogen is replenished from this source. When oxygen delivery to active muscle is insufficient, lactic acid, the reduced form of pyruvic acid is produced. Lactic acid does not accumulate in muscle except during severe sustained exercise or during extreme muscular effort, when an oxygen deficiency may occur. (Under ordinary conditions muscular activity is supported by an adequate supply of oxygen.) About 10 times as much energy is liberated during aerobic metabolism when pyruvate is oxidized by way of the mitochondria than is obtained during the anaerobic glycolysis of glucose. Muscles can contract for some time, however, without an immediate supply of oxygen.

During the recovery phase, oxidation occurs and carbon dioxide is formed. If lactic acid is produced, it is carried by the blood to the liver and resynthesized to liver glycogen (Fig. 6.14).

The energy produced in the breakdown of these chemical compounds can be measured in calories, since heat is a form of energy. There are two phases of heat production in contracting muscles: (1) *initial heat* of contraction and (2) *recovery heat*. Initial heat develops quickly, and since the process is nonoxidative, this phase of heat production comes largely from the breakdown of organic phosphates rather than from the breakdown of glycogen. Recovery heat develops more slowly after relaxation and is produced primarily from oxidative phosphorylation and the resynthesis of ATP and CP.

The mechanical efficiency of muscle compares favorably with the mechanical efficiency of gas engines, which is only about 25 percent. Muscles have been estimated to convert 25 to 30 percent of the energy they produce into movement (work); the rest is given off as heat. In an engine, the energy given off as heat is largely a loss, but in the body, the heat produced helps to maintain a body temperature of about 98.6°F, or 37°C. A whole series of

FIGURE 6·14

Energy metabolism in muscle.

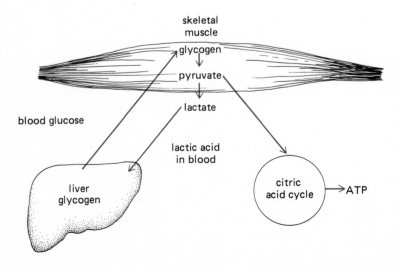

events, to be discussed later, is concerned with the maintenance and control of body temperature, but one factor should be mentioned here. As one becomes cold, the muscles tense and finally break into uncontrolled contractions, called *shivering*. These rapid contractions produce heat and help to warm the body. On a very hot day muscles are apt to be relaxed and flaccid.

Oxygen debt   Human beings are entirely dependent upon a steady intake of oxygen. About 200 to 300 cubic centimeters (cm^3) of oxygen per minute (min) are required when the body is at rest, and many times this amount is needed for vigorous exertion. The trained athlete is capable of providing 4 or 5 l of oxygen per minute to body tissues. Oxygen requirements for vigorous exercise, however, may be 16 to 29 l min or higher, depending upon the degree of exertion. Thus, during extreme exertion oxygen intake will be considerably less than the oxygen requirement of the muscles. We must, therefore, go into debt for oxygen. Runners breathe heavily as they rest after the race, until their oxygen debts are repaid. This prolonged high-oxygen requirement is due to the metabolism of lactic acid and resynthesis of ATP and CP in muscles.

It is fortunate that the energy necessary for the immediate contraction of muscle is supplied by nonoxidative reactions. However, we could not perform strenuous exercise for more than 1 or 2 min if we were deprived of oxygen. We have seen that oxygen is necessary in the recovery phase of muscle activity. If there is not enough oxygen available, lactic acid accumulates after strenuous muscular activity and is then transported to the liver, where enzyme systems convert the greater portion of it to glycogen. Oxidative energy is necessary also to rebuild organic phosphates. We can go into debt for oxygen, but the debt must be paid before normal exercise is resumed.

When we engage in moderate exercise, such as walking or working in the garden, we may not incur an oxygen debt. Physiological adjustments are such that a balance is struck between oxygen needs and energy requirements whenever possible. The heart and breathing rates become adjusted to exercise, and one reaches a condition described as a *steady state*. An athlete in training can perform moderate exercise without having additional lactic acid appear in the blood.

THE MOTOR UNIT AND TYPES OF MUSCLE CONTRACTION

Every muscle fiber within a skeletal muscle receives innervation from a motor neuron; however, a single axon of a motor nerve may innervate many muscle fibers. In order to understand muscle-nerve physiology, we must examine the concept of the *motor unit* (Fig. 6.15). A motor unit consists of the axon of a motor neuron, or nerve fiber, and the individual muscle fibers innervated by the terminal axonal filaments of this axon. Skeletal muscle fibers are thus innervated *in groups* by the axonal branches of a single motor neuron, and because of their common innervation, these fibers then respond as a unit to a nerve impulse. Muscles involved with very fine control, such as eye muscles, usually have motor units consisting of only a few muscle fibers per axon, while motor units of the large muscles of the back may have as many as 100 to 150 muscle fibers per axon. The muscle fibers of a motor unit are not all localized in one part of the muscle, but are spread out through

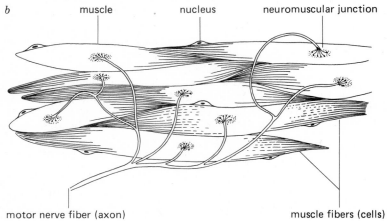

muscle   nucleus   neuromuscular junction

motor nerve fiber (axon)   muscle fibers (cells)

FIGURE 6·15

a Motor end plates in striated muscle. (*Courtesy of Dr. Mac V. Edds.*) *b* A motor unit.

the body of the muscle. This arrangement tends to equalize the strength of contraction throughout the muscle. Muscular contraction involves the activity of one or more motor units. We shall describe how motor units react to various conditions in the following sections.

GENERAL RESPONSE TO STIMULATION   A great deal may be learned about muscle contraction by observing the response of muscle under various conditions. A simple method of studying an isolated muscle is shown in Fig. 6.16. In this arrangement a muscle is removed from the animal and attached to a recording lever. A simultaneous recording of time and lever displacement is made. The muscle may be stimulated to contract in various ways; however, in our example we shall use electric current.

As the muscle is stimulated by the electric current, it contracts, but there is a brief period of perhaps 0.01 s from the time the stimulus is received until the muscle starts to contract. This is termed the *latent period*. In

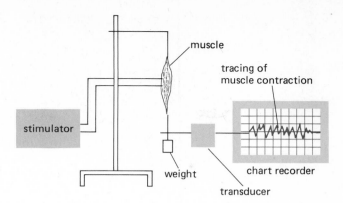

FIGURE 6·16

An apparatus for recording the contraction of an isolated muscle.

mammalian muscle, this period averages about 1 millisecond (ms), that is, 0.001 s, but different muscles may have widely varying latent periods. The *contraction phase* is represented on the recording by the upward motion of the writing lever, and this period lasts about 0.04 s.

The downward tracing represents the *relaxation phase* and lasts approximately 0.05 s. The muscle does work by contracting, but the relaxation phase is important too. Relaxation protects the muscle against fatigue, and is, therefore, not regarded as a passive stage but rather as an active process.

Single stimulus A single stimulus of a muscle preparation initiates a single response called a *twitch*. This type of contraction would not normally occur in intact muscle, but it is useful in illustrating different phases of contraction and relaxation (Fig. 6.17).

Series of stimuli Suppose we now introduce a series of stimuli of constant strength into the muscle preparation. We must allow time for the muscle to contract and relax, but this will take only about 0.1 s. If we repeat the stimuli slowly at 0.5-s intervals, we find that the muscle makes a series of

FIGURE 6·17

The phases of a simple muscle contraction, a muscle "twitch": *a* the initiation of the stimulus; *b* beginning of contraction; *a-b* latent period, about 0.01 s; *b-c* contraction phase, about 0.04 s; *c-d* relaxation phase, about 0.05 s. Each division of the time scale below represents 0.01 s.

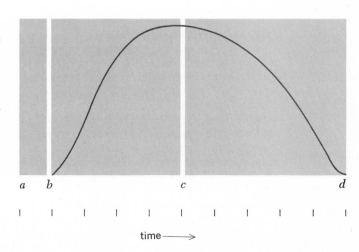

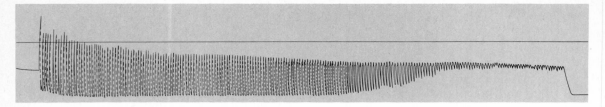

tracings as it contracts and relaxes. The first few tracings may show an increasing height with each contraction. This is the so-called *staircase phenomenon,* or *treppe,* which is associated with increasing efficiency of muscular contraction.

The muscle, having reached the peak of its performance, contracts and relaxes regularly in response to a series of regularly spaced stimuli. After numerous contractions, the muscle does not relax completely, and therefore, the tracings become shorter. The lever is not lifted so high, and with the shortening of the muscle the lever does not go so far down. This type of shortening of muscle is called *contracture* rather than *contraction.*

Fatigue As *fatigue* sets in, we see in the gradual shortening of the isolated muscle a loss of efficiency. The response time is slower, the latent period increases slightly, the time necessary for contraction increases, and the relaxation period becomes slower and slower. A study of the relaxation phase deserves special attention. In fatigue it is important to observe not only that the muscle fails to contract but also that it fails to relax (Fig. 6.18).

Summation When the rate of stimulation is increased to such an extent that the second stimulus is received by the muscle while it is still contracting in response to the first stimulus, a *summation* effect is noted. The muscle then contracts to a greater extent than it would in response to a single stimulus.

Tetanus If the muscle is stimulated at a rate of 20 or 30 stimuli per second, there is time for only partial relaxation between stimuli, and the muscle maintains a state of contraction called *incomplete tetanus* (see Fig. 6.19). Stimulation at an increased rate of 35 to 50 stimuli per second should result in complete tetanus.

A single muscular contraction, or twitch, would be of little use in coordinated movement. Those muscles that enable us to stand up, walk, run, or bend over can maintain a state of physiological tetanus for long periods of time with comparatively little fatigue. The condition of tetanus may apply to only a portion of a muscle rather than to the entire muscle. Tension in posture muscles is maintained by a series of stimuli to groups of nerve endings in the parts of the muscles involved. Flexing the biceps slowly is a coordinated contraction involving many muscle fibers and is much slower than a muscle twitch. This action may therefore be considered as a tetanus of those fibers that are in a state of contraction.

Refractory period A muscle can be experimentally stimulated at a rate so rapid that it fails to respond to each stimulus. A brief period follows each contraction in which muscle is nonirritable and, therefore, will not react to

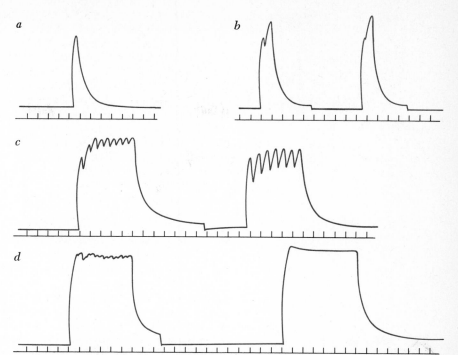

FIGURE 6·19

A tracing, or myogram, of muscle contractions: *a* a simple muscle twitch following a single stimulus; *b* summation, when two stimuli are introduced in rapid succession; *c* incomplete tetanus; *d* complete tetanus.

stimuli. This period of about 0.001 s in skeletal muscle is known as the *refractory period*. Since it is of such short duration, it probably does not limit activity under normal conditions. The stimuli that result in tetanus of skeletal muscle are evidently spaced so that they do not fall within the brief refractory period. The refractory period in heart muscles extends practically throughout the systolic contraction phase. Furthermore, a relative refractory period lasts well into diastole, the relaxing period of the heartbeat. Therefore, tetanus does not occur in cardiac tissue. Visceral muscle has the longest refractory period of any muscle tissue. Recent investigations indicate that the refractory period in muscles should be interpreted as being directly dependent upon the refractory period in the nerves that innervate them.

TYPES OF CONTRACTION   Muscles contract more efficiently if they have an appropriate load. When a load is first placed on isolated muscle and the muscle stimulated, a slight extension occurs; this slight stretching produces tension, and the muscle responds by shortening. The muscle functions more efficiently after it has adjusted to the load. The type of contraction exhibited when a muscle becomes shorter and thicker is called an *isotonic contraction*. The contraction of the biceps brachii as the lower arm is bent on the upper is an example. The tension remains essentially constant, but the muscle shortens, performing work.

In another type of contraction, the tension increases with the load, but the muscle does not shorten. This type is called *isometric*, from the Greek words that mean "the same measurement." If one carries a bucket of water with the arm extended, the muscles of the arm and trunk support the additional weight, but there is little or no change in the length of the muscles involved, only an increase in tension. In isometric contraction, energy is utilized and heat is produced, as it is in isotonic contraction. The "fixed" postural muscles of the trunk and neck undergo isometric contraction.

LENGTH-TENSION CONCEPT   A stretched muscle is somewhat similar in elasticity to a stretched rubber band. As force is applied to the muscle, it offers resistance, but if the force is increased, the muscle lengthens. If stretched beyond a certain point, however, the tension of the muscle decreases to a stage where no tension can be measured. The sliding-filament theory may explain this effect. If the muscle stretches so much that the thick and thin filaments no longer overlap, the myosin bridges cannot react with actin molecules of the thin filaments to produce contraction, and thus no tension is produced.

TONUS   Generally speaking, healthy skeletal muscles are never completely relaxed. This state of mild contraction is referred to as *tonus,* or *tone.* Tonus is a resistance to stretch. Apparently, some groups of fibers maintain a slight degree of contraction, while others are more relaxed so that fatigue does not readily occur in the intact resting muscle.

RED AND WHITE MUSCLE

Myoglobin is a muscle protein whose exact function remains unclear. This protein is found in close association with an abundance of mitochondria in red skeletal muscle. It has been shown that exercise-training will increase muscle myoglobin concentration remarkably, and it has been suggested that myoglobin may increase the availability of oxygen in muscle metabolism because it appears to facilitate oxygen diffusion. The mechanism involved can be illustrated very simply. If we measure the diffusion velocity of oxygen in an ordinary aqueous solution and then add myoglobin to the solution, we find that the oxygen diffusion velocity is greatly increased. Thus, inside muscle fibers myoglobin may be expected to speed the delivery of oxygen to the mitochondria.

Muscles containing a large amount of myoglobin are called *red muscles,* and those with a small amount are referred to as *white,* or *phasic, muscles.* An intermediate type of red muscle is also recognized. Such muscle has a high rate of oxidative phosphorylation and contracts swiftly, but in many respects it is intermediate between the more common red type and the white. The more common type of red muscle fiber is slower acting, but is also slower to show fatigue. Muscles of the legs and trunk that hold the body erect are of this type. Such muscles must necessarily resist fatigue for long periods of time.

The so-called white muscles rely largely on anaerobic glycolysis as their source of ATP. They are fast-acting but fatigue readily. They have large fibers and few mitochondria, and their blood supply is limited. They are often associated with muscles of precise movement such as those in the muscles of the eye. Since muscles may perform different types of activity, such as sustained contraction or swift movement, all three kinds of fibers may be incorporated into a single muscle.

SMOOTH MUSCLE

Smooth muscle is found in the walls of hollow organs such as the blood vessels, the digestive tract, the respiratory airways, the uterus, and the urinary bladder. It is found also in the internal eye muscles and in the erector pili muscles of the skin. Individual fibers may be found in such organs as the spleen and, in fact, in almost all internal organs.

Smooth muscle, as the name indicates, is *nonstriated*. It is not composed of the myofibrils that give skeletal muscle their striated appearance. Instead, myosin and actin filaments are found throughout the length of the fiber, but these are not arranged in a regular sequence as they are in skeletal muscle. Through the use of the electron microscope at high magnifications cross bridges have been discovered in smooth muscle. Troponin and tropomyosin are also present in smooth muscle.

Contraction is believed to be similar to that occurring in skeletal muscle. A release of calcium ions probably induces contraction, although the sarcoplasmic reticulum of smooth muscle is poorly developed and transverse tubules are apparently lacking. The surface membrane of the cell may function in some respects, however, as a sarcoplasmic reticulum, storing and releasing calcium. Smooth muscle is innervated by the autonomic nervous system, and its action is, therefore, involuntary.

TYPES OF SMOOTH MUSCLE   There are two kinds of smooth-muscle tissue: a *single-unit* type composed of thin sheets of circular bands, such as that found in the wall of the intestine, and a *multiunit* type composed of individual fibers, such as that found in the intrinsic muscles of the eyes. Most smooth muscle is of the single-unit type. This type is found in a number of involuntary sphincter muscles such as the pyloric sphincter at the distal end of the stomach.

Single-unit smooth muscle   This type is so named because its cell masses contract as a single unit. Thin flat sheets, or circular bands, of muscle are composed of spindle-shaped cells, each with a large nucleus located in the middle of the cell. These cells are about 6 μm in diameter and 50 to 100 μm in length. The cells usually lie longitudinally, with their slender ends fitted in against the thick middle portions of adjacent cells (Fig. 6.20).

FIGURE 6·20

Single-unit smooth muscle.

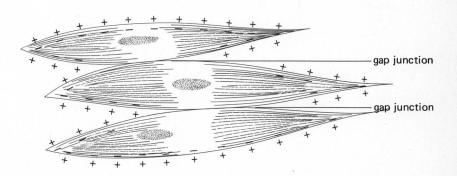

gap junction

gap junction

smooth muscle cells
showing gap junctions

Some thin strands of connective tissue lie between muscle cells, but the muscle cell membranes are closely associated with *gap junctions* formed between membranes. Gap junctions permit ions to pass easily between adjacent cells. In contraction, a wave of depolarization spreads over the muscle membrane, passing from one cell to another. The cells that receive the initial stimulus are termed *pacemaker* cells.

Pacemakers give rise to the *pacemaker potential,* which is followed by a slow depolarization of the cell's membrane potential. The resting potential of smooth muscle varies considerably but is considered to be around −60 mV, with the inside of the cell being negative and the outside, positive. Slow depolarization of the resting potential produces an action potential, which is followed by rhythmic contraction of the whole muscle. This slow depolarization is caused by a continuously decreasing permeability of the membrane to potassium. As the rate at which potassium diffuses out of the cell decreases, the magnitude of the negative resting potential decreases (goes to zero). When depolarization is sufficient, the threshold potential is reached and an action potential is produced. Membrane permeability to potassium is highest during the repolarization phase of the action potential and steadily decreases as repolarization occurs; but it continues to decrease after maximum repolarization, and this leads to production of another action potential. Smooth muscle is controlled by the autonomic nervous system, certain neurohumors, and hormones, and, unlike skeletal muscle, may be not only excited by these substances but also inhibited. Excitation and inhibition may be accomplished by altering the permeability of the cell membranes to potassium or sodium. An increase in permeability to potassium causes a more negative membrane potential, a state of *hyperpolarization,* and prevents an action potential. A decrease in permeability to potassium or an increase in permeability to sodium has the opposite effect.

Rhythmic waves of muscular contraction and relaxation such as those seen in the stomach and intestine are referred to as *peristalsis.* Visceral muscle contracts and reflexes much more slowly than skeletal muscle, and a single contraction may require several seconds or even minutes. Smooth muscle does not have specialized nerve endings or structures equivalent to the motor end plates of skeletal muscle.

Single-unit smooth-muscle tension   Single-unit smooth muscle, unlike skeletal muscle, is able to maintain a mild state of contraction for considerable periods of time even when it has been experimentally isolated from its nerve supply. The visceral muscles of hollow organs such as the stomach, urinary bladder, and uterus maintain tension even though subjected to considerable extension, a characteristic necessary to the proper functioning of these muscles. For example, the urinary bladder may be greatly extended but still able to void urine because the bladder muscles retain their tension even when stretched. The muscles also exert tension when they are greatly shortened, so that urine may be voided when the bladder is nearly empty. The ability of smooth muscle to shorten or stretch and still maintain effective tension is an example of its plasticity. Stretching may cause depolarization and contraction, thus limiting the stretch reaction. Good tone in visceral muscle is, of course, very important in maintaining a good state of health.

Multiunit smooth muscle This type of smooth muscle is found in the iris and ciliary muscles of the eye, the nictitating membranes of vertebrates, the walls of blood vessels, and the piloerector muscles attached to hairs. Individual muscle fibers may be innervated by more than one nerve fiber. The muscle fibers may contract more rapidly than single-unit smooth muscle, and they do not exhibit peristalsis. A stimulus of threshold level is followed by membrane depolarization, a rather long latent period, and then a fairly rapid contraction. In piloerector muscles, the fibers are arranged longitudinally to form a very small muscle in the skin attached to the base of the hair. These are the erector muscles that cause the hair to stand on end in reaction to the stimuli of fear or cold.

CARDIAC MUSCLE

The muscular wall of the heart is composed of specialized *cardiac muscle* tissue, which is considered to be a type of muscle intermediate in structure between smooth and skeletal muscle. Cardiac tissue consists of a branching network of fibers, giving strength to the musculature. The cells are elongate with large nuclei. Heart muscle is finely striated and involved in involuntary movement.

Cardiac muscle resembles skeletal muscle in possessing thin actin filaments, thick myosin filaments, and an actin-myosin sliding-filament contractile system. The regulatory protein troponin is also present. Sarcomeres are plainly marked by prominent Z lines. The sarcoplasmic reticulum is evident in cardiac muscle, and transverse tubules invaginate from the surface membrane in the region of the Z lines.

Like visceral muscle, heart muscle contracts in a rhythmic manner, with the action potential spreading from one cell to another through gap junctions. At the ends of cardiac muscle cells, thickened areas commonly known as *intercalated disks* are found. Under the electron microscope, the intercalated disks appear as irregular structures that provide firm support for the attachment of myofibrils. Dense, heavily pigmented regions of the disk resemble desmosomes (Fig. 6.21).

Heart muscle has an inherent ability to contract rhythmically throughout life. Nerve impulses to the heart regulate its function to some extent, but heart muscle is not entirely dependent upon its nerve supply. The heart of a frog or turtle may continue to beat even though it is removed from the body, and a small piece of cardiac tissue, given proper care, will continue to pulsate for some time. This *automaticity* is caused by the characteristic slow depolarization toward a threshold potential level of cardiac muscle. When threshold potential is reached, it is followed by contraction. During contraction, the muscle will not respond to additional stimuli for the duration of its long, absolute refractory period, which may last from 1 to 5 s. Therefore cardiac muscle does not undergo summation, and complete tetanus is not possible.

Following initiation of contraction, a slow repolarization of cardiac muscle membrane begins, but the contraction phase is completed and the muscle is in the relaxation phase before the cell membranes completely repolarize. The long refractory period provides time for the chambers of the heart to fill with blood before the next contraction phase starts.

Similarities in sarcomere structure and function in skeletal and cardiac

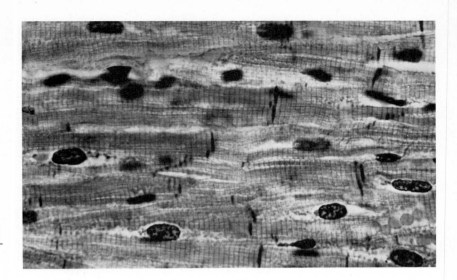

FIGURE 6·21
Cardiac tissue, showing striations, nuclei, and intercalated disks. (*Courtesy of General Biological Supply House, Inc., Chicago.*)

muscle seem to indicate that the mechanisms of contraction are closely related. In cardiac muscle, action potentials are thought to cause the release of calcium ions, presumably from the sarcoplasmic reticulum. The actin-myosin system then releases energy for contraction, as it does in skeletal muscle. However, cardiac muscle has not been studied as extensively as skeletal muscle, and there may be considerable differences in the mechanism of contraction. The effect of neurotransmitters will be considered in Chap. 16. The hormone norepinephrine acts as an excitatory transmitter to the cardiac muscles, while acetylcholine is inhibitory.

Numerous large mitochondria, or sarcosomes, are contained in cardiac tissue, a finding consistent with its need for ATP as a source of energy. Myoglobin is also present. It is concerned with oxygen supply. Heart muscle requires an abundant supply of oxygen. If this supply is cut off for even a minute, cells will die. The action of the heart will be considered further in Chap. 15.

CLINICAL ASPECTS

Myasthenia gravis Myasthenia gravis is a disorder of the skeletal muscles, whose first symptoms include a sagging of the facial muscles, particularly in the eyelids, which droop. Deterioration may progress to muscles of the larynx, pharynx, and chest, making breathing difficult. The disability seems to affect the neuromuscular junction. There may be some interference with acetylcholine transmission or with cholinesterase, an enzyme that normally breaks down acetylcholine. Recent research has focused on the development of antibodies that may inhibit the action of cholinesterase at the neuromuscular junction.

Myasthenia gravis may affect people of various ages, but it is more common in women than in men and most frequently occurs after thirty years of age. Muscular fatigue and general weakness are also characteristics of the disorder.

Muscular dystrophy   There are several types of muscular dystrophy, disorders characterized by progressive muscular degeneration. The most common type was first described by Duchenne and bears his name. Duchenne's dystrophy is a severe, disabling inherited disease. The genes that cause this disorder are sex-linked and recessive, and the disease occurs in young male children. Females are carriers and may transmit the genetic disease, but they do not develop it themselves.

Duchenne's dystrophy is characterized by progressive weakening of the pelvic and femoral muscles, which makes walking difficult. In this stage, it is often noted that an affected child will get up from the floor by bracing his arms against the thighs for support. There is extensive degeneration of muscle cells and deposition of fat in the muscles. As the disease progresses, shoulder and arm muscles may be weakened, but facial muscles are seldom involved.

Strains and sprains   Muscles or their tendons are sometimes stretched or strained beyond their normal range and become sore. Athletes and others sometimes suffer a painful condition called "charley horse," or "pulled muscle," in which muscle fibers are strained or torn.

A *sprain*, however, is an injury to a ligament. Ankles and wrists are the most common site of sprains. A bandage may be applied to give support to the injured joint.

Uncontrolled muscle contractions   Muscle *spasm* refers to a painful sustained muscle contraction that is difficult to relax. Massaging the injured muscle to increase the flow of blood may prove helpful. Muscle *cramps* are painful contractions that may be caused by overworking a muscle or by exposure to cold. "Swimmer's cramp," for example, seems to be caused by overexertion in cold water. A cramp in a leg muscle may be relieved by using opposing muscles to extend the leg.

SUMMARY

1 Skeletal muscle is striated and voluntary. It is composed of cylindrical muscle fibers, each of which is enclosed in a sheath called the sarcolemma. The protoplasm of a fiber is the sarcoplasm. Within the sarcoplasm are threadlike myofibrils. The myofibril is composed of still smaller filaments described as thin actin filaments and thick myosin filaments. Contraction occurs when the two sets of filaments slide over each other and produce a shortening of the fiber.

Two additional proteins are associated with actin in the thin filament. These are tropomyosin and troponin. They appear to be concerned with preventing actin from combining with myosin.

2 Myosin molecules are placed so that they lie in opposite directions within the thick filament; that is, the heads point in one direction in one half of the filament and in the opposite direction in the other half of the filament. Cross bridges extend out and contact reactive sites on the thin filaments.

3 The sarcoplasmic reticulum is a part of the internal membrane system of muscle. There are also transverse tubules which develop from a direct

infolding of the surface membrane. The transverse tubules apparently carry an electric signal to the myofibrils located deep inside the fiber.

4 When a nerve impulse reaches the neuromuscular junction in skeletal muscle, a transmitter substance, acetylcholine, is released and an endplate potential develops. Depolarization of the muscle-fiber membrane then occurs, and an action potential spreads over and through the fiber, resulting in contraction.

5 Groups of muscle fibers are innervated by the same axon, forming a motor unit. These fibers contract as a unit.

6 After stimulation, muscle contracts following a brief latent period. After the contraction phase, the relaxation phase follows. During a short period after each stimulation, the muscle is refractory to additional stimulation.

7 The immediate source of energy in contraction of muscle is the breakdown of ATP. Creatine phosphate is another high-energy phosphate compound capable of releasing energy. Much of this energy is used in the resynthesis of ATP. Glycolysis, through several steps, results in the formation of pyruvic acid. This is an anaerobic phase of energy metabolism, and it releases a small amount of energy.

8 Muscle contraction is characterized as isotonic or isometric. In isotonic contraction, the muscle thickens and shortens. This is the type of contraction involved in body movement. In isometric contraction, the tension increases with the load, but the muscle does not shorten. The muscles concerned with maintaining posture often exhibit this type of contraction.

9 There are three kinds of muscle tissue: smooth, cardiac, and skeletal. Smooth muscle may be of the single-unit type, which contracts as a unit, or it may be of the multiunit type, which consists of individual fibers, each with its own motor nerve ending. Smooth muscle is nonstriated and involuntary. The term visceral muscle applies to the smooth muscle found in the walls of hollow organs of the viscera, where it contracts in rhythmic waves, an action described as peristalsis. Such organs include the stomach, intestine, and urinary bladder. Multiunit smooth muscle is found in the intrinsic muscles of the eye, in the walls of blood vessels, and in the pilomotor muscles. All smooth muscle exhibits a mild state of contraction called tonus. The action of smooth muscle is regulated by the autonomic nervous system, neurohumors, and hormones.

10 Cardiac muscle is found in the muscular walls of the heart. It is striated and involuntary and contains intercalated disks. Heart muscle has developed a high degree of automaticity, which enables it to pulsate rhythmically. Since it maintains a long refractory period, it does not respond to summation and there is no complete tetanus.

QUESTIONS

1 Discuss the structure and function of smooth muscle, describing differences between the two types.
2 Why does a state of complete tetanus not occur in cardiac muscle?
3 Describe the sarcoplasmic reticulum. Suggest a possible function for transverse tubules.
4 How do the myosin cross bridges function?
5 What is a motor unit?

6 What is meant by threshold membrane potential?

7 How might you obtain complete tetanus in isolated muscle?

8 Describe the chemical events that occur during contraction in skeletal muscle.

9 What is meant by oxygen debt?

10 Does shivering have any value? If so, what is gained?

11 Distinguish between isotonic and isometric contraction.

12 What is the advantage of having many mitochondria associated with muscle fibers?

13 What is the chemical difference between rigor mortis and normal contraction?

SUGGESTED READING

Ashley, C. C.: Calcium and the Activation of Skeletal Muscle, *Endeavour,* **30** (109): 18–25 (1971).

Axelsson, J.: Catecholamine Functions, *Annu. Rev. Physiol.,* **33:**1–30 (1971).

Breckenridge, Bruce McL.: Cyclic AMP and Drug Actions, *Annu. Rev. Pharmacol.,* **10:**19–34 (1970).

Cohen, C.: The Protein Switch of Muscle Contraction, *Sci. Am.,* **233:**36–45 (1975).

Ebashi, S.: Excitation-Contraction Coupling, *Annu. Rev. Physiol.,* **38:**293–313 (1976).

Gibbs, C. L., W. F. H. M. Mommaerts, and N. V. Ricchiutti: Energetics of Cardiac Contraction, *J. Physiol.,* **191:**25–46 (1967).

Hurwitz, L., and A. Suria: The Link Between Agonist Action and Response in Smooth Muscle, *Annu. Rev. Pharmacol.,* **11:**303–326 (1971).

Huxley, H. E.: The Mechanism of Muscular Contraction, *Sci. Am.,* **213:**18–27 (1965).

———: The Mechanism of Muscular Contraction, *Science,* **164:**1356–1366 (1969).

Katz, Bernard: "Nerve, Muscle and Synapse," McGraw-Hill Book Company, New York, 1966.

Merton, P. A.: How We Control the Contraction of Our Muscles, *Sci. Am.,* **226:**30–37 (1972).

Murray, J. M., and A. Weber: The Cooperative Action of Muscle Proteins, *Sci. Am.,* **230:**59–71 (1974).

Sandow, A.: Skeletal Muscle, *Annu. Rev. Physiol.,* **32:**87–138 (1970).

Van Winkle, W. B., and A. Scharz: Ions and Inotropy, *Annu. Rev. Physiol.,* **38:**247–272 (1976).

7

SKELETAL MUSCLES: MOVEMENT AND ANATOMY

CONSIDERATIONS FOR STUDY

We shall now take up the study of the skeletal muscles: their names, locations, and how they work. We shall cover the following topics:

1. Movement: how skeletal muscles, acting together with the bones of the skeleton and joints produce movement
2. Types of muscles: how muscles combine different types of motion to produce coordinated movements
3. The names, locations, and actions of the most important muscles in the body

MOVEMENT OF THE BODY: SKELETAL MUSCLE FUNCTION

In skeletal muscles, the conversion of chemical energy into mechanical energy produces contraction and relaxation. A skeletal muscle may be considered as an organ of the muscular system. It has a nerve supply, an adequate distribution of blood and lymphatic vessels, and a considerable amount of connective tissue. In this chapter we shall consider the function of only skeletal muscles; we shall take up cardiac muscle and smooth-muscle functions and structure in later chapters.

As you recall from Chap. 6, a muscle is composed of bundles of microscopic fibers, each bundle separated by connective tissue. The word "fiber" is often used to denote whole bundles of fibers; each bundle represents a small strand within a large muscle. In this sense, the term "fiber" refers to the visible strands. Striated muscles account for a large portion of body weight. They constitute around 35 percent of the body weight in women and a little over 40 percent in men. Skeletal muscle tissue is red or pink.

Over 400 muscles compose the organs of the human skeletal muscle system. Muscles must have connections with the circulatory and nervous systems to function. One function of skeletal muscles is to enable the skeletal structure to move. Groups of muscles are attached to bone by tendons. When muscles contract, they apply force to their attachments and to the skeletal structure. The *attachment* of a muscle to bone may be a direct attachment, in which the connective tissue surrounding muscle fibers appears to fuse directly with the periosteum covering the bone, or it may be a tendinous attachment. *Tendons* are ordinarily long cords of white fibrous connective tissue. When they are flattened into a broad sheet, they are called *aponeuroses*. Each muscle is covered with a thin connective tissue sheath called the *fascia*. There is often considerable blending of these connective tissues to strengthen the attachment.

Attachments
  Origin
  Insertion

ATTACHMENTS: ORIGIN AND INSERTION Ordinarily the more stationary attachment of a muscle is called its *origin*, and the more movable attachment is called its *insertion*. In most cases the origin is located on the more fixed part of the skeletal structure, while the insertion is found on the more movable part. It is often the case, especially in the appendages, that the origin is proximal and the insertion is distal.

Levers
  1st class
  2d class
  3d class

LEVERS OF THE BODY The work accomplished by the skeletal muscles is done through a system of levers. The bones of the skeleton are the *levers*, the joints are the *fulcrums*, and the contraction of the muscles exerts force on the levers so that a weight can be lifted, a part of the body can be moved, or resistance can be overcome. This concept does not apply equally well to all muscles, however.

Levers are divided into three classes by the relative position of the fulcrum, the force applied, and the resistance to be overcome or balanced (Fig. 7.1). We shall consider only levers of the third class here, since these are the common type in body mechanics. In levers of the third class, force is applied across a joint: that is, between the weight and the fulcrum. A common example is the application of force from the biceps and the brachialis muscles, which are located anteriorly in the forearm. The elbow is the fulcrum, and the weight is at the hand. The force of contraction is applied across the elbow joint in flexing the arm. In this type of leverage,

FIGURE 7·1
The three classes of levers

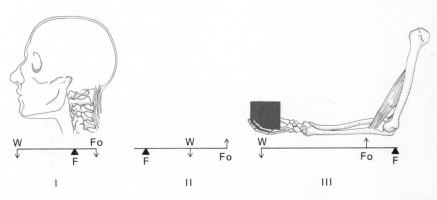

I   II   III

force is sacrificed in order to obtain quickness of movement and motion over a greater area.

COORDINATION OF MUSCULAR ACTIVITY   Ordinarily muscles do not act singly, but in groups. Even the simplest act requires *muscular coordination*. Taking a step forward is an example of very complicated muscular coordination that involves not only the muscles of the legs and feet but those of the trunk as well. Muscles are usually placed so that they operate as *antagonistic groups*. Thus, as one group of muscles contracts, the opposing group relaxes. A common example is the flexing of the arm at the elbow by the contraction of the biceps and the brachialis muscles. These muscles are placed so that their attachment is across the elbow joint anteriorly. They are inserted in the bones of the forearm just distal to the elbow. The opposing muscle is the triceps, located on the posterior side of the arm; its tendon is inserted in the olecranon process of the ulna. When the biceps and the brachialis muscles contract, the triceps relaxes. When the triceps contracts, the biceps and the brachialis relax, and the arm is extended at the elbow.

Types of movement
  Flexion
  Extension
  Abduction
  Adduction
  Circumduction
  Supination
  Pronation
  Inversion
  Eversion

TYPES OF MOVEMENT   The muscles described on the following pages have been selected largely because they illustrate a principle of contraction or leverage or because they perform an important function. Though it is difficult to say that one muscle is more important than another, certain ones are *prime movers* and offer better illustrations of muscle performance than others. For this reason, certain muscles may be described in detail, while others are mentioned only incidentally. It is not the purpose of this text to give undue emphasis to the exact origin and insertion of all muscles. The reader is referred to one of the larger works on human anatomy for more detailed descriptions.

It is misleading in many cases to mention only the most evident function of a muscle, since movement depends upon the coordinated activity of a number of muscles. One function of the biceps brachii, for example, is to flex the lower arm upon the upper. If opposing muscles are fixed, however, the biceps can be used to turn the hand to a palm-upward position. Since the heads of the biceps arise from the scapula, this muscle also helps draw the upper arm forward or pull the shoulder forward.

Muscles that bend a body part at a joint are referred to as *flexors,* and this type of movement is termed *flexion.* Muscles that extend a body part are called *extensors,* and this sort of motion is termed *extension.* The antagonistic action of muscle groups requires reciprocal innervation; that is, as nervous impulses direct one group of muscles to contract, nervous impulses cause the opposing group to relax.

The muscles concerned with maintaining posture are good examples of *fixators.* Such muscles must maintain body position against the pull of gravity. Muscles located around joints often act as fixators or stabilizers, thus freeing other muscles to act as prime movers. Muscles that help the prime mover by contracting in unison with it are termed *synergists.*

In a discussion of body movements a number of terms are useful. Movement of a part away from the midline of the body is *abduction,* whereas movement toward the body or toward the midline is *adduction.* For example, movement of the arm away from the body is abduction; while movement of the leg toward the midline is adduction. Rotating movements that hold the arm or leg straight are *circumduction.* For example, swinging the arm or leg in a cone-shaped arc is circumduction. Rotation of the hand so that the palm is upward is termed *supination;* moving the palm downward is *pronation.* The foot can be turned inward so that the sole is toward the midline, a movement called *inversion. Eversion* of the foot is a much more limited movement in which the sole is turned slightly outward.

MUSCLES OF THE HEAD, FACE, AND NECK

Muscles of the head, face, and neck
  Epicranius
  Orbicularis oculi
  Levator palpebrae superioris
  Corrugator
  Orbicularis oris
  Zygomaticus
  Buccinator
  Masseter
  Temporalis
  Pterygoideus medialis
  Platysma
  Intrinsic muscles of the tongue
  Extrinsic muscles of the tongue
  Sternocleidomastoideus
  Splenius capitis

Facial muscles alter facial expression and produce mastication movements. While the facial muscles themselves may not present an especially attractive appearance (Fig. 7.2), beauty being only skin deep, their mechanism of movement is interesting. The names, attachments, actions, and innervation of these muscles are summarized in Table 7.1. The principal muscle of the scalp is the *epicranius, or occipitofrontalis, muscle.* It comprises two sets of muscles: paired muscles at the back of the head that arise from tendinous attachments to the occipital bone and paired frontal muscles attached to the skin at the eyebrows. Between these two sets of muscles is a large epicranial aponeurosis, a fibrous membrane that extends across the top of the cranium. The occipitalis tends to fix and tighten the aponeurosis. With the aponeurosis fixed, the frontalis wrinkles the forehead horizontally, raises the eyebrows, and helps raise the eyelids. It plays a part in facial expressions denoting perplexity, surprise, horror, or visual concentration. When the lower attachment is fixed, the frontalis pulls the scalp forward.

The principal movements of the face center about the eyes and mouth, where there are circular, or sphincter, muscles. The *orbicularis oculi* circles the eye. The orbicularis oculi has numerous functions, since different parts of this muscle can be brought into play. Shortening the medial and lateral fibers produces a wink or prolonged closure of the eyelid; strong contraction of the entire muscle results in squinting or lowering of the eyebrows. The *levator palpebrae superioris* pulls the eyelid up. This muscle lies within the orbit just under the roof and is inserted into the eyelid. A small muscle arising on the frontal bone above the nasal bones and inserted into the skin of the eyebrow is called the *corrugator.* Its contraction wrinkles the forehead perpendicularly, as in frowning. Its action is closely associated with that of the orbicularis oculi.

Six muscles are located within the orbit and are attached to the eyeball. These muscles move the eyeball and will be discussed later in connection with the eye.

The lips, lower jaw, and tongue are other movable parts of the face. The *orbicularis oris,* another sphincter muscle, closes the mouth opening. Acting with other muscles the orbicularis oris purses the lips, as in whistling, or draws the lower lip upward, as in pouting. Several other muscles attach to the outer margins of this sphincter, and these muscles open the lips by their combined contraction. Among them is the *zygomaticus,* which extends from the zygomatic bone downward to the corner of the mouth, where the right and left muscles act together to pull the corners of the mouth upward into a smile. Contracting further, these muscles raise the cheek and lower eyelid, producing the little wrinkles at the outer corner of the eye so commonly associated with mirth and a kindly disposition.

The *buccinator muscle* has been called the trumpeter's muscle because it draws the corners of the mouth back and flattens the lips against the teeth. The cheeks are kept taut by its contraction, an essential movement in the correct playing of the trumpet or similar brass instruments. The buccinator

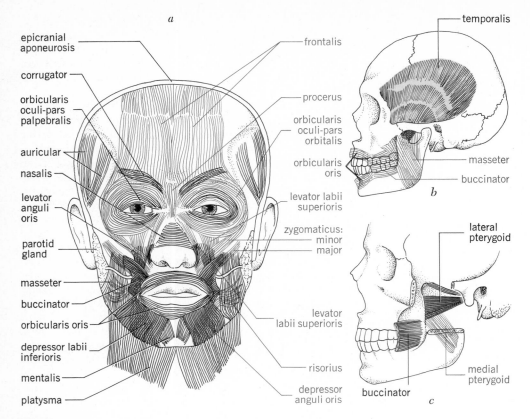

a

epicranial
aponeurosis

corrugator

orbicularis
oculi-pars
palpebralis

auricular

nasalis

levator
anguli
oris

parotid
gland

masseter

buccinator

orbicularis oris

depressor labii
inferioris

mentalis

platysma

frontalis

procerus

orbicularis
oculi-pars
orbitalis

orbicularis
oris

levator labii
superioris

zygomaticus:
minor
major

levator
labii superioris

risorius

depressor
anguli oris

temporalis

masseter

buccinator

b

lateral
pterygoid

medial
pterygoid

buccinator

c

FIGURE 7·2

a Muscles of the head and face (anterior view); *b* and *c* muscles of mastication (lateral view).

has a more important function during chewing; it cooperates with muscles of the tongue to keep food between the teeth. The effectiveness of this cooperation is shown by the infrequency with which either the tongue or the cheek are bitten.

The lower jaw moves upward with force in chewing food. One of the several muscles responsible for this action is the *masseter,* which extends downward from the zygomatic bone to the angle of the lower jaw, where it attaches to the outer surface. The muscle can be observed in action if you watch someone chewing or can be felt if you place a hand over the angle of the jaw and close the jaws tightly. Cooperating with the masseter to elevate the jaw are the *temporalis* and *pterygoideus medialis* (*internal pterygoid*) muscles (Fig. 7.2*b* and *c*).

Muscles that move the hyoid bone are listed in Table 7.2 (see Fig. 7.3).

The downward sag of the mouth in grief and exhaustion is caused by the contraction of the *platysma,* which inserts from below upon the corners of the mouth. This is an exceedingly thin superficial muscle arising over a wide area at about the level of the second rib. It forms a thin sheet of muscle anterolaterally over the neck and is inserted partly into the connective tissue over the mandible and partly into the skin of the cheek and at the corner of the mouth. During extreme exertion, for example, in a runner at a track meet, the platysma tenses, causing the lips to set and the skin of the neck to wrinkle vertically. Since the platysma is attached to the mandible, it may aid in depressing the jaw, particularly in yawning (Table 7.1).

TABLE 7·1

Principal muscles of the head and face

Muscle	Origin	Insertion	Action	Nerve
Epicranius	Epicranial aponeurosis	Skin of forehead	Elevates eyebrows, wrinkles skin of forehead horizontally	Facial
Orbicularis oculi	Medial aspect of orbit	Skin around eyes	Closes eyelids	Facial
Levator palpebrae superioris	Roof of orbit	Skin of upper eyelid	Raises upper eyelid	Oculomotor
Corrugator	Arch of frontal bone	Skin of forehead	Draws eyebrows toward each other	Facial
Orbicularis oris	Muscles about mouth	In muscles surrounding mouth	Closes and purses the lips	Facial
Zygomaticus	Zygoma	Skin at angle of mouth	Raises corner of mouth upward and backward	Facial
Buccinator	Alveolar processes of mandible and maxilla	Orbicularis oris	Compresses cheek	Facial
Masseter	Zygomatic process and zygomatic arch	Coronoid process, angle and ramus of mandible	Raises jaw and holds it tight	Trigeminal, mandibular branch
Temporalis	Temporal fossa and fascia	Coronoid process of mandible	Raises and retracts jaw	Trigeminal, mandibular branch
Pterygoideus medialis	Maxilla and pterygoid fossa	Ramus and angle of mandible	Aids in closing jaw	Trigeminal, mandibular branch
Platysma	Fasciae of upper thorax	Mandible and skin at angle of mouth	Depresses mandible and corners of mouth	Facial

Several muscles move the tongue. These include four pairs of muscles within the tongue, the *intrinsic muscles,* and three pairs of muscles having their origin outside and below the tongue, the *extrinsic muscles.* These muscles coordinate the exceedingly intricate movements of the tongue during mastication and swallowing and highly specialized movements of the tongue and lips necessary for speech.

The location and actions of certain neck muscles are summarized in Table 7.2. The *sternocleidomastoideus,* as indicated by its name, arises from two heads, one from the top of the sternum (the manubrium) and the other from the clavicle (Latin, *cleido-,* clavicle) (Fig. 7.3). The muscle is inserted by a short tendon on the mastoid processes of the temporal bone. The sterno-

TABLE 7·2

Muscles of the neck

Muscle	Origin	Insertion	Action	Nerve
Sternocleidomastoideus	Upper portion of manubrium and medial third of clavicle	Mastoid process	Draws head toward shoulder on same side; acting together they flex the neck and extend the head	Cervical 2 and 3; spinal accessory XI
Splenius capitis	First to fourth thoracic vertebrae and seventh cervical vertebra	Occipital bone and mastoid process	Pulls head back	Cervical

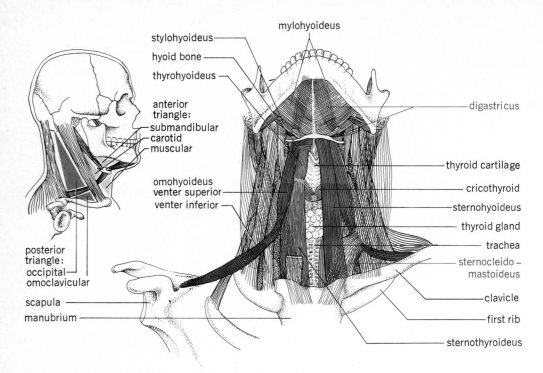

Labels on figure:
mylohyoideus
stylohyoideus
hyoid bone
thyrohyoideus
anterior triangle:
submandibular
carotid
muscular
omohyoideus venter superior
venter inferior
posterior triangle:
occipital
omoclavicular
scapula
manubrium
digastricus
thyroid cartilage
cricothyroid
sternohyoideus
thyroid gland
trachea
sternocleido-mastoideus
clavicle
first rib
sternothyroideus

FIGURE 7·3
Superficial muscles of the neck. The anterior and posterior triangles may be subdivided into smaller triangles.

cleidomastoideus muscles form the anterior triangle of the neck. Acting individually, they rotate the head toward the opposite side. When acting together, they bend the head forward, but only against increased resistance, as when you lie on your back and raise your head. This is the muscle commonly affected when one is said to have a "wry neck" or a stiff neck. The trapezius muscle may be affected also in this condition (Table 7.4).

The *splenius capitis* is one of a group of deeper, dorsal, cervical muscles. Arising from the dorsal spinal processes of the first to fourth thoracic vertebrae and from the seventh cervical vertebra, it is inserted on the occipital bone and on the mastoid process of the temporal bone underneath the insertion of the sternocleidomastoid muscle. When right and left muscles act in unison, the head is pulled back (extended). When they act singly, the head is rotated or inclined toward the muscle attached on the right or left side. Since this is one of the deeper muscles, it is not illustrated.

MUSCLES OF THE THORAX AND ANTERIOR MUSCLES OF THE SHOULDER GIRDLE

Muscles of the thorax
 Pectoralis major
 Pectoralis minor
 Serratus anterior
 External intercostals
 Internal intercostals
 Quadratus lumborum
 Diaphragm

The major muscles of the thorax and anterior shoulder girdle are shown in Fig. 7.4. The *pectoralis major* is the large triangular muscle that covers the upper part of the chest anteriorly (Table 7.3). It arises from attachments to the clavicle, the sternum, and the cartilages of the upper six ribs. The lower part of the muscle arises from the aponeurosis of the external oblique muscle. From this broad area of attachment, the muscle fibers converge and are inserted by a short flat tendon into the lateral margin of the intertubercular groove of the humerus. When the arm is raised, the pectoralis major, in cooperation with other muscles, pulls the arm down toward the chest. It also draws the arm across the chest and rotates it inward. With the arm

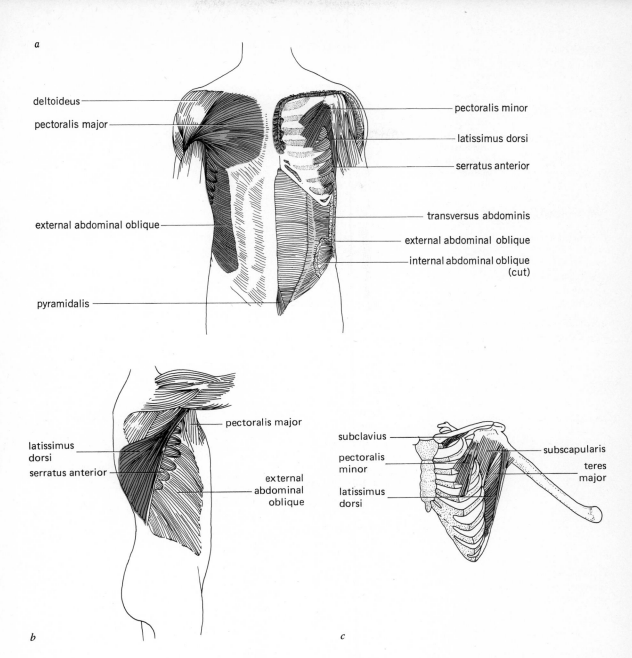

a

deltoideus

pectoralis major

external abdominal oblique

pyramidalis

pectoralis minor

latissimus dorsi

serratus anterior

transversus abdominis

external abdominal oblique

internal abdominal oblique
(cut)

latissimus
dorsi

serratus anterior

pectoralis major

external
abdominal
oblique

subclavius

pectoralis
minor

latissimus
dorsi

subscapularis

teres
major

b

c

FIGURE 7·4

Muscles of the pectoral region
and anterior abdominal wall.

fixed, as in climbing, it pulls the chest upward. The pectoralis major is
therefore used in forced inspiration or in artificial respiration.

The *pectoralis minor* lies directly underneath the pectoralis major. It is
flat, thin, and more slender than the pectoralis major. It arises from the
upper borders of the third, fourth, and fifth ribs and from their outer surfaces
close to the costal cartilages. The fiber bundles extend diagonally upward to
insert on the coracoid process of the scapula. With the scapula fixed, the

TABLE 7·3
Muscles of the thorax

Muscle	Origin	Insertion	Action	Nerve
Pectoralis major	Sternal half of clavicle, sternum, costal cartilages, aponeurosis of external abdominal oblique	Crest of greater tubercle of humerus	Flexes, abducts, and medially rotates humerus, draws body upward in climbing	Lateral and medial pectoral
Pectoralis minor	Anterior surface of ribs 3–5	Coracoid process of clavicle	Draws scapula down and forward and elevates ribs	Medial pectoral
Serratus anterior	Ribs 1–9	Inner surface of scapula	Draws and rotates scapula forward and down	Thoracic
External intercostal	Posterior tubercles of ribs	Anterior costal cartilage	Elevate ribs	Intercostal
Internal intercostal	Sternum	Angle of ribs	Depress ribs	Intercostal
Quadratus lumborum			Fixes lowest ribs	First thoracic, first lumbar
Diaphragm	Inner surfaces of xiphoid process	Central tendon	Expands and contracts thoracic cage	Phrenic

pectoralis minor elevates the ribs and aids inspiration, especially forced inspiration. If the ribs are fixed, the pectoralis minor pulls downward and forward on the scapula.

The *serratus anterior* lies beneath the pectoralis major and the pectoralis minor. It arises from the first to ninth ribs by slender digitations and is inserted on the inner surface of the scapula at the vertebral border. Its contraction moves the scapula forward and downward, rotating it somewhat as the arm is abducted.

The intercostal muscles are located between the ribs; each consists of two muscles that fill in each intercostal space (Fig. 7.5). The *external intercostal* is the thicker, outer sheet of muscle, and the *internal intercostal* forms the thinner, inner sheet. There are 44 of these muscles occupying 11 intercostal spaces on each side. The external intercostal muscles extend from the

FIGURE 7·5
Intercostal muscles and the pectoralis major.

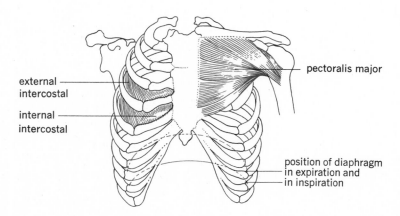

external intercostal

internal intercostal

pectoralis major

position of diaphragm in expiration and in inspiration

posterior tubercles of the ribs to the anterior costal cartilages. The fibers run diagonally from the lower border of one rib to the upper border of the next adjacent rib. They pull adjacent ribs toward each other and thus elevate the ribs, increasing the volume of the thoracic cavity, as, for example, during inspiration.

The internal intercostal muscles extend posteriorly from the sternum to the angle of the ribs. The fibers run downward diagonally, but in a direction opposite to that of the external intercostal fibers. When the lowest pair of ribs is held firmly by the *quadratus lumborum* muscles (see Fig. 7.20), the internal intercostal muscles depress the ribs, decreasing the volume of the thoracic cavity, as, for example, during expiration.

The *diaphragm* is a dome-shaped muscle that forms a wall between the thoracic and abdominal cavities (Fig. 7.6). It is attached around the lower circumference of the thorax to the inner surfaces of the xiphoid process and the lower six costal cartilages and is also joined by tendinous and muscular slips to the lateral surfaces of the upper lumbar vertebrae. The muscle fibers converge upward and are inserted into an aponeurosis called the *central tendon.* As the muscles contract, the diaphragm loses some of its convexity, and the capacity of the thoracic cavity is increased. Supported by other muscles, which act on the ribs, the elevation of the ribs and the lowering of the diaphragm expand the thoracic cavity for inspiration. To make a forced expiration, as in coughing, sneezing, laughing, or crying, a deep inspiration must precede the act. Muscles that depress the ribs and others that act to compress the abdomen then support a forcible expulsion of air during expiration. The diaphragm also plays a role in the movements of vomiting, defecation, micturition and in childbirth.

The diaphragm arches over the liver on the right side of the body and over the stomach on the left. It has three large openings: the vena cava passes through one opening in the right half of the central tendon; the esophageal opening located posterior to the central tendon on the left transmits the esophagus and branches of the vagus nerves; and a more posterior opening permits the passage of the aorta and the thoracic duct.

FIGURE 7·6

Anterior view of the diaphragm.

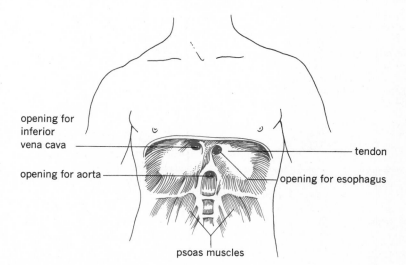

opening for inferior vena cava

opening for aorta

tendon

opening for esophagus

psoas muscles

Characteristics of the more important muscles of the thorax are summarized in Table 7.3.

MUSCLES OF THE BACK AND POSTERIOR MUSCLES OF THE SHOULDER GIRDLE

Muscles of the back
  Trapezius
  Levator scapulae
  Rhomboideus major
  Rhomboideus minor
  Serratus anterior
  Latissimus dorsi
  Erector spinae (sacrospinalis)

Major muscles of the back and posterior shoulder girdle are shown in Fig. 7.7. Their characteristics are summarized in Table 7.4. The right and left *trapezius muscles* are roughly trapezoid in shape. They arise from a flat aponeurosis that extends down the dorsal midline from the occipital bone; they also attach to dorsal vertebral spines as far down as the twelfth thoracic. The upper part of the muscle is inserted in the clavicle; the median and lower parts insert on the acromion process and spine of the scapula. The trapezius is a large muscle that is placed over other muscles of the shoulder. Contraction of the trapezius pulls the scapula toward the vertebral column and upward. It helps the arms to lift or the shoulders to carry a load, since it braces the shoulders. It also plays a part in shrugging the shoulders.

The *levator scapulae* is located posteriorly along the side of the neck. It arises from tendons from the atlas and axis and from the third and fourth cervical vertebrae. It is inserted on the medial border of the scapula, and it raises the scapula and draws it medially.

The *rhomboideus major* arises from the second to the fifth thoracic vertebrae and is inserted along the medial border of the scapula below the spine.

The *rhomboideus minor* arises from the seventh cervical vertebra and

TABLE 7·4
Muscles of the back

Muscle	Origin	Insertion	Action	Nerve
Trapezius	External occipital protuberance, superior nuchal line, nuchal ligament, and spinous process of cervical 7–12	Anterior border of scapular spine, acromion process, lateral third of posterior clavicle	Adducts, rotates, and elevates scapula	Spinal accessory XI and cervical 3, 4
Levator scapulae	Transverse processes of cervical 1–4	Medial border above spine of scapula	Draws scapula medially and depresses shoulder	Dorsal scapular
Rhomboideus major	Spinous process of thoracic 2–5 and supraspinous ligament	Medial border below spine of scapula	Adducts and laterally rotates scapula	Dorsal scapular
Rhomboideus minor	Spinous processes of last cervical and first thoracic vertebrae	Medial margin of scapula at origin of spine	Adducts and laterally rotates scapula	Dorsal scapular
Serratus anterior	Lateral surface of upper eight ribs	Ventral surface of medial border of scapula	Draws the scapula forward, rotates scapula as in abduction of arm	Long thoracic
Latissimus dorsi	Spinous processes of lower six thoracic vertebrae, thoracolumbar fasciae, crest of ilium	Intertubercular groove of humerus	Adducts, extends, rotates arm medially, draws shoulder down and backward	Thoracodorsal
Erector spinae	Posterior iliac crest, posterior sacrum, spines of lumbar vertebrae	Ribs and upper vertebrae	Pulls spinal column dorsad, extends spinal column	Spinal nerves

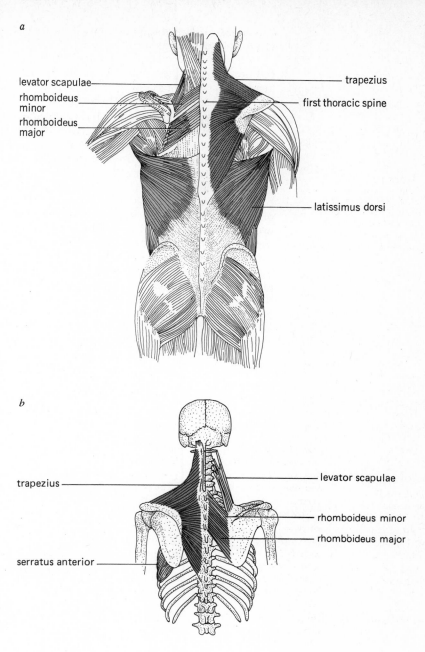

a

levator scapulae

rhomboideus minor

rhomboideus major

trapezius

first thoracic spine

latissimus dorsi

b

trapezius

serratus anterior

levator scapulae

rhomboideus minor

rhomboideus major

FIGURE 7·7
Muscles of the back and posterior muscles of the shoulder girdle.

from the first thoracic vertebra. It is inserted on the medial border of the scapula at the base of the spine.

Both rhomboideus muscles adduct the scapula and rotate it laterally.

The *serratus anterior* arises from the lateral surface of the upper eight ribs and is inserted on the ventral surface of the scapula along the median border. It is a thin muscle that lies along the side and back of the thorax. It draws the scapula forward and rotates it.

Other muscles, such as the splenial muscles, are attached at the back of the head and to the upper vertebrae. They help hold up and extend the head.

The *latissimus dorsi* muscle has a very extensive origin—from an aponeurosis attached to the spinous processes of the lower six thoracic vertebrae and all the lumbar vertebrae and from the dorsal surface of the sacrum and the posterior part of the crest of the ilium (Fig. 7.8). There is also some attachment to the lower three or four ribs. From this broad area of origin the muscle tapers upward to a tendinous insertion in the intertubercular groove of the humerus. The latissimus dorsi draws the arm downward and backward when the trunk is fixed. It also rotates the arm inward. This is one of the principal muscles involved in swimming strokes and in bringing the arm forcibly downward, as in striking a blow. When both arms are fixed above the head, the contraction of the latissimus dorsi draws the trunk forward; it is thus used in climbing.

The muscles of the back are usually divided into two classes: *superficial* and *deep*. The trapezius and the latissimus dorsi are superficial muscles of the outer layer. We shall consider only one of the muscles from the deeper layer: the erector spinae.

The *erector spinae (sacrospinalis)* is a long, deep muscle attached by a strong aponeurosis to the posterior part of the iliac crest, the posterior surface of the sacrum, and the spines of all the lumbar vertebrae. Tracing the muscle anteriorly, we find that it divides into several long muscles, which are inserted on the ribs and upper vertebrae at various levels in a consistent pattern in which each set of fibers has its origin posterior to the insertion of an adjacent set. With such a system of overlapping fibers, the erector spinae can exert a strong pull dorsad (toward the back) on the

FIGURE 7·8

The latissimus dorsi and deeper muscles of the shoulder.

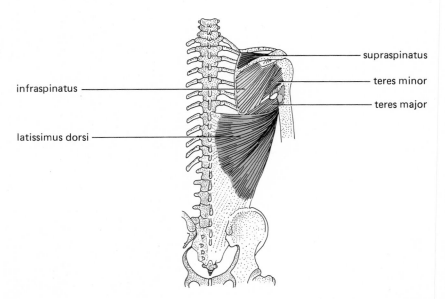

supraspinatus

teres minor

teres major

infraspinatus

latissimus dorsi

vertebral column. The erector spinae is an important posture muscle, since when both sides act together it acts as an extensor for the spinal column. When acting on one side only, it pulls the spinal column toward that side. In advanced pregnancy the erector spinae balances the weight of the fetus by bending the spinal column backward. The erector spinae is a deep muscle of the back and is thus not illustrated here.

MUSCLES OF THE SHOULDER, ARM, AND HAND

Muscles of the shoulder and upper arm
Deltoideus
Biceps brachii
Brachialis
Coracobrachialis
Brachioradialis
Triceps brachii

MUSCLES OF THE SHOULDER AND UPPER ARM   Muscles of the shoulder and upper arm are shown in Fig. 7.9. The *deltoideus,* also shown in Fig. 7.4, is a short, thick muscle located above the articulation of the arm at the shoulder. It arises from the distal third of the clavicle, the acromion process, and spine of the scapula and is inserted by a stout tendon into the deltoid tuberosity of the humerus. Its contraction raises the arm to a horizontal position or somewhat higher. When the arm is so raised, the contracted muscle can be felt as a hard bunch of muscle above the shoulder joint. There are five other muscles of the shoulder. The reader is cautioned that the deltoideus is not the only important muscle in the group.

The *biceps brachii* is a well-known muscle of the anterior part of the upper arm (Fig. 7.10). It arises from two points of attachment. The short head arises from the coracoid process of the scapula. The long head has its origin on the supraglenoid tuberosity, a bony process of the scapula just above the articulation of the humerus at the shoulder. The tendon of the long head passes over the head of the humerus and into the intertubercular groove; the tendon is protected by the synovial membrane of the joint. The two heads unite about 7 cm above the elbow joint and form the body of the biceps brachii. The muscle is inserted on the tuberosity of the radius. Its action is to flex the arm at the elbow and to rotate the forearm outward, a movement called supination. The movement of the muscle in supination is most evident when the arm is flexed. When the forearm is fixed, as in climbing or in chinning oneself on a horizontal bar, the body is pulled upward by the flexing of the arm.

The *brachialis* assists the biceps to flex the arm at the elbow. It arises from the anterior surface of the lower three-fifths of the humerus, passes anteriorly across the elbow joint, and is inserted into the tuberosity of the ulna. It is purely a flexor of the forearm.

The *coracobrachialis* has its principal origin on the tip of the coracoid process of the scapula and is inserted on the medial surface of the humerus somewhat proximal to the middle of the shaft. It is a relatively small muscle that helps flex and adduct the humerus.

The *brachioradialis* arises from the lateral ridge at the distal third of the humerus and is inserted by a long, flat tendon laterally on the radius at the base of the styloid process. A superficial muscle located above the radius on the forearm, the brachioradialis is primarily a flexor of the forearm but also helps rotate it. Although it is a forearm muscle, it is described here with other upper-arm muscles that are also flexors of the forearm (see Fig. 7.12*c*).

The *triceps brachii* is the principal extensor of the forearm. It arises from three heads: the long head arises from the scapula just below the glenoid fossa; the lateral head arises from the posterior side of the shaft of the humerus, proximally; and the medial head arises from a fleshy attachment on the posterior side of the humerus distally from the radial groove. The

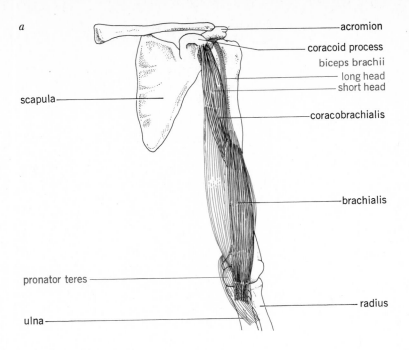

- *a*
- acromion
- coracoid process
- biceps brachii
- long head
- short head
- coracobrachialis
- scapula
- brachialis
- pronator teres
- radius
- ulna

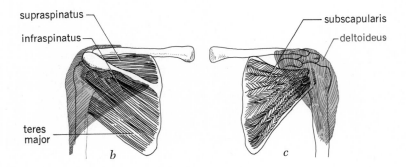

- supraspinatus
- infraspinatus
- subscapularis
- deltoideus
- teres major
- *b*
- *c*

FIGURE 7·9

Muscles of the arm and shoulder: *a* and *c* anterior view; *b* posterior view.

three heads unite to form a large muscle on the posterior side of the arm. It is inserted by a strong tendon on the olecranon process of the ulna. This muscle is the antagonist of the biceps and brachialis and extends the arm at the elbow (Fig. 7.11). The characteristics of these muscles are summarized in Table 7.5.

MUSCLES OF THE FOREARM AND HAND A number of flexors and extensors of the wrist and hand are located in the forearm. Such muscles have their origins on the humerus and on the radius and ulna and are inserted by long tendons on bones of the wrist, hand, and fingers. The flexors are on the anterior side of the arm when it is viewed in the standard anatomical position. The extensors are on the posterior side of the arm; some of their tendons are seen standing out on the back of the hand if the fingers are

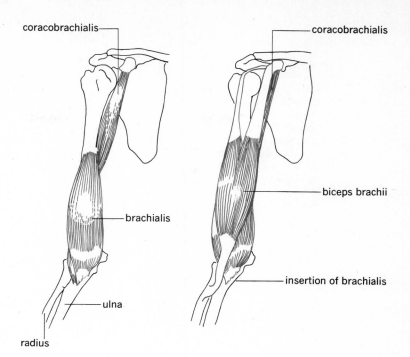

coracobrachialis

coracobrachialis

biceps brachii

brachialis

insertion of brachialis

ulna

radius

FIGURE 7·10

Superficial muscles of the arm
(anterior view).

FIGURE 7·11

The triceps brachii.

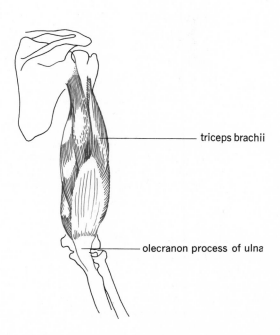

triceps brachii

olecranon process of ulna

TABLE 7·5
Muscles of the shoulder and upper arm

Muscle	Origin	Insertion	Action	Nerve
Deltoideus	Anterior surface, lateral clavicle, acromion process, and spine of scapula	Deltoid tubercle of humerus	Abducts arm; aids in flexion, extension, and adduction	Circum-flex
Biceps brachii	Long head, supra-glenoid tubercle; short head, coracoid process of scapula	Tuberosity of radius and aponeurosis	Flexes forearm and arm, and supinates hand	Musculo-cutane-ous
Brachialis	Distal two-thirds of front of humerus	Coronoid process of ulna	Flexes forearm	Musculo-cutane-ous and radial
Coracobrachialis	Coracoid process of scapula	Middle third of hu-merus	Flexes and adducts arm	Musculo-cutane-ous
Brachioradialis	Lateral ridge of distal humerus	Styloid process of ra-dius	Flexes and rotates forearm	Radial
Triceps brachii	Long head, infraglenoid tubercle; lateral head, proximal portion of shaft of humerus; me-dial head, distal half of shaft of humerus	Olecranon process of ulna	Extends arm and fore-arm	Radial

extended to the greatest degree. Some muscles of the hand spread the fingers apart or bring them together. Some of these muscles flex and extend the distal joints of the fingers. Other muscles move the thumb.

Anterior muscles of the fore-arm
Pronator teres
Flexor carpi radialis
Palmaris longus
Flexor carpi ulnaris

ANTERIOR MUSCLES OF THE FOREARM   Among the superficial muscles on the anterior aspect of the forearm are the pronator teres, the flexor carpi radialis, the palmaris longus, and the flexor carpi ulnaris (Fig. 7.12). This group of flexors and pronators forms a large part of the musculature at the base of the forearm. The *pronator teres* is a strong pronator of the hand. The *flexor carpi radialis* is located laterally along the pronator teres. It is essentially a flexor of the hand, and weakly abducts the hand toward the radius. The *palmaris longus* is a long slender muscle inserted in the palmar aponeurosis. It is a flexor of the hand, and it also tightens the palmar aponeurosis. The *flexor carpi ulnaris* is inserted on the pisiform wrist bone and the palmar fascia. It is a flexor of the hand, and adducts the hand toward the ulna (Fig. 7.12*a*, *b*, and *c*). The characteristics of these muscles are summarized in Table 7.6.

Posterior muscles of the fore-arm
Anconeus
Extensor digitorum (commu-nis)
Extensor digiti minimi
Extensor carpi ulnaris
Extensor carpi radialis longus
Extensor carpi radialis brevis

POSTERIOR MUSCLES OF THE FOREARM   Several superficial muscles are located on the posterior aspect of the forearm. These are the anconeus, the extensor digitorum (communis), the extensor digiti minimi, the extensor carpi ulnaris, the extensor carpi radialis longus, and the extensor carpi radialis brevis (Fig. 7.13).

Numerous muscles are necessary to produce all the intricate movements

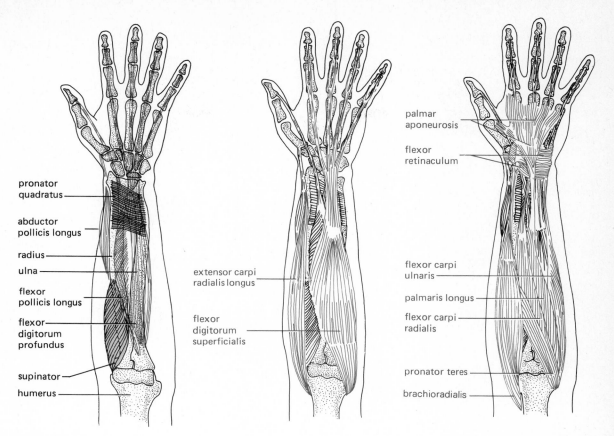

pronator
quadratus

abductor
pollicis longus

radius

ulna

flexor
pollicis longus

flexor
digitorum
profundus

supinator

humerus

extensor carpi
radialis longus

flexor
digitorum
superficialis

palmar
aponeurosis

flexor
retinaculum

flexor carpi
ulnaris

palmaris longus

flexor carpi
radialis

pronator teres

brachioradialis

FIGURE 7·12

Muscles of the anterior forearm.

TABLE 7·6

Anterior muscles of the forearm

Muscle	Origin	Insertion	Action	Nerve
Pronator teres	Medial epicondyle of humerus and coronoid process of ulna	Lateral surface of radius	Pronates hand	Median
Flexor carpi radialis	Medial epicondyle of humerus	Base of second metacarpal	Flexes wrist and aids in abducting it	Median
Palmaris longus	Medial epicondyle of humerus	Palmar aponeurosis and flexor retinaculum	Flexes hand	Median
Flexor carpi ulnaris	Medial epicondyle of humerus, olecranon process, and medial border of ulna	Pisiform, hamate, and fifth metacarpal	Flexes and adducts wrist	Ulna

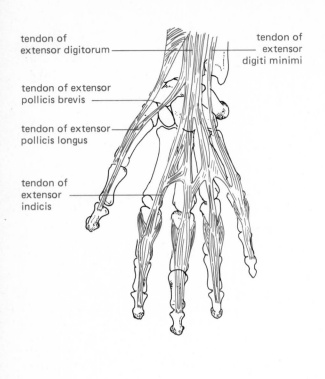

tendon of
extensor digitorum

tendon of
extensor
digiti minimi

tendon of extensor
pollicis brevis

tendon of extensor
pollicis longus

tendon of
extensor
indicis

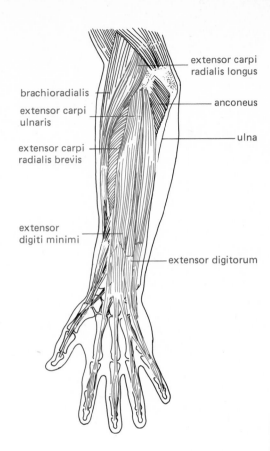

extensor carpi
radialis longus

brachioradialis

anconeus

extensor carpi
ulnaris

ulna

extensor carpi
radialis brevis

extensor
digiti minimi

extensor digitorum

FIGURE 7·13

Muscles and tendons of the
anterolateral forearm (posterior
view).

that the hand makes. As the names indicate, most of these muscles are
primarily extensors. The *anconeus* is a small triangular muscle located at the
proximal end of the forearm. It arises on the lateral epicondyle of the
humerus and is inserted along the proximal part of the ulna and to the
olecranon process of the humerus. It acts with the triceps brachii to extend
the forearm.

The *extensor digitorum* (*communis*) is primarily an extensor of the fingers.
The tendons of this muscle are seen on the back of the hand when the fingers
are fully extended. The *extensor digiti minimi* is a slender muscle whose
tendon is attached at the base of the fifth digit. It extends the little finger and
also helps abduct this finger.

The *extensor carpi ulnaris* is located along the ulna in the forearm. It
extends the hand at the wrist and helps draw the hand laterally toward the
ulna. The *extensor carpi radialis longus* and the *extensor carpi radialis brevis*
(Latin, *carpi*, wrist) have essentially the same function. They extend and
abduct the hand (Fig. 7.13). The characteristics of these muscles are summarized in Table 7.7. Muscles located entirely within the hand are the
lumbricales and the interrossei, which are discussed later in this chapter.

TABLE 7·7
Posterior muscles of the forearm

Muscle	Origin	Insertion	Action	Nerve
Anconeus	Lateral epicondyle of humerus	Posterior surface of ulna, olecranon process	Extends forearm	Radial
Extensor digitorum	Lateral epicondyle of humerus	Into distal phalanx by four tendons	Extends fingers and wrist joint	Radial
Extensor digiti minimi	Lateral epicondyle of humerus	Extensor expansion of little finger on dorsum of first phalanx	Extends little finger	Radial
Extensor carpi ulnaris	Lateral epicondyle of humerus and posterior border of ulna	Fifth metacarpal	Extends and adducts hand	Radial
Extensor carpi radialis longus	Lateral supracondylar ridge of humerus and lateral intermuscular septum	Second metacarpal	Extends and abducts hand	Radial
Extensor carpi radialis brevis	Lateral epicondyle of humerus	Third metacarpal	Extends and abducts wrist	Radial

Deeper muscles of the forearm
  Supinator
  Abductor pollicis longus
  Extensor pollicis brevis
  Extensor pollicis longus

DEEPER MUSCLES OF THE FOREARM Somewhat deeper muscles of the forearm are the supinator, the abductor pollicis longus, the extensor pollicis brevis, and the extensor pollicis longus. The *supinator* turns the palm upward. The *abductor pollicis longus* is located just beneath the supinator and is inserted on the first metacarpal. Its primary function is to abduct the thumb. The *extensor pollicis brevis* is closely allied to the abductor pollicis longus but is inserted on the proximal phalanx of the thumb. It helps extend this phalanx. The *extensor pollicis longus* is inserted on the distal phalanx and extends the distal phalanx (Fig. 7.14). The characteristics of these muscles are summarized in Table 7.8.

The thenar group
  Flexor pollicis brevis
  Abductor pollicis brevis
  Adductor pollicis
  Opponens pollicis

THE THENAR GROUP The thumb is well supplied with a number of short muscles that provide flexibility to this important digit. The thenar group forms the thenar eminence, a bulge of short muscles controlling movement of the thumb. These muscles are the *flexor pollicis brevis*, the *abductor pollicis brevis*, the *adductor pollicis*, and the *opponens pollicis* (Fig. 7.15). The thenar muscles are inserted on the proximal phalanx of the thumb. Their primary functions are indicated by their names. The opponens pollicis brings the thumb in opposition to the fingers. While this is being done, the first metacarpal is rotated medially toward the palm. The characteristics of these muscles are summarized in Table 7.9.

The use of the terms *abduction* and *adduction* may be confusing when they are applied to the musculature of the hand. In this case, these terms apply to the median line of the hand rather than to the median plane of the body. The median line of the hand is located in the third finger.

Posterior forearm muscles that flex the fingers
  Flexor digitorum superficialis
  Flexor digitorum profundus

POSTERIOR FOREARM MUSCLES THAT FLEX THE FINGERS An intermediate muscle, the flexor digitorum superficialis, and a deep muscle, the flexor digitorum profundus, provide strength for flexing the fingers. Arising from

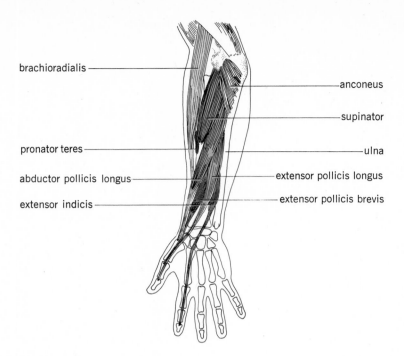

brachioradialis

anconeus

supinator

pronator teres

ulna

abductor pollicis longus

extensor pollicis longus

extensor indicis

extensor pollicis brevis

FIGURE 7·14

Deeper muscles of the forearm.

attachments on the humerus, radius, and ulna, the tendons of the *flexor digitorum superficialis* are inserted on the middle phalanges of all four fingers. The *flexor digitorum profundus* arises from the ulna and the interosseous membrane, and its four tendons insert on the distal phalanges of each finger. Since the position of this muscle is deeper than that of the flexor digitorum superficialis, the tendons of the latter divide to permit the passage of these tendons to the distal phalanges. The primary function of the flexor digitorum profundus is to flex the distal phalanges and help flex the other finger joints. The characteristics of these muscles are summarized in Table 7.10.

TABLE 7·8

Deeper muscles of the forearm

Name	Origin	Insertion	Action	Nerve
Supinator	Lateral epicondyle of humerus, supinator crest of ulna	Proximal third of radius	Supinates hand	Radial
Abductor pollicis longus	Posterior surface of ulna and radius, and interosseous membrane	First metacarpal	Abducts thumb and wrist	Radial
Extensor pollicis brevis	Middle third of radius and interosseous membrane	Proximal phalanx of thumb	Extends first phalanx of thumb and abducts hand	Radial
Extensor pollicis longus	Middle third of ulna and adjacent interosseus membrane	Distal phalanx of thumb	Extends distal phalanx of thumb and abducts hand	Radial

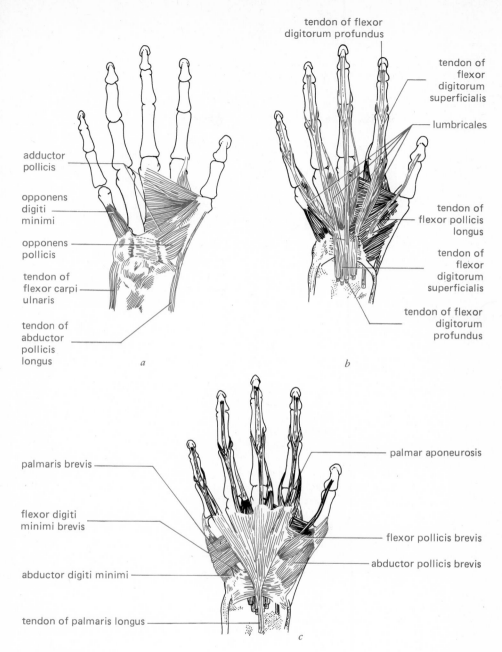

tendon of flexor
digitorum profundus

tendon of
flexor
digitorum
superficialis

lumbricales

tendon of
flexor pollicis
longus

tendon of
flexor
digitorum
superficialis

tendon of flexor
digitorum
profundus

adductor
pollicis

opponens
digiti
minimi

opponens
pollicis

tendon of
flexor carpi
ulnaris

tendon of
abductor
pollicis
longus

a

b

palmar aponeurosis

palmaris brevis

flexor digiti
minimi brevis

flexor pollicis brevis

abductor pollicis brevis

abductor digiti minimi

tendon of palmaris longus

c

FIGURE 7·15
Muscles of the palm of the
hand.

TABLE 7·9
The thenar group

Name	Origin	Insertion	Action	Nerve
Flexor pollicis brevis	Flexor retinaculum and trapezium	First phalanx of thumb	Flexes thumb	Median
Abductor pollicis brevis	Flexor retinaculum, scaphoid, and trapezium	First phalanx of thumb	Abducts thumb and aids in flexion	Mediun
Adductor pollicis	Capitate, second, and third metacarpals	First phalanx of thumb	Adducts thumb and aids in opposition	Ulnar
Opponens pollicis	Flexor retinaculum and trapezium	Lateral border of first metacarpal	Opposes thumb to fingers and abducts, flexes, and rotates first metacarpal	Median

Muscles of the hand
  Abductor digiti minimi
  Flexor digiti minimi brevis
  Opponens digiti minimi
  Lumbricales
  Interosseous
    Palmar
    Dorsal

MUSCLES OF THE HAND The *hypothenar eminence*, located on the medial side of the palm proximal to the little finger, is composed of three short muscles. These are the *abductor digiti minimi*, primarily an abductor of the little finger, the *flexor digiti minimi brevis*, a flexor of the proximal phalanx of the fifth digit, and the *opponens digiti minimi*, which enables the little finger to oppose the thumb (Fig. 7.15a and c).

There are two groups of muscles within the hand itself, the lumbricales and the interosseous muscles. The *lumbricales* are deep muscles of the palm composed of four small muscle slips that arise from the tendons of the flexor digitorum profundus muscle and are inserted by slender tendons into the tendons of the extensor digitorum muscle (Fig. 7.15b). The lumbricales help flex the fingers at the metacarpophalangeal joints while extending the interphalangeal joints. The *interosseous muscles* lie between the metacarpal bones. This group is broken into two sets: palmar and dorsal. The *palmar* set consists of three small muscles attached to the bases of all the fingers except the middle, or third, finger. Their principal action is to adduct the index, fourth, and little fingers toward the middle finger (Fig. 7.16). They also help flex the proximal phalanges of these fingers and extend the middle and distal joints. The *dorsal* interosseous set is composed of four small muscles that arise between the metacarpal bones (Fig. 7.17). The tendons of these muscles are extended to the second, third, and fourth fingers. The middle finger has two tendons, one on either side. The dorsal interossei abduct the fingers involved, flex the proximal phalanges, and extend the middle and distal joints. Characteristics of these muscles are summarized in Table 7.11.

TABLE 7·10
Posterior forearm muscles that flex the fingers

Name	Origin	Insertion	Action	Nerve
Flexor digitorum superficialis	Medial epicondyle of humerus and coronoid process of ulna	Second phalanges of fingers	Flexes the phalanx of each finger	Median
Flexor digitorum profundus	Anterior and medial aspects of ulna and interosseous membrane	Distal phalanges of fingers	Flexes distal phalanges and wrist	Median and ulnar

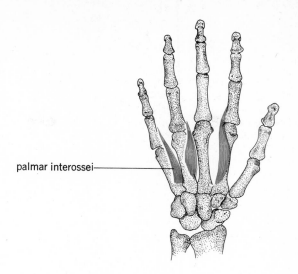

FIGURE 7·16

The palmar interossei.

This discussion has considered some of the principal muscles of the arm and hand. For a complete description of all muscles of the arm and hand, the reader is referred to larger works on human anatomy, some of which are listed at the end of this chapter.

MUSCLES OF THE ABDOMINAL WALL

Anterolateral abdominal muscles

    External oblique
    Pyramidalis
    Internal oblique
    Transversus abdominis
    Rectus abdominis

FIGURE 7·17

Deep muscles of the hand (posterior aspect); the dorsal interossei.

ANTEROLATERAL ABDOMINAL MUSCLES Important anterolateral abdominal muscles are shown in Fig. 7.18. The *external oblique* is the most superficial of the muscles. It is a broad, thin sheet of muscle arising from the outer surfaces of the lower eight ribs. From these points of origin, the fibers pass obliquely across the abdomen. The upper and middle portions of the muscle are inserted in a broad aponeurosis that covers the anterior part of the abdomen. The lower fibers are inserted directly into the crest of the ilium and, through an aponeurosis, into the *inguinal ligament (Poupart's liga-*

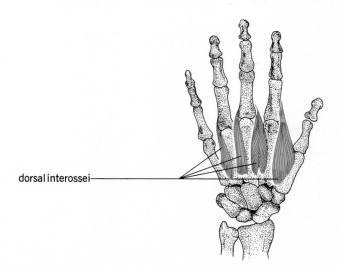

dorsal interossei——

TABLE 7·11
Muscles of the hand

Muscle	Origin	Insertion	Action	Nerve
Abductor digiti minimi	Pisiform and tendon of carpi ulnaris	First phalanx of little finger	Abducts little finger	Ulnar
Flexor digiti minimi brevis	Flexor retinaculum and hook of hamate	First phalanx of little finger	Flexes little finger	Ulnar
Opponens digiti minimi	Flexor retinaculum and hook of hamate	Fifth metacarpal	Draws fifth metacarpal toward palm	Ulnar
Lumbricales	Tendons of flexor digitorum profundus	Extensor expansion distal to metacarpophalangeal joint	Flex metacarpophalangeal joints and extend interphalangeal joints	Lateral two by median, medial two by ulnar
Interosseous palmar	Palmar surface of metacarpals of digits 2, 4, and 5	Base of proximal phalanx and expanded extensor tendon of same finger as origin	Adducts digits 2, 4, and 5 toward the midline and aids in extension of fingers	Ulnar
Interosseous dorsal	Metacarpals	Second, third, fourth digits	Abducts digits 2, 3, 4, flexes proximal phalanges, and extends middle and distal joints	Ulnar

FIGURE 7·18
Muscles of the anterolateral abdominal wall.

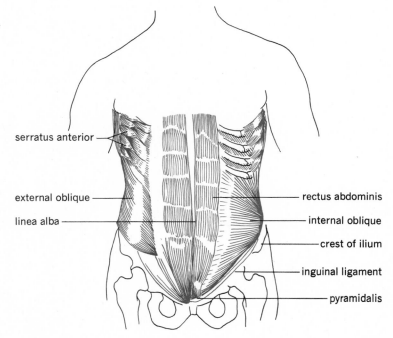

serratus anterior

external oblique

linea alba

rectus abdominis

internal oblique

crest of ilium

inguinal ligament

pyramidalis

ment). The inguinal ligament is not a true ligament; it is rather a tendinous band that extends from the anterior superior iliac spine to the pubic tubercle. It is formed by the folded aponeurosis of the lower part of the external oblique muscle.

An opening in the aponeurosis of the external oblique muscle, called the *subcutaneous inguinal ring,* is located just above and laterad to the crest of the pubic bone. This is an external opening of the inguinal canal, which transmits the spermatic cord in the male and the round ligament of the uterus in the female. The opening is somewhat larger in the male than in the female.

The *linea alba* is a narrow tendinous sheath that extends anteriorly along the middle of the abdomen from the xiphoid process of the sternum to the symphysis of the pubic bones. It is formed by the aponeuroses of the oblique and transversus muscles of both sides uniting in the midline. The umbilicus lies in the linea alba a little below the middle. The *pyramidalis muscle* arises from the pubis and is inserted in the linea alba. Its contraction pulls the linea alba and increases its tension (Fig. 7.18).

The *internal oblique* muscle lies directly under the external oblique muscle, but its fibers run approximately at right angles to those of the muscle above it. It arises from the inguinal ligament, iliac crest, and deep fascia in the lumbar region. Spreading upward, the fibers insert on the costal cartilages of the lower three ribs, into the aponeurosis of the linea alba, and onto the crest of the pubis.

The *transversus abdominis* (see Fig. 7.33) is the innermost of the flat muscles composing the abdominal wall. It arises from deep lumbar fasciae of the back and lower ribs, the iliac crest, and the inguinal ligament. Most fibers cross the abdomen horizontally and are inserted by an aponeurosis into the linea alba and onto the crest of the pubis. The abdominal wall is greatly strengthened by the three-layer arrangement of the external and internal oblique and the transversus muscles. The fibers of these muscles run in three directions. (See Fig. 7.4).

The *rectus abdominis* is a long, flat, paired muscle that extends along the anterior side of the body from the xiphoid process and the cartilages of the fifth, sixth, and seventh ribs to the crest of the pubis. It lies in a sheath formed by the aponeuroses of the oblique and transversus muscles. The medial border lies adjacent to the linea alba. Each muscle is crossed by three or four tendinous bands, and these rectangular divisions of the muscle may be observed in individuals with well-developed musculature (Fig. 7.18). The characteristics of these muscles are summarized in Table 7.12.

ACTION OF THE ANTERIOR ABDOMINAL MUSCLES The abdominal muscles compress the abdomen, and assisted by the descent of the diaphragm, they help produce the movements involved in vomiting, micturition, and defecation and in childbirth. They support the abdominal viscera and aid in expiration. When only one set of muscles contracts, the trunk is flexed toward that side. When both sets act together, they bend the body forward and thus are useful in climbing and jumping.

The rectus abdominis muscle contracts powerfully to raise the thorax and shoulders when one lies on the back, or it flexes the pelvis on the trunk when the thorax is fixed. Contraction of the recti muscles protects against body blows and helps protect the abdominal viscera from injury.

TABLE 7·12
Anterolateral abdominal muscles

Muscle	Origin	Insertion	Action	Nerve
External oblique	External surface of lower eight ribs	Anterior half of iliac crest and linea alba	Compresses abdomen, rotates and flexes vertebral column, assists in forced expiration	Intercostals 8–12, iliohypogastric, ilioinguinal
Pyramidalis	Pubis and anterior pubic ligament	Linea alba	Tenses linea alba	Subcostal, branch of twelfth thoracic
Internal oblique	Lateral half of inguinal ligament, anterior iliac crest, and lumbodorsal fascia	Lower four ribs, linea alba, and by tendinous sheath, to pubis	Compresses abdomen, rotates and flexes vertebral column, assists in forced expiration	Intercostals 8–12, iliohypogastric, ilioinguinal
Transversus abdominis	Lateral third of inguinal ligament, anterior iliac crest, and lumbodorsal fascia	Linea alba, and by tendon to pubis	Compresses abdomen and depresses ribs, assists in forced expiration	Intercostals 7–12, iliohypogastric, ilioinguinal
Rectus abdominis	Pubic symphysis and crest of pubis	Xiphoid process of sternum and cartilages of ribs 5–7	Tenses abdominal wall and flexes vertebral column	Intercostals 7–12

Posterior abdominal muscles
Psoas minor
Psoas major ⎤
Iliacus   ⎦ Iliopsoas
Quadratus lumborum

POSTERIOR ABDOMINAL MUSCLES AND MUSCLES OF THE PELVIC REGION Important muscles of the posterior abdomen and pelvic region are shown in Fig. 7.19. The *psoas minor* and *psoas major* form a portion of the posterior wall of the abdomen, but the psoas major and the iliacus are of greater importance since they move the trunk and the thigh. They may move the trunk when the thigh is fixed, or they may move the thigh when the trunk is fixed.

FIGURE 7·19

Muscles of the posterior abdominal and pelvic regions.

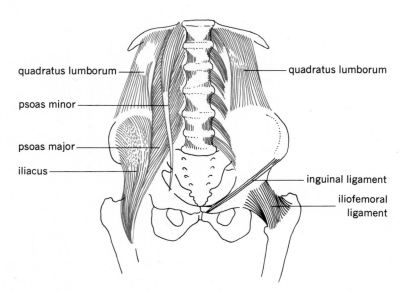

quadratus lumborum

psoas minor

psoas major

iliacus

quadratus lumborum

inguinal ligament

iliofemoral ligament

TABLE 7·13
Posterior abdominal muscles and muscles of the pelvic region

Muscles	Origin	Insertion	Action	Nerve
Psoas minor	Bodies of twelfth thoracic and first lumbar vertebrae	Pectineal line on ilium and iliac fascia	Flexes lumbar vertebral column	First lumbar
Psoas major	Transverse processes of lumbar vertebrae	Lesser trochanter of femur, with iliacus	Flexes and medially rotates thigh	Second and third lumbar
Illiacus	Iliac fossa and lateral margin of sacrum	Lesser trochanter of femur with psoas major	Flexes and medially rotates thigh	Femoral
Iliopsoas	Crest of ilium and iliac fossa	Lesser trochanter of femur	Flexes thigh, rotates thigh on pelvis, rotates femur outward, and helps flex trunk on thigh	Femoral, second and third lumbar
Quadratus lumborum	Iliolumbar ligament and crest of ilium	Lower border of twelfth rib, transverse process of upper four lumbar vertebrae	Depresses twelfth rib inferiorly and flexes trunk to same side	Twelfth thoracic and first three lumbar

The *iliopsoas* is a muscle composed of two closely related muscles, the *iliacus* and the psoas major. The iliacus arises from the crest of the ilium and from the iliac fossa. Its fibers pass under the inguinal ligament and converge toward their insertion on the lesser trochanter of the femur. The psoas major arises from the transverse processes, the bodies, and the intervertebral disks of lumbar vertebrae. It is a much longer muscle than the iliacus, but it too passes under the inguinal ligament, and both muscles have a common tendinous attachment on the lesser trochanter of the femur. The iliacus and psoas major act together to flex the thigh on the pelvis when the pelvis is fixed. They also rotate the femur outward. If the femur is fixed, however, as when you bend forward from a standing position, the iliopsoas muscle helps flex the trunk on the thigh. In the standing position its tension helps to prevent the trunk from tilting too far backward on the pelvis. A powerful antagonist to the iliopsoas is the gluteus maximus muscle on the posterior side of the hip joint.

The *quadratus lumborum* is a broad muscle forming the greater part of the posterior wall of the abdomen. Its origin is on the crest of the ilium, and it extends forward to its insertion on the inferior border of the twelfth rib. It flexes the vertebral column laterally and fixes the eleventh and twelfth ribs during forced expiration. The characteristics of these muscles are summarized in Table 7.13.

MUSCLES OF THE PELVIC FLOOR

Muscles of the pelvic floor
  Levator ani
  Coccygeus

Two muscles form the pelvic diaphragm; they are the paired *levator ani* and *coccygeus muscles* (Fig. 7.20). These muscles support the abdominal and pelvic viscera. The *levator ani* is a thin sheet of muscle arising from the body of the pubis and from the obturator fascia back to the spine of the ischium. The fibers extend downward posteriorly and are inserted along the median raphe (seam) with fibers from the muscle on the opposite side. The posterior fibers insert on the coccyx. The anal canal passes through in the midline, and

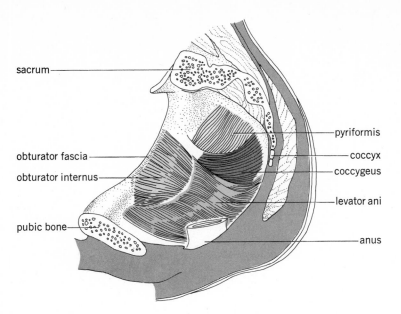

sacrum

obturator fascia

obturator internus

pubic bone

pyriformis

coccyx

coccygeus

levator ani

anus

FIGURE 7·20

Muscles of the male pelvic floor, viewed from above.

the urethra passes more anteriorly. In the female the vagina penetrates the muscle, anterior to the anal canal. The levator ani resists the downward pressure of the thoracic diaphragm. Acting together, the diaphragm and the levator ani constrict the lower part of the rectum and pull it forward.

The *coccygeus* may be considered to be essentially the posterior part of the levator ani. It arises from the spine of the ischium and is inserted on the sides of the upper coccygeal and lower sacral vertebrae. It helps support the pelvic and abdominal viscera and flexes the coccyx slightly.

The muscles of the pelvic floor and associated structures of the pelvic outlet are termed the *perineum*. Obstetricians, however, limit the term to the region between the anus and the vulva.

Included among several muscles of the perineum is a voluntary sphincter muscle around the lower part of the anal canal and around the anus. Its contraction closes the anus. There is also a sphincter muscle around the urethra. Its action is to compress the urethra after micturition. In the male the sphincter urethrae surrounds the urethra in the region between the prostate gland and the base of the penis. The female urethra is surrounded by its sphincter as it leaves the pelvis and passes through the perineum.

POSTERIOR MUSCLES OF THE HIP: THE GLUTEAL MUSCLES

The gluteal muscles
  Gluteus maximus
  Gluteus medius
  Gluteus minimus

The *gluteus maximus* is the largest and most superficial of the gluteal muscles of the posterior hip (Fig. 7.21.) Its fibers are coarse and covered with a thick layer of fat. Since this muscle is used to sit and to maintain an erect posture, it is much better developed in humans than in quadrupeds. The gluteus maximus arises from the outer surface of the ilium and the sacrum. It is inserted on the gluteal ridge of the femur and into a heavy sheath of connective tissue called the *fascia lata*. The muscles of the thigh are covered by this dense sheath. It extends downward from the hip to the knee joint, where it blends with the capsular ligament around the knee.

The actions of the gluteus maximus are complex. When the pelvis is fixed,

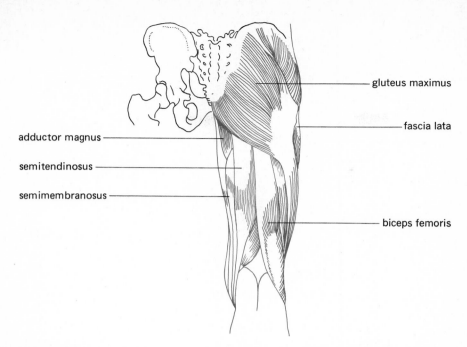

gluteus maximus

fascia lata

adductor magnus

semitendinosus

semimembranosus

biceps femoris

FIGURE 7·21
Posterior muscles of the hip
and thigh.

it extends the thigh. It is used, therefore, in jumping or walking upstairs, but
it plays little part in ordinary walking. As an antagonist to the iliopsoas, it
pulls the body erect after bending forward. This movement is made with the
femur fixed and the pelvis as the movable part. The gluteus maximus also
supports the pelvis and trunk on the femur and strengthens the knee joint.
Other gluteal muscles are the *gluteus medius* and the *gluteus minimus*. The
characteristics of these muscles are summarized in Table 7.14.

MUSCLES OF THE THIGH

ANTERIOR MUSCLES OF THE THIGH   The *quadriceps femoris* is a very large
muscle of the anterior part of the thigh. It arises by four heads and differ-
entiates into four muscles: the *rectus femoris,* the *vastus lateralis,* the *vastus
medialis,* and the *vastus intermedius* (Fig. 7.22). The rectus femoris is the
only muscle of the group attached to the pelvis, and it flexes the thigh on the

TABLE 7·14
Muscles of the posterior hip: gluteal muscle

Muscle	Origin	Insertion	Action	Nerve
Gluteus maximus	Upper portion of ilium, the sacrum, and coccyx	Gluteal tuberosity and iliotibial tract	Principal extensor and powerful rotator of thigh	Inferior gluteal
Gluteus medius	Middle portion of ilium	Oblique ridge on greater trochanter of femur	Abducts and medially rotates thigh	Superior gluteal
Gluteus minimis	Lower portion of ilium	Greater trochanter and capsule of hip joint	Abducts and medially rotates thigh	Superior gluteal

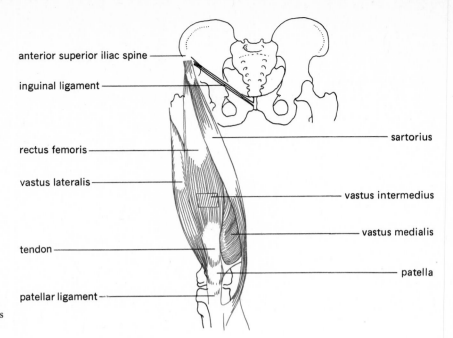

anterior superior iliac spine ——————

inguinal ligament ——————

rectus femoris ——————

vastus lateralis ——————

tendon ——————

patellar ligament ——————

sartorius

vastus intermedius

vastus medialis

patella

FIGURE 7·22
Anterior muscles of the thigh;
the quadriceps femoris muscles
and the sartorius.

Anterior muscles of the thigh
Quadriceps   Rectus femoris
femoris   Vastus lateralis
     Vastus medialis
     Vastus inter-
      medius
Sartorius

pelvis. Its origin is at the ileum, to which it is connected by two tendons. The other three muscles arise along the femur, and all four have a common insertion into the tendon that passes across the knee joint and is attached to the tibia. The tendon below the knee joint becomes the patellar ligament. The patella, or kneecap, develops in it as a sesamoid bone. The quadriceps femoris is a powerful extensor of the leg at the knee. It enables us to rise from a squatting position, for example, in setting-up exercises. It also supplies most of the power in kicking a football or in the leg kick in swimming. The quadriceps femoris is said to be three times as powerful as the flexors that oppose it.

The *sartorius* is a long, slender muscle that runs diagonally across the quadriceps femoris from its origin on the anterior, superior spine of the ilium to the medial side of the knee, where it is inserted on the upper part of the tibia near the tuberosity. It takes its name from the Latin *sartor* (tailor) and refers to the action of the muscle when the thigh is pulled into a cross-legged position formerly assumed by tailors. This is the longest muscle in the body. Although it is an anterior muscle of the thigh, because it is attached to the medial side of the knee, it can flex the leg at the knee. It is the only anterior muscle that flexes the leg. The characteristics of these muscles are summarized in Table 7.15.

Medial muscles of the thigh
Gracilis
Adductor longus
Tensor fasciae latae

MEDIAL MUSCLES OF THE THIGH   On the medial side of the thigh there are several adductor muscles. They arise from the front of the pelvis and attach at different levels to the femur. They draw the thigh inward or pull the legs toward each other. The *gracilis* (see Fig. 7.24) and the *adductor longus* (Fig. 7.23) are two of this group of muscles. One of the functions of the *tensor fasciae latae* is to assist in abducting the thigh. The characteristics of these muscles are summarized in Table 7.16.

TABLE 7·15
Anterior muscles of the thigh

Name	Origin	Insertion	Action	Nerve
Quadriceps femoris				
Rectus femoris	Anterior inferior iliac spine and upper margin of acetabulum	Tibial tuberosity	Extends leg and flexes thigh	Femoral
Vastus lateralis	Intertrochanteric line and linea aspera of femur	Tibial tuberosity	Extends leg	Femoral
Vastus medialis	Intertrochanteric line and linea aspera of femur	Tibial tuberosity	Extends leg	Femoral
Vastus intermedius	Upper shaft of femur	Tibial tuberosity	Extends leg	Femoral
Sartorius	Anterior superior iliac spine	Medial margin of tibial tuberosity	Flexes both thigh and leg	Femoral

Posterior muscles of the thigh
  Biceps femoris
  Semitendinosus
  Semimembranosus

POSTERIOR MUSCLES OF THE THIGH   The biceps femoris, the semitendinosus, and the semimembranosus muscles are three muscles of the posterior aspect of the thigh (Fig. 7.24; see also Fig. 7.21). They are commonly referred to as "hamstring muscles," because butchers use the tendons to hang up hams. The *biceps femoris* arises from two heads. The long head arises from the ischial tuberosity, and the short head arises from the middle and distal parts

FIGURE 7·23
Adductor muscles of the thigh (anterior view).

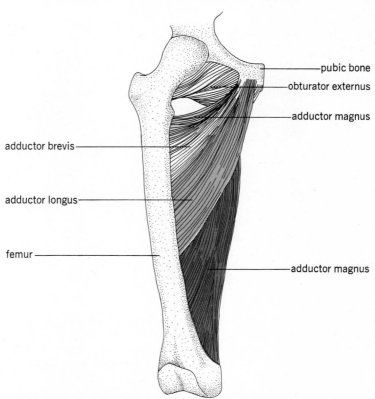

pubic bone
obturator externus
adductor magnus
adductor brevis
adductor longus
femur
adductor magnus

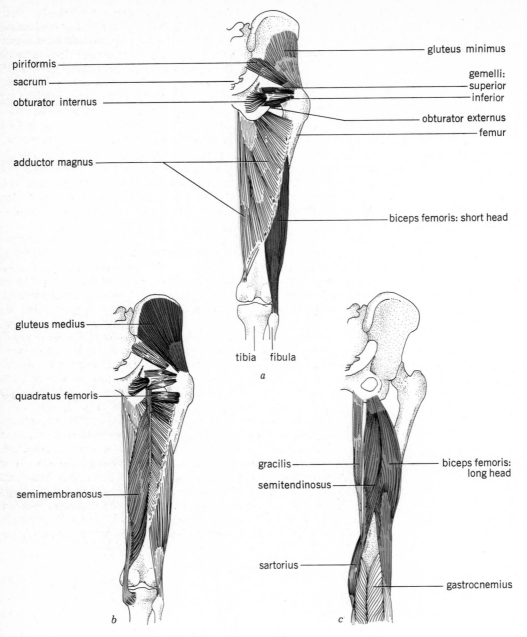

piriformis

sacrum

obturator internus

adductor magnus

gluteus minimus

gemelli:
superior
inferior

obturator externus

femur

biceps femoris: short head

tibia   fibula

a

gluteus medius

quadratus femoris

semimembranosus

gracilis

semitendinosus

biceps femoris:
long head

sartorius

gastrocnemius

b

c

FIGURE 7·24

Deeper posterior muscles of
the hip and thigh, *a*, *b*, and *c*.

of the femur. The muscle is inserted on the head of the fibula. The *semitendinosus* and the *semimembranosus* muscles arise with the long head of the
biceps femoris from the ischial tuberosity. The long tendons of these muscles
pass in back of the knee joint and outline the popliteal space, which lies
between them. The tendon of the biceps lies on the outer side. The tendons

TABLE 7·16
Medial muscles of the thigh

Muscle	Origin	Insertion	Action	Nerve
Gracilis	Inferior pubis near symphysis	Upper portion of tibia	Adducts, aids flexion, and medially rotates leg	Obturator
Adductor longus	Between pubic rami near symphysis	Middle third of linea aspera	Adducts, flexes, and medially rotates thigh	Obturator
Tensor fasciae latae	Iliac crest	Illiotibial tract	Tenses fascia lata	Superior gluteal

of the semitendinosus and the semimembranosus muscles lie on the medial side and are inserted on the proximal part of the tibia. The characteristics of these muscles are summarized in Table 7.17.

This group of muscles flexes the leg at the knee and extends the thigh at the hip joint. The posterior thigh muscles extend across two joints, the hip and the knee joints. Their antagonist, the rectus femoris, also extends across these two joints, but on the opposite side. It should be noted that the hip and knee joints flex in opposite directions.

MUSCLES OF THE LEG

Although the word "leg" commonly refers to the entire lower limb, the anatomical term "leg" refers only to the portion of the limb between the knee and ankle. Muscles of the leg affect the positioning of the foot. Since the foot is normally at right angles to the leg, *flexion* denotes raising the foot above its normal position. To *extend* the foot means to lower it from its resting position. The foot is *inverted* by raising the medial border and turning the sole inward; it is *everted* by turning the foot outward, but this is a more limited movement.

Anterior muscles of the leg
  Tibialis anterior
  Extensor digitorum longus
  Extensor hallucis longus

ANTERIOR MUSCLES OF THE LEG The anterior muscles of the leg are the tibialis anterior and the extensor muscles that extend the toes (Fig. 7.25). The *tibialis anterior* arises from the lateral condyle and from the proximal half of the shaft of the tibia. The muscle lies along the outer side of the tibia.

TABLE 7·17
Posterior muscles of the thigh

Muscle	Origin	Insertion	Action	Nerve
Biceps femoris, long head and short head	Long head: ischial tuberosity; short head: linea aspera of femur	Head of fibula and lateral condyle of tibia	Flexes leg and rotates leg laterally; long head extends thigh	Sciatic
Semitendinosus	Ischial tuberosity, fused with long head of biceps	Upper medial part of tibia	Extends thigh and flexes leg	Sciatic
Semimembranosus	Ischial tuberosity	Medial condyle of tibia	Extends thigh, flexes and medially rotates leg	Sciatic

Its tendon crosses to the inner side of the ankle and is inserted on the base of the first metatarsal bone and on the median cuneiform bones. The tibialis anterior flexes and inverts the foot. The extensor muscles of the anterior portion of the leg are the extensor digitorum longus and the extensor hallucis longus. (Latin, *hallucis*, great toe). The *extensor digitorum longus* (Fig. 7.26) arises from the lateral condyle of the tibia and from the upper anterior surface of the fibula. It lies laterad to the tibialis anterior and terminates in a tendon that divides into four slender tendons, each of which is inserted on the dorsal surface of one of the four lesser toes.

The *extensor hallucis longus* arises from the middle anterior portion of the fibula (Fig. 7.27). It is one of the deeper muscles of the anterior portion of the leg, but its tendon appears near the surface on the dorsal side of the ankle. The tendon here lies between the tendons of the extensor digitorum longus and the tibialis anterior. It then passes across the dorsal portion of the foot to its insertion on the dorsal surface at the base of the distal phalanx of the great toe. The characteristics of these muscles are summarized in Table 7.18.

The extensor muscles of the leg extend the toes by drawing them up dorsally. They also help flex the foot upon the leg.

Lateral muscles of the leg
  Peroneus longus
  Peroneus brevis

LATERAL MUSCLES OF THE LEG The peroneal muscles occupy a lateral position on the leg. The *peroneus longus* (see Fig. 7.25) arises from the lateral condyle of the tibia and from the upper part of the shaft of the fibula. The muscle terminates in a long tendon at the ankle. The tendon, which is enclosed in a common tendon sheath with the tendon of the peroneus brevis, passes behind the lateral malleolus of the fibula. It passes obliquely across the sole of the foot in another tendon sheath and is inserted on the first cuneiform bone and on the base of the first metatarsal bone. The *peroneus brevis* arises from the lower lateral part of the shaft of the fibula (Fig. 7.26a). Its tendon passes behind the lateral malleolus and then forward to insert on the dorsal surface of the fifth metatarsal bone. These peroneal muscles extend the foot and evert it. The peroneus longus and the tibialis anterior of the anterior aspect of the leg offer strong support for both the longitudinal and transverse arches of the foot. The characteristics of these muscles are summarized in Table 7.19.

TABLE 7·18
Anterior muscles of the leg

Muscle	Origin	Insertion	Action	Nerve
Tibialis anterior	Upper half of tibia and interosseous membrane	First cuneiform and first metatarsal	Dorsally flexes and inverts foot	Deep peroneal
Extensor digitorum longus	Tibia, proximal three-fourths of fibula and interosseous membrane	Tendons to middle and terminal phalanges of four lateral toes by extensor expansion	Extends toes	Deep peroneal
Extensor hallucis longus	Middle half of fibula and interosseous membrane	Distal phalanx of great toe	Extends great toe	Deep peroneal

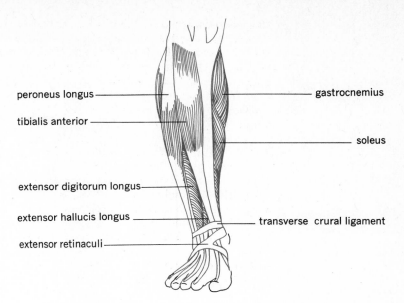

peroneus longus

tibialis anterior

extensor digitorum longus

extensor hallucis longus

extensor retinaculi

gastrocnemius

soleus

transverse crural ligament

FIGURE 7·25
Anterolateral muscles of
the leg.

Posterior muscles of the leg
Gastrocnemius ⎤ Triceps
Soleus ⎦ surae
Flexor digitorum longus
Flexor hallucis longus
Tibialis posterior

POSTERIOR MUSCLES OF THE LEG The superficial muscles on the posterior side of the leg are the gastrocnemius and the soleus. Deeper muscles include the flexor digitorum longus and the flexor hallucis longus. The tibialis posterior is the deepest of the posterior muscles.

The *gastrocnemius* and *soleus* are the muscles of the calf of the leg (Figs. 7.27 and 7.28; see also Fig. 7.26*a*). The gastrocnemius arises by two heads from areas just above the medial and lateral condyles of the femur. The soleus, which lies beneath the gastrocnemius, arises from the proximal posterior portions of both the tibia and fibula. The calcaneus tendon (Achilles), a common tendon for both muscles, is a strong tendon inserted on the calcaneus (heel bone). These muscles extend the foot, enabling us, when standing, to rise up on our toes. The gastrocnemius, since it is attached above the knee joint, also helps flex the leg at the knee. The gastrocnemius and the soleus together compose the *triceps surae* muscle. The two heads of the gastrocnemius and the proximal attachment of the soleus are the source of this name (Latin, *sura*, calf of the leg).

The *flexor digitorum longus* has its origin along the posterior side of the

TABLE 7·19
Lateral muscles of the leg

Muscle	Origin	Insertion	Action	Nerve
Peroneus longus	Upper two-thirds of fibula and intermuscular septua	First cuneiform and first metatarsal	Everts and aids in plantar flexion	Superficial peroneal
Peroneus brevis	Lower two-thirds of fibula	Fifth metatarsal	Everts and abducts foot and aids in plantar flexion	Superficial peroneal

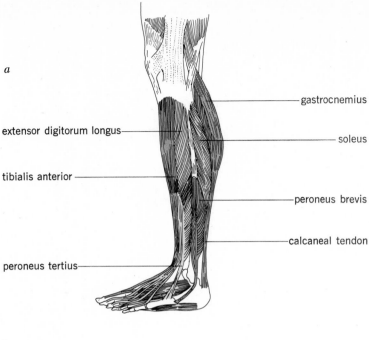

a

extensor digitorum longus—

tibialis anterior —

peroneus tertius—

———— gastrocnemius

———— soleus

————peroneus brevis

————calcaneal tendon

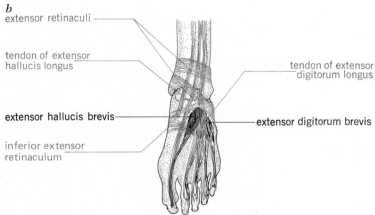

b

extensor retinaculi —

tendon of extensor
hallucis longus —

extensor hallucis brevis—

inferior extensor
retinaculum

tendon of extensor
digitorum longus

——extensor digitorum brevis

FIGURE 7·26

Deeper muscles of the leg: *a*
lateral view; *b* extensor tendons of the foot (dorsal aspect).

shaft of the tibia (Fig. 7.29). It extends downward on the medial and
posterior side of the leg. A strong tendon passes behind the medial malleolus
of the tibia and passes transversely under the arch of the foot, where it
divides into four tendons, each inserted on the ventral surface of one of the
lesser toes (see Fig. 7.26*b*). The contraction of this muscle flexes the four
lesser toes and inverts the foot. It helps support the longitudinal arch of the
foot.

The *flexor hallucis longus* (see Fig. 7.29) arises from the lower posterior
surface of the shaft of the fibula. The muscle extends downward on the
lateral, posterior side of the leg. Its long tendon crosses over to the medial

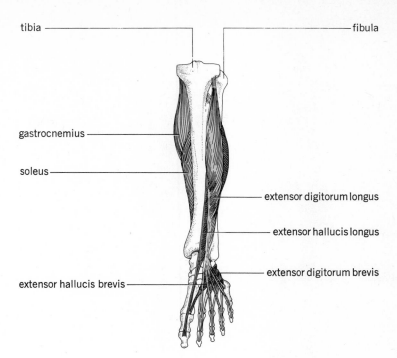

tibia — — fibula

gastrocnemius —

soleus —

extensor digitorum longus

extensor hallucis longus

extensor digitorum brevis

extensor hallucis brevis —

FIGURE 7·27
Muscles of the calf of the leg;
the gastrocnemius and soleus
(anterior view)

side posteriorly above the heel. It passes under the side of the talus on the medial side of the ankle and under the arch of the foot, and is inserted on the distal phalanx of the big toe. This muscle flexes the big toes and inverts the foot. It supports the longitudinal arch of the foot, especially when the weight is on the toes.

The *tibialis posterior* (Fig. 7.29) arises from the upper part of the shaft of both the tibia and fibula and also from an aponeurosis between these two bones. The muscle extends downward just posterior to the tibia. Its tendon passes obliquely to the medial side just above the ankle and then under the medial malleolus of the tibia and forward under the arch of the foot, where it is inserted on several tarsal and metatarsal bones. The tibialis posterior acts with the tibialis anterior to invert the foot. It helps maintain the arch of the foot. Since the medial malleolus acts as a pulley, the muscle extends the foot and thus acts as an antagonist to the tibialis anterior. The characteristics of these muscles are summarized in Table 7.20.

MUSCLES OF THE FOOT

Muscles of the foot
  Extensor digitorum brevis
  Flexor digitorum brevis
  Flexor hallucis brevis
  Lumbricales

The foot is specialized to support the body and is not therefore as flexible as the hand. The musculature in general resembles that of the hand. The muscles of the leg that control movements of the foot by their tendons are called *extrinsic muscles* (Fig. 7.29). *Intrinsic muscles* are those located within the foot itself (Fig. 7.30). On the dorsum (top) of the foot the *extensor digitorum brevis* helps the long tendons of the extrinsic muscles to extend the four medial toes. This muscle does not have a homologue in the hand (see Fig. 7.27).

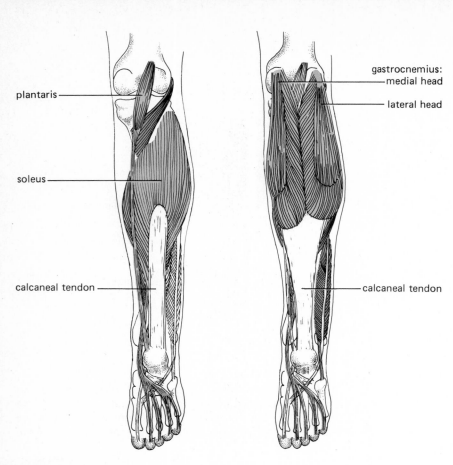

plantaris

soleus

calcaneal tendon

gastrocnemius:
medial head

lateral head

calcaneal tendon

FIGURE 7·28
Muscles of the calf of the leg
(posterior view).

TABLE 7·20
Posterior muscles of the leg

Muscle	Origin	Insertion	Action	Nerve
Gastrocnemius	Medial and lateral con-dyles of femur	With soleus into calca-neus via calcaneal ten-don	Flexes leg and plantar, flexes foot	Tibial
Flexor digitorum longus	Middle half of tibia	By four tendons into distal phalanges of lat-eral four toes	Flexes lateral four toes	Tibial
Flexor hallucis longus	Distal two-thirds of fib-ula and intermuscular septum	Distal phalanx of great toe	Flexes great toe	Tibial
Tibialis posterior	Interosseous membrane and tibia and fibula on either side	Navicular, with slips to cuneiform; cuboid; metatarsals 2, 3, and 4	Adducts and inverts foot, aids in plantar flexion	Tibial
Soleus	Upper third of fibula and soleal line of tibia	With gastrocnemius into calcaneus via cal-caneal tendon	Flexes foot	Tibial

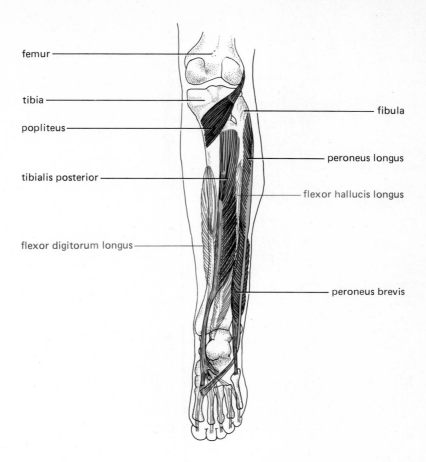

femur

tibia

popliteus

tibialis posterior

flexor digitorum longus

fibula

peroneus longus

flexor hallucis longus

peroneus brevis

FIGURE 7·29
Deeper posterior muscles of
the leg.

The bottom of the foot is the *plantar surface* (Fig. 7.31). The plantar aponeurosis is a heavy sheath of connective tissue under the skin in the sole of the foot that supports and strengthens the foot. The *flexor digitorum brevis* lies just above the plantar aponeurosis. It flexes the four lateral toes and helps support the arch of the foot. The *flexor hallucis brevis* flexes the great toe. Other intrinsic muscles of the foot such as the *lumbricales* are concerned with various movements (Fig. 7.32). Some help support the arches. Tendons in the ankle (Fig. 7.33) and foot are generally covered with synovial sheaths. A number of sesamoid bones form in the tendons of the foot, and numerous bursae protect the muscles and joints of the foot. The characteristics of these muscles are summarized in Table 7.21.

CLINICAL ASPECTS   *Hernias*   The muscular wall of the body is not intact. Many openings permit the passage of arteries, veins, nerves, and other structures. These areas, which must be kept open, constitute weak places in the body wall where a hernia may occur. *Hernia*, or *rupture*, is the protrusion of a part of the viscera through the body wall.

An *abdominal hernia* is the protrusion of a portion of the intestine, mesentery, or peritoneum through an opening in the abdominal wall.

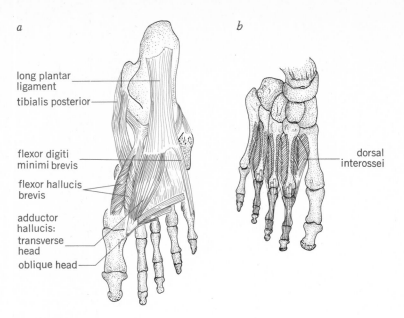

a *b*

long plantar ligament

tibialis posterior

flexor digiti minimi brevis

flexor hallucis brevis

adductor hallucis: transverse head

oblique head

dorsal interossei

FIGURE 7·30
Intrinsic muscles of the foot: *a* plantar view; *b* the dorsal interossei.

Weakening of the muscular wall and lack of tone in muscles are conditions conducive to hernia. Heavy lifting or coughing increases abdominal pressure and may cause a hernia.

A *diaphragmatic hernia* usually occurs at the opening in the diaphragm that permits the esophagus to enter the stomach. This foramen is called the *esophageal hiatus.* In *hiatus hernia,* a portion of the stomach pushes upward through the esophageal opening.

The inguinal canal is another opening in the body wall. It extends from the internal inguinal ring, an opening in the fascia of the transversus muscle, to the subcutaneous inguinal ring, which lies in the aponeurosis of the

FIGURE 7·31
Muscles of the sole of the foot (plantar surface), second layer. [*After Barry J. Anson (ed.), "Morris' Human Anatomy," 12th ed., McGraw-Hill Book Company, New York, 1966.*]

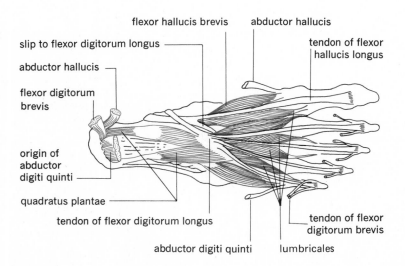

flexor hallucis brevis

abductor hallucis

slip to flexor digitorum longus

tendon of flexor hallucis longus

abductor hallucis

flexor digitorum brevis

origin of abductor digiti quinti

quadratus plantae

tendon of flexor digitorum longus

abductor digiti quinti

lumbricales

tendon of flexor digitorum brevis

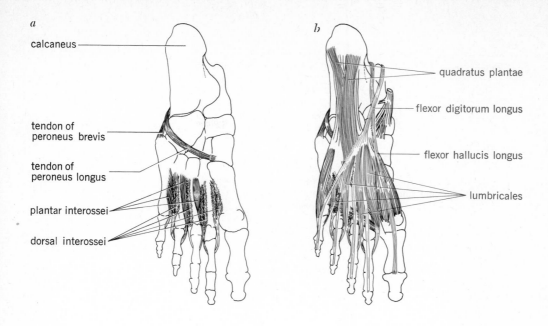

a

calcaneus

tendon of
peroneus brevis

tendon of
peroneus longus

plantar interossei

dorsal interossei

b

quadratus plantae

flexor digitorum longus

flexor hallucis longus

lumbricales

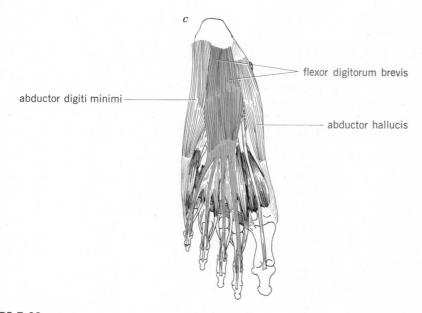

c

flexor digitorum brevis

abductor digiti minimi

abductor hallucis

FIGURE 7·32

Deep muscles of the sole of
the foot (plantar view).

external oblique muscle, just above the pubic tubercle. The inguinal canal is
about 1½ in long and lies just above the inguinal ligament and parallel to it.
A part of the intestine is occasionally forced through the inguinal canal to
form an *inguinal hernia*. This is a common type of hernia in the male
(Fig. 7.33).

TABLE 7·21
Muscles of the foot

Muscle	Origin	Insertion	Action	Nerve
Extensor digitorum brevis	Calcaneus	Base of great toe; tendons to second, third, and fourth toes	Helps extend four medial toes	Peroneal
Flexor digitorum brevis	Calcaneus and plantar aponeurosis	By four tendons into middle phalanx of lateral four toes	Flexes lateral four toes	Medial plantar
Flexor hallucis brevis	Navicular and cuneiform bones, and long plantar ligament	Two tendons to medial and lateral sides of proximal phalanx of great toe	Flexes great toe	Medial and lateral plantar
Lumbricales	Tendons of flexor digitorum longus	Extensor expansion over lateral four toes	Aids in flexion of toes	Medial and lateral plantar
Quadratus plantae	Calcaneus and long plantar ligament	Into tendons of flexor digitorum longus	Supports action of tendon of flexor digitorum longus	Lateral plantar

A *femoral hernia* occurs at the femoral ring, the internal entrance to the femoral canal, located below the inguinal ligament. The femoral sheath covers the femoral artery and vein where they pass from the body cavity to the leg. On the medial side of the femoral vein, the sheath also covers the femoral canal. A pouch of the peritoneum may pass through this canal to form a femoral hernia. Femoral hernias are not a common type, although they are more frequent in women than in men.

An *umbilical hernia* occurs at the umbilical ring (navel). This type more commonly occurs in infants, although it may occur in adults.

FIGURE 7·33

Lower abdominal structures and the external appearance of inguinal and femoral hernias.

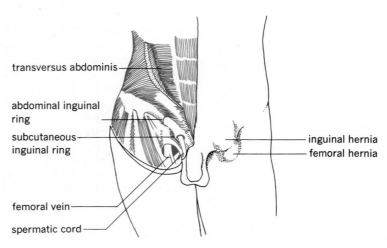

transversus abdominis

abdominal inguinal ring

subcutaneous inguinal ring

femoral vein

spermatic cord

inguinal hernia
femoral hernia

SUMMARY

1 One of the main functions of the skeletal muscles is to enable the skeletal structure to move.

2 Skeletal muscles are ordinarily attached to bones by tendons. Tendons sometimes appear as flat sheets of tissue called aponeuroses. Fasciae are connective tissue sheaths that cover muscles and related structures. Synovial bursae are sacs located between tendons or around joints that prevent damage to structures that must rub against one another.

3 The more fixed attachment of a muscle is usually its origin, and the more movable attachment is its insertion.

4 There are three classes of levers. Body movement is mostly accomplished by third-class levers.

5 Muscles of the head and face include muscles that arrange facial expression and produce mastication. The orbicularis oculi muscles around the eyes and the orbicularis oris muscle around the mouth are voluntary sphincter muscles. The masseter, temporal, and pterygoid muscles are muscles of mastication.

6 Muscles of the neck include the platysma and the sternocleidomastoideus.

7 The pectoralis major is the large anterior muscle of the thorax; the pectoralis minor lies directly underneath it. The external and internal intercostal muscles occupy the spaces between the ribs and produce respiratory movements of the thorax.

8 The deltoideus is located above the articulation of the arm at the shoulder. It raises the arm to a horizontal position. Flexors of the arm include the biceps brachii and the brachialis. The triceps brachii extends the arm at the elbow. Muscles of the forearm are flexors and extensors for the wrist and hand.

9 The diaphragm is a voluntary skeletal muscle that forms the floor of the thoracic cavity. It is dome-shaped and moves up and down as one breathes. Muscles of the abdominal wall are the external and internal oblique muscles, the transversus abdominis, and the rectus abdominis. The linea alba is a tendinous sheath extending along a midventral line from the xiphoid process of the sternum to the symphysis pubis.

10 Two prominent superficial muscles of the back are the trapezius and the latissimus dorsi. Posterior shoulder muscles that move the scapulae are: the levator scapulae, rhomboideus major, rhomboideus minor, and the serratus anterior. A deep muscle of the back is the erector spinae.

11 Superficial muscles located on the anterior aspect of the forearm are the pronator teres, the flexor carpi radialis, the palmaris longus, and the flexor carpi ulnaris.

12 On the posterior aspect of the forearm are the anconeus, the extensor digitorum, the extensor digiti minimi, the extensor carpi ulnaris, the extensor carpi radialis longus, and the extensor carpi radialis brevis.

13 Deeper muscles of the forearm are the supinator, the abductor pollicis longus, the extensor pollicis brevis, and the extensor pollicis longus.

14 The thenar muscles are the flexor pollicis brevis, the abductor pollicis brevis, the adductor pollicis, and the opponens pollicis.

15 The flexor digitorum superficialis and the flexor digitorum profundus help flex the fingers.

16 The muscles of the hypothenar eminence are the abductor digiti minimi, the flexor digiti minimi, and the opponens digiti minimi.

17 There are two groups of muscles within the hand itself: the lumbricales and the interosseous sets.

18 The iliopsoas muscle flexes the thigh on the pelvis when the pelvis is fixed; it flexes the trunk on the thigh when the femur is fixed. It is an antagonist of the gluteus maximus, the large muscle of the buttock.

19 The levator ani and coccygeus muscles form the pelvic diaphragm. Sphincter muscles constrict the anus and the urethra.

20 Posterior muscles of the hip are the gluteal muscles. They extend the thigh and pull the body erect.

21 The large muscle of the anterior part of the thigh is the quadriceps femoris. It extends the leg at the knee. The sartorius muscles enable you to cross your legs.

22 Adductor muscles of the medial thigh, such as the gracilis, pull the thigh inward. The tensor fasciae latae helps abduct the thigh.

23 The biceps femoris, the semitendinosus, and the semimembranosus muscles are three muscles located on the posterior aspect of the thigh. They flex the leg at the knee and extend the thigh at the hip.

24 Anterior muscles of the leg include the tibialis anterior that flexes and inverts the foot. Extensor muscles located anteriorly on the leg are the extensor digitorum longus and extensor hallucis longus.

25 The peroneal muscles occupy a lateral position on the leg. They extend the foot and evert it.

26 The gastrocnemius and soleus are muscles of the calf of the leg. Deeper muscles include the flexor digitorum longus and the flexor hallucis longus. The tibialis posterior, the deepest of the posterior muscles of the leg, acts with the tibialis anterior to invert the foot and helps support the longitudinal arch.

27 On the dorsum (top) of the foot, the extensor digitorum brevis joins the tendons of extrinsic muscles to extend the four medial toes. On the plantar surface, the flexor digitorum brevis flexes the four lesser toes; the flexor hallucis flexes the great toe.

QUESTIONS

1 Discuss the mechanics of body movement as an example of force applied to third-class levers.

2 Which muscles of the face are responsible for chewing, smiling, and frowning?

3 Name some of the antagonistic muscles of the appendages and describe their action.

4 The human hand is extremely versatile. Name and describe some of the muscles that provide this versatility.

5 What is the function of a pronator muscle? An adductor? An opponens?

6 Draw the location of some extensors of the wrist and fingers.

7 Which muscles compose the thenar group?

8 Name two sets of muscles located within the hand itself.

9 What is meant by synergistic muscles? Name some.

10 Explain the action of the diaphragm. What structures pass through its openings?

11 Locate three areas in the abdominal body wall where hernias may occur. What is a hernia?

12 What muscles form the floor of the pelvis?

13 Explain the action of the iliopsoas and the gluteus maximus muscles when the femur is fixed and when the trunk is fixed.

14 Why are the back muscles usually strong, while the ventral or abdominal muscles are often weak?

15 Discuss the muscles of the leg and foot relative to the support of the longitudinal arch of the foot.

16 How do the abdominal muscles support breathing?

17 Name three muscles that move the scapula.

18 List the muscles of the face, the shoulder, and the thigh and describe their functions. Which of these muscles support each other, and which are antagonistic?

SUGGESTED READING

Anson, B. J. (ed.): "Morris' Human Anatomy," 12th ed., McGraw-Hill Book Company, New York, 1966.

Crouch, J. E.: "Functional Human Anatomy," 2d ed., Lea & Febiger, Philadelphia, 1972.

Gray, H.: "Anatomy of the Human Body," 29th ed., Lea & Febiger, Philadelphia, 1973.

Schlossberg, L., and G. D. Zuidema: "The Johns Hopkins Atlas of Human Functional Anatomy," The Johns Hopkins Press, Baltimore, 1977.

UNIT THREE

CONTROLLING SYSTEMS

In Unit Two, we discussed three of the four basic tissue types: epithelial tissue, connective tissue, and muscle tissue. In Unit Three, we take up the structure and function of the remaining basic tissue: nerve tissue.

In Chap. 8, we shall first describe the structure of nerve tissue—its basic unit the neuron—and its specialization for the functions of excitability and conduction. We shall then turn to the events that make up the nerve impulse, a series of electrochemical reactions in which a stimulus is transformed into a type of energy that may be transmitted from one neuron to another. Finally, we shall discuss the reflex, one of the basic functional units of the nervous system. In a reflex, sensory information is obtained, integrated in one of the higher centers of the nervous system, and then translated into motor action.

In Chap. 9, we shall describe the basic organization and the primary functions of the central nervous system, which comprises the brain, the spinal cord, and the spinal and cranial nerves. The central nervous system is the first major division of the body's nervous system.

In Chap. 10, we shall move to the next major division of the nervous system: the autonomic nervous system. The autonomic nervous system is concerned with visceral reactions—those functions of the viscera that are regulated without conscious control.

Finally, in Chaps. 11 and 12, we shall turn to the special senses, the third division of the nervous system. In Chap. 11, we shall discuss vision and briefly describe the specialized sensory receptors involved in this sense and their connections to the central nervous system. In Chap. 12, we shall describe audition, cutaneous sensation, taste, and olfaction.

The nervous system forms one of the two major control systems of the body. We shall take up the endocrine system, the other important controller, in Unit Four.

BASIC NEUROPHYSIOLOGY

CONSIDERATIONS FOR STUDY

In this chapter, we introduce the structure and functions of neurons, or nerve cells, the fourth type of tissue. Topics covered include:

1 The structure of the neuron: its division into cell body, axon, and dendrite and the specialized functions of each of its parts
2 The synapse: how it is organized and how neurotransmitters affect it
3 The nerve impulse: the basic events in its generation and the combination of nerve impulses to produce different effects at synapses, or neuron junctions
4 The nerve: its composition
5 The spinal cord and spinal nerves: their basic structure and organization
6 Reflex activity: the participation of spinal nerves in one basic type of organized neural activity to form reflex arcs and to create conditioned reflexes

Human beings are large organisms of great complexity. While some other animal species may have relatively simple nervous systems to coordinate their metabolic activities, our nervous system is exceedingly intricate and is adapted to the needs of a highly developed organism. The human species has gone far beyond many other animal species in the development of the brain. Such complicated functions as those involved in consciousness, abstract thought, memory, and the interpretation of emotions are some of the activities of the human brain.

Nervous tissue, whose fundamental unit is the *neuron*, is ectodermal in origin. Functionally, the neuron has two outstanding characteristics: excitability and conductivity. *Excitability* means that it is capable of reacting to stimuli; *conductivity* refers to its ability to transmit electrochemical impulses to other neurons or muscle cells.

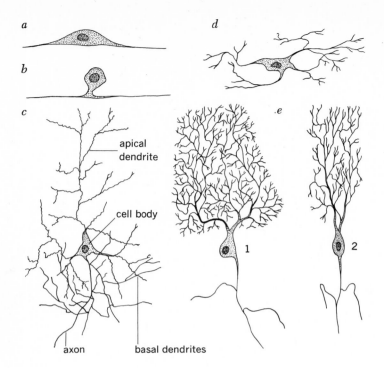

a

b

d

c

apical
dendrite

cell body

.e

1

2

axon   basal dendrites

FIGURE 8·1

Various types of nerve cells:
a embryonic bipolar nerve
cell; *b* mature ganglion cell;
c a pyramidal cell (motor)
from the cerebral cortex; *d* a
multipolar nerve cell, golgi
type II, interneuron; *e* purkinje cells from the cerebellum, *1* frontal view, *2* lateral
view.

THE NEURON

The structural and functional unit of nervous tissue is the nerve cell, or
neuron. The neuron is a microscopic grayish cell from which specialized
fibers radiate. There are various types of neurons (Fig 8.1).

STRUCTURE   The *cell body* of a neuron contains a well-defined nucleus and
nucleolus. Within the cytoplasm are small masses of a deep-staining material called *Nissl substance,* which is composed of nucleoproteins, largely
RNA, associated with the cell's endoplasmic reticulum. The Nissl substance
is concerned with the neuron's physiological activity. Neuron cytoplasm
also contains microtubules and microfilaments believed to be responsible for
the migration of substances from the cell body to other parts of the cell.
Such structures may also be involved in *exocytosis,* the expulsion of substances from the cell body.

Overall structures of some common neurons are illustrated in Figs. 8.2
and 8.3.

Neuron cell bodies vary considerably in size, ranging roughly from 5 to
50 μm in diameter. The cell body is a vital part of the neuron. If the cell
body is destroyed by mechanical injury or disease, the processes of the
neuron also die.

Dendrites (from the Greek *dendron,* tree) are the branches of a neuron
that conduct nerve impulses *toward* the cell body. Different types of
neurons have different numbers of dendrites. The neurons that conduct
motor impulses through the spinal cord usually possess several short dendrites, each with treelike branches, while some neurons found in the brain
have one long dendrite with few branches. Dendrites contain Nissl substance
and are often considered to be extensions of the cell body.

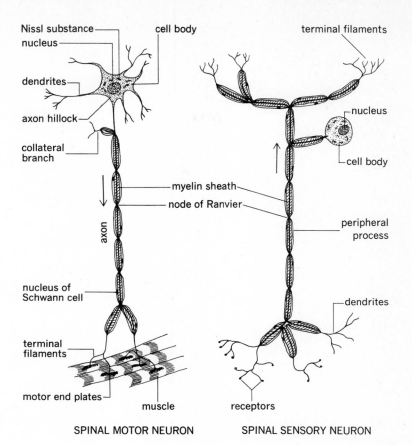

FIGURE 8·2

Two common types of neurons, a spinal motor neuron and a spinal sensory neuron.

Nissl substance — cell body
nucleus
dendrites
axon hillock
collateral branch
axon
myelin sheath
node of Ranvier
nucleus of Schwann cell
terminal filaments
motor end plates
muscle

SPINAL MOTOR NEURON

terminal filaments
nucleus
cell body
peripheral process
dendrites
receptors

SPINAL SENSORY NEURON

FIGURE 8·3

Motor nerve cells from spinal cord smear. The numerous small cell bodies are neuroglial cells. (*Courtesy of General Biological Supply House, Inc., Chicago.*)

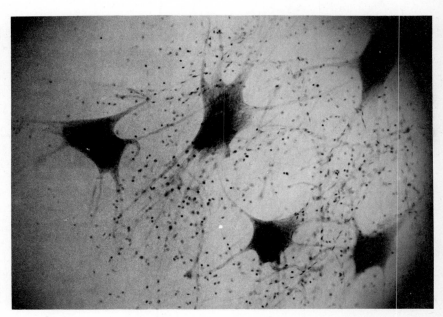

The dendrites and the cell membrane are the receptive parts of the neuron. Specialized peripheral receptors may act as transducers of specific stimuli for some sensory neurons. Such receptors are described in more detail in Chaps. 11 and 12.

The *axon* is the process of the neuron that conducts an impulse *away from* the cell body. The area of the cell from which the axon arises is termed the *axon hillock*. The part of the axon just distal to the hillock is the *initial segment*. Axons do not contain Nissl substance, which may also be absent from the axon hillock. The axon of a neuron may be very short, as in some brain neurons, or it may be 2 or 3 ft long, as in some spinal neurons whose axons extend to the muscles of the fingers or toes. Axons terminate in fine branches called *axonal terminal filaments,* or *telodendria.* The tips of these terminal branches appear to be highly specialized to transmit the nerve impulse to succeeding neurons or to a muscle effector. The difference between axon and dendrite is based primarily on difference in function, and not on minor differences in structure.

The myelin sheath Axons of some neurons are covered with a *myelin sheath* formed by the plasma membrane of Schwann cells. The fiber, enclosed in its own membrane, the *axolemma,* first becomes embedded in Schwann cells, which then are tightly wrapped around the axon in many layers. This wrapping is the result of the growth of the membrane rather than the rotation of the Schwann cells or the axon. Schwann cell membranes are always double, since the original invagination brings the two sides of the plasma membrane together, leaving a cleft (Figs. 8.4 and 8.5). The Schwann cell membrane is a phospholipid bilayer as are most cell membranes.

A myelinated axon has a segmented appearance because successive myelin sheaths are interrupted at sites termed the *nodes of Ranvier.* A single Schwann cell forms the myelin sheath between two nodes. The Schwann cell terminates at the node, and the node itself is covered only by a thin

FIGURE 8·4

a Myelinated axon, anterolateral view; *b* development of myelin sheath from Schwann cell membrane.

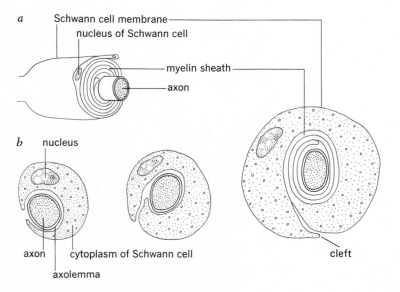

a Schwann cell membrane
nucleus of Schwann cell
myelin sheath
axon

b nucleus
axon
cytoplasm of Schwann cell
axolemma
cleft

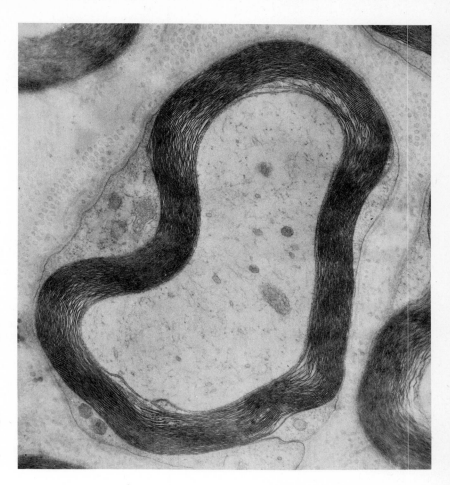

FIGURE 8·5
Electron micrograph of myelin sheath. (*From James A. Freeman, "Cellular Fine Structure," McGraw-Hill Book Company, New York, 1964.*)

basement membrane. The nodes represent relay stations for the propagation of a nerve impulse along the axon. The impulse "jumps" from one node to another in an action described as *saltatory conduction*.

Unmyelinated axons are also embedded in Schwann cells, but they are not completely enclosed, and the Schwann cell membranes do not coil around them. Several axons of this type may be embedded in a single Schwann cell, but the axolemmas remain separate (Fig. 8.6).

Nerve impulses are conducted more rapidly in myelinated fibers than in nonmyelinated ones.

STRUCTURAL AND FUNCTIONAL CLASSIFICATIONS OF NEURONS   Neurons may be classified by structure into two basic types: *bipolar cells,* which have two processes, and *multipolar cells,* which develop numerous processes. Some spinal neurons are bipolar, but multipolar cells are commonly found in the brain and also in the spinal cord.

Neurons may be classified into three types by function: (1) *motor neurons,* or *efferent neurons,* which convey impulses to muscles or glands; (2) *sensory,*

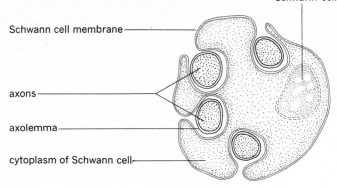

FIGURE 8·6
Unmyelinated axons embedded in Schwann cell membrane.

Schwann cell nucleus

Schwann cell membrane

axons

axolemma

cytoplasm of Schwann cell

or *afferent, neurons,* which transmit stimuli inward from various receptors; and (3) *interneurons,* which are found between other neurons and convey impulses from one neuron to another.

THE SYNAPSE AND NEURON MEMBRANE POTENTIALS

As we mentioned in our discussion of skeletal muscle in Chap. 6, the *synapse* is an area of close association between two neurons at which information is transmitted from one neuron to the next. Many of the phenomena that occur at the neuromuscular junction between a neuron and a motor end plate also occur at the synapse (Fig. 8.7). However, synaptic functions are much more highly developed, complex, and sophisticated within the nervous system. Synapses determine: (1) the circuitry, or pattern of communication, between neurons; (2) the intensity of communication between particular neurons; and (3) the inhibition or enhancement of communication. You should be aware, however, that neurophysiology and neurochemistry are presently intensely active research areas. Thus, in addition to the type of synaptic communication we shall discuss, there are now known to be simple electrical synapses, not involving neurotransmitters, and, within the brain, neurotransmitter substances which are not necessarily associated with synapses.

The neuron, as a distinct structural unit, must have a method of transmitting a nerve impulse to another neuron. Nerve impulses are usually

FIGURE 8·7
The anatomy of synapses.

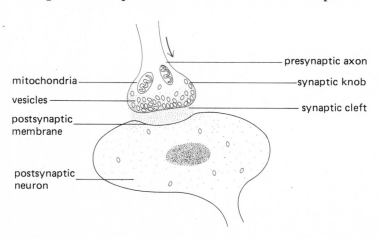

mitochondria

vesicles

postsynaptic membrane

postsynaptic neuron

presynaptic axon

synaptic knob

synaptic cleft

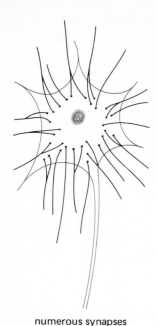

numerous synapses
on a nerve cell body

FIGURE 8·8
Nerve cell with multiple synapses.

transmitted at the *synapse,* an area where the distal ends of terminal filaments of the axon come into close morphological and physiological relationship with the dendrites and cell bodies of succeeding neurons (Fig. 8.7). The ends of the terminal filaments are enlarged into *synaptic knobs,* and are closely applied to the cell membrane of the next neuron. Electron micrographs reveal a space of about 100 to 200 Å, a *synaptic cleft,* between the two neurons. Synaptic knobs contain mitochondria and synaptic vesicles (Fig. 8.8).

TRANSMITTER SUBSTANCES Synaptic vesicles in the terminal filaments of the axon contain a *transmitter substance.* Several substances are known to act as neural transmitters. Acetylcholine seems to be the most common neural transmitter found outside the brain. It is found in high concentrations at neuromuscular junctions (see Chap. 6) and also in some parts of the brain, especially in the brainstem. Norepinephrine is a transmitter found in parts of the autonomic nervous system and in parts of the brain. Serotonin, another transmitter, is found in the brain, although its precise function is not well understood. Other transmitters in the brain are γ-aminobutyric acid (GABA) and dopamine. These transmitters will be considered later as we study the brain.

Neurotransmitters work in the following way. A neurotransmitter must be released in a controlled manner. Release is almost always keyed to depolarization of the *presynaptic neuron membrane* in the vicinity of the intracellular storage sites of the chemical. Such depolarization occurs as it does in muscle cells (see Chap. 6). Once released, the neurotransmitter attaches to a specific binding site on a receptor molecule in the *postsynaptic neuron membrane.* If the neurotransmitter is *excitatory,* the postsynaptic neuron membrane becomes increasingly permeable to sodium, and depolarization of the membrane, which may or may not reach the threshold potential, occurs. If the neurotransmitter is *inhibitory,* it causes an increased permeability to potassium in the postsynaptic membrane, and hyperpolarization occurs. Production of an action potential in a hyperpolarized neuron is much more difficult. Unlike the neuromuscular junction in which one connection is made at each motor end plate, the synaptic neuron has many synaptic connections on its dendrites as well as on its soma. The combined stimulation of *many* synapses is required to depolarize the postsynaptic neuron membrane enough to produce a threshold potential (Fig. 8.8).

A chemical agent found in the synaptic cleft destroys the transmitter substance as soon as it has interacted with the postsynaptic neuron. *Acetylcholinesterase* is an enzyme that hydrolyzes acetylcholine in neuronal synapses as well as in neuromuscular junctions and inactivates it in a fraction of a second. The enzyme *o-methyl transferase* removes the transmitter norepinephrine from the synaptic region.

THE NERVE IMPULSE

Once the neurotransmitter has done its job at the synapse, the changes it causes in membrane permeability of the postsynaptic neuron lead to generation of the *nerve impulse* in this neuron. The generation of nerve impulses is more complicated than the production of action potentials in muscle cells, which we discussed in Chap. 6. We consider these differences here, but you should remember that membrane potentials are basically similar in *all* cells.

THRESHOLD STIMULI   A weak stimulus may not produce an action potential in a postsynaptic neuron. In this case, a local depolarization may occur in the neuron membrane, but the stimulus is not strong enough to reach *threshold level* and cause the neuron's membrane to depolarize completely. It may take several stimuli of increasing strength to produce a *threshold potential,* which may be perhaps 10 mV more positive than that of the resting membrane potential. Since we consider -70 mV to be an average resting potential, the threshold potential may be around -60 mV. The threshold *stimulus,* therefore, is one that produces a membrane depolarization of at least this amount.

THE ALL-OR-NONE RESPONSE   A stimulus strong enough to depolarize a neuron to the threshold potential will produce an *action potential.* The events taking place in the action potential have been summarized in Chap. 6. Increasing the strength of the stimulus does not produce a larger action potential. The transmission of a nerve impulse by the postsynaptic neuron is said to be *all or none,* therefore.

The all-or-nothing principle, however, does not apply to all parts of the neuron. The receptor parts, that is, the dendrites and the cell membrane, react to transmitters with changes in membrane potential that may not produce an action potential. Once the entire neuron has been stimulated sufficiently to produce an action potential, this potential is transmitted in a uniform manner along the entire axon of the neuron.

Theoretically, the propagated action potential can travel in either direction along the axon, but it is not possible for an impulse to traverse a synapse from the postsynaptic neuron cell body or dendrites toward the axonal terminals of the presynaptic neuron because neurons do not produce transmitter substances in their cell bodies or dendrites that can carry impulses across the synaptic cleft.

EXCITATORY AND INHIBITORY POSTSYNAPTIC POTENTIALS   An excitatory postsynaptic potential (EPSP) brings the postsynaptic neuron closer to threshold potential, but one EPSP may not cause an action potential. The movement of potassium ions out of the neuron and the movement of sodium ions into the neuron may cause only a mild, temporary depolarization of the neuron, but such depolarization eventually may facilitate the development of an action potential.

EPSPs may be added together, a process called *summation.* A series of two or more subthreshold stimuli, each producing an EPSP, may summate and initiate depolarization. The number of synapses on the postsynaptic neuron's dendrites and soma and whether or not they are near each other affect summation.

An inhibitory postsynaptic potential (IPSP) changes the permeability of the postsynaptic membrane so that more potassium ions flow out. The internal potential of the neuron moves more to the negative side, making depolarization more difficult to attain. This increase in membrane permeability does not affect sodium ions. If a series of inhibitory impulses drives the interior voltage of the cell 10 mV further toward the negative, it becomes that much more difficult for the neuron to be stimulated to threshold potential and thus to generate an action potential. The IPSP also is capable of summation. This is illustrated in Fig. 8.9.

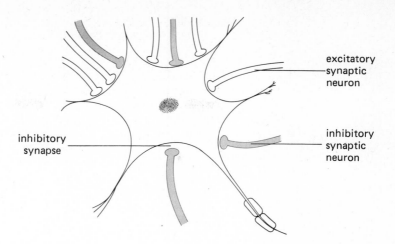

excitatory
synaptic
neuron

inhibitory
synaptic
neuron

inhibitory
synapse

FIGURE 8·9
Inhibitory synapse.

TYPES OF SUMMATION   Two types of summation are recognized. If two or more presynaptic terminals, or end knobs, are activated at the same time, their synapses, of course, are separated by a space, but their simultaneous activity may sum to produce a stronger EPSP, a phenomenon referred to as *spatial summation*. If the stimulation is limited to one presynaptic terminal end knob, but this terminal filament fires several times, the consequent effect on the postsynaptic neuron is an example of *temporal summation*. In this case the several stimuli are distributed over time to produce a larger EPSP than a single stimulus.

REFRACTORY PERIOD   A postsynaptic neuron may be stimulated experimentally at a very rapid rate by controlled electric stimuli. The neuron responds by initiating impulses at a rate that corresponds to that of the stimulus. There is an upper limit, however, to the rate at which successive nerve impulses may be conducted. As the rate of electric stimuli is increased, the rate of nerve impulses fails to increase beyond a certain amount. The passage of the nerve impulse requires time for the depolarization characteristic of the action potential. For 0.05 ms after transmitting an impulse, the axon is unable to conduct another impulse. This is referred to as the *absolute refractory period*.

The membrane potential of the postsynaptic neuron is gradually restored during a *relative refractory period*. During this period the response of the neuron to repeated stimulation is below normal. Stronger stimuli are required to initiate an action potential, and potentials may develop slowly and be weaker than usual. At the end of the relative refractory period the neuron's resting membrane potential is restored sufficiently to conduct impulses in a normal fashion.

Although the refractory period places a limit on the frequency at which nerve impulses can be conducted, this frequency is well above that required for normal physiological needs.

SYNAPTIC DELAY   If a nerve impulse is transmitted across a synapse, there is a synaptic delay of about 0.05 ms. This small time lapse accounts for events occurring at the synapse. These include the release of the transmitter substance, its diffusion into the synaptic cleft, and time for its effect to be produced on the postsynaptic neuron.

FATIGUE   Nerve tissue shows a very rapid recovery after the passage of a nerve impulse, and it is difficult to demonstrate fatigue (exhaustion of energy reserves) in neurons. However, the synapse *is* readily susceptible to fatigue. Continued stimulation of a presynaptic neuron may very well exhaust its supply of neurotransmitter. When this happens, the response of the postsynaptic neuron becomes weak and ineffective. Synaptic fatigue may protect the nervous system. If certain neurons are overstimulated, exhaustion of the synaptic transmitter substance may bring fatigue.

SPEED OF IMPULSE TRANSMISSION   The impulse that travels along a nerve was once thought to be an electric current. It was therefore assumed that it would move along a nerve at the same speed that electricity is conducted by a wire. Later research indicated, however, that the nerve impulse travels much more slowly. In myelinated nerves of mammals it travels at about the same speed as a bullet from a revolver, the maximum velocity being about 90 to 100 m/s. The speed of the nerve impulse is much slower in cold-blooded animals. In the frog, at room temperature, the maximum velocity is about 30 m/s. Conduction rates vary with the temperature, the kind of animal involved, the diameter of the fiber, and whether the fiber is myelinated or nonmyelinated. In general, the largest fibers record the highest conduction rates and the greatest action potential. Conversely, smaller fibers record slower conduction rates and lower potentials. Conduction in a myelinated fiber is about 10 times as rapid as in a nonmyelinated fiber of the same diameter because the action potential jumps from one node of Ranvier to the next. Even within a single nerve, a cathode-ray oscillograph reveals several different action potentials, for the nerve is composed of neuron axons of various sizes, and the rate of conduction varies with the size of the fiber.

CONVERGENCE OF NERVE IMPULSES   Sensory receptors initiate sensory nerve impulses, which travel in toward the central nervous system, a pathway containing many individual neurons. Such impulses converge on centers in the brain. Motor impulses arise from various parts of the brain. When these impulses are directed toward a muscle, they too travel through a pathway that passes through ventral root neurons of the spinal cord to cause contraction of muscles. The spinal motor neuron that passes through the ventral root of the spinal cord represents the *final common pathway* over which all impulses must pass to reach the muscle. Figure 8.10 illustrates the convergence of nerve impulses. Since a strong stimulus evokes a more immediate and more vigorous response than a weak stimulus, the convergence of stimuli from various sources may exert a stronger effect on a muscle than the same stimuli would if they arrived through different neurons. An example is the tensing of muscles in reaction to pain. It may be difficult to visualize a neural network with thousands of presynaptic terminals converging on a few postsynaptic cell bodies, but this is a common pattern of convergence in the central nervous system. The ability of postsynaptic neurons to react to a summation of EPSPs makes convergence a powerful organizing principle in the nervous system.

convergence

FIGURE 8·10
Convergence.

DIVERGENCE OF NERVE IMPULSES Instead of following a final common pathway, a neuron may make connections through its terminal filaments with many muscle fibers, a neuron in the brain, in another example, may synapse with many other neurons. The number of synaptic connections that can be made in this way are almost infinite. This pattern of neural organization is called *divergence* and means a spreading of impulses and information (Fig. 8.11).

FIGURE 8·11
Divergence.

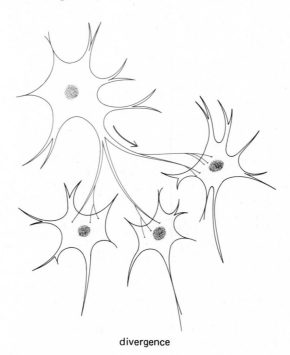

divergence

NERVES

A neuron is a single nerve cell. The term "nerve" applies to *bundles* of sensory or motor neuron axons. These bundles of axons are separated from each other and held in place in the nerve by connective tissue, which also contains fat cells. Capillaries of blood vessels are present also. A cross section of a nerve is shown in Fig. 8.12. Dense fibrous connective tissue, the *perineurium,* surrounds each nerve bundle, or fascicle. Individual neurons are embedded in a loose fibrous connective tissue that contains reticular fibers and is called the *endoneurium.*

Large peripheral nerves contain neurons and their axons, which may be of varying diameters. The axons may be both myelinated and unmyelinated. Smaller nerves may be composed entirely of myelinated axons, or they may be entirely unmyelinated. When a nerve branches, the axons of certain neurons are directed into each nerve branch. Though some nerves may contain only motor neurons or only sensory neurons, most nerves are composed of both motor and sensory neurons and are called *mixed nerves.*

DEGENERATION AND REGENERATION OF NERVES   Portions of neurons cannot survive if the fibers are separated from their cell bodies. Therefore, if a nerve is cut, these axis fibers degenerate. The neurilemma surrounding the severed fibers undergoes extensive degenerative changes. For regeneration to occur, the developing fiber must grow from the cell body into the old neurilemma of the severed portion. The neurilemma appears to be an essential factor in regeneration of nerve fibers. Regeneration does not occur in the nerve tracts of the brain or spinal cord, but peripheral nerves may regenerate at least partially if the neurilemma is not completely severed.

Cell bodies sometimes fail to survive if their axon is injured. Since mature nerve cells are unable to divide by mitosis, neurons cannot be replaced. A fairly large number of brain cells apparently die during the normal lifetime of an individual. In old people, the cumulative loss may eventually impair motor and sensory efficiency.

FIGURE 8·12

A portion of a nerve in cross section, greatly enlarged.

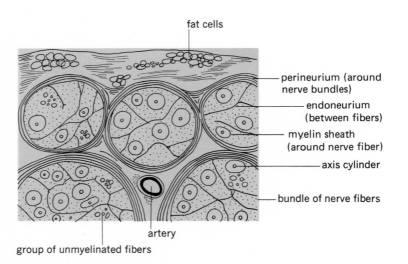

fat cells

perineurium (around nerve bundles)

endoneurium (between fibers)

myelin sheath (around nerve fiber)

axis cylinder

bundle of nerve fibers

artery

group of unmyelinated fibers

NERVE GROWTH FACTOR The *nerve growth factor* (NGF) hormone was discovered in 1948, but most information about it has emerged in the last 10 years. NGF is a protein, related in structure to insulin. It is produced primarily in the submandibular (submaxillary) salivary glands. Its production is regulated by circulating levels of male sex hormones, the androgens, which are present in significant levels in all individuals, male and female. In the developing embryonic nervous system, an excess of NGF produces hyperplasia (increased cell number). In the mature nervous system, an excess of NGF has no effect on cell number; however, it does cause hypertrophy (increased cell size) of existing cells. Normal amounts of NGF are necessary for the proper development and maintenance of the nervous system. The hormone binds to specific receptor sites found on nerve cell membrane surfaces. It stimulates both protein synthesis and energy-yielding processes, including an effect on the production of *tubulin.* NGF may, therefore, play a role in maintenance of cell body microtubules and, thus, cell structure.

ARRANGEMENT OF SPINAL CORD NERVE TRACTS: WHITE AND GRAY MATTER

A cross section of the spinal cord reveals an interior portion of gray matter shaped roughly like the letter H, surrounded by white matter (Fig. 8.13). The white matter is composed of the axonal fibers of neurons. Nearly all are myelinated, but these neurons have no neurilemma. They are grouped into great columns, which are concerned with the transmission of nerve impulses either up or down the cord. Some columns are ascending pathways; others are descending pathways. A cross section reveals that the cord is oval in outline with a dorsal median septum and a ventral median fissure. Although the gray substance of the cord is usually portrayed in cross section, it is important to realize that it consists of several columns composed largely of the cell bodies of neurons and nonmyelinated fibers. The gray matter also contains *interneurons,* or connecting neurons, which form synaptic connections between the dorsal and ventral roots or between the right and left

FIGURE 8·13
Cross section of the spinal cord, illustrating the essential structures of the reflex arc.

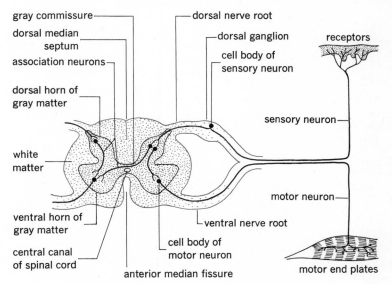

sides of the cord. Many of these neurons make connections vertically between different levels of the cord. The slender dorsal columns of gray matter contain numerous fibers of the dorsal spinal nerves. The cell bodies of some of the afferent ascending neurons are located there also; their fibers travel in the white matter of the cord.

SPINAL NERVE STRUCTURE

Each spinal nerve has two roots close to the spinal cord: a *dorsal afferent,* or sensory, *root* and a *ventral efferent,* or motor, *root* (Figs. 8.14 and 8.15). Diagrams usually portray spinal nerve roots together with a cross section of the cord. In such diagrams the motor roots usually are shown in a ventral position. To orient yourself, think of the section as being taken from an individual lying on his stomach. If the section is taken from the spinal cord in a vertical position, then the ventral motor root is anterior.

The dorsal afferent root is composed of sensory neurons, which bring impulses in toward the central nervous system. The cell bodies of these neurons are located outside the cord in the ganglion of the dorsal root (Fig 8.14).

A *ganglion* consists of a group of nerve cells usually located outside the brain or spinal cord. However, the basal ganglia are located in the basal portion of the brain itself. Afferent cell bodies of most ganglion neurons arise from embryonic neural crest cells as bipolar nerve cells. The two processes later become fused at the cell body to form a unipolar or pseudo-unipolar cell, and the single process divides into two branches. One fiber grows outward to a sensory receptor located, for example, in the skin, in a muscle, or in a tendon. The other fiber grows inward toward the cord through the dorsal root of the spinal nerve. The fiber entering the cord may not make an immediate connection with an interneuron or motor neuron. Certain of these fibers may extend to upper or lower levels of the cord before making these connections.

The cell bodies of the neurons composing the ventral root are located in the ventral columns of gray matter of the cord. Their fibers extend outward, usually terminating in a muscle or gland.

The sensory and motor roots merge as they pass outward through openings between the vertebrae, forming a *mixed nerve* (one in which there are both sensory and motor fibers). All spinal nerves are mixed nerves.

Spinal nerves divide into two large branches just outside the vertebral column. The dorsal branches innervate the muscles and skin of the posterior portions of the neck and trunk. The ventral branches are distributed to the ventral part of the body wall and to the extremities.

FIGURE 8·14
Longitudinal section through a dorsal root ganglion of the spinal cord.

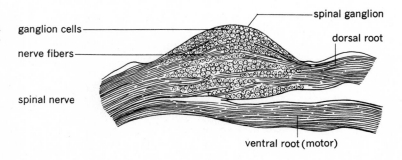

ganglion cells

nerve fibers

spinal nerve

spinal ganglion

dorsal root

ventral root (motor)

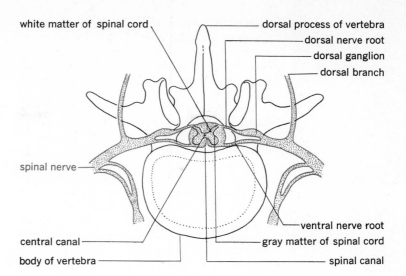

white matter of spinal cord
dorsal process of vertebra
dorsal nerve root
dorsal ganglion
dorsal branch

spinal nerve

ventral nerve root
gray matter of spinal cord
central canal
body of vertebra
spinal canal

FIGURE 8·15

Cross section of the spinal cord within a thoracic vertebra.

REFLEX ACTIVITY

The spinal nerves and the spinal cord form a connecting pathway for sensory and motor nervous impulses, which are coordinated into *reflex movements*. The sudden retraction of the hand or foot following a pain stimulus, the protective closure of the eyelids when an object comes close to the eye, or the coordinated muscular movements that enable us to keep our balance are common examples of reflex movements. We are usually more or less aware of reflex movements involving the voluntary muscles, but there are other reflexes, involving smooth muscles, that are not brought to our attention. Stimuli may affect the heart, stomach, blood vessels, or any other organ reflexly so that automatic adjustments to varying conditions may be made constantly.

THE REFLEX ARC   Simple reflexes are controlled by a functional unit of the nervous system termed the *reflex arc*. The structural elements of a reflex arc include a sensory receptor, a sensory neuron, usually interneurons within the spinal cord, and a motor neuron. Sensory stimuli are received by specialized structures called *receptors*. Sensory receptors are of many kinds and include structures in the skin for the reception of stimuli arising from change in temperature, touch (tactile sense), and pressure, as well as receptors for the special senses located in the eye, ear, and nose and on the tongue. Receptors for the sense of pain are the fine nerve endings themselves.

The sensory neuron carries the nerve impulses inward toward the brain or cord. Association neurons may connect directly with motor neurons on the same side of the cord, or they can make multiple connections with some axons crossing over to the opposite side. The motor neurons carry impulses to muscles. A diagrammatic representation of the structural elements of a reflex arc is shown in Fig. 8.16. Though reflex arcs are commonly described as though only one sensory neuron and one motor neuron were involved, it is doubtful whether any reflex arc is as simple as that. A number of neurons probably participate in a reflex circuit.

Simple reflexes are those that do not involve centers in the brain. Usually such reflex arcs are simple nervous circuits through the cord. The reaction to a pinprick, winking the eye when some object comes near it, and the knee jerk from a sudden blow to the patellar tendon are all simple reflexes. The reaction seems to be almost immediate, but actually conduction is slower over the reflex arc than it is over a single axon of the same length, because in the reflex pathway the nerve impulse is slowed somewhat as it passes across the synapses.

The strength of the stimulus eliciting a reflex response may vary. A weak stimulus, such as touching a hair, may elicit only a moderate response, while a severe pain stimulus may cause a pronounced muscular reaction, involving many muscles. Each sensory neuron is potentially in contact with many motor neurons, and if the stimulus is strong enough, there may be considerable spread or diffusion of the elicited nerve impulse to motor neurons.

The *stretch reflex,* or *myotatic reflex,* is produced by a simple reflex arc, yet this reflex is a very important postural act. In skeletal muscles some specialized muscle fibers have a *proprioceptive,* or internal feedback, function. Such fibers are sensory receptors that interpret the degree of stretch in muscles. They are mechanoreceptors termed *muscle spindles.* The stretch reflex arc involves only the receptor in the muscle spindle, an afferent sensory neuron, and a motor neuron that returns the impulse to the same muscle from which it originated. The knee jerk (Fig. 8.16) is often cited as a common example. When the patellar tendon is tapped, the *quadriceps femoris* muscle is stretched. The stretching of the muscle stimulates a proprioceptive receptor, touching off an afferent nerve impulse. When the impulse returns to the muscle, contraction occurs, and the leg jerks forward.

The stretch reflex is an example of negative feedback. It represents a controlled balance between strength of the motor stimulus, which determines the extent of contraction, and afferent sensitivity, which limits the degree of contraction. The net result of this mechanism is to enable the muscle to maintain a constant length and acquire greater smoothness in its operation.

The smooth operation of postural muscles is much more complex than it appears at first, and stretch reflexes are quite important in the regulation of activity of these muscles. The extensor muscles of the legs may appear to be the important ones for standing, but the leg flexors are also involved and are partially contracted. The muscles of the trunk and legs are in a "fixed" state of contraction, which is maintained to work against the force of gravity. Somewhat different afferent pathways involving interneurons stimulate antagonistic muscles to operate smoothly in a reciprocal fashion.

Reciprocal innervation permits one muscle to contract while its antagonist is inhibited, as seen, for example, in the leg muscles during walking. Resistance to stretch prevents undue extension of muscles. Many reflex stimuli eventually reach the level of consciousness. After we have jerked a finger away from a hot iron, we are aware of pain, and we may consider how badly the finger is burned. When we consciously direct our muscular movements, we have involved the highest level of the brain, the cerebral cortex.

CONDITIONED REFLEXES   Many reflexes are conditioned by experience. The early work of Pavlov, the Russian physiologist, demonstrated that the salivary response in dogs can be conditioned to occur following various stimuli, such as the ringing of a bell, the flash of a light, or the sight of

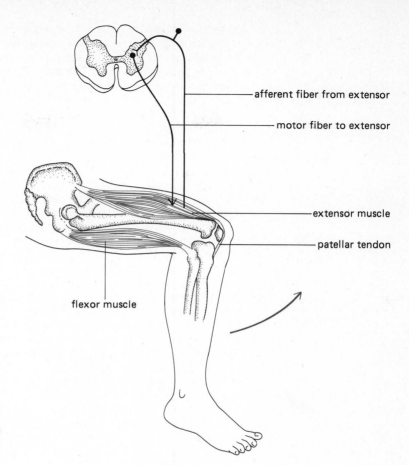

afferent fiber from extensor

motor fiber to extensor

extensor muscle

patellar tendon

flexor muscle

FIGURE 8·16
The patellar tendon reflex (knee jerk). The patellar tendon of the extensor muscle is attached to the tibia below the knee.

objects of different sizes and shapes. If a bell is rung as the dog is fed, the animal soon associates the ringing of the bell with the presence of food. Since salivation is a normal reaction associated with hunger and anticipation of eating, the dog can be conditioned so that its salivation will flow at the sound of the bell.

The conditioned reflex may be the basis for the formation of certain habitual responses. The voiding of urine by babies and young children is controlled, for example, by a spinal reflex arc. When the urinary bladder becomes distended to a certain degree, a reflex arc that results in the emptying of the bladder is set off. As the child matures, however, he learns to control voiding. Cerebral control is superimposed over the simple reflex, which becomes a conditioned reflex. It appears that simple reflexes may be conditioned by experience; these experiences, good or bad, may be made the basis of habit formation.

SUMMARY

1 The structural and functional unit of the nervous system is the neuron. Structural parts of a neuron are the cell body, dendrites, axon, and terminal filaments, or telodendria. The fiber may be myelinated, that is,

have a sheath derived from the Schwann cell membrane coiled around the fiber. The Schwann cell terminates at the node of Ranvier, so that there is a Schwann cell with its myelin sheath between nodes.

2 The synapse is the area where the enlarged ends of terminal filaments come into close contact with the dendrites and cell bodies of succeeding neurons. A synaptic cleft lies between synaptic knobs or presynaptic terminals and the postsynaptic membrane. The presynaptic terminals contain mitochondria and vesicles. The vesicles are thought to contain chemical neurotransmitter substances. Transmission at the synapse between neurons is similar in most respects to transmission at the myoneural junction.

3 Axons and peripheral processes exhibit an all-or-nothing type of response. If the stimulus produces a threshold potential in the postsynaptic neuron, the impulse is conducted along the axon, and the response will be the same no matter what the strength of the stimulus. The nerve impulse, being self-propagating, must derive its energy from metabolic processes within the cell itself.

4 At the synapse the type of transmitter substance may cause either an excitatory postsynaptic potential (EPSP) or an inhibitory postsynaptic potential (IPSP) to develop. The IPSP causes the postsynaptic neuron's internal potential to move farther to the negative, therefore making depolarization of the neuron more difficult. Both EPSPs and IPSPs may develop a summation effect on the postsynaptic membrane.

5 Neurons exhibit a refractory period during which time they are unable to conduct another impulse. Synaptic fatigue occurs when the supply of stored neurotransmitter is exhausted.

6 Nerves are composed of bundles of neuronal fibers. Most nerves are "mixed" in that they contain both sensory and motor fibers. Spinal nerves are composed of the dorsal root sensory fibers and the ventral root motor fibers, so named for their position of entry and exit from the spinal cord.

7 Neurons are classified as motor, or efferent, neurons; sensory, or afferent, neurons; and interneurons.

8 A simple reflex arc usually includes the sensory receptor; the sensory, or afferent, neuron; association neurons within the spinal cord; and the motor, or efferent, neuron. The stretch reflex does not involve interneurons, however. A conditioned reflex is one that is conditioned by experience.

QUESTIONS

1 What are some of the functions of the nervous system?
2 Why is there a ganglion on the dorsal sensory root of a spinal nerve but not on the ventral motor root?
3 Distinguish between a neuron and a nerve.
4 Describe the development of the myelin sheath.
5 Describe the presynaptic axonal terminal and discuss transmission of a nerve impulse across the synaptic cleft. Why does the impulse move in only one direction? What is meant by spatial and temporal summation?
6 Outline the pathway of a simple reflex arc.
7 A man is slightly injured. He makes an involuntary exclamation and steps back, away from danger. Trace the nerve impulses involved in these reflex activities over various pathways.

8 Explain how a reflex act may be conditioned.
9 Is there an upper limit to the rate at which nerve impulses can be conducted? Why?
10 Give an example of the principle of convergence of nerve impulses. Give an example of divergence.
11 Discuss the physiological mechanisms that produce an excitatory postsynaptic potential (EPSP). How is an inhibitory postsynaptic potential (IPSP) developed? What is the physiological advantage of an IPSP?
12 Why do myelinated fibers conduct nerve impulses at a greater velocity than nonmyelinated fibers?
13 What are some of the important functions of the synapse? How is fatigue in the nervous system related to the synapse?

SUGGESTED READING

Axelrod, J.: Neurotransmitters, *Sci. Am.,* **230:**59–71 (1974).

Bradshaw, R. A., R. Hogue-Angeletti, and W. A. Frazier: Nerve Growth Factor and Insulin: Evidence of Similarities in Structure, Function, and Mechanism of Action, *Recent Prog. in Horm. Res.,* **30:**575–595 (1974).

Bray, Dennis: The fibrillar proteins of nerve cells, *Endeavour,* **33:**131–136 (1974).

DeRobertis, E. D. P.: Molecular Biology of Synaptic Receptors, *Science,* **171:**963–971 (1971).

Glunn, I. M., and S. J. D. Karlish: The Sodium Pump, *Annu. Rev. Physiol.,* **37:**13–55 (1975).

Hebb, C.: The Central Nervous System at the Cellular Level: Identity of Transmitter Agents, *Annu. Rev. Physiol.,* **32:**165–192 (1970).

Katz, B.: "Nerve, Muscle and Synapse," McGraw-Hill Book Company, New York, 1966.

Porter, K. R., and M. A. Bonneville: "An Introduction to the Fine Structure of Cells and Tissues," 4th ed., Lea & Febiger, Philadelphia, 1973.

Stevens, C. F.: The Neuron, *Sci. Am.,* **241:**54–65 (1979).

9

THE CENTRAL NERVOUS SYSTEM AND THE SPINAL NERVES

CONSIDERATIONS FOR STUDY

In Chap. 9 we shall consider the anatomy and physiology of the central nervous system and the spinal nerves. Topics for study include:

1 The basic divisions of the brain: its anatomical and functional organization
2 The divisions and functions of the cerebrum: the highest center of neural organization
3 The functions of the brainstem and its parts
4 The cranial nerves: their function and organization
5 The control of the spinal nerves: the role of the spinal tracts and spinal nerves in controlling muscular function
6 Clinical aspects of brain and spinal cord function: injuries, disorders, and diseases of the central nervous system

The human brain is an organ of great complexity. It guides our movements, interprets the senses, and, in the greatest achievement of all, enables us to think. It is the seat of consciousness, the guide of our emotions, and the keeper of our memories. The brain and spinal cord make up the *central nervous system*, whereas the cranial nerves and spinal nerves constitute the *peripheral nervous system*, which may be separated into *somatic* and *autonomic divisions*.

276

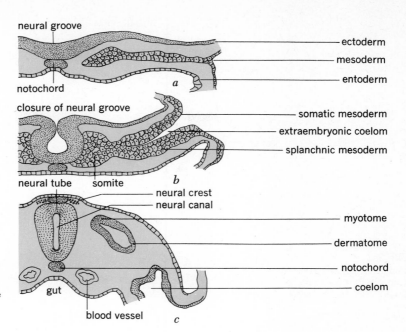

neural groove

ectoderm
mesoderm
entoderm

notochord

a

closure of neural groove

somatic mesoderm
extraembryonic coelom
splanchnic mesoderm

neural tube   somite

b

neural crest
neural canal

myotome

dermatome

notochord

coelom

gut

blood vessel   *c*

FIGURE 9·1

Development of the neural tube and neural crest (transverse section): *a* neural groove open; *b-c* gradual closure to form the neural tube.

EMBRYOLOGY OF THE NERVOUS SYSTEM

Central nervous system
  The brain and spinal cord
Peripheral nervous system
  Somatic nervous system
  Autonomic nervous system

The nervous system is first seen in the embryo as a median depression in the dorsal ectoderm called the *neural groove.* As the groove deepens, the sides grow up and fuse dorsally, forming a long, hollow neural tube that will become the brain and spinal cord (Fig. 9.1). The walls of this tube are filled with nerve cell bodies and fibers. While it is forming, there is a differentiation of the dorsal edge of the fold on either side into the neural crests. The nerve cell bodies of the crest later become the dorsal ganglia of the spinal nerves.

    The anterior end of the neural tube enlarges and later is constricted to form the three primary divisions of the brain: *forebrain, midbrain,* and *hindbrain* (Figs. 9.2 and 9.3). The posterior part of the neural tube becomes the spinal cord. The ventricles, or cavities, of the developing brain are relatively large, but they become proportionately smaller as the

FIGURE 9·2

Diagram illustrating the early development of nervous and digestive systems.

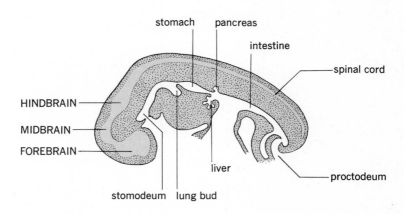

stomach   pancreas

intestine

spinal cord

HINDBRAIN

MIDBRAIN

FOREBRAIN

stomodeum   lung bud

liver

proctodeum

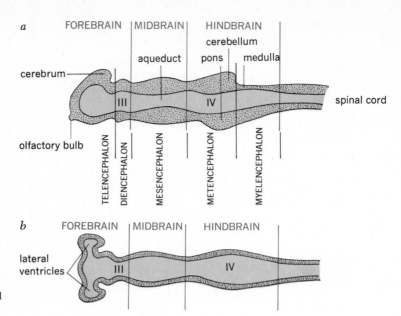

FIGURE 9·3
Diagrammatic representation
of an early embryonic brain:
a sagittal section; *b* horizontal
section.

brain grows larger. The skull forms around the embryonic brain, and the
vertebrae grow around the spinal cord, forming a protecting structure of
bone around this portion of the central nervous system.

BASIC DIVISIONS

The *forebrain,* or *cerebrum,* may be divided into two parts: an anterior part,
the *telencephalon,* and a posterior part, the *diencephalon.* The *midbrain,*
or *mesencephalon,* becomes a part of the *brainstem,* as does the hindbrain.
The hindbrain is also divided into two parts: an anterior part, the
metencephalon, and a posterior part, the *myelencephalon.* The structures

FIGURE 9·4
Diagram showing the principal
structures developed from the
three original divisions of the
brain.

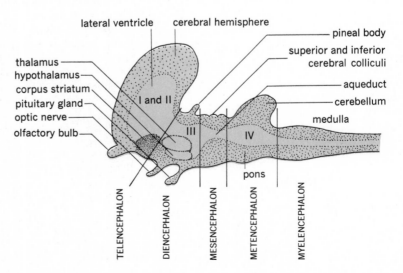

TABLE 9·1
Basic divisions of the brain

Primary divisions	Secondary divisions	Structures
Forebrain (cerebrum)	Telencephalon	Olfactory bulbs Cerebral cortex Lateral ventricles
	Diencephalon	Epithalamus, pineal body Thalamus Hypothalamus Stalk of pituitary gland Mammillary bodies Greater part of third ventricle
Midbrain	Mesencephalon	Colliculi Cerebral aqueduct Cerebral peduncles Substantia nigra Red nuclei
Hindbrain	Metencephalon	Cerebellum Pons Fourth ventricle
	Myelencephalon	Medulla Pyramidal tracts

associated with these various parts of the brain will be taken up in detail later (Fig. 9.4).

Basic divisions of the brain and their principal structures are shown in Table 9.1.

SUPPORT STRUCTURES OF THE CENTRAL NERVOUS SYSTEM

Meninges
  Dura mater
  Arachnoid
  Pia mater

MENINGES The brain and spinal cord are protected by the *meningeal membranes*. These are three in number and consist of a tough, fibrous outer layer, the *dura mater;* a delicate intermediate membrane, the *arachnoid;* and an inner vascular layer, the *pia mater* (Fig. 9.5).

The dura mater within the cranium is composed of two layers, which adhere very closely except as they are separated by venous sinuses. The outer layer fuses with the cranial bones and is termed the *internal periosteum.* The inner, or meningeal, layer of the dura mater lines the vertebral canal, which extends downward over the spinal cord. Here, the epidural space intervenes between the dura and the bony canal, a condition not present in the cranial cavity. The epidural space contains fat tissue and numerous fine veins.

Folds of the meningeal dura form the *falx cerebri,* the *falx cerebelli,* the *tentorium cerebelli,* and the *diaphragma sellae.* The falx cerebri is usually described as a sickle-shaped fold located between the two cerebral lobes in the longitudinal cerebral fissure. The falx cerebelli is a small triangular fold formed between the cerebellar hemispheres. It is attached above to the tentorium cerebelli, which forms a roof over it. The tentorium cerebelli covers the cerebellum and lies under the occipital lobes of the

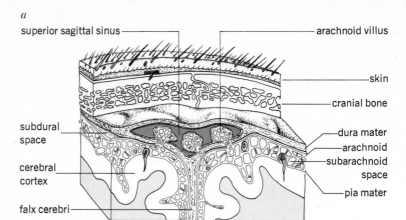

a

superior sagittal sinus

arachnoid villus

skin

cranial bone

subdural space

dura mater

arachnoid

subarachnoid space

cerebral cortex

pia mater

falx cerebri

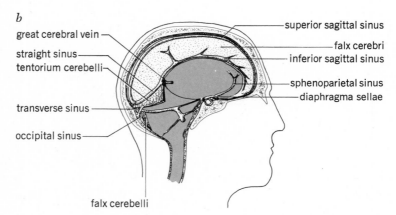

b

great cerebral vein

straight sinus

tentorium cerebelli

transverse sinus

occipital sinus

superior sagittal sinus

falx cerebri

inferior sagittal sinus

sphenoparietal sinus

diaphragma sellae

falx cerebelli

FIGURE 9·5

a A cranial section of the brain showing the meningeal membranes. Note the arachnoid villi. *b* Lateral view illustrating the venous sinuses and the folds of the cranial dura mater.

cerebrum. It joins the falx cerebri along the median line, and together they form a venous sinus called the *straight sinus*. The diaphragma sellae covers the sella turcica and partially covers the hypophysis, or pituitary gland. There is a small circular opening in the diaphragma sellae which permits the stalk of the hypophysis to pass through.

The delicate arachnoid membrane lines the dura mater and extends downward over the spinal cord. It does not ordinarily follow the *sulci* depressions in the brain's surface closely, as does the pia mater, and so there is a *subarachnoid space* at each depression between the convolutions of the brain. The subarachnoid spaces are filled with cerebrospinal fluid, a lymphlike fluid containing a few white cells that protects the brain and cord from mechanical injuries.

The pia mater contains a dense network of blood vessels. It is applied very closely to the brain surface, following down into the sulci and fissures. It is also closely applied to the spinal cord (Fig. 9.6).

VENTRICLES AND CEREBROSPINAL FLUID The *ventricles,* or cavities, of the brain communicate with each other and are continuous with the central

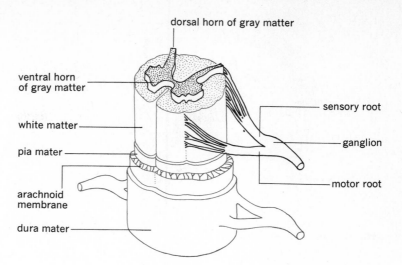

dorsal horn of gray matter

ventral horn
of gray matter

white matter

pia mater

arachnoid
membrane

dura mater

sensory root

ganglion

motor root

FIGURE 9·6

The meninges covering the
spinal cord.

Ventricles
  Lateral
  Third
  Interventricular foramen
  Fourth
  Cerebral aqueduct
  Central canal

canal of the spinal cord. They are lined by a membrane called the *ependyma*. Ependymal cells are derived from embryonic neural-tube epithelium. The ventricles, the central canal of the cord, and the subarachnoid spaces are filled with cerebrospinal fluid. The largest cavities are the *lateral ventricles* (embryonic first and second ventricles) of the cerebral hemispheres. Each lateral ventricle communicates with the third ventricle of the diencephalon by way of an *interventricular foramen* (foramen of Monro). The *third ventricle* connects with the fourth through the *cerebral aqueduct* (aqueduct of Sylvius), which traverses the midbrain. The *fourth ventricle* is continuous with the *central canal* of the spinal cord.

In each ventricle a modification of the ependymal lining covers a specialized area where a network of capillaries from the pia mater projects into the ventricle. These capillaries form the *choroid plexuses,* the source of the cerebrospinal fluid.

The cerebrospinal fluid filters out through the membrane and circulates slowly through the ventricles (Fig. 9.7). Probably the greater amount of fluid is derived from the choroid plexuses of the lateral ventricles. The fluid flows posteriorly through the interventricular foramen into the third ventricle, then through the cerebral aqueduct and into the fourth ventricle. The fourth ventricle has three openings in its roof that permit the fluid to flow into the subarachnoid spaces around the cerebellum and the medulla. There are also two openings at the tip of lateral recesses (foramina of Luschka) and a median dorsal aperture (foramen of Magendie). The fluid then flows slowly down through the subarachnoid space, covering the spinal cord, and also enters the central canal of the cord. The direction of flow in the subarachnoid space of the cord appears to be downward on the posterior or dorsal side and upward on the anterior or ventral side. It flows over the cerebrum and is returned to the blood through *arachnoid villi,* which project into a venous sinus, the *superior sagittal sinus.* A number of factors may influence the circulation of the cerebrospinal fluid: (1) differential blood pressures at the choroid plexuses

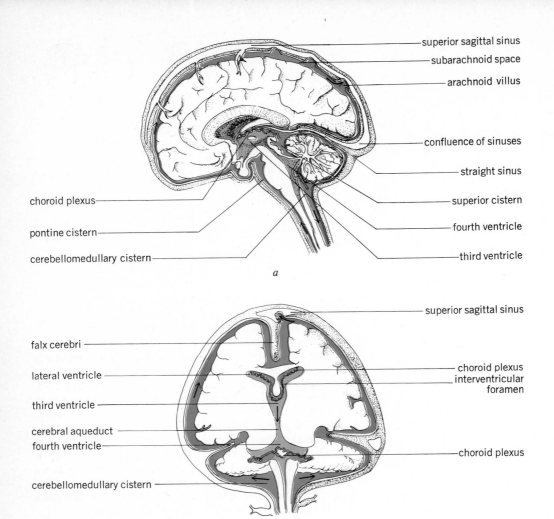

superior sagittal sinus
subarachnoid space
arachnoid villus
confluence of sinuses
straight sinus
superior cistern
fourth ventricle
third ventricle

choroid plexus
pontine cistern
cerebellomedullary cistern

a

superior sagittal sinus

falx cerebri
lateral ventricle
third ventricle
cerebral aqueduct
fourth ventricle
cerebellomedullary cistern

choroid plexus
interventricular foramen
choroid plexus

b

FIGURE 9·7

Cerebral fluid circulation; *a* sagittal section; *b* frontal section.

Neuroglia
  Astrocytes
  Oligodendrocytes
  Microgliocytes

and at the arachnoid villi, (2) osmotic pressure, (3) variation in venous pressure, and (4) changes in body position.

NEUROGLIA   For a long time neuroglial cells, or *neuroglia,* have been regarded as composing the connective tissue of the brain and spinal cord, but there is considerable doubt that they function primarily as supporting tissue. In structure or function they do not closely resemble other types of connective tissue.

There are three principal types of neuroglial cells, namely, *astrocytes, oligodendrocytes,* and *microgliocytes.* These cells vary in structure and function. Astrocytes and oligodendrocytes arise from neural tube cells and are therefore ectodermal in origin, but microgliocytes arise from the mesoderm.

Astrocytes are associated with the neurons and blood vessels of the central nervous system (Fig. 9.8). They form a membrane around the

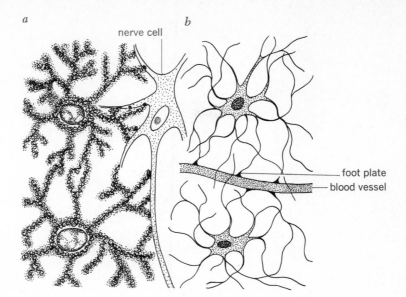

a

b

nerve cell

foot plate
blood vessel

FIGURE 9·8

Two types of neuroglial cells: *a* protoplasmic astrocytes of gray matter; *b* fibrous astrocytes, typical of white matter.

small blood vessels, especially under the pia mater. There are two kinds of astrocytes: protoplasmic and fibrous. The *protoplasmic astrocytes* have larger cytoplasmic processes than the fibrous astrocytes, are more irregularly branched, and are found predominantly in the gray matter. *Fibrous astrocytes* with long thin processes are found for the most part in white matter.

A peculiar feature of both fibrous and protoplasmic astrocytes is the development of enlargements, or *foot plates,* along blood vessels and neurons. Glial cells also form delicate membranes around the blood vessels of the brain and spinal cord, thus separating blood vessels from nervous tissue. This perivascular glial membrane was once thought to be the structural element of the blood-brain barrier, but recent research seems to indicate that the perivascular barrier is not continuous. The capillary endothelium appears to be the only continuous barrier between the blood and the neurons. Astrocytic glial cells exhibit an action potential with a duration many times longer than that of nerve cells. When stimulated electrically, they react with a slow contraction that lasts several minutes.

Fibrous astrocytes form glial scar tissue after an injury to the brain or spinal cord. Neuroglial cells are commonly involved in primary tumors of the central nervous system. Such tumors are called *gliomas.*

Oligodendrocytes are found abundantly in the central nervous system (Fig. 9.9*a*). They are smaller than astrocytes, and ordinarily only their small round nuclei are seen. Cells of this type are commonly clustered around the large cell bodies of neurons, where they are called *satellite cells.* In the white matter of the central nervous system, oligodendrocytes are found in rows along nerve fibers; they function somewhat similarly to the Schwann cells of the peripheral nervous system, by producing myelin. They are also associated with capillaries of the central nervous system.

In general, there are about 10 times as many glial cells as there are

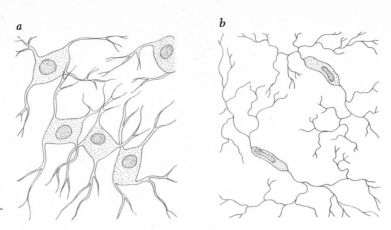

a   b

FIGURE 9·9

Neuroglial cells: *a* oligodendrocytes; *b* microgliocytes.

neurons, but the physiological functions of neuroglia are not well defined. There are indications that some may regulate the neuron's nutrition, and it has also been proposed that some have an excretory function. They seem to stick to nerve tissue, and this has given rise to the name *glial* (from the Greek, glue).

Microgliocytes have very small, irregularly shaped cell bodies with two or more processes (Fig. 9.9*b*). The processes are finely branched with short, spinelike projections. They have no foot plates and are active, motile, phagocytic cells somewhat like the reticuloendothelial cells of some organs of the body.

CEREBRUM

Cerebrum (forebrain)
    Telencephalon
    Diencephalon

Telencephalon
    Cerebral cortex
    White matter
    Lateral ventricle

The human *cerebrum*, or *forebrain*, has had a long developmental history. From meager beginnings in the lower vetebrates, it has developed into a large, dominant part of the brain. The cerebrum's connections to other parts of the brain are shown in Fig. 9.10.

TELENCEPHALON   The cerebrum grows until it completely covers the brainstem dorsally. The *telencephalon* comprises the two *cerebral* hemispheres. The two hemispheres are separated medially by the great longitudinal fissure. *Fissures* are deep depressions on the brain surface. The numerous lesser depressions of the surface are called *sulci*. These separate the elevations, which are termed *convolutions*. The surface of the cerebrum has a covering, or *cerebral cortex*, of gray matter, which is composed of millions of unmyelinated nerve cell somas and axons.

Beneath the cortex the inside of the brain appears white because this area of the brain is made up largely of myelinated axons. These fibers usually lie in tracts. Some connect the cortex with the spinal cord, while others extend between different parts of the brain itself. The axons that comprise the white matter are usually classified as *association fibers, projection fibers,* and *commissural fibers.* Association fibers connect convolutions, or gyri, within the same hemisphere, whereas projection fibers extend from the cortex to lower structures in the central nervous system. Projection fibers may carry efferent impulses to lower motor neurons, or they may be afferent fibers traveling from the brainstem to sensory areas of the cortex.

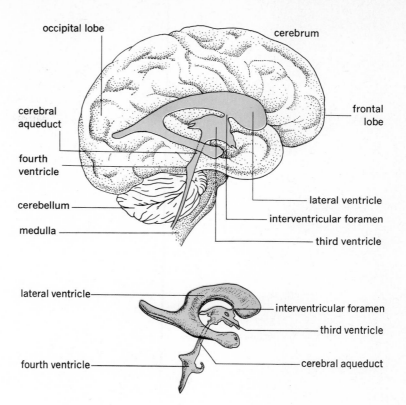

occipital lobe

cerebrum

cerebral aqueduct

fourth ventricle

cerebellum

medulla

frontal lobe

lateral ventricle

interventricular foramen

third ventricle

lateral ventricle

interventricular foramen

third ventricle

fourth ventricle

cerebral aqueduct

FIGURE 9·10

The cerebrum with ventricles of the brain in lateral view.

Commissural fibers extend between corresponding parts of the two hemispheres. The largest *commisssure* connecting the right and left hemispheres of the cerebrum is called the *corpus callosum*. The fiber bundles of this commissure surround the lateral ventricles.

Nervous activity within the cerebral cortex results in conscious thought. It provides for the higher intelligence of humans and for all that may be inferred under such general terms as reasoning ability, good judgment, memory, and willpower. We become aware of the outside world through the interpretations of the special senses of sight, hearing, taste, smell, and touch. We initiate voluntary muscular responses and direct our voluntary body movements through the activity of the cerebral cortex. Certain emotions—feelings of charity, appreciation of beauty, a desire to do right—are thought to be functions of the cerebrum; more primitive emotions such as rage may be, in part, the expression of activity of a more primitive part of the brain, the hypothalamus. Many acts that are ordinarily reflex may be dominated or controlled by the cerebrum, as when we cough voluntarily or hold our breath for a time.

Lobes
Frontal
Parietal
Temporal
Occipital
(Insula)

LOBES OF THE CEREBRAL CORTEX Fissures divide the cerebral cortex into anatomical areas called *lobes*. They are the frontal, parietal, temporal, and occipital lobes (Fig. 9.11*a*). The *insula* is sometimes considered a lobe. The lobes, with the exception of the insula, lie under cranial bones of the same name. Thus the *frontal lobe* lies below the frontal bone and is

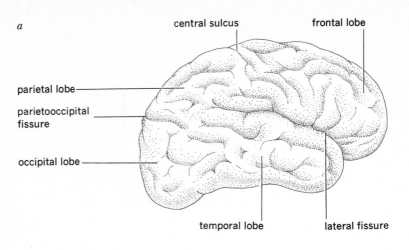

a

central sulcus   frontal lobe

parietal lobe

parietooccipital
fissure

occipital lobe

temporal lobe   lateral fissure

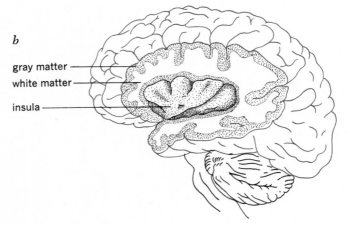

b

gray matter

white matter

insula

FIGURE 9·11

a The lobes of the cerebrum
in lateral view. *b* Section of
the left cerebral hemisphere
cut away to expose the insula,
an area of gray matter covering the corpus striatum.

separated from the parietal lobe by the *central sulcus* (the Rolandic
fissure). The *parietal lobe* is therefore posterior to the frontal lobe. The
temporal lobe lies on the side of the brain below the *lateral cerebral fissure*
(the Sylvian fissure). The *occipital lobe* is demarcated by the parietooccipital
fissure. The *insula* is an area of gray matter located under the
lateral cerebral fissure and can be seen only when the adjacent portions of
the frontal and temporal lobes are raised (Fig. 9.11*b*).

LOCALIZATION OF FUNCTION It is not possible to localize many functions of
the cerebrum. No one can point out the area that governs willpower or to
any one part of the brain that is the seat of music appreciation. Still, a
number of cerebral functions are thought to originate in localized areas.

Motor areas
  Primary
  Broca's
  Precentral

Motor areas The *motor*, or efferent, area of the cortex lies just anterior to
the central area (Fig. 9.12). Motor impulses arise in the neurons of this
area. These are voluntary impulses that cause contractions of voluntary
muscles. It can be demonstrated experimentally that when certain parts of
the motor area are stimulated, certain groups of muscles or individual

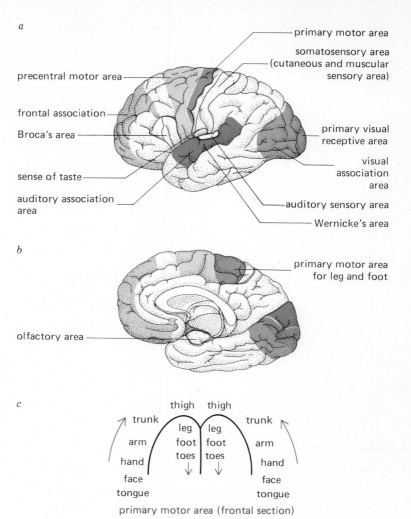

a

primary motor area

somatosensory area
(cutaneous and muscular
sensory area)

precentral motor area

frontal association

Broca's area

primary visual
receptive area

visual
association
area

sense of taste

auditory association
area

auditory sensory area

Wernicke's area

b

primary motor area
for leg and foot

olfactory area

c

thigh   thigh

trunk   trunk
        leg   leg
arm   foot foot   arm
        toes toes
hand   hand

face   face

tongue   tongue

primary motor area (frontal section)

FIGURE 9·12

Localized functional areas of the cerebral cortex: *a* left cerebral hemisphere, lateral view; *b* medial view; *c* the primary motor area, frontal section. The labels indicate the part of the body governed from motor areas that lie between the cerebral hemispheres and laterally along the central sulcus. The lateral motor area of the brain is inverted with respect to the body.

muscles respond. The area governing the thigh lies at the upper part of the motor area, while the muscles of the face and tongue are controlled from the lower part. It is interesting that a proportionately greater area is devoted to governing the muscles of the hand and tongue than to the muscles of the trunk. The motor speech area is concerned with the coordination of all muscles of the face, tongue, and throat necessary for speaking. It is commonly called *Broca's area*. In right-handed persons, the speech area is developed in the left hemisphere.

Not only is the motor area inverted with respect to the body part it controls, but the motor area of the right cerebral hemisphere governs the movement of muscles of the left side of the body and vice versa. The reason for this is that the nerve tracts from the motor area cross over either at the base of the medulla or at various levels in the spinal cord (Fig. 9.13). In right-handed persons, it would appear that the coordinating centers of the left cerebral lobe are somewhat better developed.

A knowledge of the location and function of the motor area has been of

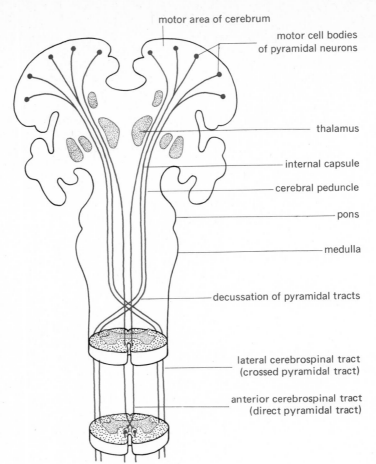

motor area of cerebrum

motor cell bodies
of pyramidal neurons

thalamus

internal capsule

cerebral peduncle

pons

medulla

decussation of pyramidal tracts

lateral cerebrospinal tract
(crossed pyramidal tract)

anterior cerebrospinal tract
(direct pyramidal tract)

FIGURE 9·13

Diagram illustrating the crossing of the descending pyramidal tracts. Neurons of the direct pyramidal tract may cross at any spinal level, where they synapse with appropriate peripheral motor neurons.

great value in the diagnosis of many types of brain injuries. Motor paralysis of a given part can be caused by injury or pressure involving the motor area of the brain. *Cerebral hemorrhage,* often referred to as a "stroke" or *apoplexy,* which usually occurs deep within the brain, has the same effect as a surface injury. For example, if a blood clot located below the motor area of the right cerebral hemisphere prevents nervous impulses from reaching the muscles, a motor paralysis of the left side of the body results.

The precentral motor area is not as specific as we are often led to believe. The instrument ordinarily used for experimental exploration is a stimulating electrode. If the upper part of the motor area is stimulated, muscular movements occur in the lower extremity on the opposite side of the body. But if we stimulate the middle or lower part of the motor area with adequate stimuli, lower-extremity movements occur also. It is found that in the area for trunk movement, the muscles of the trunk have the lowest threshold, that is, are most easily stimulated, but, evidently, various regions of the precentral motor area are not specific.

It is generally agreed that the cortex is composed of five to seven layers of neurons, but the nonspecific effect of cortical stimulation apparently is not caused by diffusion of impulses at the cortical level. Electrical stimulation of the surface area can produce very different motor effects from those produced by normal nerve stimuli originating in this area. The sequence of neurons or pattern of distribution may be very different in the normal functioning organism.

Stimulation of the cortex of unanesthetized monkeys with implanted electrodes elicits motor effects in all areas of the cortex. Such experimentation suggests that probably all areas of the cortex are actually *sensorimotor;* that is, they are not exclusively sensory or exclusively motor. The region directly anterior to the central sulcus is predominantly motor, or efferent, whereas the region posterior to the central sulcus is predominantly sensory, or afferent. Even areas such as the visual cortex contain motor fibers, however. Motor fibers in localized sensory areas produce a motor response in muscles associated with the activity of the sense organ involved. For example, stimulation in the visual area may evoke eye and head movements, but only the precentral motor area controls movement of the body as a whole.

In the classical scheme, voluntary movement is said to originate in the motor area of the cerebrum. Recent experimental evidence makes us consider the idea that other areas influence the motor area and perhaps some direct motor impulses normally originate outside the motor area. The motor cortex receives sensory stimuli from a wide variety of receptors, as well as from the eyes, ears, and a great many other sources. Since coordination plays such an important part in all bodily movement, lower subcortical coordinating areas may be largely responsible for routine movement. Consider that, in walking, the position of the leg muscles varies from second to second. To maintain essential muscular coordination, sensory feedback impulses must be sent continuously from the muscles to the brain. Even cortical zones formerly considered remote from the motor area, such as the parietal and occipital zones, may have some influence on spatial orientation. The premotor area neurons coordinate certain muscular activity, and the frontal zone plays some part in programming movements.

Primary sensory areas
  Cutaneous
  Visual
  Auditory
  Olfactory

Primary sensory areas Posterior to the central sulcus lies a sensory area where interpretations such as touch, pain, temperature, pressure, and muscle sense are made. This area is called the *cutaneous sensory* (somaesthetic) *area.* It is thought that the subdivisions of this area correspond approximately to those of the motor area just anterior to the central sulcus. The right sensory area interprets sensations received from the left side of the body and vice versa.

The *visual interpreting area* is located at the back of the brain in the occipital lobes. Nervous impulses arising in the retina are conveyed to the interpreting area by nerve tracts. This sensory area enables us to interpret what we see. A closely associated function is the ability to understand written words or symbols. When this function fails, the individual is said to suffer from a type of *aphasia* in which the patient is unable to read language. The patient's eyes may be able to see and follow the printed words, but the ability to interpret the symbols is lost.

The *auditory areas* are concerned with the interpretation of the sense of hearing and are located in the temporal lobes. Each area receives nervous

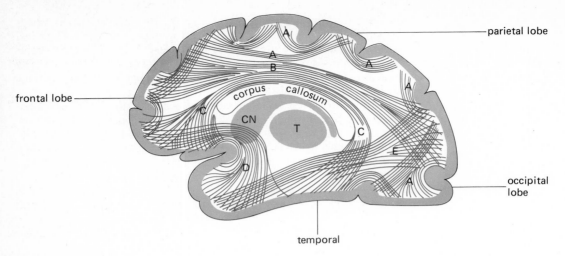

parietal lobe

frontal lobe

corpus callosum

occipital lobe

temporal

FIGURE 9·14
Association fibers of the brain.
(A) Short fibers, (B) superior
longitudinal fasciculus, (C) cin-
gulum, (CN) caudate nucleus,
(D) uncinate fasciculus,
(E) association fibers, and
(T) thalamus.

impulses from both ears. Injury to closely related areas may produce a type of aphasia in which the individual loses understanding of the meaning of words. Although hearing may remain unimpaired, the meaning of words is not recognized. An area located in the temporal lobe close to the auditory area is known as *Wernicke's area.* Injuries to this area result in failure to understand spoken language.

Interpreting areas for the senses of taste and smell have not been definitely located. A cortical area associated with the sense of taste may be located at the base of the central sulcus where sensory impulses from the face and tongue are interpreted.

Located ventrally below the frontal lobes of the cerebrum are the two *olfactory bulbs,* where important synaptic connections are made between the fibers of the olfactory interpreting areas of the brain. The interpreting area for the sense of smell is thought to lie deep within the cerebrum in a region known as the limbic system. This is part of the primitive cortex known as the *rhinencephalon.*

Association areas
   Common integrative

Association areas *Association areas* extend deeply throughout the brain. These are large delicate tracts connecting motor and sensory areas of the cortex (Fig. 9.14). Since such areas do not give specific responses when stimulated, it is difficult to interpret their exact functions, but they are known to link various types of information coming from one region of the brain to those coming from another area. It is believed, therefore, that the association areas enable us to coordinate information: that is, to think and reason. A *common integrative area* located posteriorly to the lower part of the cutaneous sensory area functions in coordinating impulses from all the senses. Such integration of sensory information enables us to react to what we see, hear, taste, smell, or feel.

EMBRYONIC DEVELOPMENT OF THE CEREBRUM   A brief consideration of the development of the embryonic cerebrum may help us understand the relationships between parts of the cortex. The older part, from an evolutionary standpoint, is the *paleocortex,* which includes the *corpus striatum,*

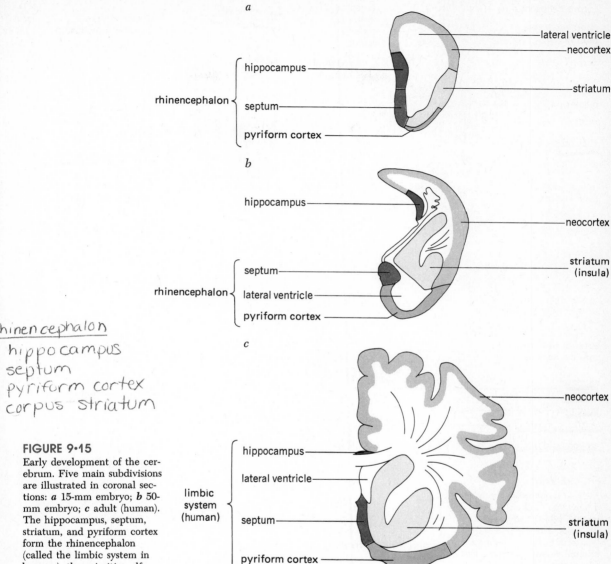

a

rhinencephalon
{
  hippocampus
  septum
  pyriform cortex
}

lateral ventricle
neocortex
striatum

b

hippocampus

rhinencephalon
{
  septum
  lateral ventricle
  pyriform cortex
}

neocortex
striatum (insula)

c

neocortex

limbic system (human)
{
  hippocampus
  lateral ventricle
  septum
  pyriform cortex
}

striatum (insula)

Rhinencephalon
 hippocampus
 septum
 pyriform cortex
 corpus striatum

FIGURE 9·15
Early development of the cerebrum. Five main subdivisions are illustrated in coronal sections: *a* 15-mm embryo; *b* 50-mm embryo; *c* adult (human). The hippocampus, septum, striatum, and pyriform cortex form the rhinencephalon (called the limbic system in humans), the primitive olfactory part of the brain. The striatum and the cortex of gray matter above it form the insula. The older structures of the rhinencephalon become involved in functions other than olfactory in the human being and now form part of the limbic system. (*After W. J. S. Krieg, "Functional Neuroanatomy," 2d ed., McGraw-Hill Book Company, New York, 1953.*)

the *hippocampus*, the *septum*, and the *pyriform cortex* (Fig. 9.15). These four structures form the *rhinencephalon* of more primitive vertebrate brains, which is part of the *limbic system* in man. The corpus striatum and the gray matter above it form the insula. The rhinencephalon was the first part of the cortex to develop in evolution, and it was primarily a center for olfactory interpretation, but in humans it has been superseded by the neocortex, or new cortex, although it still retains some olfactory connections.

The *limbic system* is sometimes expanded to include the thalamic nuclei, the hypothalamus, the amygdaloid nuclei, and several tracts running through the midbrain. It is also referred to by some neuroanatomists as the

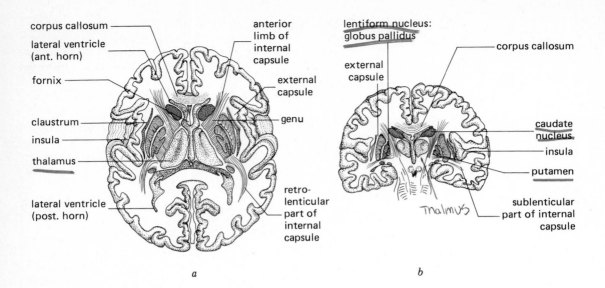

corpus callosum

lateral ventricle
(ant. horn)

fornix

claustrum

insula

thalamus

lateral ventricle
(post. horn)

anterior
limb of
internal
capsule

external
capsule

genu

retro-
lenticular
part of
internal
capsule

a

lentiform nucleus:
globus pallidus

external
capsule

corpus callosum

caudate
nucleus

insula

putamen

sublenticular
part of internal
capsule

Thalmus

b

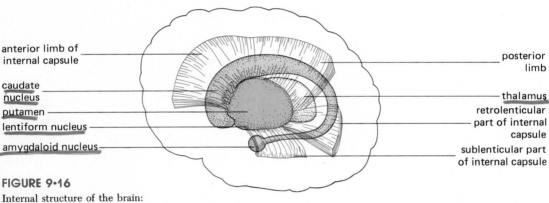

anterior limb of
internal capsule

caudate
nucleus

putamen

lentiform nucleus

amygdaloid nucleus

posterior
limb

thalamus

retrolenticular
part of internal
capsule

sublenticular part
of internal capsule

c

FIGURE 9·16

Internal structure of the brain: *a* horizontal section; *b* coronal section; *c* basal ganglia and thalamus. The caudate and lentiform nuclei, including the internal capsule, compose the corpus striatum. The internal capsule is an important nerve tract, probably best seen in a horizontal section of the brain. It is V-shaped, with anterior and posterior limbs meeting at the apex, which is called the *genu.* The internal capsule is composed of motor and sensory fibers from all parts of the cerebral cortex. Cerebral hemorrhage, often called a "stroke," commonly occurs at the internal capsule.

limbic-midbrain system. In animals, this region has been shown to affect sexual and aggressive behavior.

The phylogenetically newer portion of the brain is the *neocortex*. The more highly developed neocortex has grown posteriorly and has covered over the paleocortex. It includes the motor areas and the sensory association areas, with the exception of the olfactory sense.

The *basal ganglia* comprise areas of gray matter deep within the white matter in the lower part of the telencephalon. They include the *caudate nucleus,* the *putamen,* and the *globus pallidus.* The latter two form the *lentiform* (lens-shaped) *nucleus.* The caudate nucleus has a thickened head portion and a long slender tail, at the end of which is an enlargement called the *amygdala* or amygdaloid body (Fig. 9.16*a, b,* and *c*). Not much is known about the function of the amygdala except that it is closely associated with the hypothalamus. Experimental injury to the amygdala in animals appears to produce emotional responses. The basal ganglia are not easily reached for experimentation without injury to other parts of the brain, so their functions are not clearly understood; but they have elaborate connections with the thalamus, hypothalamus, and reticular system, which we shall discuss later. The basal ganglia are thought to help coordinate movements of skeletal muscles, in walking, for example. There is also some indication that the basal ganglia inhibit impulses from the motor cortex to the skeletal muscles. Malfunction of some of the basal ganglia give rise to the muscular tremors characteristic of Parkinson's disease.

Diencephalon
  Epithalamus
  Thalamus
  Hypothalamus
  Pineal body
  Hypophysis
  Optic chiasma
  Mammillary bodies

THE DIENCEPHALON   The *diencephalon* is the posterior part of the forebrain. The lines of demarcation are not distinct, since many structures lie partly in the telencephalon and partly in the diencephalon (see Fig. 9.4). Thus the choroid plexus of this region is continuous with that of the telencephalon, since the greater part of the third ventricle is the cavity of the diencephalon, but a part of it is the cavity of the telencephalon. The thalamic region may be divided into an upper portion, or *epithalamus;* an intermediate portion, the *thalamus* proper; and a lower portion, the *hypothalamus.* From the epithalamus arises an outgrowth that becomes the *pineal body.* This structure apparently functioned in prehistoric reptiles as a median, light-sensitive third eye; in more modern vertebrates it is usually considered to be an endocrine gland, although its function is not well understood. The pineal body contains an enzyme that catalyzes the reaction in which serotonin becomes melatonin. The most striking example of the action of melatonin is its effect on the melanocytes of the frog, where it causes bleaching of the frog skin. In humans, the pineal body appears to influence body rhythms, or the "biological clock" of physiological activities through an indirect response to light. See Chap. 13 for further discussion of the pineal body.

The *thalamus* is the largest of the number of areas of gray matter deep within the lower part of the brain. It is an important relay center for both motor and sensory impulses and has extensive cortical connections. It has been likened to a telephone switchboard where messages are received and relayed. In its evolutionary development, it has retained synaptic connections with the optic nerves and is sometimes referred to as the optic thalamus. All the senses, however, except olfactory fibers have thalamic connections.

The *hypothalamus* is one of several structures located in the basal portion of the diencephalon. It is situated below the thalamus and close to the base of the stalk (infundibulum) of the hypophysis (pituitary). Various autonomic functions are controlled from centers in the hypothalamus. A body-temperature regulatory center is located here, as is a center concerned with thirst and water metabolism. A feeding and hunger center has also been identified. The hypothalamus is a regulatory center for both the sympathetic and parasympathetic divisions of the autonomic nervous system. "Sham" rage in animals has long been associated with stimulation at the hypothalamic level. The hypothalamus is also closely associated with various endocrine activities of the hypophysis. It produces hormones and releasing factors itself that regulate and control the activities of the hypophysis. These will be considered later in Chap. 13.

In addition to the infundibulum and *hypophysis* the *optic chiasma* and the *mammillary bodies* are visible at this level in a ventral view of the brain. The optic chiasma is formed here by the crossing-over of some of the fibers of the optic nerve. The mammillary bodies do not have clearly defined functions, but since they have synaptic connections with the hypothalamus, their activities may be connected with similar functions.

LEARNING AND MEMORY PROCESSES

Memory is the ability to recall events or information about previous experiences that may have happened only a few minutes ago or many years ago. There seem to be both a short-term memory, which functions in tasks such as remembering a telephone number until one has dialed it, and a long-term memory of facts and events, which may last for the lifetime of the individual.

The cerebral cortex was originally thought to be the seat of learning, thinking, and memory. A few years ago, however, experimental studies proved that injury to the cortex, except for motor and sensory areas, produces little change in behavior. In humans, the language area of the cortex is an exception, since it is easily injured. Language abilities have been located in the left cerebral hemisphere in over 90 percent of the human population. This does not mean that the right hemisphere is not used at all in language. The right hemisphere probably provides some complementary information regarding shape and form, but does not, by itself, enable an individual to write or speak a language. Furthermore, the right hemisphere controls the recognition of objects, faces, and spatial dimensions. It is also concerned with the appreciation of music.

Learning and memory are closely related. Neurons of the brain must be able in some way to record and store information. The neural theory of memory is referred to as *memory trace*. Memory traces may be formed during learning. For example, if a laboratory animal learns how to obtain food or other rewards by associating the reward with certain responses, a memory trace may have been laid down.

Neural circuits in the brain have been proposed as an explanation for memory trace formation. *Neuronal loops* with electrical activity passing around a closed circuit of neurons and synapses may provide a means of storing information for short-term memory. Evidence from individuals who have been under deep anesthesia, in a coma, or who have undergone electroconvulsive shock therapy indicates that these conditions, which temporarily affect brain activity, affect short-term memory but do not suppress long-term memory. The hypothesis that short-term memory is produced by neuronal circuits has gained some credence.

The mechanism of memory, however, is still largely unknown. If this function can be localized, it may possibly be associated with the hippocampus, an area of the *limbic system,* and with the reticular system, but it is more likely, considering the complexity of neuronal connections, that it is not localized.

Another line of investigation concentrates on the nature of conduction across the synapse. Learning may increase the conductivity of a certain group of synapses, but it is more likely that the postsynaptic membrane involved in such synapses may increase their sensitivity to the transmitter substance. Such sensitivity might very well increase during the learning experience and decrease during a period of forgetting. Some very interesting experiments performed by Penfield (see Suggested Reading at the end of this chapter) seem to indicate that memories can be recalled in certain cases by stimulating an area of the temporal lobe below the auditory area. Penfield calls this region the *interpretive cortex*. These experiments have been performed during surgical operations on the brain under local anesthesia.

Electrodes are introduced into this area, and with very mild electrical stimulation, patients sometimes recall past events quite vividly and in great detail. Penfield did not obtain similar results from stimulating any other part of the brain. The memory record may not be stored in the interpretive area itself but in some deeper area related to it.

It is not surprising that many investigators have turned to research on ribonucleic acid (RNA) and protein metabolism in their search for the process that involves memory storage. A molecular theory for long-term memory is an attractive hypothesis but very difficult to establish on an experimental basis. Such theories hold that memory is stored in the configuration of certain molecules, especially RNA and protein. Learning, in itself, may very well involve changes in the structure of the RNA molecule and its metabolism within brain cells, but this concept awaits further investigation.

It is known that brain cells are lost with aging. This loss in itself may eventually reduce mental ability. The aged person sometimes becomes forgetful or is unable to recall recent events although earlier events may be remembered quite well. Other factors related to loss of memory in the elderly may include a reduced blood supply to the brain or a reduction in the oxygenation of brain cells. It should be pointed out that memory loss does not automatically occur with aging, but may be the result of certain pathological processes that affect some individuals.

The fact that some learning and some memories are retained for a lifetime seems to indicate that long-term memory involves some permanent change in the brain. Most body cells change many times during the life of the individual. New neurons do not appear, however, after age six.

If there are permanent changes in the brain, then how does forgetting take place? Possibly the learned impression was not deep enough or was not reinforced by additional information. Possibly we are unable to make the proper associations to induce recall. The mechanisms involved in forgetting are not well understood.

CHEMICAL COMPOUNDS AFFECTING BRAIN FUNCTION

The study of the chemical mechanisms of the brain has assumed a position of great importance in present-day research. Studies of excitatory transmitters, inhibitory transmitters, tranquilizers, energizers, and drugs that produce hallucinations have been in the forefront of investigation. A more recent classification of brain-affecting chemicals includes neuroregulators and neuromodulators in addition to neurotransmitters. In this classification, neuromodulators do not act as transmitters at synapses but are present in physiological fluids and are capable of modifying the activity of neurons. The term "neuroregulator" is applied to both neurotransmitters and neuromodulators.

CHEMICAL TRANSMITTERS Neurotransmitters are chemical substances released by the terminal filaments of axons that bind to specific receptors located on the surface of postsynaptic neurons. Neurotransmitters may be classified as either excitatory or inhibitory. Norepinephrine and acetylcholine are recognized as synaptic neurotransmitters. Acetylcholine is a transmitter that functions at neuromuscular junctions and at the synapses of cholinergic neurons. Norepinephrine is also a brain transmitter.

Serotonin (5-hydroxytryptamine), discovered in 1947, is present in brain tissue. Together with norepinephrine, it is found in considerable amounts in the hypothalamus. Experimentally administered serotonin apparently does not cross the blood-brain barrier, but the amino acid precursor, 5-hydroxytryptophan does. The *substantia nigra* is a layer of gray matter embedded in the cerebral peduncles. In a cross section through the midbrain, this dark-colored layer lies anterior or ventral to the *red nucleus* on either side (Fig. 9.17). The substantia nigra contains a dark pigment, which is probably some form of melanin formed from dopa or dopamine by polymerization.

The tremors of Parkinson's disease appear to involve the extrapyramidal pathways and are relieved in many cases by the administration of L-dopa. Dopamine does not cross the blood-brain barrier, but L-dopa readily enters the brain.

Several neuromodulatory drugs are widely used in the treatment of certain types of mental and emotional illnesses. Such drugs have a quieting effect and relieve some anxiety states and emotional imbalances. These drugs are commonly known as *tranquilizers*. These drugs seem to affect subcortical areas of the brain, such as the midbrain reticular formation, the limbic system, the hypothalamus, and certain nuclei associated with the primitive olfactory area, now thought to be concerned with emotion. Mild tranquilizers (Valium and Librium) apparently depress the limbic system. Stronger tranquilizers, phenothiazines, used in the treatment of psychotic patients depress the activity of catecholamines (norepinephrine, dopamine, serotonin) in the brain, reticular system, and basal ganglia.

The effects of long-term use of tranquilizing drugs are not known. A long period of experimentation will be necessary to determine their value. The human nervous system is well adapted to withstand ordinary emotional emergencies and anxieties. Tranquilizing drugs should not be used routinely to avoid cares and worries that are part of everyday life.

FIGURE 9·17

A section through the midbrain revealing the red nuclei and the substantia nigra. These structures are part of an intrinsic system involving the basal ganglia and the thalamus. They function as an intricate relay system of short neurons linking various basal ganglia and in carrying impulses between the cerebral cortex, basal ganglia, and the spinal cord. [*Redrawn from J. Parsons Schaeffer (ed.), "Morris' Human Anatomy," 12 ed., McGraw-Hill Book Company, New York, 1966.*]

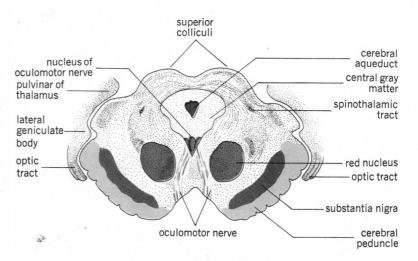

It is difficult to study neurotransmitters in the brain because many of these transmitters have an inhibitory effect. One method is to determine which transmitter affects receptor sites on a postsynaptic neuron. An inhibitory transmitter causes hyperpolarization of the postsynaptic neuron, thus making it more difficult for an excitatory neuron to produce an action potential in the postsynaptic neuron. Gamma-aminobutyric acid (GABA) and glycine appear to be inhibitory transmitters. Dopamine is an inhibitory brain neurotransmitter.

Gamma-aminobutyric acid is considered to be the most important inhibitory neurotransmitter agent of the cerebral cortex (Table 9.2). *Glycine*, an amino acid, on the other hand, is more effective as an inhibitor in the brainstem and spinal cord and has relatively weak effects in the brain.

Dopamine is primarily an inhibitory neurotransmitter in the brain. It is

TABLE 9·2
Neurotransmitters or neuroregulators

Chemical substance	Source of secretion	Activity
Acetylcholine	Neuron endings; myoneural junctions; all preganglionic autonomic and all postganglionic parasympathetic endings. Brain postganglionic fibers to sweat glands. (See autonomic nervous system.)	Causes impulse to be transmitted to skeletal muscle at myoneural junctions and to cholinergic neurons.
Norepinephrine	Most postganglionic sympathetic neuron endings; spinal cord, parts of brain, reticular system, adrenal medulla *pons*	Sympathetic response
Epinephrine	Brain cells, hypothalamus?, adrenal medulla *Cortex*	Sympathomimetic, neural regulator
Dopamine	Brain neurons, limbic system, basal ganglia, hypothalamus	Inhibitory neurotransmitter in the brain. L-dopa affords some relief in Parkinson's disease. Metabolized by monoamine oxidase (MAO)
Serotonin	Brain, brainstem, and spinal cord	Induces sleep; a strong vasoconstrictor; inhibitory neuroregulator in brain. Metabolized by monoamine oxidase (MAO)
GABA Gamma-aminobutyric acid	Brain and spinal cord, purkinje cells of cerebellum	Inhibitory neuroregulator
Glycine	Spinal cord	Inhibitory neuroregulator
Glutamic acid	Brain and spinal cord	Excitatory? neuroregulator
Endorphins enkephalins	Adenohypophysis, brain	Analgesic, may suppress pain response
Substance P	Dorsal sensory neurons and spinal cord	Transmits pain stimuli

amino acids {

Histamine → hypothalmus.

present in the limbic system, basal ganglia, and hypothalamus. In its formation, the amino acid tyrosine is acted upon by the enzyme tyrosine hydroxylase and converted to 3,4-dihydroxyphenylalanine, commonly called *dopa*. Dopa is converted to dopamine by the action of the enzyme dopa decarboxylase.

$$ATP \xrightarrow[\text{cyclase}]{\text{adenylate}} \text{cyclic AMP}$$

$$\text{dopa} \xrightarrow[\text{decarboxylase}]{\text{dopa}} \text{dopamine} \xrightarrow[\beta\text{-oxidase}]{\text{dopamine}} \text{norepinephrine}$$

Dopamine nerve tracts involve the *substantia nigra, the caudate nucleus, and the putamen of the corpus striatum.* The caudate nucleus and putamen are basal ganglia and are included in the extrapyramidal motor system.

Endorphins (endogenous morphine-like substances) are recently discovered peptide transmitters that have an opiate-like effect in suppressing pain perception. Found in the brain and hypophysis and in some peripheral areas, they may represent a built-in mechanism for reducing awareness of pain. Substance P, another peptide, is currently being investigated as a transmitter of pain impulses to the brain. Endorphins may produce their pain-reducing effect by suppressing the release of substance P.

ENERGIZERS   Many drugs function as energizers. Such drugs increase neural activity. Some, such as the amphetamines (dexedrins), have chemical structures that closely resemble that of norepinephrine. This type of energizer competes with norepinephrine for receptor sites. Such energizers also inhibit the action of monoamine oxidase (MAO), an enzyme which ordinarily destroys the neurotransmitter or blocks its repeated uptake by neurons. Energizers make a person feel wide awake, but they do not create energy; they simply stimulate the body to use stored energy. Abuse of these drugs eventually results in poor health, for the energy debt must be paid.

HALLUCINOGENS   The hallucinogen lysergic acid diethylamide (LSD) produces hallucinations and sometimes severe mental disturbances even when administered in minute amounts. LSD appears to act as an antagonist to serotonin, the brain neurotransmitter. *Marijuana* is a milder hallucinogen than LSD, but its use may cause a loss of coordination and sensory distortion. In large doses, it may produce hallucinations.

BRAIN WAVES

The active brain produces electrical impulses in steady rhythm. These *brain waves* can be recorded by an instrument called an electroencephalograph. Electrodes are placed on the scalp, and the amplified electrical impulses of the brain are recorded on an electroencephalogram (EEG), in which brain waves are translated into a series of wavy lines or tracings (Fig. 9.18). The resting rate of impulses recorded from the cerebral cortex of a person who is awake, but whose eyes are closed and who is not unduly excited, is around 10 to 12 per second; these are called *alpha waves*. It is thought that they arise in the reticular formation and then stimulate the cerebral cortex to produce consciousness.

FIGURE 9·18

a Normal EEG. This illustrates the effects of eye opening (EO) and eye closing (EC) on the alpha waves which were taken with parietal to occipital lobe linkages on the left and right sides, respectively. *b₁* Abnormal EEG. Note that the sine wave activity seen about equally on left and right sides in *a* is sharply reduced on the left side here and very slow irregular waves are recorded, whereas the waves from the right hemisphere *b₂* are normal. The reduced normal activity and abnormal slow waves resulted from occlusion of the middle cerebral artery on the left side. (*Courtesy of Hartford Hospital, Hartford, Conn.*)

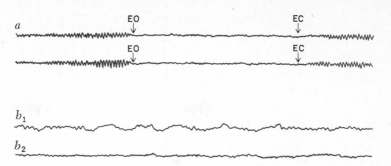

In the alert state, greater cerebral activity occurs, and *beta waves* are recorded at a frequency that varies considerably but is often in the range of 16 to 30 hertz (Hz). The origin of brain waves is not known with certainty, but they may be the result of current flow in the mass of pyramidal cell dendrites of the cerebral cortex. The rhythmic pattern may originate in the thalamus, where a brain pacemaker has been discovered, but precise location of the source will require further study.

Waves of very low frequency, fewer than 4 or 5 per second, occur during sleep. These are *delta waves*, characteristic of restful sleep. Dreaming accompanied by rapid eye movements (REM), produces a period of fast waves resembling those of the alert state.

BRAINSTEM

Brainstem
　Midbrain
　Hindbrain

Great nerve tracts that connect the spinal cord with higher synaptic levels in the cerebrum constitute the white matter of the *brainstem*, which is composed of the midbrain and the hindbrain. The gray matter consists of numerous ganglia and nuclei, which are interwoven into the white matter. Some of these are the ganglia of the cranial nerves, since all the cranial nerves arise from the brainstem except the first pair, the olfactory nerves. Large motor and sensory tracts pass through this region, and many reflex centers are located here. The brainstem includes structures of the midbrain and the hindbrain, including the pons and medulla but excluding the cerebellum.

Midbrain
　Cerebral peduncles
　Cerebral aqueduct
　Red nuclei
　Tectum
　Superior colliculi
　Inferior colliculi

THE MIDBRAIN (MESENCEPHALON)　The *midbrain* is the upper portion of the brainstem (Fig. 9.19). In more primitive animals, the optic lobes develop in this region, but in more advanced species and in humans, the midbrain is covered over by the cerebrum and loses most of its optic tract connections. The greater part of the midbrain consists of nerve tracts that carry impulses between the cerebrum and the cerebellum, the medulla, and the spinal cord. The anterior part is composed largely of two great nerve tracts, the *cerebral peduncles*, or *crura cerebri*. The cerebral aqueduct extends from the third ventricle to the fourth and traverses the midbrain. Between the aqueduct and the cerebral penduncles lie the *red nuclei* (see Fig. 9.17), two masses of gray matter connected by nerve tracts with the cortex of the cerebrum, the thalamus, the cerebellum, and the spinal cord. The red nuclei, through their connections, are thought to be concerned with muscle tone and with delicate or skilled movements. The nuclei of the oculomotor and trochlear cranial nerves are also located in the midbrain. Posterior to the cerebral

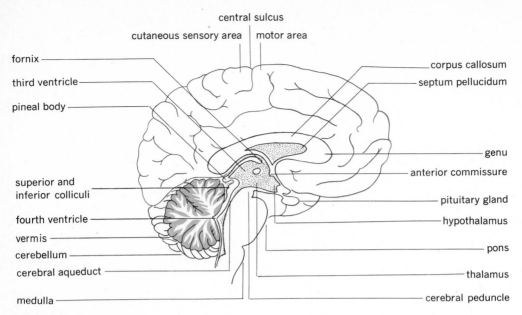

central sulcus

cutaneous sensory area | motor area

fornix

third ventricle

pineal body

superior and
inferior colliculi

fourth ventricle

vermis

cerebellum

cerebral aqueduct

medulla

corpus callosum

septum pellucidum

genu

anterior commissure

pituitary gland

hypothalamus

pons

thalamus

cerebral peduncle

FIGURE 9·19
Sagittal section through the
brain.

aqueduct is an area called the *tectum.* Within this dorsal area are four
rounded structures called the *superior* and *inferior colliculi* (the *corpora
quadrigemina*). The colliculi contain the optic lobes in the brains of lower
vertebrates, but in the human brain they are subdivided into four structures.
The superior colliculi still retain some connections with the optic tract and
are a center for visual reflexes in most mammals, and possibly in humans.
The inferior colliculi have become auditory in function and may therefore
be considered as an auditory reflex center in mammals, including humans.

Metencephalon
  Cerebellum
  Pons
  Fourth ventricle

THE HIND BRAIN *The metencephalon* The *metencephalon,* the first section of
the hindbrain, lies just below the midbrain. Its most obvious structures are
the *cerebellum,* which lies posterior to the brainstem, and a bridge of nerve
tracts, called the *pons,* which extends across the anterior part of the
brainstem. The *fourth ventricle* is the cavity of this region. All ventricles
contain choroid plexuses (Fig. 9.20).

Cerebellum
  Hemispheres
  Vermis
  Arbor vitae
  Cerebellar peduncles

CEREBELLUM Two *cerebellar hemispheres* are located below the occipital
lobes of the cerebrum. Between the two hemispheres is an area called the
vermis. The cerebellum has a cortex of gray matter, but it differs from that
of the cerebrum in certain respects. It is not convoluted in the same manner
as the cortex but instead is arranged as a series of layers. Within the cortex
are the large *purkinje cells.* These efferent inhibitory neurons with their
remarkable branching dendrites are found only in the cortex of the cerebellum. They are large, flattened, fanlike cells arranged in layers. Leading to
them are the elongate fibers of climbing cells and the many-branched mossy
fibers that represent the input to the purkinje cells. The output is through
the long axons of purkinje cells that lead to areas of gray matter deep within
the cerebellum. Three other kinds of cerebellar cells are classed as interneurons: *basket cells, stellate cells,* and *golgi cells,* all inhibitory in function.

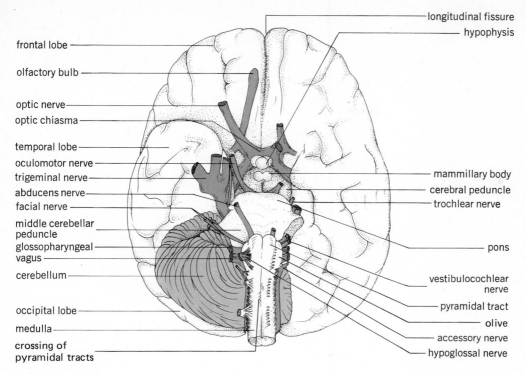

frontal lobe

olfactory bulb

optic nerve

optic chiasma

temporal lobe

oculomotor nerve

trigeminal nerve

abducens nerve

facial nerve

middle cerebellar
peduncle

glossopharyngeal

vagus

cerebellum

occipital lobe

medulla

crossing of
pyramidal tracts

longitudinal fissure

hypophysis

mammillary body

cerebral peduncle

trochlear nerve

pons

vestibulocochlear
nerve

pyramidal tract

olive

accessory nerve

hypoglossal nerve

FIGURE 9·20

Ventral view of the brain including the bases of cranial
nerves.

In a deep layer below the purkinje cells are minute excitatory neurons called *granule cells.* These types of cells are mentioned largely to indicate the complexity of cells in the cerebellum; you should not think that purkinje cells are the only type present (Fig. 9.21).

The interior of the cerebellum is largely composed of white matter, although the gray matter of the cortex descends deeply into the white matter and elaborates into an inverted treelike pattern of branching. The branching pattern, as seen in sections of the interior, gives rise to the name *arbor vitae,* or "tree of life," as a descriptive name for this area. The cerebellum contains four pairs of nuclei, the largest of which are the dentate nuclei. Three great nerve tracts, the *cerebellar peduncles,* connect the cerebellum with the midbrain, thalamus, pons, medulla, and spinal cord.

The cerebellum has been called the "secretary" to the cerebrum. It does not initiate motor responses but coordinates muscular movements so that action will be smooth and efficient instead of jerky and uncoordinated. The cerebellum is also concerned with maintaining equilibrium. It is connected by nerve fibers with the semicircular canals of the inner ear, which are also concerned with equilibrium. The cerebellum directs muscular coordination that keeps the body balanced in various positions. It also coordinates impulses received from the ear, the eye, and the tactile sense.

The cerebellum helps to maintain muscle tone. Birds whose cerebellum has been removed allow their wings to droop and are never able to fly. Mammals whose cerebellum has been removed exhibit a peculiar, uncoordinated walking gait. All cerebellar functions are below the level of con-

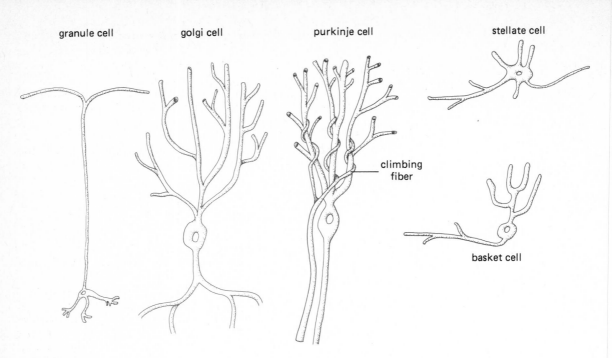

granule cell   golgi cell   purkinje cell   stellate cell

climbing
fiber

basket cell

FIGURE 9·21
Cells found in the cortex of
the cerebellum.

scious activity; sensory impulses received in the cerebellum do not produce sensations.

Localization of function in the cerebellum does not appear to be as definite as it is in the cerebrum. Each cerebellar hemisphere controls the coordination of movement of the appendages on the same side of the body, and the vermis is said to control the coordination of the trunk musculature. Projection areas in the cerebellum have been localized in mammals for the sense of touch and the proprioceptive (body position) sense. Such sensations seem to be received middorsally on the cerebellar cortex. Auditory and visual stimuli are received in an area located at about the middle of the dorsal aspect. Equilibrium seems to be controlled from two centers, one in the anterior and one in the posterior cortical area of the cerebellum. It seems significant that the tactile area and the proprioceptive areas from the same muscles coincide and that the auditory and visual areas overlap. The vermis is considered to control *saccades,* the rapid eye movements used in scanning an object.

The *pons* is composed primarily of horizontal nerve tracts that connect the two hemispheres of the cerebellum anteriorly and of vertical tracts that connect the cerebrum with the medulla. The pons also contains nuclei of gray matter, including the nuclei of the trigeminal (Vth cranial), abducent (VIth cranial), and the motor nucleus of the facial (VIIth cranial) nerves.

THE RETICULAR FORMATION   A diffuse mixture of gray matter spread throughout the white matter of the brainstem is known as the *reticular formation.* The connections made by the reticular formation are not well defined from an anatomical standpoint. It has been established that the reticular formation receives fibers from the cerebral cortex, the hypothalamus, the cerebellum, and the spinal cord. It also receives collateral branches from the basal ganglia and the limbic system. Acting as a reflex center, the reticular formation facilitates muscular coordination.

This region of the brainstem also contains the nuclei of cranial nerves associated with such vital activities as breathing and cardiovascular control.

The reticular activating system (RAS) The creation of consciousness involves the *reticular activating system,* a part of the reticular formation. Impulses flowing through the upper part of this system reach the thalamus and are distributed to various areas of the cerebral cortex. Stimulation of the upper part of the RAS in a sleeping animal causes almost immediate arousal, while injury to this area produces unconsciousness. The lower part of the reticular activating system cooperates with a sleep area in the pons to control sleep.

Sleep A slowing of external stimuli, a relaxation of the mental state, and some chemical changes in the sleep center of the pons produce a sense of drowsiness and allow the creation of the periodic state of diminished consciousness known as sleep. A person falling asleep goes from periods of light sleep to those of deep sleep in which breathing becomes slower and deeper and the pulse rate decreases. During sleep there may be muscular movements of the limbs and rapid eye movements (REM), which are associated with dreams. Unusual or extreme loss of sleep may be detrimental to health. A sleep factor has been obtained from the spinal fluid of laboratory animals, but its chemical structure has not been determined. An understanding of the functions of the reticular formation in its relation to sleep, arousal, and consciousness is proving to be of great importance in human physiology.

Myelencephalon
Medulla
Pyramidal tracts
Olive

Myelencephalon The base of the brainstem, the *myelencephalon,* is formed by the medulla (see Fig. 9.20). The *medulla* is continuous with the spinal cord but does not have the same internal structure. Though the nerve tracts are continuous, some are larger and better defined in the medulla. Some of the fibers in the *pyramidal tracts* cross to the opposite side in the medulla forming the decussation of the pyramidal tracts. The continuous gray matter of the cord is broken up into groups of nuclei in the medulla. These include the nuclei of the IXth, Xth, XIth, and XIIth cranial nerves. The central canal of the cord is continuous anteriorly through the medulla, where it opens into the lower part of the fourth ventricle. The medulla, however, is not just the upper part of the spinal cord. It contains vital reflex centers, such as the cardiac inhibitory center, which acts through the vagus nerve to slow the heart rate; the vasoconstrictor center, which is responsible for the constriction of peripheral blood vessels and the consequent rise of arterial pressure; and a respiratory center, which provides the nervous stimulus for regular respiratory movements. The medulla also controls a number of common reflex activities, such as laughing, coughing, and sneezing, and many of the activities of the digestive tract.

A rounded enlargement on either side of the medulla just below the pons is called the *olive.* The inferior olivary nucleus gives rise to the climbing cell fibers of the cerebellum.

Twelve pairs of nerves arise from the brain within the cranial cavity. These are termed the *cranial nerves*. They are numbered from I to XII and are also named (Table 9.3). The cranial nerves are essentially like spinal nerves, but they are more highly specialized. Though some are mixed nerves, some cranial nerves have lost one branch or the other and are, for the most part, motor or sensory nerves.

TABLE 9·3
Cranial nerves I through XII

No.	Name	Type	Function
I	Olfactory	Sensory	Sense of smell
II	Optic	Sensory	Sense of vision
III	Oculomotor	Motor	Innervates extrinsic muscles of eyes except lateral rectus and superior oblique. Also innervates the levator palpebrae superioris muscles
			Parasympathetic fibers to sphincter of pupil and to ciliary muscle of lens
IV	Trochlear	Motor	Superior oblique muscle of eye
V	*Trigeminal*	Mixed	
	Ophthalmic branch	Sensory	Anterior surface of eye; nasal mucous membrane; skin of forehead
	Maxillary branch	Sensory	Teeth of upper jaw; skin area of face above mouth and below eyes
	Mandibular branch	Mixed	Sensory to teeth of lower jaw, skin, lower lip, and tongue.
			Motor to muscles of mastication Muscle sense
VI	Abducens	Motor	Lateral rectus muscle
VII	Facial	*Mixed*	
		Motor	Muscles of the face
		Sensory	Taste, anterior two-thirds of tongue
			Parasympathetic fibers to lacrimal, submandibular, and sublingual glands
VIII	*Vestibulocochlear*		
	Vestibular branch	Sensory	Hearing
	Cochlear branch	Sensory	Equilibrium
IX	Glossopharyngeal	Mixed	Motor to swallowing muscles of pharynx
			Sensory to pressure receptors in carotid sinuses. Parasympathetic fibers to parotid salivary glands
X	Vagus	Mixed	Motor to muscles of pharynx, larynx; thoracic and abdominal viscera
			Sensory to mucous membranes of respiratory and digestive tracts
XI	*Accessory*	Motor	
	Cranial portion		Pharyngeal and laryngeal muscles
	Spinal portion		Sternomastoid and trapezius muscles
XII	Hypoglossal	Motor	Muscles of the tongue

(Acoustic)

On Old Olympus' Towering Tops A Fin Viewed Greece, Venus And Hermes.

OLFACTORY NERVE (I) The *olfactory nerve* arises from sensory receptors located in the upper part of the mucous membrane that lines the nasal cavity. Separate fibers grow inward through the cribriform plate of the ethmoid bone to the olfactory bulbs, where they make synaptic connections with the secondary neurons leading inward to the olfactory-interpreting centers of the brain. The olfactory nerve is purely sensory and carries nervous impulses that are translated into sensations of odor.

The olfactory nerve is peculiar in that the individual fibers grow inward. Therefore, the olfactory nerve has no sensory ganglion. The fibers, furthermore, are unmyelinated.

OPTIC NERVE (II) The *optic nerve* is a sensory nerve concerned with the sense of light. It arises from ganglion cells located in the retina of the eye, and its fibers form the optic tract, which leads back to the lateral geniculate body of the thalamus. From there, sensory impulses are conveyed by secondary neurons to the visual interpreting area in the occipital lobe of the cerebrum. In a ventral view of the brain (see Fig. 9.20) the optic nerves form a structure something like the letter X. This structure is the *optic chiasma*, where the fibers from the inner half of the retina of either eye cross over and go to the opposite side of the brain. The crossing of some optic nerve fibers probably produces better coordination of responses between the eye and the brain.

OCULOMOTOR, TROCHLEAR, AND ABDUCENT NERVES (III, IV, VI) The *oculomotor*, the *trochlear*, and the *abducent* nerves are essentially motor nerves that innervate the muscles that move the eyeball. The oculomotor and trochlear nerves arise from nuclei of gray matter located beneath the cerebral aqueduct in the midbrain. The trochlear nerve nucleus is posterior to the nucleus of the oculomotor nerve. The nucleus of the abducent nerve is in the lower part of the pons, beneath the fourth ventricle. The oculomotor and trochlear nerves emerge near the anterior border of the pons; the abducent nerve emerges at the lower border of the pons in the fissure between the pons and the medulla. The oculomotor nerve also carries fibers of the parasympathetic system to the circular muscle of the iris and to the ciliary muscle of the eye. A few sensory fibers may travel in nerves that serve the extrinsic eye muscles, but the presence of sensory receptors in these eye muscles has not been adequately demonstrated as yet.

TRIGEMINAL NERVE (V) The *trigeminal* is a mixed nerve with both motor and sensory nuclei. The motor nucleus and the sensory root are located in the pons laterad and below the fourth ventricle. The sensory nuclei are located in a large *semilunar ganglion* (gasserian ganglion) outside the brainstem. There are three large branches of the trigeminal nerve: the ophthalmic, the maxillary, and the mandibular.

The *ophthalmic branch* is a sensory branch and carries impulses originating in the surface of the eye, in the lacrimal gland, and from the nose and forehead.

The *maxillary branch* is also a sensory branch, and its fibers are broadly distributed. Among the structures supplied by this branch are the teeth and gums of the upper jaw, the upper lip, and the cheek.

The *mandibular branch* is a mixed nerve. It has many small branches. Some of these nerves supply the teeth and gums of the lower jaw, the chin, the lower lip, and the tongue. It is a motor nerve innervating the muscles concerned with mastication, and it also contains sensory fibers from proprioceptors in muscle tendons of these structures (see Table 9.2).

FACIAL NERVE (VII) The *facial nerve* is a mixed nerve. Its motor nucleus lies in the lower part of the pons, and fibers are supplied to the muscles of the face and forehead. The sensory branch is very small. Its fibers arise from the geniculate ganglion, which is located in the temporal bone, and are distributed to the anterior two-thirds of the tongue, from which they convey sensations of taste. The motor branch also carries fibers of the parasympathetic system to the sublingual and submaxillary salivary glands. Parasympathetic fibers stimulate vasodilatation and secretion in these glands. Fibers concerned with taste sensation and parasympathetic fibers pass through the tympanic cavity in the *chorda tympani* branch.

VESTIBULOCOCHLEAR NERVE (VIII) The *vestibulocochlear* nerve (acoustic) is a sensory nerve concerned with hearing and equilibrium. It is composed of two nerves of different origin and function. The *cochlear nerve* carries auditory impulses, and its ganglion lies in the cochlea. From receptors in the *spiral organ of Corti,* auditory impulses are conveyed inward to the medulla. The impulses cross over to the opposite side and pass upward through the pons and midbrain over a series of neurons to the auditory-interpreting area in the temporal lobe. A few efferent fibers are also present.

The *vestibular nerve* arises in the vestibular ganglion of that portion of the ear associated with the semicircular canals. It enters the medulla but has important connections with the cerebellum. The vestibular nerve is concerned with maintaining equilibrium. Its functions are discussed in Chap. 12 in connection with the ear.

GLOSSOPHARYNGEAL NERVE (IX) The *glossopharyngeal nerve* arises from the medulla and supplies the tongue and pharynx. It is a mixed nerve; the motor fibers are distributed to muscles of the pharynx, while the sensory fibers are supplied to the tonsils, the mucous membranes of the pharynx, and the posterior part of the tongue. Taste stimuli originate from receptors located in the large papillae at the back of the tongue. This nerve also carries fibers of the craniosacral system. Secretory and vasodilator fibers are distributed to the parotid salivary gland.

VAGUS NERVE (X) The *vagus nerve* is the longest cranial nerve. Its pathway lies from the medulla through the neck and thorax to the abdomen. It is a mixed nerve: sensory branches convey impulses from the mucous membranes lining the respiratory and digestive tracts; voluntary motor fibers are distributed to certain muscles of the pharynx and larynx.

The right and left vagus nerves send branches to the cardia and pulmonary plexuses. Above the stomach they unite to form the esophageal plexus. Branches supplying the abdominal viscera arise below the esophageal plexus and contain involuntary fibers from both vagus nerves.

The vagus nerve is also one of the principal nerves of the craniosacral (parasympathetic) division of the autonomic nervous system. It carries inhibitory fibers to the heart and secretory fibers to the gastric glands and pancreas, as well as vasodilator fibers to the abdominal viscera. Autonomic fibers are also supplied to the bronchial tubes, esophagus, stomach, pancreas, gallbladder, small intestine, and ascending colon.

ACCESSORY NERVE (XI) The *accessory nerve*, a motor nerve, is composed of two branches, a cranial portion and a spinal portion. The *cranial branch* arises from a nucleus in the medulla and emerges from the side of the medulla just below the roots of the vagus. The *spinal branch* arises from the spinal cord in the upper cervical region and ascends, passing upward through the foramen magnum. It then turns and descends beside the vagus. The cranial portion is accessory to the vagus and supplies most of the pharyngeal and laryngeal muscles. The spinal portion innervates the sternomastoid and trapezius muscles.

HYPOGLOSSAL NERVE (XII) The *hypoglossal nerve* is a motor nerve distributed to the muscles of the tongue. It arises from the medulla. Injury to this nerve causes difficulty in speaking or swallowing.

THE SPINAL CORD

The *spinal cord* is that portion of the central nervous system within the vertebral canal. It is continuous with the base of the brain anteriorly; posteriorly it tapers to a threadlike strand below the second lumbar vertebra. Cervical and lumbar enlargements are seen in the regions in which large nerves to the appendages enter or leave the cord.

In the early development of the fetus, the spinal cord extends the length of the spinal canal, but as the fetus grows, the vertebral column grows in length at a greater rate than the spinal cord. Hence the cord is drawn forward in the vertebral canal, and the roots of lumbar, sacral, and coccygeal nerves travel down the spinal canal to reach their normal places of exit. The canal below the second lumbar vertebra then contains the threadlike strand of the cord surrounded by lumbar, sacral, and coccygeal nerves. This taillike group of nerves is called by a descriptive name, the *cauda equina* (Fig. 9.22).

The cord is suspended rather loosely in the spinal canal. Since its diameter is considerably less than that of the canal, the vertebral column can be moved freely without injury to the cord.

Ascending tracts
  Fasciculus gracilis
  Fasciculus cuneatus
  Cerebellar tracts
  Spinothalamic tracts

CONDUCTION PATHWAYS OF THE SPINAL CORD *Ascending tracts* The vertical neurons of the cord are arranged in orderly bundles. Many of these tracts have been carefully investigated, and their origin, termination, and function recorded. Two of the large posterior ascending tracts are the *fasciculus gracilis* and the *fasciculus cuneatus* (Fig. 9.23). The cell bodies of the neurons composing these tracts lie in the dorsal ganglia of spinal sensory nerves, and their fibers extend upward to the medulla, ending in the *nucleus gracilis* or the *nucleus cuneatus*. (A nucleus, in this sense, is a group of nerve cell bodies lying within the central nervous system.) Other neurons connect the nuclei and the thalamus; a third set of neurons conveys impulses from the thalamus to the sensory-interpreting areas of the cerebral cortex. This is the pathway through which position of the skeletal muscles (voluntary proprioceptive sense) is interpreted and sensations of touch are received.

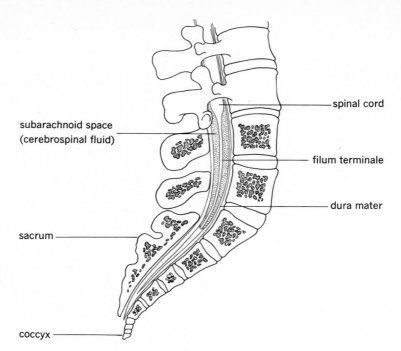

FIGURE 9·22
Longitudinal section through the lower part of the vertebral column and the spinal canal, showing the relationship between the spinal cord and the meninges in this region. (Diagrammatic)

spinal cord

subarachnoid space (cerebrospinal fluid)

filum terminale

dura mater

sacrum

coccyx

Large tracts in the lateral part of the cord are direct *cerebellar tracts*. They are concerned with muscular coordination. Sensory impulses that convey unconscious proprioceptive sensations travel upward to the cerebellum through these tracts.

Spinothalamic tracts lie in lateral and ventral portions of the white matter. These pathways convey impulses to the thalamus and then to the

TABLE 9·4
Principal ascending tracts of the spinal cord

Name	Location in spinal cord	Origin	Terminal endings in brain	Functional sensation
Fasciculus gracilis Fasciculus cuneatus	Posterior white columns	In spinal ganglia on same side	Medulla	Touch, pressure, conscious muscle sense, motion
Direct cerebellar	Lateral white columns	Neuromuscular receptors	Cerebellum	Unconscious muscle sense, muscular coordination
Lateral spinothalamic	Lateral white columns	Cell bodies in posterior gray columns of opposite side	Thalamus	Pain and temperature sense on opposite side
Anterior spinothalamic	Anterior white columns	As above	Thalamus	Touch and pressure

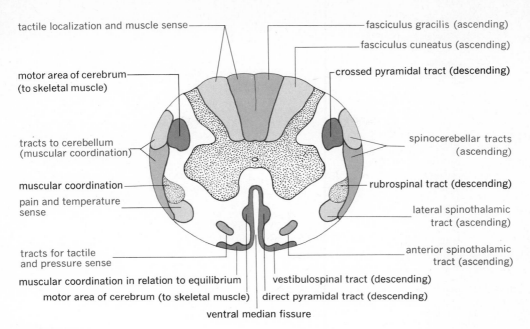

tactile localization and muscle sense

motor area of cerebrum (to skeletal muscle)

tracts to cerebellum (muscular coordination)

muscular coordination

pain and temperature sense

tracts for tactile and pressure sense

muscular coordination in relation to equilibrium

motor area of cerebrum (to skeletal muscle)

ventral median fissure

fasciculus gracilis (ascending)

fasciculus cuneatus (ascending)

crossed pyramidal tract (descending)

spinocerebellar tracts (ascending)

rubrospinal tract (descending)

lateral spinothalamic tract (ascending)

anterior spinothalamic tract (ascending)

vestibulospinal tract (descending)

direct pyramidal tract (descending)

FIGURE 9·23

Diagrammatic cross section of the spinal cord, indicating the general location of some of the principal nerve tracts. The motor area of the cerebrum gives rise to both the crossed and the direct pyramidal tracts.

Descending tracts
Crossed pyramidal
Direct pyramidal
Rubrospinal
Vestibulospinal

cerebral cortex, where they may be interpreted as pain, temperature, pressure, touch, and muscle sense.

Features of the ascending tracts are summarized in Table 9.4.

Descending tracts The neurons which make up the pyramidal tracts have their origin in the motor area of the cerebral cortex. Most of their axons cross over to the opposite side of the brain in the medulla and descend in the lateral part of the cord, giving rise to the *crossed pyramidal tracts* (*lateral corticospinal tracts*). The fibers that do not cross in the medulla form two smaller ventral columns called the *direct pyramidal tracts* (*ventral corticospinal tracts*). The neurons of both crossed and direct tracts make synaptic connections with the motor nerve roots of spinal nerves at various levels. The neurons of the direct pyramidal tract cross in the anterior white commissure just before they make a synaptic connection with a spinal nerve root. The direct pyramidal tracts form the pathways of voluntary motor impulses to the muscles of the trunk; the crossed pyramidal tract supplies muscles of the arms and legs.

Neurons of the pyramidal tract do not synapse until they descend to a peripheral spinal motor nerve outlet at a certain spinal level. Their fibers extend from the cerebral cortex to the spinal motor nerve cell or to short association neurons before they synapse. The *extrapyramidal system* involves a greater area of the cortex, including the supplementary motor area. The neurons of this system may synapse several times at subcortical levels before reaching a spinal motor outlet. The subcortical levels include the basal ganglia, the red nucleus, and the reticular formation. Some corticothalamic and corticohypothalamic fibers are included also. The extrapy-

TABLE 9·5
Principal descending tracts of the spinal cord

Tract	Location in spinal cord	Origin in brain	Terminal endings	Motor function
Pyramidal tracts				
Crossed pyramidal tract	Lateral white columns	Voluntary motor areas	Anterior gray or anterolateral columns of spinal cord	Voluntary motor impulses, especially to muscles of the arms and legs
Direct pyramidal tract	Anterior or ventral columns	Voluntary motor areas	Anterior gray or anterolateral columns of spinal cord	Mainly to muscles of the trunk
Extrapyramidal tracts				
Rubrospinal	Lateral white columns	Red nucleus	Anterior gray or anterolateral columns of spinal cord	Muscular coordination, postural control
Vestibulospinal	Anterior or ventral columns	Vestibular nucleus of VIIIth cranial nerve in medulla	Anterior gray or anterolateral columns of spinal cord	Maintaining equilibrium

ramidal system supports the complex voluntary movements mediated by the pyramidal system. Probably the best explanation of its function is that a balance between the two systems permits a refinement of muscular movement. The tremors associated with Parkinson's disease are usually considered to involve pathology in certain areas of the basal ganglia, but they may also involve fibers that connect the cortex with other subcortical areas.

The *rubrospinal tracts* descend through the lateral part of the cord. The cell bodies of these neurons are located in the *red nucleus* of the midbrain. Their fibers cross immediately and descend to various levels of the cord where they make connections with spinal motor nerve roots. Since the red nucleus has both cerebral and cerebellar connections, much voluntary muscular control may be transferred to voluntary muscular coordination through these connections. The rubrospinal tract has been shown experimentally to be concerned with reflexes that aid in righting the body and with the tone of muscles affecting posture.

The *vestibulospinal tract* originates from the nucleus of the vestibular (VIIIth cranial) nerve in the medulla, and the neurons end at various levels near the origin of the spinal motor nerve roots. Since they receive impulses from the semicircular canals of the ear, their function is to adjust muscular coordination in relation to maintaining equilibrium.

Features of the descending tracts are summarized in Table 9.5.

SPINAL NERVES

Thirty-one pairs of *spinal nerves* arise from the spinal cord. They are grouped as follows: cervical, 8 pairs; thoracic, 12 pairs; lumbar, 5 pairs; sacral, 5 pairs; coccygeal, 1 pair (Fig. 10.1).

There are eight pairs of cervical nerves rather than seven, because the first pair arises from the medulla and emerges above the atlas. The remaining spinal nerves arise from the cord and emerge through openings between the vertebrae.

We have learned that spinal nerves are composed of a dorsal root and a ventral root at the spinal cord. The two roots join to form the mixed spinal nerve which emerges from the vertebral canal. Soon after leaving the canal the spinal nerve divides into two branches, a dorsal ramus and a ventral ramus. These rami are called the posterior and anterior primary divisions of the nerve. The posterior branch supplies a certain segment of skin area of the back (a dermatome) and also a segment of back muscles (Fig. 9.24). The ventral branch in the thoracic region supplies the muscles and skin of the thoracic and abdominal regions. Other ventral primary divisions form the great plexuses: cervical, brachial, lumbar, and sacral.

Dermatomes supplied by adjacent spinal nerves overlap so injury to a single spinal root may be compensated by active nerves of adjacent segments.

SPINAL PLEXUSES Nerves leaving the spinal cord form complex interlacing networks in the cervical, brachial, and lumbosacral regions. Such networks are called *plexuses.* Motor branches of the first four cervical nerves form the *cervical plexus.* Peripheral branches innervate muscles of the neck and shoulder. The phrenic nerve, which is distributed to the diaphragm, arises from the plexus.

The *brachial plexus* is a large plexus of the neck and shoulder composed of motor branches of the Vth to VIIIth cervical nerves and the Ist thoracic nerve. Among the nerves that supply the arm are the axillary, median, radial, and ulnar nerves.

The lumbosacral plexus consists of branches from all the lumbar nerves and the Ist to IIId sacral nerves. This is a very large network, with nerves extending to the lower extremities. Included among several nerves are the femoral nerve and the very large sciatic nerve. Inflammation and injury of the latter nerve may cause a neuralgic condition called *sciatica.*

CLINICAL ASPECTS OF BRAIN AND SPINAL CORD FUNCTION

DIAGNOSTIC TECHNIQUES *Reflex testing* A physical examination commonly involves the testing of *muscular reflexes.* The most common test is that of the *patellar reflex,* or *knee jerk.* The patellar tendon is tapped, an action that usually produces a mild kicking movement. Failure to generate a knee jerk may indicate injury to some of the nerves emanating from the lumbar region of the spinal cord. An exaggerated kick, on the other hand, may indicate corticospinal tract injury.

Another reflex test often performed is the *calcaneal* (Achilles) *tendon reflex,* or ankle jerk test. The normal reaction is a slight extension of the foot. Failure to obtain this reflex may indicate injury to the muscles or nerves of the leg or an injury to the lower part of the spinal cord.

The *Babinski reflex* may be elicited by stroking the outer edge of the sole of the foot. In babies, such action causes the great toe, and often all the toes, to extend. The presence of this reflex is considered normal in babies, but indicates injury to the corticospinal tract in older children and adults. A positive Babinski sign usually indicates injury in the upper part of the cord,

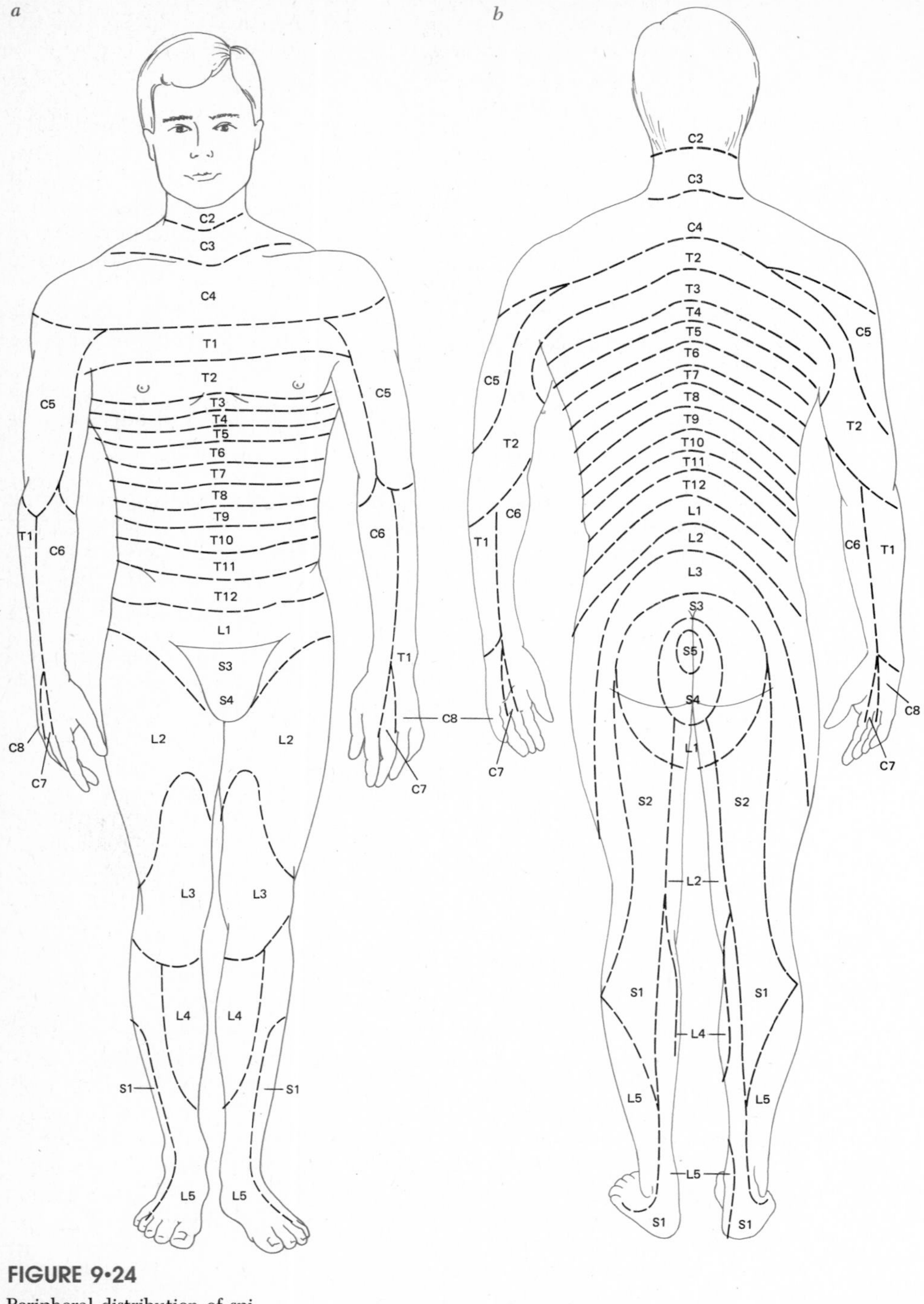

FIGURE 9·24

Peripheral distribution of spinal nerves; *a* anterior surface, *b* posterior surface.

as in paralysis following a stroke. The *plantar reflex* is a normal reflex causing contraction of the toes after stroking the sole of the foot.

In the *abdominal reflex,* the abdominal muscles contract when the side of the abdomen is stroked. This is the normal response. The absence of this reflex may indicate injury to the corticospinal tract or to reflex centers in the spinal cord.

Other reflexes are also used in the diagnosis of nervous system abnormalities, but the ones described above are those most commonly tested.

Electroencephalogram (EEG)   An EEG is used to detect brain malfunctions that give rise to abnormal brain waves. Mechanical injuries to the brain, as well as brain tumors and epilepsy, may be diagnosed in this way. Two kinds of epilepsy, petit mal and grand mal, produce quite different types of brain waves. *Petit mal,* a form of epilepsy in which a person may lose consciousness for a few seconds, is recorded on the EEG as a slow rhythm with a spike, usually alternating with a broader, dome-shaped pattern. In *grand mal* epilepsy, the person loses consciousness and motor control and falls into convulsions. The EEG indicates great neural excitement and a confusion of strong impulses. There is a wild synchronous discharge of impulses between neurons of the brain. Petit mal is characterized by a very brief loss of consciousness and only minor muscular twitching. Sometimes the only sign of a petit mal seizure is a sudden cessation of activity; the patient stops and stares as though daydreaming.

Epilepsy does not affect intelligence. Most epileptics are able to control seizures by the use of drugs that depress synchronous nervous system activity.

Lumbar puncture   Diseases that attack the central nervous system or the meninges (meningitis) can alter the composition or increase the amount of spinal fluid. Diagnosis and treatment of such diseases can be made easier by analyzing a sample of the spinal fluid. The sample is obtained by a procedure called a *lumbar puncture.* Since the spinal cord ends at the first or second vertebra and since the dura and arachnoid membranes extend below this point, a needle may be inserted between the vertebrae and into the subarachnoid space of this region and cerebrospinal fluid may be withdrawn. Spinal anesthetics are introduced into the subarachnoid space in the same region by lumbar puncture.

INJURIES TO SPINAL CORD   The spinal cord is the pathway for impulses between the body and the brain. Injury to the ascending tracts causes lack of coordination, since spinal reflexes are disturbed. Walking becomes uncoordinated, movements are jerky, and maintaining balance may be difficult. *Locomotor ataxia (tabes dorsalis)* is an injury of this sort caused by the degeneration of some of the large ascending dorsal tracts of the cord.

The nerve fibers of the cord cannot regenerate after an injury. Degeneration of fibers proceeds upward, away from the neuron cell bodies, in the ascending tracts, and downward, away from the neuron cell bodies, in descending motor tracts.

Severing a motor pathway produces muscular paralysis. The paralysis may be of two types, depending upon which tracts are severed and where the injury occurs. If the injury occurs in the pyramidal tract or if the motor neurons of a spinal nerve are injured, the paralysis may be complete and of the *flaccid type*. In this type of injury, muscle tone fails and considerable atrophy of the muscles occurs. Certain extrapyramidal injuries in the brain or cord cause spastic paralysis, a condition in which there is an exaggerated tonicity and uncoordinated reflexes.

DISORDERS AND DISEASES OF THE CENTRAL NERVOUS SYSTEM *Abnormal states of unconsciousness* COMA A coma is a state of unconsciousness from which the person cannot be aroused. It is usually associated with injuries to the brain or with extreme bodily injuries that deprive the brain of oxygen or nutrition. Coma may be caused by a brain tumor, a stroke, unusual physiological disturbances, such as a diabetic crisis, or diseases such as meningitis. The reticular formation appears to be involved in producing coma no matter what the exact cause.

NARCOLEPSY This is a condition in which the individual is subject to uncontrollable attacks of deep sleep. The sleep is of short duration, usually 15 min to 1 hour, and the patient is rather easily awakened.

Neuralgia This disorder produces severe pain without evidence of any structural changes in the involved nerve. *Trigeminal neuralgia* (*tic douloureux*) is a well-known form of neuralgia characterized by severe pain on one side of the face. An attack lasts only a few seconds, but successive attacks may continue at intervals for several weeks. Middle-aged and elderly people are the usual victims of this condition, whose cause is unknown.

Neuritis Inflammation or lesions in the peripheral nerves cause a condition termed *neuritis*. Neuritis may be acute or chronic, degenerative or inflammatory, and painful or paralytic. Sciatica is a neuritis of the sciatic nerve.

Shingles This painful condition is caused by the virus herpes zoster. The disease affects the sensory ganglia of peripheral nerves and is characterized by intense itching and severe pain. It usually subsides in a few weeks.

Multiple sclerosis Multiple sclerosis is one of the most common chronic diseases of the central nervous system. It is characterized by degeneration of the white matter (myelin sheaths) in the brain and spinal cord. The cause is unknown, and, to date, there is no cure. The disease often strikes the optic nerves first. Later, multiple lesions appear in the myelin sheaths of axons of the brainstem and spinal cord. When these areas are involved, tendon reflexes are exaggerated, while abdominal reflexes are absent. The Babinski reflex is positive.

Cerebral palsy This disorder is a neurological defect of the brain's motor neurons. The cerebral motor areas, the deeper cerebral nuclei, or the cerebellum may be affected. Usually all these areas are affected to a certain extent. The physical signs are usually partial facial paralysis and partial paralysis of the leg muscles with poorly coordinated movement. Cerebral palsy is often accompanied by mental retardation. This is not a progressive disease, but once the damage has been done there is no reversal.

Poliomyelitis This disease, also called infantile paralysis, is caused by a virus. It may affect the motor neurons to skeletal muscles and cause paralysis, a type called acute anterior poliomyelitis, or it may attack the neurons that supply the muscles concerned with breathing, a form called bulbar poliomyelitis. These neurons are located in the breathing center of the medulla. Vaccination establishes immunity, but today many preschool children are not vaccinated. There are two ways of acquiring immunity: (1) by vaccination with the dead virus (the Salk vaccine) or (2) by live attenuated virus taken orally (the Sabin method). All children should be immunized against this serious disease.

Parkinson's disease This disease is becoming more common in the general population, largely because many people live to an older age, and Parkinson's disease is an affliction of the middle-aged and the elderly. Parkinson's disease involves the basal ganglia. The disease progresses slowly and is characterized by tremors affecting the arms and hands and, more rarely, the legs or head. Parkinsonian tremors are not too serious a problem, providing they are not widespread, but the disease also causes the muscles to be rigid, which does cause problems. Facial muscles may show rigidity, and the face may become expressionless and masklike. Fingers and toes may become deformed by continued contracture of the muscles. As rigidity spreads over the musculature of the body, walking may become difficult.

People with Parkinson's disease seem to lack the ability to produce dopamine in the brain. As we indicated earlier, dopamine is unable to cross the blood-brain barrier, but levodopa (L-dopa), a precursor of dopamine, can diffuse across this barrier and enter the brain, where it is changed chemically to dopamine. The administration of L-dopa has proven helpful to many Parkinsonian patients.

Since L-dopa is rapidly broken down in the bloodstream, large doses must be used for the drug to be effective. A decarboxylase inhibitor has been employed, therefore, to reduce L-dopa's rapid destruction. This has made possible use of smaller doses of L-dopa and has lessened the undesirable side effects of this drug.

Brain tumors These are growths that invade brain tissues. Since there is little room for expansion in the cranium, brain tumors produce pressure on other parts of the brain. Symptoms vary widely, depending upon location and size of the tumor, but may include severe headaches, dizziness, and lassitude.

Malignant tumors affecting the brain are of various kinds. If the tumor affects neuroglial cells, it is called a *glioma*. If the meninges are the site of the tumor, it is termed a *meningioma*. The latter kind is usually benign. Some brain tumors can be successfully removed by surgery, especially if they are small and close to the surface, but brain surgery is always a delicate undertaking.

Headache The headache, one of the common human ailments, is attributed to pain in the head, but it is usually a symptom of some malfunction elsewhere in the body. Headaches may be caused by unusual fatigue, eyestrain, indigestion, sinus trouble, too much alcohol, or strong emotional tension. However, not much is known about the actual source of the pain. The brain itself is considered to be insensitive to pain. If unusual pressure is exerted on the brain's blood vessels, though, pain receptors in their walls may respond and produce pain. Pain receptors in the meninges and venous sinuses also may be involved. *Migraine headaches* are severe headaches, and are of several kinds. Older people are more often affected, and such headaches are often accompanied by gastrointestinal disturbance. One type of migraine is characterized by visual disturbances in which wavy lines or flashes of light are seen. These disturbances cause temporary, localized blindness, apparently in the retina but more likely arising from the brain. Persons subject to this type of migraine are often abnormally sensitive to light. The head may feel sore for days following a migraine attack.

SUMMARY

1 The brain in its early embryonic development forms three primary divisions: forebrain, midbrain, and hindbrain.

2 The forebrain, or cerebrum, is divided into two secondary divisions: the telencephalon and the diencephalon. The hindbrain also is divided into two secondary divisions: the metencephalon and the myelencephalon. The midbrain, or mesencephalon, is the connecting link between the forebrain and the hindbrain.

3 The brain and spinal cord are covered by three meningeal membranes: dura mater, arachnoid, and pia mater.

4 The cavities of the brain are called ventricles. The lateral ventricles communicate with the third ventricle through the interventricular foramens. The third and fourth ventricles are connected by way of the cerebral aqueduct. The fourth ventricle is continuous with the central canal of the spinal cord. The ventricles and the central canal contain cerebrospinal fluid.

5 Neuroglial cells are closely associated with the neurons and blood vessels of the central nervous system. There are three principal types: astrocytes, oligodendrocytes, and microgliocytes.

6 The cerebrum is divided into two cerebral hemispheres. Each hemisphere is divided into lobes: frontal, parietal, temporal, and occipital lobes. The central sulcus separates the frontal and parietal lobes. The lateral cerebral fissure demarcates the temporal lobe. The occipital lobe lies posterior to the parietooccipital fissure. The insula is sometimes considered as a lobe of the cerebrum.

7 Some, but not all, functions of the cerebrum may be localized. The motor area lies just anterior to the central sulcus. It is inverted with respect to the area of the body that any given portion controls. The right motor area governs the left side of the body and vice versa, because the pyramidal tracts, arising on one side of the brain, cross over to the opposite side. The decussation occurs in the medulla or at the spinal level where the spinal nerve emerges. The cutaneous sensory area and the interpreting area for the proprioceptive sense lie directly posterior to the central sulcus. The visual cortex is located in the occipital lobes. Auditory impulses are interpreted in the temporal lobes.

8 There are many neurotransmitters, which may be either excitatory or inhibitory (refer to Table 9.2).

9 The diencephalon contains the thalamus, which lies in the lateral walls of the diencephalon surrounding the third ventricle. The hypothalamus

is located in the basal portion of the diencephalon. The thalamus is described as a relay center for nerve impulses, while several autonomic functions are regulated by the hypothalamus. Among them are regulatory centers for the control of body temperature and water metabolism. The hypothalamus also produces hormones and releasing factors.

10 The medulla lies at the base of the brainstem. It contains a number of vital centers, such as the cardiac inhibitory center, the respiratory center, and the vasoconstrictor center.

11 The reticular activating system, located in the brainstem, is thought to play a role in arousal from sleep and in maintaining a state of consciousness.

12 The cerebellum coordinates muscular movements and maintains equilibrium. The large cortical cells with branching dendrites are purkinje cells. The pons connects the two cerebellar hemispheres anteriorly, and the cerebellum also contains vertical tracts that connect the cerebrum with the medulla.

13 There are 12 pairs of cranial nerves.

14 Ascending tracts in the spinal cord carry impulses concerned with touch, pressure, pain, temperature, and the proprioceptive sense. Descending tracts, such as the pyramidal tracts, transmit motor impulses to muscles of the trunk and the appendages.

15 There are 31 pairs of spinal nerves. Spinal nerves form the cervical, brachial, and lumbosacral plexuses.

QUESTIONS

1 Trace the flow of the cerebrospinal fluid through the ventricles and cavities of the brain and spinal cord.
2 List some functions of the cerebrum.
3 What is meant by cerebral localization?
4 An injury affecting the upper part of the motor area on the left side may cause motor paralysis of what part of the body? Why?
5 Locate the sensory-interpreting areas of the brain.
6 Name some structures found in the diencephalon.
7 Discuss the functions of the thalamus and hypothalamus.
8 What do neurotransmitters do? Describe the difference in action between an excitatory and an inhibitory neurotransmitter.
9 What does the cerebellum control?
10 Describe the different kinds of neuroglial cells and discuss their function.
11 In what ways do cranial nerves differ from spinal nerves?
12 A man is seated at a desk, writing. He hears the door open and gets to his feet. Discuss the brain centers and nervous pathways involved.
13 What function is ascribed to Broca's area?
14 Name some of the structures included in the rhinencephalon.
15 Discuss some of the ideas concerning memory.
16 Large, delicate nerve tracts that connect various brain areas and lie within the brain have what function?
17 What is meant by memory trace?
18 Name some kinds of cells found in the cerebellum.
19 What are some of the functions of the basal ganglia?
20 What is the significance of the Babinski reflex?

SUGGESTED READING

Andersson, B.: Thirst and Brain Control of Water Balance, *Am. Sci.*, **59**(4):408–413 (1971).

Barkus, J. D., et al.: Behavioral Neurochemistry: Neuroregulators and Behavioral States, *Science*, **200**:964–973 (1978).

Deutsch, J. A.: The Cholinergic Synapse and the Site of Memory, *Science*, **174**:788–794 (1971).

Eccles, J. C.: "The Understanding of the Brain," McGraw-Hill Book Company, New York, 1977.

Frenk, H., B. C. McCarty, and J. C. Liebeskind: Different Brain Areas Mediate the Analgesic and Epileptic Properties of Enkephalin, *Science*, **200**:335–337 (1978).

Geschwind, N.: The Organization of Language and the Brain, *Science*, **170**:940–944 (1970).

Gorden, B.: The Superior Colliculus of the Brain, *Sci. Am.*, **227**:72–82 (1972).

Greenough, W. T.: Experimental Modification of the Developing Brain, *Am. Sci.*, **63**(1):37–46 (1975).

Iversen, L. L.: The Chemistry of the Brain, *Sci. Am.*, **241**:134–149 (1979).

Katz, B.: Quantal Mechanism of Neural Transmitter Release, *Science*, **173**:123–126 (1971).

Lasansky, A.: Nervous Function at the Cellular Level: Glia, *Annu. Rev. Physiol.*, **33**:241–256 (1971).

Llinas, R. R.: The Cortex of the Cerebellum, *Sci. Am.*, **232**:56–71 (1975).

Luria, A. R.: The Functional Organization of the Brain, *Sci. Am.*, **222**:66–78 (1970).

Merton, P. A.: How We Control the Contraction of Our Muscles, *Sci. Am.*, **226**:30–37 (1972).

Nathanson, J. A., and P. Greengard: "Second Messengers" in the Brain, *Sci. Am.*, **237**:108–119 (1977).

Neale, J. H., et al.: Enkephalin Containing Neurons Visualized in Spinal Cord Cultures, *Science*, **201**:467–469 (1978).

Pappenheimer, J. R.: The Sleep Factor, *Sci. Am.*, **235**:24–29 (1976).

Penfield, W.: The Interpretive Cortex, *Science*, **129**:1719–1725 (1959).

Segal, D. S., et al.: B- Endorphin: Endogenous Opiate or Neuroleptic, *Science*, **198**:411–413 (1977).

Snyder, S. H., and J. P. Bennett: Neurotransmitter Receptors in the Brain: Biochemical Identification, *Annu. Rev. Physiol.*, **38**:153–175 (1976).

Yates, F. E., S. M. Russel, and J. W. Maran: Brain-Adenohypophyseal Communication in Mammals, *Annu. Rev. Physiol.*, **33**:393–444 (1971).

10

THE AUTONOMIC NERVOUS SYSTEM

CONSIDERATIONS FOR STUDY

In this chapter, we take up the anatomy and physiology of the autonomic nervous system **(ANS). Topics to be covered include:**

1 **The sympathetic division and its organization**
2 **The parasympathetic division and its organization**
3 **The physiology of the ANS: how the two divisions work, both together and in opposition, to regulate visceral functions**
4 **Chemical transmitters in the ANS and their effects**

The autonomic nervous system (ANS) provides the key to homeostasis. Through very complex feedback systems and reflex activities, the ANS helps to regulate and control the normal balanced functioning of organs and organ systems. Emotional states are supported by very extensive bodily changes. A state of fear can induce the desire to run, but we cannot run far unless physiological adjustments support the effort. Anger can prepare us to fight, but it will not be an efficient battle unless our circulatory systems make necessary adjustments to provide strength and endurance for the contest. The situation need not be a highly emotional one; work and exercise are supported by the same physical adjustments.

The autonomic nervous system plays a large role in regulation of the internal environment of the body with regard to temperature and body fluids. A good example is the adjustment the body makes to a marked change in the temperature of the surrounding medium. As the room temperature rises, the sweat glands are stimulated to secrete by fibers of the sympathetic division of the ANS, and the surface of the skin becomes moist. If the humidity of the air is low enough, evaporation tends to cool the surface. At the same time peripheral vasodilatation of the arterioles and capillaries of the skin permits a greater volume of blood to be brought to the body's

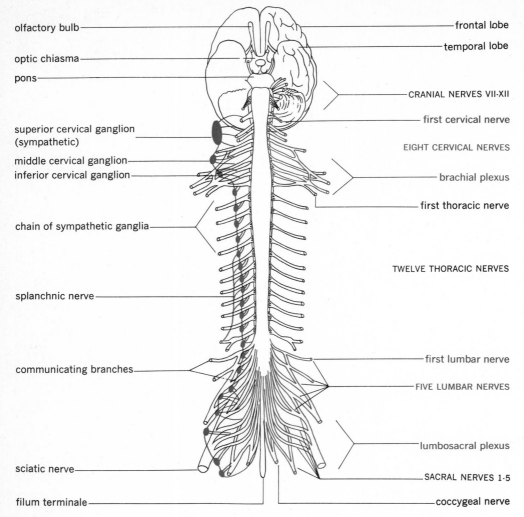

olfactory bulb

optic chiasma

pons

superior cervical ganglion (sympathetic)

middle cervical ganglion

inferior cervical ganglion

chain of sympathetic ganglia

splanchnic nerve

communicating branches

sciatic nerve

filum terminale

frontal lobe

temporal lobe

CRANIAL NERVES VII-XII

first cervical nerve

EIGHT CERVICAL NERVES

brachial plexus

first thoracic nerve

TWELVE THORACIC NERVES

first lumbar nerve

FIVE LUMBAR NERVES

lumbosacral plexus

SACRAL NERVES 1-5

coccygeal nerve

FIGURE 10·1

Ventral view of the brain and spinal cord. Cranial and spinal nerve roots are shown, including the chain of sympathetic ganglia, on one side only. [*Redrawn from J. Parsons Schaeffer (ed.), "Morris' Human Anatomy," 12th ed., McGraw-Hill Book Company, New York, 1966.*]

surface. Skeletal muscles tend to relax when the body is exposed to high temperatures, reducing production of heat from contraction.

When the body is exposed to cold, the skin surface is nearly dry, and the surface arterioles constrict to keep more blood away from the cool surface and thus conserve heat. Skeletal muscles increase their tone, and shivering, which produces more heat from the repeated contraction of muscles, may begin.

Physical adjustments to an emergency are largely controlled by the autonomic nervous system. Preparations to strengthen the body for a critical situation include acceleration and strengthening of the heartbeat, raising the blood pressure, release of glucose from the liver, and the secretion of a small amount of epinephrine by the adrenal glands. Breathing is made easier by the relaxation of muscles in the bronchial tubes. During an emergency,

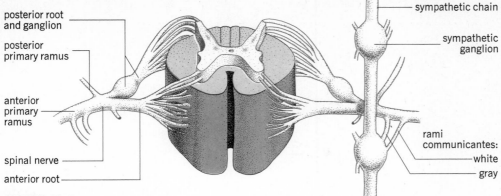

posterior root
and ganglion

posterior
primary ramus

anterior
primary
ramus

spinal nerve

anterior root

sympathetic chain

sympathetic
ganglion

rami
communicantes:
white
gray

FIGURE 10·2

Segment of spinal cord showing nerve roots and connection
with sympathetic chain (anterior view).

digestion can wait, and the activity of the digestive system is therefore depressed; the blood supply is largely diverted from the digestive system to the skeletal muscles. These effects are obtained mainly by the stimulation of one division of the autonomic nervous system: the sympathetic, or thoraco-lumbar, portion.

The autonomic system is divided somewhat artificially into a *thoraco-lumbar,* or *sympathetic,* portion and a *craniosacral,* or *parasympathetic,* part. The thoracolumbar division is composed of a chain of ganglia and nerves that extends on either side of the spinal cord from the cervical region through the thoracic and lumbar regions (Fig. 10.1). Throughout the tho-racic and lumbar regions, each ganglion is connected to a spinal nerve by a communicating branch (Fig. 10.2). Fibers extend upward to the head from the superior cervical ganglion; they also extend downward from the sacral ganglia, thus increasing the distribution of sympathetic fibers.

The craniosacral, or parasympathetic, division is associated with certain cranial and sacral nerves. The terms *thoracolumbar* and *craniosacral* are well suited to anatomical considerations; the terms *sympathetic* and *para-sympathetic* seem better suited to discussions of the physiological aspects of the autonomic nervous system.

SYMPATHETIC, OR THORACOLUMBAR, DIVISION

In the sympathetic division of the autonomic nervous system, motor im-pulses are conveyed over two visceral efferent neurons instead of the one seen in somatic nerves (Fig. 10.3). For example, in somatic efferent nerves there is only one neuron between the central nervous system and the skeletal muscle. In the autonomic nervous system, however, there are two neurons between the central nervous system and the effector, which may be smooth muscle, cardiac muscle, or glandular tissue. The first neuron, called the *preganglionic neuron,* extends from the central nervous system to a ganglion where it synapses with a second neuron. The second neuron, termed the *postganglionic neuron,* extends from the ganglion to the effector (Fig. 10.4). A synaptic connection between the two neurons is ordinarily made in a ganglion of the thoracolumbar chain, although this is not necessarily so. A preganglionic neuron has its cell body in the intermediolateral column of gray matter of the spinal cord. Its axon extends, ordinarily, from the cell

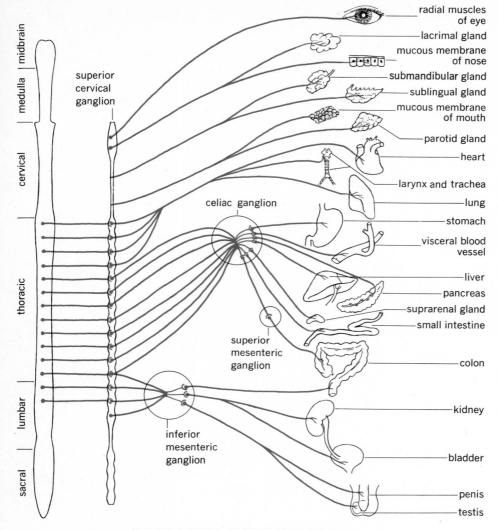

radial muscles of eye
lacrimal gland
mucous membrane of nose
submandibular gland
sublingual gland
mucous membrane of mouth
parotid gland
heart
larynx and trachea
lung
stomach
visceral blood vessel
liver
pancreas
suprarenal gland
small intestine
colon
kidney
bladder
penis
testis

superior cervical ganglion

celiac ganglion

superior mesenteric ganglion

inferior mesenteric ganglion

midbrain
medulla
cervical
thoracic
lumbar
sacral

THORACOLUMBAR (SYMPATHETIC) SYSTEM

FIGURE 10·3

The thoracolumbar system. *(Adapted from "Blakiston's New Gould Medical Dictionary," 2d ed., McGraw-Hill Book Company, New York, 1956.)*

body to the autonomic ganglion outside the cord. The postganglionic neuron has its cell body in the ganglion, and its axon extends to a visceral muscle. Some preganglionic fibers may extend through the autonomic ganglion to a collateral ganglion. In these cases, a short postganglionic fiber is supplied to the organ. These pathways are summarized in Table 10.1.

PREGANGLIONIC NEURON   The cell body of a sympathetic preganglionic neuron is smaller than that of a motor neuron of the central nervous system, and the particles of its Nissl substance are finer and more rounded. Its axon emerges from the spinal cord as a part of the motor root of a spinal nerve

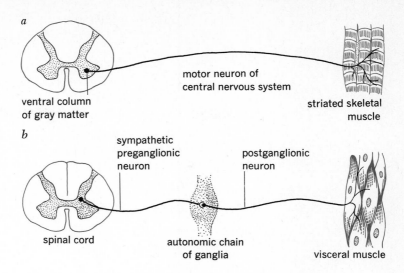

a

ventral column
of gray matter

motor neuron of
central nervous system

striated skeletal
muscle

b

sympathetic
preganglionic
neuron

postganglionic
neuron

spinal cord

autonomic chain
of ganglia

visceral muscle

FIGURE 10·4

Common types of motor pathways: *a* spinal peripheral motor neuron with no synapse between the spinal cord and skeletal muscle; *b* typical sympathetic pathway; *c* the sympathetic preganglionic neuron, in this case, passes through the chain of ganglia to a collateral ganglion; the shorter postganglionic neuron leads to the organ supplied.

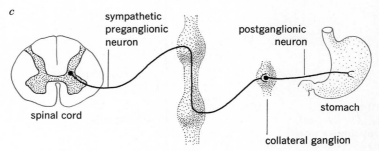

c

sympathetic
preganglionic
neuron

postganglionic
neuron

spinal cord

stomach

collateral ganglion

but soon leaves it to enter an autonomic ganglion. The majority of preganglionic axons are myelinated. A group of myelinated fibers presents a white appearance, and so the group traveling to the sympathetic ganglion is called the *white branch,* or *white ramus communicans.* When a preganglionic neuron enters a sympathetic ganglion, it makes a synaptic connection with many postganglionic neurons. This arrangement is significant, since it provides for the rapid, widespread response characteristic of the sympathetic system (Fig 10.5).

The thoracic and the first two lumbar nerves are connected with the autonomic chain of ganglia by a white rami; hence the name *thoracolumbar*

TABLE 10·1
Pathways of the preganglionic neuron

1 May synapse in the vertebral chain of ganglia
2 May pass through a ganglion to another ganglion in the chain before synapsing
3 May pass through a ganglion and proceed to a collateral ganglion to synapse

The preganglionic neuron always synapses in a ganglion or passes through a ganglion of the sympathetic chain

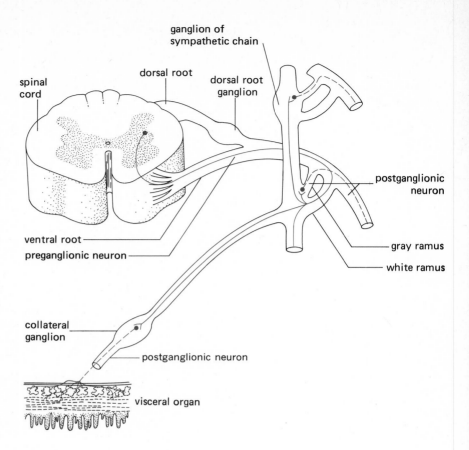

spinal cord

dorsal root

ganglion of sympathetic chain

dorsal root ganglion

postganglionic neuron

gray ramus

white ramus

ventral root

preganglionic neuron

collateral ganglion

postganglionic neuron

visceral organ

FIGURE 10·5

A cross section of the spinal cord with spinal nerve roots and sympathetic ganglia.

for this division of the ANS. Cervical ganglia are supplied by preganglionic fibers that extend upward from the thoracic nerves through the lateral chains of ganglia. The lower lumbar and sacral ganglia are supplied by fibers that extend downward from the thoracic and lumbar nerves.

POSTGANGLIONIC NEURON   A postganglionic neuron of the thoracolumbar division has its cell body in a sympathetic chain ganglion or in a collateral ganglion. Its axon extends to involuntary muscle tissue or to glandular cells; thus the cell bodies of the autonomic chain ganglia are entirely motor. Postganglionic fibers may take two courses that extend beyond the lateral ganglia. They may proceed inward by way of a visceral branch to terminate in the muscles of the viscera, or they may rejoin the spinal nerve by way of a gray root, usually called the *gray ramus communicans,* and terminate in the involuntary muscles of the peripheral region, such as the muscles in the walls of the blood vessels or those in sweat glands of the skin. Since these postganglionic fibers are not myelinated, the connection appears gray in contrast with the white branch of myelinated preganglionic fibers.

Both white and gray rami are attached to the thoracic and upper lumbar spinal nerves, but the cervical, lower lumbar, sacral, and coccygeal nerves have no white rami since no preganglionic sympathetic fibers originate in these nerves. However, each spinal nerve has a gray branch and therefore receives sympathetic postganglionic fibers.

Sympathetic plexuses
  Cardiac
  Celiac
  Hypogastric

SYMPATHETIC PLEXUSES   The great plexuses of the autonomic system are the cardiac; the celiac, or solar; and the hypogastric plexuses. Though these plexuses are regarded as essentially sympathetic, they also receive fibers from the parasympathetic system. A nerve plexus, as we have said previously, is a network of interlacing nerves.

The *cardiac plexus* lies under the arch of the aorta just above the heart. It receives branches from the cervical sympathetic ganglia and from the right and left vagal nerves, i.e., nerves of the parasympathetic division, and regulates the action of the heart.

The *celiac, or solar, plexus* is the largest network of cells and fibers of the autonomic system. It lies behind the stomach and is associated with the aorta and the celiac arteries. The ganglia of this plexus receive the splanchnic nerves from the sympathetic system and branches of the vagus

FIGURE 10·6

The chain of ganglia and the larger plexuses of the sympathetic system (lateral view).

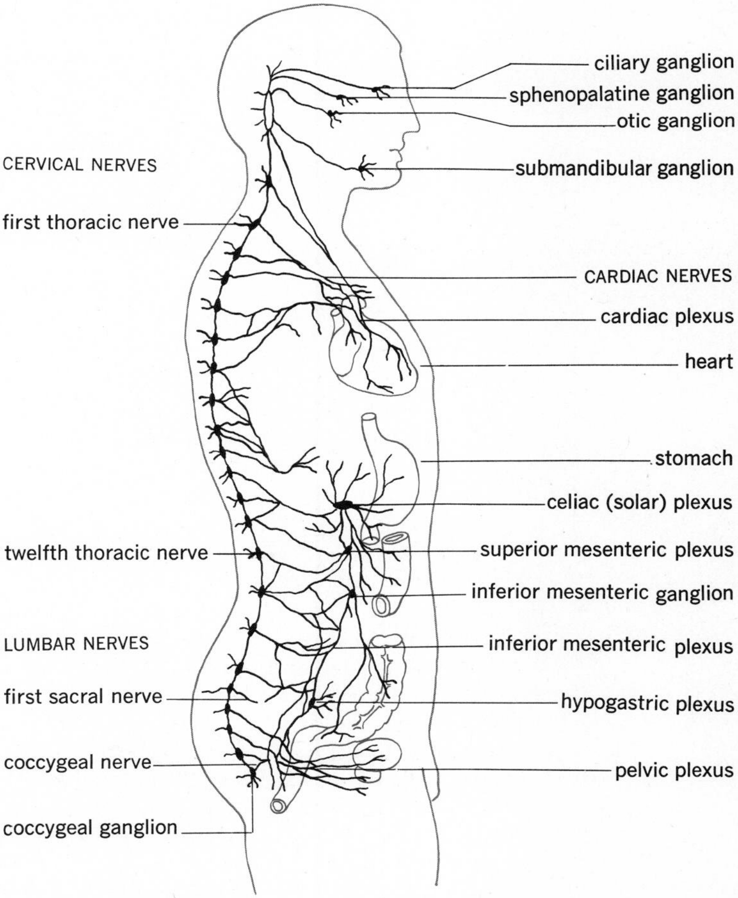

CERVICAL NERVES

first thoracic nerve

twelfth thoracic nerve

LUMBAR NERVES

first sacral nerve

coccygeal nerve

coccygeal ganglion

ciliary ganglion
sphenopalatine ganglion
otic ganglion
submandibular ganglion
CARDIAC NERVES
cardiac plexus
heart
stomach
celiac (solar) plexus
superior mesenteric plexus
inferior mesenteric ganglion
inferior mesenteric plexus
hypogastric plexus
pelvic plexus

from the parasympathetic system. A blow to this region may slow the heart, reduce the flow of blood to the head, and depress the breathing mechanism.

The *hypogastric plexus* forms a connection between the celiac plexus above and the two pelvic plexuses below. It is located in front of the fifth lumbar vertebra and continues downward in front of the sacrum, forming the right and left pelvic plexuses. These plexuses supply the organs and blood vessels of the pelvis (Fig. 10.6).

PARASYMPATHETIC, OR CRANIOSACRAL, DIVISION

The parasympathetic, or craniosacral, division of the ANS is associated with certain cranial and sacral nerves in which autonomic fibers are incorporated, hence the name *craniosacral division* (Fig. 10.7). The oculomotor (IIId

FIGURE 10·7

The craniosacral system. (*Adapted from "Blakiston's New Gould Medical Dictionary," 2d ed., McGraw-Hill Book Company, New York, 1956.*)

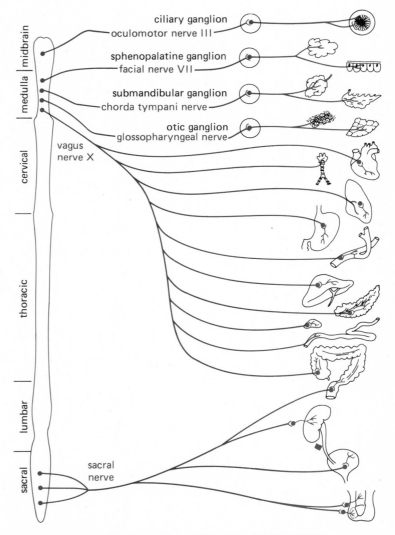

CRANIOSACRAL (PARASYMPATHETIC) SYSTEM

cranial) nerve, which arises in the midbrain, innervates certain voluntary muscles that move the eyeball; in addition, it carries parasympathetic fibers to involuntary muscles within the eyeball. Preganglionic fibers are distributed to the ciliary ganglion located behind the eyeball. Postganglionic fibers arising in the ganglion extend to the ciliary muscles of the eye and to the sphincter of the pupil.

The facial (VIIth cranial), glossopharyngeal (IXth cranial), vagus (Xth cranial), and accessory (XIth cranial) nerves constitute a group of cranial nerves arising from the medulla. Since they also contain parasympathetic fibers, they are considered to be a part of the craniosacral division. The vagus supplies the viscera of the thorax and abdomen; this may be why no parasympathetic fibers arise from the thoracic or lumbar regions of the cord.

The sacral portion of this system is identified with certain sacral nerves that carry parasympathetic fibers to the pelvic viscera.

FIGURE 10·8

Various types of autonomic pathways: *a* parasympathetic pathway with a synapse in a collateral ganglion; *b* typical parasympathetic pathway, with a short postganglionic neuron within the organ supplied; *c* sympathetic pathway to visceral muscle, showing a synapse in the sympathetic ganglion; *d* typical synapse of a sympathetic preganglionic neuron; the postganglionic fiber reaches the sweat gland through a peripheral nerve.

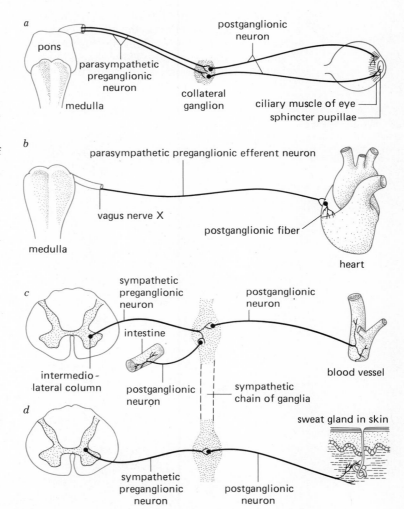

PREGANGLIONIC AND POSTGANGLIONIC FIBERS   A parasympathetic preganglionic fiber typically extends from its nucleus in the brain or sacral region of the spinal cord to the organ supplied. The postganglionic fiber is often a very short fiber located within the organ itself. The preganglionic fiber may end in a *collateral ganglion*. For example, preganglionic fibers extend to the ciliary ganglion of the eye. The postganglionic fibers in this case are longer than those incorporated within the innervated organ (Fig. 10.8).

Parasympathetic cranial nerves
  Facial
  Glossopharyngeal
  Vagus
  Accessory

CRANIAL NERVES CARRYING PARASYMPATHETIC FIBERS   It has been mentioned that four cranial nerves, which arise from the medulla, carry autonomic fibers. These are, therefore, a part of the craniosacral system. These nerves are the facial, glossopharyngeal, vagus, and accessory nerves.

The *facial nerve* includes parasympathetic fibers that cause the lacrimal gland and the sublingual and submandibular salivary glands to secrete. The lacrimal gland is supplied with postganglionic fibers from the sphenopalatine ganglion. Sublingual and submandibular salivary glands receive postganglionic fibers that arise in the submandibular ganglion (Fig. 10.9).

Preganglionic fibers in the *glossopharyngeal nerve* extend outward to the *otic ganglion*. Postganglionic fibers arise in the otic ganglion and supply the parotid salivary glands. These glands, including the lacrimal, have both sympathetic and parasympathetic innervation. They derive their sympa-

FIGURE 10·9
Autonomic innervation of the salivary glands. Afferent neurons of cranial nerves V, VII, and IX are shown by dotted lines.

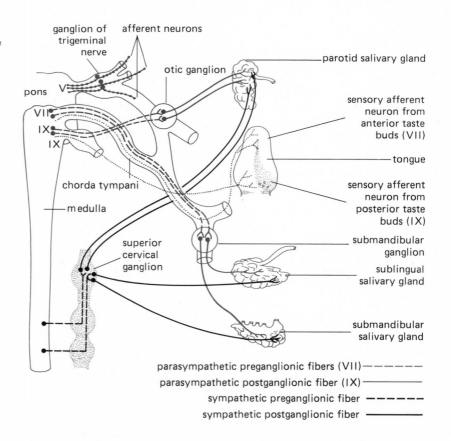

parasympathetic preganglionic fibers (VII) ————
parasympathetic postganglionic fiber (IX) ————
sympathetic preganglionic fiber — — — — —
sympathetic postganglionic fiber ————

thetic innervation from the superior cervical sympathetic ganglion and the carotid plexuses. The functions of these two sets of nerves is not clear. Both parasympathetic and sympathetic secretory fibers are present, but the parasympathetic fibers seem to be dominant.

The *vagus nerve* contains both motor and visceral afferent axons. Long preganglionic motor axons extend through the vagus to the organ supplied. Very short postganglionic neurons make connections with the long preganglionic within the organ. The vagus supplies motor axons to the larynx, trachea, bronchioles, heart, esophagus, stomach, small intestine, and some parts of the large intestine. Stimulation of the vagus inhibits action of the heart, causing the heart rate to slow or to stop. Branches of the vagus act as accelerator nerves to the muscles of the wall of the digestive tract. Peristalsis is increased by parasympathetic stimulation. Parasympathetic fibers to the glands of the digestive tract have a regulatory function on secretion, but food content of the stomach or intestine and hormones circulating in the blood can also stimulate secretion (Fig. 10.10).

Parasympathetic fibers from both the right and left vagus nerves enter the great plexuses of the sympathetic system. There is, however, a definite parasympathetic nerve supply to such organs as the pancreas, liver, and kidneys. Nervous stimulation of these organs is, for the most part, merely regulatory. Hormones in the blood normally cause the pancreas and liver to secrete, but stimulation of the vagus increases the flow of pancreatic juice.

The autonomic nervous system is considered to be entirely motor by some neurologists. Other neurologists consider visceral afferent neurons to be part of this system. The broader classification is followed here. Visceral afferent neurons carry nerve impulses from the organs of the body to the

FIGURE 10·10

Distribution of parasympathetic neurons to the viscera by the vagus nerves.

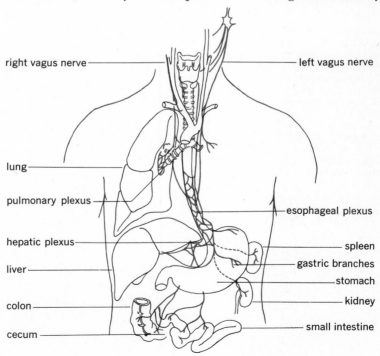

right vagus nerve — left vagus nerve

lung

pulmonary plexus — esophageal plexus

hepatic plexus — spleen

— gastric branches

liver — stomach

— kidney

colon — small intestine

cecum

central nervous system where they are interpreted as pain. Visceral afferent impulses from the stomach may be interpreted as hunger.

A part of the *accessory nerve* contains visceral motor and cardiac inhibitory axons.

THE SACRAL AUTONOMICS The sacral portion of the craniosacral system is composed of preganglionic fibers incorporated in the second, third, and fourth sacral nerves. The fibers extend to the pelvic plexuses, where they enter into close relationship with neurons of the sympathetic system. Parasympathetic presympathetic neurons innervate the urogenital organs and the distal part of the colon. Postganglionic neurons are considered to arise in the organs supplied or in small ganglia located close by. Such parasympathetic neurons send motor impulses to the muscles of the distal two-thirds of the colon, to the rectum, and to the urinary bladder. They carry vasodilator impulses to the penis and clitoris. Inhibitory impulses pass to the internal sphincter muscle of the bladder and to the internal sphincter of the anus.

Enteric plexuses
Myenteric
Submucous

PARASYMPATHETIC PLEXUSES (ENTERIC PLEXUSES) The digestive tube has its own intrinsic nerve supply in the *enteric plexuses*. These plexuses consist of the *myenteric plexus*, located between the longitudinal and circular muscles, and the *submucous plexus*, located under the mucous layer in the submucosa (Fig. 10.11). This part of the nervous system extends the entire length of the digestive tube. It may be assumed that parasympathetic fibers which enter the walls of the digestive tract are preganglionic fibers that make synaptic

FIGURE 10·11

Enteric plexuses: *a* a longitudinal section through the wall of the small intestine, illustrating the location of the myenteric and submucous plexuses (these intestinal plexuses are considered to act as postganglionic neurons of the parasympathetic system and may act in this capacity for the sympathetic system also); *b* surface view of the myenteric plexus; *c* surface view of the submucous plexus.

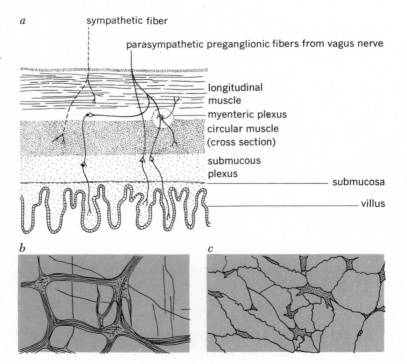

connections with neurons of the enteric system. Sympathetic fibers entering the muscular wall, however, are postganglionic fibers, and until recently have been considered to terminate on the muscles they supply without making synaptic connections at the intestinal plexus.

New evidence suggests that sympathetic postganglionic neurons may not end directly on intestinal muscle. It now appears that sympathetic postganglionic neurons may synapse on the ganglion cells of the submucous and myenteric plexuses along with the parasympathetic neurons.

The enteric plexuses maintain rhythmic peristaltic movements along the digestive tract. Peristalsis is maintained if both sympathetic and parasympathetic nerve supplies are cut, but the nerves of the autonomic system modulate such activity (see Table 10.2).

PHYSIOLOGY OF THE AUTONOMIC NERVOUS SYSTEM

SYMPATHETIC AND PARASYMPATHETIC RELATIONSHIPS The parasympathetic system functions as an antagonist of the sympathetic system when an organ is innervated by both systems. If the sympathetic system is the accelerator system, as in the heart, for example, then the parasympathetic system is the inhibitor. The function of the parasympathetic system in this case is to slow the accelerated heart and thus restore normal heart rate. Even though the parasympathetic system acts as an inhibitor, it does not ordinarily depress the heart rate below normal unless it is unduly stimulated by the action of drugs or by pressure on a nerve.

An example of sympathetic-parasympathetic interaction may be seen in the oculomotor nerve. If this nerve is cut, the pupil dilates. The parasympathetic fibers within the oculomotor nerve carry nervous impulses that cause the pupil to constrict. Cutting the nerve destroys the balance between parasympathetic and sympathetic innervation. The sympathetic nervous impulses from the superior cervical ganglion then cause the pupil to dilate. The "drops" placed in the eye during an optical examination act by blocking the parasympathetic nerve impulses, thus causing the pupil to dilate.

Effects of the autonomic nervous system are usually conditioned by other factors such as the presence of hormones in the bloodstream or by circulatory effects. The secretion of a gland may be depressed by the stimulation of an inhibitor nerve or by vasoconstriction of blood vessels supplying the gland, thus limiting its blood supply.

Although the sympathetic system is an accelerator to the heart, the situation is reversed in the digestive tract. Here the sympathetic nerves depress peristalsis and the secretion of digestive glands during emotional excitement, while the parasympathetic system accelerates digestive activity. Parasympathetic stimulation during emotional stress may cause increased production of hydrochloric acid in the stomach, a condition that may lead to the development of stomach ulcers.

When we speak of the sympathetic and parasympathetic nerves as being *antagonistic,* we mean this in the sense of antagonistic muscles. The nerves from the sympathetic and parasympathetic systems may produce opposite effects, but they work together to provide smooth adjustments to many physiological conditions.

Sympathetic and parasympathetic effects are not always clearly antagonistic. The accommodation reflex of the eye in which the lens and iris are adjusted to facilitate clear vision appears to be controlled primarily by the parasympathetic system, and to a lesser extent by the sympathetic system. The two sets of muscles of the iris work together in a synergistic relationship that causes them to contract or dilate the pupil smoothly. The pupil also dilates, however, in response to an emotional state such as fear or pain through stimulation by the sympathetic system.

Some effects of the autonomic system are summarized in Table 10.2.

NEGATIVE FEEDBACK SYSTEMS OF THE ANS   The autonomic nervous system provides numerous examples of physiological mechanisms that regulate the maintenance of homeostasis. Such control mechanisms are examples of the principle of *negative feedback*, which we discussed in Chap. 3.

Body-heat regulation   The sweat glands are innervated by neurons of the sympathetic system, which are carried by nerves of the peripheral nervous system. Although the usual response of the eccrine sweat glands to high temperature is an increase in sweating, these glands do not respond in this

TABLE 10·2
Some effects of the autonomic nervous system

Structure	Sympathetic	Parasympathetic
Eye		
Iris	Dilatation of the pupil	Contraction of the pupil
Ciliary muscle	Relaxation of ciliary muscles, accommodation for distance vision	Contraction of ciliary muscles, accommodation for close-up vision
Bronchial tubes	Dilatation	Constriction
Heart	Accelerates and strengthens actions	Depresses and slows action
Blood vessels		
Coronary arteries	Dilatation	Constriction
Abdominal and pelvic viscera	Constriction	
Peripheral blood vessels	Constriction	
External genitalia	Constriction of blood vessels	Dilatation of blood vessels, erection
Stomach		
Muscles	Depresses activity	Increases activity
Glands	Alters secretion	Increases secretion
Liver	Stimulates glycogenolysis	
Visceral muscle of intestine	Depresses peristalsis	Increases peristalsis
Adrenal medulla	Secretion of epinephrine	
Adrenergic fibers	Most of the sympathetic postganglionic fibers; among the exceptions are those to the sweat glands	
Cholinergic fibers	All preganglionic fibers; all parasympathetic postganglionic fibers; motor nerve fibers to voluntary muscle; postganglionic sympathetic fibers to the sweat glands	

way during a fever. In the latter case, the toxins produced by disease apparently affect the heat regulatory center in the hypothalamus, "setting the thermostat" higher than normal. The heat regulatory mechanism is altered in several ways, actions that cause heat to be conserved and therefore result in a rise in temperature. Sweating in response to a strong emotional situation may also involve the apocrine sweat glands and may be most evident on the forehead, in the palms of the hands, the soles of the feet, and in the axillae of the arms. Regulation of sweat gland activity involves the integrated action of the hypothalamus, centers in the brainstem, and directly, hormone secretion.

In general, the hypothalamus acts as a coordinating center for the autonomic nervous system, helping to regulate the functions of the viscera.

Circulatory effects The sympathetic system greatly influences vasoconstriction through neurons, but other factors play their part as well. Some of these factors are epinephrine, norepinephrine, oxygen tension, carbon dioxide tension, and temperature. Sympathetic neurons that innervate the blood vessels show tonic, or continuous, activity. If their activity is increased, greater constriction of the blood vessels is produced. If activity is decreased, vasodilatation occurs. Relaxation of smooth muscle in the walls of blood vessels is probably accomplished by autoregulation. Most blood vessels (except capillaries) apparently receive only sympathetic innervation and are regulated by sympathetic tone (Fig. 10.12).

Vasoconstriction may be localized or general. In an emergency calling for quick action, general vasoconstriction causes a rise in blood pressure. At the same time, vasoconstriction may reduce the flow of blood to the digestive tract. Muscular exercise requires an increased flow of blood to the skeletal muscles and, therefore, vasodilatation of the blood vessels supplying them. Coronary arteries supplying the heart muscle are also dilated. The action of the sympathetic system is supported by release of epinephrine and norepinephrine from the adrenal medulla. These hormones tend to have a more prolonged effect on receptive tissue than the direct effect of sympathetic innervation.

There is little, if any, neural control of vasoconstriction or vasodilatation in the coronary arteries. The rate of flow depends largely upon the arterial blood pressure and the amount of resistance offered by the arteries themselves. Local metabolic control systems play a major part.

FIGURE 10·12
Effect of stimulation by sympathetic neurons on blood vessels.

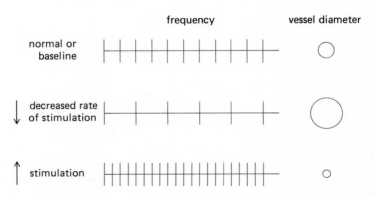

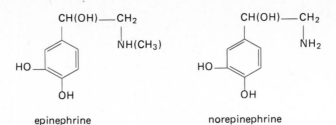

FIGURE 10·13
The chemical structure of epinephrine and norepinephrine.

epinephrine

norepinephrine

CHEMICAL TRANSMITTERS OF THE ANS   *Adrenergic neurons*   The terminal filaments of most sympathetic postganglionic neurons produce *norepinephrine*
(Fig. 10.13), and such neurons are classified as *adrenergic*. Sympathetic
neurons to the blood vessels of skeletal muscles, the skin, and sweat glands
are exceptions since they produce acetylcholine and are therefore *cholinergic*. Such postganglionic fibers enter peripheral nerves to reach blood vessels
and other structures.

The effects of norepinephrine, as well as those of epinephrine, are
general and widespread. These chemical substances are carried by the
bloodstream and may affect organs remote from their place of origin.
Norepinephrine is broken down by the enzyme catechol-*o*-methyl transferase, but most of it is reabsorbed by the nerve endings. The sympathetic
ganglia and the medullary portion of the adrenal glands, both of which are
involved in secretion of epinephrine and norepinephrine, have the same
embryonic origin. They both arise from neural crest cells (Fig. 9.1). Epinephrine is produced by the medullary portion of the adrenal glands, which
secrete 80 percent epinephrine and 20 percent norepinephrine.

ALPHA AND BETA RECEPTORS   It is commonly recognized that epinephrine
and norepinephrine sometimes have different effects and sometimes similiar
effects on body tissues. To explain these seemingly contradictory results, it
has been suggested that there are two kinds of adrenergic receptors: *alpha*
and *beta*. It has been proposed that norepinephrine binds only to alpha
receptors, while epinephrine exerts its effect on both alpha and beta sites.
This concept appears to be too simple, however, to explain the problem.
Such receptors are now referred to as "patterns of forces," indicating some
of the difficulties involved in interpreting their functions. In addition to
alpha and beta receptors, inhibitors known as *alpha* and *beta blockers* block
the effectiveness of one receptor, while allowing transmitters full access to
the other one. Alpha and beta receptors also have been shown to be
involved in the release of hormones. As an example, growth hormone release
is said to be stimulated by alpha receptors and inhibited by beta receptors.

Alpha receptors are thought to be concerned with vasoconstriction,
dilatation of the pupil by muscles of the iris, and relaxation of the muscles of
the intestine. Beta receptors are considered to be involved in vasodilatation,
stimulation of the heart rate, and strengthening of heart contraction. At the
present time, we must consider that the action of these receptors is poorly
understood and very complex.

Both types of receptors are intimately associated with adenyl cyclase on

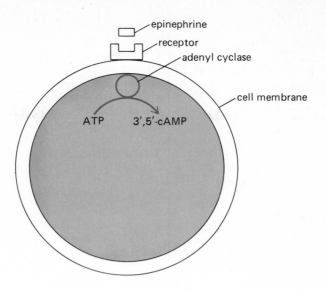

FIGURE 10·14
Epinephrine receptor on cell membrane.

the inner surface of the cell membrane. When activated, this enzyme forms 3′,5′ cyclic AMP from ATP (Fig. 10.14).

Cholinergic fibers Parasympathetic fibers also produce a chemical transmitter substance. In this case the substance is *acetylcholine*, which is promptly converted to choline and acetic acid by the action of an enzyme called *cholinesterase*. Since acetylcholine does not remain in its most active state for any great length of time, it is probable that its effects are entirely local. Unlike norepinephrine, it is probably not carried by the bloodstream.

All autonomic preganglionic fibers, whether sympathetic or parasympathetic, have been shown to release a cholinergic substance, probably identical to acetylcholine. The transmission of nervous impulses across the synapse between the preganglionic and postganglionic neurons of the ANS is accomplished, therefore, by the production of acetylcholine.

As we have indicated, postganglionic sympathetic neurons that innervate the sweat glands and the smooth muscles of the skin are cholinergic. The axons of such neurons are carried by peripheral nerves. Voluntary motor neurons that innervate skeletal muscles are also cholinergic. If we classify neurons on the basis of the chemical transmitter substances that they produce, the division of the autonomic system into sympathetic and parasympathetic appears somewhat artificial.

VISCERAL AFFERENT NEURONS

Some afferent neurons arise in the viscera and are associated with the autonomic nervous system. Such neurons do not give rise to sensation in the ordinary sense. The sense of touch or localized pain is not recognized internally, as it is externally in the skin and body wall. The viscera are relatively insensitive to cutting, burning, or crushing pain.

The receptors of visceral afferent neurons are located in the viscera. Large, medullated fibers extend through the autonomic ganglion and the white ramus to the cell bodies of neurons located in the ganglion of the dorsal nerve root (Fig. 10.5). (Visceral afferent neurons do not synapse in the autonomic ganglion.) Each neuron extends from the viscera to the spinal cord. From cell bodies located in the ganglia of the dorsal spinal nerve roots, visceral afferent processes enter the dorsal horn of gray matter of the cord along with the processes of sensory spinal nerves. They may synapse with interneurons or with ascending tracts in the spinal cord. Visceral afferent impulses also can reach the medulla by way of the vagus and other cranial nerves. Even though the vagus nerve may not carry recognizable pain impulses, visceral afferent stimuli are involved in the reflex that causes the pupil of the eye to dilate when the vagus nerve is stimulated in the abdominal region.

CLINICAL ASPECTS

Drugs affecting the sympathetic system Certain drugs affect the functioning of the autonomic nervous system. Most drugs that affect the sympathetic system are administered to alleviate *hypertension*. Such drugs control the sympathetic effect on vasoconstriction. If these drugs fail to eliminate hypertension, a *sympathectomy* may be performed. This surgical operation involves the removal of sympathetic ganglia from the thoracic and lumbar ganglia chain. Blood pressure then is reduced, but the patient may experience difficulty in raising the blood pressure during stress or following sudden changes in position, e.g., after rising quickly from a reclining position.

Some drugs that oppose sympathetic effects of the autonomic nervous system and are therefore useful in reducing hypertension caused by vasoconstriction are reserpine, methyldopa, and hydralazine.

Reserpine increases vasodilatation and slows heart rate.

Methyldopa interferes with the synthesis of norepinephrine and is a vasodilator.

Hydralazine depresses the vasoconstrictor center in the medulla, thereby partially blocking sympathetic effects.

Drugs affecting the parasympathetic system The drug *atropine* acts as an antagonist to acetylcholine of the parasympathetic system. It is used primarily to dilate the pupil in eye examination. *Pilocarpine* acts as a mimetic of acetylcholine, the transmitter of the parasympathetic system. It is a physiological antagonist to atropine. Pilocarpine nitrate is a miotic that causes contraction of the pupil and is used in the treatment of glaucoma.

Visceral pain The exact way in which pain arises is not well understood. It may be caused by certain kinds of muscular contraction in visceral muscle or by pressure within a hollow organ. Internal pain is not readily localized; it is often "referred" to some surface area. One explanation of such *referred pain* is that since the visceral afferent neuron enters the gray matter of the cord in close association with a spinal sensory nerve, the pain is interpreted as arising in the surface area supplied by that spinal nerve. Thus pain arising in the gallbladder is felt as if it were coming from the surface area above the organ. Cardiac pain may be felt in the scapular area or, as in *angina pectoris*, can radiate down the left arm. A reflex arc may be established between visceral afferents and spinal motor neurons, which would account for the tensing of muscles over an area of internal pain.

Though referred pain is often described as an acute, sharp, stabbing pain, a direct type of visceral pain is described as a dull pain. It is usually not well localized, but the individual is aware of general discomfort. Afferent visceral neurons carry the impulses that give rise to the direct type of pain. Such impulses enter the spinal cord, but little is known about the mechanism by which awareness of such pain is transmitted.

Autonomic functions, biofeedback, and meditation   Animals subjected to operant conditioning using reward or punishment have been able to learn to increase or decrease heart rate, blood pressure, intestinal muscle contractions, and a number of other visceral responses controlled by the autonomic system. Human subjects have also had some success in modifying blood pressure and heart rate by the training method known as *biofeedback*. This method is in the early stages of investigation, but physiologists may have to reconsider their belief that the autonomic nervous system is purely involuntary.

The effects of various kinds of prolonged meditation do not differ physiologically from other methods of relaxation. Their functional effects, mediated by the autonomic nervous system, may be beneficial. Like most methods of relaxation, they produce restful sleep and a general sense of well-being.

SUMMARY

1 The autonomic nervous system is divided into a thoracolumbar, or sympathetic, division, and a craniosacral, or parasympathetic, division.

2 The thoracolumbar division is characterized by a chain of ganglia on either side of the spinal cord in the thoracic and lumbar regions. A preganglionic neuron extends from the spinal cord to one of the ganglia of the lateral chain. A postganglionic neuron extends from the autonomic ganglion to the organ supplied.

3 The great plexuses of the autonomic system are the cardiac, celiac, and hypogastric plexuses.

4 The craniosacral division is associated with certain cranial and sacral nerves. In this division, preganglionic fibers are carried in cranial or sacral nerves. Postganglionic fibers arise in collateral ganglia or within the organ supplied.

5 Cranial nerves that carry parasympathetic fibers are the oculomotor (IIId cranial), facial (VIIth cranial), glossopharyngeal (IXth cranial), vagus (Xth cranial), and accessory (XIth cranial) nerves.

6 The enteric plexuses form the intrinsic nerves of the intestine. The intestine also receives postganglionic sympathetic fibers and preganglionic parasympathetic fibers. The enteric system neurons are considered to act as postganglionic neurons of the parasympathetic system.

7 The hypothalamus is a center for autonomic regulation.

8 Sympathetic fibers that release norepinephrine at their terminal filaments are said to be adrenergic. Sympathetic postganglionic fibers are adrenergic except those innervating sweat glands, the smooth muscles of the skin, and the blood vessels of skeletal muscles.

9 Parasympathetic fibers release acetylcholine as a chemical transmitter substance, and, like motor neurons of peripheral nerves, they are cholinergic. Sympathetic preganglionic fibers also are cholinergic.

10 Visceral afferent fibers carry sensory impulses from the viscera, which

are interpreted as pain. Often such impulses are interpreted as referred pain, which arises in a certain area but seems to come from some other area. The spinal visceral afferent axons pass through the autonomic ganglia, but the cell bodies are located in ganglia of the dorsal spinal nerve roots along with the cell bodies of spinal sensory nerves. Visceral afferent neurons do not synapse until they reach the dorsal column of gray matter of the spinal cord.

QUESTIONS

1 List some of the physiological adjustments mediated by the autonomic system that take place in climbing stairs.
2 Discuss the autonomic effects of worry or a severe fright.
3 Why is it advantageous to be happy and to have pleasant surroundings at mealtime?
4 If the oculomotor nerve is severed, why does the pupil of the eye dilate?
5 Explain how it is possible to stimulate the vagus nerve and yet observe that the heart rate slows or stops altogether.
6 Discuss the role of the autonomic system in the control of body temperature.
7 By what pathway do sympathetic fibers reach the peripheral blood vessels and the sweat glands?
8 Why does the hair seem to stand on end when one is frightened?
9 Explain the chemical mediation of nerve impulses at the synapse.
10 What is meant by referred pain?
11 Explain negative feedback.
12 How does the negative feedback mechanism contribute to homeostasis?
13 Why is acetylcholine of little use as a drug?
14 What is a sympathectomy?

SUGGESTED READING

Axelrod, J.: Noradrenaline: Fate and Control of Its Biosynthesis, *Science*, **173**:598–600 (1971).
Axelsson, J.: Catecholamine Functions, *Annu. Rev. Physiol.*, **33**:1–30 (1971).
Cannon, W. B.: "Bodily Changes in Pain, Hunger, Fear and Rage," Appleton-Century-Crofts, Inc., New York, 1929.
Davidoff, R. A.: Gamma-aminobutyric Acid Antagonism and Presynaptic Inhibition in the Frog Spinal Cord., *Science*, **175**:331–333 (1972).
Dicara, Leo V.: Learning in the Autonomic Nervous System, *Sci. Am.*, **222**(1):30–39 (1970).
Hardman, J. G., G. A. Robinson, and E. W. Sutherland: Cyclic Nucleotides, *Annu. Rev. Physiol.*, **33**:331–336 (1971).
Michaels, R. R., et al.: Evaluation of Transcendental Meditation as a Method of Reducing Stress, *Science*, **192**:1242–1244 (1976).
Von Euler, U. S.: Adrenergic Neurotransmitter Functions, *Science*, **173**:202–205 (1971).
Wallace, R. K., and Herbert Benson: The Physiology of Meditation, *Sci. Am.*, **226**:85–90 (1972).

11

SPECIAL SENSES I: VISION

CONSIDERATIONS FOR STUDY

In this chapter, we take up vision, one of the special senses. We discuss this sense first because it is a highly developed and important one and because its anatomy and physiology are illustrative of concepts involved in all the special senses. Topics include:

1 The anatomy of the eye: its accessory structures, the muscles involved in eye movement and focusing, the structure of the eyeball
2 A brief account of the embryology of the eye
3 The rods and cones, the eye's receptor cells: their structure and function under different conditions
4 The mechanism of vision: the optical properties of the eye
5 The physiology of vision: the translation of the stimuli received by the visual receptors into nerve impulses and the coding of such impulses to produce vision
6 Color vision: the three-pigment theory of color vision

THE SPECIAL SENSES

The sense organs provide awareness of external stimuli, permitting us to react to forces impinging upon the body or arising within the body. Whatever we are able to learn about the world around us depends upon our recognition of various forces arising from an area outside our own bodies. It is not by accident that the highly developed sensory organs are located at the anterior end of the body and close to the brain. This has been the course of development in the evolutionary history of all organisms. Ordinarily sense organs are adapted to react to only one kind of stimulus. The eye receives light waves; the ear is stimulated by sound waves. *Perception* is the recognition of sensory or afferent stimuli by the brain. Perception requires that the sensory unit be intact and functioning.

THE SENSORY UNIT   The receptor reacts to only one primary type of stimulus, i.e. it responds to a certain *modality* of stimulation. If we define a motor unit as a motor neuron and its terminal filaments, then we should define a *sensory unit* as a sensory neuron with its receptors. The receptors constitute a *sensory field*. The sensory, or afferent, neuron makes connections with interneurons, and eventually the nerve impulse will reach interpreting centers in the brain. Receptors may be divided into various types by their structures, but their function is to react to certain types of stimuli, which they are adapted to receive. The rod and cone cells of the eye or the specialized nerve endings in the skin are examples of sensory receptors.

Intensity   The afferent neuron and its pathway are able to convey more than just the primary sensation. We sense the *intensity* of stimuli. Perception of stimulus intensity involves the number of sensory units stimulated and the frequency with which they fire. The intensity of sound waves is interpreted as *loudness,* the intensity of light waves as *brightness.* In most cases we are able to determine the *location* of the stimuli, as in those coming from the skin, or the *direction,* as in sound and light stimuli.

PROJECTION OF SENSATION   The interpretation of sensation occurs in more or less localized areas of the brain, but these sensations are projected to their *source.* If a finger is cut, the sensation of pain is projected to that particular finger. Our visual perceptions are directed to the object rather than to the area of interpretation in the brain. Sometimes the sensation is projected within the body, as when internal pain or referred pain is experienced or when the sensations of hunger or thirst are projected from organs of the digestive tract.

THE ACTION OF SENSORY RECEPTORS   Sensory receptors are biological transducers. In a physical sense, they receive energy in a particular form (light, sound, heat waves) and convert it into another form of energy, the nerve impulse. Mechanical stimulation of the receptor core in pacinian corpuscles initiates a weak local electric current, called the *generator potential.* This generator current starts the *action potential.* The nerve ending in the core is the transducer. The generator potential does not travel along the neuron; it only starts the action potential, which is then propagated along the fiber. The neuron is myelinated but the core is not. It has been determined that the action potential is initiated at the first node of Ranvier, which is located at the beginning of the neuron's myelin sheath.

Pressure on the pacinian corpuscle causes a wave of depolarization to pass along the core membrane, creating the generator potential. This potential is not an all-or-none response and is not self-propagating. It is a graded response, but if it is of sufficient strength, a propagated action potential is initiated over the afferent neuron when it reaches the low threshold level of the fiber's first node.

CLASSIFICATION OF RECEPTORS   Receptors are commonly classified into three groups: exteroceptors, proprioceptors, and interoceptors (Table 11.1). The

TABLE 11·1
Sensory receptors

Classification	Location	Source of stimulus
Exteroceptors	Skin, tongue, nose, ear, eye	External
Proprioceptors	Muscle spindles, tendons, joints	Stretch reflexes, postural reflexes, position in space
Interoceptors	The viscera	Endings of the visceral afferent nervous system

sense organs that receive stimuli from the exterior, such as the eye, ear, and skin, are *exteroceptors*. Exteroceptors are the receptors of the somatic sensory, or somatic afferent, system. *Proprioceptors* are located in muscle spindles and tendons and around joints. They are important in the interpretation of the position of any part of the body, in muscular coordination, in postural reflexes, and in determining the degree of stretch in muscle or tendon. Proprioceptors are the receptors for muscle sense, or proprioception. Sensations arising from stimuli to proprioceptors are ordinarily below the level of consciousness, but they can reach the level of consciousness. *Interoceptors* are the receptors of the visceral afferent system. They are located in the viscera, primarily in the organs of the respiratory, digestive, and reproductive systems. Stimuli arising from these receptors are ordinarily below the level of consciousness, although visceral pain can reach the conscious level. The visceral afferent system is closely related to autonomic functions and was discussed in Chap. 10.

THE EYE AND VISION

Visual receptors react to the physical stimulus of light. The eye looks out upon the world and by a very intricate mechanism reports its observations to the brain. The exact means by which receptors convert light radiations into nerve impulses and the interpretation of these impulses by the brain are still among the most perplexing problems in physiology.

ANATOMY OF THE EYE

ACCESSORY STRUCTURES   Before we take up the action of the eye itself, let us consider the accessory structures surrounding it. The *eyeball* lies in a bony orbit, which protects it on all sides except the anterior. It is cushioned by a padding of fat that lines the bony orbit.

A protective bony *brow ridge* above the eyes, together with the eyebrows, helps to ward off objects falling into the eyes from above. The eyebrows also direct perspiration away and tend to shade the eyes from the direct rays of the sun.

The *eyelids* are movable folds of skin that form a protective curtain over the anterior surface of the eyeball. The upper lid is the larger and is more freely movable. It is pulled up by the *levator palpebrae superioris* muscle. The sphincter muscles, the *orbicularis oculi*, close the eyelids. The *eyelashes* are stiff, curved, protective hairs, placed along the margin of each lid. Sebaceous glands are associated with the eyelashes. Posterior to the bases of the eyelashes are the tarsal, or meibomian, glands. They secrete an oily fluid that tends to retain tears and prevents the eyelids from sticking together.

The *conjunctiva* is a thin mucous membrane covering the anterior surface of the eyeball and lining the eyelids. It is transparent, since light must pass through it to reach the receptors of the eye. The portion between the eyeball and the lining of the eyelids is folded and loose to permit free movement of the lids and the eyeball. At the inner angle of the eye there is a fold of the conjunctiva called the *plica semilunaris*. It is crescent-shaped and is commonly known as the *third eyelid*, or *nictitating membrane*. Deep in the inner angle of the eye is a fleshy mound known as the *caruncle*. In the embryo this structure develops as a separate portion of the lower lid and has glands similar to those found in the eyelid. It may even bear an eyelash or two in a few cases. The caruncle is an island in a space formed at the inner angle of the eye. The eyelids form the boundaries of this conjunctival space, which holds a small amount of tear fluid and is referred to as the *lacrimal lake*. The caruncle forms part of the yellowish secretion that tends to gather in this area, especially overnight.

The conjunctiva is well supplied with blood vessels and nerve endings. Irritation of this membrane can cause the blood vessels to become plainly visible. The pain felt when a foreign body gets into the eye indicates that there are abundant sensory nerve endings in the conjunctiva.

LACRIMAL STRUCTURES   The *lacrimal gland* produces tears (Figure 11.1). It is an elongate compound gland located in a depression of the frontal bone above the outer angle of the eye. Several ducts carry the lacrimal secretion from the gland to the conjunctiva of the upper eyelid. Ordinarily just enough fluid is secreted to keep the surfaces moist. Much of this secretion is lost by evaporation, and the remainder flows toward the inner angle of the eye where it is drained off through two minute openings, called the *puncta lacrimalia*. The puncta are the openings of two *lacrimal canals*, which carry the secretion to the *lacrimal sac*. An extension of the lacrimal sac, the *nasolacrimal duct*, leads the secretion down into the nasal passageway (Fig. 11.1).

Lacrimal secretion contains various salts in aqueous solution and a small amount of mucin. It is slightly antiseptic, since it contains the enzyme lysozyme, which has a bactericidal action. The principal function of tears,

FIGURE 11·1
The lacrimal structures of the eye.

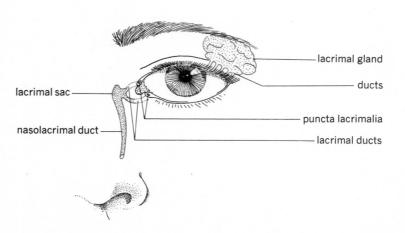

lacrimal gland

ducts

lacrimal sac

puncta lacrimalia

nasolacrimal duct

lacrimal ducts

however, is to moisten the surface of the eyes. Irritation of the conjunctiva or the nasal membranes increases the flow of tears, as do emotional states. The lacrimal glands are not fully functional at birth. A young baby's eyes should be protected therefore against intense light, dust, and the drying effect of wind.

EXTRINSIC EYE MUSCLES AND THEIR INNERVATION   Six extrinsic muscles are attached to the eyeball (Table 11.2 and Fig. 11.2). Four of them, the *recti muscles,* arise from a ring tendon (Zinn's ring) at the back of the orbit. They are about 40 mm long and are inserted on the eyeball in positions corresponding to their names. The *superior rectus* is attached to the superior surface of the eyeball and turns the eye upward. The *inferior rectus* is attached underneath the eyeball and turns the eyeball downward. The *lateral rectus* is attached to the outer side; the *medial rectus* is attached to the inner side. The latter two muscles turn the eyeball to the lateral and medial sides, as their names indicate. There are also two *oblique* muscles, which help rotate the eyeball. The *inferior oblique muscle* arises anteriorly from the floor of the orbit near the fossa for the lacrimal sac. It passes backward laterally between the inferior rectus tendon and the floor of the orbit, inserting posteriorly on the eyeball under the lateral rectus. This is the shortest of the ocular muscles.

The *superior oblique* is the longest and most slender of the ocular muscles. Arising from the apex of the orbit (the edge of the sphenoid bone in the region of the aperture that transmits the optic nerve), it passes forward on the upper nasal side between the superior and medial rectus muscles. Above the eye on the medial side, it passes through a loop of cartilage commonly called the *trochlea.* It becomes tendinous at the trochlea and turns downward and backward, passing under the superior rectus to insert on the posterior aspect of the eyeball. The oblique muscles probably do not act independently but work with the rectus muscles to roll the eyes.

TABLE 11·2

Extrinsic muscles of the eye

Muscle	Origin	Insertion	Action	Nerve
Inferior oblique	Anterior orbital floor	Between insertion of superior and lateral recti muscles	Rotates eyeball upward and laterally	Oculomotor
Rectus; inferior, lateral, medial, superior	Common tendinous ring around optic foramen	By tendons into sclera at their respective positions	Medial, rotates eyeball medially; lateral, rotates eyeball laterally; superior, rotates eyeball upward and medially; inferior, rotates eyeball downward and medially	Inferior, medial, and superior by oculomotor, lateral by abducens
Superior oblique	Root of orbital cavity, medial to optic foramen by a narrow tendon	A narrow tendon; passes through trochlear pulley and inserts between rectus superior and rectus lateralis	Rotates eyeball downward and laterally	Trochlear

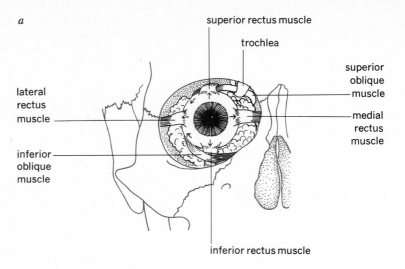

a

superior rectus muscle

trochlea

superior oblique muscle

medial rectus muscle

lateral rectus muscle

inferior oblique muscle

inferior rectus muscle

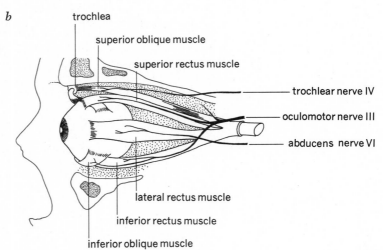

b

trochlea

superior oblique muscle

superior rectus muscle

trochlear nerve IV

oculomotor nerve III

abducens nerve VI

lateral rectus muscle

inferior rectus muscle

inferior oblique muscle

FIGURE 11·2
Extrinsic muscles of the eye:
a superior view; *b* lateral
view.

Innervation of the extrinsic eye muscles   The *oculomotor nerve* innervates all the rectus muscles with the exception of the lateral rectus, which is innervated by the *abducens nerve*. The inferior oblique muscle is also innervated by the oculomotor nerve. The *trochlear nerve* supplies the superior oblique muscle.

The eyeball is suspended in the orbit by the extrinsic muscles and by a complex series of ligaments and other connective tissues. One of the most important of these suspensory structures is the *fascia bulbi*, or Tenon's capsule. This is a serous sac that is attached to the eyeball anteriorly in a region around the cornea but does not cover the cornea. Posteriorly it envelops the eyeball completely so that the eye muscles, the nerves of the eye, and the blood vessels pass through it. The eyeball rests and moves on these structures, which are further supported by a padding of fat within the orbit.

STRUCTURES OF THE EYEBALL The eyeball is nearly spherical, but the distance from front to back is a little greater because the cornea, which appears as an arc of smaller dimension, is located on the anterior surface of the larger sphere (Fig. 11.3). The eyeball is remarkably constant in size in either sex, measuring about 24 mm in its anterior-posterior diameter, but it is slightly smaller in the female (0.5 mm in all diameters). The eye may appear large or small because of its position in the orbitals. If it protrudes somewhat so that more of the white shows than would normally be exposed, the eyeball appears to be abnormally large. The eye may also appear deep-set and small or sunken after a long illness, when considerable body fat has been used up.

The wall of the eyeball is composed of three layers. The *outer coat* is a fibrous tunic consisting of the sclera and cornea. The *intermediate layer* is a highly vascular, pigmented tunic comprising the choroid, a muscular ciliary body, and the iris. The *inner layer* is the retina. The *refracting media* of the eye are the aqueous humor, the lens, and the vitreous body.

The outer layer The *sclera* is the white coat of the eye. It covers the fibrous tissue. This opaque covering helps maintain the shape of the eyeball and protects the more delicate structures from injury. The anterior surface is covered by the conjunctiva, and small blood vessels can be seen through this layer in living organisms. The sclera itself is not abundantly supplied with blood vessels. At the junction of the sclera and cornea, however, there is an important venous sinus, the *sinus venosus* of the sclera (canal of Schlemm), which encircles the base of the cornea. This thin-walled sinus at the filtration angle of the eye provides an outlet through which the aqueous humor flows into the venous system of the eyeball (Fig. 11.3).

The *cornea* is the transparent part of the fibrous tunic. It is a portion of a sphere, superimposed upon the anterior surface of a larger sphere, the eyeball. It represents about one-sixth of the eyeball's total area. The cornea proper is composed of modified fibrous tissue, which during development

FIGURE 11·3
Horizontal section through the eyeball.

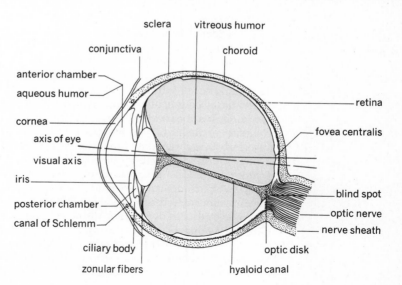

becomes clear and presents a fairly homogeneous appearance. Over the outer surface there is an epithelial layer that is continuous with the epithelial layer of the conjunctiva. The embryonic cornea contains capillaries, but they are lost in the functional eye. The cornea contains no blood vessels except around its border; it is nourished by lymph, which diffuses slowly through the tissue. Oxygen is absorbed directly from the atmosphere. The corneal epithelium has an abundant nerve supply of fine unmyelinated nerve endings from the ciliary nerves, making it highly sensitive. Injury to the fibrous portion of the cornea heals by the formation of opaque tissue.

The intermediate layer   The intermediate layer, or vascular tunic, of the eyeball is composed of the choroid, the ciliary body, and the iris. The *choroid* is a dark-brown membrane that lines the sclera. It is highly vascular, since it is concerned with maintaining the nutrition of the retina. There are numerous irregular-shaped pigment cells that, with the blood vessels, give the membrane its dark color. It therefore darkens the interior of the eye, absorbing light rays and preventing their reflection. The optic nerve passes through the choroid at the back of the eyeball. The membrane extends forward to the ciliary body. Human eyes do not shine in the dark because they lack the reflecting layer (*tapetum*) in the choroid that is a characteristic structure in the eyes of most animals.

The *ciliary body* may be divided into three regions: the ciliary ring, the ciliary processes, and the ciliary muscle. The ciliary ring is a darkened area about 4 mm wide and continuous with the choroid. The ciliary processes are folds or ridges arranged as the meridians of a sphere. About 70 processes radiate out from behind the iris and taper toward the ciliary ring. They provide attachment for the zonular fibers (suspensory ligament) of the lens and for the ciliary muscles (Fig. 11.4).

The ciliary muscle is composed largely of fibers running in a meridional direction, but there is also a small band of circular muscle. When the meridional fibers contract, they pull forward on the ciliary ring and the ciliary processes, thus slackening the zonular fibers. The lens then becomes more convex for accommodation to close work, such as reading. Some recent evidence indicates that the circular muscles may play a greater part as antagonists of the meridional fibers than had been formerly thought.

The *iris* is the most anterior portion of the vascular tunic. It is the highly colored part of the eye. Human irises are blue, gray, green, or brown, depending on the distribution and the amount of pigment. The iris of an albino has no pigment and appears pink. The circular opening in the iris is the pupil. It appears black because it opens into the cornea and the anterior chamber of the eye. The iris has two layers. A colorless layer is located anteriorly and a blue layer is located posteriorly. The blue layer is usually present in all eyes, but in those persons with dark eyes, a dark pigment, *melanin*, develops in the anterior layer and obscures the blue color.

The inheritance of eye color is a more complex topic than can be carefully considered here. In general, blue eye color is a recessive trait, and brown eye color is dominant. Albinism is also a recessive character. If both parents have pure blue or gray eyes, their children generally will have eyes the same color as the parents, since these colors indicate presence of homozygous recessive genes. Brown eye color may be determined by either a homozygous (*BB*) gene pair or a heterozygous (*Bb*) gene pair. Parents who

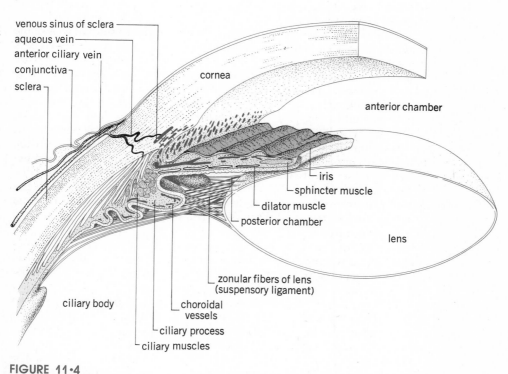

a

conjunctiva

sclera

sinus venosus of sclera (canal of Schlemm)

cornea

pupil

iris

ciliary body

ring of ciliary muscle

retina

fovea centralis

optic nerve

blood vessels of choroid

retinal blood vessel

b

venous sinus of sclera

aqueous vein

anterior ciliary vein

conjunctiva

sclera

cornea

anterior chamber

iris

sphincter muscle

dilator muscle

posterior chamber

lens

zonular fibers of lens (suspensory ligament)

ciliary body

choroidal vessels

ciliary process

ciliary muscles

FIGURE 11·4
a Section through eyeball showing ciliary muscle; *b* detailed section through anterior part of the eye (greatly enlarged).

are heterozygous for brown eyes may, therefore, have blue-eyed children (*bb*). Persons with blue eyes have a homozygous recessive gene pair for this trait. There is more variability in eye color than these simple illustrations indicate, however, and probably more than one gene is involved in the final determination.

The iris is dark-colored on its posterior surface, and it contains two antagonistic sets of muscles. The *dilator muscles* are radial muscles; their contraction dilates the pupil. They derive their innervation from the sympathetic chain of ganglia of the autonomic nervous system. Preganglionic fibers pass upward from the first and second thoracic levels to the superior cervical ganglion. Postganglionic fibers in the ophthalmic branch of the Vth cranial nerve enter the eyeball in the ciliary nerve to the dilator muscles of the pupil.

A circular muscle, the *sphincter pupillae,* surrounds the pupil, and its contraction causes the pupil to constrict. Parasympathetic fibers are carried by the oculomotor nerve to the sphincter muscle, which constricts the pupil. Preganglionic fibers arise in the midbrain and extend outward to the ciliary ganglion located behind the eyeball. Postganglionic fibers in the short ciliary nerve enter the eyeball and innervate the ciliary muscle and the sphincter of the pupil.

The inner layer   The *retina* is the inner layer, or nervous tunic, of the eye. The optical portion forms the lining of the posterior part of the eyeball, extending forward to the ciliary ring. At the back of the eye the retina is about 0.4 mm thick, becoming thinner as it extends forward. Within this thin, delicate membrane the receptors for the sense of sight are located.

This portion of the eye forms originally as a part of the lateral wall of the brain. In the early embryo an infolding of the neural tube is followed by a lateral outpocketing in the region of the eye so that the receptor cells are located at the back of the retina next to the pigment layer. They face *away* from the source of light; this type of retina, found in all vertebrates, is called an *inverted retina.*

The retina, although very thin, contains several layers of neurons, which may be separated by suitable histological methods (Fig. 11.5). Actually, 10 layers may be differentiated, of which we shall mention only a few. At the back of the retina is the *pigment layer,* composed of low columnar epithelium containing great amounts of pigment. The next layer contains the *receptor cells.* These cells are sensitive to light. Two types of receptors are found: *rod* and *cone cells,* so named because of their shapes. An external limiting membrane extends through the receptor cell area at the base of rod cell and cone cell bodies.

The *outer nuclear layer* contains the nuclei of the receptor cells (outer layers refer to those layers farther away from the center of the eye). The outer nuclear layer comprises a mass of dark-staining nuclei located adjacent to the rod-and-cone layer.

Adjoining the outer nuclear layer is the *outer synaptic layer,* or *outer plexiform layer.* It consists of fibers and synaptic endings of receptor cells as they connect with the dendrites of bipolar cells of the next layer. It may be identified as a lighter-colored area.

Bipolar cells form the next large area near the middle of the retina. The bipolar layer contains, in addition, some *horizontal cells* and some *amacrine cells.* Bipolar cells have large round nuclei surrounded by rather thin layers of cytoplasm. They form a connecting link between the receptor cells and the inner synaptic layer. The horizontal cells appear to be association cells that relay impulses to bipolar cells nearby, but their lateral connections are not extensive, and no direct connection between horizontal cells and

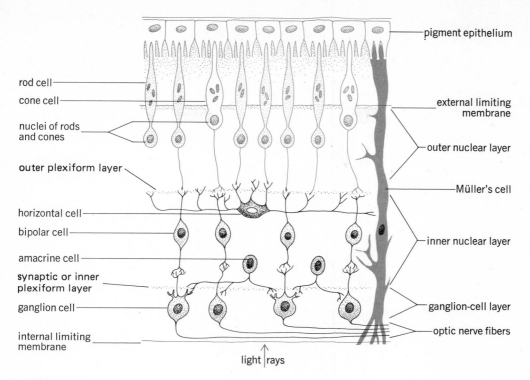

rod cell

cone cell

nuclei of rods
and cones

outer plexiform layer

horizontal cell

bipolar cell

amacrine cell

**synaptic or inner
plexiform layer**

ganglion cell

internal limiting
membrane

light rays

pigment epithelium

external limiting
membrane

outer nuclear layer

Müller's cell

inner nuclear layer

ganglion-cell layer

optic nerve fibers

FIGURE 11·5

Cell types and layers of the
retina (diagrammatic and simplified).

ganglion cells has been found. Horizontal cells have a small oval nucleus and
a rather large amount of cytoplasm.

The amacrine cells, which lie close to the ganglion cells and synapse in
the inner synaptic (plexiform) layer, are now the subject of investigation.
The function of these cells is not clear, but they synapse, at least in part, on
ganglion cells. In the primate retina, bipolar axons synapse with both
ganglion cell dendrites and processes of amacrine cells. Amacrine cells are
probably more important in lateral transmission than are horizontal cells,
since their lateral processes extend much farther. Amacrine cells may also
participate in feedback loops to bipolar neurons, and they may play some
part in the organization of the visual input to ganglion cells whose axons
form the optic nerve.

The nuclei of bipolar, longitudinal, amacrine, and Müller's cells form the
inner nuclear layer of the retina. Adjoining this layer is the *inner synaptic,*
or *plexiform, layer.*

The inner cell layer is composed of large *ganglion cells,* which synapse
with bipolar and amacrine cell processes, whose axons make up the fibers of
the optic nerve.

Müller's cells are considered to be modified neuroglial cells. The fibers
extend throughout the retina, curving around ganglion cells and spreading
out to form a supporting framework for both the internal and external
limiting membranes. They envelop rod and cone cells, perhaps taking the
place of myelin as an insulating layer. They may be more than passive
supporting cells, since they exhibit a slow change of potential associated

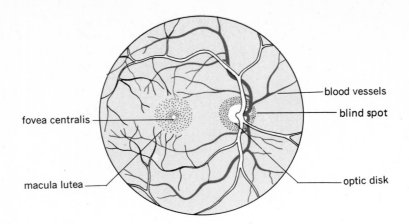

FIGURE 11·6
Posterior portion of the right
retina as viewed from the
front.

with excitation of the rod and cone cells that they surround. Their exact
function has not been determined, but they have been shown to contain
glycogen and oxidative enzymes.

The blind spot   Since the neurons that form the optic nerve arise on the
anterior side of the retina, which faces the lens, they must turn and pass
back through the retina in order to carry impulses to the brain. The optic
nerve is formed by the union of these optic neurons, which all pass through
the retina at one spot. Since all receptors are pushed aside to permit the exit
of the optic nerve fibers, there is a blind spot in each eye at this point
(Fig. 11.6).

As an experiment to prove the existence of the blind spot, draw a small
circle and a small cross 2 in apart on a 3 × 5 in card. Now close the left eye
and, staring at the cross, move the card from a position about 6 in away from
the eye to a position either a little closer or a few inches farther away. As
you maneuver the card so that the circle's image falls on the blind spot, it
will not be seen, since there are no visual receptors there (Fig. 11.7).

The blind spot is not located in the exact center of the posterior part of
the eye; instead, it is on the inner side of a median line passing through the
posterior pole of the sphere. Blood vessels enter the eye at the center of the
optic disk or blind spot. The retinal artery then divides into four main
branches, which, with numerous smaller branches, cover the retina. The
retinal veins follow the path of the arteries and exit beside the retinal artery
in the optic disk. Blood vessels of the retina can be observed under proper
illumination. A physician can look into the depths of the eye as if the eye
were a window and make observations about the condition of the blood
vessels. In some disease states, the blood vessels change in appearance; they
may be notched where one crosses another or even partially obliterated.

FIGURE 11·7
Diagram for determination of
the blind spot of the retina.
With the left eye closed or
covered, gaze steadily at the X
while moving the book from a
distance about 12 in. in front
of the face slowly toward the
face. Note that at a certain
distance, the image of the dot
becomes invisible. To check
the other eye, close or cover
the right eye and gaze steadily
at the dot, noticing that the X
becomes invisible when the
book is advanced a certain distance toward the face.

The refracting media The *anterior chamber* of the eye lies immediately posterior to the cornea. It is filled with the *aqueous humor.* The *posterior chamber* is a small area located on either side between the lens and the iris.

The *lens* is not a part of the choroid coat, but it is directly attached to the ciliary muscles by zonular fibers. The *lens* is a clear, transparent tissue located posterior to the pupil and iris. It lies between the anterior chamber and the vitreous body and is slightly yellow in older people. The adult lens measures 9 mm in diameter and is about 4 mm thick. It is a biconvex lens, but somewhat more convex on the posterior side. In the fetal eye, the lens is almost spherical, and even in children, the lens is strongly convex; but in adults past middle age, there is usually a gradual and partial loss of convexity and elasticity. The human lens is able to change its degree of convexity during accommodation for near and far vision, a characteristic not found in other mammals.

The lens is enclosed by a clear, elastic capsule. Just underneath the capsule on the anterior surface only is a single layer of cells that constitutes the lens epithelium. A study of the structure of the lens itself reveals that it is a tissue composed of specialized *lens fibers.* The fibers appear to be long, slender, and flat, but in cross section they appear as flattened hexagons. The fibers lie along meridional lines, but no fiber is long enough to reach from pole to pole. The lens fibers are arranged in concentric lamellae like layers of onion tissue.

The *vitreous body* is a transparent, jellylike substance that fills the cavity of the eyeball posterior to the lens. An anterior cavity in the vitreous body accommodates the lens. The vitreous body is largely water, and its refractive index is close to that of water. A lymph canal passes through it from the optic disk to the region of the lens. This is called the *hyaloid artery* in the embryonic eye, where it provides nourishment of the lens during its development. The vitreous body helps maintain the shape of the eyeball and supports the retina.

Refractive index of eye tissue In order to reach the retina, light rays must pass through the cornea, the aqueous humor, the lens, and the vitreous body. *Refraction,* or the bending of light rays, varies for each of these tissues and fluids. Refraction occurs because light rays travel at different speeds as they pass through media that vary in density; for example, a stick lying obliquely and half-submerged in water appears to be bent.

The curved *cornea* is the first refracting tissue. Light waves pass from the air through the cornea and aqueous humor and enter the pupil. The refractive power of the cornea is often underrated. Its refractive power causes rays of light to converge considerably before they reach the crystalline lens. Light passes next through the aqueous humor, which has a refractive power that is about the same as water.

When light waves strike the curved surfaces of the lens they are further bent or refracted. The refractive index of the air is 1.00, whereas the refractive index of the lens is 1.40. The refractive indices of other parts of the eye are as follows: cornea, 1.38; aqueous humor, 1.33; vitreous humor, 1.34. Normally, light rays are directed to focus on the central focusing spot of the retina.

FORMATION AND DEVELOPMENT OF FLUIDS WITHIN THE EYEBALL The intraocular fluid keeps the eyeball distended and firm and is maintained at an average pressure of 19 mmHg, the range being about 15 to 25 mmHg. This fluid includes the aqueous humor and the vitreous body.

The aqueous humor is formed from the inner folds of the ciliary processes. These processes are infoldings of the choroid and are arranged in a circle around the lens. The aqueous humor flows forward through the zonular fibers, through the pupil, and into the anterior chamber of the eye. It also diffuses into the gelatinous consistency of the vitreous body, but since there is little exchange of fluid there, it cannot be said to flow into it.

The outlet for the intraocular fluid is in the angle between the cornea and the iris where the sinus venosus of the sclera (canal of Schlemm) drains the fluid eventually into the veins of the sclera (Fig. 11.4). The sinus venosus drains into thin-walled aqueous veins, which normally carry aqueous humor rather than blood. Projections from endothelial cells lining the canal of Schlemm have been observed. They increase the surface area, and it has been suggested that they may enhance the exchange of fluid in these cells.

EMBRYONIC DEVELOPMENT OF THE EYES An evagination of the lateral wall of the forebrain is called the *optic vesicle,* a structure that forms as early as $3\frac{1}{2}$ to 4 weeks in the human embryo. The optic vesicle forms a two-layered cup, the optic cup, as it grows out to meet the surface ectoderm. Just above the optic cup, the surface ectoderm thickens and invaginates into the cup as a hollow lens vesicle. Later it becomes pinched off from the surface ectoderm to form the lens of the eye (Fig. 11.8). The lens is nearly spherical when it is first formed, but it gradually flattens and becomes biconvex. The cells multiply until the cavity of the lens is completely filled. The growing lens is supplied with blood from the hyaloid artery, which normally disappears before birth, leaving only a remnant called the hyaloid canal.

The layer of the optic cup closest to the lens becomes the neural retina, and the deepest layer forms the pigment back of it. The optic nerve fibers

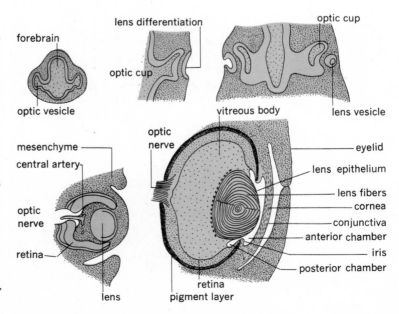

FIGURE 11·8

The development of the eye in the mammalian embryo. The lens is enclosed in an external noncellular capsule (not shown). Epithelial cells elongate and form lens-fiber cells. [*After Lewis, in Barry J. Anson (ed.), "Morris' Human Anatomy," 12th ed., McGraw-Hill Book Company, New York, 1966.*]

form the ganglion cells in the retina and grow back through the optic stalk as the optic nerve.

The eyes of the 6-week embryo are indicated by the optic vesicle outlined under a thin covering of ectoderm. They are located far around on the side of the head. With the development of the face at 7 weeks, the eyes are found in a forward position, although still somewhat far apart. The formation of eyelids at this time emphasizes the outlines of the eyes. The eyelids meet and fuse during the ninth week and remain closed until the seventh month of intrauterine life. Unlike many mammals, the human infant is born with the eyes open.

The eyeball continues to grow after birth and reaches its adult size at about the age of puberty.

VISUAL RECEPTOR CELLS: RODS AND CONES

The photoreceptor cells of the retina are transducers adapted to receive light energy and to transform them, by a photochemical process, into nerve impulses that can be interpreted by the brain. Rod cells provide vision in the dark or in dim light; cone cells are responsible for color vision in daylight or under adequate lighting conditions (Fig. 11.9).

There are roughly 10 times as many rods as cones in the human retina. It is estimated that there are approximately 1 million cone cells and 10 million rod cells. The structural organization of these receptors is interesting, since electron microscopy indicates that they are derived from cilia and retain some of the characteristic structures of these organelles. Rod and cone cells have essentially the same basic structure, but rod cells are typically longer and more slender than cone cells. Rod cells are generally 50 to 60 μm long in the retinas of various animals and about 2 μm wide. Cone cells, being shorter than rod cells, have a somewhat conical appearance and are about three times wider, 6 to 7 μm wide. The basic structure of rod and cone cells

FIGURE 11·9
Receptor cells of the retina, greatly enlarged; rod cell at left; cone cell at right.

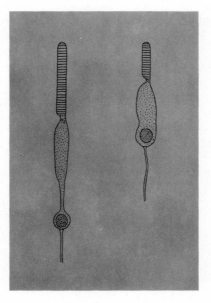

consists of an outer segment, a slender connecting stalk, and a basal portion. Rod cells are more numerous toward the periphery of the retina, whereas cone cells are concentrated in the fovea centralis, the area of most acute vision.

While the structures through which light must pass have a high degree of transparency, they are not completely transparent. One investigator, using blue-green light under test conditions, found that in order to see a light, the eye must receive from 54 to 148 quanta at the cornea of the dark-adapted eye. Because of reflection and absorption, of the 54 quanta falling on the cornea, only 26 arrive at the retina. Here again only 5 quanta are estimated to be absorbed by the rods, and there are indications that an individual rod cell may be stimulated by 1 quantum of light, the smallest unit of energy. Such results indicate that the visual receptors approach the ultimate in sensitivity.

ROD CELLS   The rod cell, like other cell types, is contained within a thin plasma membrane. The outer segment of the rod cell is characterized by laminated foldings of the endoplasmic reticulum, or ergastoplasm. This is a smooth, folded membrane (without ribosomes), and the visual pigments concerned with the biochemical aspects of vision are believed to be located in or on the endoplasmic reticulum.

The central connecting stalk of the rod cell has the structural characteristics of a cilium in its embryonic development, with nine pairs of filaments and a basal body that resembles a centriole. A second centriole at right angles to the first has been observed in some electron micrographs. The connecting cilium has an off-center appearance. This is caused by the

FIGURE 11·10

A rod cell, showing detail based on electron micrographs.

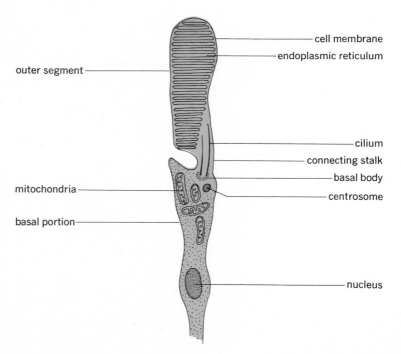

one-sided enlargement of the outer segment during its embryonic development (Figs. 11.10 and 11.11).

The basal portion of the rod cell contains numerous elongate mitochondria, the energy source of the cell. The rod cell nucleus lies just beyond the mitochondrial area.

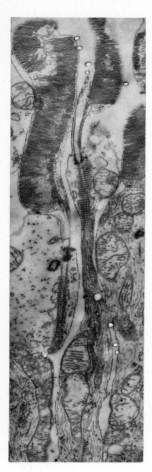

FIGURE 11·11
Electron micrograph of a rod cell. (*From James A. Freeman, "Cellular Fine Structure," McGraw-Hill Book Company, New York, 1964.*)

Rhodopsin and vision Rod cells contain a reddish pigment called *rhodopsin,* or visual purple, in the outer segments. Rhodopsin is a colored aldehyde of 11-*cis*-vitamin A plus the protein opsin. The energy of light waves causes the isomerization of rhodopsin, resulting in a rapid change in the physical form of the structure as all-*trans*-retinal. *Retinal* is a carotenoid pigment closely related chemically to beta-carotene, while the *opsins* are colorless proteins chemically bound to a *chromophore* (color-bearing) *group.* In a brief explanation of *cis-trans isomerism,* we note that when a double bond exists between two carbon atoms, their position is fixed in regard to each other. If additional carbon atoms, such as two carboxyl groups, are added on the same side of the double bond, this represents a cis position. If located on opposite sides of the double bond, it is a trans position. The cis molecule bends at the numbered position, whereas the all-trans molecule is generally in a more or less straight line. This distinction is pertinent here relative to the formation of 11-*cis*-rhodopsin, since only the cis forms of vitamin A and retinal can take part in this reaction (Fig. 11.12).

Energy derived from exposure to light causes rhodopsin to bleach and to be converted to another chemical form called lumi-rhodopsin, but this form is highly unstable and it almost immediately changes into meta-rhodopsin. This compound is also unstable, and in a matter of seconds it forms all-*trans*-retinal plus opsin. The cis and trans forms of retinal can each be converted into corresponding types of vitamin A. The reverse reactions also take place, but these are relatively slow reactions. The cycle is completed when 11-*cis*-retinal plus opsin is converted to rhodopsin. Rhodopsin is stable as long as it remains in the dark. When it is exposed to light, it bleaches, causing excitation of the rod cells. Subsequently the ganglion cells of the optic nerves are stimulated.

Adenylate cyclase is present in the photoreceptors of the retina. This enzyme is inactivated by light, and this inactivation is found to be proportional to the bleaching of rhodopsin (Miller). Evidently adenylate cyclase and cyclic AMP are involved in photoreception, but their exact functions have not been determined.

Dark adaptation *Dark adaptation* is the process by which the eye accommodates to darkness. When we enter a dark room, for example, our pupils dilate, but a change in the retina also occurs so that rod cells increase in sensitivity. It requires at least half an hour to accomplish dark adaptation. Most of us have had the experience of going into a darkened theater and finding that we were hardly able to locate our seats. Later we find that there seems to be plenty of light available and we can see fairly well. Rhodopsin is regenerated in the dark so more of this visual pigment is available. *Some* light is necessary for vision, since rhodopsin must be stimulated by light energy to transform this energy into nerve impulses.

It is often advantageous to have the eyes dark-adapted, as, for example, under war conditions. One way to accomplish this is to have persons sit in

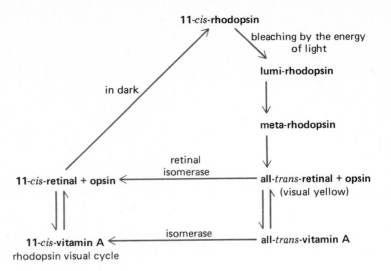

11-*cis*-rhodopsin

bleaching by the energy of light

lumi-rhodopsin

meta-rhodopsin

in dark

retinal isomerase

11-*cis*-retinal + opsin ← all-*trans*-retinal + opsin (visual yellow)

11-*cis*-vitamin A ← isomerase ← all-*trans*-vitamin A

rhodopsin visual cycle

FIGURE 11·12
The development and break-down of rhodopsin.

a totally dark room for 30 min. There remains the problem of transferring them to the scene of operations without exposure to light. It was discovered that by using goggles with red glass to act as filters it was possible to have good cone vision while permitting the rods to become dark-adapted. The red filters screen out the light rays that stimulate the rods but do not greatly affect the light rays stimulating the cones.

Rod cells are not concerned with color vision. Looking at a flower garden in the deep twilight is much like observing a photographic image on black-and-white film. Light-blue flowers are seen as various shades of gray, and the deep red flowers appear to be black. With adequate illumination, the yellow-red end of the spectrum appears the brightest, but with decreasing illumination, the blue-green end of the spectrum is more readily discerned by the dark-adapted eye. This effect, known as the *purkinje effect*, is explained as the result of the ability of the retina to shift to rod vision in the process of dark adaptation. After sundown, as daylight decreases, red objects become darker in color, but light-blue objects remain clearly blue for a while before gradually resolving to gray as twilight deepens.

Nocturnal animals presumably have good vision in the dark. Some of the nocturnal animals, such as bats, have retinas that are said to contain only rods. The retinas of rats contain a few cone cells among the rods, whereas some owls have a fair distribution of cone cells and are able to see very well in daylight. The eye of a human is essentially adapted for vision during the day but functions reasonably well in the dark if there is some weak light available.

Since the concentration of rods is greater around the periphery rather than in the center of focus for vision, there is an advantage to looking a little to one side of an object when trying to locate an indistinct object in the dark.

Light adaptation *Light adaptation* is accommodation to normal or bright light. It is a somewhat misleading term, since human eyes are ordinarily adapted to daylight vision. It is best observed when we suddenly turn on the lights after being in the dark for some time or when we emerge from a dark theater into bright sunlight. The light appears too bright for our eyes, but after a few minutes the eyes adapt themselves to the new condition. The excess rhodopsin stored in the dark-adapted retina is bleached rapidly, the pupil constricts somewhat, and we find that we can see quite well again, even in bright sunlight. Dark adaptation is a slow process, but light adaptation is fairly rapid.

CONE CELLS   The cone cells provide acute detailed vision and perception of color. It appears to be well established now that there are three different light-sensitive pigments located in three different kinds of cone cells. One pigment exhibits its greatest sensitivity in blue-light wavelengths, one in the spectrum for green light, and one in sensing red-light wavelengths. The cone pigments have been given the following names: *iodopsin* for blue, *chlorocruovin* for green, and *erythrocruovin* for red. It has become possible experimentally to pass a beam of light through a single retinal cone cell. When the spectra from several different cone cells are plotted, they fall into the three groups indicated above, but cone cells of these groups cannot be distinguished by their appearance, for they contain no colored pigment. We shall discuss color vision later in this chapter.

Most cone cells are found in the *area centralis,* the region around the visual axis located 2 to 2.5 mm laterad from the optic disk, or blind spot (Fig. 11.6). The human area centralis is a pigmented area, corresponding in general to the area called the *macula lutea,* or *yellow spot.* Though the area centralis is about 6 mm in diameter, the most evident yellow pigment area is only about 3 mm in diameter. Structurally, this area is differentiated by an increase in the number of ganglion cells, which are arranged here into more than one row. The color of the pigmentation is described as canary yellow or as about the color of grapefruit. This pigment is a carotenoid, related to the xanthophyll of leaves.

The *fovea centralis* is a bowl-shaped depression in the middle of the area centralis. It is a tiny pit, about 1.5 mm in diameter and about 0.24 mm deep. In the center of the depression is the *foveola,* an additional concavity measuring ony 0.35 mm in diameter. The fovea centralis is specialized in many ways to provide the most acute vision possible (Fig. 11.13).

The concentration of cone cells increases inward from the periphery of the area centralis until, in the foveola and a surrounding area about 0.5 mm in diameter, rod cells are completely excluded. The periphery of the fovea, however, and even the rim of the inner slope of the depression contain both rods and cones. The cone cells of the fovea are longer, much more slender, and more closely packed than elsewhere in the retina. Larger blood vessels pass around this area, for although there are small blood vessels present in the fovea centralis, even capillaries are not found in the foveola. The retina is thinner in the fovea. At the depth of the pit, in the foveola, most layers of neurons either have been pushed aside or are absent altogether. Here the ratio of cone to ganglion cell approaches 1:1. It has been suggested that the slope of the fovea acts to magnify the image projected on the foveal cones.

Everything that we see clearly is projected on this very small area of the fovea centralis. The most detailed observations are made by the projection of a minute image on the foveola. The size of the retinal image varies with the size of the object and its distance from the eye. Since the fovea centralis

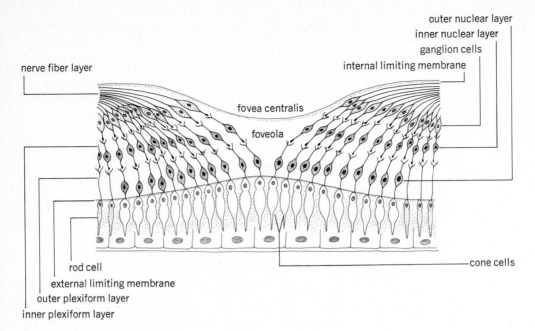

Labels on figure:
- nerve fiber layer
- outer nuclear layer
- inner nuclear layer
- ganglion cells
- internal limiting membrane
- fovea centralis
- foveola
- rod cell
- external limiting membrane
- outer plexiform layer
- inner plexiform layer
- cone cells

FIGURE 11·13
Horizontal section of the retina through the macula lutea and fovea centralis.

is only about 1.5 mm in diameter and the foveola is only 0.35 mm in diameter, most retinal images to be clearly seen should be less than 1 mm in size. The retinal image may be much more minute; in fact, two points can be distinguished when their retinal images are only 0.003 mm apart. The term *retinal image* is commonly used, but it is more accurate to think of vision as an energy effect produced by stimulation of certain receptors on the retina. A visible image is not created on the retina.

MECHANISM OF VISION

FUNCTION OF THE LENS The camera has been used many times to illustrate the optical mechanism involved in vision, and it provides a fair working model. Both eye and camera have a biconvex lens to bring light rays to a focus. The camera has a diaphragm to control the amount of light admitted; the iris performs this function in the eye. Light rays enter the dark interior of the camera and focus on sensitized material on the film or plate, Light rays also enter the eye and pass through its dark interior, focusing upon the light-sensitive retina.

A *biconvex lens* converges light rays, bringing them to a focal point behind the lens. The distance to the focal point varies according to the curvature of the surfaces of the lens and the refractive index of the glass used in its construction. A biconvex lens is used in a magnifying glass; the light rays are brought to a focus as a bright spot of light. Heat rays also are brought to a focus, and such a lens is sometimes called a "burning glass."

A *convex lens* focuses parallel rays at its focal point. For practical purposes, the rays of light from an object 20 ft away are considered to be parallel, since they diverge so little. But the object to be observed or photographed is often closer than 20 ft. In this case the eye and the camera have different ways of focusing the diverging light rays from a nearby object.

The eye can change the convexity of its lens. The lens becomes more convex if it is necessary to focus strongly divergent rays on the retina, while the distance between the lens and the retina remains the same. The glass camera lens cannot change its shape in order to focus on some object only a few feet away. The focusing type of camera moves the lens forward toward the object in this case, in order to increase the distance between the lens and the film.

Accommodation by the lens The lens accommodates, or adjusts its convexity, to bring into focus both near and distant objects. A relatively flat lens may be able to focus parallel light rays and so accommodate for distant vision, but to focus the diverging rays of nearby objects requires a more convex lens. The ciliary muscles are attached anteriorly in the region around the cornea; posteriorly they blend into the choroid coat. The zonular fibers, supporting the lens, are continuous with the capsule of the lens and extend out radially to make connection with the ringlike ciliary body. The ciliary muscles therefore exert tension only indirectly on the zonular fibers. When the ciliary muscles are relaxed, the ciliary ring is thin and the zonular fibers are tightened. Tension on the capsule of the lens causes the lens to become less convex so that it can accommodate for far vision. In accommodation for near vision, the ciliary muscles contract, pulling the ciliary body and choroid forward toward the cornea. The zonular fibers slacken, and the lens becomes as convex as its elasticity will permit. The greatest change in the convexity of the lens is at its anterior surface: there is not much change in the posterior surface. Looking at some distant object is restful to the eyes, since most of the ciliary muscles are relaxed, while hours of close work require these muscles to maintain a more or less constant tension.

Accommodation reflex While looking at a distant object, bring the finger into the field of vision about 12 in from the eyes. The finger is seen, but the image is blurred. Now a quick change of focus brings the image of the finger into sharp focus, but the distant object is not seen sharply. When the image is not focused sharply on the retina for near vision, sensory impulses pass over the optic nerve to the brain, and motor impulses are transmitted by autonomic fibers in the oculomotor nerve to the ciliary muscles, which contract to make the lens more convex, thereby bringing the image into sharper focus. This mechanism is termed the *accommodation reflex*. A negative feedback mechanism enables the lens to adjust automatically to secure the best focus. The visual cortex is necessarily involved in accommodation.

Diopters The refractive power of an optic lens is measured in *diopters*. A convex lens focusing an image at 1 m (focal length) has a refractive power of 1 diopter. A stronger convex lens focusing at 0.5 m has a greater refractive power of $+2$ diopters. Concave lenses focus the image in front of the lens. By the same rating system employed for measuring the power of convex lenses, concave lenses are given a negative number. A concave lens focusing an image at 0.5 m would have a refracting power of -2 diopters.

INVERSION OF THE IMAGE If we look at the back of a camera that focuses the image on a ground-glass plate, we observe that the image is inverted. This is one of the properties of a convex lens; the rays passing through it are refracted to such an extent that they cross. This means that light rays from the top of the object are projected at the bottom of the image and vice versa. The image is also reversed from left to right. Slides provide another example of inversion; they must be placed in the projector upside down if the picture on the screen is to be right side up.

The image formed on the retina is inverted by the biconvex lens of the eye. We do not see objects, partly because we learn by experience to distinguish them in proper relation to their surroundings. We should keep in mind, however, that the retinal image is not seen; nervous impulses arising from the visual receptors pass back to the visual area of the brain and are then interpreted as an image.

The following is an interesting experiment that illustrates the inversion of the retinal image. Take a card and punch a pinhole in it. Now take the pin and hold it close to the eye, between the eye and the pinhole in the card, while looking at some bright light through the pinhole. When the shadow of the pin falls upon the retina, the image formed by the shadow is seen but the pin appears to be upside down. The brain apparently interprets all retinal images as having been inverted. In this case the shadow is not inverted on the retina, and so the pin is seen upside down.

THE LIGHT REFLEX We have noted that the pupil dilates in response to sympathetic stimulation in the emotional states of fear and pain. The pupil also reacts to light with the *light reflex*. We have observed that the pupil constricts when exposed to bright light and dilates in the dark or in dim light. The most active part of the light reflex is the swift protective closing of the pupil when the eye is exposed to bright light. Constriction of the pupil is mediated through the parasympathetic system. The pupil also plays an important part in accommodation. The radial muscles of the iris contract somewhat for distant vision, and the pupil therefore dilates to a certain extent. For near vision, the sphincter muscle constricts the pupil. Camera fans know that for best results with a close-up, the photographer must use good lighting and reduce the aperture of the lens. The center of the lens is optically the best part of the lens for close work. The action of the iris in constricting the pupil, when a printed page is brought closer to the eyes, for example, permits light rays to pass through the center of the lens, producing a clearer image. These rays can then be brought into a sharp focus at the fovea centralis.

CONVERGENCE The eyes appear to look straight ahead at a distant object, but if an object is brought close, the eyes converge markedly until, with the object close to the face, the eyes appear "cross-eyed." Have someone hold his finger vertically at arm's length and then bring it close to the face, keeping the eyes focused on it as long as possible. You will note the extreme convergence of the eyes as the finger approaches the face. Convergence is a necessary corollary of accommodation, since it keeps the retinal image focused on corresponding points in each eye, an essential part of binocular vision.

An object may be brought so close to the eye that it is impossible to obtain a sharp focus. The *near point* therefore is defined as the closest position at which an object can be clearly seen, with the lens completely accommodated. The near point is very close in children (8.8 cm at ten years of age). As the age increases, the distance of the near point from the eye increases. The increase is gradual in young persons, but between the ages of forty and fifty there is a marked change and some individuals may find that they are becoming farsighted.

Parallel rays from a distant object are focused sharply by the normal eye with the ciliary muscles relaxed, or without accommodation. Theoretically the *far point* is at infinity, but practically an object at a distance of 20 ft gives off rays so nearly parallel that the normal eye can focus them without accommodation. This distance is taken as the far point for most practical purposes.

THE PHYSIOLOGY OF VISION

The processing of visual information has been difficult to determine. The rods and cones are receptors of light waves, but action potentials do not originate with these cells. The receptors are biological transducers, and they initiate only receptor potentials. It is not clear as yet whether bipolar cells produce action potentials, but certainly their graded potentials are strong enough to initiate action potentials in the ganglion cells. Such action potentials are then transmitted by the optic nerves. There is evidently considerable coding of the visual input by ganglion cells before impulses leave the retina. Synaptic connections in the geniculate bodies of the thalamus continue the sorting and refining process before information reaches the visual cortex.

Processing of information begins when the eye scans the object with a series of rapid movements. The series of retinal images provides an estimate of form, size, lines, and color. Rapid eye movements (REM) constitute the *saccade*. The individual is unaware of these movements as the eye fixes on various details. Rapid movements of the eyes during dreaming are well documented. In reading a line of print, the eye also jumps along from one point of fixation to another several times in a second.

Let us take the scanning of a green plastic bucket as an example. By a series of rapid movements, the eye will fix in succession on the contour, the slope of the sides, height, width, and circular top as an indication of size, and recognition. Probably some details will be overlooked, for example, whether or not it has a bail and the nature of its color and composition. Obviously this series of retinal images in a fraction of a second is quite different from a single picture taken with a camera. The series of retinal images enables the viewer to form a picture of the object that he may recognize as a bucket.

The eye is even more sensitive to objects that move. In fact, there is evidence that in some animals the retina responds selectively to the direction of movement; some ganglion cells apparently respond to movement in one direction but give no response when the movement is in the opposite direction. The element of contrast is quite important in initiating visual impulses. Think of how difficult it is to observe insects, birds, or mammals with protective coloration who blend with their background as long as they remain quiet. It is very easy to see such animals when they move.

Various transmitter substances have been found in the retina. These include GABA, dopamine, and norepinehrine. Acetylcholine may also be present. It is suggested that presynaptic fibers may be cholinergic. The roles of these transmitter substances as they pertain to different kinds of cells and their synapses have not been clarified.

THE NEURAL PATHWAY   Processing of information takes place at many locations along the visual neural pathway. The axons of ganglion cells leave the retina at the optic disk and form the optic nerve, which contains nearly a million of these fibers. As the neurons gather to form the nerve at the back of the retina, they become myelinated. The nerves from each eye progress toward each other and form a partial crossing in the *optic chiasma*. Only the fibers from the inner, or nasal, half of the retina cross over to the opposite side in the chiasma. The fibers from the outer, or lateral, half of the retina remain on the same side. The fibers now form the optic tracts, which pass around the cerebral peduncles to the lateral geniculate bodies, and succeeding fibers extend back to the visual areas in the occipital lobes (Figs. 11.14 and 11.15).

Some fibers, however, do not synapse in the lateral geniculate bodies but enter the *superior colliculi* instead. These small structures in the midbrain function as optic reflex centers. Optic nerve fibers make connections with fibers of the oculomotor nerve here, so that the sphincter muscle of the pupil is stimulated to reduce the size of the pupil in response to strong light. This reaction is termed the *light reflex*. Other optic nerve connections with the ciliary nerves, which innervate the ciliary muscles, mediate the accommodation reflex for distant vision, in which the lens relaxes. More elaborate neural connections form protective reflexes, such as those that close the eyelids as an object approaches the eye and raise the arms to protect the eyes. The superior colliculi also supply impulses to the extrinsic eye muscles that move the eye.

THE VISUAL CORTEX   The visual cortex is located in the right and left occipital lobes of the cerebrum (see Chap. 9). Nerve fibers from the right

FIGURE 11·14

a Visual pathway, including the optic chiasma and the optic tract. Note that the fibers from the outer half of each retina do not cross in the chiasma. *b* Enlargement of lateral geniculate bodies.

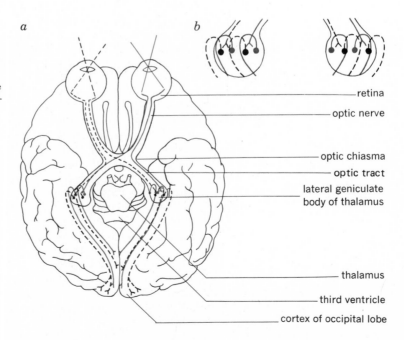

a

b

retina

optic nerve

optic chiasma

optic tract

lateral geniculate body of thalamus

thalamus

third ventricle

cortex of occipital lobe

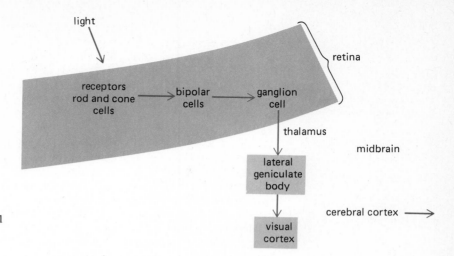

FIGURE 11·15

Effect of light falling on visual receptors. The impulse is traced to the visual cortex.

half of each retina carry impulses to the right visual area, whereas fibers from the left half of the retina bring impulses to the left visual area. This means that the optic fibers arising from the inner half of each retina must cross to the opposite side. The crossing of the fibers occurs back of the eyeballs in the *optic chiasma*. The crossing, as we have described, is incomplete, but it is thought to contribute to the correlation of eye movements and visual perception.

Visual impulses are transmitted to the visual areas of the cortex from the retina, maintaining a faithful *spatial* representation of the retinal image. Not only are right and left halves of this image correlated in the cortex, but neurons from the upper half of the retina carry impulses to the upper part of the visual cortex and impulses arising from the lower part of the retina are transmitted to the lower half of the visual area. Evidently the inverted image at the retina is transmitted directly to the visual area; impulses transmitted from the inverted retinal image do *not* arrive right side up at the visual cortex.

FIGURE 11·16

Accommodation for distant and near vision. Light rays from a distant object are practically parallel and can be focused with the ciliary muscles relaxed and with the normal convexity of the lens. Diverging rays from near objects, however, require greater convexity of the lens to focus sharply on the retina.

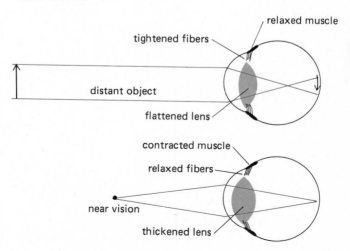

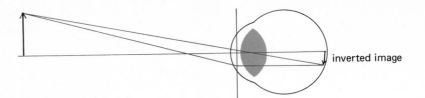

inverted image

FIGURE 11·17
The inversion of the retinal
image in the schematic eye.

Coding in the visual cortex The cerebral cortex can no longer be regarded as a passive "switchboard" where connections are made and information is passed on. There are some 10 billion neurons in the human brain, and apparently they are able to accept, reject, and interpret selectively the information supplied to them by sensory neurons.

Hubel and Wiesel, investigators at Harvard University, have directed their research on vision toward the visual cortex of the cat, rather than to the retina. They have attempted to explain how the visual cortex analyzes and interprets the retinal image by studying the responses of individual cells to light stimuli, which are presented in various shapes and patterns. Synapses in the visual system, as elsewhere, may be excitatory or inhibitory.

It has been known for some time that the ganglion cells of the retina maintain a rhythm of firing even when no external stimulus is present. Earlier investigators had shown that a small circular pattern of light evoked a maximum response from one small group of receptors in the retina, while the same stimulus suppressed firing in another retinal area until the light was turned off. The areas that reacted to the small circular spot of light were termed "on" areas; those that did not react were called "off" areas (Fig. 11.18). The researchers also determined that a typical bright round stimulus elicited "on" responses from a central spot and "off" responses from a surrounding field. Contrasting with this situation were regions where a circular spot of light produced an "off" response in the center of the retinal field and an "on" response from the surrounding field (Fig. 11.18).

The contrasting reactions to such stimuli from different retinal regions is most likely interpreted as fields of contrasting light, such as that surrounding the "edge" of an object. A weak illumination, even though it may cover the entire retina, does not produce as great a response in the ganglion cells as a small circular spot of light concentrated on a small area since weak overall illumination does not elicit contrasting responses.

FIGURE 11·18
Response of "on" and "off"
center ganglion cells.

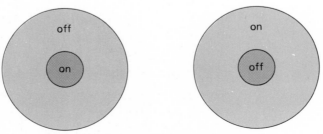

response of on and off center ganglion cells

The complex cortical cells recognize "lines" by interpreting the contrasting responses of retinal regions to a bright light. Many groups of cells also respond to movement of a narrow beam of light across them in one direction only. The cells will fire as a beam of light is moved vertically across them but not in any other direction. Another group of cells will respond to a narrow beam moving only in a horizontal direction. Thus we have the ability to recognize form and to record movement of an object in various directions.

BINOCULAR OR STEREOSCOPIC VISION   Human eyes, very fortunately, make every effort to maintain a constant visual field as the head is moved. It would be most confusing if we shook our head vigorously during conversation and found that the visual field also appeared to move from side to side. As we approach an object, we may keep the object fixed by greater and greater convergence of the eyes. *Binocular vision* is the blending of two images to produce one image that looks real and has depth. When the eyes converge upon an object, the right eye sees a little more of the right side of the object, while the left eye sees a little more of the left side. The two images are not identical, but when they are superimposed and interpreted by the visual area of the cortex, we perceive an image that has depth and unity.

The stereoscope is an instrument that gives the illusion of depth and solidity when one looks at pictures taken in a certain way. A picture taken by an ordinary camera appears flat in comparison. Two pictures made to be viewed through a *stereoscope* are taken at slightly different angles, equivalent to the angle of convergence of the eyes. Light rays, passing from the two pictures through the lenses of the stereoscope, stimulate corresponding retinal points in each eye, and the image is interpreted as having depth and reality.

Binocular vision aids in judging distance. There are, however, other factors that help us measure distance. The color of distant objects, the relative size of intervening objects, shadows, and perspective all play a part.

Corresponding points on both retinas must be stimulated simultaneously to obtain a single, binocular image. Obviously the retinal foveae are corresponding areas. If we fix our eyes on some object across the room and then bring a finger up close to the near point of the eyes, we see two fingers instead of one. If the right and left fingers are held horizontally in front of the eyes at about the distance of the near point while the eyes are fixed on some distant object, the right finger will be seen by the right eye, the left finger by the left eye. In the middle of the visual field, however, we will see a seemingly detached portion of the fingers, the part of the image that is superimposed by both eyes.

COLOR VISION

PHYSICAL CHARACTERISTICS OF LIGHT   An understanding of color vision requires some knowledge of the physical characteristics of light waves. The sensation of color vision is produced by different wavelengths of light. It should be remembered that color vision is the interpretation by the brain of certain wavelengths of light received on the retina. This fact gives rise to the saying "color exists only in the mind."

Sunlight or natural white light is a combination of all colors of the spectrum. When such a light is passed through a prism, it is separated into its component parts: red, orange, yellow, green, blue, and violet. These are the colors of the spectrum. The longest visible wavelengths, 760 Å (760 millionths of a millimeter), produce the sensation of red, while the shortest visible wavelengths, 390 Å (390 millionths of a millimeter), are interpreted as violet. Intermediate wavelengths produce the sensations corresponding to the other colors of the spectrum. (Light waves are commonly measured in angstroms; one ten-millionth of a millimeter equals one angstrom.)

An object has color because it absorbs certain wavelengths of light and reflects, or transmits, other wavelengths. The eye receives the reflected or transmitted wavelengths, which are interpreted as representing the color of the object. Objects that appear white to the eye reflect all wavelengths. Objects that appear black in sunlight absorb all wavelengths. Black appears to be a definite sensation; it is not equivalent to seeing "nothing," the sensation we have when objects are focused on the blind spot.

Since white light can be separated into the colors of the spectrum, it is also possible to combine these colors to produce white light. However, all the colors are not needed to produce white light, and all the colors of the spectrum can be formed from three primary colors: red, green, and violet. When two or more wavelengths fall upon the retina at the same time, the result is a color fusion and the sensation is of a different color than that produced by any single wavelength. Thus the wavelengths for red and green give rise to the color sensation of yellow or orange, while the fusion of red and violet is interpreted as purple.

The fusion of certain wavelengths of light should not be confused with the mixing of pigments to produce different colors of paint. These pigments are not optically pure and so reflect and absorb light waves in quite a different manner. Mixing yellow and blue pigments produces a green color; the fusion of yellow and blue wavelengths of the spectrum in the proper proportion gives rise to a white color sensation.

COMPLEMENTARY COLORS   The fusion of light waves of two colors or of combinations of colors can produce the sensation of white. Depending upon the proportions of colors used, a suitable color complement can be found for any given color, and the result of their fusion will be the sensation of white. Color pairs that produce white when fused are *complementary colors*. In the light spectrum the following color pairs are complementary to each other: red and blue-green; orange and blue; yellow and indigo blue. On a color scale, these colors are placed diametrically opposite each other. Closely related colors, upon fusion, produce a more saturated, "deeper" color, or one that produces less of the sensation of white. The fusion of red and yellow to form orange or the fusion of red and violet to produce purple are examples of this phenomenon. The pale or so-called pastel shades are examples of unsaturated colors.

It is easy and interesting to perform the following experiment, but far more difficult to explain the result. If you place a green filter over one eye and a red filter over the other eye and then look at a white area, you do not see either color. You will see instead a mixing of the two colors to produce a weak yellow or orange. Obviously the retina of one eye is stimulated by light waves in the red region of the spectrum, while the other is reacting to

TABLE 11·3

*Wavelengths of light
interpreted as color*

Wavelengths in nm	Color
435	Blue
540	Green
550	Yellow
570	Red
580	Orange

light waves in the green region. It appears that the mixing does not occur in either retina but more likely occurs in the visual cortex.

THE THREE-PIGMENT THEORY OF COLOR VISION   Color vision is thought to require three photopigments, which are incorporated into three types of cone receptors. Each type of receptor has an individual maximum sensitivity in either the blue, the green, or the yellow regions of the spectrum. "Red" receptors reach their sensitivity peak in the yellow area of the spectrum, but they extend their sensitivity well into the red region and so are efficient in sensing red.

Table 11.3 indicates the light-wave lengths, in nanometers (nm), that are interpreted as a certain color.

A range of light waves is interpreted as the same color. For example, light waves shorter than 435 nm are seen as blue-violet and violet. In an intermediate-length light-wave range between green and red, orange is found at about 580 nm and yellow at about 550 nm. (A nanometer is 0.0001 micrometer.)

The spectral sensitivity curves, as determined by several research workers employing somewhat different procedures, vary slightly, but in general, blue sensitive receptors reach their maximum sensitivity at about 435 nm; this corresponds to the blue-violet area of the spectrum (Fig. 11.19). Green-sensitive cones have a maximum sensitivity at 540 nm, which is interpreted as green although their sensing ability extends from blue through green, well into yellow, orange, and red wavelengths. As we have indicated, red-sensitive cones reach their maximum sensitivity at 570 nm in the yellow area of the spectrum, but their sensitivity extends through green and well into the red wavelengths. In color television, orange-red is a very strong color; that is, it has a dominant color reception.

Blending of colors in vision occurs when more than one type of cone cell is stimulated at the same time. The stimulation may be essentially equal or it may be unequal. If red and green receptors are stimulated equally, as they are by wavelengths in the yellow area of the spectrum, the sensation is interpreted as yellow. But if the stimulation is unequal and red-receptor cones are stimulated more strongly than green-receptor cones, then the color is interpreted as orange or orange-red.

Processing of color information apparently begins with certain retinal ganglion cells that respond strongly to light waves of one of the primary colors but are suppressed by light waves of another color. Some cells in the thalamic lateral geniculate bodies also respond in a similar manner to impulses initiated by light of different wavelengths. These cells typically operate from fields with "on" or "off" centers. Finally, such coded impulses reach the primary visual cortex, where the coded messages are given their final interpretation.

THE AFTERIMAGE   Visual sensations develop, last for a certain length of time, and then fade. If the image can be seen for a longer time than the actual exposure, it is called an *afterimage*. It is termed a *positive afterimage* if it has the same color and the same appearance as the object. If it changes from white to black or if it is produced in the complementary color, it is called a *negative afterimage.*

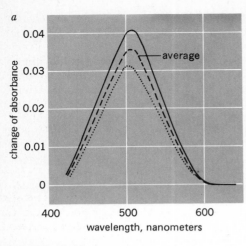

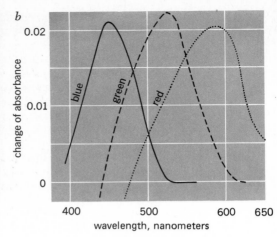

FIGURE 11·19

Spectral sensitivity of rod and cone cells: *a* Difference spectra of the visual pigment in a human rod cell in the parafoveal region. The average curve maximum at 505 nm closely resembles the difference spectrum of human rhodopsin. The pigment of the rod consists primarily or entirely of rhodopsin. *b* Difference spectra of visual pigments in single human cone cells in the parafoveal region. [*After Paul K. Brown and George Wald, Visual Pigments in Single Rods and Cones of the Human Retina, Science, **144**:45–51 (1964). Copyright, 1964, by the American Association for the Advancement of Science.*]

The persistence of the positive afterimage enables us to enjoy moving pictures; the afterimage is retained just long enough for the motion picture projector to substitute the succeeding frame in the series that makes up a film. The rapid succession of frames gives the illusion of movement to the images. When a bicycle wheel is turned rapidly, the individual images of the spokes soon become blurred and the interior of the wheel appears solid. At the same time, if you were to place your finger lightly on the spokes, you would feel each spoke pass by. Evidently the visual afterimage lasts much longer than the "afterimage" for the sense of touch.

A positive afterimage may be developed by looking at an ordinary electric light for a few seconds. Upon closing the eyes a bright image persists for some time and is seen in the same color as the original. The persistence of the afterimage appears to be caused by nerve impulses that continue to develop in the retina after the initial stimulus has passed.

A negative afterimage is thought to be caused by adaptation of the cone cells for the particular light or color involved. If we stare at a white globe over an electric light and then turn our gaze to a white wall, the afterimage is black, not white. The theory is that long exposure to white light stimulation has temporarily adapted all the receptors in the retinal region focused upon the light globe. When the eyes are turned to the white wall, a second and weaker stimulus of white falling on this area of the retina produces no response. We therefore see a black area surrounded by the white of the wall. Black is the sensation produced by a lack of retinal stimulation. (We should note that this "black" is distinct from the sensation produced by looking at a black-colored object.)

If we stare at an object with a bright red color for about 30 seconds and then look quickly at a white surface, the afterimage will appear in the complementary color, which is blue-green. It is assumed that the retinal elements sensitive to red have become adapted, and the second stimulus of white light, without the aid of the retinal red-sensitive pigment, causes the blue-green–sensitive pigment to respond.

COLOR BLINDNESS   The term *color-blind* is usually applied to those who do not have normal color vision. For the most part it is an inappropriate term, since most "color-blind" individuals see color of one sort or another. There are a very few persons who are truly color-blind; but if so, their vision is much like viewing a photograph on ordinary black-and-white film. The image is in tones of white and black—a true condition of achromatic vision.

Many persons are simply color-weak. They are poor at distinguishing different hues, and their color-perception mechanism adapts readily. While gazing steadily at some small area of color, as in attempting to identify an insect by verifying its color, the color seems to disappear. But if the color area is large and definite, these persons are not aware of any difficulty. Many are quite unaware that they do not have normal color vision.

For some, the visual spectrum is simply shortened at one end. If the shortening occurs in the longer wavelengths, the individual may not be able to distinguish a red line from the black line on a chart or graph. All appear black to him. Apparently there is also considerable individual variation in ability to distinguish the shorter wavelengths that are interpreted as violet hues.

Color blindness occurs ordinarily when one type of color receptor is deficient. If the red-receptor cones are missing or nonfunctional, the color-blind person is forced to interpret color primarily with green and blue receptors. This perception is inadequate because true color perception requires both red and green cones to provide contrast. If green receptors are deficient, the consequences are similar. The color-blind person is unable to distinguish between various shades of color and is most deficient in distinguishing between the longer wavelengths of green and red. This condition is referred to as *red-green color blindness*. A red-blind person is called a *protanope*, and a green-blind person is a *deuteranope*.

Blue color blindness is very rare and differs in its inheritance from the more typical red-green type. It is a dominant trait and is not sex-linked.

The inheritance of color blindness The inherited form of red-green color blindness is passed on as a sex-linked trait. This deficiency is determined by a gene that cannot direct the development of full color vision. The gene is recessive and is located on the X chromosome. A female must have this gene on both X chromosomes if she is to be color-blind herself. If the gene is present on only one X chromosome (the usual condition, if present at all), the woman will not be color-blind, but she may transmit the defective gene to some of her children (Fig. 11.20).

The gene for color blindness has no normal dominant counterpart in the Y chromosome of the male, so that a male who inherits an X chromosome with this defective gene will have the trait. Almost without exception, the color-blind father will transmit the gene for this *trait* to his female children, because the female children will inherit one X chromosome from the father and one from the mother.

When a female with one gene for color blindness marries a normal male, each male child has a 50 percent chance of being color-blind and each female child has a 50 percent chance of becoming a carrier for this trait. Color blindness is far commoner in men, affecting about 8 percent of the male population. It is much rarer in women, affecting only about 0.5 percent of the female population according to one estimate.

Red-green color blindness may involve a deficiency in either or both of

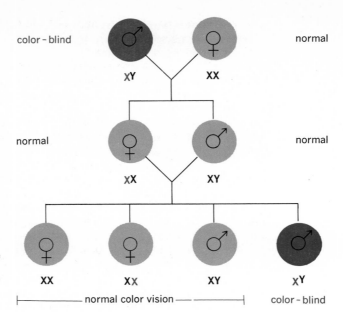

color - blind   normal

XY   XX

normal   normal

XX   XY

XX   XX   XY   XY

├────── normal color vision ──────┤   color - blind

FIGURE 11·20
Schematic design illustrating the inheritance of color blindness. The male is color-blind, and the female has normal color vision. The defect is transmitted through the daughters, but both male and female children have normal color vision. The grandsons who inherit the X-chromosome gene for color blindness will be color-blind; all other grandchildren will have normal color vision.

CLINICAL ASPECTS

two different color-sensitive pigments, each presumably controlled by a specific gene. Both genes are thought to be on the X chromosome.

DISEASES OF THE EYE *Glaucoma* When the drainage of aqueous humor fails to keep up with the amount produced by the ciliary processes, *glaucoma* occurs. In this condition, intraocular pressure may rise to such levels that severe pain is produced and the nerve fibers arising in the retina may be injured. Damage usually occurs at the optic disk, which is the weak spot in the retina where nerve fibers exit to form the optic nerve. The nerve fibers may be forced against the edge of the disk, and the continued pressure on them may cause their destruction. The retinal artery enters the eye at the optic disk, and it may be compressed by the increased pressure within the eyeball. Drugs that constrict the pupil may relieve mild cases of glaucoma, because constriction of the pupil tends to enlarge and relax the angle between the iris and the cornea, thus permitting better drainage.

The normal intraocular pressure is about 15 mmHg. This pressure reading is taken with an instrument called a *tonometer*. Abnormal pressures may reach 20 to 30 mmHg, levels at which damage to the optic nerve and retina occurs. If drugs fail to cause an opening of the drainage pathway by constricting the pupil, a surgical procedure may be used to relieve the pressure. This consists of removing a small piece of the iris which would then permit the fluid to flow forward into the sinus venosus (canal of Schlemm) (see Fig. 11.3).

Conjunctivitis Several diseases may cause inflammation of the conjunctiva and associated structures, a condition called *conjunctivitis*. *Trachoma* is a disease of the conjunctiva that affects the upper and lower eyelids. If it is not controlled, it may lead to blindness from ulceration and scarring of the cornea. A bacterial infection may also cause the common conjunctivitis termed *pinkeye*.

Keratitis *Keratitis* is an inflammation of the cornea. If untreated, it may so damage the cornea that a corneal transplant may be advisable. It is usually of bacterial origin.

Hordeolum, or sty An infection at the base of the eyelashes, usually caused by a staphylococcus type of bacteria, is called a *hordeolum*, or *sty*. If the infection invades glands along the edge of the eyelid, it is called a *chalazion*. In either case a physician may open the infected area to establish drainage, and this procedure usually will clear up the infection.

Cataract A condition in which the lens becomes opaque or milk-white in appearance, a condition termed a *cataract*. The change often occurs first in the firm nucleus, or central portion, although there are various types of cataract. Since light cannot pass through the clouded lens in adequate amounts, a cataract produces increasing blindness as the lens becomes progressively more opaque. The lens may be removed in a very delicate surgical operation, and sight may be restored by this means. When the lens is removed, of course, focusing power is lost. Strong bifocal eye glasses or a combination of contact lenses and eyeglasses are required to achieve near and far vision after cataract removal. Cyclotron-induced radiation cataracts are a new hazard in the modern world.

Night blindness Adequate amounts of vitamin A, which is supplied to the retina by the blood, are needed for good night vision. Persons who are unable to see in a dim light as well as the average person, have a defect called *night blindness*. If this condition is caused by a deficiency of vitamin A, it may be improved by adding vitamin A to the diet.

Argyll Robertson pupil An interesting abnormal condition affecting the pupillary light reflex is termed the *Argyll Robertson pupil*. This condition is characterized by the loss of the light reflex. The pupil remains small in dim light, but the accommodation reflex still persists. This reaction usually indicates the presence of tertiary syphilis in the central nervous system, and its presence has diagnostic value. The contraction of the pupil still occurs during accommodation because the disease has not affected the parasympathetic nerve fibers to the eyes.

Drug effects The pupils of the eyes dilate widely following the use of marijuana. Morphine and opium greatly constrict the pupils into pinholes. Atropine, often used in eye examinations, dilates the pupils. Pilocarpine, which is used in the treatment of glaucoma, is parasympathomimetic and constricts the pupils.

OPTICAL DEFECTS An eye that has no optical defects is said to be *emmetropic;* one with optical defects is called *ametropic*.

Astigmatism The condition *astigmatism* is usually caused by imperfect curvature of the cornea. If all the arcs of the cornea are equal, light waves passing through form a cone of light behind the cornea and focus at a single point. If there is a variation in the curvature, however, so that, for example, the vertical curvature is greater than the horizontal, two points of focus are found. The light rays from the greater curvature focus at a shorter distance than those from the lesser curvature. In astigmatism the focus of the image is not sharp, and images may appear blurred. Bright lights may appear to have shafts of light extending outward in something of a star-shaped pattern. Seen through the corrected lenses of a telescope, the stars appear as spheres, as they actually are; but with the unaided eye, the stars often appear to have shafts of light extending outward, indicating that all eyes are astigmatic to a certain extent.

In a test for astigmatism, the person looks at vertical and horizontal lines that cross or a figure with radiating lines, like the spokes of a wheel. If the vertical lines appear more distinct or darker than the horizontal lines, the inference is that the vertical curvature gives a sharper image than the horizontal. The type of astigmatism in which the vertical curvature is greater than the horizontal is the most common form. Astigmatism is corrected by wearing glasses in which cylindrical lenses are designed to equalize the vertical and horizontal curvatures of the cornea. In other words, a greater curvature of the lens along one direction compensates for the lesser curvature of the cornea in that direction and vice versa.

Presbyopia An older person may find that he or she has to hold the newspaper quite a distance from the eyes in order to read it; in such a case, the near point may be around 50 or even 75 cm. If the eyes are normal otherwise, this condition is called *presbyopia*, the farsightedness of the aged. The lens has lost so much of its elasticity that, even though the ciliary muscles contract strongly, the lens does not become convex enough for close work such as reading small print. The necessary additional refraction can be obtained by using the proper convex lenses in glasses.

Hyperopia This is a type of farsightedness in which the refractive power of the eye is not great enough or the eyeball is too short for proper near vision (Fig. 11.21a). It should not be confused with presbyopia, which is a gradual loss of accommodation as one grows older. The light rays from an object in this case are focused behind the retina, resulting in a blurring of vision unless the accommodation is strong enough to secure a sharp focus. Ordinarily, objects at a distance may be seen clearly, but the diverging rays from near objects are focused with difficulty, if at all. Reading or other detailed work may cause eyestrain. The condition may be corrected by using proper convex lenses in glasses or contact lenses.

Myopia In *myopia*, or nearsightedness, the refractive power of the eye is too great or the eyeball is too long for proper far vision (Fig. 11.21b). Light rays focus somewhere in the vitreous body, a little in front of the retina. The image, when projected back to the retina, will be blurred. To correct this condition, concave lenses are used in glasses or contact lenses.

Biconcave lenses are thin in the middle and spread the light rays that pass through. Such lenses diverge light rays before they pass through the cornea, so that the focal point is pushed farther back in the eye. The nearsighted person is handicapped in regard to distant objects, but can do close work

FIGURE 11·21

Diagram illustrating two common eye defects, hyperopia and myopia: *a* Hyperopia, a type of farsightedness. If the eyeball is too short, the rays of light focus behind the retina, as shown in the upper illustration. The correction of farsightedness is accomplished by wearing a convex lens before the eye. This lens brings the rays of light into sharp focus on the retina, as illustrated. *b* Myopia, or nearsightedness. If the eyeball is too long or if the refractive power of the eye is too great, the rays of light focus at some point in front of the retina. Nearsightedness is corrected by wearing a concave lens before the eye. This type of lens brings the rays to sharp focus on the retina, as illustrated.

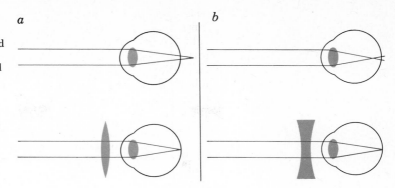

without difficulty. The eyes of those elderly persons who can read fine print without their glasses are almost certainly myopic.

Testing for visual acuity   In testing the eyes for visual acuity, a properly illuminated test chart is placed at a distance of 20 ft. The chart has rows of letters, each row in a different-sized print. At the end of each row is a number indicating the ability of the normal eye to read the type of that particular size. An individual who has a 20/20 rating (normal vision) can read the line indicated for 20 ft at a 20-ft distance. Each eye must be tested separately, of course. If the individual being tested is rated 20/40, it means that he can read only the larger letters, marked 40 ft, at 20 ft. If the individual has better-than-normal vision, he may be able to read a row of smaller letters at 20 ft—perhaps a row that the normal eye could read only at 15 ft. In this case his rating would by 20/15.

Strabismus   In some cases the eye muscles are not properly balanced, and the two eyes do not focus together normally. There may be too much convergence, as in the condition commonly called "cross eyes," or *strabismus;* then again one or both eyes may be strongly divergent (*divergent strabismus*). Surgery or exercises of eye muscles may restore muscular balance. If the two eyes do not focus on the same retinal points, the vision of one eye is usually disregarded and the individual sees only one field.

SUMMARY

1 A sensory unit is a sensory neuron with its receptors. The receptors constitute a sensory field. Sensory receptors act as biological transducers.
2 Sensory receptors are classified as exteroceptors, proprioceptors, and interoceptors.
3 The accessory structures of the eye are the eyelids, eyebrows, eyelashes, conjunctiva, and lacrimal apparatus. The conjunctiva is a thin, transparent membrane covering the anterior surface of the eyeball and lining the eyelids. The lacrimal gland produces tears. The lacrimal secretion moistens the anterior surface of the eye, and the excess fluid drains into the nasolacrimal duct.
4 There are six extrinsic muscles attached to the eyeball. They are the superior, inferior, lateral, and medial rectus muscles plus the superior and inferior oblique muscles.

5 The wall of the eyeball is composed of three layers: an outer fibrous tunic, consisting of the sclera and cornea; an intermediate vascular layer, called the choroid; and an inner nervous tunic, the retina.

6 The outer coat, or sclera, is white except for the anterior transparent portion, called the cornea. Astigmatism is a condition usually caused by imperfect curvatures of the cornea.

7 The dark-colored choroid gives rise to the ciliary body and the iris as it extends anteriorly. The muscles of the ciliary body control the degree of convexity of the lens. The iris is the colored portion of the eye, and eye color refers to the color of the iris. The pupil is the circular opening in the iris. It appears black because it opens into the dark recess of the eyeball. The posterior side of the iris is dark-colored and contains two antagonistic sets of muscles. Contraction of the radial muscles dilates the pupil, whereas contraction of the sphincter pupillae causes the pupil to constrict.

8 Parasympathetic fibers of the oculomotor nerve innervate the sphincter muscle, while sympathetic fibers innervate the radial muscles.

9 The inner nervous coat contains receptors called rod and cone cells. There are several layers of neurons in the retina, the rod and cone cells representing the deepest layer. The retina is inverted, with the receptor cells turned away from the light source. The optic disk is a blind spot because the nerve fibers that form the optic nerve are concentrated at this spot, excluding receptors.

10 Light waves must pass through the cornea, the aqueous humor, the lens, and the vitreous body before reaching the retina. The refractive index varies for each of these tissues and fluids; light waves therefore are refracted or bent in passing through them.

11 Rod cells react to low intensities of light, providing vision in dim light. Rhodopsin, or visual purple, is a photosensitive pigment associated with rod cells. Cone cells react to light of high intensity and provide acute, detailed vision. They are also the receptors for the perception of color.

12 The area centralis, or yellow spot, contains the fovea centralis, which is the area of acute vision. The fovea is a depressed area of the retina. In the center of the depression is the foveola, an area only 0.35 mm in diameter; it contains only cone cells.

13 The lens is biconvex and capable of changing its convexity in accommodation for objects at varying distances. It becomes more convex to accommodate for objects that are close to the eye.

14 The pupils respond to varying light intensities by constricting when subjected to bright light and dilating when exposed for some time to light waves of low intensity. This is the light reflex. The pupil constricts also during the accommodation of the lens for near vision and dilates to a certain extent for distant vision.

15 When the radiant energy of light waves is focused on the retina, nerve impulses are initiated in the optic nerve. The nerve pathway includes the optic chiasma, the optic tract, the lateral geniculate body of the thalamus, and, finally, the visual area of the occipital lobes. In the optic chiasma, only the nerve fibers from the inner half of each retina cross over to the opposite side. The fibers from the outer half of each retina remain on the same side as that from which they originated.

16 Color sensation is produced by different wavelengths of light stimulating

the cone cells of the retina. Visual sensation can last longer than the actual exposure time. If the sensation persists for a time after the exposure has ended, the persisting image is called an afterimage. If it has the same color as the object, it is a positive afterimage; if it is produced in the complementary color, it is a negative afterimage.

17 Color-blind persons do not have normal color vision. Inability to distinguish red and green colors is the most common type. The inherited type of red-green color blindness is passed on as a sex-linked trait. The gene for color blindness is recessive but is expressed in a male (XY) who has inherited one gene in the X chromosome, because the Y chromosome contains no dominant gene to prevent the development of the defect. If the gene is present in only one X chromosome in a female (XX), she will not be color-blind but may transmit the gene to some of her children. A color-blind female inherits the gene in both of her X chromosomes.

18 A number of diseases and disorders affect the eyes.

QUESTIONS

1 Name some of the accessory structures surrounding the eye.
2 Locate the conjunctiva and indicate its function.
3 Give some of the functions of the lacrimal structures.
4 Name and locate the muscles that move the eyeball.
5 Discuss the three layers or coats of the eyeball. Indicate the derivatives of each coat.
6 What is astigmatism?
7 Why is the vertebrate retina inverted?
8 How is a blind spot formed in the eye?
9 What is visual purple?
10 How do rod cells function?
11 Discuss dark adaptation and night vision.
12 Explain the structure and function of the fovea centralis.
13 What are the refracting media of the eye?
14 What is a cataract?
15 Compare the focusing of a camera with that of the eye.
16 Give some of the uses of a convex lens.
17 Explain the action of the lens and the ciliary body during accommodation.
18 Discuss the action of the pupil when exposed to light and during accommodation.
19 Why is convergence of the eyes necessary for accommodation?
20 How does presbyopia differ from hyperopia?
21 What structural conditions can cause the eye to be myopic? What kind of lens is used in the glasses to correct the condition?
22 Explain binocular vision.
23 Discuss color vision.
24 What is a positive afterimage? A negative one?
25 Discuss color blindness. How are certain forms inherited?

SUGGESTED READING

Bioemendal, Hans: The Vertebrate Eye Lens, *Science*, **175**:127–138 (1977).
Brindley, G. S.: Central Pathways of Vision, *Annu. Rev. Physiol.*, **32**:259–268 (1970).

Favreau, O. E., and M. C. Corballis: Negative Aftereffects in Visual Perception, *Sci. Am.*, **235**:42–48 (1976).

Glickstein, M., and A. R. Gibson: Visual Cells in the Pons of the Brain, *Sci. Am.*, **235**:90–98 (1976).

Guillery, R. W.: Visual Pathways in Albinos, *Sci. Am.*, **230**:44–54 (1974).

Heyningen, Ruth van: What Happens to the Human Lens in Cataract, *Sci. Am.*, **233**:70–81 (1975).

Hubel, David H.: The Visual Cortex of the Brain, *Sci. Am.*, **209**:54–62 (1963).

——— and T. N. Wiesel: Receptive Fields, Binocular Interaction and Functional Architecture of the Cat's Visual Cortex, *J. Physiol.*, **160**:106–154 (1962).

Miller, W. H., R. E. Gorman, and M. W. Bitensky: Cyclic Adenosine Monophosphate: Function in Photoreceptors, *Science*, **174**:295–297 (1971).

Noda, H., R. B. Freeman, Jr., and O. D. Creutzfeldt: Neuronal Correlates of Eye Movements in the Visual Cortex of the Cat, *Science*, **175**:661–663 (1972).

Noton, D., and L. Stark: Eye Movements and Visual Perception, *Sci. Am.*, **224**:35–43 (1971).

Pollen, D. A., James R. Lee, and J. H. Taylor: How Does the Striate Cortex Begin the Reconstruction of the Visual World?, *Science*, **173**:74–77 (1971).

Richards, W.: The Fortification Illusions of Migraines, *Sci. Am.*, **224**:89–96 (1971).

Rodieck, R. W.: Central Nervous System: Afferent Mechanisms, *Annu. Rev. Physiol.*, **33**:203–240 (1971).

Rushton, W. A. H.: Visual Pigments and Color Blindness, *Sci. Am.*, **232**:64–74 (1975).

Schmidt, F. O., Parveti Dev, and B. H. Smith: Electrotonic Processing of Information by Brain Cells, *Science*, **193**:114–120 (1976).

van Heyningen, R.: What Happens to the Human Lens in Cataract, *Sci. Am.*, **236**(6):70–81 (1975).

Werblin, F. S.: The Control of Sensitivity in the Retina, *Sci. Am.*, **228**:71–79 (1973).

Witkovsky, P.: Peripheral Mechanisms of Vision, *Annu. Rev. Physiol.*, **33**:257–280 (1971).

Worthen, D. M.: Endothelial Projections in Schlemm's Canal, *Science*, **175**:561–562 (1972).

Young, R. W.: Visual Cells, *Sci. Am.*, **223**:89–91 (1970).

12

SPECIAL SENSES II: AUDITION, EQUILIBRIUM, CUTANEOUS SENSATION, GUSTATION, AND OLFACTION

CONSIDERATIONS FOR STUDY

In this chapter, we discuss the other special senses: audition (hearing), equilibrium, cutaneous sensation, gustation (taste), and olfaction (smell). As we did in Chap. 11, we shall first outline the physical properties of the stimulus involved in each sensory modality, then cover the anatomy of the organs involved in reception and transmission of sensory impulses, and finally describe the integration of these impulses in centers of the central nervous system to produce perception.

Topics include:

1 The physical characteristics of sound: pitch, loudness, and timbre
2 The anatomy of the ear and its division into outer, middle, and inner parts and the structures in each
3 The pathway for conduction of sound in the ear
4 The neural pathway of audition, the physiology of hearing, and the translation of reception into perception.
5 The role of the vestibular system of the ear in equilibrium
6 Cutaneous sensation: the receptors and the neural pathways involved in the perception of pressure, temperature, and pain

7 The gustatory sense: the taste buds and the translation of sensory information into taste sensations

8 The olfactory sense: the olfactory receptors and the neural pathway involved in the perception of odor

AUDITION (HEARING)

The ear functions in two special senses: *audition*, or hearing, and *equilibrium*, the sense of position, motion, and balance. We shall take up audition first, but before we describe the anatomy of the ear, we shall briefly discuss the physical characteristics of sound, the auditory stimulus.

SOME PHYSICAL ASPECTS OF SOUND When a shot is fired or when a bell is rung, the atmosphere is disturbed by sound waves radiating away from their source. The succession of sound waves consists of alternate *condensations*, or compressions, and *rarefactions* of the air. Sound waves can be likened to waves in water and are commonly shown diagrammatically as a curved line representing a series of hills and valleys. Such sound waves can vary in frequency, a characteristic termed *pitch*, and in amplitude, a characteristic called *loudness*.

When sound waves impinge upon a stretched membrane such as the eardrum, they exert a push-and-pull effect and cause the membrane to vibrate in sympathy with the frequency of the force exerted upon it. However, sound waves generated in most cases are not simple. For example, the sound waves produced by most musical instruments are compound; they are composed of a series of vibrations, which produce *overtones* in addition to a dominant tone. The wave diagram can therefore be complex rather than a series of hills and valleys.

If a thundercloud is very distant, we may see the lightning many seconds before we hear the thunder because sound travels at a relatively slow rate (1100 ft/s) compared with the speed of light (approximately 186,000 mi/s).

The *pitch* of sound waves may vary from a low frequency of around 16 Hz (cycles per second) to very high frequencies of 20,000 Hz. There are, of course, vibrations both below and above these frequencies that are inaudible to the human ear. An example often cited is that of a whip moved gently through the air; the waves created by its movement are inaudible. But if the whip is moved swiftly through the air, a high, piercing sound is heard. Apparently many animals hear frequencies higher than the upper limit for man, but there are supersonic frequencies presumably inaudible to any vertebrate ear.

Audibility curves show that hearing is good within a range of 1000 to 3000 Hz but falls off rapidly for frequencies below or above this range. It has been observed that the range of frequencies common to the human voice falls within the range of optimum reception by the human ear.

Intensity of sound is commonly associated with loudness, but intensity and loudness are not necessarily identical. Intensity is simply the physical energy of sound transmitted through the atmosphere. The *loudness* of the sound is subjective, a matter of auditory perception. The intensity of sound can be so great that it is felt rather than heard, as in the case of heavy concussions. Such disturbance close to the ear can give rise to pain (Table 12.1).

The ear is also able to recognize *quality*, or *timbre*, in musical sounds. It

TABLE 12·1

Sound ratings in decibels

Auditory threshold	0
Watch ticking	20
Business office sounds	40
Conversation, close up	60
Loud radio	80
Automobile horn	90
Amplified music	130
Jet plane at takeoff	150

is easy to distinguish between the sound of a violin, piano, or horn, even though they may all play the same note, because each instrument produces different overtones, or partial vibrations, which make up a compound wave.

AUDITORY STRUCTURES AND PATHWAYS   Auditory receptors in the inner ear react to physical energy in the form of sound waves. The ear represents an intricate mechanism designed to transmit sound waves by mechanical means inward to these receptors. The ear consists of three primary parts: (1) the external ear, (2) the middle ear, and (3) the inner ear. The auditory pathway consists of the auditory (VIIIth cranial) nerve and the auditory interpreting center, which is located in the temporal lobe of the cerebrum.

EXTERNAL EAR   The *outer ear,* or *pinna,* is a structure designed to collect sound waves and to direct them into the external auditory meatus (Fig. 12.1*a*). The pinna is composed of elastic cartilage, covered with skin. The lobe of the ear contains vascular connective tissue rather than cartilage. The *external auditory meatus* is about an inch in length and leads inward to the tympanic membrane. The meatus is not straight; pulling upward and backward on the outer ear tends to straighten it. The skin covering the ear is extended into the meatus. It becomes thinner and more sensitive as it approaches the eardrum. Close to the external opening are stiff protective hairs and *sebaceous glands,* which produce earwax, or *cerumen.* The glands are thought to be modified sweat glands. Both the hair and the glands tend to keep foreign objects out of the meatus.

The *tympanic membrane,* or eardrum, completely covers the inner extremity of the external auditory meatus and thus separates the outer and middle parts of the ear. It is a very thin fibrous membrane covered with a thin layer of skin on the outer side and with a mucous membrane on the inner side. It lies at an angle of about 55° to the lower border of the canal. Most of the eardrum is tightly stretched to form a vibrating membrane, but a small, thin section in the upper part is not stretched and is called the *flaccid part.* It is in this area that injuries from concussion or inflammation are likely to occur.

MIDDLE EAR   An air space in the petrous portion of the temporal bone houses the *middle ear* and is continuous with the spongy bone of the mastoid process. The *auditory,* or *eustachian, tube* connects the cavity of the middle ear with the nasopharynx and helps equalize pressure on the eardrum.

Within the cavity of the middle ear is a lever system of three very small bones. The bones are the malleus, incus, and stapes (Fig. 12.1*b*). The *malleus* is shaped somewhat like a mallet, with the handle portion attached to the inner surface of the eardrum. The rounded head fits into a depression in the *incus.* The head of the malleus is held tightly in place by ligaments so that the malleus and the incus move as one. The incus is the intermediate bone in the series. The *stapes* is a stirrup-shaped bone that fits into the oval window of the vestibule. The joint between the incus and the stapes is freely movable, the stapes performing a rocking movement at the oval window. That these are, indeed, small bones is indicated by their measurements. The malleus is about 8 mm long, the incus is approximately 7 mm, and the stapes measures only 4 mm.

There are two small muscles in the middle ear. The *tensor tympani*

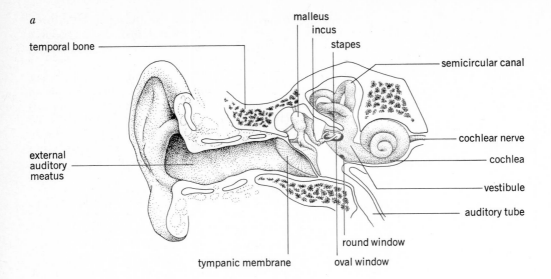

a

temporal bone

malleus
incus
stapes

semicircular canal

external
auditory
meatus

cochlear nerve

cochlea

vestibule

auditory tube

round window

tympanic membrane

oval window

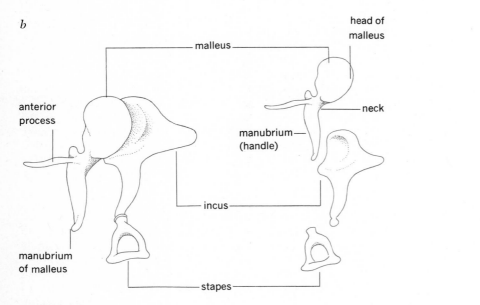

b

head of
malleus

malleus

anterior
process

neck

manubrium
(handle)

incus

manubrium
of malleus

stapes

FIGURE 12·1

a The structure of the ear, including detail of the middle ear; *b* auditory bones.

attaches by a tendon to the manubrium of the malleus. When it contracts, it tightens the eardrum. The second muscle is the *stapedius,* the smallest of all the skeletal muscles. Its tendon attaches to the neck of the stapes. It opposes the action of the stapes as the stapes is pushed inward. Contractions of both muscles have a damping effect on the vibrations of the ear bones and so attenuate and protect the delicate inner ear from intense vibrations.

A branch of the facial (VIIth cranial) nerve, the chorda tympani, passes through the cavity of the middle ear.

The *auditory tube* extends from the middle ear to the pharynx, a distance

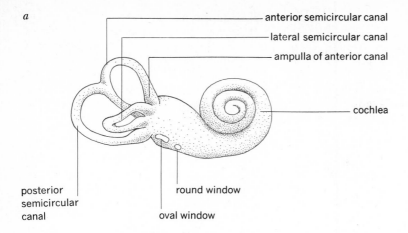

a

anterior semicircular canal

lateral semicircular canal

ampulla of anterior canal

cochlea

posterior semicircular canal

round window

oval window

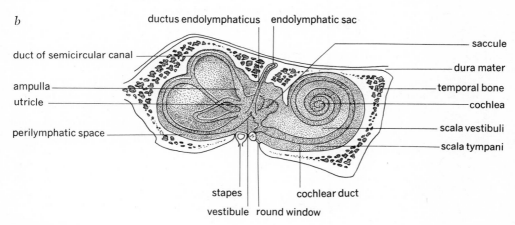

b

ductus endolymphaticus   endolymphatic sac

duct of semicircular canal

ampulla

utricle

perilymphatic space

saccule

dura mater

temporal bone

cochlea

scala vestibuli

scala tympani

stapes

cochlear duct

vestibule   round window

FIGURE 12·2

The inner ear, illustrating the membranous and osseous labyrinths; *a* right osseous labyrinth; *b* membranous labyrinth.

of about 1.5 in. It is only about 0.125 in in diameter at its narrowest point. Its function is to permit air to enter the middle ear behind the eardrum so that air pressure is equalized on both sides. It is a common experience to notice the effect of a change in air pressure on the eardrums when riding up and down hills. The lower opening of the tube is closed except during such movements as swallowing or yawning. You may help equalize pressure on the ear membrane, therefore, by swallowing or opening the mouth wide. If the auditory tube is blocked or its lower end not open freely, an inequality of air pressure may cause the eardrum to bulge inward or outward. Under these conditions the membrane cannot vibrate freely, and hearing will be impaired.

INNER EAR   The major structural divisions of the *inner ear* are the vestibule, the cochlea, and the semicircular canals (Fig. 12.2). These structures develop in the harder portions of the temporal bone and make up the *osseous* (bony) *labyrinth*. A fluid, the *perilymph*, separates the osseous labyrinth from the smaller *membranous labyrinth* within. The fluid of the membranous labyrinth is called the *endolymph*.

Vestibule The *vestibule* is a chamber in the osseous labyrinth located between the semicircular canals and the cochlea. The stapes rocks in and out at the *oval window* of the vestibule. Below this opening is the *round window*. The vestibule also has communications with the semicircular canals and the cochlea. Within the bony vestibule are two small sacs of the *membranous labyrinth*. The larger of the two is called the *utricle* and is associated with the membranous semicircular canals; the other, the saccule, is associated with the membranous duct of the cochlea.

Cochlea The *cochlea* comprises two labyrinths: an *osseous* and a *spiral labyrinth* (Fig. 12.3). A bony core, the *modiolus,* is surrounded by a thin bony shelf, the *spiral lamina.* Like the threads of a screw, the spiral lamina winds around the central modiolus. The osseous labyrinth is therefore partially divided by the spiral lamina, which projects out into the cavity. The upper end of the spiral is called the *scala vestibuli,* and the lower end is called the *scala tympani.* The scala vestibuli leads from the oval window to the apex of the cochlea. The scala tympani extends from the round window to the apex, where the two spiral passageways are confluent. The small aperture permitting confluence at the apex is called the *helicotrema.* The fluid of the osseous labyrinth is perilymph (Fig. 12.3).

The *cochlear duct (scala media)* is the *membranous labyrinth* of the cochlea (Fig. 12.4). It is separated from the scala vestibuli above by the thin vestibular membrane (of Reissner). Below, the *basilar membrane* lies between the cochlear duct and the scala tympani (Fig. 12.4). The basilar membrane, therefore, forms the floor of the cochlear duct by extending from the bony shelf of the spiral lamina. The basilar membrane actually becomes wider as the spiral approaches the narrowing apex of the cochlea. The cochlear duct is connected with the saccule in the vestibule and extends upward through the cochlea, where it lies between the scala vestibuli and the scala tympani. It ends as a blind sac at the apex. The outer wall of the cochlear duct is composed of fibrous periosteum and epithelial tissue.

FIGURE 12·3
Section through the cochlea of the inner ear.

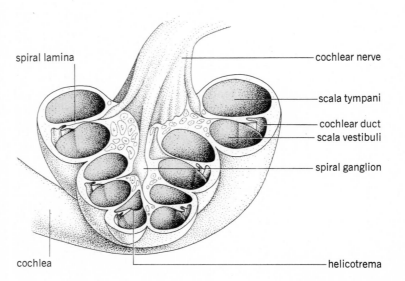

spiral lamina

cochlear nerve

scala tympani

cochlear duct
scala vestibuli

spiral ganglion

cochlea

helicotrema

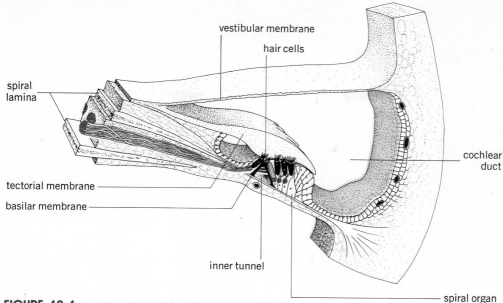

vestibular membrane

hair cells

spiral lamina

tectorial membrane

basilar membrane

cochlear duct

inner tunnel

spiral organ

FIGURE 12·4

The cochlear duct and spiral organ (organ of Corti). The figure represents a cross section of the cochlear duct and spiral organ in one turn of the cochlea. The hair cells are thought to be stimulated by movement of the tectorial membrane, and nerve impulses are transmitted to the spiral ganglion by nerve fibers.

When viewed in vertical section, the cochlea has a spiral that turns $2\frac{1}{2}$ times. Each whorl is divided into three ducts: the scala vestibuli, the cochlear duct, and the scala tympani.

The *spiral organ* (*of Corti*) contains the auditory receptors. It lies on the basilar membrane and is composed of supporting cells and hair cells. The *supporting cells* are tall epithelial cells of four types; they are attached to the basilar membrane, and their free surfaces form a reticular membrane over the spiral organ. The *hair cells* are neuroepithelial cells, with hairlike processes that extend through openings in the reticular membrane into the endolymph of the cochlear duct. The hair cells are receptors and they are arranged in outer and inner rows. Above the spiral organ lies the *tectorial membrane*, which is attached to the spiral lamina and extends over the hair cells. Though it usually appears well above the spiral organ in prepared sections, in the living state apparently it rests on the hair cells and transmits stimuli to them.

Semicircular canals The semicircular canals, the third division of the inner ear, will be described later when we discuss the sense of equilibrium, since these structures are involved with equilibrium rather than with audition.

THE PHYSIOLOGY OF AUDITION

THE AUDITORY NEURAL PATHWAY The axons of the *cochlear nerve*, which is a branch of the vestibulo-cochlear (VIIIth cranial) nerve, arise from bipolar nerve cell bodies located in the *spiral ganglion*. The spiral ganglion, in turn, is located in the modiolus. In the early embryo, a fiber grows inward from each cell body and terminates about the bases of hair cells, while the other fiber grows in the opposite direction as a part of the cochlear nerve pathway.

Sensory afferent neurons converge first in the superior olivary nuclei, which are located on either side of the brainstem. The medial superior olive is thought to compare inputs from right and left ears. Both excitatory and inhibitory cells are present in the olivary complex, a greater number being excitatory.

The cochlear nerve fibers terminate in nuclei in the medulla. Many fibers cross over to the opposite side in the medulla and then ascend in lateral tracts until the auditory-interpreting center in the temporal lobe of the cerebrum is reached. There are many neural relays in the auditory pathway. Among the more important nuclei are the inferior colliculi of the midbrain and the medial geniculate bodies of the thalamus.

The spoken word presumably stimulates word-understanding areas in both temporal lobes, but in the majority of individuals, the left area appears to be better developed. There is evidence also that the right side of the brain receives noise and other nonverbal stimuli whereas the left side of the brain receives the greater amount of verbal stimuli. (See R. Cohn's article in Suggested Reading.)

REVIEW OF CONDUCTION Let us review the series of structures set in motion by the initial impact of sound waves upon the ear membrane.

The impact of sound waves causes the stretched membrane to vibrate. The energy of these vibrations is carried across the air space of the middle ear by the lever system formed by the three ear bones. The rocking motion of the stapes sets up waves in the fluid of the inner ear. The lever system does not change the pressure in the inner ear, however. Since the area of the tympanic membrane is some 20 times larger than that of the base of the stapes, the pressure exerted on the fluid within the cochlea is about 20 times that exerted on the tympanic membrane by the impact of sound waves. Since a fluid has greater inertia than air, however, the greater pressure exerted on the footplate of the stapes is esential to equalize the strength of the vibrations between air and fluid.

THEORIES OF HEARING The first theory of hearing, called the *resonance theory,* was developed by the great physicist Helmholtz. He compared the crosswise fibers of the basilar membrane with the strings of a piano. In the basilar membrane, the shortest fibers are at the base of the cochlea, the longest at the top, or apex. Helmholtz knew that stretched strings, such as those of the piano, vibrate in sympathy with sound waves that strike them. He reasoned that the shorter fibers of the basilar membrane might vibrate in sympathy with sound waves of high frequency, while the longer fibers might respond to sound waves of low frequency. Though subsequent investigation has indicated that this may be true in a general way, the fibers of the basilar membrane are not piano strings and there is little evidence that they are stretched. Such fibers are also bound together in the membrane.

The inner-ear fluid also affects the vibration of the basilar membrane fibers. Waves in the fluid are responsible for the vibrations of the membrane. The longer fibers near the helicotrema are farthest away from the stapes footplate at the oval window where the waves are initiated, and these fibers therefore vibrate under a greater fluid "load." The shorter fibers, which vibrate at high frequencies, are near the oval and round windows, where the slight "load" favors their rapid vibration.

It is generally accepted today that certain areas of the basilar membrane vibrate in sympathy with sound waves of given frequency and that a change in area corresponds to a change of frequency. Thus the area that vibrates most in response to high frequencies would be the shorter fibers at the lower end of the basilar membrane, whereas lower frequencies would affect the fibers of the upper part (Fig. 12.5). This is, in general, an elementary explanation of the theory of hearing known as the *place theory*. This theory assumes that the hair cells in the spiral organ immediately above the vibrating area are stimulated and that they initiate nerve impulses. These impulses ultimately reach a specific cortical area of the temporal lobe, which interprets the impulse as sound of a certain pitch.

The *traveling-wave theory*, another theory of hearing, is essentially a modified place theory. Traveling waves of fluid have been shown to affect the entire basilar membrane, but as the stimulus frequency setting up the waves increases, the area of maximal vibrations moves toward the lower part of the basilar membrane. As the frequency decreases, the maximum moves toward the upper part. Thus, the area of maximal vibrations shifts its position along the membrane according to the frequency of the sound stimulus. Transmission of impulses from the area of maximal vibration determines the pitch perceived by the hearer. The traveling-wave theory is considered a good one by many researchers today.

TRANSFORMATION OF THE AUDITORY STIMULUS The activity of the spiral organ causes changes of electric potential, which can be accurately recorded by electrical devices. The cochlea may be compared in some ways to a telephone.

FIGURE 12·5

The cochlea and the basilar membrane: *a* the cochlea extended to show relative positions of the scalae; *b* the basilar membrane extended to show areas affected by sound waves of low and high frequency.

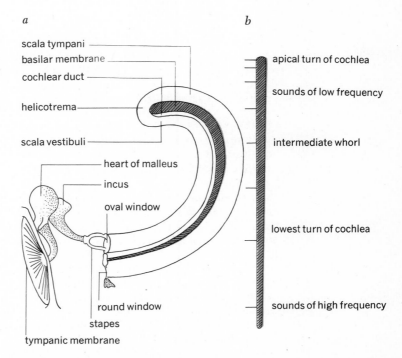

a

scala tympani
basilar membrane
cochlear duct
helicotrema
scala vestibuli
heart of malleus
incus
oval window
round window
stapes
tympanic membrane

b

apical turn of cochlea
sounds of low frequency
intermediate whorl
lowest turn of cochlea
sounds of high frequency

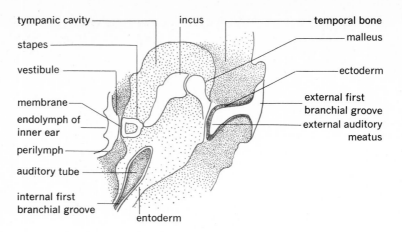

tympanic cavity
stapes
vestibule
membrane
endolymph of inner ear
perilymph
auditory tube
internal first branchial groove

incus
temporal bone
malleus
ectoderm
external first branchial groove
external auditory meatus

entoderm

FIGURE 12·6

The embryonic ear. (*After G. S. Dodds.*)

Development of the ear The early formation of the ear is indicated by a thickening of the ectoderm on either side of the neural groove, which deepens to form the auditory pit. The auditory pit is present in the embryo of 4 weeks. It deepens and becomes a closed sac—the auditory, or otic, vesicle. Later, outgrowths from the vesicle give rise to the membranous labyrinths of the cochlea, the vestibule, and the semicircular canals of the inner ear.

The first pharyngeal pouch forms the tympanic cavity of the middle ear and the auditory tube, whereas the first branchial groove forms the external auditory canal. The partition separating the two acquires an inner layer of mesoderm and becomes the tympanic membrane (Fig. 12.6).

EQUILIBRIUM

The ear is involved not only in hearing but in the sense of *equilibrium*. This sensory modality involves the *vestibular apparatus*, which is composed of the semicircular canals of the inner ear, the membranous structures of the vestibule, and the fluids that fill them.

VESTIBULAR APPARATUS The *semicircular canals* are deeply embedded in the temporal bone. Three *osseous canals*, about 1 mm in diameter, are hollowed out of bone and lined with connective tissue. The three canals occupy a superior-posterior position with reference to the cochlea. They are located in planes that are approximately at right angles to each other. The anterior and posterior canals are nearly vertical: their medial extremities join and enter the vestibule as a single cavity. The lateral canal lies in a horizontal plane. The canals communicate with the vestibule by five openings rather than six because of the fusion of two of the ducts (Fig. 12.2*a*).

Within the osseous canals are the *membranous canals*. They follow the general contour of the osseous canals but are much smaller, occupying hardly a third of the space within the bony canals. They are supported by strands of connective tissue and by fluids. The *perilymph* is the fluid of the osseous canals, while the *endolymph* is the fluid found within the membranous canals (Fig. 12.7).

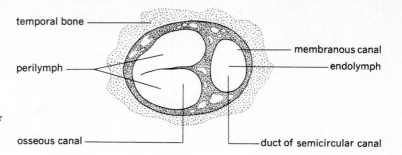

FIGURE 12·7

Cross section of a semicircular canal, showing the semicircular duct and adjacent perilymph spaces.

temporal bone

perilymph

osseous canal

membranous canal

endolymph

duct of semicircular canal

Each osseous canal has an enlargement near one end, called the *ampulla*. A crest of hair cells called the *crista ampullaris* lies within this structure. The hair cells are end organs of the vestibular nerve. Rising above the crista and hair cells is a wedge-shaped gelatinous structure, the *cupula*. Hair cells extend upward into the cupula. Movement of the cupula in response to movement of the endolymph is thought to stimulate the hair cells (Fig. 12.8)

The *membranous labyrinth* of the vestibule does not follow the outline of the osseous structure. The vestibule contains two membranous sacs, the utricle and the saccule. The *utricle* is the larger. It lies in the superior-posterior portion of the vestibule and is closely associated with the semicircular canals, which open into it. It also communicates with the saccule through a very fine duct. On the inner surface of a small outpocketing is a thickened oval area called the *macula*. It contains cells of two types, supporting cells and hair cells. A gelatinous substance that contains minute calcareous concretions called *statoconia*, or *otoconia* is found on the surface of the macula. These are similar to the larger "ear stones," or otoliths, of certain invertebrates and fish. The macula is about 3 mm long, and since it is oval in shape, its width is a little less than its length (Fig. 12.9).

The *saccule* is a small oval sac located in the lower anterior recess of the vestibule. It contains a macula similar to that located in the utricle. The saccule connects with the cochlear duct through a very slender tube. Another tiny duct from the posterior part of the saccule joins with a duct from the utricle to form the *ductus endolymphaticus*, which ends as a blind sac under the dura mater on the inner surface of the temporal bone (see Fig. 12.2*b*).

The cells of the cristae and the maculae are supplied by branches of the vestibular nerve, which in turn is a branch of the vestibulocochlear (VIIIth cranial) nerve. The pathway of the vestibular nerve leads back through the pons to nuclei in the cerebellum.

The utricle is considered to be affected by gravity (Fig. 12.10). Therefore a change of position (for example, tilting away from a horizontal plane) affects the macula in the utricle.

The function of the saccule is less well known. It is closely associated with the utricle and may function in proprioception, particularly in helping maintain the normal upright position of the head.

MAINTENANCE OF EQUILIBRIUM   Many experiments have indicated that the vestibular apparatus plays an essential part in equilibrium, postural reflexes, and righting reflexes. Even in standing quietly, the maculae send out impulses because the pull of gravity affects the jellylike endolymph in which

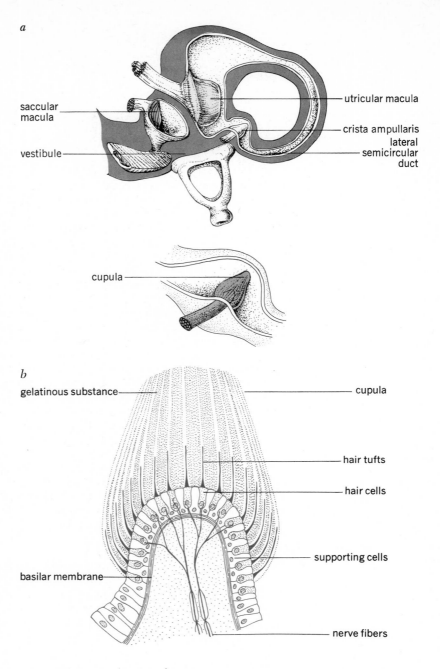

a

saccular macula

vestibule

utricular macula

crista ampullaris

lateral semicircular duct

cupula

b

gelatinous substance

cupula

hair tufts

hair cells

supporting cells

basilar membrane

nerve fibers

FIGURE 12·8

a Vestibular system, showing cristae in the semicircular ducts and the maculae in the utricle and saccule;
b enlargement of the crista ampullaris and cupula of the semicircular canal.

the "hair" cells are embedded. A change in posture, such as bending over, bends the hair cells of the maculae to a greater extent (Fig. 12.10).

Maintaining balance while in motion involves the fluid endolymph and the cristae of the semicircular canals. The hair cells in the cristae indicate changes in the rate of motion. In acceleration, because of inertia, the

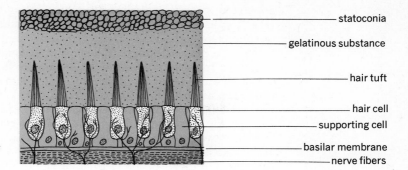

statoconia

gelatinous substance

hair tuft

hair cell

supporting cell

basilar membrane

nerve fibers

FIGURE 12·9

Section through the macula of the utricle.

endolymph tends to lag behind the motion of the head and the hair cells are bent accordingly. The hair cells slowly return to their original position as the inertia of the endolymph is slowly overcome and the fluid moves to a new position.

The hair cells of the cristae are activated in a sudden change of position, such as turning the head or moving in a direct line. A familiar example of response to acceleration in a direct line mediated by the semicircular canals is the *placing reaction* of a cat. If a cat is blindfolded and lowered rapidly head down, as if it were jumping down from some high place, the forelegs extend and the toes spread—a position assumed for making a safe landing. Another example is the *righting reflex,* in which the cat is able to right itself and land on its feet when dropped from an inverted position. The vestibular

FIGURE 12·10

The utricular macula, showing how a change of position affects the hair cells. *a* Upright; *b* bending over.

a

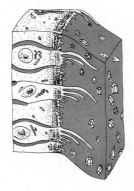

utricular macula

statoconia

hair cells

b

statoconia fall with gravity

apparatus is concerned primarily with righting the head, so the righting reflex also involves proprioceptors of the neck and body that coordinate the orientation of both head and body in maintaining the equilibrium necessary to secure a proper landing. The righting reflex is not present in humans.

During rotation of the head, acceleration effects from the semicircular canals may affect muscles of the neck, trunk, and appendages as well as the extrinsic muscles of the eyes. As a person seated in a swivel chair is rotated, the eyes tend to fix on some object until the head is turned away; then the eyes quickly move back and focus on a new object. There is a slow phase of this eye movement, in which the eyes turn in the opposite direction as they attempt to hold their focus on the same object as the head turns. The eyes then move in the direction of rotation in a quick phase. This reflex movement of the eyes is called *nystagmus*. This response has been observed in vertebrate animals from fish to mammals.

One interesting aspect of nystagmus is that it is an acceleration effect and is therefore not caused by whirling motion alone. If a person or animal is rotated at a uniform speed for some period of time, nystagmus disappears. When the rotation is stopped suddenly, however, postrotatory nystagmus is observed. In this type of nystagmus, the quick movement is in the opposite direction from that of the previous rotation. If a person is rotated in a vertical position with the head erect, the horizontal canals are affected and the resultant nystagmus is horizontal. If the head is in such a position that the vertical canals are primarily affected, the nystagmus is vertical: that is, the eyes move up and down. The inertia of the endolymph determines the direction of nystagmus. In accelerated turning to the right, the endolymph in the right horizontal canal moves toward the ampulla, while the hair cells of the crista move toward the utricle. In other words, the endolymph, because of its inertia, moves in the opposite direction from the turning of the head. In the left horizontal canal, the movement of the endolymph is in the direction away from the ampulla, while the movement of the hair cells is toward the ampulla.

Movement of the endolymph deflects the hair cells of the crista and initiates impulses in the vestibular nerve. The nerve pathway involved in nystagmus leads from the semicircular canals to the medulla, where secondary neurons ascend and make synaptic connections within the nuclei of the IIId, IVth, and VIth cranial nerves, which are motor nerves to the extrinsic eye muscles and which also connect with the cerebellum. Nystagmus is therefore a vestibular reflex.

The movement of the cupula is correlated with observed nystagmus, and experimental evidence indicates that nystagmus is an effect caused by deviation of the cupula, which lasts until the cupula returns to its normal position. Dohlman, using a technique developed by Steinhausen, was able to make the cupula visible within the ampulla in living fish (see Suggested Reading) by introducing dyes, such as india ink, into the ampulla. A drop of oil was then placed in the semicircular canal and was shown to move toward the ampulla during acceleration. At the same time the cupula swung inward toward the utricle as a result of movement of the endolymph within the canal and ampulla. The cupula swung back in the opposite direction when the rotation stopped.

Past pointing after rotation is a similar reflex. After being rotated, a blindfolded subject is unable to bring the arm straight down in a vertical plane. Instead he points his finger a little to one side, in the direction that he has been turned. This effect is linked with postrotatory nystagmus, because the subject points in the same direction as the slow movement of the eyes. Vestibular reflexes from this type of rotation produce widespread motor stimuli to the muscles of the neck, trunk, and appendages; all motor impulses are concerned with maintaining balance during rotation or regaining a normal position after rotation.

The effects of prolonged whirling, such as dizziness and staggering, are well known. Sometimes such stimulation of the horizontal canals affects the autonomic nervous system, causing nausea, pallor, sweating, and vomiting.

The action of the vestibular apparatus plays an important part in enabling the body to maintain its balance. A precise coordination exists between the proprioceptive sense organs located in muscles and tendons and the higher centers of the brain. The eyes also help to maintain equilibrium, and the cerebellum is a center for coordination of movements involved in maintaining balance.

CUTANEOUS SENSATIONS

Cutaneous sensations include those of touch, pressure, heat, cold, and pain. There are several types of receptors involved in cutaneous sensation, and we shall describe the structure and function of each receptor (Fig. 12.11).

SIMPLE NERVE ENDINGS Invading practically all tissues are simple nerve endings. Such structures have been demonstrated in such tissues as oral epithelium, mucous and serous membranes, and skin. Nerve fibers penetrate the basal layers in bundles, breaking up into individual unmedullated fibers in the outer layers as they pass between cells. The nerve endings may be simple or somewhat enlarged. Since similar endings are found in the cornea and in the teeth, it is assumed that naked nerve endings are primarily pain receptors.

Various types of stimuli can excite pain receptors. In general, an energy stimulus of extreme degree, i.e., one that threatens to cause damage to tissues, such as increasing the amount of heat or pressure, is interpreted as pain.

Free sensory nerve endings occur also in connective tissue and in muscles and tendons, and there is a nerve network around the base of each hair. Nerve endings associated with hair follicles are concerned with touch.

ENCAPSULATED SENSORY NERVE ENDINGS Another type of sensory receptor is characterized by a connective tissue capsule that surrounds the nerve ending. The capsule varies considerably in shape and thickness in the

FIGURE 12·11

Various types of receptors: *a* free nerve endings (pain); *b* end bulb of Krause (cold); *c* end organ of Ruffini (warmth); *d* Meissner's corpuscle (touch); *e* pacinian corpuscle (pressure). (*After Wendell J. S. Krieg, "Functional Neuroanatomy." McGraw-Hill Book Company, New York, 1953*).

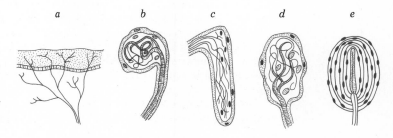

various types of receptors. Such receptors relay touch, pressure, and temperature stimuli.

While touch is commonly mediated by the simple nerve endings associated with hair follicles, long, encapsulated structures called *Meissner's corpuscles* also react to tactile stimuli. These receptors have a rather thin covering, and within the capsule are thin, transverse connective tissue plates that divide the corpuscle into minute compartments. A nerve fiber enters the capsule and branches into the transverse compartments. Meissner's corpuscles are found in connective tissue elevations of the skin, called *papillae.* Though present in the skin throughout the body, they are most numerous on the fingertips, the palm of the hand, and the sole of the foot. They are also present in the tip of the tongue, the lips, the nipples, the glans penis, and the clitoris. Recent work seems to indicate that they are ordinarily present in groups of two or three. A touch-sensitive spot in the skin ordinarily represents an area above several Meissner's corpuscles.

Located deeply in subcutaneous tissues and elsewhere throughout the body are oval structures called *pacinian corpuscles.* These receptors react to pressure stimuli. Pacinian corpuscles are laminated; that is, the capsule is composed of layers resembling an onion. A nerve fiber grows into the capsule's central cavity. Pacinian corpuscles are also found in the mesentery, in the connective tissue of the pancreas and other organs, and along blood vessels. Those located along blood vessels apparently function in the control of blood pressure. Cholinesterase has been found in the core of the pacinian corpuscle, indicating that transmisson of the impulse generated by the receptor involves acetylcholine.

Located in the outer portion of the dermis, in the tip of the tongue, and in the cornea of the eye are receptors known as the *end bulbs of Krause.* These are assumed to be receptors for cold, and they underlie spots in the skin sensitive to cold. Although they vary in shape, the commonest type is an oval capsule with small, branching nerve fibers in its cavity. These receptors ordinarily react to temperatures below the normal temperature of the skin or to temperatures lower than those needed to stimulate receptors for warmth.

The receptors for warmth are thought to be elongate, encapsulated structures located deep in the dermis. This type of receptor is called the *end organs of Ruffini.* The long, cylindrical capsules are filled with fine nerve fibers. The end organs of Ruffini are not as abundant as the end bulbs of Krause, and warm spots on the skin are not as abundant as cold spots. Ordinarily these receptors react to temperatures above normal skin temperature.

It has become evident that each type of cutaneous sensation, for the most part, has its own receptors. These receptors are not evenly distributed over the surface of the body. The fingertips are highly specialized for the sense of touch but are not so efficient in indicating temperature. The wrist, for example, gives a better estimate of temperature than can be obtained from the fingertips.

Though there are two types of receptors for temperature, each type actually reacts only to degrees of heat. The receptors for warmth are stimulated by a degree of heat that is greater than normal skin temperature. The cold receptors react to a degree of heat less than normal skin temperature and below that required to stimulate the receptors for warmth.

Although much is known about the regulation of body temperature, there is a lack of conclusive evidence regarding the identity and function of the receptors concerned with the temperature sense.

THE NEURAL PATHWAY FOR CUTANEOUS SENSATION   Touch impulses travel through several pathways in the spinal cord. Impulses for both touch and pressure travel in the anterior spinothalamic tract, crossing over directly to the opposite side upon entry to the cord. From synaptic endings in the thalamus, the impulses travel through the internal capsule to the cutaneous sensory area of the cerebral cortex, which lies posterior to the central sulcus. Other routes for touch impulses lie in the large posterior tracts of the cord, the fasciculus gracilis and fasciculus cuneatus (see Fig. 9.23). These tracts lead to nuclei of the same name in the medulla. The fibers synapse in the medulla, and succeeding fibers cross to opposite sides. They join the fibers ascending from the anterior spinothalamic tracts and pass upward to the thalamus, where again they synapse and a final set of fibers leads to the cutaneous sensory area in the cerebral cortex. Touch and pain impulses from the head and face are carried by fibers in the trigeminal nerves.

The pathway for pain and temperature impulses from receptors in the skin of the trunk and appendages lies in the lateral spinothalamic tract. It is probably entirely a crossed tract ascending to the medulla and the thalamus.

PAIN   The sensation of pain is difficult to explain. Simple nerve endings are abundant in the skin and widely present in various organs. There are no specialized encapsulated end organs for the initiation of pain impulses. Nerve impulses from simple nerve endings arrive at the thalamus by way of the spinothalamic tract, and from the thalamus they are distributed to the cerebral cortex. Most pain fibers entering the thalamus from the skin cross over to the opposite side of the brain.

Stimulation of pain receptors in the skin produces the most direct and clear-cut sensation. Pain from a skin cut, bruise, or burn is easily located and identified. Pain from the gastrointestinal tract, however, is more likely to be caused by inflammation, stretching, or spasm of smooth muscle. Such pain is often referred to some other region rather than to its source, a phenomenon known as *referred pain*. Pain in skeletal muscle and in heart muscle is usually caused by ischemia (reduced blood supply). Such pain may also be referred to another area of the body.

Electrical stimulation of the cortex and lesions in the parietal lobes seldom produce pain. The sensation of pain may arise from a variety of causes, and its function is to warn of possible injury or disease.

The pain threshold is considered to be about the same for everyone, but some people are more tolerant of pain than others. Pain may be forgotten during stressful situations involving intense excitement, anger, or fear. Mild pain may be disregarded if a person is distracted or if forgetting is suggested. The pain threshold may be raised by anaesthetics, and pain may be relieved by analgesic drugs. Endorphins apparently are capable of reducing pain. It is thought that this effect may be accomplished by suppressing the release of a pain transmitter called substance P.

GUSTATION (TASTE)

The senses of *gustation* (taste) and *olfaction* (smell) are called the "chemical senses" because the stimuli that elicit these sensations are chemical. The receptors for the sense of taste are stimulated by chemical substances in solution. Although many taste sensations are closely correlated with odors, the taste receptors seem to be more similar in structure to the cutaneous senses discussed in the preceding pages.

The sense of taste plays an important part in the selection and enjoyment of food. If the nose is closed so that we are unable to benefit from our olfactory sense, we must depend upon taste, which is limited to only four sensations: sweet, sour, salty, and bitter, or some combination of these. In order to taste, the tongue must be moist enough to permit the substances to be tasted to go into solution. We cannot taste sugar or salt on a perfectly dry tongue.

The gustatory sense involves receptors in the tongue (Fig. 12.12). Certain areas of the tongue react more strongly to one of the four taste sensations than to the others. The posterior part of the tongue reacts to bitter stimuli. The lateral edges respond to sour and salt stimuli but are most sensitive to sour stimuli. The tip of the tongue responds to all four sensations of taste but

FIGURE 12·12
a The tongue, illustrating papillae and primary taste centers; *b* vertical section through a taste bud.

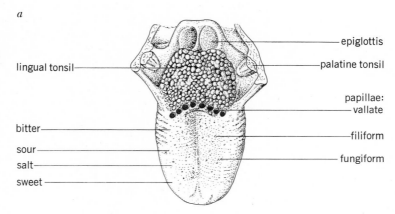

a

lingual tonsil

epiglottis
palatine tonsil

papillae:
vallate

bitter
sour
salt
sweet

filiform

fungiform

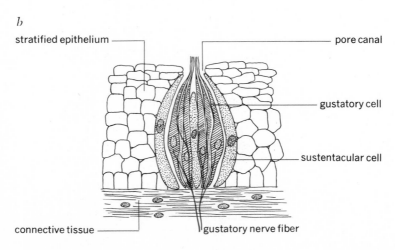

b

stratified epithelium

pore canal

gustatory cell

sustentacular cell

connective tissue

gustatory nerve fiber

is most sensitive to sweet and salt stimuli. The middle portion of the tongue's surface does not react to any great extent to taste stimuli.

GUSTATORY RECEPTORS   The receptors for the sense of taste are located in *taste buds*. These are minute oval bodies composed of supporting cells and elongated sensory cells. The sensory cells have hairlike processes that converge at the *taste pore*. Through the opening at the taste pore, the sensory cells make contact with chemical stimuli. The sensory cells and their processes within the taste bud are protected from the frictions of rough food surfaces against the tongue. Taste cells degenerate constantly and are replaced by cells that migrate into the taste buds.

It is thought that all receptor cells for the sense of taste are similar in structure but that they respond to different concentrations of chemical substances. Specific reactive sites probably bind individual molecules of a chemical substance. One theory suggests that after the substance binds to the receptor cell a change occurs in the receptor cell membrane, making it more permeable to the inward movement of ions. Depolarization of the membrane would initiate an action potential to stimulate nerves concerned with the sense of taste.

One group of neurons from taste bud receptor cells may respond vigorously to the stimulus of one chemical, while another group may respond only weakly to this stimulus. The situation may be reversed in response to a different chemical. This type of response may give some indication of how taste discrimination is developed.

The taste buds are located in small, rounded elevations on the tongue, called *papillae*, of which there are various types. The papillae are responsible for the rough appearance of the upper surface of the tongue. The largest are the *vallate papillae*, which form a V-shaped row at the back of the tongue. While taste buds are found primarily on the tongue's upper surface, they are also distributed to a limited extent over the palate, the epiglottis, and the back part of the mouth, especially during embryonic development and in young children. The number of taste buds decreases as the individual becomes older. Elderly persons usually lose some of the acuteness of the sense of taste exhibited by children. Mucus secreted during colds and catarrhal conditions tends to cover the taste buds and interfere with their normal function.

THE NEURAL PATHWAY FOR GUSTATION   The two cranial nerves involved in transmitting impulses from the taste buds are the facial (VIIIth cranial) and the glossopharyngeal (IXth cranial). A branch of the facial nerve supplies receptors on the anterior two-thirds of the tongue, and the glossopharyngeal nerve supplies the posterior one-third. Taste buds in the posterior part of the mouth (pharynx) and around the larynx are supplied by a few fibers of the vagus (Xth cranial) nerve.

The neural pathway for impulses from the taste buds leads to gray matter in the medulla, where most of the fibers terminate. The second part of the relay leads from the medulla to a nucleus in the thalamus. Neurons of the third set lead from the thalamus to the lower part of the cutaneous sensory area in the cerebral cortex. The sensory-interpreting area for taste is now thought to be located there in association with the sensory-interpreting area for the face as a whole and close to the sensory area for the tongue.

OLFACTION (SMELL)

Olfaction, or the sense of smell, is closely correlated with the sense of taste, in that the sensation of taste of many substances is largely supplemented by their odor. When we taste orange juice with the nose closed, it is merely sweet or acid, but we do not taste the "orange" flavor. Orange juice is a highly volatile substance, and to classify it accurately the sense of smell must also be used.

The olfactory sense is also a chemical sense stimulated by volatile substances. The substances must be soluble in water, at least to some extent, since the olfactory cell processes project into the mucous covering of the olfactory epithelium. Probably olfactory substances must also be soluble in lipids or at least be subject to active transport through the lipid layer of the receptor-cell membrane.

There are several theories regarding the mechanism of olfaction. Such theories are generally based upon chemical or physical principles. A recent proposal is a stereochemical theory based on the shape of molecules. It hypothesizes an olfactory membrane with receptor sites approximately the same shape as the molecules that it is designed to receive.

The olfactory receptors occupy an area about the size of a postage stamp in the upper part of each nostril. It is a rounded yellowish area containing vitamin A, located on the upper part of the superior nasal conchae and the nasal septum just opposite. Since this area is high in the nasal passageway, it is somewhat above the common pathway of air passing through the nose in ordinary respiration. When we wish to smell something whose odor is not very strong, we sniff at it, thus drawing the air containing the volatile substance higher into the nasal passageway and over the olfactory area (Fig. 12.13).

The posterior part of the nasal passageway leads down into the throat by way of the posterior nares. The posterior nares are closed during swallowing, but at other times odors of foods may pass up through them to reach the olfactory area.

The greater part of the nasal passageway is innervated by branches of the trigeminal (Vth cranial) nerve. These sensory branches are concerned with pain, temperature, and tactile stimuli within the nostrils. Some substances do not produce an olfactory effect but rather produce an irritating or tactile

FIGURE 12·13
The olfactory area in the nasal passageway.

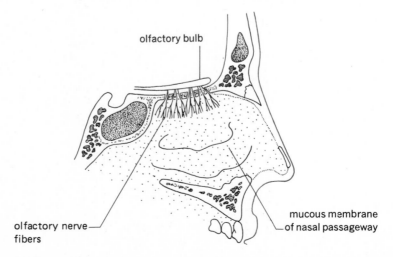

olfactory bulb

olfactory nerve fibers

mucous membrane of nasal passageway

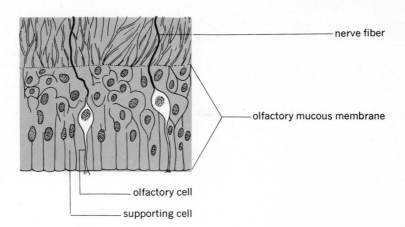

nerve fiber

olfactory mucous membrane

olfactory cell

supporting cell

FIGURE 12·14
Olfactory cells and supporting cells as seen through the olfactory mucous membrane (vertical section).

effect. Substances such as pepper, camphor, phenol, ammonia, ether, and chloroform produce other sensations than olfactory ones since they also stimulate the trigeminal nerve. Chloroform, in addition, gives rise to a sweetish taste because it stimulates the taste buds.

OLFACTORY RECEPTORS The *olfactory receptors* are primitive bipolar cells in which the distal part, located in the nasal mucosa, is the receptor and the basal part is extended inward as an axon (Fig. 12.14). The bipolar receptor cell is long and slender, with a nucleus at its base. Distally, arising from the cell body, there is a tuft of hairlike processes (dendrites), 2 to 12 in number, that project into the mucus that covers the olfactory membrane. Volatile substances are thought to stimulate the receptor cells by coming in contact with these hairlike processes. The receptor cells lie in columnar epithelium. The yellowish color of the olfactory area is caused by the pigmentation of the columnar cells. It seems likely that the excitation of the receptor cell for the sense of smell is similar to that for the sense of taste.

THE OLFACTORY NEURAL PATHWAY The olfactory nerve is composed of unmyelinated axons that extend inward from receptor cells to neurons within the *olfactory bulbs*, which are a primitive part of the cortex (Fig. 12.15). These synaptic relay centers lie beneath the frontal lobes of the cerebrum and on either side of the crista galli of the ethmoid bone. Bundles of fibers from the nasal mucosa pass through openings in the cribriform plate of the ethmoid and enter the olfactory bulbs. There, they make contact with the dendrites of *mitral cells*, the impulse first passing through areas of intricate synaptic branching called *glomeruli*. The glomeruli seem to function as relays for impulses directed to various parts of the brain. Neurons of the mitral cells form the olfactory tract (Fig. 12.15).

The olfactory tract Neurons of the olfactory bulbs lead inward to form the *olfactory tract*. Some neurons pass over to the opposite side in the anterior commissure. The olfactory tract passes along the ventral side of each frontal lobe. The posterior connections of the olfactory tract have not been fully worked out, and the location of the interpreting area for olfaction remains in doubt. The tract itself contains some gray matter dorsally, and the interpreting area may be located either in the tract itself or in areas of the

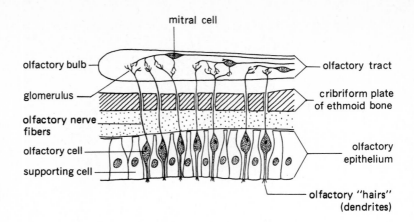

mitral cell

olfactory bulb — olfactory tract

glomerulus — cribriform plate of ethmoid bone

olfactory nerve fibers

olfactory cell — olfactory epithelium

supporting cell

olfactory "hairs" (dendrites)

FIGURE 12·15
Olfactory cells and their synaptic connections with mitral cells in the olfactory bulb.

brain close to the olfactory bulb and tract. The olfactory sense differs from others in having no known direct connections with the thalamus. There is, however, an indirect connection with the hippocampus, located in the inferior horn of the lateral ventricle. Some olfactory tract fibers enter the pyriform cortex, an area in the lower part of the temporal lobe and below the hippocampus in the rhinencephalon. Such fibers have been traced also to some of the basal nuclei, such as the amygdaloid and caudate nuclei, as well as to the mammillary bodies. Since the olfactory connections extend to so many areas deep within the brain, it has been difficult to determine an exact interpreting center for the olfactory sense.

The olfactory ability of human beings is not as highly developed as that of most animals. The ability of hunting animals to track their prey by scent requires olfactory ability of a high order. Humans do have the ability to recognize a great variety of odors, both good and bad. The florist may be able to identify many different flowers by their individual odors, as the chemist can recognize a great variety of volatile substances. The variety of recognizable odors is not easily classified and put into definite categories as are the sensations of taste.

OLFACTORY PERCEPTION   The perception of smell develops quickly, but it also adapts quickly. After we smell a flower a while, the odor seems to vanish. This condition is not caused by fatigue of the receptor mechanism, for we can readily smell some other odor. It illustrates the property of *adaptation* and explains why we can so easily become accustomed to prevalent odors. When we walk into a room where someone is smoking, we will probably be quite aware of the odor of burning tobacco. But after remaining in the room for some time, we probably will be accustomed to the odor and generally unaware of the condition of the air. Persons have been asphyxiated by gas while they were asleep, apparently unaware of their danger.

CLINICAL ASPECTS   AUDITORY DEFECTS   *Transmission deafness*   Deafness is common among the elderly, but people of any age may suffer from an impairment of hearing. One type, called *transmission deafness*, occurs when the external canal and tympanic membrane fail to receive sound waves or as a result of failure of the middle ear bones to transmit vibrations to the inner ear. Frequently this condition is caused by a middle ear infection following a severe cold, tonsillitis, or a sinus infection. In severe infections (*otitis media*) the membranes of the middle ear become inflamed and pus may be produced behind the eardrum.

Nerve deafness   This type of deafness is attributed to damage in the spiral organ by sounds of high intensity or continued exposure to loud noise. Some factory workers, aircraft workers, or musicians playing in bands that use high amplification, may become "hard of hearing" through *nerve deafness*.

One of the hazards of modern times lies in exposing the ear to loud sounds. Continuous exposure to highly amplified music, for example, may cause partial deafness. Mechanics who work around jet engines may become nearly deaf even though their ears are partially protected. The firing of heavy artillery also poses a hazard to the ears.

Ménière's disease   A syndrome termed *Ménière's disease* involves head noises, dizziness, and increasing deafness. The head noises are entirely subjective. The exact cause of the disease has not been determined. An excess of fluid in the endolymphatic system appears to affect the semicircular canals and causes degeneration of the hair cells in the spiral organ.

Tinnitus   Tinnitus is a condition involving head noises, usually described as a ringing, hissing, or roaring in the ears. Tinnitus may indicate disease of the middle ear or inner ear or damage to the cochlear nerve.

Testing hearing   Hearing can be tested in several ways, for example, listening to a watch tick and recording the distance from the ear that sound can be heard, testing with a tuning fork placed in various positions, or testing with an electric audiometer that will test the entire range of hearing frequencies.

Since the eardrum is a delicate membrane and subject to injury, persons should be advised against cleaning wax out of the ears with hard objects, such as matches, pencils, or hairpins. Wax must be gently washed out, if necessary, with clean warm water, preferably by a physician.

Ear infections   Although it is essential for the equalization of pressure, the *auditory tube* may also become a liability because it offers a passageway by which infections may invade the middle ear. Infections of the throat can cause inflammation of the tube and middle ear. The tube is shorter and more horizontal in children, providing somewhat more ready access for invading organisms. An earache ordinarily signals an active infection behind the eardrum. Such infections should be brought to the attention of a physician without delay. If uncontrolled, the infection may invade the spongy bone of the mastoid process, perhaps making a surgical operation necessary to clean out the infected area in this bony process. The middle ear has a rather thin, bony roof. The infection can spread upward and invade the meninges that lie just above. During infection the eardrum may bulge outward because of the accumulation of material behind it. A physician can pierce the eardrum in order to relieve the pressure and permit drainage through the external auditory canal. If the eardrum is not pierced, it may rupture and thus permit drainage. This is more likely to occur in children, and it produces "running ears." Small ruptured areas usually heal, but larger ruptures may persist as perforated eardrums and cause impaired hearing. Persons with openings through the eardrums should be advised against diving or swimming underwater lest infections reach the middle ear.

EQUILIBRIUM DISTURBANCES   *Acceleration disturbances*   The vestibular apparatus and the vestibular nerve are involved in acceleration disturbances. Loss of equilibrium, dizziness, and nausea may indicate disease of the vestibular apparatus. The patient is rotated in a *Barany chair* with the head held in different positions to determine whether the semicircular canals are functioning normally.

Motion sickness   While not regarded as a disease, *motion sickness* is a fairly common experience, especially for children. It may be experienced while traveling in an automobile, a plane, a train, a boat, or even on a merry-go-round. It may cause dizziness, headache, nausea, and vomiting, but it is a temporary and self-limiting illness. Certain antimotion sickness drugs may suppress its effects. Motion sickness of various kinds, including seasickness, is thought to be caused by continuous stimulation of the maculae rather than by a disturbance of receptors in the ampullae of the semicircular canals.

CUTANEOUS DISORDERS   Disorders of the cutaneous senses accompany a number of diseases and injuries. Following brain injuries, such as stroke, there may be impairment of the senses of touch, temperature, or pressure, depending upon where the injury occurs. A sensation of numbness or tingling may indicate that some nerve fibers are responding in the affected area. An injury in a peripheral nerve, in the spinothalamic tract, in the brainstem, or in the thalamus itself may also cause cutaneous disturbances. Injuries causing derangements of cutaneous sensation are difficult to locate since the neural pathways involved may be injured at many points.

Pain and acupuncture   *Acupuncture* has come into prominence in the Western world recently, although it has been practiced for centuries in China. In acupuncture, needles are inserted through certain selected areas of the skin and rotated. In a short time, pain is relieved as though an anaesthetic had been applied. An entirely satisfactory explanation of this result remains to be found, although several theories have been advanced. Endorphins may be involved.

OLFACTORY DISORDERS   Olfactory disorders are often temporary. A deep cold or other infection of the upper respiratory tract may cause a temporary inability to smell food. The sense of taste is usually impaired also under these conditions. A permanent loss of the olfactory sense may cause long-lasting taste deficits. These causes may indicate olfactory nerve damage or injury to olfactory-interpreting areas in the brain.

SUMMARY

1 The ear consists of three primary parts: the external, middle, and internal ear. The external ear leads inward from the pinna, or auricle, through the external auditory meatus to the tympanic membrane.

2 The middle ear is a cavity containing a lever system of three ear bones:

the malleus, incus, and stapes. The auditory tube extends from the pharynx to the middle ear. It permits air to enter behind the eardrum to equalize pressure on either side.

3 The major structures of the inner ear are the vestibule, the cochlea, and the semicircular canals. The vestibule contains an oval window and a round window. Within the bony vestibule are two small sacs, the utricle and the saccule. The utricle is associated with the semicircular canals; the saccule apparently is more closely associated with the membranous duct of the cochlea.

4 The cochlea is coiled into $2\frac{1}{2}$ turns and consists of an osseous and a membranous spiral labyrinth. The modiolus is the bony core of the cochlea. There are cavities, called scalae, in the cochlea. The scala vestibuli is continuous with the vestibule; the scala tympani leads to the round window at its base. These two passageways are confluent at the apex of the cochlea by a small aperture, the helicotrema.

5 The cochlear duct, or scala media, is the membranous labyrinth of the cochlea. It contains endolymph. The spiral organ lies on the basilar membrane of the scala media. Receptors for the sense of hearing are the hair cells located in the spiral organ. The cochlear nerve is a branch of the vestibulocochlear nerve.

6 The place theory of hearing holds that certain areas of the basilar membrane vibrate in sympathy with sound waves of given frequency. High frequencies affect the narrow, lower part of the basilar membrane, whereas low frequencies affect the wider, upper part of the membrane.

7 The semicircular canals are three in number. The anterior and posterior cànals are nearly vertical, whereas the lateral canal is horizontal. The osseous canals are hollowed out of bone, but within each osseous canal is a much smaller membranous canal, or duct. Each canal has an enlargement at one end, called an ampulla. Within the ampulla is the crista ampullaris, a crest of hair cell receptors. The cupula rises above the crista. Movement of the endolymph causes movement of the cupula and stimulates the hair cells. The hair cells are the end organ receptors of the vestibular nerve axons. The pathway of the vestibular nerve leads to the cerebellum.

8 The semicircular canals function in maintaining equilibrium principally when there is a sudden change of position, either in a direct line or during rotation. When a person is rotated and then stopped suddenly, a postrotatory nystagmus is observed. The eyes move back and forth, the quick movement being in the opposite direction from the previous rotation. Whirling can cause dizziness and involve the autonomic nervous system, causing nausea and sweating. Motion sickness, such as seasickness, is thought to be caused by continuous stimulation of the maculae rather than by disturbance of receptors in the semicircular canals. Equilibrium involves not only the semicircular canals and màculae but (1) proprioceptive organs in muscles and tendons, (2) the eyes, and (3) the cerebellum as the center for muscular coordination.

9 Encapsulated sensory nerve endings are described as Meissner's corpuscles, pacinian corpuscles, end bulbs of Krause, and end organs of Ruffini.

10 Taste buds contain chemoreceptors for the sense of taste. The taste buds are located in papillae on the tongue. The facial and glossopharyngeal nerves transmit taste receptor impulses.

11 The olfactory area is a small area in the upper part of each nostril that

contains receptor cells for the olfactory sense. The olfactory nerve is composed of axons from receptor cells that pass inward through openings in the cribriform plate of the ethmoid bone. The fibers synapse in the olfactory bulb. A second set of neurons leads inward toward the brain, forming the olfactory tract. A great variety of odors can be distinguished, but adaptation to specific odors occurs quickly.

12 Abnormal stimulation of any sensory receptor may be interpreted as pain.

13 Some disorders related to the special senses are discussed.

QUESTIONS

1 Discuss the function of the auditory tube and its relation to the middle ear.
2 List the structures involved in transmitting the energy of sound waves from the ear membrane to the spiral organ.
3 Describe the auditory pathway.
4 Where are the receptors for the sense of hearing located?
5 Is the narrowest part of the basilar membrane at the top or bottom of the cochlea? The shortest fibers may be expected to react to sounds of which frequencies?
6 What structures compose the osseous labyrinth? The membranous labyrinth? The perilymph is the fluid of which labyrinth?
7 Discuss the structure and function of the utricle and saccule.
8 Explain the reaction of the semicircular canals to acceleration.
9 List the receptors located in the skin and describe their structure and function.
10 Why is it that if the nasal passageway is closed, the sense of taste seems limited?
11 Where on the tongue are the taste receptors located?
12 Does the tongue respond to stimuli other than those that stimulate the sense of taste?
13 Just where is the olfactory area located in the nasal passageway?
14 Describe the appearance of olfactory receptors.
15 Trace the pathway of a nerve impulse from a tactile receptor to the interpreting center in the brain.
16 Explain the process of adaptation to odors.

SUGGESTED READING

Buser, P.: Higher Functions of the Nervous System, *Annu. Rev. Physiol.*, **38:**217–245 (1976).

Cohn, R.: Differential Cerebral Processing of Noise and Verbal Stimuli, *Science*, **172:**599–601 (1971).

Dohlman, B.: Some Practical and Theoretical Points in Labyrinthology, *Proc. R. Soc. Med.*, **28:**1371–1380 (1953).

Eldredge, D. H., and J. D. Miller: Physiology of Hearing, *Annu. Rev. Physiol.*, **33:**281–310 (1971).

Galambos, R., and H. Davis: The Response of Single Nerve Fibers to Acoustic Stimulation, *J. Neurophysiol.*, **6:**39–57 (1943).

Goldberg, Jay M., and César Fernández: Vestibular Mechanisms, *Annu. Rev. Physiol.*, **37:**129–162 (1975).

Lim, Robert K. S.: Pain, **32:**269–288 (1970).

Lynn, Bruce: Somatosensory Receptors and their CNS Connections, *Annu. Rev. Physiol.,* **37:**105–127 (1975).

Lowenstein, W. R.: Biological Transducers, *Sci. Am.,* **203:**99–108 (1960).

———: Mechano-electric Transduction in the Pacinian Corpuscle, Initiation of Sensory Impulses in Mechanoreceptors, in "Handbook of Sensory Physiology," vol. 1, pp. 269–299. Springer-Verlag, New York, 1971.

McCutcheon, N. B., and J. Saunders: Human Taste Papilla Stimulation: Stability of Quality Judgments over Time, *Science,* **175:**214–216 (1972).

Melzack, Ronald, and Patrick D. Wall: Pain Mechanisms: A New Theory, *Science,* **150:**971–979 (1965).

Rosenzweig, Mark R.: Sensory Receptors, *Cold Spring Harbor Symp. Quant. Biol.,* **30** (1965).

Wever, E. G., and O. W. Bray: The Nature of Acoustic Response: The Relation between Sound Frequency and Frequency of Impulses in the Auditory Nerve, *J. Exp. Psychol.,* **13:**373–387 (1930).

13

THE ENDOCRINE SYSTEM

CONSIDERATIONS FOR STUDY

We have studied the regulatory functions of the nervous system, one of the body's two major regulatory systems. We come now to the second main regulatory system: the endocrine system. In discussing the functions of the endocrine system, we include the following topics:

1 The hypophysis or pituitary gland, the master gland of the endocrine system
2 The functions of the important hormones of the hypophysis' anterior lobe
3 The function of the hormone secreted by the hypophysis' intermediate lobe
4 The functions of the hormones of the hypophysis' posterior lobe
5 The functions of the thyroid gland hormones, thyroxine and thyrocalcitonin
6 The functions of the parathyroid gland hormone
7 The functions of the pancreatic hormones, insulin and glucagon
8 The functions of hormones secreted by the two parts of the adrenal glands
9 The function of the gonadal sex hormones
10 The function of the prostaglandins, hormones found in various tissues of the body
11 The function of the thymus hormone, erythropoietin

The regulatory functions of the endocrine system and those of the central nervous system are closely intertwined. The hypothalamus, a brain center, plays a significant part in controlling the secretion of certain hormones, while the endocrine system directly regulates many nervous functions.

Neurons send impulses selectively to certain tissues or organs, whereas hormones are carried by the blood to all tissues. However, only certain tissues or cells are keyed to accept certain hormones. Those tissues or organs that are designed to accept a specific hormone are called *target tissues* or *organs.* Target tissue cells must have a receptor either on the cell membrane or within the cytoplasm for that particular hormone.

The body has two kinds of glands: exocrine and endocrine. *Exocrine* glands, such as the salivary glands, produce a secretion that is carried to a specific region of the body through ducts. The larger digestive glands, sweat glands, and mammary glands are exocrine glands, otherwise known as glands of external secretion.

Endocrine glands are glands of internal secretion. Such glands have no ducts, and their products, called *hormones,* are absorbed and carried by the bloodstream to target organs, which are often remote from the site of origin. The secretions of the endocrine glands are produced in minute amounts (micromilligrams or smaller quantities), and they are constantly broken down so that large amounts do not accumulate. Hormones have specific actions on target cells and tissues, and they are chemical regulators (Fig. 13.1).

Most hormones are derived chemically from either peptides, amines, amino acids, or steroids. The larger hormone molecules such as insulin, parathyroid hormone, and growth hormone, are usually proteins made up of amino acid chains. Epinephrine and norepinephrine are amines derived from individual amino acids. Amines contain carbon, hydrogen, and nitrogen but do not contain oxygen. Well-known peptide-derivative hormones are antidiuretic hormone, oxytocin, gastrin, and calcitonin. Peptide hormones are produced by ribosomes on cellular rough endoplasmic reticulum. A few hormones are glycoproteins, a term that means that they possess a carbohydrate moiety, or part, in addition to a protein moiety. Chorionic gonadotrophins, thyroid-stimulating hormone (TSH), and luteinizing hormones are in this group. The steroids are nonprotein hormones. These include the sex hormones (the testosterones, the estrogens, and the progesterones) and the adrenocorticoids. Prostaglandins are known chemically as cyclic fatty acids.

Most hormones enter the bloodstream, where they are bound to plasma proteins, although some free, or unbound, hormone molecules are found in the blood. Hormone concentration in the blood is maintained at a steady state since some is removed by target tissues, some portion is inactivated in the liver, and some is excreted by the kidneys. Negative feedback mechanisms regulate the production of most hormones.

If the cell is not a target cell for a particular hormone, that hormone passes through the cell membrane in either direction so that its concentration within the cell does not increase. Hormone concentration does increase in target cells because the hormone molecules are bound to receptors in such cell's cytoplasm and thus very few molecules are free to leave.

Hormones may infiltrate many tissues without causing any effect since they react only with specific protein receptors in target tissues that bind the hormone. Protein hormones, one major type, bind to receptors on the cell membrane. This requires a second messenger to introduce the hormone to

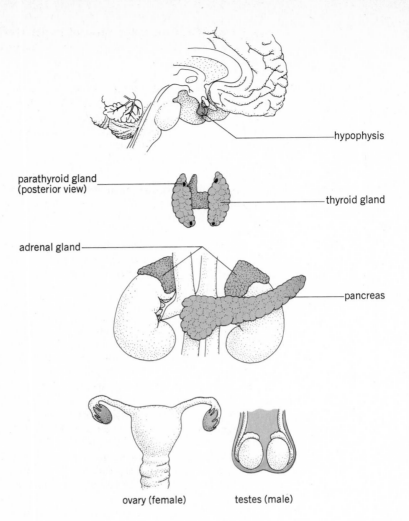

hypophysis

parathyroid gland
(posterior view)

thyroid gland

adrenal gland

pancreas

ovary (female)   testes (male)

FIGURE 13·1
Various glands performing endocrine functions.

the cytoplasm inside the cell. A common second messenger is cyclic adenosine monophosphate (AMP). When a hormone is bound to a cell receptor,
the enzyme adenyl cyclase is activated. This enzyme moves into the
cytoplasm of the cell, where it converts ATP to cyclic AMP. Cyclic AMP
then supplies the energy to direct the metabolism of the cell.

Steroid hormones, the other major kind, are derived from the chemical
precursor cholesterol. Such hormones are small, lipid-soluble molecules that
pass through the cell membrane and into the cytoplasm. They also pass out
again through the cell membrane and into the blood. The receptors for such
hormones are located in the cellular cytoplasm.

Even though a hormone may be produced in minute amounts—nearly 0.5
million sheep brains were required to produce 5 mg of growth-hormone
inhibitory factor somatostatin—a deficiency or an overproduction may
cause striking changes in the physiology and morphology of an individual.
Hormones pass into the bloodstream where they combine with plasma
proteins, which probably delay their breakdown. Hormones then bind

readily to specific receptor sites on target tissues. It is likely that most hormones break down or bind to receptor sites in a matter of minutes.

FUNCTIONS OF HORMONES

Most hormones have specific functions, such as the role of insulin in carbohydrate metabolism or the action of glucagon in influencing the liver to convert glycogen to glucose. The hormones of the thyroid gland accelerate oxidative enzyme systems. Many hormones play an important part in the regulation of growth and metabolism. The role of the sex hormones in the development and maintenance of the reproductive system is well known. Some effects of hormones are less obvious, but they contribute indirectly to personality, intellectual attainment, and physical ability.

Research workers have opened up many avenues of investigation in the field of endocrinology. The search for hormone receptors and receptor sites on target organ tissues has been one topic. For example, research has been conducted to find cellular receptor sites that bind insulin. Releasing factors, inhibitory releasing factors, and hormones have been extensively studied and identified. In the forefront of these investigations have been studies of the role of cyclic AMP as a mediator of intracellular activity attributed to hormones.

A second nucleotide, guanosine 3'5'-monophosphate (cyclic GMP), may prove to be another intracellular mediator. Cyclic GMP may be an antagonist to cyclic AMP in bidirectionally controlled systems such as those involved in muscular contraction and relaxation or those that respond to the action of certain hormones. It has been suggested that cyclic GMP influences cholinergic responses, especially to the heart and brain.

ENDOCRINE BALANCE   Probably no endocrine gland within the body can function adequately without being affected by the secretions of other glands. A delicate balance between the activities of the endocrine glands provides, in a normal individual, for normal growth and metabolism. If the activity of one of the endocrine glands is disturbed by disease or injury, other endocrine glands may be adversely affected. The amount of hormone produced by a gland is normally regulated by a negative feedback mechanism. If the blood hormone level is too low, the endocrine gland secretes additional hormone; if the blood level is too high, secretion of this particular hormone is inhibited.

Some endocrine glands may act independently, but others do not appear to be able to do so. It would be helpful if the functions of all the endocrine glands could be discussed at the same time in order to stress their interrelationships, but for the sake of clarity, we must consider them one at a time. Among the glands to be considered are the hypophysis, the thyroid and the parathyroids, the adrenals, and the gonads. The hypophysis is located just below the brain, the thyroid and parathyroids develop anteriorly in the neck region, the adrenals are located on top of the kidneys, and the gonads are male and female sex glands.

HORMONES OF THE HYPOTHALAMUS

The relationship between the nervous system and the endocrine system is most clearly developed in the close association between the hypothalamus and the *hypophysis,* or pituitary. Small blood vessels in the lower portion of the hypothalamus join and extend down the *stalk (infundibulum)* of the hypophysis, forming the hypothalamic-hypophyseal portal system, which supplies the adenohypophysis (Fig. 13.2).

The hypothalamus is the source of hypophyseal hormones, or *releasing factors*, which are considered to be *neurohormones* rather than neurotransmitters. These neurohormones are released into the bloodstream, where they travel only a short distance through the hypophyseal portal veins into the *anterior lobe* of the hypophysis, the *adenohypophysis*. There is, presumably, a releasing factor for each hormone, although all such factors have not yet been identified. We may speak, for example, of growth-hormone releasing factor or of thyrotrophin releasing factor, and so on. The hypothalamus also produces inhibitory releasing factors, such as melanocyte inhibiting factor (MIF). The actions of these hormones are summarized in Table 13.1.

Negative feedback regulation of hormonal secretion is well demonstrated by the suppression of releasing hormones. The hypothalamus, as we have said, produces releasing factors for several hormones. For example, the hypothalamic releasing factor stimulates the secretion of thyroid-stimulating hormone (TSH) by the anterior lobe of the pituitary. TSH, in turn, controls the secretion of thyroxine by the thyroid gland. If the blood level of thyroxine rises above normal, the secretion of TSH releasing factor by the hypothalamus is suppressed. On the other hand, if the blood level of thyroxine is lowered, more releasing factor for TSH is secreted, and down the line, more thyroxine is secreted.

HYPOPHYSIS (PITUITARY)

The *hypophysis*, or *pituitary gland*, has been called the master gland of all the endocrines because, through its *trophic* hormones, it exerts a regulatory effect over the activity of other endocrine glands. The term "pituitary" is somewhat unfortunate, since this term originally referred to a mucous

FIGURE 13·2

Lateral view of the hypophysis illustrating the hypophyseal portal system.

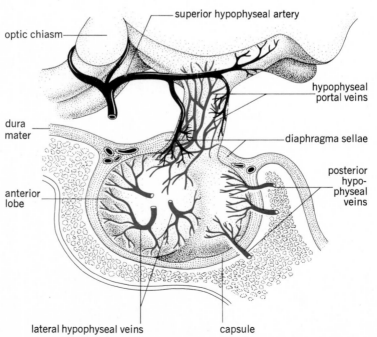

optic chiasm

superior hypophyseal artery

hypophyseal portal veins

dura mater

diaphragma sellae

posterior hypophyseal veins

anterior lobe

lateral hypophyseal veins

capsule

TABLE 13·1

Functions of the hypothalamic hormones

Hypothalamic hormones	Function	Target organ or tissue
Somatotrophin release factor (SRF) [Growth-hormone releasing hormone (GRH)]	Increases secretion of growth hormone. Increases growth	Adenohypophysis
Somatostatin, growth-hormone release inhibitory factor (GIF) (SRIF)	Inhibits growth-hormone secretion	Adenohypophysis
Follicle-stimulating hormone releasing factor (FSH-RF)	Stimulates development of seminiferous tubules in male	Adenohypophysis
Luteinizing release factor (LH-RF)	Promotes release of LH. Development of corpus luteum of ovary.	Ovaries
	Stimulates interstitial cells of testes to secrete testosterone (ICSH)	Testes
Thyrotrophin release factor (TRF)	Release of prolactin in some species	Mammary glands
Thyrotrophin releasing hormone (TRH)	Release of thyroid-stimulating hormone (TSH)	Regulates thyroid function
Prolactin release factor (PRF)	Stimulates release	Mammary glands
Prolactin inhibiting factor (PRF)	Inhibits release	Mammary glands
Corticotrophin release factor (CRF) or Corticotrophin releasing hormone (CRH)	Stimulates secretion of ACTH	Adrenal cortex
Melanocyte releasing factor (MRF)	Secretion of MSH	Melanocytes
Melanocyte inhibiting factor (MIF)	Inhibits MSH secretion	Melanocytes

secretion, or phlegm. The term "hypophysis" refers to the location and development of the gland under the brain. Both are in common use, but hypophysis is favored in medical and scientific research. The term "trophic" is used by some endocrinologists to indicate those hormones that have specific endocrine glands as their target organs. The functions and target organs of the hypophyseal hormones are summarized in Table 13.2.

ANTERIOR LOBE (ADENOHYPOPHYSIS) The hypophysis is about 1 cm in diameter along its anteroposterior axis. It lies well protected in a depression of the sphenoid bone called the *sella turcica*. The most prominent parts of the gland are the anterior lobe, the posterior lobe, and the stalk, or infundibulum. There is also an intermediate lobe, which varies greatly in different species. The anterior and posterior lobes have different embryonic origins.

TABLE 13·2
Hormones of the hypophysis

Hormone	Target organ	Functions
Adenohypophysis		
Growth hormones (GH) (somatotrophin)	Various tissues, skeletal, muscle	Growth of bones and other tissues
Thyroid-stimulating hormone (TSH) (thyrotrophin)	Thyroid gland	Stimulates the secretion of thyroxine
Thyrocalcitonin (TCT)		Hypoglycemia
Adrenocorticotrophic hormone (ACTH)	Adrenal cortex	Secretion of corticoids
Gonadotrophic hormones:		
Male:		
Follicle-stimulating hormone (FSH) plus luteinizing hormone (LH)	Seminiferous tubules	Production of sperm
Luteinizing hormone (LH) [interstitial cell–stimulating hormone (ICSH)]	Interstitial cells of testes	Secretion of testosterone
Female:		
Follicle-stimulating hormone (FSH)	Follicles of ovaries	Maturation of ovarian follicles
Luteinizing hormone (LH)	Interstitial cells of ovaries	Stimulates secretion of estrogen and progesterone. Formation of corpus luteum
Prolactin (PRL) (mammotrophin)	Mammary glands	Production of milk in activated glands, if estrogen and progesterone are present
Intermediate Part (pars intermedia)		
Melanocyte-stimulating hormone (MSH)	Melanophores	Darkening of skin by pigmentation
Neurohypophysis		
Oxytocin	Uterus	Stimulates contraction
Antidiuretic hormone (ADH) (vasopressin)	Kidney	Retention and reabsorption of water

The *anterior lobe* (*pars distalis*) arises as a diverticulum of Rathke's pouch, which in the embryo is a backward and upward extension of the oral epithelium. The ventral portion of the pouch thickens and later gives rise to glandular cells, which can be differentiated by appropriate stains. The anterior lobe formed in this way is also called the *adenohypophysis*, meaning glandular portion of the hypophysis (Fig. 13.3).

Hormone-producing cells The anterior lobe of the hypophysis contains at least three kinds of glandular cells. These are classified as acidophils, basophils, or

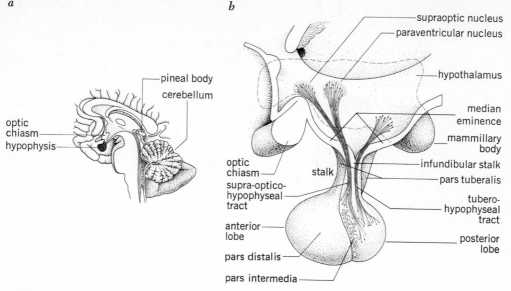

a

b

pineal body
cerebellum
optic chiasm
hypophysis

supraoptic nucleus
paraventricular nucleus
hypothalamus
median eminence
mammillary body
optic chiasm
stalk
supra-optico-hypophyseal tract
anterior lobe
pars distalis
pars intermedia
infundibular stalk
pars tuberalis
tubero-hypophyseal tract
posterior lobe

FIGURE 13·3

Hypophysis and pineal body: *a* sagittal section of the brain; *b* sagittal section of hypophysis illustrating the supraopticohypophyseal and the tuberohypophyseal tracts between the hypothalamus and the posterior lobe.

chromophobes depending on their reaction to certain laboratory stains. Two of these types take stains readily, and these are called *chromophil* cells. Chromophil cells that show an affinity for acid stains are termed *acidophils;* those that stain with basic dyes are termed *basophils. Chromophobes,* the third type of cells, are small cells that take stains very poorly.

Acidophils produce two hormones: growth hormone (GH) and prolactin (PRL). Basophils release four hormones: thyroid-stimulating hormone (TSH), adrenocorticotrophic hormone (ACTH), follicle-stimulating hormone (FSH), and luteinizing hormone (LH).

Growth hormone   The process of growth is dependent upon a number of factors that determine the course of metabolism. Genetic and nutritional factors exercise important limiting influences on growth, while hormonal factors can regulate or modify the process. The growth hormone (GH), also called somatotrophin, secreted by acidophilic cells in the adenohypophysis is not solely responsible for growth, but it does influence its rate.

Basically, growth occurs when protein accumulates in the tissues of the body. When growth stops or when weight loss occurs, an increase in the excretion of nitrogen is noted. The growth hormone stimulates an increase in amino acid and lipid utilization and the conversion of additional amounts of glucose from glycogen in the liver. Raising blood glucose level to very high concentrations may lead to diabetes, and this effect is referred to as the *diabetogenic effect* of the growth hormone. Under normal circumstances, this does not occur, however.

Growth hormone is difficult to study since no single target organ is involved. In young animals, bones, cartilage, and muscles are stimulated to grow by GH. Humans are not affected by growth hormones obtained from most other species, but do respond to human growth hormone and to growth hormones from some other primate species.

Several growth factors, which are not clearly defined at present, may be involved. Peptides called *somatomedins* promote growth, and they are thought to be essential for the growth of bone. More research is necessary in this field.

Negative feedback mechanisms may influence GH secretion, but a hormone from the hypothalamus is in more direct control. This hormone is a peptide known as *somatostatin*, or growth-hormone inhibiting factor (GIF and GHIF). It is known also as somatotrophic hormone release-inhibiting factor (SRIF). Aside from inhibiting secretion of growth hormone, somatostatin also inhibits the release of thyrotrophin release factor (TRF).

Prolactin The development of the mammary glands (breast glands of the female) is stimulated by ovarian hormones, but milk production is controlled by the hormone *prolactin*, which is secreted by the adenohypophysis. Prolactin is also a *luteotrophic hormone* in rodents (mouse and rat) and is so named because in these species it has an effect on the development of the corpus luteum of the ovary. Stimulation of the corpus luteum causes it to secrete the hormone progesterone. The luteotrophic effect does not occur in the human being.

A prolactin release-inhibiting hormone (PIH) is secreted by the hypothalamus. PIH inhibits the release of prolactin by the adenohypophysis. PIH appears to override prolactin secretion except at the time of lactation, at which time suckling may initiate the suppression of PIH and the release of prolactin. There is also a prolactin-release factor (PRF) which stimulates release.

Thyrotrophin (TSH) The adenohypophysis also produces hormones whose function is to regulate hormone secretion by other endocrine glands (Fig. 13.4). The adenohypophysis produces *thyrotrophin* (TSH), a thyroid-stimulating hormone that maintains the normal functioning of the thyroid gland. TSH increases the blood circulation through the thyroid gland, and

FIGURE 13·4
Some of the regulatory effects of the adenohypophyseal gland.

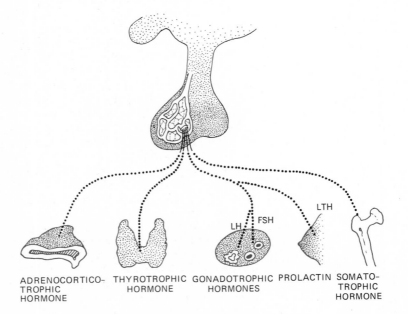

ADRENOCORTICO-TROPHIC HORMONE   THYROTROPHIC HORMONE   GONADOTROPHIC HORMONES   PROLACTIN   SOMATO-TROPHIC HORMONE

LH FSH   LTH

this, in turn, increases the production of the thyroid hormones thyroxine (T_4) and triiodothyronine (T_3).

The hypothalamus secretes a thyrotrophin-releasing hormone (TRH) which stimulates the adenohypophysis to release TSH.

Corticotrophin (*ACTH*) Corticotrophin, or adrenocorticotrophic hormone, is secreted by the anterior lobe of the hypophysis. This hormone influences the development and function of the adrenal cortex, especially in the secretion of the glucocorticoids.

Corticotrophin releasing hormone from the hypothalamus stimulates the secretion of ACTH.

A single injection of ACTH into experimental animals causes a rapid depletion of the normally high ascorbic acid concentration of the adrenal cortex. ACTH reduces the cholesterol concentration of the adrenal cortex more slowly as cholesterol is converted into adrenal glucocorticoids. The blood-serum cholesterol is also reduced by the administration of ACTH over a period of several days. Cholesterol is a source of precursors in the biosynthesis of adrenocortical hormones as well as in that of the sex hormones.

Follicle-stimulating hormone (*FSH*) *and luteinizing hormone* (*LH*) Introduction of adenohypophyseal extracts produces marked enlargement of the ovaries of both normal female rats and females of other mammalian species and those who have had their hypophyses removed. By careful experimental procedures, investigators have been able to differentiate two gonadotropic hormones secreted by the adenohypophysis. One is the *follicle-stimulating hormone* (FSH). In the female, FSH stimulates development of the follicle cells surrounding each maturing ovum in the ovary (Fig. 13.4). In the male, FSH stimulates development of the seminiferous tubules in the testes, which produce the male sex cells, or sperm. The second hormone is the *luteinizing hormone* (LH). In the female, LH stimulates the development of corpora lutea in the ovary. Actually the combined actions of both LH and FSH are essential to the development of the ovum and to ovulation. In the male, LH stimulates the interstitial cells of the testes to secrete testosterone.

The corpora lutea are the yellow bodies of the ovary. After the ovum ruptures through the surface of the ovary (ovulation), the follicular cells, which formerly surrounded and nourished the ovum, change in appearance and function. They form a yellow glandular body called the *corpus luteum,* which then begins to function as an endocrine gland in its own right (see the section on gonadal hormones in this chapter; see also Chap. 21).

LH is thought to stimulate the development of the ovarian follicle before ovulation as well as to stimulate ovulation and to provide for the development of the corpus luteum. LH also stimulates the secretion of estrogen by the ovaries and the secretion of progesterone by the corpus luteum.

In the male, LH is sometimes called the *interstitial cell–stimulating hormone* (ICSH) because it stimulates the development of the small interstitial cells of the testes. These cells are located in groups surrounded by the sperm-producing cells (seminiferous tubule cells). The interstitial cells secrete the male sex hormone. The dual names LH and ICSH arose when it was thought that they were two separate hormones; now it is known that LH and ICSH are identical in both sexes.

The secretions of FSH and LH are regulated by hypothalamic releasing factors. The releasing factors are *follicle-stimulating hormone release* factor (FSH-RF) and *luteinizing hormone release factor* (LH-RF). Inhibitory factors secreted by the hypothalamus apparently depress the secretion of gonadotrophic hormones until puberty.

Melanocyte-stimulating hormone (MSH) Melanocytes are pigment cells, most evident in lower vertebrates that darken their skin by expansion of these cells. In humans, the pigment melanin not only is found in static melanocytes but also is distributed freely in the outer layer of the skin. The *melanocyte-stimulating hormone (MSH)* is found in the adenohypophysis or in the intermediate lobe of the hypophysis in those vertebrates who have one. If it comes from the intermediate lobe, MSH is often called *intermedin*. The intermediate lobe is poorly developed in humans, and the function of MSH is obscure. MSH is generally considered to have some effect on melanin production and thus on darkening the skin. MSH release-inhibiting factor (MIF) and MSH release factor (MRF) are produced in the hypothalamus.

POSTERIOR LOBE (NEUROHYPOPHYSIS) The posterior lobe of the hypophysis, or neurohypophysis, arises as a downgrowth of the diencephalon from the region that will later become the floor of the third ventricle. The stalk by which the neurohypophysis is attached is called the *infundibulum*. The base of the infundibulum is located behind the optic chiasma. The *neurohypophysis (pars nervosa)* has intimate connections with the hypothalamus through the infundibulum (Fig. 13.3). The cells that compose the neurohypophysis are modified neuroglial cells amidst numerous nerve fibers; the cells are called *pituicytes*. There seems to be no doubt that the secretions of the posterior lobe arise in the hypothalamus. Droplets of secretion pass down the nonmyelinated axons that connect the two structures and are stored in the neurohypophysis. The neurohypophysis differs from the adenohypophysis, which secretes its own hormones. The two principal hormones found in the neurohypophysis are *oxytocin* and *antidiuretic hormone* (ADH).

Oxytocin *Oxytocin* induces marked contractions in the uterus, especially in the pregnant uterus. Its function in the male, if any, is not known. Oxytocin also causes the release of milk in the mammary glands of the nursing mother. The sucking stimulus initiates nerve impulses that are directed to the hypothalamus. These afferent impulses to the hypothalamus cause the neurohypophysis to release oxytocin into the bloodstream. In the mammary gland, numerous myoepithelial cells surrounding alveoli are stimulated to contract, ejecting milk into the mammary ducts. The infant is then able to remove milk from the breast by sucking. The synthesis and secretion of milk are governed by various hormones, among them, prolactin. Oxytocin affects the ejection of milk after it has been formed by glandular cells of the mammary glands.

Antidiuretic hormone (ADH) Early experiments with the *antidiuretic hormone* were performed on dogs. In these studies, the principal effect of this hormone was a rise in blood pressure; hence, the hormone was called "vasopressin." This hormone has little effect on general blood pressure in humans, however, although it may have some regional effects. ADH may be involved, for example, in contraction of the smooth muscle in the arterioles of the vasa recta surrounding the kidney tubules. ADH may also exert a mild pressor effect on the general circulation following severe hemorrhage.

The most evident function of ADH is the depression of urine flow from the kidneys. This function gives the hormone its name: antidiuretic hormone. ADH decreases urine flow by increasing the permeability of cell membranes in the distal convoluted tubules and collecting ducts so that more water is removed from the filtrate that passes through these tubules. ADH thus enables the kidney to reabsorb and conserve water.

CLINICAL ASPECTS: HYPOPHYSEAL HORMONES

Effects of hypophysectomy Removal of the entire gland (*hypophysectomy*) in a young mammal produces a series of degenerative changes but not necessarily death. Many early experiments with hypophysectomy resulted in extensive damage to the brain, making the true effects more difficult to interpret. Most effects that follow total hypophysectomy can be obtained by removal of the adenohypophysis alone.

Removal of the adenohypophysis of a young mammal prevents the animal from growing and developing into a normal adult. The gonads do not develop, so the animal remains sexually immature. Metabolism is depressed, resulting in increased accumulation of fat. Thyroid and adrenal abnormalities occur. Injection of adenohypophyseal extract tends to promote growth and development of the gonads in hypophysectomized animals.

When normal young animals are treated with extract of the adenohypophysis, gigantism, overgrowth of skull bones, large feet, and early sexual maturity are observed. The thyroid gland and adrenal cortex enlarge, with evidence of hypersecretion. The external genitalia may become larger than normal.

Removal of the hypophysis in experimental animals is followed by a great reduction in the size and activity of the adrenal cortex. Enough activity remains, however, to maintain life. The adrenal medulla undergoes no change in size, and apparently its function is unaffected in hypophysectomized animals. When ACTH is given to normal animals, it induces hypertrophy of the adrenal cortex and stimulates hyperfunction of that gland.

When the neurohypophysis of an animal is removed without injuring the hypothalamus, the kidneys begin to excrete great amounts of water (polyuria). This condition is called *diabetes insipidus*, and a similar disease, in which an unusual amount of urine of low specific gravity is excreted, occurs in humans. In diabetes insipidus, the urine does not contain glucose, as it does in diabetes mellitus, and the amount of urine excreted is much greater. Diabetes insipidus is considered to be caused by a neurohypophyseal deficiency, and the condition can be alleviated by continued administration of ADH. Injury to the hypothalamus or to the nerve pathway in the infundibulum also causes polyuria to develop. Maintenance of a functional nerve pathway between the hypothalamus and the neurohypophysis is essential to the production of ADH.

Growth hormone abnormalities Skeletal growth may be stimulated by daily injections of the growth hormone into young animals, and rats and dogs have been stimulated to grow to unusual proportions by this method. One growth hormone abnormality that affects humans is related to an abnormal increase in the length of the long bones of the arms and legs, producing individuals with heights of 7 to 8 ft. This condition is called *gigantism*, since the body proportions are different from those of normally tall persons. Gigantism is caused by a hypersecretion of GH. Growth in height can occur only so long as the zones of growth in the long bones remain open, so this condition only affects children. Another condition resulting from hypersecretion of GH affects adults and is called *acromegaly*. Overgrowth of bone in acromegaly is especially noticeable in the brow ridges, the lower jaw, and the hands and feet, all structures containing much cartilage. The nose and lips thicken, the face loses its intelligent expression, and there may be mental regression. Tumors of the adenohypophysis are usually found to be responsible for hypersecretion of GH in these cases (Fig. 13.5).

Hyposecretion of GH in infants and children produces *pituitary dwarfism*. Pituitary dwarfs, unlike cretins, have normal facial features, normal intelligence, and good body proportions. They apparently do not obtain enough growth hormone to permit normal growth. Since the gonadotrophic hormones are also lacking in this condition, pituitary dwarfs remain sexually infantile (Fig. 13.6).

THYROID GLAND

The greater part of the *thyroid gland* arises from the floor of the pharynx. Its connection with the pharynx is soon lost, and the gland develops into a bilobed structure with a connecting portion anteriorly across the trachea just below the larynx (Fig. 13.7). The thyroid grows during childhood and reaches its normal adult size at puberty. The thyroid is the largest of the glands that are entirely endocrine in function. It is larger in the female than in the male, but its size is affected by a number of factors. It enlarges, for example, if normal quantities of iodine are not available. There is a greater incidence of abnormal conditions of the thyroid in the female than in the male.

FIGURE 13·5
Acromegaly. Patient exhibiting coarse facial features. (*From E. Cheraskin and L. Langley, "Dynamics of Oral Diagnosis," The Year Book Medical Publishers, Inc., Chicago, 1956.*)

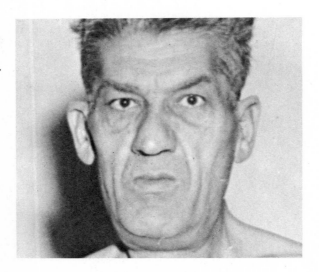

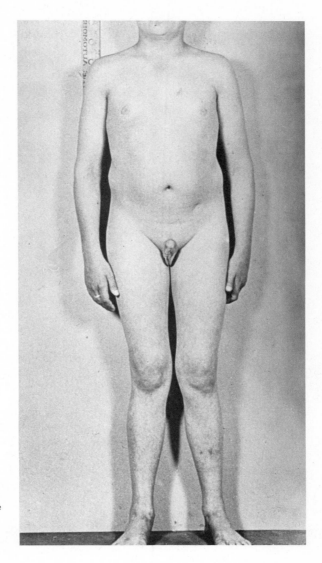

Histological preparations of thyroid tissue show it to be composed of
spherical sacs called *vesicles*, or *follicles*. The walls of the vesicles are
composed of a single layer of cuboidal epithelial cells. Within the vesicle is
a viscid colloidal fluid, which contains iodine. The colloid stores the thyroid
hormones (Fig. 13.8).

The thyroid gland receives an abundant blood supply from the thyroid

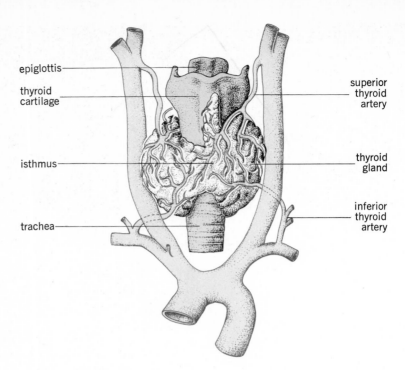

epiglottis

thyroid
cartilage

isthmus

trachea

superior
thyroid
artery

thyroid
gland

inferior
thyroid
artery

FIGURE 13·7

The thyroid gland, anterior
view.

arteries, which branch from the external carotid and subclavian arteries. The
arteries anastomose freely, and much of the blood volume appears to pass
more or less directly into the thyroid veins. Each vesicle is surrounded by a
dense capillary network.

THYROID HORMONES   The major thyroid hormone is *thyroxine*, or *tetra-
iodothyronine* (T_4). *Triiodothyronine* (T_3), which is physiologically very
active but is present in the blood in only very small amounts, is also found. A
third thyroid hormone, quite different in chemical nature from the
thyronines, is *thyrocalcitonin*, which will be discussed later.

Iodine forms a part of thyronine structure. There are several steps in the
synthesis of these two thyroid hormones. Ionic iodide is first converted to
elemental iodine. This is an oxidative reaction catalyzed by a tissue peroxi-
dase. Elemental iodine is then combined with the amino acid tyrosine by
enzymatic action to form monoiodotyrosine. Some tyrosine molecules
combine with two atoms of iodine to form diiodotyrosine. A coupling
enzyme aids in combining the mono- and diiodo- forms into the hormone
triiodothyronine. Two molecules of diiodotyrosine form tetraiodothyro-
nine, or thyroxine.

Large vesicles of the thyroid gland contain the glycoprotein *thyroglobulin,* which is secreted by cells of the vesicles. The thyroid hormones are
stored in the colloid of the vesicles, chemically attached to thyroglobulin.
Before the hormones can enter the bloodstream, they must be released from
this large globulin, since it cannot enter the blood. In the blood, the thyroid
hormones are carried by thyroid-binding globulin (TBG) and are released by
this protein as they enter cell membranes.

418   THE ENDOCRINE SYSTEM

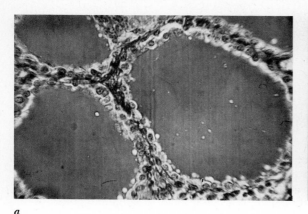
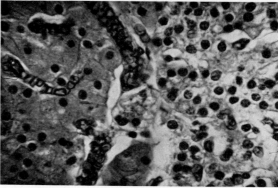

a

b

FIGURE 13·8

a Normal thyroid tissue from a guinea pig. The large vesicles filled with colloidal fluid are typical of thyroid tissue.
b Parathyroid tissue, human. This tissue is quite different in appearance from that of the thyroid. (*Courtesy of Dr. W. M. Copenhaver.*)

Thyroxine The most outstanding effect of thyroxine, the major thyroid hormone, is its ability to accelerate metabolism. The effect appears to work at the cellular level and is probably caused by the hormone's action in oxidative enzyme systems of the cell. The rate of energy exchange is affected by thyroxine. Energy exchange is accelerated in hyperthyroidism and is reduced in hypothyroidism. Growth is also affected by thyroxine. Hypothyroidism in the young results in retarded growth. Metabolism in patients with low basal metabolism rates can be greatly accelerated by the administration of thyroid extract.

The normal functions of the thyroid are merely regulatory. As we have seen, the adenohypophysis releases thyrotrophin (TSH), which is the chief regulator of thyroid function. An increase in the metabolic rate inhibits the secretion of TSH by a negative feedback mechanism that includes the hypothalamus. Much can be learned concerning the functions of the thyroid, however, by considering conditions of abnormal function.

Thyrocalcitonin One of the more recently discovered hormones is the polypeptide *thyrocalcitonin,* produced by interstitial, or parafollicular, "C" cells of the thyroid gland. The interstitial cells have an interesting biological history. They arise from the epithelial tissue of embryonic branchial pouches which develop into the postbranchial, or ultimobranchial, bodies of lower vertebrates (Fig. 13.13). In the human thyroid as well as in that of most other mammals, cells from the embryonic ultimobranchial bodies are incorporated into the thyroid gland as interstitial cells. The hormone produced by these cells is not found in the colloidal material within the thyroid vesicles. Calcitonin has also been found in the parathyroid and thymus, but its primary source is the thyroid.

Thyrocalcitonin regulates blood calcium level. It cooperates with parathyroid hormone, which is also calcemic in function. A negative feedback mechanism maintains a precise blood calcium level. Skeletal structures supply a great reserve source of calcium, and thyrocalcitonin plays a major part in the remodeling of bones as it stimulates bones to retain calcium. Thyrocalcitonin also increases sodium and chloride excretion and acts as a diuretic when administered in large amounts.

Protein-bound iodine All the thyroid hormones can take up iodine. The iodine is incorporated into thyroid-binding globulin as a plasma protein, and is then referred to as *protein-bound iodine* (PBI). Since PBI concentration can be measured, it is often used to indicate the level of thyroid hormone secretion. The test for PBI concentration in the blood plasma has its limitations, however. It measures the amount of circulating thyroid hormone fairly accurately, but the measurement may also include other iodine-containing proteins.

CLINICAL ASPECTS: THYROID HORMONES

Abnormal conditions of the thyroid gland are responsible for three well-known pathological states: cretinism, myxedema, and exophthalmic goiter. The first two conditions are examples of *hypothyroidism,* or a deficiency in thyroid hormone secretion. The last condition, exophthalmic goiter, is an example of hyperthyroidism, or an oversecretion in thyroid hormone. The functioning of the thyroid gland may also be affected by iodine deficiency, a condition that produces goiter.

Goiter Many types of thyroid gland enlargement may be called *goiter.* A simple uncomplicated goiter is usually caused by a lack of iodine. Simple goiters do not necessarily result in any endocrine disturbance, such as a change in the rate of metabolism, although the presence of a goiter indicates a low rate of thyroid hormone production. The incidence of simple goiter is much higher in mountainous or desert regions where the drinking water lacks iodine. Areas remote from the sea are also apt to have iodine-deficient water. Deficiencies in iodine are mostly overcome today by the widespread use of iodized table salt. Goiter occurs more frequently in women than in men. Adolescents and pregnant women should be especially careful to obtain an adequate supply of iodine in their food and drinking water. The incidence of simple goiter has become relatively low now that its cause is understood.

Cretinism Hypothyroidism manifests itself in many ways. It is probably more accurate to limit the term *cretin* to those individuals whose extreme hypothyroidism is caused by failure of the thyroid to develop in intrauterine life. When this happens, the brain of the fetus is retarded in its development, and the infant shows signs of hypothyroidism early in life. The child fails to grow properly, the tongue thickens and often protrudes, the bridge of the nose fails to develop (saddle nose), and mental retardation is present. The treatment of this type of hypothyroidism may be unsatisfactory, largely because the developing brain of the fetus has already been seriously injured. Administration of thyroid extract usually produces good physical growth if treatment is begun early in infancy, but mental development may still be retarded. There are, however, many less severe cases of hypothyroidism in infants and children that do yield to treatment, and such individuals improve greatly. Thyroid extract is one of the few hormones that can be taken by way of mouth. The earlier the treatment of hypothyroidism is begun, the better are the chances of normal development (Fig. 13.9).

Myxedema In some forms of hypothyroidism, the skin becomes thick and puffy, resembling an edema or swelling; the term *myxedema* was, therefore,

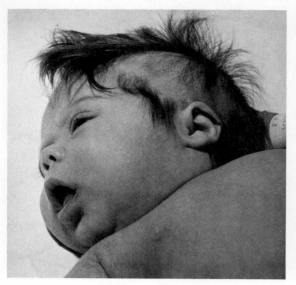
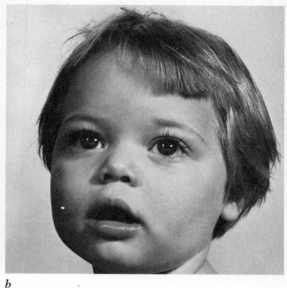

a   *b*

FIGURE 13·9

a A ten-week-old baby showing typical features of severe
hypothyroidism. Note especially the fragile hair, which is
falling out, and the thickened
facial features. *b* The same
child after 16 months of treatment. (*Courtesy of Dr. Milton
S. Grossman.*)

suggested as descriptive of this condition. The condition is not a true edema,
however, although there is an accumulation of a semifluid albuminous
substance in the skin. Some types of hypothyroidism in children may be
described as juvenile myxedema. Not all forms of adult hypothyroidism are
myxedematous, especially in the beginning. A low basal metabolic rate is
the most consistent finding in myxedematous hypothyroidism. The patient
feels cold, because a low rate of metabolism results in decreased heat
production. The physical changes and mental retardation associated with
hypothyroidism develop slowly. The skin thickens and becomes dry, and the
face loses its normal intelligent expression. If the patient is an adult, growth
cannot be retarded, but otherwise the symptoms of myxedema resemble
those of cretinism (Fig. 13.10).

Cases of myxedema respond to treatment with thyroid extract, often with
dramatic results. Within a few days the patient may appear alert and show
improved speech and a higher basal metabolic rate. The facial expression
slowly improves as the individual returns to normal mental and physical
health.

Hyperthyroidism When the thyroid gland is overactive and produces excessive
amounts of thyroid hormones, the various conditions that develop may be
described generally as *toxic goiters*. In toxic goiters, unlike simple goiter, the
thyroid may be only slightly enlarged, if enlarged at all. One form of toxic
goiter is caused by an adenomatous growth (or tumor). The causal factors in
this case would appear to be within the gland itself. In the case of exophthalmic goiter, the thyroid is probably overactive because it is stimulated
from some other source. Experimental evidence points to an unknown
factor, probably associated with thyrotrophic hormone.

Exophthalmic goiter, or Graves' disease, is characterized by moderate
enlargement of the thyroid, protrusion of the eyeballs (exophthalmos), and a

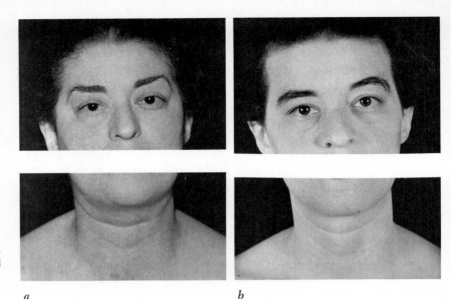

a   b

high basal metabolic rate. Ocular signs are not always present, but when the
eyes are affected, the protruding eyeballs show more sclera than usual and
the patient has a tense, frightened appearance (Fig. 13.11). The patient is
apt to be nervous and to have a high pulse rate. Since the basal metabolism
is greatly increased, the hyperthyroid patient may eat a great deal and still
lose weight. The patient will produce considerable body heat from height-

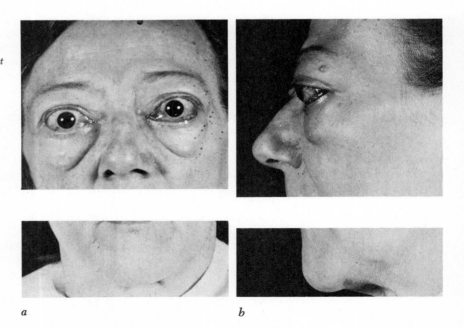

a   b

ened metabolic activity, feel warm, and perspire readily. Muscular weakness may result in tremor, especially evident in the fingers. It was originally thought that Graves' disease was caused by a lack of thyroid hormone inhibition or by an excessively high level of TSH. A gamma globulin that stimulates the thyroid much like TSH but acts more slowly and over a longer period of time has recently been discovered. This factor is called *long-acting thyroid stimulator* (LATS). The evidence of LATS may account for some features of Graves' disease, but it does not seem to explain exophthalmos, which may involve other factors.

PARATHYROID GLANDS

Partially embedded on the posterior surface of the thyroid gland are four small yellowish or brown bodies known as the *parathyroid glands* (Fig. 13.12). The parathyroid glands are closely related to the thyroid in point of origin, for they arise in the embryo from the entoderm of the third and fourth pharyngeal pouches (Fig. 13.13). The cells are found together in a compact mass, so that parathyroid tissue does not closely resemble thyroid tissue (see Fig. 13.8). Neither are they closely related from a functional standpoint. The parathyroid glands are the smallest of the compact endocrine glands, being only about the size of a cherry seed. The parathyroid hormone, which is primarily produced by these glands, is concerned with the metabolism of calcium and phosphorus.

PARATHYROID HORMONE (PTH) *Parathyroid hormone (PTH)* is a polypeptide protein. This hormone plays an important role in regulating homeostasis of calcium and phosphate levels of the blood. We have noted that calcium ions

FIGURE 13·12
Thyroid and parathyroid glands, posterior view.

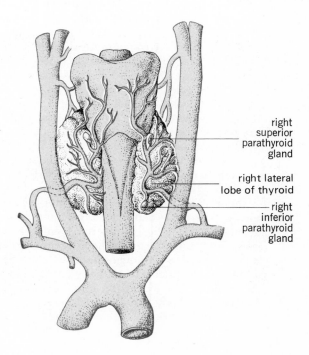

right superior parathyroid gland

right lateral lobe of thyroid

right inferior parathyroid gland

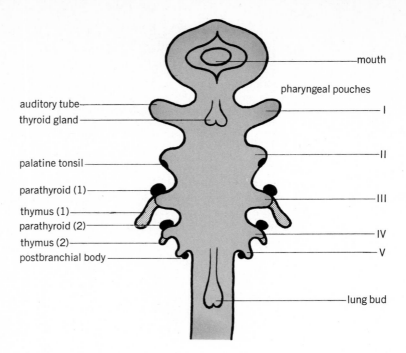

auditory tube
thyroid gland
palatine tonsil
parathyroid (1)
thymus (1)
parathyroid (2)
thymus (2)
postbranchial body

mouth
pharyngeal pouches
I
II
III
IV
V
lung bud

FIGURE 13·13

Derivatives of the embryonic pharynx, shown diagrammatically.

are essential for muscular contraction, for transmission of the nerve impulse, and for the clotting of the blood.

The parathyroid glands are an exception to the general rule of hormonal control in that they are not controlled by any known secretion of the adenohypophysis. When blood calcium ion level becomes low, the parathyroids are stimulated, apparently directly, to increase secretion of PTH. If the calcium ion concentration of the blood plasma increases above normal, secretion of the parathyroid gland is inhibited.

The blood calcium ion level may be increased in several ways: (1) by increased absorption of calcium from the intestine with the aid of vitamin D, (2) by increased reabsorption of calcium by kidney tubules, or (3) by increased release of calcium stored in bones. As bone structure is broken down by osteoclast cells to release calcium, phosphate is also released, but at the same time, the kidneys are stimulated to increase the excretion of phosphate so that a balance is maintained. PTH acts upon bone to stimulate osteoclast activity.

Almost all (about 99 percent) calcium stored in the body is found in the skeleton, where it is deposited as calcium phosphates and carbonates. The calcium ion level of the blood serum is only about 10 mg/100 ml. Calcium is present in protoplasm and in the extracellular fluid in minute amounts. It is absorbed from foods through the intestinal wall and eliminated chiefly by way of the intestinal tract, although a small amount appears in the urine.

Phosphorus, largely in the form of phosphates, is widely distributed throughout the body, but about 80 percent is found in the skeleton. Phosphates occur in various combinations in the blood, protoplasm, and extracellular fluid. They form a part of adenosine triphosphate, phosphocreatine, phospholipids, phosphoproteins, and many other organic compounds. Phosphates are obtained from foods and are eliminated largely by the kidneys.

CLINICAL ASPECTS: PARATHYROID HORMONES

Removal of the parathyroid glands from experimental animals produces striking effects, especially in those of the order *Carnivora*. Muscular tremors occur after a few days, and these progress toward tetany and death. The tremors are accompanied by a sharp drop in the blood calcium level, as well as a rise in blood phosphorus level. The tetany can be alleviated by the injection of an extract of the parathyroid glands containing the hormone, by administration of calcium, or by using certain drugs that enable the body to maintain the calcium level of the blood.

Hypoparathyroidism Parathyroid insufficiency in humans is not often of the acute type. As we have noted, complete extirpation of the parathyroids in animals usually produces tetany and death. In humans these glands are seldom destroyed completely, and tetany produced by hypoparathyroidism is more commonly of a chronic or latent type. Hyperexcitability of the peripheral neuromuscular system is a common feature of tetany. The muscles of the face contract when the facial nerve is stimulated by tapping, giving the face a tense, sad, or crying expression. Pressure on the arm can cause the fingers to fold together in a characteristic fashion. The blood calcium level is low, and the blood phosphate concentration is high. Using parathyroid hormone for long-term treatment has not proven satisfactory, though it will cause a rise in the serum blood calcium level and a reduction in blood phosphate level through urinary excretion. Massive doses of calciferol (vitamin D_2) are used to increase calcium absorption from the intestine, thus maintaining blood calcium level. Another sterol used in treatment of hypoparathyroidism is dihydrotachysterol (A.T.10), which is produced by irradiating ergosterol. The action of this sterol resembles the action of parathyroid extract in reducing blood phosphate level through increasing urinary excretion. Dihydrotachysterol also causes a moderate rise in blood calcium level by stimulating increased absorption of calcium from the intestine. These effects vary with the amount of secretion available from the patient's parathyroid glands, since this compound has a tendency to decrease the activity of the parathyroids.

Hyperparathyroidism A greater-than-normal secretion of PTH is commonly associated with a tumor or growth on one or more of the parathyroid glands. Hyperparathyroidism results in a loss of muscle tone and a disturbance in calcium and phosphate metabolism. The blood calcium level is high and is partially maintained by the withdrawal of calcium from the skeleton. The skeleton may become greatly weakened through loss of calcium, and spontaneous fractures may occur. The bones often become greatly deformed, and cysts are seen in fibrous tissue. The excess blood calcium may be deposited in tissues other than bone, especially as stones in the kidneys. The blood phosphate level is also high, and the excretion of both calcium and phosphate by the kidneys is increased. Surgical removal of excess parathy-

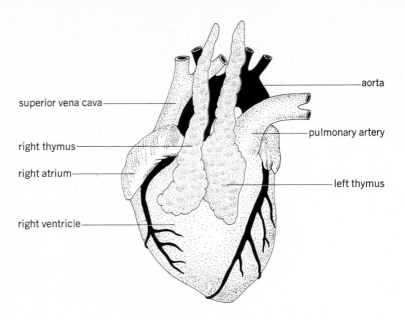

superior vena cava

aorta

right thymus

pulmonary artery

right atrium

left thymus

right ventricle

FIGURE 13·14
The thymus gland in a young adult.

roid tissue appears to be the best treatment for hyperparathyroidism. Parathyroid function is usually stable, and hypoparathyroid and hyperparathyroid conditions are rarely primary disturbances.

THYMUS

In the upper part of the thorax above the heart is a bilobed structure called the *thymus* (Fig. 13.14). The thymus is pink and consists of lymphoid tissue. The thymus of meat animals is the organ called the "throat sweetbread." Relatively large in young children, the thymus reaches its greatest size at puberty and then regresses as the lymphoid tissue is largely replaced by adipose and fibrous tissue. Numerous attempts to discover a thymus secretion have been made, and recently a hormonelike substance called *thymopoietin (thymin)* has been reported. Thymopoietin is thought to be concerned with the development of lymphocytes. The thymus does produce an abundance of lymphocytes called *T cells*. Such cells provide immunity against many viruses and other organisms which might be injurious. T cells are thought to provide immunity to some types of cancer. The T cell is also the immunological defense that resists tissue transplants. Lymphocytes and immunological problems will be discussed in Chap. 14.

THE PINEAL GLAND

The pineal gland was originally a photoreceptor, forming a median, or third, eye in prehistoric reptiles (Fig. 13.3*a*). Often regarded as a vestigial structure, it now appears to be a true endocrine gland, which produces the hormone *melatonin*.

The clearest picture of melatonin's function is found in those animals that possess active melanophores. Whereas the hypophyseal hormone melanocyte-stimulating hormone (MSH) causes the skin of these animals to darken, melatonin causes the melanophores to gather and the color of the skin to lighten. In these lower vertebrates, the pineal gland responds to light waves received at the retina. Nerve impulses are probably carried to the gland by sympathetic nerves, causing the release of the melatonin. The function of the pineal in humans, however, remains largely obscure. Several chemically active substances have been obtained from the pineal gland. These include serotonin, norepinephrine, and histamine. However, only melatonin is regarded as being secreted there.

PANCREAS

The pancreas, as we shall see later, is a gland that secretes both hormones and digestive enzymes. It is a slender, tapering gland located below and somewhat behind the stomach (see Fig. 19.12*a* and *b*).

INSULIN   The pancreas produces the hormone *insulin,* which is concerned with carbohydrate metabolism. The hormone is not produced by the large glandular cells that secrete enzymes but by groups of very small cells, called the *islet cells of Langerhans.* There are two kinds of islet cells: *alpha* and *beta* (Fig. 13.15*a* and *b*). The beta cells secrete insulin. Alpha cells secrete

FIGURE 13·15

The pancreas, the endocrine portion. *a* Beta cells of the islets of Langerhans produce insulin; alpha cells produce glucagon. These hormones are absorbed directly into the bloodstream. *b* Photomicrograph of islet cells, which are located in the lighter areas. (*Courtesy of L. Kelemes, M.D.*)

a

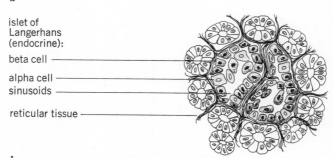

islet of Langerhans (endocrine):

beta cell ——————

alpha cell ——————
sinusoids ——————

reticular tissue ——————

b

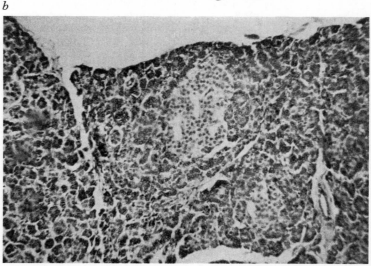

another hormone called *glucagon,* which will be considered later in this chapter. Insulin is not dispersed by way of the pancreatic duct; it is absorbed and distributed by the bloodstream, as are all hormones. Insulin is a protein composed of two polypeptide chains held together by two disulfide linkages. One chain comprises 21 amino acids; the other contains 30 amino acids.

The action of insulin   Most cells utilize considerable amounts of glucose in their metabolic activities. Those cells most dependent upon glucose are neurons, in general, adipose cells, and especially brain cells. Glucose enters cells by facilitated diffusion down a steep concentration gradient. The control of the blood glucose level is regulated by a balance between several hormones. Insulin contributes to the maintenance of blood glucose level by facilitating the entry of glucose into certain cells, such as muscle cells. After a meal, the blood glucose level rises as glucose is absorbed from the digestive tract, and the beta cells of the pancreas begin to secrete more insulin. Under the influence of insulin, glucose is readily taken up by the liver and converted to glycogen. Actively contracting muscle cells also take up considerable amounts of glucose, and resting muscles store glucose as muscle glycogen.

Fasting reduces the concentration of blood glucose, and as a result glucose uptake by cells is limited. Brain cells are an exception since their ability to utilize glucose is independent of insulin. Reduction of blood glucose causes an increase in the breakdown of glycogen to glucose. At the same time, protein catabolism releases amino acids into the bloodstream. Under these conditions, the liver forms glucose from certain amino acids in gluconeogenesis. Insulin secretion therefore is reduced by a reduction in blood sugar concentration.

Other endocrine glands and carbohydrate metabolism   The pancreas is not the only endocrine gland involved with the maintenance of normal carbohydrate metabolism. The thyroid, adrenals, and hypophysis are also involved.

Thyroid imbalance produces profound changes in metabolism in general. When linked with insulin deficiencies, the disturbance in carbohydrate metabolism is marked. Hyperthyroidism increases general metabolic activities; when it accompanies diabetes, the effects of this condition are much more severe.

Excess epinephrine secretion by the adrenal glands may cause hyperglycemia and glycosuria. The conversion of liver and muscle glycogen is accelerated. The hyperglycemia caused by the injection of epinephrine is a direct countereffect to the hypoglycemia that follows the injection of insulin. While epinephrine plays a part in the normal regulation of blood sugar, it should not be inferred that it has any direct relation to the basic causes of diabetes.

As we have indicated, the action of insulin in carbohydrate metabolism is complex. It involves not only insulin but growth hormone, epinephrine, and glucagon. Such hormones as thyroxine and cortisol supplement the action of insulin in a diabetic condition, but they do not cause diabetes. Thyroxin regulates the metabolic rate, and cortisol stimulates the liver to convert glycogen to glucose.

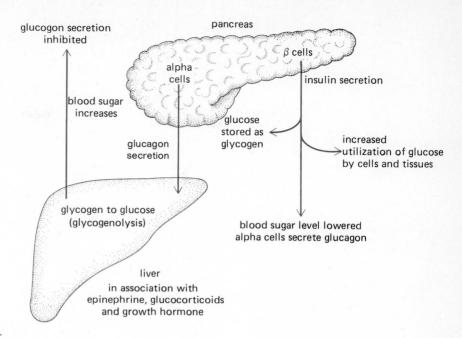

glucogon secretion inhibited

pancreas

β cells

alpha cells

insulin secretion

blood sugar increases

glucose stored as glycogen

increased utilization of glucose by cells and tissues

glucagon secretion

glycogen to glucose (glycogenolysis)

blood sugar level lowered alpha cells secrete glucagon

liver

in association with epinephrine, glucocorticoids and growth hormone

FIGURE 13·16

The effects of insulin and glucagon.

GLUCAGON   The hormone *glucagon*, which is produced by the alpha cells of the pancreas, regulates blood glucose level in coordination with insulin secretion. Glucagon stimulates the liver to convert glycogen to glucose and to release it into the blood stream. You will recall that insulin reduces the blood glucose level. If the blood glucose level is lowered, glucagon secretion will cause it to rise. In other words, a low blood glucose level causes an increased secretion of glucagon, whereas a high blood glucose level depresses glucagon secretion. Fasting, therefore, stimulates glucagon secretion, which causes a rise in the blood glucose level by accelerating the conversion of liver glycogen to glucose through the glycogenolytic pathway. Glucose then enters the bloodstream. The blood glucose level rises rapidly following an injection of glucagon. An increase in gluconeogenesis is also seen, although this effect may be caused by a depletion of glycogen (Fig. 13.16).

CLINICAL ASPECTS: DIABETES MELLITUS

When the pancreas fails to produce enough insulin, a serious derangement of carbohydrate metabolism occurs; this condition is known as *diabetes mellitus*. In the absence of insulin, tissues are unable to utilize glucose, and the blood glucose level may rise to twice the normal value. As expressed in milligrams per 100 milliliters, the blood glucose level in diabetics may rise from a normal of 80 to 120 mg/100 ml to 200 or 300 mg/100 ml. It is generally considered that the blood glucose "threshold," or the level at which glucose begins to appear in the urine, is around 100 mg/100 ml. High levels of blood glucose result in the condition called *hyperglycemia*. As the glucose level of the blood rises above normal, the kidneys begin to excrete this substance. Since the kidneys do not ordinarily excrete glucose, a chemical test for glucose in the urine is routinely used in the diagnosis of diabetes. The kidneys do excrete glucose after a meal rich in carbohydrates, but this produces only a temporary *glycosuria* in a normal individual.

The glucose reserves of the body are rapidly depleted in diabetes. Proteins are then utilized to supply glucose to starved tissues, and fat metabolism is also disrupted. The oxidation of fats is increased, resulting in the accumulation of certain organic acids (ketone substances) in the blood and urine. One of the breakdown products of fatty acids is acetoacetic acid. This is a toxic substance and may be responsible for the development of diabetic coma in extreme untreated cases.

Early research on diabetes In early studies of diabetes, researchers removed the pancreas from dogs and discovered that there was a great increase in the production of urine, that the urine contained sugar, and that these animals were unable to live more than a month without the pancreas. They determined also that the digestive portion of the pancreas was not responsible for these conditions. Investigators tied off the ducts of the pancreas in dogs, causing the glandular cells producing digestive enzymes to degenerate and leaving only the islands of Langerhans still active. The dogs lived and did not develop diabetes. Many attempts were made to extract a hormone from the pancreas, but these attempts met with failure until 1922. In that year Banting and Best prepared an extract of pancreatic tissue, free of digestive enzymes, from islet cells. When this extract was injected into depancreatized dogs, it reduced their diabetic symptoms. This outstanding work was done in the laboratory of J. J. R. Macleod in Toronto, Canada.

Injectable insulin Once insulin had been identified as the substance involved in diabetes, efforts were made to produce an insulin that could be used on a regular basis by diabetics. Insulin is a complex protein. It can be prepared in crystalline form with salts of zinc, nickel, cadmium, or cobalt. Most forms must be injected hypodermically, since, if taken orally, insulin is readily digested before it can be absorbed. Highly soluble insulin must be injected subcutaneously two or three times daily by the diabetic. It is absorbed by the blood rather rapidly, causing the available concentration in the blood to fluctuate and the blood glucose level to vary. The normal secretion of islet cells is apparently produced steadily and in minute amounts.

A search was made for less soluble forms of insulin, which would be absorbed less rapidly by the blood. It was found that insulin precipitated with certain protein derivatives is absorbed slowly and therefore remains effective over a longer period of time. Protamine zinc insulin is one of the long-acting forms. Protamine is a fish-sperm protein; the addition of zinc renders the solution more stable. NPH (Neutral Protamine Hagedorn) insulin is another long-acting form developed more recently. Since these insulin preparations are not highly soluble, a single injection in the morning may be sufficient for 24 h if the diabetic condition is mild.

Effects of insulin When insulin is injected, there is an almost immediate drop in the blood glucose level. Part of the glucose is converted to glycogen and stored in the liver and muscles; part is oxidized, with a consequent release of energy; and part undergoes synthesis into fat. Insulin does not cure diabetes, but it does regulate carbohydrate metabolism to such an extent that diabetics can live essentially normal lives. With a lowering of the blood glucose level, glycosuria tends to disappear, the fat metabolism becomes more normal, and the presence of abnormal amounts of acids in the blood and urine is reduced. Diabetics usually have increased thirst, which is allayed only by drinking considerable amounts of water. A large amount of water is necessary to provide for the filtration of the quantity of glucose and salts excreted. Insulin, by reducing the glycosuria, also reduces diuresis.

Development of diabetes Experimental work indicates that there is a difference in tolerance to diabetes between the carnivorous and herbivorous animals investigated. Almost all the early investigators used dogs or cats as experimental animals and found that they could not live without the pancreas for more than a few weeks. On the other hand, herbivorous animals, such as rabbits, sheep, goats, and monkeys, may survive without their pancreas for several months. These animals gradually lose weight and eventually die from causes directly or indirectly related to loss of insulin. From such studies, it may be inferred that insulin, like other hormones, is merely a regulator of physiological activities. Without the regulator, the level of these physiological activities may vary in different animals, but all activity does not necessarily cease.

Hyperinsulinism Hyperinsulinism is a condition in which the islet cells secrete too much insulin. Hyperinsulinism is accompanied by hypoglycemia (low blood glucose), and the central nervous system seems to be primarily affected, probably from the effects of glucose starvation. Tumors of the islet cells appear to be a common cause of excess insulin production.

Insulin shock Repeated injections of insulin resulting in extreme hypoglycemia to the extent of causing convulsive seizures have been used in treating certain types of mental cases. Considerable mental improvement has been reported in many cases, although the treatment is not without physical danger. Various stages of insulin shock can usually be terminated by the administration of glucose.

ADRENAL GLANDS

The *adrenal gland* is actually two glands in one. It is composed of an outer portion, or *cortex*, and an inner portion, or *medulla* (Figure 13.17). The two parts are distinct in origin and function. The adrenal glands of sharks and similar fish have a separate internal gland corresponding to the cortex, which lies between the kidneys, and a double row of bodies close to the sympathetic ganglionic chains, which are homologous to the tissue of the medulla.

The cortical portion of the human adrenal arises in the embryo from mesoderm of the same general region that produces the gonads. The medulla, on the other hand, is formed by the infiltration of neural crest ectodermal cells, which have a common origin with cells of the sympathetic ganglionic chain. These strands of cells, which invade the mass of cortical tissue, stain dark brown with chromic acid and are called *chromaffin cells*. Eventually they form the entire middle, or medullary, portion of the gland.

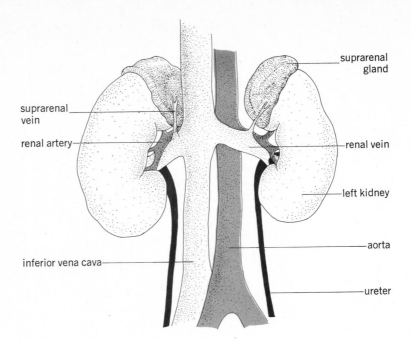

FIGURE 13·17
Adrenal glands (suprarenal glands) *in situ*.

suprarenal gland

suprarenal vein

renal artery

renal vein

left kidney

aorta

ureter

inferior vena cava

Preganglionic cholinergic fibers supply the medullary tissue by way of the splanchnic nerves. There are no true postganglionic fibers; the medullary cells themselves probably act in this capacity.

In the human the glands lie on top of the kidneys, so they are called *suprarenal glands* by some authorities, but in most mammals the adrenals are separate from the kidneys, and therefore *adrenal* is the better term. In a section of a fresh adrenal gland, the cortex is yellow, while the medulla is a red-brown color. Each adrenal is enclosed by a capsule of fibrous connective tissue (Fig. 13.18).

Microscopic examination reveals that the cortex is differentiated into three zones, while the medulla consists chiefly of chromaffin cells arranged

FIGURE 13·18
Human adrenal gland in schematic section to show cortex and medulla.

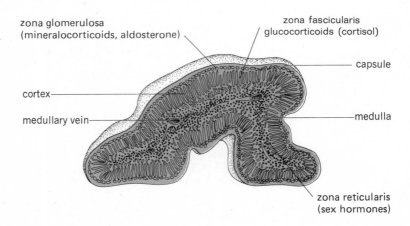

zona glomerulosa (mineralocorticoids, aldosterone)

zona fascicularis glucocorticoids (cortisol)

capsule

cortex

medulla

medullary vein

zona reticularis (sex hormones)

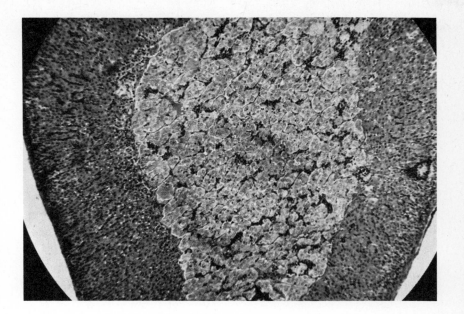

FIGURE 13·19
Photomicrographic section
through mouse adrenal gland
to show differentiation between cortex and medulla.
(*Courtesy of Edgar P. Jayne.*)

in irregular masses (Fig. 13.19). The three zones differ in appearance and function. The zona glomerulosa lies directly under the capsule of connective tissue that covers the gland. The cells of this outer zone are arranged in bundles and secrete mineralocorticoids. The zona fasciculars is the intermediate zone. It consists of cords of cells in long straight rows. These cells secrete glucocorticoids. The inner zone, the zona reticularis, has the appearance of a network of cords that branch and reunite. This inner zone of cells secretes androgens. The medulla receives a most abundant blood supply—a significant factor in view of its function. The cells of the cortex secrete hormones that are distinct from the hormones secreted by the medullary tissue. The functions of these hormones are summarized in Table 13.3.

HORMONES OF THE ADRENAL MEDULLA   *Epinephrine and norepinephrine*   Two closely related hormones are produced by the adrenal medulla. They are *epinephrine*, also known as adrenaline, and *norepinephrine*, or noradrenaline. Both hormones are catecholamines derived from the precursor, dopamine. The two hormones have similar, but not identical, effects. Epinephrine is produced much more abundantly than its counterpart norepinephrine. Release of the hormones is secured directly by stimulation of the splanchnic nerves and not by other hormones. Norepinephrine, it may be recalled, is the principal neurotransmitter in the sympathetic nervous system. Epinephrine has been shown to increase the level of cyclic AMP in many tissues. It does this by activating adenyl cyclase, which, in turn, catalyzes the synthesis of cyclic AMP. Cyclic AMP causes many inactive cellular enzymes to convert to an active form.

The action of epinephrine is almost identical with effects produced by stimulating the sympathetic nervous system. Blood pressure rises, largely because the smooth muscles in the walls of the arterioles are stimulated to

TABLE 13·3

Hormones of the adrenal glands

General classification	Hormone	Functional effects
Medullary hormones	Epinephrine	Increases blood supply to lungs and muscles, increases cardiac output, raises blood sugar level
	Norepinephrine	Causes peripheral vasoconstriction; an adrenergic neurotransmitter
Cortical hormones		
Mineralocorticoids	Aldosterone	Increases sodium reabsorption by kidney tubules, increases potassium in urine, increases water retention, increases blood sodium level
Glucocorticoids	Cortisol (hydrocortisone)	Stimulates gluconeogenesis, increases protein breakdown, increases blood glucose and amino acids in blood plasma, anti-inflammatory response
	Corticosterone	
	Cortisone	
Adrenal sex hormones	Androgens	None in normal individual
	Estrogens	None in normal individual

constrict; peristaltic movements of the intestine are inhibited; the pupil dilates; and bronchial muscle relaxes. The similarity of action between the hormone and the sympathetic system is not so surprising when we recall that during fetal development, the modified ganglion cells of the adrenal medulla migrate from the mass of tissue that forms the sympathetic ganglia. The action of epinephrine, therefore, is said to be *sympathomimetic;* that is, it mimics the action of the sympathetic system.

Cannon and his associates proposed that epinephrine helps an organism deal with an emergency. A considerable body of data has been compiled in support of this theory. Though the evidence is largely indirect, it seems to indicate that many of the effects produced by epinephrine in an emergency situation are beneficial to the organism. Still it is doubtful that epinephrine is essential for survival in an emergency. The theory, therefore, does not necessarily explain the function of the adrenal medulla. There is still some question about the importance of the adrenal medulla in enabling an animal to resist shock and stress, although this concept is seemingly well established.

Differential effects When epinephrine is injected into the veins, a sudden rise in blood pressure, which is of short duration, is noted. The effect of epinephrine on human blood pressure is much more striking than the effect produced in laboratory animals. Though many factors contribute to a rise in blood pressure, the most evident factor here is the constriction of peripheral arterioles. The great splanchnic capillary bed is also affected by epinephrine-directed vasoconstriction.

Not all parts of the circulatory system are affected in the same way. The coronary arterial system, for example, is not constricted. The concentration of the epinephrine solution injected and the site of injection—a vein, the abdominal cavity, or an artery—affect the outcome. In a normal laboratory animal, epinephrine increases the blood supply to the lungs and to the skeletal muscles.

Norepinephrine is primarily a peripheral vasoconstrictor; this hormone has little effect on heart rate but can cause a sharp rise in blood pressure. Natural epinephrine, as secreted by the adrenal medulla, contains norepinephrine in small amounts. Cannon and his associates stated that when sympathetic postganglionic fibers are stimulated, a substance resembling epinephrine, but not identical with it, is liberated at the neuromuscular junction. They called the substance *sympathin*. Sympathin is now known as norepinephrine, which is the adrenergic substance released at all postganglionic sympathetic axonal terminals.

The injection or increased secretion of epinephrine increases oxygen comsumption and respiration rate, since it increases the basal metabolic rate. Epinephrine causes blood glucose level to rise through the reduction of stored liver glycogen to glucose. Muscle glycogen is also mobilized, but it appears in the blood as lactic acid rather than glucose. The lactic acid is carried to the liver, where it is resynthesized to glycogen. It has been suggested that during starvation, when liver glycogen stores have been depleted, the administration of epinephrine may increase liver glycogen through this pathway. Cigarette smoking is said to cause an increased secretion of epinephrine.

THE ADRENAL CORTEX   The functions of the hormones of the adrenal cortex are more complex than those of the medullary portion. These hormones affect many organs and tissues in various ways. Unlike the adrenal medullas, the adrenal cortices are essential to life; the effects caused by their removal are very severe and invariably fatal unless essential hormones are replaced.

The hormones secreted by the adrenal cortices may be divided into three groups: mineralocorticoids, glucocorticoids, and the sex hormones (androgens and estrogens). As we have noted, mineralocorticoids are secreted by cells of the zona glomerulosa, a layer of circular groups of cells placed directly below the connective tissue capsule (Figs. 13.18 and 13.19).

Mineralocorticoids   *Aldosterone* is the principal mineralocorticoid. It plays a large role in the regulation of electrolyte balance and is especially active in promoting the reabsorption of sodium by the renal tubules and the increased absorption of sodium by the intestinal epithelium. At the same time, aldosterone stimulates potassium excretion by the kidneys. Water loss from the kidneys is decreased when aldosterone level is high, causing a consequent increase in the volume of the blood plasma. The mineralocorticoids are important also in regulating tissue water balance.

Glucocorticoids   The more important glucocorticoids are *cortisol* (*hydrocortisone*), *corticosterone*, and *cortisone*. Cortisol is by far the most active of the three hormones, and as much as 95 percent of glucocorticoid effectiveness has been attributed to it. Cortisone is the least effective. Many functions have been assigned to these steroids. Among those that are well established by research investigation are: (1) the promotion of glycogen synthesis in the liver, which is in part related to an increase in gluconeogenesis; (2) a redistribution of fat deposits; (3) an anti-inflammatory response; and (4) an increased resistance to injury and to physical "stress."

The sex hormones The adrenal cortex normally produces sex hormones only in trace amounts. Only under very abnormal conditions do these sex steroids have a noticeable effect. Visible effects, if any, appear to be produced mainly by androgens, and are usually seen in females and young boys.

CLINICAL ASPECTS: ADRENAL HORMONES

Medullary tumor The adrenal medulla is remarkably stable in its function, and abnormalities are rare. Insufficient secretion of epinephrine is not a serious problem. Occasionally a tumor of the adrenal medulla produces an excess secretion. The most common effect is a sudden, paroxysmal rise in the blood pressure to readings above 200 mmHg, with a return to fairly normal readings between attacks. Persistent hypertension is not characteristic of this condition, although it has been reported associated with tumors of the adrenal medulla. Surgical removal of the tumor usually results in a return to normal blood pressure levels.

Epinephrine and allergies Epinephrine dilates the bronchioles and so may be used in the treatment of bronchial asthma. It is also effective in treating other types of allergy, such as hives, presumably because it regulates the permeability of capillaries and the retention of salts and water in the tissues.

Adrenal response to stress Inflammation, injury, and stress involve the hypothalamus and a hormone from the adenohypophysis as well as the adrenocortical hormones. The adenohypophyseal hormone is ACTH. Pain from an injury or the neural response to stress sets off a hormonal chain that begins in the hypothalamus. The hypothalamus, then, produces an ACTH-release factor that causes the adenohypophysis to release ACTH, which, in turn, stimulates the adrenal cortex to secrete its hormones.

Adrenocortical insufficiency renders most organisms much more susceptible to the stress of disturbing conditions. Such insufficiency also makes the organism more susceptible to infections and to the onset of fatigue. When the corticoids are not secreted in sufficient amounts, blood potassium level is increased, while the excretion of water by the kidneys is decreased. Under these conditions the fluid content of the tissue cells increases as blood volume decreases. Blood pressure falls as the blood volume decreases, and the kidneys are unable to function properly with a reduced blood supply. If these conditions go unchecked, eventually a state of shock is produced.

Addison's disease A well-known form of adrenocortical insufficiency is called *Addison's disease* after an English physician who, in 1855, described a condition associated with the destruction of the cortex of the adrenal glands. The condition is characterized by extreme lassitude, muscular weakness, weak heart action, lowered blood pressure, and digestive disturbances accompanied by loss of weight. Especially characteristic is a peculiar increase in the pigmentation of the skin and mucous membranes of the mouth (Fig. 13.20). The pigmentation change is due to an increase in melanin, the normal pigment of the skin. The additional pigmentation may

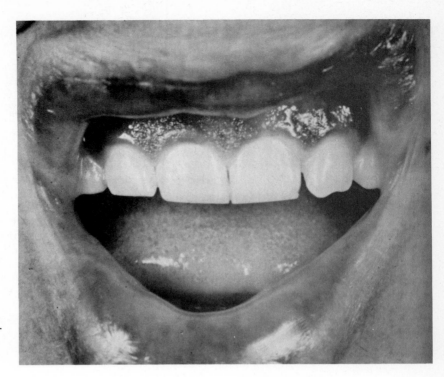

FIGURE 13·20
Pigmentation of gums and lips (oral buccal mucosa) as a characteristic of Addison's disease. (*Courtesy of Robert Gittler, M.D.*)

be evenly distributed, or it may be in blotches. It is greatest in those areas that normally contain pigment, such as around the nipples.

Deterioration of the adrenal cortex is commonly found in Addison's disease. Often the cortex is destroyed by tuberculous lesions. Addison's disease is treated with salt therapy (to replace salts lost through inefficient reabsorption) and cortical steroids. A great deal of progress has been made in treatment of this disease, which was regarded as fatal only a few years ago. The outlook now is greatly improved, but the disease is still serious.

Hyperactivity of the adrenal cortex One occasionally hears of children who attain considerable sexual development by the time they are two or three years old. Such cases are usually associated with abnormal activity of the adrenal cortex, sometimes caused by a tumor. Young girls with the adrenogenital syndrome develop rapidly before puberty, usually with some degree of masculinization. Pubic and axillary hair may develop at an early age, and the distribution of body hair is diffuse or more like the masculine type. Some growth of facial hair may occur, especially on the upper lip and chin. The clitoris usually becomes greatly enlarged, but the ovaries and uterus remain small and menstruation does not start.

Young boys with adrenal hyperactivity appear much older than their actual age. They grow rapidly during the early years and show considerable muscular development. Pubic hair develops, even though the child may be only three years old. The penis becomes greatly enlarged, but the testes remain small and apparently do not produce mature sperm.

Hyperactivity of the adrenal cortex in the female after puberty is characterized by virilism, or masculinization. A great amount of body hair, with masculine distribution, often grows. Facial hair develops strongly, especially on the chin. The "bearded lady" in a circus side show may be suffering from abnormal growth of the adrenal cortex. The body build of affected females tends toward the masculine type, and the voice is deeper than normal. The effects of this condition in the adult male are not well known.

Another type of hyperfunction of the adrenal cortices produces the condition known as *Cushing's disease.* This disease is characterized by redistribution of fat, often with great obesity, muscular weakness, skeletal weakness, and high blood pressure.

THE GONADS

The gonads are the reproductive organs that produce the sex cells. The ovaries of the female produce ova, or eggs; the testes of the male produce sperm, or spermatozoa. The gonads also function as endocrine glands by secreting the sex hormones. These hormones exercise control over the development of the secondary sexual characteristics and over the normal production and maturation of the sex cells.

FEMALE SEX HORMONES   Hormones are secreted by various parts of the female reproductive system.

A group of estrogenic substances is secreted by the ovary. They are commonly called the *estrogens. Estradiol* is the most potent of the several naturally occurring estrogens. The estrogenic hormones are so named because they produce *estrus,* the mating state in female mammals with cyclic receptive periods.

The corpus luteum (Fig. 13.21) produces a progestational hormone called *progesterone. Gestation* means pregnancy, and progesterone influences changes in the lining of the uterus that favor the implementation of a fertilized ovum, therefore permitting pregnancy to develop.

The placenta also secretes estrogens and progesterone. In addition, it produces a gonadotrophic hormone called *chorionic gonadotrophin.* The chorion is the fetal membrane that forms the embryonic portion of the placenta.

Functions of estrogenic substances   The estrogens are responsible for many of the changes that occur in the female at puberty. These changes include the development of the internal and external genitalia and the appearance of the secondary sexual characteristics. The development of the breasts, the pubic and axillary hair, and the feminine body contours, with the broadening of the pelvis, are secondary sexual characteristics. The maturing female also undergoes a change of voice. The change is not as striking as in the male, but most women lose the high thin voice of little girls and assume a lower voice of better quality. Menstruation—the cyclic flushing of the degenerating lining of the uterus—begins with the attainment of puberty.

Estrogens are also found in high concentration in the follicular cells surrounding the developing ovum, and these hormones play a large role in the menstrual cycle, which is outlined below. Estrogenic substances are not produced exclusively by the ovary; they can be recovered from the urine of

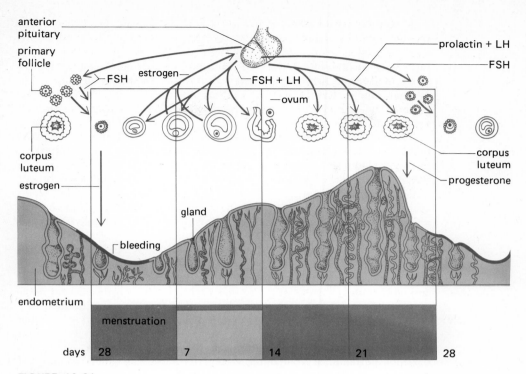

anterior
pituitary

primary
follicle

prolactin + LH

FSH

FSH

estrogen

FSH + LH

ovum

corpus
luteum

corpus
luteum

estrogen

progesterone

gland

bleeding

endometrium

menstruation

days   28   7   14   21   28

FIGURE 13·21

Role of hormones in the menstrual cycle.

both male and female. The urine of pregnant women contains more estrogen
than the urine of nonpregnant females, because the placenta probably
secretes estrogens during pregnancy. The urine of a stallion is also a good
source of estrogen, and estrogenic substances have also been isolated from
plant sources.

Sex hormones in menstrual cycle and pregnancy   The human menstrual cycle is
essentially a lunar-month, or 28-day, cycle in which menstruation lasts 4 or
5 days. After menstruation, there follows a period in which the lining of the
uterus is regenerated and a new ovum and its surrounding follicle mature.
This is sometimes called the *recovery*, or *preovulatory*, *period*. It lasts until
around the middle of the cycle. Ovulation, or the rupture of a mature ovum
from the ovary, usually occurs at approximately the thirteenth to fifteenth
day of the 28-day cycle. The cells of the ruptured follicle that formerly
surround the ovum change into the yellowish cells of the *corpus luteum* (or
yellow body) of the ovary, a structure that secretes progesterone. The lining
of the uterus is soft, thick, and ready for implantation by a fertilized ovum.
If the ovum is fertilized and implanted, the corpus luteum remains active
during the early weeks of pregnancy before it begins to degenerate. If a
fertilized ovum is not implanted during this time, changes occur in the
lining of the uterus that lead to menstruation. The latter part of the period
between ovulation and menstruation is called the *premenstrual period*. The
corpus luteum undergoes degeneration during the latter part of the premenstrual period, progesterone secretion slows, and the lining of the uterus

degenerates. Menstruation follows at the end of the cycle (Fig. 13.21) when the degenerate lining is flushed out, along with sloughed off blood vessels and a small amount of blood.

During the preovulatory period of the menstrual cycle, the estrogen level of the blood rises; this is correlated with the growth and maturation of the ovarian follicles. The lining of the uterus also develops rapidly during this period. The adenohypophyseal hormones FSH and LH stimulate the development of the follicles.

The hormone secreted by the corpus luteum is progesterone. Estrogen is required to build up the lining of the uterus after menstruation, but progesterone stimulates the maintenance of the lining, once it is developed. In animal experiments, progesterone seems to play no part in the development of the secondary sexual characteristics, but it does stimulate further development of the glandular tissue of the breasts after the tubules have been developed under the influence of estrogen and the adenohypophyseal hormones.

Progesterone plays an important part in pregnancy, since it helps maintain the secretory lining of the uterus after implantation. As we have seen, during the early part of pregnancy the corpus luteum remains functional and continues to secrete progesterone. Ovulation is suppressed, and menstruation does not occur, since the lining of the uterus is maintained in a state of proliferation and at a high degree of activity. Progesterone is used in birth control pills for this reason. Progesterone also appears to be essential for the development of the maternal portion of the placenta.

Progesterone is metabolized to an inactive substance called *pregnandiol* and excreted in this form in the urine. Pregnandiol can be converted readily to progesterone in the laboratory. It is of interest that the bull and the human male also excrete pregnandiol or a closely related substance. In males, progesterone is probably secreted by the testes or the adrenal cortices.

A hormone called *relaxin* is secreted by the corpus luteum, especially during pregnancy, and by the placenta. The corpus luteum of pregnancy probably contributes the greatest amount. The effects of relaxin vary in different species. It relaxes the symphysis pubis in some species but not in others. In the human female it softens and relaxes the cervix of the uterus before childbirth. Experiments have shown that tissues must first be exposed to estrogen before relaxin can be effective.

PLACENTAL HORMONES The placenta provides an exchange of food materials and oxygen between the blood of the mother and the blood of the fetus. The placenta also functions as a temporary endocrine gland, producing estrogens, progesterone, and chorionic gonadotrophin. The secretion of estrogens by the placenta causes a very considerable rise in the blood estrogen level during pregnancy. The production of estrogens during pregnancy is much greater than that seen during the normal monthly cycle. The high estrogen level inhibits the production of FSH; consequently, follicular development in the ovary is suppressed during pregnancy.

Progesterone also is secreted in much greater amounts by the placenta. It produces changes in the lining of the uterus favorable to the developing fetus. Estrogens and progesterone help develop the breasts for lactation.

Chorionic gonadotrophin prevents the decline of the corpus luteum during early pregnancy by stimulating it to enlarge and become more active. It also stimulates the corpus luteum to produce estrogens and progesterone in large quantities. During the third or fourth month of pregnancy, the secretion of chorionic gonadotrophin is suppressed and the corpus luteum begins to deteriorate. By this time the placenta has begun to secrete estrogen and progesterone in adequate amounts, so that the degeneration of the corpus luteum can proceed without loss of necessary hormones.

Chorionic gonadotrophin is obtained from pregnancy urine (P.U.) and from placentas. Unlike the hypophyseal gonadotrophins, human chorionic gonadotrophin does not stimulate the growth of ovarian follicles. When injected into immature mice, rats, or isolated rabbits, it causes marked changes in the ovary, where it has a predominantly luteinizing effect. Mature follicles may erupt, hemorrhage spots may occur in unruptured follicles, and atretic corpora lutea may be formed from the unruptured follicles. Changes in the ovary of immature or nonpregnant laboratory animals after injection with pregnancy urine are the basis for various pregnancy tests. These tests on laboratory animals give accurate indications even when there is a pregnancy of only 2 weeks and before there are any outward signs.

More rapid pregnancy tests are also performed with certain species of African and South American toads or native male frogs. Tests of this nature take little time (about 2 h), and each animal can be used again. Female toads extrude their eggs if there is a positive test for chorionic gonadotrophin; with male toads or frogs a positive test is determined upon finding seminal discharge in their urine.

There are now direct chemical tests that determine the presence of chorionic gonadotrophin. Immunological tests for antigen-antibody reaction are also used to detect pregnancy.

Human chorionic gonadotrophin is used clinically to stimulate the descent of the testes if they have not come down into the scrotum. The interstitial cells of the testes also are stimulated.

MALE SEX HORMONES   Several steroids that have masculinizing effects can be obtained from urine. These substances are called *androgens*. At first it was thought that these substances were male sex hormones, but they are now considered to be breakdown products of the hormone that retain some androgenic activity. A hormone substance called *testosterone* has been isolated from the testes. It possesses great androgenic activity and is thought to be the male sex hormone.

Testosterone is produced by the interstitial cells of the testes. The cells that secrete the hormone are located in the tissue between the sperm-producing follicles. The interstitial cells are stimulated by ICSH from the adenohypophysis. ICSH appears to be more effective when FSH is present also (Fig. 13.22).

The male sex hormone is responsible for the development of the secondary sexual characteristics of the male, such as the beard, the pubic and axillary hair, the male body contour, and the deep voice. The male sex hormone influences the development of the primary sex organs and internal

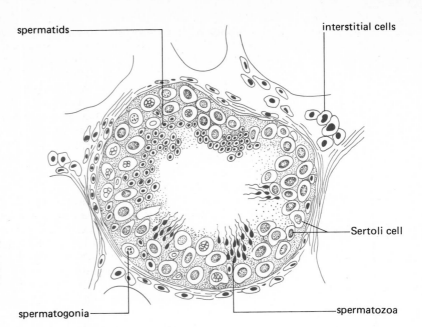

spermatids

interstitial cells

Sertoli cell

spermatogonia

spermatozoa

FIGURE 13·22

Cross section through a seminiferous (convoluted) tubule of the testis. Note the location of the interstitial cells, which produce the hormone testosterone.

sex organs such as the prostate gland and the seminal vesicles. Gonadotrophic hormones from the adenohypophysis are also essential to normal development.

Much of our knowledge of the action of the male sex hormone has been derived from the castration of animals. It has been known for centuries that castration of meat animals makes them more docile and easier to fatten. The meat is of better quality, and the muscles are not so tough. Castration of a young cockerel causes the bird to grow larger than a normal male because the closure of the zones of growth in the long bones is delayed. The castrate (or capon) then becomes a large and desirable bird. The loss of the sex hormone is especially evident in the appearance of the head. The large comb and wattles of the rooster fail to grow; instead the head tends to resemble that of the female, with a low pale comb.

Castration before puberty in human males prevents the expression of the secondary sexual characteristics. The voice does not change, facial hair growth is scanty, and the skin remains pale and soft. Eunuchs (or castrates) commonly grow tall, with long arms and legs, narrow shoulders, and a wide pelvis. Late castrates may not show much change in outward appearance. There is usually no change in the voice. The seminal vesicles and prostate gland gradually regress, however.

OTHER SOURCES OF ANDROGENS AND ESTROGENS   The adrenal cortex, as we have seen, produces both androgens and estrogens. We have noted that tumors of the cortex often produce masculinizing effects. The ovaries also produce androgens, and the testes, estrogens. Androgens and estrogens can be removed from the urine of either sex, but estrogens predominate in the female and androgens are more abundant in the male.

REJUVENATION It is almost an axiom that the introduction of a hormone will fail to stimulate the gland that normally secretes that hormone because of the negative feedback mechanisms that control secretion of most hormones. The introduced hormone very often has a depressant, rather than a stimulating, effect on the gland involved. It is unlikely that injections of androgen or testicular grafts or implantations will produce anything more than a transient beneficial effect on aging men. The power of suggestion is great, however, and claims which some have made of improved muscular strength and sexual vigor after treatment appear to be largely psychological rather than physical.

PROSTAGLANDINS

Prostaglandins are hormonelike substances found in various tissues of the body. These substances have a wide spectrum of physiological functions. They are largely regulatory, affecting the action of smooth muscle, blood pressure, glandular secretion, and the female reproductive system, especially contraction of the uterus. They are probably associated with cyclic AMP in their regulatory activities.

Prostaglandins were found originally in semen from the prostate gland, from which they derive their name, but they are now known to be secreted by many tissues.

Numerous regulatory actions of prostaglandins have been reported. Prostaglandins (1) stimulate the contraction of smooth muscle, especially in the uterus; (2) regulate renal blood vessels through the renin-angiotensin system; (3) lower blood pressure by inhibiting the reabsorption of sodium and by increasing the secretion of water in the renal tubules; (4) reduce inflammation by stimulating the response of white cells (leucocytes) in an injured area. Aspirin is said to inhibit prostaglandin synthesis.

OTHER HORMONES

We shall discuss kidney hormones and the renin-angiotensin system in Chap. 18. *Erythropoietin* is considered in Chap. 14.

Numerous hormones are associated with various digestive processes. The functions of the following hormones will be considered in Chap. 19: enterogastrone, gastrin, cholecystokinin, secretin, and enterocrinin.

SUMMARY

A summary of the functions and target organs of principal hormones described in this chapter is provided in Table 13.4.

1 Endocrine glands are glands of internal secretion. They are ductless glands, and their secretions are called hormones.
2 Some of the hormones, or factors, secreted by the hypothalamus act as release factors or release-inhibiting factors. These are: somatotrophin release factor; somatostatin, growth hormone release-inhibiting factor; luteinizing release factor; prolactin release-inhibiting factor; prolactin release factor; thyrotrophin release factor; corticotrophin release factor; melanocyte release-inhibiting factor and melanocyte hormone release factor.
3 The hypophysis is composed of two lobes, a posterior lobe (neurohy-

TABLE 13·4
Survey of principal hormones

Gland	Hormone	Functional effects
Thyroid	Thyroxine, T_4, and triiodothyronine, T_3	Accelerates metabolism, promotes normal growth
	Thyrocalcitonin	Regulates blood calcium level
Parathyroid	Parathyroid hormone (PH)	Raises blood calcium levels by stimulating calcium reabsorption from bone
Adrenal glands (see Table 13.3)		
Pancreas	Insulin	Lowers blood glucose level; part of the sugar is stored as glycogen in liver and muscles
	Glucagon	Causes a rise in blood glucose by accelerating the conversion of liver glycogen to glucose
Kidneys	Erythropoietin	Stimulates the production of erythroblasts
Gonads Female: ovaries	Estrogens	Development of female reproductive system and secondary sexual characteristics; development of ovum
	Progesterone	Produced by the corpus luteum, it is concerned with the maintenance of the uterine lining, preparing it for implantation. Also development and maintenance of the placenta; development of mammary alveoli
Male: testes	Testosterone	Development of male reproductive system and secondary sexual characteristics; development of sperm
Pineal	Melatonin	Function in humans unknown
Thymus	Thymopoietin	The immune system?
Hormones of the digestive tract (see Table 19.3)		

pophysis) and an anterior lobe (adenohypophysis). The adenohypophysis produces six hormones: growth hormone, two gonadotrophic hormones, adrenocorticotrophic hormone, thyrotrophic hormone, and prolactin. The gonadotrophic hormones are follicle-stimulating hormone (FSH) and luteinizing hormone (LH). ACTH stimulates the adrenal cortex; the thyrotrophic hormone regulates the secretion of the thyroid gland. The neurohypophysis stores two hormones, antidiuretic hormone and oxytocin. The antidiuretic hormone promotes water reabsorption by the kidneys. Oxytocin stimulates especially the smooth muscle of the pregnant uterus.

4 The thyroid gland secretes several iodine-containing hormones, of which thyroxine is the most active. The thyroid also secretes thyrocalcitonin.

5 The parathyroid glands secrete parathormone, a hormone concerned with regulating calcium metabolism.

6 The thymus and the pineal gland are now thought to be endocrine glands.

7 The pancreas, aside from secreting digestive enzymes, also produces the hormones insulin and glucagon. Insulin reduces blood glucose level, whereas glucagon increases blood glucose level. Insulin enables cell membranes to absorb glucose. If cells are unable to absorb and utilize glucose, a condition called diabetes develops. If the blood glucose level rises beyond a certain point, the kidneys begin to excrete glucose. A chemical test for sugar in the urine is a routine method for detecting diabetes.

8 The adrenal glands are composed of a medulla and a cortex. The medulla secretes epinephrine and norepinephrine. These hormones exert a pressor effect on the heart, blood pressure, and smooth muscle. Norepinephrine is essentially a peripheral vasoconstrictor.

9 The adrenal cortex secretes many hormones, which are essential for life. The more important active corticosteroids are as follows: the glucocorticoids cortisone and cortisol, and the mineralocorticoid aldosterone. Glucocorticoids regulate carbohydrate metabolism; mineralocorticoids affect salt concentration and water balance.

10 The adrenal cortex also secretes sex hormones. Hyperactivity of the adrenal cortex may cause premature sexual development in children.

11 The gonads produce sex hormones. Female sex hormones are estrogens and progesterone. The placenta also secretes these hormones as well as chorionic gonadotrophin. Estrogens stimulate the development of the secondary sexual characteristics of the female and prepare the ovum for ovulation. Progesterone, which is secreted by the corpus luteum, maintains the lining of the uterus until after ovulation during the regular menstrual cycle. During pregnancy it maintains the lining of the uterus after implantation. Ovulation and menstruation are suppressed during pregnancy by special hormonal conditions.

12 Chorionic gonadotrophin prevents the decline of the corpus luteum during the first 3 or 4 months of pregnancy. Chorionic gonadotrophin excreted in the urine of the pregnant female forms the basis for various pregnancy tests.

13 The male sex hormones are androgens, which are secreted by the interstitial cells of the testes. Androgens may also be produced by the adrenal cortex, which is also a source of estrogens in the male.

1 Name and locate the endocrine glands.
2 Describe a set of experiments that will prove that a certain gland produces a specific hormone.
3 What is meant by endocrine balance?
4 Why is the incidence of goiter higher in some geographical regions than in others?
5 Discuss various conditions that can arise from abnormalities of the thyroid gland.
6 Explain the action of the parathyroid gland in the regulation of calcium and phosphorous metabolism.
7 Describe the differences in structure and function of the adrenal medulla and of the adrenal cortex.
8 Discuss the practical uses of adrenal hormones.
9 Explain the function of the hypophysis as the master gland of all the endocrine glands.
10 Distinguish between the origin and function of the two major lobes of the hypophysis.
11 List the hormones of the hypophysis and explain their functions.
12 Describe the changes that occur as a result of male castration. Which hormones are concerned?
13 Discuss the function of the female sex hormone in relation to the menstrual cycle.
14 Devise a test for determining an early pregnancy.
15 Is it true that androgenic substances can be recovered from the urine of either sex? Where might these substances arise in the female?
16 Discuss the function of the pancreas as a gland of internal secretion.
17 Explain the role of insulin in carbohydrate metabolism.
18 How does glucagon function?
19 What are prostaglandins?
20 Does the placenta produce any hormones? What are their functions?
21 Describe the role the hypothalamus plays in regulating hormone level. Select a hormone secreted by the female gonads and trace control of its secretion to the hypothalamus and the hypophysis.

SUGGESTED READING

Axelrod, J.: Comparative Biochemistry of the Pineal Gland, *Am. Zool.*, **10:**259–267 (1970).

Cannon, W. B.: "Bodily Changes in Pain, Hunger, Fear and Rage," Appleton-Century-Crofts, Inc., New York, 1929.

Copp, D. H.: Endocrine Regulation of Calcium Metabolism, *Annu. Rev. Physiol.*, **32:**61–86 (1970).

Frieden, E., and H. Lipner: "Biochemical Endocrinology of the Vertebrates," Prentice-Hall, Inc., Englewood Cliffs, N.J., 1971.

Gillie, R. B.: Endemic Goiter, *Sci. Am.*, **224:**92–101 (1971).

Green, R.: "Human Hormones," McGraw-Hill Book Company, New York, 1970.

Guillemin, R., and R. Burgus: The Hormones of the Hypothalamus, *Sci. Am.*, **227:**24–33 (1973).

Hollander, C. S., et al.: Thyrotropin-releasing Hormone: Evidence for Thyroid Response to Intravenous Injection in Man, *Science*, **175:**209–210 (1972).

Kolata, G. B.: Hormone Receptors: How Are They Regulated?, *Science,* **196:**747–748 (1977).

Mitnick, M., and S. Reichlin: Thyrotropin-releasing Hormone: Biosynthesis by Rat Hypothalamic Fragments in Vitro, *Science,* **172:**1241–1243 (1971).

Mulrow, P.: The Adrenal Cortex, *Annu. Rev. Physiol.,* **34:**409–424 (1972).

Netter, F.: CIBA Collection of Medical Illustrations, vol. 4, "Endocrine System and Metabolic Diseases," CIBA Corp., Newark, N.J., 1965.

O'Malley, Bert W., and W. T. Schrader: The Receptors of Steroid Hormones, *Sci. Am.,* **234:**32–43 (1976).

Orci, L., K. H. Gabbay, and W. J. Malaisse: Pancreatic Beta-Cell Web: Its Possible Role in Insulin Secretion, *Science,* **175:**1128–1130 (1972).

Pastan, Ira: Cyclic AMP, *Sci. Am.,* **227:**97–105 (1972).

Pike, J. E.: Prostaglandins, *Sci. Am.,* **225:**84–92 (1971).

Schally, A. V., et al.: Gonad-releasing Hormone: One Polypeptide Regulates Secretion of Luteinizing and Follicle-stimulating Hormones, *Science,* **173:**1036–1038 (1971).

Yates, F. E., S. M. Russell, and J. W. Maran: Brain-Adenohypophyseal Communication in Mammals, *Annu. Rev. Physiol.,* **33:**393–444 (1971).

THE CIRCULATORY SYSTEM

A study of the blood is considered in Chapter 14. This discussion is followed by a description of the heart and general circulation in Chapter 15. Chapter 16 is concerned with the physiology of circulation.

14

THE BLOOD

In an organism as large as the human one, transporting oxygen, food, and chemical elements to tissues that are relatively far away from their source of supply and relieving distant cells of their waste products pose a problem. It would be impossible to supply their needs by diffusion or any carrier mechanism that depends upon proximity. The blood and circulatory system have developed, therefore, as a transport mechanism. In this chapter, we take up the following topics:

1 The development of the blood and its composition
2 The formed elements of the blood: red and white blood cells
3 The characteristics of the erythrocytes, or red blood cells
4 The characteristics of leukocytes, or white cells
5 The functions of the granular leukocytes: neutrophils, eosinophils, and basophils
6 The clotting mechanism of the blood
7 Blood typing
8 The immune system: cellular immunity (T cells) and humoral immunity (B cells)
9 The function of the spleen as an accessory organ of the circulatory system
10 Clinical aspects: some diseases of the blood

The *blood* transports oxygen and nutrient fluids through the circulatory system to cells that are often at a considerable distance from the heart, lungs, or digestive tract. This arrangement also helps tissues to rid themselves of waste materials even though the tissues are located in the feet or hands, relatively far away from the kidneys. The blood is commonly considered to be a liquid tissue; that is, in blood the intercellular structure is liquid rather than fibrous or more or less solid. The *circulatory system* comprises a series of tubes (arteries, veins, capillaries) through which blood is carried to the tissues. The heart acts as a pump to supply the motive force to the blood traveling through the circulatory system.

blood island

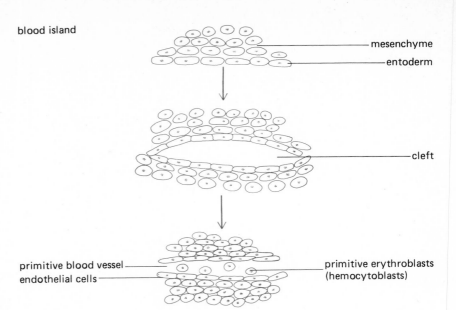

mesenchyme

entoderm

cleft

primitive blood vessel

endothelial cells

primitive erythroblasts
(hemocytoblasts)

FIGURE 14·1
Development of blood islands.

DEVELOPMENT OF BLOOD

The embryo develops its own blood and is not dependent upon the mother for the formation of blood cells. The first indication of an embryonic blood supply is the formation of clusters of blood cells in mesodermal blood islands of the yolk sac (Fig. 14.1). Centers for blood formation also produce endothelial tubes similar to those that line blood vessels. There are various sites of blood formation; the early yolk-sac center is soon replaced by centers in the body mesenchyme and later by those in the liver and spleen. When bone is formed, the red marrow of bone becomes the principal blood-forming center.

The mesenchyme is a tissue derived from mesoderm; it has great potential for producing other tissues such as connective tissues, bone, muscle, reticuloendothelial tissue, and blood. Wandering mesenchyme cells form the blood islands and produce various kinds of blood cells. The first recognizable blood cell from which other types of blood cells are thought to originate is called a *hemocytoblast*. The first corpuscles to appear in the embryonic blood are not identical with mature *erythrocytes*, or red blood cells. Hemocytoblasts retain their nucleus and more closely resemble the immature red cell, or *erythroblast*. The hemocytoblast is thought to be the parent cell from which the red cells and various kinds of white cells arise. Very large cells called *megakaryocytes* may arise from undifferentiated mesenchyme cells. These cells give rise to the blood platelets through fragmentation of their cytoplasm. The hormone *erythropoietin*, which has now been isolated in pure form, stimulates red-cell formation.

FUNCTIONS OF BLOOD

Most of us know that the blood carries oxygen, carbon dioxide, nutritive elements, and waste materials, but we are commonly unaware that the blood also has many other functions. A brief list of some of the blood's evident functions is given here.

1 Oxygen is transported from the lungs to the tissues by the blood.
2 Carbon dioxide, a product of cellular metabolism, is transported from the tissues to the lungs via the blood.
3 Nutrient materials are absorbed from the intestine and carried to the tissues by the blood.
4 Many organic breakdown products of metabolism (urea, uric acid, creatinine, purine wastes) are carried by the blood to the kidneys for excretion.
5 Hormones are distributed throughout the body by the blood.
6 The blood flows from the deeper, warmer parts of the body to the extremities, thus distributing heat more evenly to all parts of the body. When surface blood vessels in the skin dilate, more blood comes to the body's surface, thus promoting heat loss; when surface vessels constrict, blood is kept away from the surface, thus reducing heat loss. The blood, therefore, plays an important role in the regulation of body temperature.
7 The blood plays an important part in maintaining the acid-base balance of the tissues. Most tissues, including the blood, are slightly alkaline. The pH of arterial blood is between 7.35 and 7.45.
8 The blood volume and the fluid content of the tissues are kept in a hemostatic balance through continuous exchanges of fluid and solutes. The capillary wall acts as a selectively permeable membrane, permitting a constant filtration into the tissues of water, molecules, and other solutes. Small molecules, such as those of oxygen, glucose, or amino acids, pass through the capillary wall readily, but larger protein molecules pass through very slowly, if at all. Filtration, the movement of water and dissolved substances out of the bloodstream, is aided by capillary blood pressure.

The blood also has a number of proteins in colloidal state that tend to attract fluid from the tissues into the bloodstream and hold it there. Food proteins are broken down in digestion to amino acids and absorbed in this form. Amino acids enter into the formation of plasma protein, including albumin and globulin. Plasma protein plays an important part in building up osmotic pressure of the blood. Abnormal conditions such as increased permeability, increased capillary pressure, or decreased plasma-protein content of the blood may permit excessive filtration of fluid into the tissues. The tissues swell and literally become waterlogged, a condition known as *edema*. After severe loss of blood, water moves from the tissues into the bloodstream, and the volume of the blood may be quickly restored in this manner. The blood in this case is able to draw on a water reserve normally held in the tissues.
9 The clotting ability of the blood, which reduces bleeding, has been of obvious survival value. The mechanism of clot formation will be discussed later in the chapter.
10 The blood plays an important part in protecting the body from bacteria and other organisms that can cause disease or other abnormal conditions. Some kinds of white blood cells protect the body by ingesting bacteria or other foreign matter appearing in the bloodstream.

BLOOD VOLUME

The amount of blood in the body has been measured in various ways. The volume of blood may be expected to vary with the size of the individual. Blood volume has been estimated at $\frac{1}{20}$ to $\frac{1}{13}$ of the body weight. The blood volume of a man of average size is about 5 l.

The liquid portion of circulating blood is called the *plasma*. It is a straw-colored fluid, very complex chemically, that contains a wide variety of substances (Table 14.1). The red blood cells, white blood cells, and platelets float in this liquid medium. In this respect the blood may be regarded as a liquid tissue; it contains cells, but its intracellular substance is a changing liquid rather than some more substantial building material.

Most functions of the blood previously mentioned directly involve the plasma. Even though the blood is continuously engaged in transporting absorbed food products and receiving the waste products of cell metabolism, its chemical content is fairly constant. Plasma is about 92 percent water; the remaining 8 percent of materials in solution make blood thicker than water, as the saying goes. Its specific gravity is greater than 1, more nearly 1.025 as an average. Its chemical balance is largely self-adjusting, in much the same way as water balance is maintained between the blood and the tissues. As some chemical elements in the plasma are taken up by cells for their metabolism, more of these same elements are absorbed by the plasma from food sources. The chemical balance is thus maintained. Waste products of metabolism are steadily absorbed by the plasma and are just as regularly removed by excretory organs. The buffering action of the blood helps to preserve its chemical balance.

SALT AND WATER BALANCE   The plasma and tissue fluids are the *internal environment* of the body. The ions in plasma are mostly chloride, bicarbonate, phosphate, sulfate, sodium, calcium, potassium, and magnesium. In order to maintain a proper acid-base equilibrium, these salts must be present and maintained in proportion to one another. Physiologists have known since the experiments of Sydney Ringer around 1885 that a salt balance is maintained in the blood. Physiological (normal) salt solutions are used to maintain the internal environment of experimental animals during demonstrations, surgical procedures, and various other laboratory procedures.

TABLE 14·1
Constituents of blood

Plasma

Water, 92 percent
Solids, 8 percent
Inorganic chemicals: sodium, calcium, potassium, magnesium, chloride, bicarbonate, phosphate, sulfate
Organic chemicals
  Proteins: serum albumin, serum globulin, fibrinogen
  Nonprotein nitrogenous substances: urea, uric acid, creatine, creatinine, ammonium salts, amino acids
  Nonnitrogenous substances: glucose, fats, cholesterol
Hormones
Gases: oxygen, carbon dioxide, nitrogen

Cells

Erythrocytes
Leukocytes
(Platelets)

Through their buffering action, the salts of the plasma help maintain an acid-base balance between the blood and the tissues; these salts are also involved in maintaining blood-tissue water balance.

PLASMA PROTEINS Organic substances in plasma include such proteins as serum albumin, serum globulin, and fibrinogen. These proteins are not food proteins; they are not directly absorbed from food sources, and they are not being transported to the tissues for their use. Fibrinogen is formed in the liver, and most other plasma proteins are thought to be formed there also. The plasma proteins, present in a colloidal-sol state, exert considerable osmotic pressure, which is a factor in the maintenance of the water balance between the blood and the tissues and in regulating the blood volume. The proteins give viscosity, or thickness, to the plasma, a factor in the maintenance and regulation of blood pressure. Serum globulin is involved in antibody formation—the reaction of the blood to toxins formed by bacteria or to foreign proteins introduced into the blood. Fibrinogen is essential to the clotting mechanism.

OTHER PLASMA ELEMENTS Nonprotein nitrogenous substances found in the blood include urea, uric acid, creatine, creatinine, and ammonium salts. These substances represent breakdown products of protein metabolism and are carried by the blood to the organs of excretion.

Protein foods are reduced to amino acids during the process of digestion and are absorbed as such. Amino acids, the building blocks for all proteins found in the body, are therefore present in the blood plasma.

Glucose, fats, and cholesterol are nonnitrogenous substances found in plasma. A considerable amount of glucose is absorbed from the blood and stored as glycogen in liver and muscle tissue. Though the role of glucose in nutrition is well recognized, it also acts as a physiological constant in the plasma. The glucose level of the blood is fairly constant at an average concentration of about 0.1 mg percent (80 to 120 mg/100 ml). A reduction in the blood glucose level may cause weakness, fainting, or more serious consequences. The kidneys excrete glucose if the glucose level becomes too high.

Fats are also carried by the plasma, as well as several fatlike substances, such as cholesterol and the phospholipids. The blood plasma contains hormones, the secretions of ductless glands. It also contains the chemical substances concerned with the clotting of the blood.

Only a small amount of carbon dioxide is carried in solution in plasma, even though it is continuously produced as a waste product of metabolism and is constantly absorbed by the blood. Carbon dioxide forms carbonic acid; in this form it is buffered by hemoglobin and salts, such as sodium phosphate, which remove carbonic acid from the plasma by combining with it. Hemoglobin is one of the chief carriers of both oxygen and carbon dioxide. Nitrogen is carried in the plasma as an inert gas.

THE FORMED ELEMENTS

ERYTHROCYTES (RED BLOOD CELLS) Carried along in the plasma are red-colored cells so numerous that the plasma itself appears red. These cells are called *erythrocytes,* or red blood cells. As they arise in the red marrow of the bone, they are nucleated and are called *erythroblasts,* but shortly before

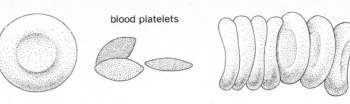

blood platelets

erythrocyte   rouleau formation

FIGURE 14·2

Red blood cells and blood platelets.

entering the bloodstream, they usually lose their nuclei and become highly specialized cells, or *red corpuscles*. Erythrocyte size is fairly constant, with a diameter of about 7.5 to 8.6 μm and a thickness of about 2 μm. The erythrocyte is a biconcave disk, thinner in the middle than at the edge (Fig. 14.2). When observed singly under a microscope by transmitted light, erythrocytes are only faintly colored and, seen individually, appear slightly yellowish in color.

A hormone called *erythropoietin* controls the production of erythrocytes. It is a glycoprotein, generally thought to be produced by the kidneys, although the liver and perhaps other tissues may produce small amounts. It stimulates the production of erythroblasts in the early stages (stem cells) as they develop in red bone marrow, a process called *erythropoiesis*.

The number of erythrocytes per cubic millimeter of blood is usually stated as an average of 5 million erythrocytes/mm^3 for men and 4.5 million/mm^3 for women. It is well known, however, that active young men frequently have more than 5 million erythrocytes/mm^3. A more accurate estimate, therefore, would be 5.45 million cells/mm^3 for men and 4.75 million cells/mm^3 for women. The red blood cell count is higher in a newborn infant than it is in older children or adults. Ascending to high altitudes increases the number of red cells in the bloodstream. Other factors causing an increase in the number of red cells are concerned with muscular exercise and with a rise in the environmental temperature. Under these conditions additional red cells can be discharged from the spleen and other reservoirs.

BLOOD-CELL COUNTING

THE HEMOCYTOMETER   It is possible to make a fairly accurate estimate of the number of erythrocytes per cubic millimeter by actually counting a limited number of cells as they are spread out on a ruled microscope slide called a *hemocytometer*.

THE HEMATOCRIT   The *hematocrit* is the percentage of erythrocytes per 100 milliliters of whole blood by volume after a blood sample has been centrifuged. The reading is taken in a special graduated tube. The average hematocrit for a man is 47.0 ($\pm$7.0) percent, and for women 42.0 ($\pm$5.0) percent. If a man has a hematocrit of 47 percent, it means he has 47 ml erythrocytes for every 53 ml plasma in 100 ml whole blood. If you are interested in this technique, you may find detailed instructions in laboratory manuals and in clinical textbooks.

DIFFERENTIAL COUNT   A differential count of blood cells indicates the proportion of each kind of white cell present. Ordinarily, the number of each kind in 100 cells is counted. After counting 100 or 200 cells, the percentage of each type of white cell is determined. The percentage is, of course, variable.

Following is a normal differential count as it might appear:

Neutrophils	68%
Eosinophils	3%
Basophils	1%
Lymphocytes	24%
Monocytes	4%
Total	100%

Hemoglobin Erythrocytes derive their color from a complex protein called *hemoglobin.* This substance is composed of the pigment *heme,* which contains iron, and the protein *globin.* Hemoglobin attracts oxygen molecules and holds them in a loose chemical combination known as *oxyhemoglobin.* Hemoglobin is said, therefore, to have a chemical affinity for oxygen. The structure of the hemoglobin molecule has been successfully analyzed by x-ray diffraction and chemical methods. It consists of four folded polypeptide chains. The four chains form the globin, or protein, part of the molecule. In addition there are four atoms of iron, each associated with a heme group. The heme group gives the cells their red color and their oxygen-combining ability. The iron atoms of hemoglobin are bivalent, or in the ferrous state. It has been stated that one erythrocyte contains approximately 280 million molecules of hemoglobin (Perutz).

As the blood passes through a capillary network in the thin air sacs of the lungs, oxygen enters into a loose chemical combination with hemoglobin to form oxyhemoglobin, and in this form, it is carried to the tissues. There, as the blood passes through tissue capillaries, the oxyhemoglobin loses oxygen to the tissues. Hemoglobin in this state is referred to as *deoxyhemoglobin.* Arterial blood, after passing through the lungs, is a somewhat brighter red than that found in the veins, but venous blood is never blue. Erythrocytes not only carry oxygen to the tissues, but they also help indirectly to carry carbon dioxide away from the tissues (Fig. 14.3).

The number of erythrocytes in the circulation remains fairly constant. The average life of a red blood cell is about 120 days, but such cells are constantly being replaced. The maintenance of a constant number is yet another example of homeostasis. If the tissues fail to obtain enough oxygen (hypoxia), erythropoietin is produced and carried to the red bone marrow by the blood. The red bone marrow is then stimulated to produce more erythrocytes, and thus more carriers of hemoglobin. This is another example of a negative feedback system. In this one, the supply of oxygen to tissues is increased as oxygen is utilized and becomes deficient. Other factors, such as some hormones and vitamin B_{12}, may influence the production of erythrocytes.

LEUKOCYTES (WHITE BLOOD CELLS) Leukocytes, or white blood cells, are nucleated and somewhat variable in size and shape (Table 14.2). They are far less numerous than erythrocytes, numbering 5000 to 9000/mm³. If 7000 white blood cells per cubic millimeter is taken as an average, red blood cells outnumber white blood cells by about 700:1. Some kinds of leukocytes, especially neutrophils and monocytes, exhibit ameboid movement and are

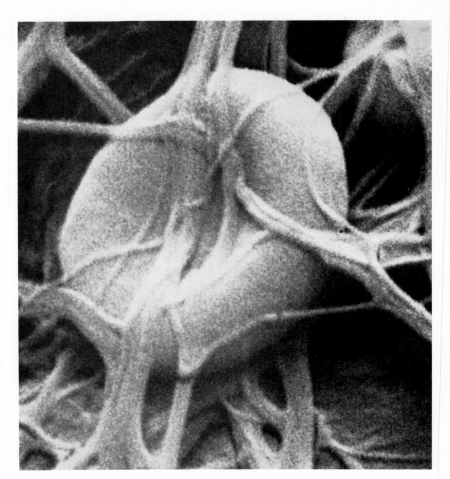

FIGURE 14·3
Scanning electron micrograph of an erythrocyte enmeshed in fibrin. Original magnification, 10,000. [*Courtesy Emil Bernstein and Eila Kairinen, Gillette Company Research Institute, Rockville, Md. Cover photo, Science, 173 (Aug. 27, 1971). Copyright, 1971, by the American Association for the Advancement of Science.*]

actively *phagocytic* (cell eating). Their life span is probably only a few days.

There are two major groups of white cells. The first group includes those cells that have granules in the cytoplasm and possess a nucleus of two or three lobes. They are called *granular,* or *polymorphonuclear, leukocytes.* According to some workers, this is the only group properly referred to as leukocytes. Granular leukocytes arise from bone marrow cells called myeloblasts.

The second group includes those cells that do not have granules in the cytoplasm and in which the nucleus is more or less spherical in shape. These are the *nongranular white cells;* this grouping includes *lymphocytes* and *monocytes.*

Granular leukocytes NEUTROPHILS The most abundant type of white cell is the neutrophil. These cells constitute 65 to 70 percent of the total number of white cells. Averaging about 10 μm in diameter, they are somewhat larger than erythrocytes. The nucleus appears to vary with age. The younger mature cells have a bilobed or trilobed nucleus with thick connections

TABLE 14·2

Classification of blood cells

Name	Normal no.	Diameter, μm	Description
Erythrocytes	4.75–5.5 million/mm^3	7.5–8.6	Biconcave disk, non-nucleated
Leukocytes	5000–9000/mm^3		
Granular			
Neutrophils	Approx. 65% total leukocytes	10–12	Fine granules in cytoplasm usually stain to lavender; two- or three-lobed nucleus
Eosinophils	2–4% total leukocytes	12	Large cytoplasmic granules usually stain bright eosin red; nucleus usually bi-lobed
Basophils	0.2–0.5% total leukocytes	10	Large cytoplasmic granules usually stain deep, dull blue or purple; thick bi-lobed nucleus
Nongranular			
Lymphocytes	20–25% total leukocytes		
Large		12	A thin rim of nongranular cytoplasm, usually stains light blue; large globular nucleus stains dark blue or purple
Small		9	
Monocytes	5–7% total leukocytes	15–20	Resemble large lymphocytes, but with relatively more cytoplasm staining light blue; nucleus large and deeply indented

between the lobes, whereas older cells have more lobes in the nucleus with thin connections between the lobes. The granules of the cytoplasm are very fine, and when stained with Wright's or some similar stain, they take both acid and basic stains and appear as a lavender or lilac color (Fig. 14.4).

Neutrophils are active ameboid cells and are also phagocytes. They can push their way between cells in the wall of the smallest blood vessels and move about through the tissues. *Phagocytosis* refers to the ability of such cells to ingest bacteria or other foreign bodies. They contain protein-digesting enzymes that digest most of the material they engulf. Since they are capable of ingesting other cells, they are called *phagocytes*. The phagocytic activity of neutrophils is important in helping to rid the body of injurious bacteria. The effectiveness of antibodies depends upon their ability to slow or immobilize infectious particles, thereby making phagocytosis more effective.

The movement of leukocytes through the capillary wall is called *dia-*

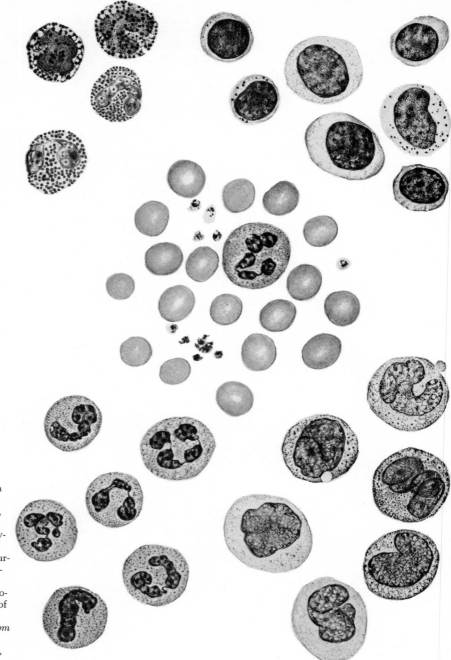

FIGURE 14·4

Cells from smear preparation of normal human blood, Wright's stain. In the center, adult red cells and a few blood platelets around a polymorphonuclear neutrophil. Upper left, two basophils (purple granules) and two eosinophils (red granules). Upper right, large and small lymphocytes. Lower right, a group of six monocytes. Lower left, a group of six neutrophils. (*From Roy O. Greep, "Histology," McGraw-Hill Book Company, New York, 1973.*)

pedesis. When passing through a capillary wall, a leukocyte pushes out a pseudopodium as in ameboid movement. It then squeezes its way between cells in the direction of injured tissue or an area of inflammation. The process by which the neutrophil is attracted to the site of the injury is called *chemotaxis*. Injured tissue is known to release chemotaxic substances such as *leucotaxine*, which attract neutrophils. The neutrophil works its way toward the injury with diapedesic movement. At the site, it engulfs bacteria and debris by phagocytic action (Fig. 14.5).

EOSINOPHILS   Although they resemble neutrophils, eosinophils are slightly larger and the eosinophil nucleus is usually bilobed. The granules in the cytoplasm are larger and stain a bright red with acid dyes such as eosin. These cells make up only 2 to 4 percent of the total number of white cells in the blood, but in the tissues they can congregate in considerable numbers. According to some workers, eosinophils are considered to be less active and not so highly phagocytic as neutrophils.

BASOPHILS   Basophils have a bilobed nucleus, and their large cytoplasmic granules stain a deep blue with basic stains such as methylene blue or Wright's stain, which contains methylene blue. Basophils constitute only 0.5 percent of the white cells of the blood. Not much is known about the motility or phagocytic activity of basophils, but they are considered to be less active than eosinophils. Basophils are said to contain histamine and a heparinlike substance. Histamine dilates capillaries and often permits a fluid to move through the capillary wall into the tissues; heparin is an anticoagulant. Tissue basophils apparently become the mast cells of the tissues. The large granules of mast cells are thought to store enzymes.

Nongranular leukocytes   LYMPHOCYTES   Lymphocytes are considered to be of two distinct sizes: small and large. *Small lymphocytes* are the smallest of the white cells and constitute 20 to 25 percent of all leukocytes. The small

FIGURE 14·5

Neutrophils moving toward an area of injured tissue by chemotaxis.

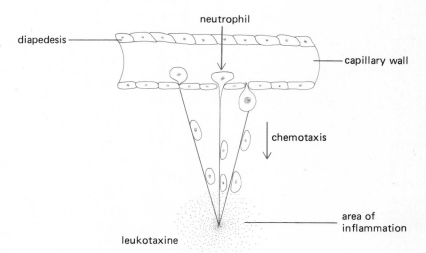

lymphocyte nucleus is comparatively large and spherical and is surrounded with a thin covering of cytoplasm. The cytoplasm is nongranular and stains a light blue, while the nucleus is dyed a much deeper blue or purple with Wright's stain. *Large lymphocytes* represent only about 3 percent of the total of white cells. A large lymphocyte contains more cytoplasm, and its nucleus is large, oval, and indented. Lymphocytes are present in bone marrow and in lymphoid tissues such as thymus, spleen, and lymph nodes.

Lymphocytes have their origin as stem cells, or lymphoblasts, in bone marrow. The lymphoblast has a large spherical nucleus and a rather small amount of cytoplasm.

When lymphoblasts leave the bone marrow, they follow two pathways. Some migrate to the thymus; these cells and their progeny are known as *T cells*. A second group moves to other lymphoid tissues, where they are called *B cells*.

In the thymus the T cells become capable of producing antibodies, and such cells become part of the immune system. The site at which the B cells are sensitized to produce antibodies is not known.

Until recently there was no way to recognize any physical difference between these two cell types, but evidence collected with the scanning electron microscope has shown that B cells have some 150 projections protruding from their surface whereas T cells have an almost smooth surface with very few projections. The projections indicate a difference in function.

Using blood samples, it has been discovered that about 80 percent of lymphocytes are T cells and the remaining 20 percent are B cells. In some illnesses, however, such as lymphocytic leukemia, B cells may reproduce rapidly and become the more numerous type.

It would be unwise to separate the functions of B and T cells too strictly. In many ways they act together to protect the body from disease and foreign antigens. We shall return to problems of immunity later in the chapter.

Plasma cells are derived from large lymphocytes, probably from B cells that have been stimulated by antigens. The cytoplasm of plasma cells is filled with an endoplasmic reticulum covered with ribosomes. Plasma cells produce immunoglobulins that are antibodies. These cells are usually found in lymphoid tissues rather than in the circulating blood and lymph (Fig. 14.6).

MONOCYTES   Research seems to indicate that monocytes develop from lymphocytes. Closely resembling large lymphocytes in appearance, monocytes are about 15 μm in diameter, with a large, deeply indented nucleus. The nucleus stains deep blue or purple with Wright's stain. Monocytes have relatively more cytoplasm than do large lymphocytes. There are compara-

FIGURE 14·6
a A megakaryocyte, diagrammatic; *b* a plasma cell, diagrammatic.

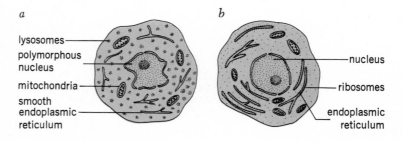

a

lysosomes
polymorphous nucleus
mitochondria
smooth endoplasmic reticulum

b

nucleus
ribosomes
endoplasmic reticulum

tively few monocytes in the blood—about 5 percent of the total white cell count.

Monocytes may develop into *macrophages*. Monocytes, which are present in the blood, and macrophages, which are found in tissues, are somewhat different in function, but they appear to be morphologically similar. Monocytes move into connective tissues from the blood, whereas macrophages are wandering cells. Attracted to inflammation and injured tissues by chemotaxis, macrophages are very active phagocytes, ingesting old blood cells, bacteria, and foreign particles. Macrophages are scavengers, and their cell bodies have abundant lysosomes that enable them to digest ingested organic matter. Monocytes have rather weak digestive powers. Both monocytes and macrophages, however, can engulf old, worn-out neutrophils, mast cells, antigens, and particles of tissue as they clean up an area of inflammation or infection after the initial stages have been passed and recovery is in progress.

Leukocytes in inflammation and disease   The white blood cells found in the bloodstream are usually in transit from their place of origin to their destination in the tissues. Unlike the red blood cells, which must remain within a closed circulatory system, the white blood cells, as we have seen, are able to pass through capillary walls and move through tissues.

The protective action of white cells can be illustrated by following the sequence of the events following the lodging of a splinter under the skin. Some initial injury to surrounding tissues occurs, and a minor infection is caused by bacteria introduced on the splinter. Toxic products released by the bacteria tend to destroy tissues locally, and blood vessels in the immediate vicinity are affected. They tend to dilate, thus bringing more blood to the affected area. Therefore the area reddens somewhat and feels warm, since the temperature of the blood is warmer than that of the normal skin surface. White blood cells congregate in great numbers in the infected area. The white cells are actively phagocytic and rapidly ingest bacteria, as well as particles of tissue cells. Some of the phagocytes are destroyed by toxic substances liberated by the bacteria. The breakdown of phagocytic cells liberates their digestive enzymes.

The festering that often occurs around the base of a splinter is called *suppuration*, or the formation of pus. The exudate from blood vessels, dead tissue cells, bacteria, and living and dead leukocytes forms a more or less liquid material called *pus*. In the normal course of events the splinter becomes loosened and can be easily withdrawn. The white cells clean up the infected area, and regeneration of the skin and tissue cells restores the area to normal.

PLATELETS (THROMBOCYTES)   Minute, granular, disk-shaped objects in the blood are called *platelets* (thrombocytes). They arise from fragmentation of giant cells in the red marrow of bone. These giant cells are derived from cells of the reticuloendothelial system and are called *megakaryocytes* (Fig. 14.6). Even though they arise by fragmentation of a large cell, platelets are not irregular in shape; they are normally round disks. Platelets are much smaller than erythrocytes, have no nucleus, and average around 25,000 to 400,000/mm^3. Blood platelets contain a number of chemical substances; probably the most important are the phospholipid cephalin, a prothrombin activator similar to tissue thromboplastin, and serotonin, a vasoconstrictor substance. Blood platelets tend to stick together if they are near a vascular injury and so are able to plug small breaks in capillaries. Platelets play an important part in blood clotting. Their life span is short, probably 3 to 10 days.

The body would be unable to maintain hemostasis effectively if it were not provided with physiological mechanisms to prevent loss of blood in case of injury to a blood vessel. Two mechanisms protect the body against the consequences of vascular injury: vascular spasm and blood clotting.

MAINTAINING HEMOSTASIS

VASCULAR SPASM AND THE PLATELET PLUG   Injury to a blood vessel first causes *vascular spasm*. The injured blood vessel contracts, bringing its opposing surfaces closer together. Platelet aggregation occurs almost immediately over crushed or cut surfaces. Platelets tend to stick to exposed tissues such as the collagenous layers lying under the endothelial lining of blood vessels. When these layers are exposed by an injury, platelets stick to them. As the stuck platelets disintegrate, they release ADP, and this attracts other platelets to the injured area. Vasoconstriction is further enhanced by the release of serotonin and epinephrine from the growing aggregate of platelets. Finally a soft *platelet plug* develops at the site of injury. This plug helps control bleeding. Vascular spasm and the formation of a platelet plug mark the first stage in maintaining homeostasis (Fig. 14.7).

THE CLOTTING MECHANISM   After the first stage, the clotting mechanism, an elaborate process that protects against bleeding in injury, starts. Twelve or more chemical substances are involved in the stepwise series of reactions that comprise the clotting process. An international committee has designated names of the clotting factors involved; these are shown in Table 14.3. Most clotting factors are synthesized in the liver, but calcium is supplied from the store of plasma calcium ions and platelets in the bone marrow.

FIGURE 14·7

Hemostatic mechanisms in the formation of platelet plug.

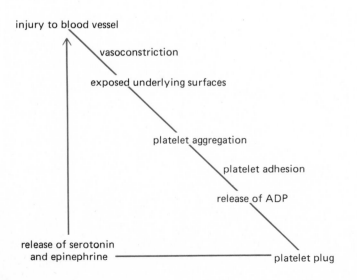

TABLE 14·3

Factors involved in blood clotting

Factor	Chemical substances
Factor I	Fibrinogen, a globulin
Factor II	Prothrombin, an albumin
Factor III	Thromboplastin, a phospholipid from platelets and from tissues
Factor IV	Calcium
Factor V	Labile factor (accelerator globulin, ACG)
Factor VI	Now considered to be identical with factor V
Factor VII	Serum prothrombin conversion accelerator (SPCA)
Factor VIII	Antihemophilic factor (AHF), antihemophilic globulin (AHG), platelet cofactor I
Factor IX	Plasma thromboplastin component (PTC), platelet cofactor II
Factor X	Stuart-Prower factor, a globulin, enhances thromboplastin
Factor XI	Plasma thromboplastin antecedent (PTA)
Factor XII	Hageman factor, a contact factor, starts reaction
Factor XIII	Fibrin-stabilizing factor
Platelet factor	A phospholipid, cephalin

Two systems are involved in blood clotting: an intrinsic system and an extrinsic system. These two systems are described below and illustrated in Fig. 14.8.

Intrinsic system Internal clotting, which occurs when blood vessels are injured or broken, involves the intrinsic system. First, platelets break down, releasing platelet factor. Several factors then act, and prothrombin-activator substance, which converts inactive prothrombin to active thrombin, develops. Finally, thrombin interacts with soluble fibrinogen to form insoluble fibrin. Fibrin provides an insoluble framework in which red blood cells and platelets are caught, forming a clot.

Extrinsic system Blood clots from injuries on the surface of the skin are formed by the extrinsic system. When surface blood vessels are broken and tissue cells are injured, a substance called tissue thromboplastin is liberated. This substance must be activated by calcium ions and several clotting factors, such as the labile factor V (accelerator globulin), factor VII (SPCA), and factor X (the Stuart-Prower factor), for it to convert prothrombin to thrombin. Thrombin, as in the intrinsic system, interacts with fibrinogen to produce fibrin and a clot (Table 14.3).

Three phases of clotting The clotting process follows three phases in the extrinsic system:

Phase 1 Injured tissue releases thromboplastin.
Phase 2 Thromboplastin is activated by interaction with calcium ions, the labile factor (accelerator globulin), and several other factors and converts inactive prothrombin to active thrombin.
Phase 3 Thrombin interacts with fibrinogen. Fibrinogen comes out of solution as insoluble threads of fibrin. Fibrin forms a network that provides a support for blood cells and platelets, leading to formation of a clot.

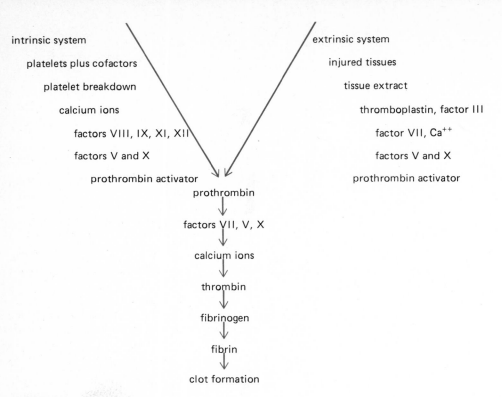

intrinsic system   extrinsic system

platelets plus cofactors   injured tissues

platelet breakdown   tissue extract

calcium ions   thromboplastin, factor III

factors VIII, IX, XI, XII   factor VII, Ca^{++}

factors V and X   factors V and X

prothrombin activator   prothrombin activator

prothrombin

↓

factors VII, V, X

↓

calcium ions

↓

thrombin

↓

fibrinogen

↓

fibrin

↓

clot formation

FIGURE 14·8

Some differences between intrinsic and extrinsic blood clotting.

The clotting process is somewhat different in the intrinsic system because it starts with platelet breakdown and the release of platelet factor (see Fig. 14.8). The platelets appear to release a substance similar to tissue thromboplastin, but this substance apparently requires several factors for activation.

ANTICOAGULANTS Blood does not ordinarily clot inside blood vessels because it contains chemical substances that prevent clotting. Blood will clot when it is removed from blood vessels unless anticoagulants are added to it. Calcium ions are essential to the clotting process, and if they are removed from the blood, it will not clot. When sodium oxalate, an anticoagulant, is added to a sample of blood, it reacts with calcium to form insoluble calcium oxalate, which precipitates. Citrates can also be used as anticoagulants.

In circulating blood, *heparin* is an antiprothrombic substance; that is, it blocks the change of prothrombin to thrombin. It also decreases the production of thromboplastin. Heparin is secreted by *mast cells* in the connective tissue of various organs, and it is a natural anticoagulant.

INTERNAL CLOTTING Internal clotting is called *thrombosis*. A clot, or *thrombus,* may form in some blood vessel of the arm or leg and do comparatively little harm, but if it blocks the blood supply to the brain, the heart (coronary thrombosis), or lungs, it can be very serious. Clots are thought to form over plaques of cholesterol-containing deposits on the linings of blood

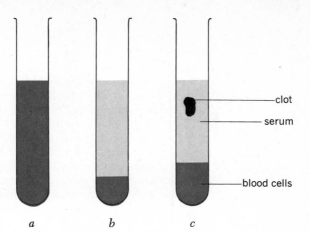

FIGURE 14·9
The formation of a clot when blood is drawn into a container and allowed to stand: *a* whole blood; *b* the formation of fibrin cells; *c* the separation of the serum, the clot, and the blood cells.

— clot
— serum
— blood cells

a   *b*   *c*

vessels. Cholesterol plaques present a rough surface to which blood cells and platelets can adhere.

Blood clots are sometimes dissolved by a plasma protein called *fibrinolysin,* or *plasmin*. This substance is present in the inactive form in the plasma, where it is known as *profibrinolysin* or *plasminogen*. The inactive lysin must be activated before it can act as a digestive enzyme to dissolve a clot. A high plasmin content in the uterus ordinarily prevents the blood in the menstrual flow from clotting.

Dicoumarol is an anticoagulant drug used to prevent internal clotting. It depresses the production of prothrombin by the liver.

SERUM   When a sample of blood is drawn into a container and allowed to stand, a clot forms and the cells and clot separate from a yellowish fluid, the blood *serum*. Serum differs from plasma because the chemical substances concerned with clotting have interacted. In serum, fibrinogen has been precipitated as fibrin, and a clot has been formed (Fig. 14.9).

CLOT RETRACTION   After a clot forms, whether in a test tube or in the body, it then retracts and becomes more firm. It may actually harden and dry if it is a skin-surface clot. Platelets are responsible for clot retraction. They produce pseudopodia that contract and pull the fibrin strands closer together, thus forcing out the serum and leaving a hardened, dry mass of blood cells.

BLOOD TRANSFUSIONS AND STORING BLOOD

Blood transfusions were first used on a large scale in World War I. Large-scale use prompted the development of methods for storing and preserving blood. It was found that sodium citrate was a safe anticoagulant and that prompt refrigeration made it possible to keep whole blood for 5 to 7 days. Blood banks were established at various army bases and later in many hospitals.

A better whole-blood preservative was introduced in 1944. This consisted of sodium citrate, citric acid, and dextrose. With this preservative, refrigerated whole blood could be preserved and kept in good condition for 21 days. More efficient techniques of preservation have continued to lengthen the time that blood can be kept in usable condition.

During the greater part of World War II, blood plasma was used very extensively as a substitute for whole blood in transfusions. Plasma can be preserved, either frozen or dried, for long periods, and it has far less bulk than whole blood. Dried lyophilized plasma remains effective for years and can be readied for use simply by redissolving it in distilled water. Since a great part of plasma's effectiveness in combating shock caused by blood loss is due to the serum albumin it contains, several agencies concentrated on fractionating the plasma proteins, with particular emphasis on obtaining serum albumin. Although there were definite advantages in reduced shipping bulk for these newer preparations, physicians concluded at about this time that, in general, whole blood was the most satisfactory material for transfusion.

BLOOD GROUPING

THE ABO SYSTEM   The widespread transfusion of whole blood caused an increased interest in the problem of blood grouping. It was discovered that it was not safe to transfuse blood between members of different species. Blood from two human beings was often found to be incompatible. When two blood samples are incompatible upon mixing, the red blood cells *agglutinate*, or clump together.

Little was known about why some bloods were compatible while others were incompatible until the research of Karl Landsteiner, around 1900, revealed that there are four human blood groups. Landsteiner's system divided all blood into A, B, AB, and O groups, and showed that these blood groups are genetically determined by mendelian inheritance.

Agglutination occurs when the antigen of red blood cells belonging to a certain group interacts with its specific antibody, which is found in the plasma of blood of another group. Obviously, both antigen and antibody cannot be present in the same blood, for the red blood cells would then be agglutinated. The letters specifying the four blood groups also label the antigen on red blood cells of blood of that type. The red blood cells of group A have antigen A; those of group B have antigen B; those of group AB have antigen AB; those of group O have no antigens.

Blood plasma contains antibodies that will agglutinate red blood cells with the appropriate antigen. These antibodies normally occur in blood and are designated *alpha* and *beta*, or *anti-A* and *anti-B*. They are also called *isoagglutinins*. The plasma of blood of group A contains anti-B antibodies, while the plasma of group B contains anti-A antibodies. Group AB plasma contains no antibodies; the serum of group O contains both anti-A and anti-B antibodies. The antigens and antibodies of the four ABO blood groups are summarized in Table 14.4.

From this discussion we can see the absolute necessity of determining the blood group before making a transfusion. It is always desirable to use blood from a donor of the same blood type as the recipient, but if such blood is not available, other types can be used, as long as the red blood cells of the donor are not agglutinated by the plasma of the recipient. For practical purposes, the plasma of the donor is so diluted as it mixes with the total blood volume of the recipient that the plasma antibodies do not ordinarily cause any reaction. For this reason, blood plasma, which is usually pooled,

TABLE 14·4
Blood groups and corresponding antigens and antibodies

Blood group or phenotype	Antigens of red cells	Antibody in plasma or serum	Genotype
A	A	Anti-B	AA or Ao
B	B	Anti-A	BB or Bo
AB	AB	None	AB
O	None	Anti-A and anti-B	oo

can be safely used without typing. The donor's red blood cells, however, may easily be agglutinated by the antibodies in the recipient's plasma; these clots pose a real threat to the recipient. Blood groups must be carefully typed and matched before transfusion, therefore.

To determine blood group, the red blood cells are tested with the serum of group A and of group B (Fig. 14.10).

The reaction of the blood when red blood cells of the donor are mixed with plasma or serum of the recipient is shown in Fig. 14.11. The four donor blood groups are tested against the four recipient groups. All tests should be observed with a microscope to determine agglutination since extensive rouleau formation, the stacking of erythrocytes, might otherwise be interpreted as agglutination.

Cross matching To ensure safety, donor and recipient blood is *cross-matched* before transfusion. A suspension of red cells from the donor is mixed with

FIGURE 14·10
Determination of the four blood groups. The serums of group A and group B are tested for reaction against the unknown cells of the four groups: A, B, AB, and O. Four large drops of group A serum and four large drops of group B serum are placed on microscope slides. A small drop of blood of the unknown type is added to each drop of serum. Agglutination occurs only when the antigens of the red cells react with the corresponding antibodies in the serum. Refer to Table 14.4.

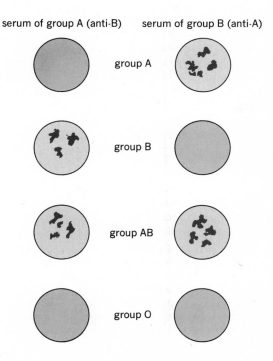

serum of group A (anti-B)   serum of group B (anti-A)

group A

group B

group AB

group O

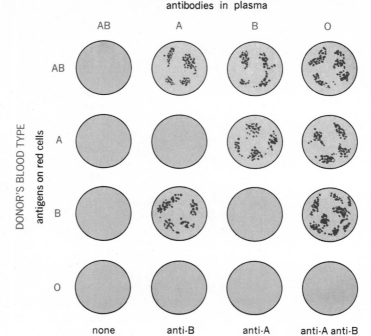

RECIPIENT'S BLOOD TYPE

antibodies in plasma

DONOR'S BLOOD TYPE
antigens on red cells

FIGURE 14·11

The reaction occurring when antigens on red blood cells of the donor's blood type are introduced into the plasma of the recipient's blood type. Agglutination indicates a positive reaction. Persons with type O blood are called universal donors, whereas persons with type AB blood are termed universal recipients. However, all blood should be carefully typed and tested before transfusion.

defibrinated serum from the recipient, and the mixture is observed through a microscope to determine whether agglutination has occurred. In another test, the erythrocytes of the recipient are cross-matched with the serum of the donor. If no agglutination occurs in either test, the donor blood may be safely transfused.

In the early stages of our knowledge of blood typing and whole-blood transfusions, type O blood was in greatest demand because it has no antigens on the red cells. In the general white population, about 45 percent have type O blood. Individuals with type A blood represent about 41 percent of the population, while type B individuals total about 10 percent and type AB only 4 percent. In the American black population the distribution of blood types is approximately: O, 48 percent; A, 26 percent; B, 23 percent; AB, 3 percent.

THE Rh FACTOR   In addition to the A-B antigens, several other antigens are known to be associated with erythrocytes. One of these is commonly called the *Rh factor*, because the antibody was originally developed in the plasma of rabbits and guinea pigs by injecting them with the red blood cells of *Macacus rhesus*, the rhesus monkey. The Rh factor has a more complicated inheritance than the ABO factors. Three genes are involved, and there are eight possible *genotypes*. Dominance is indicated by the symbols *D, C,* and *E* and recessives by the symbols *c, d,* and *e.* Two phenotypes, Rh+ and Rh−, are produced by combinations of these three genes. Those individuals who inherit a *D* gene will be Rh+ and those with a *d* gene are Rh−.

Possible genetic combinations are as follows:

	CDE		cdE
Rh +	cDE	Rh —	Cde
	cDe		CdE
	CDe		cde

Rh + individuals have the Rh antigen on their red blood cells and no anti-Rh antibodies in their plasma. Rh — persons lack both Rh antigens *and* anti-Rh antibodies. However, the Rh — individual may develop anti-Rh antibodies, if Rh + erythrocytes are mixed with that person's plasma. About 85 percent of the white population are Rh +. The dark-skinned races are largely Rh +; about 93 percent of American blacks are Rh +.

Incompatible Rh antigens may be introduced in two possible ways. The transfusion of Rh + blood to any Rh — person can cause a transfusion reaction in which anti-Rh antibodies are developed. If an Rh — woman becomes pregnant by an Rh + man, incompatible antigens may also be introduced into both fetal and maternal circulations during pregnancy, if the fetus is Rh +. When blood is typed properly for transfusion, the Rh factor is typed along with the ABO groups and blood compatible with all the recipient's antigens is given.

OTHER FACTORS There are many antigens on human red blood cells in addition to those already mentioned. The M, N, and MN factors are well known. Blood factors, with the exception of the ABO groups, are detected by immunization and by the production of specific antibodies against the antigens. The M and N factors are detected by mixing rabbit plasma from animals immunized with human red blood cells and observing agglutination.

The ABO blood groups, with subgroups determined by factors M and N, and the Rh factor are often used in the field of legal medicine in attempts to prove parental relationships. For example, neither antigen A nor B could be present in a child's blood unless it were present in the blood of one or both of the parents, nor would it be possible for type AB parents to have type O children. The genotypes for MN factors are MM, NN, or MN. If a child has blood factor MM, for example, the father cannot be NN, and the parents must be MM or MN. Blood group typing has its limitations, but it is often useful in establishing an unchangeable means of identification.

IMMUNITY AND IMMUNE SYSTEMS

We could not live without self-protection against disease. This form of protection, known as *immunity*, includes all physiological processes that enable the body to recognize foreign materials and to erect defenses against them. Defense mechanisms may be nonspecific or specific. *Nonspecific immune mechanisms* involve barriers to the invasion of disease organisms; these barriers include the skin, mucous membranes, or ciliated membranes. *Specific immune mechanisms* are of two kinds: *cellular-mediated immunity*, derived from sensitized or activated thymic lymphocytes (T cells), and *humoral immunity*, derived from blood antibodies produced by B cells.

The body's *immune system* is a line of defense against invading organisms. This system is, in general, successful in warding off disease, but it does not make us immune to all invading organisms. The body has several lines of defense. The first is the skin. As long as the skin remains unbroken, it forms a barrier against most microorganisms. The second line of defense is formed by the digestive tract with its soft mucous linings and digestive enzymes. The mucous linings entrap injurious organisms, and digestive enzymes destroy them.

The third line of defense is formed by the immune system. We have discussed leukocytes and their role in phagocytosis of foreign organisms and debris. We have also considered the activity of lymphocytes and plasma cells in defense. Monocytes and macrophages also take part in phagocytic activity.

CELLULAR IMMUNITY   The stem cell lymphocytes originate in the bone marrow of the fetus. Some of these cells later appear in the thymus gland as lymphoblasts. In the thymus, lymphocytes multiply and become sensitized, immunocompetent, cells that can resist disease organisms (Fig. 14.12). These cells appear also in other lymphoid tissues such as the lymph nodes and the spleen. In these tissues, germinal centers produce clones of sensitized cells when stimulated by foreign substances. (Clones are different populations of lymphocytic cells that develop from a single cell. Each clone responds to a single antigen and to it alone.)

T cells   Lymphocytes that develop in the thymus are called *T cells*. Such cells may live for many years, retaining their sensitivity to a specific antigen throughout their lives. T cells act in immune responses that do not involve antibody production, but they do release chemical substances (lympho-toxins) that inhibit the invading antigen. Macrophages are then attracted by chemotaxis to the antigenic area, which they clean by phagocytosis. Macrophages are the cells that resist and destroy transplanted tissues as they do any other foreign tissue or organism. T cells appear to act more specifically against viruses, fungi, and parasites. They differ from B cells in that they migrate to an invasion site to release their cytotoxic substances.

HUMORAL IMMUNITY   Humoral immunity is produced by lymphocytes that remain in lymphoid tissues and produce antibodies that circulate in the blood stream.

B cells   Lymphocytes known as *B cells* apparently migrate to lymph nodes and other lymphoid tissues when they leave the fetal bone marrow. In the chicken, such cells migrate to the *bursa of Fabricius,* an organ of the digestive tract, where they become antibody-producing cells. The human body has no such bursa, but these lymphocytes are apparently able to produce antibody somewhere in the tissues. However, the name "B cell" (for bursa) is retained for these cells.

When stimulated by an antigen, B cells may be transformed into *plasma cells,* which secrete immunoglobulin, or antibodies. Antibodies enter the bloodstream to attack foreign objects (Fig. 14.13). We therefore use the term "humoral immunity" in speaking of this type of immune reaction. Antibodies produced by B cells become blood proteins called *gamma globulins,* or *immune globulins (Ig).* There exist several kinds of gamma globulins, designated by the letters G, A, M, D, and E. IgG and IgM are two gamma globulins that provide the greatest amount of specific immunity against

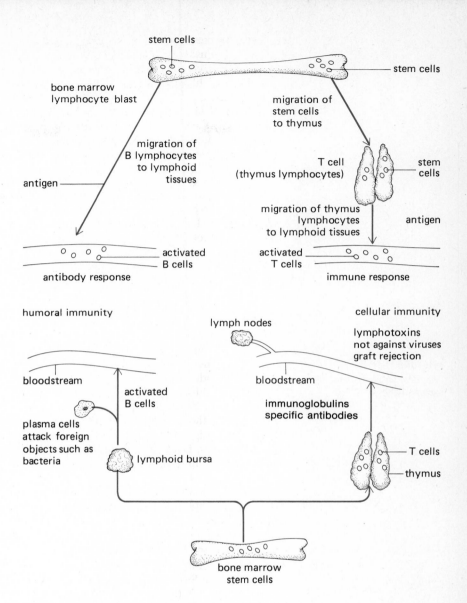

FIGURE 14·12

Migration of B cells and T cells.

infectious bacteria. IgE is concerned with certain kinds of allergy, whereas IgA antibodies are produced by lymphoid tissues in the gastrointestinal tract, the respiratory tract, and the genitourinary tract and are largely confined to these areas. The function of IgD has not been determined with certainty.

The structure of IgG immunoglobulin   IgG is one of the best known of the five types of B-cell antibodies and may be used to illustrate their function. IgG is a large protein molecule with a Y-shaped structure composed of two pairs of light and heavy polypeptide chains. The binding site is at the distal end of

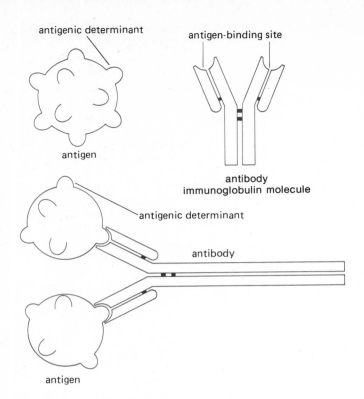

antigenic determinant

antigen

antigen-binding site

antibody
immunoglobulin molecule

antigenic determinant

antibody

antigen

FIGURE 14·13
Antigen, antibody, and antigen-antibody coupling.

each arm of the Y. Antibodies have at least two binding sites that fit exactly with the antigen, giving it specificity. The ability of the immune system to produce antibodies that bind to literally thousands of antigens preserves us from infections of various sorts.

Active immunity *Active immunity* to a particular disease is acquired by having the disease. During recovery, the body has prepared antibodies against the disease antigens. When an antigen is first introduced, a modest buildup of antibody occurs, but if the same antigen is introduced a second time (during reinfection, for example), there will be a tremendous outpouring of its specific antibody. This phenomenon illustrates what has been called the *memory* factor of the immune system. Immune "memory" calls up a greater resistance to a later infection and thus enables the body to ward off recurrent infection. This type of resistance to infection, which results from actual contact with bacterial organisms, their toxins, or other antigens, is active immunity.

For some potentially fatal diseases such as smallpox, typhoid fever, and rabies, vaccines and toxoids are used to establish an artificial active immunity to protect the individual against an initial infection. *Vaccines* are prepared from bacteria or viruses that have been *attenuated,* or weakened, so that they cause few or no ill effects, but the presence of such agents still stimulates the immune system to produce antibodies against them. *Toxoids* are toxins that have been rendered harmless, but whose presence still stimulates the development of antibodies against the toxin.

Passive immunity   Passive immunity is a short-term type of artificial immunity used to protect an individual who has been or may be directly exposed to a disease. Passive immunity is acquired by injections of already developed antibodies. The antibodies are usually obtained by injecting an animal with a toxin or with human serum taken from a person who has recently recovered from the disease. The animal thus develops active immunity to the disease, and its serum can be used as an *antitoxin* to provide passive immunity to humans. Antitoxins can also be obtained from bacterial cultures. We now can acquire protection against such diseases as diphtheria, scarlet fever, mumps, tetanus, and many others through passive immunity.

COMPLEMENT   The general term *complement* stands for the assembly of forces that fight inflammation. It is activated by antigen-antibody complexes and involves the plasma proteins. Complement causes greater vasodilatation, increased capillary permeability, and neutrophil migration toward the inflamed area, drawn there by chemotaxis. In all, complement is an organized response to control inflammation.

Complement also operates by direct attack on invading cells. These are identified by antibodies and killed by complement proteins. Only immunoglobulins M (IgM) and G (IgG) bind complement in a process known as *complement fixation*. After antibody has recognized the invading cell and attached to its surface, complement fixes to the cell's surface and makes a lesion from which the cellular contents leak, destroying the cell. These activities take place in the plasma and are therefore considered to be humoral reactions.

INTERFERON   Interferon is a protein produced by cells as they react to a viral infection. An invasion of a virus causes the cell's ribonucleic acid (RNA) to produce *interferon*, which then enters the circulation. Other cells are then influenced to produce the antiviral substance. All viruses cause the synthesis of the same kind of interferon, which acts against other viruses. Interferon is not, therefore, virus-specific. It is species-specific, however, as was shown by Derek Burke who found that interferon made in chick cells was active in other chick cells but not in duck cells (see Suggested Reading). Human interferon must be used, therefore, to combat human viral infections, and human cells produce very little interferon.

Interferon itself does not apparently directly attack the virus but induces cell proteins to do so. An antiviral substance would prove valuable if it could be used to increase resistance to viruses such as those associated with the common cold, and great interest is shown in finding a means to synthesize interferon. Interferon is being tested as an anticancer substance, but even if successful it would not be available for several years.

RADIATION AND CARCINOGENS AND THEIR EFFECT ON THE BLOOD

RADIATION HAZARDS   One of the hazards of modern living is the possibility of exposure to atomic radiation and fallout. The blood-forming tissues are very susceptible to radiant energy but show considerable ability to recover. Soon after exposure to radiation, leukocytes decrease in number. The clotting mechanism of the blood is adversely affected, causing a tendency toward bleeding. Anemias occur later on because radiation injures the tissues that form erythrocytes.

Strontium 90, a radioactive fallout product, is absorbed like calcium and deposited in bone and bone marrow. Radioactive fallout contaminates the ground and enters the food chain through plants. When individuals eat plant foods directly or drink milk, strontium 90 is incorporated into human tissues. It may then affect the blood-forming tissues or cause cancer to develop. Strontium 90 emits beta rays for a very long time; it has a physical half-life of 28 years. (Half-life is the period of time required for a given quantity of an element to lose half its radioactivity. A half-life of 28 years means that the element will lose half its radioactivity by the end of that time.)

Most experimental evidence regarding injurious effects of radiation has been obtained from experimentation on animals. Shielding a portion of the body from excessive radiation gives the animal a better chance for recovery. Shielding the spleen, in particular, has been shown to help recovery.

Bone marrow transplants have proved to be successful in animals, and eventually this technique may be used to save the lives of persons who have been exposed to excessive amounts of radiation.

Carcinogens often affect the blood-forming tissues and are thought to be carried by the blood to other organs and tissues where cancers may become localized.

THE SPLEEN AND BLOOD SUPPLY

The spleen, which is an important accessory of the circulatory system, is a highly vascular organ located to the left of the stomach and somewhat behind it. It lies below the diaphragm and directly above the left kidney. It is covered by the peritoneum, a membrane lining the abdominal cavity. Though it can change size, it is ordinarily 5 or 6 in long and about 4 in wide. It is a soft, pliable organ of dark purplish color (Fig. 14.14).

The internal structure of the spleen is rather complex. Just beneath the serous coat is a connective tissue capsule, from which a fibrous framework (the trabeculae) projects into the interior. The capsule and the internal fibrous structure contain smooth muscle, which accounts for the ability of the organ to contract. The area between the fibrous framework is filled with splenic pulp, which is composed of loose reticular tissue that supports small arteries and veins and contains red and white blood cells in great numbers. A complicated network of blood spaces, the splenic sinuses, is found in the splenic pulp. Here also are lymphatic nodules composed of lymphoid tissue and concerned with the development of lymphocytes. Large phagocytic white cells of the reticuloendothelial system (macrophages) are found in great numbers in the splenic pulp. They function in the removal from the blood of old or agglutinated red blood cells, pieces of red blood cells, and foreign matter.

FUNCTIONS OF THE SPLEEN Numerous functions are ascribed to the spleen, but three are of primary importance. The spleen: (1) holds a reserve blood supply; (2) destroys old or agglutinated red blood cells and platelets; and (3) forms lymphocytes.

As an accessory organ of the circulatory system, the spleen can direct a considerable amount of reserve blood into the arteries in response to an emergency, thus increasing the oxygen-carrying capacity of the blood. Stimulated by impulses from the sympathetic nervous system and by epinephrine from the adrenal glands, the spleen is strongly activated in

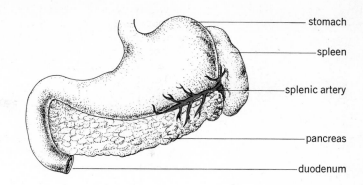

a

stomach

spleen

splenic artery

pancreas

duodenum

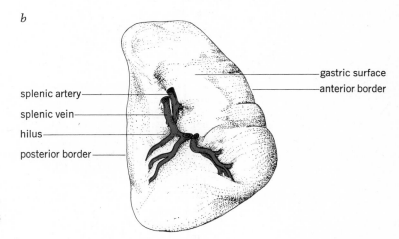

b

gastric surface
anterior border

splenic artery

splenic vein

hilus

posterior border

FIGURE 14·14

a The spleen, anterior view of stomach, pancreas, and spleen; *b* the spleen visceral surface.

emotional states and in times of physical stress. When there is a loss of blood, the spleen helps not only to make up the volume of the blood but also to supply additional red blood cells.

Although the spleen is not essential to life and well-being, it helps organisms to meet emergencies. Exercise and an increase in external temperature cause the spleen to decrease its volume of blood. Animals whose spleens have been removed appear to have less resistance to the toxic effects of disease.

The large phagocytic cells of the spleen, liver, and other tissues apparently do not ingest normal, healthy erythrocytes; they are most useful in taking old red blood cells, fragments, and agglutinated masses of cells out of circulation. The spleen has been called the "graveyard" for old platelets.

In the embryo, the spleen forms both red blood cells and lymphocytes. After birth, erythrocytes are formed in the red marrow of bone, but the spleen continues to be an important organ for the manufacture of lymphocytes and monocytes throughout life.

Other reservoirs of blood are found in the body. The liver, pulmonary blood vessels, venous plexuses of the skin, and the mesenteric blood vessels all hold a considerable blood supply.

CLINICAL ASPECTS: BLOOD CONDITIONS AND DISEASES

INFLAMMATION The biological events associated with *inflammation* are very complicated. Inflammation is a reaction to injury, often caused by such agents as bacteria. Phagocytic cells have been observed to stick to the inner surface of the capillary wall at the site of trauma. Soon these cells begin to move through the capillary wall by diapedesis. The polymorphonuclear cells in the inflamed area are predominantly neutrophils, but lymphocytes and monocytes are also present in great numbers. Tissue macrophages increase in number in response to inflammation and move about through the injured area.

Sometimes inflammation may be deep-seated, as in the case of an infected appendix. The white-cell count may rise remarkably as the protective mechanisms of the body attempt to control the infection. In an abscess around the root of a tooth, in tonsillitis, in appendicitis, and in many infectious diseases, the white-cell count may rise to 13,000, 18,000, or even higher than 30,000 white cells/mm^3. This condition is called *leukocytosis*. A rise in the white-cell count gives the physician excellent corroborative evidence that there is an infection within the body. Often the severity of the infection is indicated by the number of leucocytes.

A rise in the white-cell count does not mean ordinarily that the various kinds of white cells retain their normal numerical relation to each other. Inflammation and certain infectious diseases, especially those involving the round or coccal species of bacteria, cause the number of neutrophilic cells to increase. Some chronic infections cause an increase in the number of lymphocytes, and some cases of asthma, certain skin diseases, and some infections by parasitic roundworms (*Trichinella*) cause an increase in the number of eosinophils (eosinophilia). The eosinophil count rises sharply in reaction to blood parasites such as *Trichinella* in the condition called *trichinosis*, which is the result of eating a minute, encysted roundworm in raw or undercooked pork.

Chronic diseases, such as tuberculosis, may cause the number of monocytes to rise. These cells are slow to multiply at the start of an infection but may increase in numbers over a period of time.

ANEMIA Any condition of the blood in which there is a serious reduction of the number of erythrocytes or of the amount of hemoglobin is known as an *anemia.* Often both the red-cell count and the percentage of hemoglobin are reduced. The typical patient with anemia appears pale and weak and has a loss of energy because the oxygen-carrying capacity of the blood is reduced by either the loss of hemoglobin or a reduction in erythrocyte number, which causes an indirect decrease in available hemoglobin.

There are many kinds of anemia. Anemia may be caused by inability of the body to manufacture enough hemoglobin or by failure of the bone marrow to produce enough erythrocytes. The first type of anemia is caused by an inadequate amount of iron in the diet or by the faulty absorption and utilization of iron in the production of hemoglobin. Iron absorbed from the intestine combines with a protein to form a compound known as *ferritin*. It is then found in the liver, bone marrow, and in minute amounts in other tissues. In some kinds of anemia the number of red cells may be normal, but the percentage of hemoglobin is greatly reduced. Occasional anemia is caused by an abnormal rate of erythrocyte destruction either by chronic bleeding or by some hemolytic substance in the blood.

Pernicious anemia   The failure of the red marrow of bone to produce erythrocytes causes a very low red blood cell count, but individual cells may be normal or large and contain normal or even above normal amounts of hemoglobin. *Pernicious anemia* is this type of anemia. The discovery of an *antianemic factor* in liver has benefited victims of pernicious anemia and of other, similar forms. Understanding the relation between the erythrocyte maturation factor and vitamin B_{12} has also helped physicians interpret the factors involved in this disease. (See discussions of folic acid and vitamin B_{12} in Chap. 20.)

Sickle-cell anemia   Normal adult hemoglobin (Hb^A) is composed of four polypeptide chains and four heme groups. The four polypeptide chains comprise two alpha chains and two beta chains. The number and location of the amino acids composing each chain have been determined; each alpha chain contains 140 amino acids, and each beta chain consists of 146.

Sickle-cell anemia is caused by the presence of an abnormal hemoglobin (Hb^S), which is associated with a defective gene. When two defective genes are present (homozygous recessive condition), the victim's erythrocytes exhibit a peculiar thickened, elongate appearance called "sickling" (Figs. 14.15 and 14.16). The abnormal cells are not strictly half-moon–shaped. Sometimes, such cells are misshapen, with protruding strands that entangle other blood cells and tend to block small blood vessels in the tissues of affected persons. The abnormal cells are fragile and are therefore rapidly destroyed, giving rise to the condition called *sickle-cell anemia*.

Individuals heterozygous for the sickle-cell gene, those with one defective gene and one normal gene ($H^A H^S$), may show some sickling of their erythrocytes, but these individuals are essentially normal and do not have sickle-cell anemia. They are said to carry the *sickle-cell trait*. About 60 percent of their hemoglobin is normal Hb^A, while 40 percent is Hb^S. Sickled cells are not very efficient oxygen carriers, although individuals with sickle-cell trait do not suffer any great deficiency.

The abnormal sickle-cell hemoglobin can be identified by electrophoresis. A clinical test for sickle-cell trait is also available. It has been discovered, in addition, that sickled cells regain their normal appearance, temporarily, under reduced pressures.

It is of considerable interest to geneticists, physiologists, and biochemists that the only chemical difference between the Hb^A and Hb^S is the substitution of the amino acid valine for glutamic acid at position 6 in the beta

FIGURE 14·15
Distorted blood cells found to occur in sickle-cell anemia.

a
homozygous

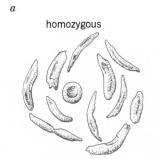

b
heterozygous

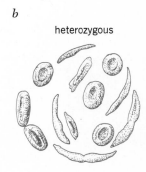

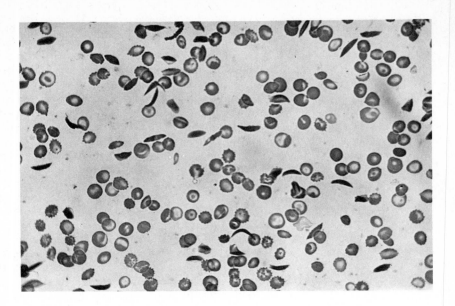

chain. In the heterozygous individual the substitution produces sickle-cell trait, and in the homozygous individual it produces the very serious condition of sickle-cell anemia.

Hb^S occurs almost exclusively in blacks of African descent. Various surveys indicate that American blacks, as a group, have about two-thirds African ancestry and one-third Caucasian ancestry. It has been estimated that the sickle-cell trait is present in around 8 percent of American blacks. If this estimate is correct, the homozygous condition would be comparatively rare, perhaps as low as 0.2 percent.

The incidence of sickle-cell trait in the population of certain areas in Africa is relatively high, perhaps as great as 20 percent. It has been shown that in areas where subtertian malaria is endemic there is a much higher incidence of Hb^S than in regions free of malaria. It appears that the sickle-cell hemoglobin gives individuals greater resistance to malaria. In areas free of malaria there is, of course, no selective advantage for having this trait.

Thalassemia   *Thalassemia* is an anemia similar to sickle-cell anemia. In thalassemia the amount of hemoglobin is reduced and the red blood cells tend to be smaller than normal and pale. Thalassemia is also genetically determined. This condition is sometimes called "Mediterranean anemia" because it occurs largely among peoples who live in areas around the Mediterranean Sea and, of course, among people who have migrated from this area. There are two types: *thalassemia minor*, the heterozygous form that has few symptoms, and *thalassemia major*, the homozygous severe form that attacks children. Thalassemia major is also called Cooley's anemia.

OTHER BLOOD DISEASES   *Polycythemia*   The blood condition known as *polycythemia* is characterized by the production of a greater than normal number of erythrocytes. The erythrocyte-forming areas become overactive.

As a result, the viscosity of the blood increases and blood pressure rises. The heart may become overloaded, and there is also a danger of blood clots.

Hemophilia   In the peculiar condition known as *hemophilia*, the blood clots very slowly, and internal bleeding is especially difficult to control. The defect is inherited as a sex-linked, recessive character in the same way that color blindness is inherited. Women are not ordinarily affected. If a normal woman marries a man who is a hemophiliac, their sons will not be affected, but their daughters may inherit the defective gene and transmit the condition to some of their own sons. Thus the defect skips a generation before it is actively expressed in males (see Fig. 11.20).

It was formerly thought that the antihemophilic globulin, factor VIII (AHF), was not present in the blood of hemophiliacs or was found in amounts too small to be effective. It is now known that hemophiliacs do produce antihemophilic globulin (factor VIII), but it is biologically inactive. Investigations with rabbit and human antiserum indicate that all hemophiliacs possess antigen that reacts with human factor VIII. Thus it is possible to determine those persons who have an immunologically active factor VIII. Those who have an antigen that biologically inactivates this factor are hemophiliacs.

Von Willebrand's disease   Another condition in which slow and protracted bleeding is characteristic is *Von Willebrand's disease*. Persons suffering from this disease show diminished biological activity of factor VIII and some factor VIII antigen, but their condition is not determined by the same genetic mechanism as hemophilia. In Von Willebrand's disease, the gene is not sex-linked, but is rather a dominant, autosomal gene, so both men and women may have the disease. Von Willebrand's disease is usually considered to be less severe than hemophilia.

Leukopenia, leukocytosis, lymphocytosis   Some diseases, especially pneumonia and typhoid fever, cause a reduction in the white-cell count. Certain drugs also appear to reduce the number of white cells. Destruction or degeneration of red bone marrow also causes a reduction of white as well as red blood cells. *Leukopenia* means a reduction of white cells below normal. Reduction in the number of white cells lowers resistance to infection.

The number of white cells normally rises as the result of infection; this increase is called *leukocytosis*. There is, however, an abnormal condition in which white cells are produced in tremendous numbers without any known infection. This condition is called *leukemia* (cancer of the blood). In one type of leukemia, the red bone marrow, for some unknown reason, produces white cells at an uncontrolled rate. In another type, lymphoid tissues are overactive in the production of lymphocytes. A deficiency in the number of red blood cells (*erythropenia*) is frequently associated with leukemia.

Infectious mononucleosis is a disease most commonly observed in young persons, in the age group of sixteen to twenty-five years. It is a viral disease, caused by a herpes virus and characterized by fever, enlargement of cervical lymph nodes, and *lymphocytosis* (overproduction of lymphocytes) with atypical lymphocytes. In many cases the upper eyelids become puffy and sag downward. The spleen is usually enlarged. Normally healthy young persons can expect to recover and resume their usual activities in 3 or 4 weeks. It is not contagious by exposure in the sense that a disease such as measles is contagious.

THE Rh FACTOR   A problem sometimes arises in pregnancy when the father of the child is Rh+ and the mother is Rh−. The blood of an Rh-positive mother has no Rh antigens, but may develop anti-Rh antibodies. If there is an occasional break in the placental barrier, which ordinarily separates the circulatory system of the mother from that of the fetus, the Rh antigen on the erythrocytes of an Rh+ fetus may be accidentally introduced into the blood of the Rh− mother. In such cases, the mother's blood would form anti-Rh antibodies just as if the woman received a direct transfusion with Rh+ blood. Rh+ genes are dominant over Rh− genes, so this situation is not uncommon.

The anti-Rh antibody formed in the mother's blood most likely may pass into the fetal blood supply, where it agglutinates the fetal red blood cells. The anti-Rh antibody formed during pregnancy tends to react more strongly with the fetal red blood cells during the later months of pregnancy, at which time it causes hemolysis of red cells. This condition is called *erythroblastosis fetalis,* and it frequently results in stillbirth.

It takes time for the mother's blood to develop anti-Rh antibodies, however, so the mother's first Rh+ child may be born without any Rh disturbance, but subsequent Rh+ fetuses are much more likely to undergo serious hemolysis (Fig. 14.17).

If the fetus is too young to survive a Caesarean section delivery when hemolysis occurs, it may be given transfusions of group O Rh− blood which is injected into the peritoneal cavity. This is a supportive procedure that may help the fetus to live until delivery.

Severe erythroblastosis fetalis in the newborn infant is often treated by

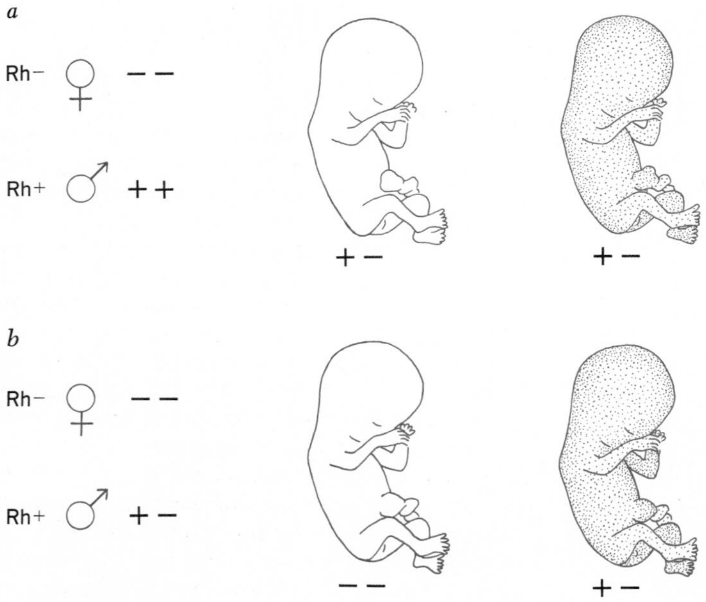

FIGURE 14·17
Possible effects of the Rh factor on the fetus: *a* Even though the father in this case is Rh+ and the mother Rh−, the first child probably will not have erythroblastosis fetalis. The second child is more likely to be affected. *b* If the father is heterozygous for the Rh factor, the fetus may be Rh−, in which case there would be no reaction with the Rh− blood of the mother. An Rh+ fetus, however, increases the possibility of an antibody reaction in the mother's blood.

FIGURE 14·18

Inheritance of the Rh factor. The serious problem arises only when the fetus is Rh+ and the mother is Rh−.

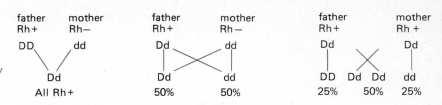

replacing the infant's Rh+ blood with Rh− blood in an *exchange transfusion*. This procedure stops the destruction of the child's red blood cells, and the symptoms of jaundice and anemia tend to disappear. The infant's blood-forming tissues will then begin to develop Rh+ blood to replace gradually the transfused Rh− blood. The antigen is not present in the Rh− blood, and the anti-Rh antibodies are largely removed during the exchange transfusion. The antibodies remaining in circulation will gradually disappear as the infant develops its own Rh+ blood.

It has been noted that anti-Rh antibody concentration usually rises in the blood of an Rh− mother about 72 hours after the birth of an Rh+ infant. Placental separation during delivery of the afterbirth probably permits some Rh+ antigen to enter the mother's blood and cause production of the anti-Rh antibody. Large doses of anti-*D* gamma globulin may be given to the mother after delivery to combat the activity of the fetal antigen. The anti-*D* antibody tends to clear the mother's blood of the *D* antigen. If this treatment is performed after each Rh-incompatible pregnancy, the chances of the mother becoming sensitized to the Rh antigen are greatly reduced.

One or both parents may be heterozygous for the Rh factor. Figure 14.18 indicates genotypes and phenotypes for this trait. Only the *D* gene is shown. The results indicate that the *D* gene is inherited in a simple mendelian pattern of inheritance.

IMMUNE SYSTEMS ABNORMALITIES *Allergy* People who are hypersensitive to some foreign substance are said to be allergic to this substance. Allergens—(IgM) agents that cause allergy—are commonly proteins. Such proteins may come from a host of substances: pollen, hair, feathers, dust, milk, eggs, wheat products, and meat. There are various kinds of allergy, but a common type is *atopic* allergy. Atopic allergy is a familial tendency to react to allergic conditions such as asthma without unusual exposure to allergens.

An allergy is an abnormal reaction to an antigen. Allergic reactions may be manifested by symptoms such as a runny nose, watery eyes, difficulty in breathing, and skin irritation and by some tissue damage caused by the release of histamine by mast cells. Apparently the body produces some antibodies against the allergen but not enough to provide immunity. A disposition toward the tendency to be allergic appears to be inherited.

In discussing humoral immunity, we mentioned that immunoglobin E (IgE) is involved in certain types of allergy. Some antigens enhance the production of IgE by plasma cells. IgE then binds to mast cells, causing the release of histamine and other chemicals. The histamine and various chemicals irritate the tissues they contact. If the respiratory tract is affected, the symptoms of "hay fever" or asthma may be produced. Antihistamines are commonly used to relieve symptoms of atopic allergy.

Some people are allergic to the stings of wasps and bees. The first sting may cause only a red, painful area at the site of the sting, but this sting sensitizes the person to the allergen. A second or subsequent sting may cause a severe allergic reaction or even death.

There is a rise in the number of eosinophils in some cases of allergy, possibly in response to toxic substances released by the allergic reaction.

Autoimmune response Occasionally the body develops antibodies that are directed against its own cells. This destructive form of allergy gives rise to *autoimmune disease*. Autoimmune disease may be caused by several factors: (1) an overdevelopment of complement, (2) a release of chemical substance from platelets, or (3) failure of the body to recognize its own tissues. The glomeruli of the kidneys are often injured or destroyed by autoimmune disease.

CANCER The search for causes of cancer has involved a host of substances that are suspected or known to be injurious and carcinogenic. Coal tars, cigarette smoke, combustion products from engine exhausts, asbestos, environmental gases, smog, and insecticides are only a few of these products. One theory concerning the cause of cancer is that it is due largely to exposure to some injurious substance in the environment. The carcinogen is not thought to cause cancer directly but induces the breakdown of normal cell-growth controls, perhaps by adversely affecting the immune response. The incidence of most cancers increases with the age of the individual. There is no cure for cancer at present, but clean air, proper diet, and early detection offer the best means of survival.

SUMMARY

1 The blood has many functions. It transports many substances, including oxygen, carbon dioxide, food materials, waste materials, and hormones. It plays an important part also in regulating body temperature and in maintaining both the acid-base balance and the water balance of the tissues. Protective functions of the blood include its clotting ability, the phagocytic activity of leukocytes, and the production of antibodies.

2 Plasma is the fluid portion of the blood. It is very complex chemically and contains the clotting components. After the blood has been permitted to clot, the amber-colored fluid is called blood serum.

3 Red blood cells are called erythrocytes. They are biconcave disks, averaging 5 million cells/mm^3 of blood. They arise as nucleated erythroblasts but ordinarily lose their nuclei by the time they enter the bloodstream. They derive their color from the oxygen-carrying compound called hemoglobin.

4 White blood cells, or leukocytes, are classified as granular and nongranular. Granular leukocytes include neutrophils, eosinophils, and basophils. Nongranular leukocytes include lymphocytes and monocytes. White cells number about 7000/mm^3 of blood.

5 Blood platelets are not cells but fragments of large cells, called megakaryocytes. They number about 250,000 to 400,000/mm^3.

6 There are two systems for blood clotting: an intrinsic system and an extrinsic system. The intrinsic system involves the release of platelet factor, several cofactors, and the development of prothrombin-activator

substance. The extrinsic system requires tissue extract, thromboplastin plus several cofactors, and prothrombin activator. Both systems require calcium ions and convert prothrombin to active thrombin. In the final stages of clot formation, thrombin reacts with fibrinogen to form insoluble fibrin, which provides the framework of the clot.

7 Blood is divided into four groups by the ABO system: A, B, AB, and O. These letters represent the antigens present on the red blood cells. The red blood cells of group A have antigen A; the red blood cells of group O have no antigens. The plasma or serum contains antibodies. The plasma of group A contains anti-B antibodies and the plasma of group B contains anti-A antibodies. Group AB plasma has no antibodies, whereas the serum of group O contains both anti-A and anti-B antibodies. Persons with group O blood have been termed universal donors; persons with group AB blood, universal recipients.

8 The Rh antigen, if introduced by transfusion or otherwise into the Rh− blood, causes the formation of anti-Rh antibodies. About 85 percent of the white population are Rh+; about 15 percent, Rh−.

9 The immune system refers to a defense mechanism against foreign agents. There are two types of immunity: cellular and humoral. Immunity may be active or passive. Active immunity may be obtained by injecting attenuated or killed organisms or by toxoids, antitoxins, or gamma globulins.

10 Interferon is a protein produced by cells for protection against virus infections. This protein is part of the body's immune system.

11 The spleen is an important organ concerned with reserve blood supply. Old erythrocytes and platelets are taken out of circulation by the spleen; lymphocytes are formed in this organ, as well as in lymph nodes. The blood-forming tissues, including the spleen, are especially sensitive to atomic radiation.

12 An allergy is a hypersensitivity reaction to foreign antigens. An autoimmune response occurs when the body develops antibody against its own cells.

QUESTIONS

1 List and explain the functions of the blood.
2 Differentiate between plasma and serum.
3 Why has the blood and tissue fluid been called the internal environment?
4 What part do erythrocytes play in the functioning of the blood?
5 Is it possible to count the red blood cells? How?
6 List the different kinds of leukocytes and give their functions.
7 What is the function of plasma cells?
8 Follow the sequence of events that results in the formation of a clot.
9 Why is internal clotting infrequent?
10 Name the four ABO blood groups and discuss blood typing. Why would some transfusions cause agglutination of the recipient's blood?
11 Discuss the Rh factor.
12 What are some of the functions of the spleen?
13 How does the thymus gland contribute to the production of antibodies?
14 Describe the function of macrophages.
15 What is chemotaxis?
16 Distinguish between cellular immunity and humoral immunity.

17 Discuss active and passive immunity.
18 What is allergy?
19 Define complement.
20 How does interferon function?

SUGGESTED READING

Adamson, J. W., and C. L. Finch: Hemoglobin Function, Oxygen Affinity and Erythropoietin, *Annu. Rev. Physiol.,* **37**:351–369 (1975).

Auerbach, R.: "Experimental Analysis of Lymphoid Differentiation in the Mammalian Thymus and Spleen," pp. 539–557, Organogenesis, Holt, Rinehart and Winston, Inc., New York, 1965.

Burke, Dereck C.: The Status of Interferon, *Sci. Am.,* **236**:42–50 (1977).

Cairns, J.: The Cancer Problem, *Sci. Am.,* **233**(5):64–78 (1975).

Capra, J. D., and A. E. Edmundson: The Antibody Combining Site, *Sci. Am.,* **236**:50–59 (1977).

Cooper, H. A., R. G. Mason, and K. M. Brinkhous: The Platelet: Membrane and Surface Reactions, *Annu. Rev. Physiol.,* **38**:501–533 (1976).

Cunningham, Bruce A.: The Structure and Function of Histocompatibility Antigens, *Sci. Am.,* **237**:96–107 (1977).

Gatti, R. A., O. Stutman, and R. A. Good: The Lymphoid System, *Annu. Rev. Physiol.,* **32**:529–546 (1970).

Jerne, Neils K.: The Immune System, *Sci. Am.,* **229**:52–60 (1973).

Kabat, D.: Gene Selection in Hemoglobin and in Antibody-synthesizing Cells, *Science,* **175**:134–140 (1972).

Lerner, R. A., and F. J. Dixon: The Human Lymphocyte as an Experimental Animal, *Sci. Am.,* **228**:82–91 (1973).

Mayer, M. M.: The Complement System, *Sci. Am.,* **229**:54–66 (1973).

Miller, W. J.: Blood Groups: Why Do They Exist?, *BioScience,* **26**:557–562 (1976).

Old, L. J.: Cancer Immunology, *Sci. Am.,* **236**(5):62–79 (1977).

Paul, W. E., and B. Benacerraf: Functional Specificity of Thymus-Dependent Lymphocytes, *Science,* **195**:1293–1300 (1977).

Perutz, M. F.: The Hemoglobin Molecule, *Sci. Am.,* **211**:64–76 (1964).

Raff, M. C.: Cell-Surface Immunology, *Sci. Am.,* **234**:30–39 (1976).

15

THE HEART, THE GENERAL CIRCULATION, AND THE LYMPHATIC SYSTEM

CONSIDERATIONS FOR STUDY

Having discussed the blood and its many functions, we now turn to the heart, which drives the blood, and the general circulation, which is the pathway through which blood is distributed. The lymph and lymphatic systems act as a medium of exchange between the blood and the tissues. In this chapter, we consider the following topics:

1 The structure and function of the heart, including the functions of the sinoatrial node, the atrioventricular node, and the neuromuscular bundle
2 The cardiac cycle
3 The general circulation: arteries, veins, and capillaries
4 The principal systemic arteries, including the aorta and its branches
5 The venous system, including the hepatic portal system
6 The fetal circulation and its modification at birth
7 The structure and function of the lymphatic system
8 Some diseases affecting the circulatory system

The heart is a muscular organ, the contractions of which force the blood to circulate through a closed system of arteries, arterioles, capillaries, and veins. The blood is conducted away from the heart through arteries, which

487

divide into smaller and smaller branches. Finally, at the periphery, the blood passes through very fine capillaries and starts its return flow into small venules. Larger and larger veins bring the blood back to the heart. The lymphatic system gathers lymph from the body tissues and returns it to the heart where it mixes with the blood and is pumped out again. Lymph enables nutrient materials to reach the cells. It also gathers cellular waste materials.

THE HEART

The heart
  Mediastinum
  Atria
  Ventricles

Path of blood
  Right atrium
  Tricuspid valve
  Right ventricle
  Pulmonary semilunar valvula
  Pulmonary artery
  Pulmonary vein
  Left atrium
  Mitral valves
  Left ventricle
  Aortic semilunar valvula
  Aorta

The heart lies in a double-walled pericardial sac located in the interpleural space, or *mediastinum*, a space bounded laterally by the lungs and extending from the backbone to the sternum dorsoventrally. Many blood vessels closely associated with the heart also lie in this area. The diaphragm forms the floor of the mediastinum as well as that of the entire thoracic cavity. Although the heart is centrally located, its tip (or apex) is inclined downward and toward the left. As the heart fills with blood and starts to contract, its beat is felt against the wall of the chest between the fifth and sixth ribs and about 3 in to the left of the midventral line.

The heart varies in size, but it is about 5 in long and 3.5 in wide in the adult male. In general, a large person can be expected to have a proportionately larger heart. The heart is often described, however, as being about the size of the closed fist.

The human heart contains four chambers. At the upper end are the right and left *atria*. The atria receive blood from veins. Their muscular walls are thin and are usually observed in a collapsed condition when hearts are preserved for study. The right and left *ventricles* form the lower part of the heart. They have thick muscular walls, which enable them to force the blood out of the heart and into the arteries. In its actual position, the right ventricle is largely anterior to the left (Fig. 15.1).

PATH OF BLOOD THROUGH THE HEART Large caval veins bring the blood back from the body; these veins empty into the *right atrium* (Fig. 15.2). The

FIGURE 15·1

The heart and associated arteries and veins in anterior view.

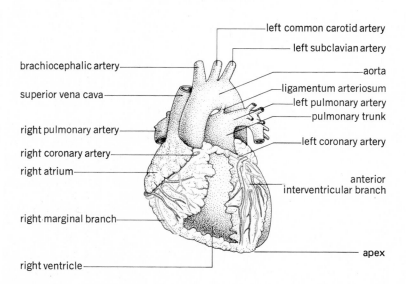

left common carotid artery
left subclavian artery
brachiocephalic artery
aorta
ligamentum arteriosum
superior vena cava
left pulmonary artery
pulmonary trunk
right pulmonary artery
left coronary artery
right coronary artery
right atrium
anterior interventricular branch
right marginal branch
apex
right ventricle

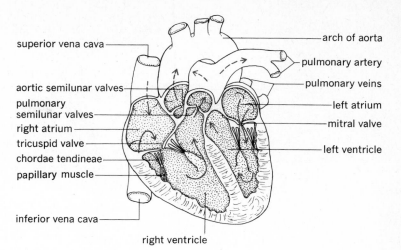

superior vena cava

aortic semilunar valves

pulmonary
semilunar valves

right atrium

tricuspid valve

chordae tendineae

papillary muscle

inferior vena cava

arch of aorta

pulmonary artery

pulmonary veins

left atrium

mitral valve

left ventricle

right ventricle

FIGURE 15·2
Chambers and valves of the heart, showing direction of blood flow, anterior view.

blood flows down into the *right ventricle,* even before the atrial walls begin to contract. Between the right atrium and the right ventricle lies a cylindrical valve, which permits the blood to pass from atrium to ventricle but closes as the ventricle starts to contract. This is the *tricuspid valve,* so called because it is composed of three cusps. Valves increase the efficiency of the heart action by preventing the blood from moving backward into the space it has just vacated. The atrioventricular valves are supported by tendinous strands below, which attach to small mounds of muscle projecting from the ventricular wall. The strands are *chordae tendineae,* and they attach to *papillary muscles.* The chordae tendineae prevent the valve from being forced back into the atrium by the pressure of the blood as the muscular ventricular walls contract. Since the tricuspid valve prevents the blood from being forced back into the atrium, the only exit for the blood is through the pulmonary artery to the lungs.

At the base of the pulmonary artery are three leaflets, somewhat half-moon–shaped, called *pulmonary semilunar valvula.* Their concave surfaces are directed upward, like cups, to hold the column of blood in the *pulmonary artery.* When contraction of muscles in the ventricular wall forces the blood upward, the valves collapse, thus offering little resistance to the passage of the blood into the pulmonary artery. As soon as this forward flow has ceased, the valves close, preventing the blood from flowing backward.

The pulmonary artery leads to the lungs, where it divides into many small branches. Arterioles finally subdivide into a capillary network traversing the walls of air sacs in the lungs. Here, in the heart-lung cycle, carbon dioxide is given off by the blood, to be exhaled, and oxygen is taken up.

The lung capillaries give rise to small veins, or *venules,* which, in turn, give rise to larger veins. Eventually, four *pulmonary veins,* two veins from each lung, carry freshly oxygenated blood back to the *left atrium.* The blood passes through the left atrium and goes on down into the left ventricle.

The thin atrial walls contract, and this action is followed by a strong contraction of the ventricles. Between the left atrium and the left ventricle lies the *mitral,* or *bicuspid, valve.* The valves located between atria and ventricles are also referred to as right and left *atrioventricular,* or *AV, valves.*

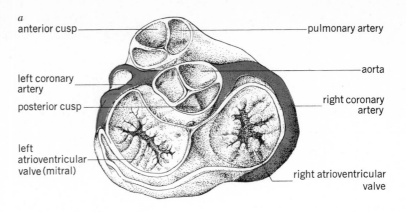

a

anterior cusp — — pulmonary artery

left coronary artery — — aorta

posterior cusp — — right coronary artery

left atrioventricular valve (mitral) —

— right atrioventricular valve

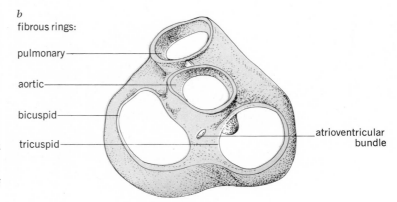

b
fibrous rings:

pulmonary —

aortic —

bicuspid —

tricuspid —

— atrioventricular bundle

FIGURE 15·3

Structure of the heart: *a* valves of the heart viewed from above; *b* fibrous framework (cardiac skeleton); this is the framework of dense collagenous and connective tissue fibers that separates the musculature of the atria from that of the ventricles.

The mitral valve closes as the left ventricle starts to contract. The blood is forced up through the only exit available, the large *aorta*, or aortic artery. At the base of the aorta three *aortic semilunar valvula* prevent backflow of a column of blood in the aorta into the left ventricle as it relaxes. Branches of the aorta distribute the blood all over the body (Fig. 15.3).

The heart repeats this action 60 to 80 times per minute throughout life.

PERICARDIUM   The heart and the bases of the great blood vessels are enclosed in a two-layered membrane called the *pericardium*. If one side of a hollow rubber ball is forced in against the other side, it forms a fair example of the shape of this double-layered membrane (Fig. 15.4). The outer layer is composed of fibrous tissue, and the more delicate, inner layer is a serous membrane. The serous membrane is closely applied to the heart, but it also lines the fibrous pericardium. The serous portion forms a closed sac, the *pericardial cavity*. The heart does not lie in a cavity, however; it is merely covered by two layers of membrane. The pericardial cavity is only a potential cavity under normal conditions; the outer serous lining is in contact with the inner serous membrane, with only a very small amount of pericardial fluid to prevent friction as the heart beats.

ENDOCARDIUM   The cavities of the heart are lined by a serous membrane called the *endocardium*. This layer is continued over the valves and chordae

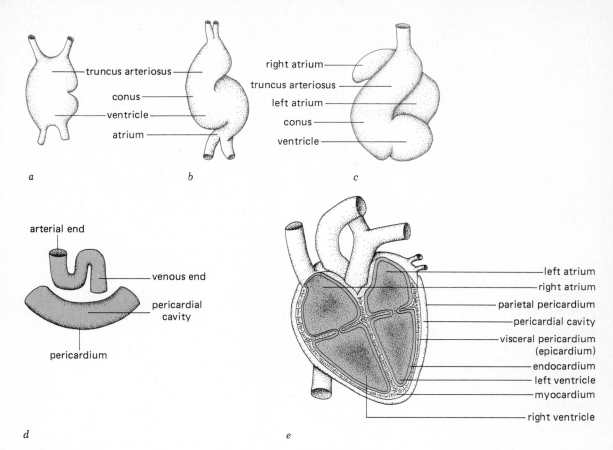

FIGURE 15·4

a-c Development of the embryonic heart. *d-e* Relationship of the developing pericardium to the embryonic heart and the fully developed heart.

tendineae and fuses with the membrane lining the large blood vessels of the heart. An inflammation of the lining membrane is termed *endocarditis*. Such inflammation is considered by some investigators to be a secondary infection commonly accompanying such diseases as scarlet fever and rheumatic fever. Since the endocardium covers the valves of the heart, as well as lining the openings guarded by the valves, the inflammation can alter the shape of the valves or their openings so that the valves no longer close tightly. In this case a small amount of blood may leak past the valves, causing a low sound in a stethoscope, called a *murmur*. If the valvular opening is constricted as a result of inflammation, the condition is referred to as *stenosis*. The same valve can be affected by both types of injury.

Fortunately, the heart ordinarily can compensate for considerable loss of efficiency from valves that do not close tightly. The heart with valvular deficiency often performs well enough to permit ordinary activity but reaches its peak load before a normal heart. Persons with well-defined heart murmurs are therefore advised not to engage in strenuous activities that might overtax their hearts.

MYOCARDIUM   The musculature of the heart is referred to as the *myocardium*. It is composed of cardiac muscle tissue, which is involuntary, finely striated, and has branching fibers. You may wish to review the discussion of cardiac muscle tissue in Chap. 6. The muscles of the heart are arranged in complex patterns of irregular whorls so that the heart contracts with a twisting, or wringing, motion. Heart muscle exhibits an all-or-nothing type of contraction. The atrial walls are relatively thin, while the ventricular walls are thick and strong. This is especially true of the left ventricle, which forces blood throughout the body. Its walls are much thicker than those of the right ventricle, which only moves the blood through the lungs.

CORONARY BLOOD VESSELS The heart muscle does not receive its oxygen and nutrition from the blood that passes through the chambers of the heart. Cardiac tissue has its own blood supply, which comes to it by way of the right and left *coronary arteries*. Any blockage of the coronary arteries is serious, therefore, because the heart muscle cannot continue its work without proper nourishment. The distribution of blood vessels throughout cardiac tissue is so complete that no other tissue is better nourished. After passing through a dense capillary network, the blood passes into a venous system that eventually empties into the coronary sinus and so into the right atrium (Fig. 15.5).

NERVE SUPPLY The nerve supply to the heart is considered to be largely regulatory. The heart of cold-blooded and warm-blooded animals, if prop-

FIGURE 15·5

Large arteries and veins associated with the heart, plus the coronary circulation: *a* The arch of the aorta and its branches. At the base of the aorta are the coronary arteries arising from aortic sinuses just above the semilunar valves. *b* The heart, posterior view, illustrating the distribution of the coronary veins. The blood is returned to the right atrium by the posterior cardiac veins, which open into the coronary sinus, but several anterior cardiac veins open directly into the right atrium.

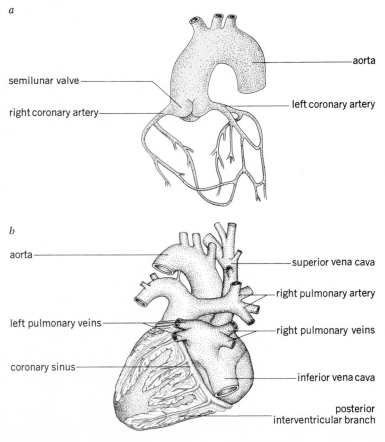

THE HEART, THE GENERAL CIRCULATION, AND THE LYMPHATIC SYSTEM

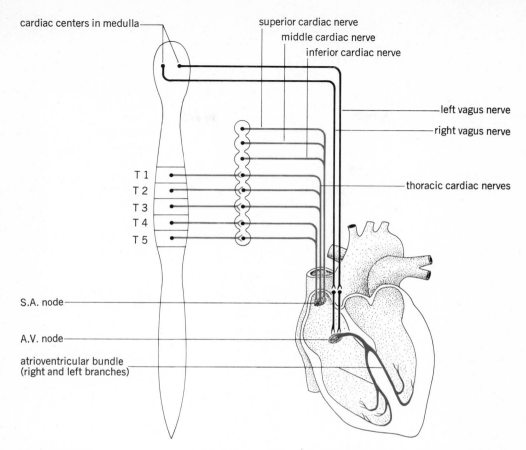

cardiac centers in medulla

superior cardiac nerve
middle cardiac nerve
inferior cardiac nerve

left vagus nerve
right vagus nerve

T 1
T 2
T 3
T 4
T 5

thoracic cardiac nerves

S.A. node

A.V. node

atrioventricular bundle
(right and left branches)

FIGURE 15·6

Autonomic innervation of the
heart.

erly prepared, may be completely removed from the body, and it will
contine to beat for some time without assistance from nerve connections.
The heartbeat is stimulated by the distension of the muscular wall as the
chambers of the heart fill with blood and by certain salts (electrolytes) in the
blood.

The heart is innervated by nerves from the sympathetic and parasympathetic systems. Sympathetic fibers reach the heart by way of the cardiac
nerves from the cervical ganglia and the upper four or five thoracic ganglia
of the sympathetic chain (Fig. 15.6). The adrenergic substance norepinephrine, released at the sympathetic postganglionic nerve endings, increases the
rate of firing at the sinoatrial node and so increases heart rate. Norepinephrine increases sodium ion conductance into the muscle cells of the ventricles,
and the force of contraction probably is increased by greater calcium ion
activity within the muscle fiber.

Parasympathetic neurons reach the heart by way of the vagus nerves.
Acetylcholine released at the endings of the parasympathetic fibers slows
heart rate by decreasing potassium loss, that is, by decreasing the rate at
which potassium ions move out of the muscle cell.

CONTRACTIONS OF THE HEART   *The sinoatrial node (SA node)*   The wave of
contraction that spreads over the atria originates in a small area of special-

ized cardiac tissue called the *sinoatrial node* (*SA node*). In the mammalian heart the SA node is located in the wall of the right atrium near the entrance of the superior vena cava.

Cardiac muscle cells have an inherent ability for rhythmic contraction and pacemaking. One group of cells excites another group through *gap junctions*. Therefore, the cells with the fastest rhythm will set the pace for all other cells. The cells with the most rapid rhythm are found in the SA node, which then becomes the pacemaker for the heart.

Depolarization of the SA node creates an action potential, and an excitation wave spreads over the atria. Specialized nerve fibers called *internodal tracts* conduct the nerve impulse from the SA node to the musculature of the right and left atria so that both atria contract at the same time. The internodal tracts also conduct the nerve impulse to the atrioventricular node.

The atrioventricular node (*AV node*)　The musculature of the atria, however, is separate and distinct from the ventricles. A layer of connective tissue around the crown of the heart separates the two sets of muscles. A strand of differentiated neuromuscular tissue in the septum (or wall), which completely separates the right and left sides of the heart in postembryonic life, affords a pathway by which excitation can reach the ventricles (see Fig. 15.3).

Located in the lower part of the right atrium and very close to the septum between the atria is another area of specialized tissue called the *atrioventricular node* (*AV node*) (Fig. 15.7). After the nerve impulse arrives at the AV node, there is a slight delay in its passage through the nerve fibers of the node. The fibers are so small that they conduct impulses very slowly. The conduction is delayed for about 0.1 s. This delay permits the ventricles to fill with blood before contracting.

The neuromuscular bundle and the purkinje network　The *neuromuscular bundle* (*bundle of His*) divides into right and left branches and is the next center in the heart to receive the nerve impulse. The neuromuscular bundle conducts

FIGURE 15·7

Cardiac conduction: sinoventricular conduction system. The heartbeat originates at the sinoatrial node, the pacemaker of the heart. It spreads rapidly over all the atrial musculature and converges on the atrioventricular node. Impulses pass down the atrioventricular bundle at about 500 cm/s; the conduction rate then slows to around 50 cm/s as impulses pass through the ventricular muscle enabling both ventricles to contract at the same time.

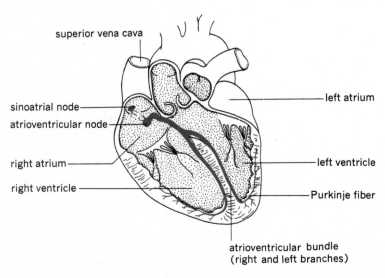

superior vena cava

sinoatrial node

atrioventricular node

right atrium

right ventricle

left atrium

left ventricle

Purkinje fiber

atrioventricular bundle
(right and left branches)

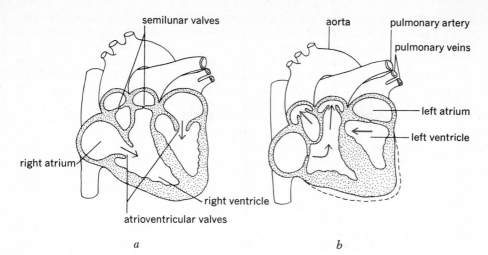

semilunar valves

aorta   pulmonary artery

pulmonary veins

left atrium

left ventricle

right atrium

right ventricle

atrioventricular valves

a   *b*

FIGURE 15·8

The heart cycle: *a* Relaxation phase, with the ventricles in diastole. The atrioventricular valves open as the ventricles fill with blood. The pulmonary and aortic semilunar valves are closed. *b* Contracting phase, showing the ventricles in systole. The atrioventricular valves are closed, and the semilunar valves are open.

impulses to the right and left ventricles, respectively. The terminal branches produce a dense network of minute filaments, the *purkinje network;* located underneath the endocardium of the ventricles.

The contraction wave   Since the muscles of the atria are the first to be stimulated, the atria are the first to contract. The wave of contraction spreads through normal cardiac tissue from SA node to AV node. But the rate of conduction through the neuromuscular bundle is about 10 times as fast as it is through nonspecialized cardiac muscle. The impulse therefore descends rapidly through the neuromuscular bundle and is distributed to all parts of the ventricular musculature by way of the purkinje network. The muscular walls of the ventricles then contract almost immediately after the atria contract. The conduction time from atria to ventricles is around 0.12 to 0.2 s in humans. The neuromuscular bundle is normally the only pathway by which impulses can be transmitted to the ventricles; it is, therefore, a very important passageway. In experiments with animals, if transmission across the neuromuscular bundle is interrupted by cutting or tying, the ventricles lose pace with the atria. The ventricles can be inhibited temporarily and prevented from contracting, but ordinarily they resume beating at a much slower rate than that maintained by the atria. Various conditions known as *heart block* can affect conduction in the human heart. A partial or complete blockage of conduction in the neuromuscular bundle slows the ventricular rhythm. If the blockage is complete, the ventricular contraction rate drops to around 35/min, while the atrial rate remains around 70/min.

THE CARDIAC CYCLE   If we look at the beating heart of an animal, we observe first that there is a regularity or rhythm to its movement. The atria ordinarily contract first, and this movement is followed by contraction of the ventricles. The contraction phase, during which blood is forced out of the chambers of the heart, is referred to as *systole* (Fig. 15.8). The term "systole" may refer to either atrial systole or ventricular systole, but when used alone it refers to the contraction of the ventricles. *Diastole* refers to the dilatation or relaxation phase of the cycle in which the chambers of the heart relax so that they may fill with blood. Unless atrial diastole is indicated, the term refs to the relaxation of the ventricles. The action of the heart is so arranged that the atria are already entering diastole by the time the ventricles begin the systolic phase. The ventricles then fill with blood, and there is a short period of *diastasis* during which the heart muscle is inactive but filling continues. This is a brief period of rest for the heart musculature. It is almost immediately ended by atrial systole as the cardiac cycle is repeated. As we mentioned in Chap. 6, cardiac tissue has a long refractory period and is not subject to summation or tetany.

During systole the pressure within the ventricles is high, higher than the pressure in the great arteries leaving the heart. As the semilunar valves are forced open, therefore, blood rushes out into these arteries. The atria are then already in their diastolic phase, the pressure within is low, and blood flows into them from the great veins. As the ventricles complete their relaxation phase, the pressure within these chambers drops below atrial pressure. The atrioventricular valves then open, and the ventricles begin to fill with blood even though the atria are still in a resting stage. Atrial systole produces a final and positive surge of blood through the atrioventricular valves, which then close as the ventricles start to contract (Fig. 15.9).

Heart sounds If you listen to a heart by placing your ear against the chest over the heart or by means of a stethoscope, two well-defined sounds may be

FIGURE 15•9

The cardiac cycle as illustrated by pressure and volume charts from the left ventricle. AV, atrioventricular; LV, left ventricle; SL, semilunar valve. Note that as the left ventricle contracts, the pressure exceeds that in the aorta.

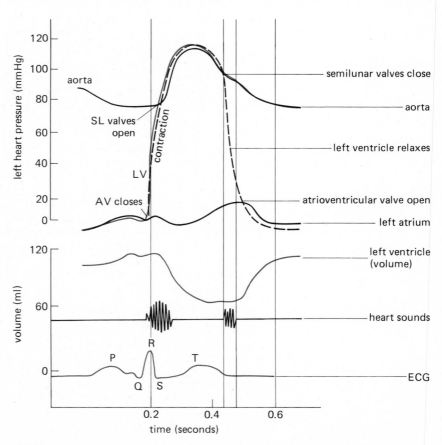

heard. The first sound is lower and of longer duration than the second sound. Though the sound may be caused by vibrations arising from various sources, it is primarily associated with the closure of the atrioventricular valves. The first and second heart sounds are commonly expressed by the syllables lub-dup.

The second sound is associated with the closure of the semilunar valves as the ventricles enter their diastolic phase. This sound is louder, sharper, and of higher pitch than the first sound. The sounds are altered in hearts with defective valves.

Heart rate So many factors influence the heart rate that statements concerning rate only apply to specified conditions. It is often said that the average heart beats about 70 beats/min. Such statements most likely refer to studies on the heart rate of young men, seated, and in a postabsorptive state. The heart rate for young women is somewhat faster, on an average, than that of young men. One method of judging the sex of the fetus during the latter part of pregnancy is to count the number of heartbeats per minute. The heart rate of the male fetus usually is around 130 to 135 beats/min, while that of the female is ordinarily 140 to 145 beats/min.

The heart rate of a person in a recumbent position commonly is 58 to 68 beats/min. A typical rate in the standing position would be around 80 beats/min. The digestion of a meal increases the strength and rate, partly because of activity of the muscles of the stomach and intestine. Muscular exercise increases the heart rate; riding a bicycle may increase the rate to 120 to 130 beats/min. Different investigators have obtained some variation in heart rate measurements. The average for young male college students with some athletic training is perhaps as follows:

Reclining position: 58 beats/min
Standing position: 78 beats/min
Immediately after light exercise: 90 beats/min

Highly trained male athletes have much lower heart rates, perhaps ranging down to 40 beats/min when in a resting position.

The following average heart rates (beats/min) were reported for a group of healthy young women: reclining, 64.7; seated, 71.8; standing, 85.1.

Emotional excitement affects the heart rate. Many people feel their hearts "race" as a result of fear. Others feel a decrease in rate and strength of heartbeat from overwhelming fear.

Age and size influence heart rate. A newborn child begins postnatal life with a heart rate of about 140 beats/min. As the child grows older and larger, heart rate slows. Typical figures are as follows:

Three years old: around 100 beats/min
Ten to twelve years old: around 90 beats/min
Young male adult: around 70 beats/min
The aged: around 75 to 80 beats/min

Heart rate is also affected by the secretions of the endocrine glands, especially the hormones of the thyroid and adrenal glands. Thyroxine, secreted by the thyroid gland, increases the heart rate, while epinephrine, secreted by the adrenal glands, increases both the rate and strength of the heartbeat.

CARDIAC OUTPUT AND STROKE VOLUME   The *stroke volume* is the quantity of blood ejected by one side of the heart per beat. Each time the heart of a resting man beats, about 80 ml of blood are ejected from each ventricle. If the heart beats 70 beats/min, the amount of blood ejected from one ventricle would be 70 beats/min × 80 ml/beat = 5600 ml/min, or 5.6 l/min. This is called the *minute volume*. This quantity is useful in estimating the capacity of the heart under various conditions.

ELECTROCARDIOGRAPH AND ELECTROCARDIOGRAM   The *electrocardiograph* is an instrument that amplifies and records the small electric potentials produced by the beating heart. It is essentially a string galvanometer hooked to a recording device. The *electrocardiogram (ECG)* is a record of differences in electric potential produced by the contraction of fibers of heart muscle. Electric currents spread over the heart and into the tissues around the heart. Some of these electric currents can be recorded from the surface of the body by electrodes placed on the arms, legs, and chest.

The standard lead system for recording on ECG consists of five elec-

FIGURE 15·10
Figure illustrating the three standard leads for the electro-cardiograph. The chest leads are only approximate.

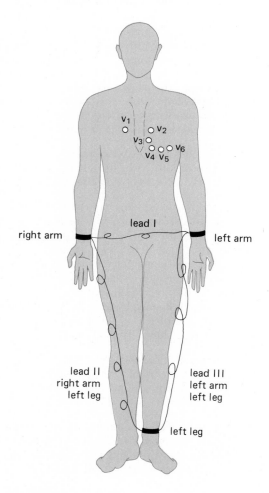

trodes: one on each arm, one on each ankle, and one on the chest. The leads from the arms to the left leg constitute *Einthoven's triangle,* a standard arrangement. The right leg electrode is inactive and serves as the ground for all leads.

The chest electrode may be placed in six different positions, each called a *lead,* as the ECG is taken. The chest leads are V_1, located in the fourth intercostal space to the right of the sternum; V_2, placed just to the left of the sternum; V_3, located below V_2 and between V_2 and V_4; V_5, placed to the left of V_4 and toward the axilla; and V_6, positioned to the left of V_5 (Fig. 15.10).

The ECG is a record of the electric events during each contraction; the events of each cycle are typically broken into P, QRS, and T waves. The P wave represents the initial excitation of the atria, as depolarization sweeps over them. The P wave indicates the contraction of the atria. The second wave, the QRS wave, occurs 0.1 or 0.2 s later and represents the depolarization of the ventricles and their contraction. The final T wave signifies repolarization of the ventricles. The letters designating the waves are arbitrary and do not refer to any part of the heart (Fig. 15.11).

The ECG provides an accurate record of heart action that is useful in the study and diagnosis of heart conditions. Interpreting the electrocardiograph requires a thorough knowledge of heart function. For example, the normally brief atrioventricular interval betweeen P and QRS waves may be lengthened in cases of partial heart block. Heart block may be caused by injury to the conduction system between the atria and the ventricles that involves the AV node and the neuromuscular bundle.

THE GENERAL CIRCULATION: ARTERIES, VEINS, AND CAPILLARIES

Circulatory system
Arteries
Veins
Capillaries

The circulatory system includes a series of tubes through which blood is pumped to all parts of the body. The heart is the motive force in this system, while arteries, veins, and capillaries form the tubes. The *arteries* in the atrial system are responsible for carrying the blood from the heart to the tissues through a series of successively smaller branches. The occasionally used term *arterial tree* refers to the arterial trunk and its treelike branching. The trunk arteries are often very large. The aorta in an adult is about 1 in in diameter at its base. Successive branches have increasingly smaller diameters, and the smallest arteries may have diameters of only 0.3 mm. The total diameter of all the branches, however, is much greater than the diameter of the trunk artery, and so the blood flows with decreasing pressure and velocity through the smaller arteries and arterioles toward the capillaries.

The *veins* of the venous system gather the blood from the tissues and return it to the heart. Starting with the smallest venules, the succeeding venous branches become larger and larger as one vein joins another. The veins bringing blood into the heart are large, trunk blood vessels, which often lie side by side with arteries of equal size.

The capillaries are minute tubes that connect the arterioles and the venules. Smaller than the diameter of a hair, many capillaries are so tiny that even red blood cells must pass through in single file. Capillaries vary in diameter from 5 to 20 μm and are ordinarily about 0.5 mm in length. The capillary wall consists of flat, endothelial cells that are only one cell layer thick (Fig. 15.12). Oxygen and food materials pass to the tissues through the capillary wall, while carbon dioxide and other breakdown products of metabolism enter the bloodstream at this point and are carried away.

FIGURE 15·11

The drawing is taken from an electrocardiogram recording the action of a normal heart. The P wave indicates the contraction of the atria; the QRS period represents the contraction of the ventricles, whereas the T wave corresponds to the relaxation of the ventricles. The P, QRS, and T waves represent one heart cycle. Irregularities in these waves may indicate an abnormal heart condition. (*Courtesy of the American Heart Association.*)

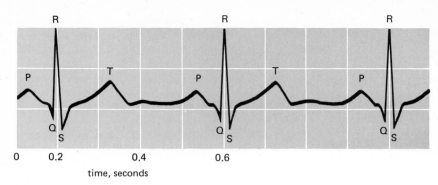

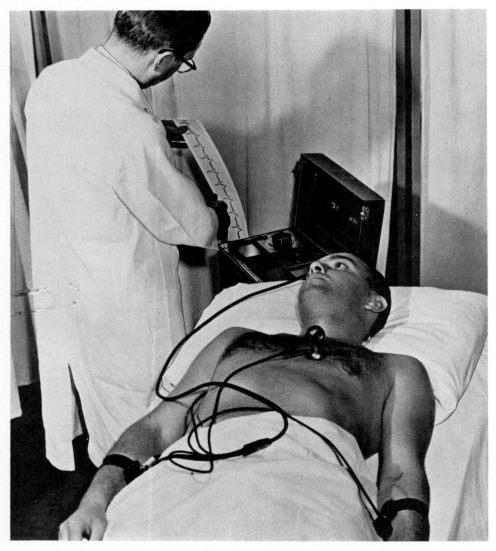

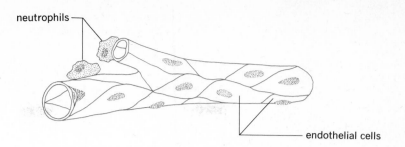

FIGURE 15·12

Portion of a blood capillary, illustrating the endothelial cells of the capillary wall. Two phagocytic neutrophils are shown outside the capillary wall.

STRUCTURE OF ARTERIES AND VEINS   The conducting vessels of the circulatory system, including those supplying the heart, develop from endothelial tubes, which persist as the endothelial linings of these structures. The *endothelial intima* provides a smooth surface over which the blood moves. The *arterial wall* consists of three layers: the endothelial lining, which contains some elastic tissue; an intermediate layer of smooth muscle and elastic tissue; and an outer layer of loose connective tissue, composed of both collagenous and elastic fibers. Large arteries, such as the aorta, possess little muscular tissue; their walls are composed largely of elastic tissue. This elastic tissue provides a tough yet resilient wall with extensibility and elasticity.

Veins have the same three layers in their walls as arteries, but since the middle layer is poorly developed, the walls of veins are thin and contain little muscle or elastic tissue. Walls of veins tend to collapse after death, while the thicker-walled arteries retain their shape (Fig. 15.13). Some veins, especially those of the extremities, have flaplike valves, which prevent backflow of the blood.

PULMONARY AND SYSTEMIC ARTERIES   There are two main divisions of the vascular system: the *pulmonary* and the *systemic*. The pulmonary circulation consists of the pulmonary artery and its branches, the capillary network

FIGURE 15·13

The structure of blood vessels: *a* Veins. Veins have thinner walls than arteries, largely because the media is thinner. *b* Artery. Note the thickness of the media, a mixture of elastic connective tissue and smooth muscle fibers.

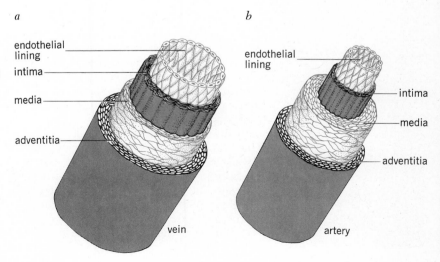

in the lungs, and the pulmonary veins. The pulmonary artery, unlike other arteries, carries venous blood from the right ventricle to the lungs. It divides into the right and left pulmonary arteries, which supply the right and left lungs, respectively. The blood, after passing through a capillary network within the lungs, flows into venules and small veins forming the pulmonary veins. There are four pulmonary veins, two from each lung. These are short veins without valves; they return oxygenated blood from the lungs to the left atrium of the heart.

The aorta and its branches supplying blood to the tissues of the body, as well as the capillaries and the veins returning blood to the heart, compose the systemic circulatory system.

PRINCIPAL SYSTEMIC ARTERIES

Aorta
  Ascending aorta and arch
    of the aorta
  Thoracic aorta
  Abdominal aorta

Ascending aorta and arch
  Coronary
  Brachiocephalic
  Common carotid
  Carotid sinus
  Carotid body
  Subclavian

THE AORTA   The largest trunk artery is called the *aorta*. It arises from the left ventricle, ascends to a position above the heart, arches to the left, and then descends behind the heart. It is referred to by different descriptive names in different locations as the ascending aorta, the arch of the aorta, and the descending aorta. The section of the descending aorta above the diaphragm is called the *thoracic aorta*, while that below it is termed the *abdominal aorta* (Fig. 15.14).

The principal branches of the aorta are listed in Table 15.1.

BRANCHES OF THE ASCENDING AORTA AND ARCH OF THE AORTA   *Right and left coronary arteries*   The first branches to arise from the ascending aorta are the *coronary arteries*. They arise from dilatations at the base of the aorta just above the semilunar valves. The dilatations are called *aortic sinuses*. The coronary arteries grow around the crown of the heart, giving off branches to the muscles of the atria and ventricles. The coronary arteries are of great importance, since they supply the heart musculature (Fig. 15.5a).

Brachiocephalic artery   The first branch off the arch of the aorta is the *brachiocephalic artery*. A large artery in diameter but only 4 to 5 cm long, it gives rise to the right common carotid and to the right subclavian arteries (Fig. 15.14).

Carotid arteries   The *left common carotid artery* branches directly from the arch of the aorta. Both right and left common carotid arteries then pass obliquely upward, covered by several muscles, especially the sternomastoid muscles. At the upper level of the larynx each artery divides into external and internal carotid arteries. The *external carotid* in its course up the side of the head gives off many branches and rapidly decreases in size. The *internal carotid* passes upward from the bifurcation of the common carotid and then turns inward and enters the cranium. It supplies the anterior part of the brain, the forehead, the nose, and the orbit of the eye. The *ophthalmic artery* is the branch of the internal carotid that enters the orbit alongside the optic nerve. Orbital branches supply the extrinsic muscles of the eyeball and accessory structures of the eye; ocular branches supply the eyeball itself. The *central artery* of the retina and the *ciliary arteries* are branches of the ocular group. Branches of the internal carotid arteries (the anterior cerebral arteries), with the aid of other arteries supplying the brain, form an interesting circle at the base of the brain called the *arterial circle (circle of*

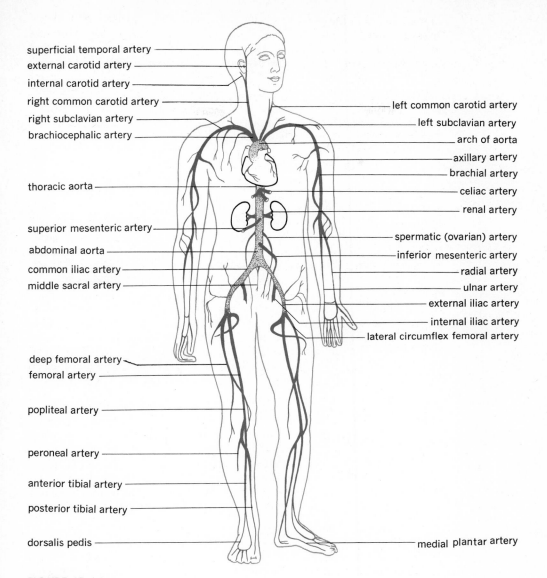

superficial temporal artery
external carotid artery
internal carotid artery
right common carotid artery
right subclavian artery
brachiocephalic artery

left common carotid artery
left subclavian artery
arch of aorta
axillary artery
brachial artery
celiac artery
renal artery

thoracic aorta

superior mesenteric artery
abdominal aorta
common iliac artery
middle sacral artery

spermatic (ovarian) artery
inferior mesenteric artery
radial artery
ulnar artery
external iliac artery
internal iliac artery
lateral circumflex femoral artery

deep femoral artery
femoral artery

popliteal artery

peroneal artery

anterior tibial artery

posterior tibial artery

dorsalis pedis

medial plantar artery

FIGURE 15·14
The principal arteries.

Willis). It is a remarkable anastomosis of arteries aiding distribution of blood to the brain (Fig. 15.15).

Carotid sinus and carotid body   A slight enlargement called the *carotid sinus* arises at the base of the internal carotid artery. In the region between the bases of the external and internal carotid arteries lies a small structure composed of epithelioid tissue termed the *carotid body* (Fig. 15.16). Similar structures in the arch of the aorta are called the *aortic bodies*. Sensory receptors of these regions play an important part in the reflex regulation of circulation and respiration.

TABLE 15·1
Branches of the aorta

Ascending aorta and arch	Thoracic aorta	Abdominal aorta
Coronary	Pericardial	Celiac
Brachiocephalic	Bronchial	Superior mesenteric
Right common carotid	Esophageal	Renal
Left common carotid	Mediastinal	Spermatic or ovarian
Right subclavian	Intercostal	Inferior mesenteric
Left subclavian		Inferior phrenic
		Lumbar
		Middle sacral

Subclavian arteries   Although the *right subclavian artery* arises as a branch of the brachiocephalic artery, the *left subclavian artery* arises directly as a branch off the arch of the aorta a little above the clavicle and then passes below it into the shoulder area. The subclavian is a good example of the custom of giving the same artery different names as it passes through different regions. The *subclavian* becomes the *axillary* as it passes through the axilla of the arm. In the area above the elbow, it is the *brachial* artery. At the elbow it divides into the *ulnar* and *radial arteries*. The ulnar is the larger and passes down the ulnar side of the forearm. The radial, looking like a direct continuation of the brachial, but much smaller, passes down the radial side of the forearm toward the thumb. The pulse is commonly taken from the radial artery at the wrist. Branches of the two arteries anastomose in the palm through the volar arches (see Fig. 15.14 and Plate 7).

BRANCHES OF THE THORACIC AORTA   The arteries of the thoracic region are numerous but small. Included in this group are the pericardial, bronchial, esophageal, mediastinal, and intercostal arteries. Briefly, the *pericardial arteries* are small arteries on the posterior surface of the pericardium. Branches of the *bronchial arteries* supply the walls of the bronchial tubes and the tissues of the lungs. The *esophageal branches* (usually four or five small branches) anastomose with other small arteries to form a network along the esophagus. The *mediastinal arteries* are numerous small branches

Thoracic aorta
  Pericardial
  Bronchial
  Esophageal
  Mediastinal
  Intercostal

FIGURE 15·15

Arteries at the base of the brain. The arterial circle (circle of Willis) is formed by the two anterior cerebral arteries (which are branches of the right and left internal carotid arteries) and the anterior communicating artery; in addition there are the two posterior cerebral arteries and the posterior communicating arteries.

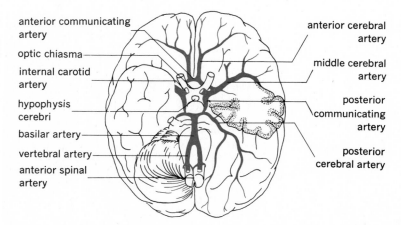

anterior communicating artery

optic chiasma

internal carotid artery

hypophysis cerebri

basilar artery

vertebral artery

anterior spinal artery

anterior cerebral artery

middle cerebral artery

posterior communicating artery

posterior cerebral artery

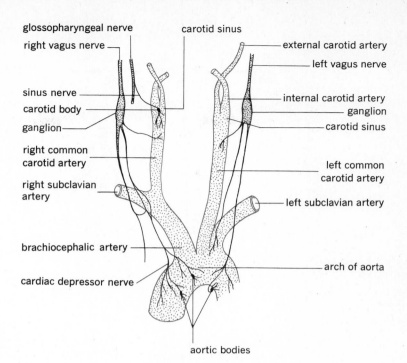

right vagus nerve ─
glossopharyngeal nerve
carotid sinus
external carotid artery
left vagus nerve

sinus nerve ─
carotid body ─
ganglion─

internal carotid artery
ganglion
carotid sinus

right common
carotid artery

left common
carotid artery

right subclavian
artery

left subclavian artery

brachiocephalic artery ─

arch of aorta

cardiac depressor nerve ─

aortic bodies

FIGURE 15·16

The branches off the arch of
the aorta and the location of
the carotid and aortic bodies.

in the posterior part of the mediastinum. They supply the lymph nodes and
areolar tissues of this region. Nine pairs of *intercostal arteries* arise from the
dorsal side of the thoracic aorta and supply the muscles and skin of the back
as well as the intercostal muscles.

Abdominal aorta
  Celiac
  Superior mesenteric
  Renal
  Spermatic
  Ovarian
  Inferior mesenteric
  Phrenic
  Lumbar
  Middle sacral

BRANCHES OF THE ABDOMINAL AORTA   The abdominal aorta diminishes
considerably in size as it progresses posteriorly, since it gives off a number of
large branches. Its branches, with few exceptions, supply the walls and
viscera of the abdominal cavity.

Celiac artery   The first large branch of the aorta below the diaphragm is the
celiac artery (Fig. 15.17). The celiac has three branches: the *left gastric,* the
splenic, and the *hepatic.* The left gastric is a small artery that passes from
left to right along the lesser curvature of the stomach. Toward the right it
forms an anastomosis with the right gastric branch of the hepatic artery. The
gastric arteries give off branches to both the anterior and posterior surfaces
of the stomach. The splenic is the largest branch of the celiac artery. It
extends behind the stomach and above the pancreas to the spleen, where its
branches enter at the hilus to supply the splenic tissue. Small branches are
given off to the pancreas. Close to the spleen, a large branch of the splenic
artery, the *left gastroepiploic,* passes around the lower or greater curvature
of the stomach and anastomoses with the *right gastroepiploic.* The epiploic
arteries give off numerous branches to the stomach and greater omentum.
The hepatic artery supplies the tissues of the liver. Among its several
branches is the *cystic artery,* which branches over the gallbladder.

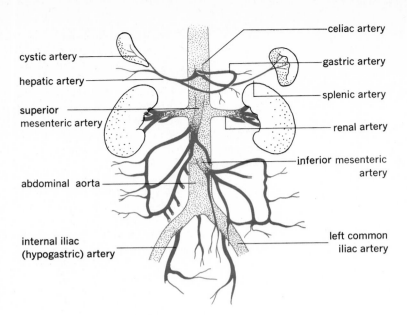

celiac artery

cystic artery

gastric artery

hepatic artery

splenic artery

superior
mesenteric artery

renal artery

inferior mesenteric
artery

abdominal aorta

internal iliac
(hypogastric) artery

left common
iliac artery

FIGURE 15·17

Principal branches of the abdominal aorta.

Superior mesenteric artery   The *superior mesenteric* is a large artery that sends branches to the greater part of the entire intestinal tract with the exception of the upper part of the duodenum and the distal part of the colon. Its branches use the mesenteric tissues for support in order to reach the intestine. It has its origin about 1.2 cm below the origin of the celiac artery.

Renal arteries   The *renal arteries* arise laterally from the aorta and are distributed to the kidneys. They are large arteries, each breaking up into several branches before entering the hilus of the kidney.

Spermatic arteries   Arising from the anterior part of the aorta below the renal arteries, the *spermatics* are long, slender arteries that course downward and pass through the inguinal canals in the male. They extend into the scrotum, and their branches terminate in the testes (See Fig. 15.14).

Ovarian arteries   The female *ovarian arteries* are homologous with the spermatic arteries of the male. They arise at a similar location and descend into the pelvis, where each is distributed to an ovary. One branch forms an anastomosis with the uterine artery.

Inferior mesenteric artery   Branching from the aorta not far above the place where the aorta itself divides to form the two common iliacs, the *inferior mesenteric* supplies the latter part of the colon and partially supplies the rectum. It is smaller than the superior mesenteric artery.

Phrenic arteries   Small arteries supplying the diaphragm are the *phrenic arteries.*

Lumbar arteries   Similar to the intercostal arteries, the *lumbar arteries* arise from the posterior surface of the aorta in the region of the first four lumbar vertebrae. There are usually four pairs of lumbar arteries, which supply the muscles and skin of the lumbar region of the back.

Middle sacral artery   A small artery, the *middle sacral artery*, originates from the posterior surface of the aorta about 1.2 cm above its bifurcation into the common iliac arteries. It descends medially along the anterior surface of the fourth and fifth vertebrae, the sacrum, and the coccyx. Terminal branches supply nodules on the anterior surface of the coccyx, which compose the coccygeal gland.

ARTERIES OF THE PELVIS AND LOWER EXTREMITIES   *Common iliac arteries*   The aorta divides at the level of the fourth lumbar vertebra and gives rise to the *common iliac arteries*. Each artery passes downward for about 5 cm, diverges, and gives rise to two branches: the internal iliac, or hypogastric, and the external iliac.

Internal iliac, or hypogastric, arteries   The *internal iliac artery* has numerous branches that supply the pelvic muscles and viscera. Branches extend also to the gluteal muscles, to the medial area of each thigh, and to the external genitalia. The *uterine artery* is an important branch in the female. It becomes the largest branch of the internal iliac during pregnancy. Pelvic viscera supplied by branches of the internal iliac include the urinary bladder, the rectum, the prostate gland in the male, and the uterus and vagina in the female.

Internal iliac arteries in fetal circulation   The internal iliac arteries are large, vital arteries in the fetal circulatory scheme, and they represent a direct extension of the common iliacs in fetal development. Each internal iliac gives off branches to the pelvic viscera and then extends up the anterior wall of the abdomen as an umbilical artery. Reaching the umbilicus (navel), each umbilical artery passes out through the umbilical cord to the placenta (see Fig. 15.21). After the umbilical cord is severed at birth, the greater part of the abdominal umbilical artery atrophies and leaves in its place a strand of fibrous tissue called the *lateral umbilical ligament*. The proximal part remains as the umbilical artery of postnatal life. The umbilical artery gives rise to the vesical arteries, which supply the urinary bladder.

External iliac artery   Extending from the base of the internal iliac about 10 cm downward, the *external iliac artery* diverges through the pelvis to the point where it passes underneath the inguinal ligament; here it enters the thigh and becomes the *femoral artery*. Like the subclavian artery, the external iliac is called by different names in different parts of its course.

Femoral artery   After passing below the inguinal ligament, the *femoral artery* lies in a compartment of the femoral sheath, with the femoral vein on its medial side and the femoral nerve on its lateral side. The blood vessels and nerve are close to the anterior surface of the thigh at this point. They are covered by some fatty tissue. Lymph nodes and lymphatics are found in this region also. The femoral artery sends small branches to the wall of the abdomen, to the external genitalia, and to the muscles of the thigh.

The *femoral artery* becomes the *popliteal artery* as it passes through the popliteal space at the back of the knee joint. Below the knee it gives rise to the *posterior tibial* and *anterior tibial arteries*. The posterior tibial gives off a large branch, the *peroneal artery*. Both arteries descend beneath the muscles of the calf of the leg. The posterior tibial arteries form the *plantar arteries* of the sole of the foot. The anterior tibial artery passes forward between the tibia and the neck of the fibula and extends downward close to the tibia. On the anterior surface of the ankle bones it becomes the *dorsalis pedis artery* and branches over the top of the foot.

THE VENOUS SYSTEM

The venous system is responsible for returning blood to the heart. The large veins that empty into the heart originate in the tiny venules that receive blood from the capillary network. Through a system of coalescence of ever-larger vessels, the large trunk veins are formed. The systemic veins are the pathway for the return flow from the systemic arteries, and they return the blood from all over the body to the heart. Since the flow of blood is from smaller vessels to larger ones, it is advantageous to study the veins in this manner—not by working out from the large vessels at the heart, as in the study of the arterial system. The following are some of the principal veins (see Fig. 15.18 and Plate 9).

Pulmonary veins

PULMONARY VEINS   The blood returns to the heart by way of the pulmonary veins. Originating in the capillary network of the alveoli, successively larger branches finally form two pulmonary veins emerging from each lung. They empty separately into the left atrium of the heart. There are no valves in these veins.

Veins of the heart
  Cardiac veins
  Coronary sinus

VEINS OF THE HEART   The return flow from the coronary arteries is taken up by the *cardiac veins*, most of which empty into a large sinus, the *coronary sinus*, which empties into the right atrium of the heart. Its opening is partially protected by a valve, consisting of a single semicircular fold of membrane. The valve of the coronary sinus prevents backflow of blood into the sinus during contraction of the right atrium (see Fig. 15.5*b*).

Veins of the head and neck
  External jugular
  Internal jugular
  Brachiocephalic
  Superior vena cava
  Azygos
  Hemiazygos

LARGER VEINS OF THE HEAD AND NECK   *External jugular veins*   Superficial veins of the scalp and the deeper veins of the face flow into the *external jugular vein*. The external jugular descends over the sternomastoid muscle, but underneath the platysma, and empties into the *subclavian vein* on either side (see Plate 10).

   The brain receives a very abundant blood supply from the internal carotid arteries, and numerous veins on the surface and within the brain gather the blood and return it through the internal jugular veins. Branches of external or surface veins can be seen, for the most part, in the sulci or gyri of the cortex. Deep within the substance of the brain are many veins that drain into large sinuses. These veins and sinuses have very thin walls and possess no valves. Veins within the brain lack a muscular coat.

Internal jugular veins   The *internal jugular vein* arises as a continuation of a brain sinus. It is a large vein that descends laterally beside the common carotid artery. At its base it joins the subclavian vein; together they form the

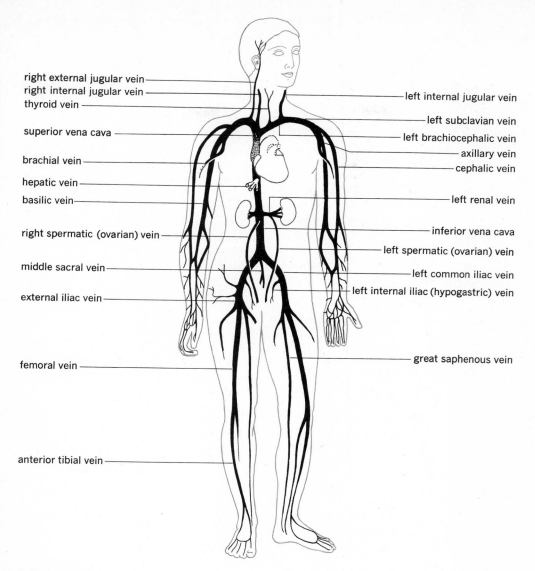

right external jugular vein
right internal jugular vein
thyroid vein

superior vena cava

brachial vein

hepatic vein

basilic vein

right spermatic (ovarian) vein

middle sacral vein

external iliac vein

femoral vein

anterior tibial vein

left internal jugular vein

left subclavian vein
left brachiocephalic vein
axillary vein
cephalic vein

left renal vein

inferior vena cava

left spermatic (ovarian) vein

left common iliac vein
left internal iliac (hypogastric) vein

great saphenous vein

FIGURE 15·18

The venous system. Only the major veins are shown.

brachiocephalic vein. In addition to its principal function of returning the blood from the brain, the internal jugular receives tributaries from the face and from the neck. The right vein is usually larger than the left, and both are much larger than the external jugular veins.

Brachiocephalic (innominate) veins The right and left *brachiocephalic veins* vary in length and position. The right brachiocephalic is almost vertical in position and appears to be an extension of the right internal jugular vein. It is only about 1 in in length. The left brachiocephalic is almost horizontal in position as it passes above the heart to join the right brachiocephalic and form the *superior vena cava.* It is 6.2 to 7 cm in length. The brachiocephalic veins receive the return flow from the internal jugular and subclavian veins, but in addition, the vertebral, deep cervical, inferior thyroid, and internal thoracic veins are tributaries.

Superior vena cava Formed by the confluence of the right and left brachiocephalic veins, the *superior vena cava* extends downward about 7 cm and empties into the right atrium of the heart. It receives the azygos vein as a tributary. The superior vena cava contains no valves.

Azygos vein The *azygos vein* originates along the dorsal wall of the abdomen as the ascending lumbar vein. It ascends through the thorax along the right side of the vertebral column and empties into the superior vena cava just above the pericardium.

Hemiazygos vein Arising from the lumbar vein and occasionally from the left renal vein, the hemiazygos vein ascends to the left of the vertebral column as high as the eighth or ninth thoracic vertebra, where it turns to the right and enters the azygos vein. There is also a small *accessory hemiazygos vein.* The hemiazygos varies in regard to its size, position, and connections (see Fig. 15.25).

Veins of the upper extremity
Cephalic vein
Basilic vein
  Radial
  Ulnar
  Brachial
  Axillary
  Subclavian

VEINS OF THE UPPER EXTREMITY There are two sets of veins in the forelimb: deep and superficial. The *deep veins* are associated with arteries and are known as the *radial, ulnar, brachial, axillary,* and *subclavian* veins. The *superficial veins* form a great network of anastomoses with each other and with the deep veins. Both the superficial and deep veins have valves, and both groups finally open into the axillary and subclavian veins.

Cephalic vein The *cephalic vein* arises in the radial network of superficial anastomoses. Above the elbow it passes along the lateral side of the biceps muscle. At the shoulder it penetrates deeply and empties into the axillary vein.

Basilic vein The *basilic vein* has its origin in a venous network on the ulnar (little finger) side of the back of the hand. It curves around to the anterior surface before it reaches the elbow and continues upward along the medial side of the biceps brachii muscle. At the shoulder it unites with the *brachial vein* to form the *axillary vein.* The axillary becomes the *subclavian vein* in the area along the clavicle. The *thoracic duct* of the lymphatic system flows into the left subclavian vein at its junction with the internal jugular. On the right, the *right lymphatic duct* enters the right subclavian vein at the same junction (see Fig. 15.25).

Veins of the lower extremity
Superficial veins
  Saphenous
Deep veins
  Tibial
  Popliteal
  Femoral

VEINS OF THE LOWER EXTREMITY The veins of the lower extremity are divided into *superficial* and *deep veins,* as are the veins of the upper extremity. The superficial veins are formed of a great network of anastomoses close to the surface, while the deep veins are covered deeply with muscles and follow the larger arterial trunks.

Superficial veins The *great saphenous vein* is the longest vein in the body. It receives branches from the sole and dorsal aspect of the foot and crosses the

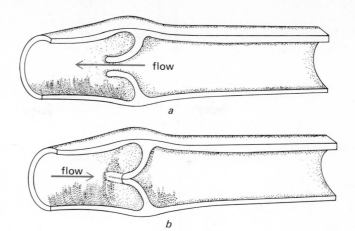

flow

a

b

FIGURE 15·19

Valves in a vein. Arrows indicate the direction of pressure. *a* Blood moving toward the heart, valve open. *b* Pressure reversed; pockets of the valve dilated, occluding passage and preventing backflow.

ankle on the medial side, just anterior to the medial malleolus. It ascends along the medial side of the leg and thigh and enters the *femoral vein* 2 cm or so below the inguinal ligament. There is also a *small saphenous vein,* which passes up the back of the leg and empties into the popliteal vein at the knee. The saphenous veins have valves (Fig. 15.19), but not as many as the deep veins of the lower extremity. The long saphenous vein is especially likely to enlarge under conditions of prolonged standing because of back pressure of the blood. Such enlarged veins are called varicose veins, which are described in the clinical aspects section at the end of the chapter.

Deep veins Below the knee there are two veins placed beside each artery. The names of the veins correspond to the names of the arteries that they accompany. The *anterior* and *posterior tibial veins* arise from veins of the foot. They ascend through the leg and drain into the *popliteal vein* at the back of the knee. The popliteal becomes the *femoral vein* as it passes through the upper two-thirds of the thigh. Just above the inguinal ligament it becomes the *external iliac vein.*

Veins of the pelvis and abdomen
Internal iliac
External iliac
Common iliac
Uterine
Inferior vena cava
Renal
Hepatic

VEINS OF THE PELVIS AND ABDOMEN The *internal iliac veins (hypogastric)* join the *external iliac veins* to form the *common iliac veins.* Tributaries to the internal iliac veins are numerous and carry the return flow from the internal iliac arteries. The *uterine veins* of the female are important tributaries to the internal iliacs. The common iliac veins unite to form the inferior vena cava at the level of the fifth lumbar vertebra.

The *inferior vena cava* is the largest vein that returns blood from the lower parts of the body. Formed by the union of the common iliac veins, it extends upward through the abdomen and thorax to the right atrium of the heart. It lies deeply protected by the viscera just anterior to the vertebral column and to the right of the aorta. Numerous small tributaries enter the inferior vena cava. For the most part these carry the return flow from smaller branches of the abdominal aorta, and their names correspond to names given to the arteries. The largest tributaries are the short renal veins, which enter at nearly right angles.

The *renal veins* return the blood from the kidneys. The left renal vein is a little higher, since the left kidney is higher than the right. It is also longer, crossing over the aorta. The left spermatic vein in the male, or the left ovarian in the female, and the left suprarenal are tributaries (Fig. 15.18). The renal circulation provides for the elimination of waste products from the blood (see Chap. 18).

The *hepatic veins* arise in the liver and flow into the inferior vena cava. There are usually two or three large veins and several small veins. They carry the return flow from the hepatic artery and the blood that comes to the liver through the portal system.

HEPATIC PORTAL SYSTEM Veins ordinarily transmit blood directly back to the heart. In certain pathways, however, the blood passes through a capillary network before returning to the heart. Such a venous pathway has capillaries at either end and is called a *portal system* (Fig. 15.20). In humans the *hepatic portal system* takes up blood returning from the spleen, stomach, and intestine. The blood enters the portal vein and passes through capillaries and sinusoids in the liver before returning to the heart. The *splenic vein* returns blood from the spleen, stomach, and pancreas. The *inferior mesenteric vein,* returning blood from the left colon, is a tributary of the splenic vein. The *superior mesenteric vein* joins the splenic to form the *portal vein.* The mesenteric blood vessels lie on a supporting tissue called *mesentery.*

The *splenic vein* arises from the union of several branches at the spleen. It passes behind the pancreas, from which it receives several pancreatic veins. The left gastroepiploic vein from the greater curvature of the stomach is a tributary. The largest vein joining the splenic is the inferior mesenteric. This vein transmits blood from the rectum and the descending colon.

The *superior mesenteric vein* is the returning pathway for blood from the small intestine and the ascending and transverse colon. It is the larger of the two mesenteric veins. Many of its tributaries correspond to branches of the superior mesenteric artery. It joins the splenic vein at the level of the pancreas, and together they form the portal vein.

FIGURE 15·20

The hepatic portal system.

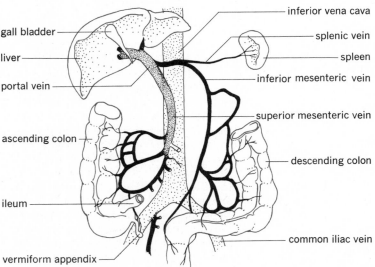

The *portal vein* is a short but large vein. It extends upward about 7 cm from its origin, near the neck of the pancreas, to the liver, where it divides into a right and left branch as it enters the liver. The left gastric vein from the lesser curvature of the stomach is a tributary, and at the liver the right branch usually receives the cystic vein from the gallbladder. Within the liver, the branches of the portal vein break up into enlarged capillarylike blood vessels called *sinusoids*. The blood leaves the liver through the sinusoids that unite to form small branches of the hepatic veins, by which blood enters the inferior vena cava (Fig. 15.20).

The portal system transports absorbed food products from the intestine to the liver. Glucose, in particular, is stored in the liver as glycogen. The portal system acts also as a reservoir for the storage of blood. The spleen has great capacity for holding blood in reserve and forms an important part of the portal storage system (see Chap. 14).

FETAL CIRCULATION

The circulatory plan of the fetus is somewhat different from that of the adult, largely because the lungs are not functional in the fetus. The *placenta* acts as an organ of respiration by providing for the exchange of O_2 and CO_2 between the maternal and fetal bloodstreams. The placenta also provides for passage of food products into the fetal circulation and for removal of waste products. The fetal portion of the placenta is a flat disk of highly vascular tissue attached to the lining of the uterus.

The *umbilical cord* extends from the placenta to the *umbilicus*, or navel, of the fetus. It contains the umbilical vein and two umbilical arteries. There is also a mucoid connective tissue in the umbilical cord, called *Wharton's jelly*.

The *umbilical vein* passes upward within the abdomen of the fetus to the undersurface of the liver. A considerable portion of the bloodstream passes through the liver. The umbilical vein gives off two or three branches that enter the liver directly. The portal vein joins a second large branch before it enters the liver. A smaller branch, the *ductus venosus*, bypasses the liver and enters the inferior vena cava. The blood that passes through the liver enters the inferior vena cava through the hepatic veins.

Upon entering the inferior vena cava, blood from the hepatic veins mixes with blood returning from the lower extremities of the fetus. The mixed blood enters the right atrium, and much of it passes through an opening in the wall between the two atria directly into the left atrium. The opening is called the *foramen ovale*. Some blood from the right atrium passes into the right ventricle and out to the lungs through the pulmonary artery, but the tissues of the lungs do not require a great amount of blood so long as the lungs are nonfunctional. Much of the blood that passes out over the left branch of the pulmonary artery flows directly into the dorsal aorta through a connecting blood vessel called the *ductus arteriosus*.

Blood from the left side of the heart is forced out into the aorta, as in the adult. The descending aorta supplies the abdominal viscera and the lower extremities, but a considerable amount of this blood is returned to the placenta by the two umbilical arteries. The umbilical arteries in the fetus are almost direct extensions of the internal iliac arteries.

The diagram of fetal circulation in Fig. 15.21 shows that the umbilical

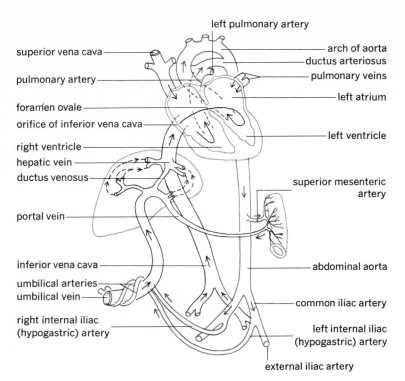

left pulmonary artery

superior vena cava

pulmonary artery

foramen ovale

orifice of inferior vena cava

right ventricle

hepatic vein

ductus venosus

portal vein

inferior vena cava

umbilical arteries

umbilical vein

right internal iliac
(hypogastric) artery

arch of aorta

ductus arteriosus

pulmonary veins

left atrium

left ventricle

superior mesenteric
artery

abdominal aorta

common iliac artery

left internal iliac
(hypogastric) artery

external iliac artery

FIGURE 15·21

Plan of the fetal circulation in a mature fetus. Arrows indicate the direction of blood flow.

vein carries the most highly oxygenated blood. Much of the food-laden blood passes through the liver, and this organ is proportionately large in the fetus. Since much of the blood from the umbilical vein goes to the left side of the heart directly, the blood passing over the arch of the aorta and to the brain is fairly rich in oxygen and food products.

MODIFICATIONS IN THE VASCULAR SYSTEM AT BIRTH   The lungs become functional at birth, and more blood begins to pass through the pulmonary arteries to the lungs. Respiration is stimulated when the placental supply of oxygen is cut off and the elimination of CO_2 through the placenta ceases. The CO_2 content of the blood rises, and the respiratory center is stimulated reflexly. The newborn infant, therefore, gasps for breath, and respiration is established. A secondary valvular fold covers the foramen ovale, incompletely at first, but it usually adheres to the septum and closes the opening completely toward the end of the first year after birth.

The ductus arteriosus degenerates and is modified into a ligament connecting the left pulmonary artery and the aorta. The umbilical vein and the ductus venosus also persist only as ligaments. Only the basal or proximal parts of the umbilical arteries remain functional. The distal parts that extend from the urinary bladder to the umbilicus degenerate within a few days after birth. In the adult, fibrous cords persist, representing the obliterated branches of these arteries.

The changes in circulation that occur at birth, while pronounced, are not revolutionary. As the fetal heart and lungs mature during fetal development, the amount of blood passing through the foramen ovale and the ductus arteriosus is regulated to meet the needs of the developing body. In the early stages of development the greater portion of the blood passes directly across the atrial septum to the left side of the heart. As the lungs mature, the foramen ovale becomes more occluded and more blood is directed to the lung tissues. At the time of birth practically all the blood entering the right side of the heart passes over the pulmonary artery to the fetus's lungs, so the changeover to respiration is not startling. It is not the same as suddenly putting a lid over a wide-open valve. The foramen ovale may be considered to balance the work of the atria. In the same way, the ductus arteriosus serves to balance the workload of the ventricles. At birth the ductus arteriosus is constricted, probably by the action of its smooth muscle. The constriction is made permanent by replacement of its cavity by connective tissue. Occasionally a congenital condition develops in which the foramen ovale or the ductus arteriosus fails to close properly. In such cases, pulmonary blood supply is inadequate, the blood is insufficently oxygenated, and a condition of *cyanosis* develops. Babies with such circulatory defects are commonly called *blue babies.*

Heart defects are not ordinarily as simple as those described above. The entrance to the pulmonary arteries may be too narrow, or the aorta may be displaced to the right, thus constricting both the aorta and the pulmonary artery, or the septum between the ventricles may not be complete, leaving an opening in the wall between the two lower chambers.

THE LYMPHATIC SYSTEM

The spaces between the cells that form body tissues are filled with tissue fluid. This tissue fluid is derived from the portion of blood plasma that filters through the walls of capillaries. A system of *lymphatic vessels* drains such fluid from extracellular spaces. When it is found within the lymphatic vessels, the fluid is called *lymph.* Since tissue fluid and lymph are essentially the same, this definition of lymph is not strictly adhered to.

Lymph is a clear fluid containing a low count of granular leukocytes and a varying number of lymphocytes. It may also contain a few erythrocytes. Lymph protein content is less than that of plasma, and the total phosphorus and calcium content is lower. Lymph contains enzymes such as amylase, maltase, protease, and lipase. It exhibits some ability to clot, but the process is slow and the clot soft.

Since the blood passing through the tissues is confined to capillaries, it normally never comes in direct contact with the cells. Lymph, on the other hand, which is outside the capillary wall, bathes the tissue cells and acts as a medium of exchange between the blood and the tissues. Nutrient materials can pass through the capillary wall and be carried by the tissue fluid to the cells. Waste materials of cell metabolism pass from the cell into the tissue fluid and then are absorbed into the bloodstream to be carried away.

Lymph flows very slowly. Its flow is aided by muscular contractions and the pressure built up by the filtration of fluid from the capillaries. Even in the larger lymphatic vessels the rate of flow is slow, and the pressure is low. Various estimates place the amount of flow in the human thoracic duct as averaging around 2 or 3 l/24 h. The pressure in lymphatic capillaries has been estimated at 1.5 to 3 mmHg, whereas pressure in blood capillaries commonly averages 15 to 30 mmHg.

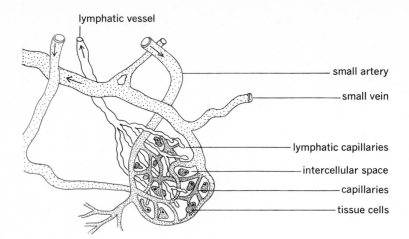

lymphatic vessel

small artery

small vein

lymphatic capillaries

intercellular space

capillaries

tissue cells

FIGURE 15·22
The relationship between the blood and lymphatic capillaries in the intercellular spaces.

Some research workers consider that the lymph is moved by a pumping action initiated at the distal end of lymphatic capillaries. As the lymph capillary fills with fluid, endothelial valve flaps close and periodic contractions move the lymph forward. Other research workers maintain that the movement of lymph largely results from the movement of body muscles and also from changes of pressure in the thorax during respiration.

The lymphatic system is a one-way, collecting system, concerned with gathering the tissue fluid, not with distributing it. The larger lymphatic vessels eventually drain into veins, so lymph enters the bloodstream and is distributed by the arterial system.

LYMPHATIC CAPILLARIES *Lymphatic capillaries* are the smallest vessels of the lymphatic system (Fig. 15.22). The lymphatic capillary somewhat resembles a blood capillary in structure. Both consist of a single layer of endothelial tissue, probably derived embryonically from venous endothelium. The terminal ends of lymphatic capillaries are closed. Lymph is absorbed from tissue spaces through *flap valves* in the delicate endothelial membrane (Fig. 15.23a and b). Lymphatic capillaries are wider and more irregular than blood capillaries. They anastomose readily and form elaborate plexuses.

LARGER LYMPHATICS Larger lymphatic vessels drain the capillary network. The walls of these vessels resemble the walls of veins in structure. There are three layers: the inner layer is composed of endothelial cells; the middle layer is largely muscular, with some fine elastic fibers; and the third, or outer, coat is largely connective tissue, with some smooth-muscle tissue. The muscle fibers in both the middle and outer coats are longitudinal and oblique. The larger lymphatics therefore are contractible, although the lymphatic capillaries are not.

The larger lymphatics have a constricted or beaded appearance caused by the presence of valves, which prevent backflow, on the inner surface (Fig. 15.24). The valves are bicuspid or tricuspid and are placed at shorter intervals than those of veins. Lymphatic capillaries do not contain valves.

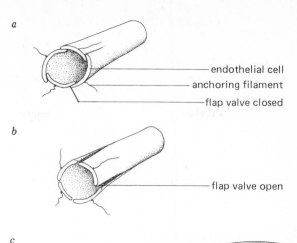

a

endothelial cell
anchoring filament
flap valve closed

b

flap valve open

c

skeletal muscle

anchoring filament

flap valve

endothelial cell

anchoring filament

skeletal muscle

FIGURE 15·23
Lymph capillary: *a* flap valve closed; *b* flap valve open (cross section); *c* longitudinal section.

FIGURE 15·24
A portion of a lymphatic vessel from cat mesentery, showing a valve. (*Courtesy of General Biological Supply House, Inc., Chicago.*)

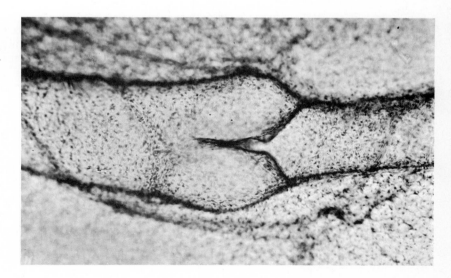

The larger lymphatics are white and pursue an irregular course, with frequent anastomoses. Unlike veins, they do not unite to form larger and larger vessels but tend to pursue individual courses. The deep lymphatics frequently form a plexus around blood vessels. The larger lymphatics are collecting vessels and flow from the capillary network toward lymph nodes. They are the afferent vessels of lymph nodes.

Functions of the lymphatics Lymph normally filters out of the blood capillaries, and most of it is resorbed through lymph capillaries. If there is any breakdown in this balance between the blood and the tissue fluid, there may be an accumulation of extracellular fluid in the tissues, resulting in *edema*.

A small amount of protein steadily filters out of blood capillaries and is returned by lymph capillaries. However, when a lymphatic malfunction such as a blockage occurs, protein (colloid) concentration may increase in the extracellular fluid until it equals the colloid concentration in the blood. In this case, the change in osmotic pressure moves water rapidly from the blood into the intracellular spaces, resulting in severe edema.

The lymphatics in the region of the small intestine have a special function in the absorption of digested fat from the intestine. The inner wall of the small intestine is lined with small processes called *villi*. Each villus contains a central lymph channel, or *lacteal*. The lacteals are continuous with the lymphatic vessels of the intestinal wall and the mesentery (see Fig. 19.17). Lymph carrying absorbed fat has a milk-white appearance and is called *chyle*. Intestinal villi and abdominal lymphatics may also appear white when filled with chyle. Chyle passes through the lymph nodes that are distributed along intestinal blood vessels and in the mesentery.

LYMPHATIC DUCTS AND CISTERNS The *cisterna chyli,* or *chyle cistern,* is a dilatation of the base of the thoracic duct that collects lymph. It is located just anterior to the body of the second lumbar vertebra. It receives several tributaries that bring lymph from the lower limbs, the intestine, the pelvic viscera, and the kidneys.

The thoracic duct arises in the cisterna chyli (Fig. 15.25). It passes upward through the diaphragm and mediastinum anterior to the vertebral column. It curves to the left, passing behind the left internal jugular vein, and opens into the left subclavian vein in the angle between the left subclavian and left internal jugular veins. The thoracic duct is about 18 in long and is provided with paired valves. It is called the *left lymphatic duct* in humans. Its tributaries drain all parts of the body except the upper right quadrant (Fig. 15.26).

The *right lymphatic duct* is very much reduced in humans. The collecting trunks that drain the upper right quadrant of the body commonly fail to unite, each entering the subclavian vein separately or in some combination. When the collecting trunks of this region do unite, they form a short right lymphatic duct, a little over 1 cm long, which enters the right subclavian vein at its junction with the right internal jugular vein.

LYMPH NODES Distributed along the course of lymphatic vessels are small bodies of lymphoid tissue called *lymph nodes* (Fig. 15.27). They are usually ovoid or bean-shaped, although the shape may vary considerably. The larger nodes exhibit a depressed area called a *hilus*. Afferent lymphatics enter the

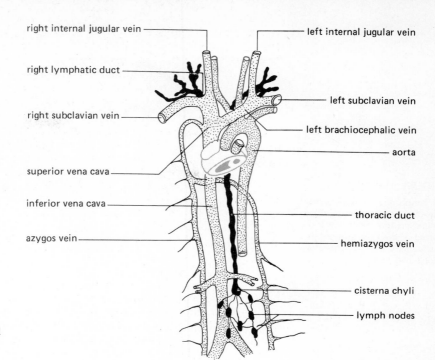

right internal jugular vein

right lymphatic duct

right subclavian vein

superior vena cava

inferior vena cava

azygos vein

left internal jugular vein

left subclavian vein

left brachiocephalic vein

aorta

thoracic duct

hemiazygos vein

cisterna chyli

lymph nodes

FIGURE 15·25

The thoracic and right lymphatic ducts. The azygos and hemiazygos veins are illustrated also.

FIGURE 15·26

Areas of the body drained by the right and left lymphatic systems. The colored area is drained by the right lymphatic duct. The rest of the body is drained by the thoracic duct.

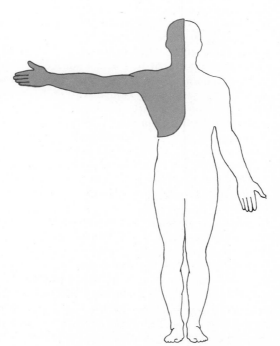

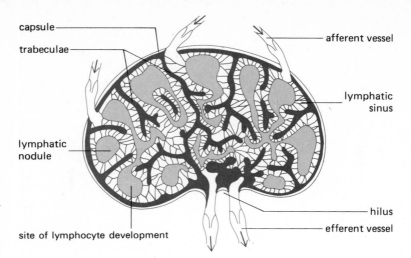

capsule

trabeculae

afferent vessel

lymphatic sinus

lymphatic nodule

site of lymphocyte development

hilus

efferent vessel

FIGURE 15·27

A lymph node. Lymphocytes develop and proliferate in germinal centers which are located in the central portions of follicles. Lymph sinuses are not blood sinuses.

node at various places, but efferent lymphatics emerge at the hilus. Blood vessels enter and also emerge at the hilus. The lymph node has an internal supporting framework of connective tissue that provides spaces called *lymph sinuses*. The lymph stream widens greatly as it passes through the node, and therefore the rate of flow is greatly reduced. The lymph filters through a maze of passageways lined with phagocytic cells. Such cells engulf bacteria or other foreign products from the lymph stream. A mechanical filtering process may also exist. Lymph nodes of the respiratory tract are often black from the filtering of carbon particles. Lymph nodes commonly become swollen and inflamed during severe bacterial infections.

The lymph nodes protect the body against the invasion of foreign substances. Lymphocytes of lymph nodes and spleen become immunologically competent cells that attack antigens introduced into the body.

Lymph nodes are so distributed that most, if not all, of the lymph passes through at least one lymph node before it returns to the bloodstream. Though lymph nodes are distributed throughout the body along lymphatic pathways, great collections of them lie in certain parts of the body. There are deep and superficial lymph nodes, just as there are deep and superficial lymphatic vessels. Groups of large nodes are located in the neck, the axillae, and the groin (see Fig. 15.28 and Plate 11). Six or seven small lymph nodes are embedded in the fat of the popliteal space behind the knee. The mesentery also supports numerous lymph nodes.

Lymphoid tissues Aside from lymph nodes, there are a number of lymphoid tissues in the body. Such tissues consist essentially of a framework of reticular tissue enclosing groups of cells, which are largely lymphocytes. Organs that are lymphoid in structure are the spleen, the thymus gland, the palatine tonsil, the pharyngeal tonsil (adenoids), and the solitary and aggregated lymph follicles. Lymphoid infiltrations are common in the mucous membrane of the intestinal tract and in the lungs. Solitary nodules are common in the intestine, and in the lower part of the small intestine these nodules form groups of lymphoid patches in the intestinal lining. These are

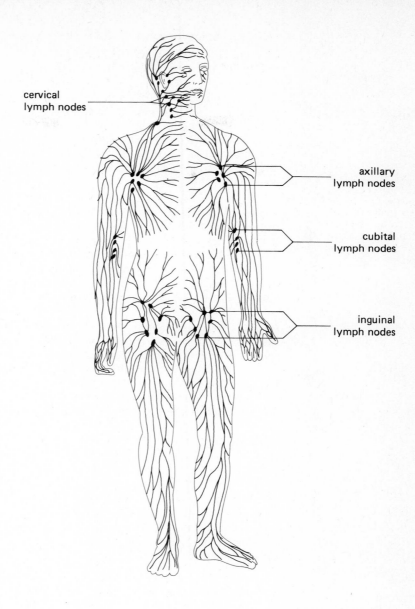

cervical
lymph nodes

axillary
lymph nodes

cubital
lymph nodes

inguinal
lymph nodes

FIGURE 15·28
The distribution of major
lymph nodes and plexuses.

aggregated nodules, or *Peyer's patches*. The solitary nodules are small, about the size of a pinhead, but the aggregated nodules may be found in patches 2 or 3 in long. They produce lymphocytes which may later become immunologically competent cells. Lymphoid tissues such as tonsils are generally larger in children than in adults.

Lymphoid tissues such as lymph nodes, spleen, and thymus give rise to lymphocytes. It has been observed that the lymphocyte count is higher after the lymph has passed through lymph nodes than it was before it entered. Lymph in the thoracic duct, having passed through many lymph nodes, can

have a lymphocyte count of 40,000 cells/mm³, whereas lymph from peripheral lymph spaces may have only a few hundred lymphocytes per cubic
millimeter.

CLINICAL ASPECTS

DISORDERS OF THE ARTERIES   Several disorders of the arteries cause many
circulatory problems. Cholesterol-containing plaques block the arteries in
a condition called *atherosclerosis.* Atherosclerosis is usually accompanied by
arteriosclerosis, a hardening of the arteries. These two conditions may cause
hypertension, or high blood pressure, and they may lead to severe heart
conditions.

Aneurysm   An *aneurysm* is a weakening of the arterial wall with subsequent
distension, or "ballooning out." The weakened areas usually occur along the
main arteries where blood pressure is high, such as in the aorta. The effects
of arteriosclerosis are considered to be the leading cause, but syphilis and
other diseases and injuries may also be involved. A large, bulging aneurysm
is dangerous, and surgery may be required to prevent its rupture. The weak
area can be replaced by synthetic tubing.

Ischemia of leg muscles   Some people experience cramping of leg muscles after
walking. This condition may be due to *ischemia,* a local reduction of blood
supply. It may also represent the effects of atherosclerosis, which have
caused the arteries to become so plugged that very little blood passes
through. The pulse below the arterial occlusion may be very weak, or there
may be no pulse at all. Diseased portions of the arteries may be replaced by
synthetic textile tubing with very good results.

Angina pectoris   When atherosclerosis affects the coronary arteries, several
types of cardiac conditions may develop. *Angina pectoris* is a painful heart
condition caused by coronary artery blockage. The angina patient often
feels a pain in the upper left chest after moderate exercise. The pain may
radiate under the left shoulder and down the left arm, and it is usually
relieved by sitting down and resting. Nitroglycerin tablets held under the
tongue and dissolved are the common medication prescribed. Other conditions may cause chest pain, so mild chest discomfort does not necessarily
indicate angina.

Coronary thrombosis   A *coronary thrombosis* is a clot in one of the coronary
arteries. This condition causes what we usually call a severe "heart attack."
Acute coronary thrombosis causes striking chest pain, with nausea, sweating,
and often collapse. Anyone suffering from these symptoms should be taken
to a hospital at once. If the thrombosis shuts off the blood supply to only a
small portion of the heart muscle, causing a small *myocardial infarction*
(death of heart muscle), the victim may survive. The infarcted portion of the
heart muscle dies and is replaced by noncontractile scar tissue. If the patient
survives the initial attack, alternate routes for arterial blood supply to the
heart then develop, and after a time the patient may be able to lead a
normal life.

DISTURBANCES OF HEART RHYTHM A strong, steady heartbeat is characteristic of a healthy heart, but in certain pathological conditions the steady rhythm of the heart beat is interrupted by irregular beats.

Atrial flutter and fibrillation Weak rapid beats of the atria, which are not coordinated with the beats of the ventricles, are called *atrial flutter*. Atrial flutter may develop into the more serious condition of *atrial fibrillation*. In atrial fibrillation, the atrial contractions fail to coordinate with those of the ventricles, so that the atria become quite ineffective in pumping blood to them.

Ventricular fibrillation *Ventricular fibrillation* is a similar condition in which the ventricles do not coordinate their contractions with the atria. This is a serious situation because it results in circulatory failure unless it can be corrected at once. It occurs occasionally following a severe coronary thrombosis and sometimes during surgical operations on the heart. Defibrillator machines may be used to shock the heart back into a normal rhythm of contraction.

The artificial pacemaker The nerve impulse is conducted to the ventricles by way of the AV node and neuromuscular bundle. If the contractions of the ventricles become uncoordinated with those of the atria, an artificial pacemaker may be used to restore normal rhythm. This is a compact battery-powered device that may be worn over the heart or implanted under the skin. The batteries provide a steady pulse to the faulty pacemaker in the heart and are generally quite successful.

OTHER HEART CONDITIONS A slow heartbeat (or pulse) is known as *bradycardia*, whereas a very rapid heart rate is called *tachycardia*. In the condition called *paroxysmal tachycardia*, a series of rapid contractions that may last for a few seconds or a few minutes is followed by a resumption of the normal heartbeat. Such attacks can usually be treated with drugs. *Congestive heart failure* occurs in diseased or damaged hearts. The failure may be one-sided (right or left heart failure), but often both sides are involved. If the right ventricle, which pumps venous blood to the lungs, weakens, edema may occur, especially in the legs, ankles, and feet. This condition was formerly known as "dropsy." The left ventricle may also weaken under these conditions, and congestive heart failure results. This kind of heart failure can be treated with the drug *digitalis*, which is derived from the leaves of a plant known as foxglove. Diuretics are also helpful in eliminating excess water and sodium.

OPEN-HEART SURGERY Many congenital and acquired defects of the heart are corrected today by open-heart surgery. Such surgery requires special techniques. It is well known that reducing the temperature of the body will reduce the oxygen requirements of the tissues. The brain can withstand reduced oxygen under these circumstances for brief periods of a few minutes. If the body temperature is lowered to 30°C, brief operations on the heart can be performed without using a heart-lung machine. Lowering the body temperature creates the condition called *hypothermia*.

More elaborate heart operations require a heart-lung machine to pump the blood and oxygenate it. A heart-lung bypass, or *extracorporeal circulation*, is established during the procedure. Hypothermia is also induced in such procedures. Blood drawn from the caval veins is oxygenated, maintained at the required temperature, filtered, and finally returned to the aorta.

VENOUS DISORDERS   *Varicose veins*   The veins of the legs must return blood to the heart against the pull of gravity. They are assisted in this by the massaging action of muscles and by venous valves, which prevent backflow. Since the veins are thin-walled, they often enlarge under pressure and present tortuous bluish lines under the skin. The condition is more common in women than in men. Surgery for severe cases requires removing (stripping) the superficial vein (the greater saphenous) and perhaps tying off some of the smaller veins. Superficial veins are most affected.

Phlebitis   *Phlebitis* refers to an inflammation of a vein. It is occasionally associated with varicose veins. The inflammation may lead to *thrombophlebitis* in which thrombin forms blood clots in the vein. This condition is dangerous since free or floating clots (emboli) may block circulation in vital organs.

LYMPHATIC SYSTEM DISORDERS   *Lymphagogues*   Many substances can act as capillary poisons, causing an increased infiltration through damaged capillary walls with accompanying loss of protein. Such substances have been called *lymphagogues of the first class*. Some examples are the following: extracts of strawberries, extracts of crayfish muscle, extracts of leech heads, and various proteins, such as histamine and peptones. *Lymphagogues of the second class* are such substances as hypertonic glucose and hypertonic NaCl. These substances, when injected into the blood, pass readily through the capillary wall into tissue spaces. An excessive accumulation of such substances in the tissue spaces attracts fluid from the capillaries by osmosis. The lymph flow, draining the tissue spaces, is greatly increased.

Edema   An abnormal accumulation of fluid in the tissue spaces is called *edema*. The accumulation may be great enough to cause a swelling of the affected part. Pressing with the finger leaves a pitted depression, which is slow to fill up. Edema may result from many causes, such as increased capillary pressure, increased capillary permeability, reduction of plasma proteins, and lymphatic obstruction. Edema is often associated with heart disease and kidney disease.

SUMMARY

1 The heart, located in the interpleural space, or mediastinum, has four chambers: two atria and two ventricles. Blood enters the right atrium, passes through the tricuspid valve, and fills the right ventricle. As the ventricle contracts, the blood is forced up into the pulmonary artery, which carries it to the lungs. Backflow is controlled by the closing of the pulmonary semilunar valvula. The blood returns to the left atrium by way of the pulmonary veins. It passes from the left atrium to the left ventricle through the mitral valve. When the left ventricle contracts, the

blood is forced up into the aorta and out over the body. Backflow in the aorta is controlled by the closure of the aortic semilunar valvula.

2 The heart is enclosed in a two-layered membrane called the pericardium. The endocardium is a serous membrane lining the cavities of the heart. The myocardium refers to the musculature of the heart. Cardiac muscle tissue is an involuntary, striated type with transverse bars.

3 The wave of contraction that spreads over the atria originates at the sinoatrial (SA) node. When the excitation reaches the atrioventricular (AV) node, it is conveyed to the ventricular musculature by way of the neuromuscular bundle of His. The neuromuscular bundle is located in the median septum of the heart.

4 Heart sounds are caused by the closing of valves. The first sound is caused primarily by the closure of the atrioventricular valves, whereas the second sound indicates the closing of the semilunar valves.

5 Systole refers to the contracting phase of the heart, usually the contraction of the ventricles. Diastole occurs when the heart muscles are relaxed and the chambers fill with blood.

6 The heart rate varies, depending upon a number of factors. The average heart rate for a young adult male, seated, is 70 beats/min. Heart rates for women are a little faster than those recorded for men. The heart rate of the female fetus is around 140 to 145 beats/min, whereas that of the average male fetus is 130 to 135 beats/min.

7 Heart muscle derives its oxygen and nutrition from the coronary arteries. The coronary arteries are the first branches off the ascending aorta. Blocking of the coronary arteries by a clot (thrombosis) or by deposition of fatty materials can lead to serious heart conditions.

8 There is an unequal branching of arteries off the arch of the aorta. On the right, the brachiocephalic artery gives rise to the right common carotid and right subclavian; on the left, the left common carotid and left subclavian arise separately and directly off the arch.

9 An enlargement at the base of the internal carotid artery is called the carotid sinus. The carotid body lies between the bases of the external and internal carotid arteries. Around the arch of the aorta lie the aortic bodies. These structures contain chemoreceptors and pressoreceptors for the reflex regulation of circulation.

10 The principal branches of the thoracic aorta are the pericardial, bronchial, esophageal, mediastinal, and intercostal.

11 The principal branches of the abdominal aorta are the celiac, superior mesenteric, renal, spermatic or ovarian, inferior mesenteric, lumbar, and middle sacral.

12 The venous system returns the blood to the heart. The pulmonary veins return freshly oxygenated blood from the lungs to the heart. The return flow from the head and brain enters the superior vena cava by way of the external and internal jugular veins, which, in turn, flow into the brachiocephalic veins.

13 The largest veins of the arm are the cephalic and basilic. They carry venous blood to the subclavian veins.

14 The azygos vein originates along the dorsal wall of the abdomen as the ascending lumbar vein. It traverses the thorax and empties into the superior vena cava. One of its tributaries is the hemiazygos vein.

15 The largest superficial vein of the lower extremity is the great saphenous

vein. The anterior and posterior tibial veins flow into the popliteal vein back of the knee joint. The popliteal vein becomes the femoral vein, a deep vein of the thigh. The femoral vein becomes the external iliac vein of the pelvic region.

16 The internal iliac veins (hypogastric) together with the external iliac veins form the common iliac veins. The uterine veins of the female are important tributaries to the internal iliac veins. The common iliac veins form the inferior vena cava.

17 The inferior vena cava is a large vein of the trunk. It lies beside the dorsal aorta. It has numerous small tributaries, but relatively few large veins enter it. The two largest tributaries are the renal veins and the hepatic vein.

18 The hepatic portal system receives the return flow of blood by way of the splenic, inferior mesenteric, and superior mesenteric veins. These veins form the portal vein, which enters the liver. The portal system returns food-laden blood from the intestine to the liver.

19 Fetal circulation differs from postnatal circulation in several respects. In the embryo and fetus, the umbilical cord extends from the placenta to the umbilicus of the fetus. It contains the umbilical vein and two umbilical arteries. The umbilical vein carries oxygen and food-laden blood from the placenta. The umbilical vein is joined by the portal vein before it enters the liver. A more or less direct channel through the liver is called the ductus venosus.

20 Within the fetal heart, the blood may pass directly from the right atrium to the left atrium through the foramen ovale, or it may enter the right ventricle and pass out over the pulmonary artery. Since the lungs are not functional in the fetus, much of this blood passes directly into the aorta by way of a connection between the left pulmonary artery and the aorta called the ductus arteriosus.

21 Lymph is the medium of exchange between the blood and the tissues. The lymphatic system consists of lymphatic capillaries, larger lymphatic vessels, the thoracic duct, the right lymphatic duct, and lymph nodes. Lymph capillaries in the villi of the intestine are called lacteals. Lymph carrying absorbed fat is called chyle. Chyle from the intestinal lymphatic vessels drains into the cisterna chyli. It then moves slowly up through the thoracic duct and empties into the left subclavian vein. The left lymphatic system drains all portions of the body except the upper right quadrant. This portion is drained by the right lymphatic duct, which is very much reduced in size.

22 Lymph nodes are generally distributed over the body and around organs. Superficial lymph nodes are well distributed, but there are also large groups under the jaw, in the axillae of the arms and in the groin. Lymph nodes function (1) to protect the body against the invasion of bacteria and other foreign substances and (2) to produce lymphocytes, which have an immunological function.

QUESTIONS

1 Locate the heart.
2 Describe the heart's gross structure.
3 Describe the minute structure of cardiac muscle.
4 Explain the function of the heart valves.

5 Give an explanation of the factors causing heart sounds.
6 Indicate some of the factors influencing the heart rate.
7 Discuss the function of capillaries.
8 Trace the circulation of the blood from the time it leaves the liver until it returns to that organ.
9 How do veins differ from arteries?
10 Name and locate some of the larger veins.
11 Trace the blood from the heart to the small intestine and back to the heart.
12 Describe fetal circulation.
13 In what way does the lymph serve as a medium of exchange?
14 Discuss the structural nature of the lymphatic system.
15 What are the functions of lymph nodes?
16 Explain the function of the lymphatic system in the absorption of fat.
17 Name some lymphoid tissues.

SUGGESTED READING

Adolph, E. F.: The Heart's Pacemaker, *Sci. Am.*, **216**:32–67 (1967).

Benditt, Earl P.: The Origin of Atherosclerosis, *Sci. Am.*, **236**(2): 74–85 (1977).

Fozzard, Harry A.: Heart: Excitation-Contraction Coupling, *Annu. Rev. Physiol.*, **39**:201–220 (1977).

Korner, P. I.: Integrative Neural Cardiovascular Control, *Physiol. Rev.*, **51**(2):312–267 (1971).

Lundgren, O., and Mats Jodal: Regional Blood Flow, *Annu. Rev. Physiol.* **37**:395–414 (1975).

Parmley, Wm. W., John V. Tyberg, and S. A. Glantz: Cardiac Dynamics, *Annu. Rev. Physiol.*, **39**:277–299 (1977).

Some standard anatomy textbooks are listed in Reference Books, at the back of this book. These books may be consulted for detailed descriptions and illustrations of the anatomy of the circulatory system.

16

PHYSIOLOGY OF CIRCULATION

The physiology of circulation involves a consideration of factors that regulate blood flow and pressure. In this chapter, we consider:

1 Regulatory factors such as vasomotor control, baroreceptors, chemoreceptors, and other control mechanisms
2 The movement of fluids through the capillary membrane: the foundation for filtration and absorption pressure
3 The measurement of blood pressure and the effects of hypertension upon health
4 The clinical control of hypertension

Circulation of the blood is determined by two variables: *blood flow* and *blood pressure*. The blood flows rapidly from the heart and through the deeper parts of the body to the extremities and returns to the heart in about 60 s. It can take various routes, and some are longer than others. For example, the heart-lung route is a much shorter circuit than the round trip between the heart and the foot.

The blood also travels at different rates in various parts of the circulatory system. The rate of flow in arteries is not uniform, being somewhat faster in larger arteries than in smaller ones. Furthermore, blood flow is faster at the crest of the systolic wave than it is during diastole.

BLOOD FLOW

BLOOD FLOW IN ARTERIES, AND THE PULSE   As the heart beats, the ventricles regularly force blood out into the aorta. With each contraction of the ventricles, a wave of increased pressure starts at the heart and travels along arteries until it is lost in the capillaries. This pressure wave is the *pulse*. It can be felt in any of the arteries that are close to the body's surface. It is

528

common practice to take the pulse from the radial artery at the wrist. The pulse rate is, of course, the same as the heart rate, around 70/min in the resting state. The pulse should not be confused with the movement of the blood itself. The pressure wave travels 5 to 8 m/s, depending upon the age of the subject. As the arteries lose their elasticity with age, the velocity of the pulse wave increases. The mean velocity of blood flow in the arteries, however, has been estimated to be only about 20 cm/s (Fig. 16.1). The blood may travel as fast as 50 cm/s in the aorta, but it moves much more slowly in the smaller arteries. It is evident, therefore, that the pulse wave felt in the artery could not be caused by the passage of the blood itself.

The elasticity of the arterial wall causes the pressure wave to travel more slowly than it would through unyielding pipes. It also permits the blood to flow in a continuous stream through the capillaries. If a larger artery is cut, the blood escapes in spurts. In the intact circulatory system, however, much of the force of ejection from the heart is absorbed by the elasticity of the arterial wall. The blood, therefore, moves steadily forward even during the diastolic phase of the heart. If the circulatory system were constructed of glass tubing, which is inelastic, the blood would move by spurts in all parts of the apparatus.

A physician or trained observer can learn a great deal about the circulatory system by simply taking the pulse. Not only may the heart rate be observed, but its strength and regularity may also be determined. Even the blood pressure can be judged by noting whether the pulse is soft or hard. A soft pulse means that arterial tension is low because the artery has retained its elasticity. A hard pulse indicates high arterial tension because the wall of the artery is hard and inelastic.

BLOOD FLOW IN VEINS   The venous system collects blood from the great capillary bed. The overall area of the venous system narrows down again as the blood is returned to the heart, so the velocity of the blood as it travels through the veins may increase, therefore, to more than 250 times its

FIGURE 16·1

Diagram illustrating the velocity of blood flow and blood pressure in various parts of the circulatory system. The total capillary bed is much greater than it appears to be in the diagram. Mean arterial pressure is usually a little less than the average of systolic plus diastolic pressures, (120 + 80)/2.

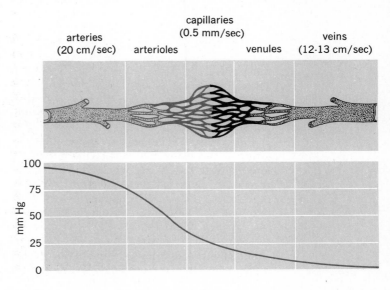

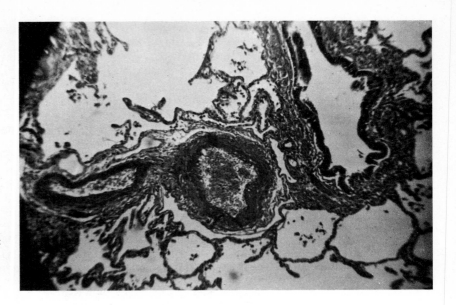

FIGURE 16·2
Cross section through an artery and a vein. The arterial
wall is much thicker than that
of the vein. Veins are often
partially collapsed, as illustrated. (*Courtesy of L.
Kelemes, M.D.*)

velocity in the capillaries. Since the venous bed is greater than the arterial bed, the velocity of blood in the veins is lower than that in the arteries. Many conditions influence the rate of blood flow in the veins. As in arteries, the blood velocity is greater in the large vessels than in the small ones. The velocity of blood flow increases during exercise and is directly proportional to the blood volume of any given vein. The beating of the heart helps to move the blood in the veins, but other factors supplement heart action. For example, the massaging action of skeletal muscles pushes the blood toward the heart. The veins of the lower extremities especially have numerous valves that prevent backflow. Muscle activity during exercise promotes the return flow of blood through the veins, but standing quietly for long periods places unusual pressure on the veins of the legs. Elevation of a limb above the level of the heart permits the blood to flow toward the heart, but the effect is not as great as might be expected. Veins tend to collapse under these conditions, thus increasing the resistance of the movement of blood within their walls (Fig. 16.2).

Inspiratory movements assist the flow of blood into the atria from the great veins. As the volume of the thorax is increased, air rushes into the lungs. The heart lies within the thoracic cavity, and the decrease in pressure within the cavity also helps fill the atria of the heart with blood.

Blood vessels, in general, are distensible, but veins are much more so than arteries. The veins of the skin, like other surface blood vessels, react to temperature; they dilate when the skin surface is warm and contract as a reaction to chilling. All the veins act as blood reservoirs, but the deep ones especially are capable of holding a considerable quantity of blood. More than half the total blood volume is contained in the veins alone.

RATE OF FLOW THROUGH CAPILLARIES   The velocity of the blood slows to about 0.5 mm/s as it passes through the capillaries. Though the diameter of a single capillary is exceedingly small, the total diameter of all capillaries

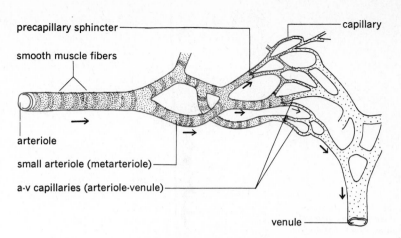

precapillary sphincter

smooth muscle fibers

arteriole

small arteriole (metarteriole)

a-v capillaries (arteriole-venule)

capillary

venule

FIGURE 16·3
Capillary pathways.

exceeds the total diameter of all arteries. Therefore, like a river flowing into a great open area, arterial blood flows slowly through the great capillary bed (Fig. 16.3).

Some capillaries provide a direct connection between the smallest arterioles and the venules; these are called *a-v capillaries* for convenience. A small volume of blood always flows through these capillaries, even during vasoconstriction. The a-v capillaries provide an important pathway for circulation between the peripheral ends of arteries and veins.

Vasoconstriction in the smallest arterioles (*metarterioles*) limits the flow into the capillaries. These small arterioles have cells of smooth muscle in their walls and thus can constrict and relax. The typical capillary branches off the smaller arteriole, often at nearly right angles to it. The flow of blood into a typical capillary is controlled by a band of smooth muscle, the *precapillary sphincter*, located at the place where it branches off the arteriole. The precapillary sphincter can constrict so that no blood passes through the capillary. In an area of tissue under observation, vasodilatation brings to view hundreds of capillaries that were not visible under conditions of normal vasoconstriction. The skin may appear pale, since surface blood vessels constrict when the body is exposed to chilling. Constriction of surface blood vessels tends to keep most blood away from the cool surface and so reduces heat loss. Such vessels may dilate locally when the skin is exposed to cold, as seen, for example, in the pink cheeks of healthy children. Surface vessels commonly expand to protect a tissue from injury by extreme cold. Surface blood vessels also dilate after exposure to heat, or after exercise in a warm room, and a greater volume of blood is brought to the cooler skin area. Sweat glands pour their secretions onto the surface of the skin. As sweat evaporates, the surface is cooled and the blood loses heat to the cooler area. Vasoconstriction and vasodilatation of blood vessels located in the skin play an important part in temperature regulation.

BLOOD PRESSURE

The heart pumps blood against the frictional resistance of blood vessels, generating *blood pressure*. Cardiac output is always an important factor in determining blood pressure. If the heart rate or the stroke volume increases, blood pressure is elevated. Conversely, a slowing of the heart rate or weakening of its beat causes blood pressure to fall. Blood pressure is highest in the arteries near the heart and diminishes as the blood passes into smaller vessels farther away from the source of pressure.

The viscosity of the blood influences its resistance to flow and thus raises blood pressure. Viscosity and resistance to flow increase as the concentration of red blood cells increases. A greater concentration of plasma proteins can also increase viscosity. Blood with high viscosity offers greater resistance to flow than blood with low viscosity.

The quantity of blood to be moved is another factor that influences blood pressure level. Though the body can withstand relatively small losses of blood without an accompanying loss of pressure, the loss of a large amount of blood causes a drop in blood pressure.

The elasticity of arterial walls also affects blood pressure. As the ventricles contract, blood is forced into the aorta under high pressure and the artery is somewhat distended. When the ventricles have completed their contraction and have entered diastole, blood pressure falls to the diastolic level. The elastic rebound of the arterial wall helps to maintain the diastolic pressure at a lower level, while moving the blood along and delivering a steady flow to the capillaries.

PRESSURE-FLOW RELATIONSHIPS IN BLOOD VESSELS To understand blood pressure and how it affects blood flow, we must first understand the factors that affect movement of fluid through cylindrical tubes. The rate at which fluid will flow through a rigid-walled cylindrical tube is equal to the difference in *hydrostatic pressure* exerted by the fluid at the two ends of the tube (ΔP) divided by the resistance of the tube to the flow of fluid through it (R_F). This may be expressed in the mathematical equation

$$\text{Flow rate} = \frac{\text{pressure difference } (\Delta P)}{\text{resistance to flow } (R_F)}$$

This equation cannot be used to compute flow rate through blood vessels, however, because blood vessels are not rigid-walled tubes. Nonetheless, blood flow in blood vessels is *proportional* to the ratio of pressure and resistance. As we stated above, the hydrostatic pressure generated by the blood is greatest in arteries near the heart and falls off in vessels farther away from the heart. This change gives direction to the blood flow. Heart rate and stroke volume determine the pressure differential in blood vessels. The question then is what factors determine resistance? Resistance to flow in blood vessels is determined by three main factors: (1) blood viscosity, (2) vessel length, and (3) vessel radius. Blood normally has a viscosity 1.025 times that of water, and the viscosity of the blood rarely changes sufficiently to influence flow rate. Vessel length is, of course, constant in each individual and therefore cannot be a factor in determining flow rate of blood. The third factor, vessel radius, can be readily altered in the vascular system. Flow is extremely sensitive to vessel radius because resistance is proportional to the reciprocal of vessel radius raised to the fourth power, as shown in this equation:

$$\text{Resistance} \propto \frac{1}{(\text{vessel radius})^4}$$

An illustration of the importance of this relationship can be seen in Fig. 16.4a. It is seen that simply doubling the radius of a tube increases flow rate 16-fold. Increasing the radius fourfold increases flow rate 256-fold. The body has made use of this important relationship in the arterioles. Fine control of smooth-muscle activity in the walls of these vessels greatly influences the rate of blood flow to a given tissue by a slightly altering vessel radius. Constriction increases resistance and reduces flow rate; dilatation decreases resistance and increases flow rate.

VASOMOTOR CONTROL OF BLOOD VESSEL DIAMETER   Any change in peripheral resistance to the flow of blood is largely a matter of change in the diameter of arterioles. The smooth muscles in the walls of these small blood vessels are supplied with vasomotor nerves. When the muscle fibers are stimulated to contract, the diameter of the blood vessel is reduced; this process is known as *vasoconstriction*. When muscle fibers relax, permitting the diameter of blood vessels to increase, the process is known as *vasodilatation* (Fig. 16.4b).

The *vasoconstrictor center* (vasomotor) is located in the medulla, in the floor of the fourth ventricle. Nervous outflow from this center travels to the arterioles through the postganglionic axons of the sympathetic system. These are nonmyelinated fibers arising from cell bodies located in the chain of sympathetic ganglia. The vasoconstrictor center maintains a constant flow of impulses to muscles of arterioles through the vasoconstrictor nerves, keeping the muscles in a state of tonic contraction. Severing the spinal cord below the medulla or severing the great splanchnic nerves to the abdominal viscera causes vasodilatation and a fall in blood pressure. Stimulating the splanchnic nerve causes increased tonic contraction in the arterioles of the abdominal viscera and a consequent rise in the blood pressure. In general, the sympathetic system is responsible for the rise in blood pressure that accompanies emotional states, such as anger and fear, and the physical adjustments attending muscular exercise.

The nervous control of vasodilatation is more obscure. In some instances vasodilatation seems to be caused largely by the inhibition of vasoconstriction. Many vasodilatation mechanisms affect only a localized area. So far as

FIGURE 16·4
a Increase in flow when the tube radius is increased to the fourth power. *b* The smooth muscle of the arteriole.

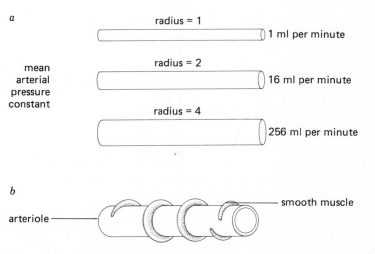

nervous control is concerned, sympathetic vasoconstriction appears to be far more important than vasodilatation.

The heart and skeletal muscles are known to receive a rapid increase in blood flow through vasodilatation. This *active hyperemia* is a local response and not the direct result of nervous stimulation. The mechanism evidently depends on local chemical changes in the blood supply to muscles in the arterioles.

Oxygen need, rather than nervous control, appears to be the most effective means of regulating blood flow to the tissues in general. Lack of oxygen is at least one factor in effecting local arteriole dilatation. Other factors are: a rise in carbon dioxide concentration, and an increase in potassium ion concentration. These factors influence the muscle of arterioles to relax.

Reflex control of sympathetic stimulation Sympathetic stimulation to most arterioles creates a mild state of contraction, or tone, in the smooth muscle since the sympathetic nerve endings release norepinephrine, a strong vasoconstrictor. If the rate of stimulation rises above a certain level, vasoconstriction occurs. If the rate of stimulation is lowered, the arteriolar muscle relaxes and vasodilatation results. Rapid vasodilatation of blood vessels in the skin causes blushing. Conversely, turning pale with fear is a result of peripheral vasoconstriction. It appears that the sympathetic nervous system is responsible for both constriction and dilatation of blood vessels except for those of the brain and heart.

Parasympathetic vasodilator neurons are found in the chorda tympani branch of the facial nerve, which supplies the blood vessels of the sublingual and submandibular salivary glands, and in the glossopharyngeal nerve, which supplies the parotid salivary gland. The sacral portion of the parasympathetic or craniosacral division is represented by axons of the pelvic nerve that supply the penis and the clitoris. Parasympathetic vasodilatation causes the erectile tissue in these organs to fill with blood. The walls of blood sinuses in the erectile tissue become turgid as the cavities fill with blood, and the whole organ becomes larger, hard, and erect.

Sympathetic nerves may also contain dilator axons. The cervical sympathetic outflow supplies vasodilator fibers to blood vessels in the mucous membrane of the mouth, the nostrils, and to surface areas of the face innervated by the trigeminal nerve. Sympathetic vasodilators to the coronary arterioles apparently have been demonstrated in dogs.

CHEMICAL CONTROL OF BLOOD VESSEL DIAMETER Some naturally secreted chemicals affect vasoconstriction and dilatation. These are:

Norepinephrine Norepinephrine is, as we have mentioned, a vasoconstrictor. Norepinephrine alters the balance of flow between sodium and potassium ions passing through the cell membranes of smooth muscle fibers. Normally potassium ions move out and a smaller number of sodium ions move inward. Under the influence of norepinephrine, however, sodium ions generally move inward and potassium ions move outward. Acetylcholine does not have this effect.

Epinephrine A hormone of the adrenal medulla, epinephrine causes vasoconstriction of the arterioles in the skin, mucous membranes, and the great

FIGURE 16·5
The development of brady-
kinin.

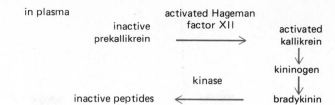

splanchnic area. In animal experimentation, however, epinephrine has also been shown to dilate arterioles of the coronary arteries. The effects produced by epinephrine closely parallel the action of norepinephrine released at the nerve endings of postganglionic fibers of the sympathetic system.

Histamine Normally released from mast cells, basophils, and blood platelets at the site of inflammation, histamine causes vasodilatation, increased permeability of capillaries, and lowering of blood pressure.

The kinins A group of polypeptide compounds, the kinins are much more active vasodilators than is histamine. Kinins are formed in the blood plasma from kallikrein. The inactive plasma enzyme prekallikrein is converted to the active enzyme kallikrein by a number of factors but primarily by the Hageman blood factor XII, which itself must be activated. Active kallikrein then reacts with another plasma protein called *kininogen,* converting it to *bradykinin,* an active vasodilator (Fig. 16.5). Bradykinin release causes a rapid drop in both systolic and diastolic blood pressure as a result of the vasodilatation it produces. Kinins are quickly inactivated by the enzyme *kinase.*

THE BARORECEPTOR MECHANISM The walls of the carotid sinuses and the base of the aortic arch contain *baroreceptors,* sensory receptors that are stimulated by a rise in blood pressure. As blood pressure rises, the walls of the arch and the sinuses are distended slightly, causing impulses from the baroreceptors to travel to the vasomotor center in the medulla. The vaso-constrictor center is depressed, and the arterial pressure is reduced. The nerves involved include a branch of the vagus, which supplies the aortic arch, and a branch of the glossopharyngeal, which supplies the carotid sinus (see Fig. 15.12). Stimulation of the medulla by baroreceptor impulses is followed by a slowing of the heart rate and vasodilatation, especially in the abdominal region supplied by the splanchnic nerves. The blood pressure consequently is lowered.

It is well known that some persons are especially sensitive to pressure over the carotid sinus. Pressure applied to the sinuses is followed by an immediate lowering of blood pressure, which may possibly cause giddiness or fainting.

When arterial blood pressure is lowered, the baroreceptors stop sending impulses, vasoconstriction occurs, and the heart rate increases. When arterial pressure increases, the baroreceptors are activated and send impulses to the vasomotor center in the medulla. This center stimulates the vagus nerves that innervate the heart. Vagal stimulation slows the heart and also decreases the strength of its contraction. A reduced output from the heart reduces the blood pressure. These baroreceptor mechanisms illustrate

Marey's law of the heart, which describes the inverse relationship between blood pressure and heart rate.

An increase in venous pressure close to the heart stimulates baroreceptors located in the walls of the caval veins and in the right atrium. Afferent neurons in the vagus nerves then carry inhibitory impulses to the vasomotor center in the medulla, which depress the activity of parasympathetic efferent impulses sent to the heart via the vagus nerves. Sympathetic nerves are stimulated, and the heart rate increases. This effect is called the *Bainbridge reflex*, and it results in an increase in stroke volume and heart rate to relieve venous pressure at the heart.

LAW OF THE HEART   It has been shown experimentally that the heart, with its nerves severed, can still respond to an increased amount of blood in the caval veins by increasing its stroke volume and thereby moving an increased amount of blood into the arterial system. Within certain limits, the strength of contraction is determined by the lengthening of ventricular muscle fibers, which increase the volume or capacity of the ventricles. This has been called the *law of the heart*, originally derived from work by Starling. The heart muscle can adapt to increased venous volume and also to increased arterial resistance and still maintain an adequate circulation. When nerves to the heart are intact, however, this intrinsic reflex mechanism ensures the ability of the heart to adjust to sudden changes in blood pressure as well as to maintain an adequate circulation.

THE CHEMORECEPTOR MECHANISM   *Chemoreceptors* located in the carotid and aortic bodies react to changes in the chemical composition of the blood by stimulating changes in blood pressure. The *carotid body* is a reddish-brown, oval structure about 5 mm long and 2.5 mm wide, located posterior to the bifurcation of the external and internal carotid arteries on either side (see Fig. 15.12). The *aortic bodies*, which are similar to the carotid bodies in structure and function, are located at the base of the right subclavian artery and in the arch of the aorta. These bodies are composed of epithelioid cells and contain chemoreceptors. The carotid bodies are innervated by a branch of the glossopharyngeal nerve; a branch of the vagus nerve supplies the aortic bodies. The carotid body chemoreceptors appear to function primarily in the regulation of respiration, but they have some circulatory effects as well. (Respiratory effects are considered in Chap. 17.) The aortic bodies may have some respiratory effects, but they are primarily concerned with regulating blood pressure. The chemoreceptors located within the carotid and aortic bodies react to a decrease of oxygen tension (concentration) in the blood or to an increase of carbon dioxide tension. They also react to a change in the acidity of the blood. As the oxygen content of the blood decreases, or as the carbon dioxide content rises, the chemoreceptors are stimulated. They send impulses to the vasoconstrictor center in the medulla, which stimulates vasoconstriction, and the blood pressure rises.

OTHER REGULATORY FACTORS, OSMOTIC PRESSURE   Many years ago, Starling recognized the importance of plasma-protein osmotic pressure in the exchange of fluids between the blood plasma and the tissue spaces. The hydrostatic pressure, or blood pressure, in the capillary and the plasma-protein osmotic pressure represent two opposing forces that regulate the exchange of fluid between the blood and the tissues.

The capillary wall is the membrane between the blood plasma and the interstitial fluid. Materials in solution pass through the capillary wall by diffusion rather than by active transport. The interstitial fluid acts as the medium of exchange between the blood and the tissues. The entrance of materials through the cell membrane may be by diffusion or active transport, in contrast to the diffusion through the highly permeable capillary wall. However, proteins (colloids) composed of large molecules diffuse through the capillary wall very slowly. As a result, water tends to move from the interstitial spaces into the capillary.

MOVEMENT OF MATERIALS THROUGH THE CAPILLARY MEMBRANE Capillary hydrostatic (blood) pressure averages 25 mmHg at the arterial end of the capillary and decreases to around 10 mmHg at the venous end. The higher pressure at the arterial end favors the movement of materials out of the capillary and into the interstitial fluid, the hydrostatic pressure of 25 mmHg being greater than the net osmotic pressure of 20(25 − 5) mmHg (Fig. 16.6). At the venous end of the capillary, the hydrostatic pressure is only 10 mmHg and the osmotic pressure remains at 20 mmHg; the flow here is reversed and favors fluid movement into the capillary. Actually, the movement of fluid in and out throughout the length of the capillary tends to maintain a balance between the amount of filtration and absorption. A small excess filtration is taken up by the lymphatic system. It is difficult to obtain accurate capillary and tissue pressures by direct methods, and so the above numbers are averages.

It is estimated that about 90 percent of the fluid that filters out at the arteriole end of the capillary will be reabsorbed by venous capillaries. Only about 10 percent is absorbed by lymphatic capillaries. In edema, the tissue spaces become filled with fluid and a much greater amount of fluid passes out of the capillaries than is reabsorbed. This results in swelling of the tissues.

FIGURE 16·6

The pressures that determine the flow of fluid between the capillary and the tissue spaces. The larger red arrows indicate the direction of blood flow. All pressures are recorded in mmHg.

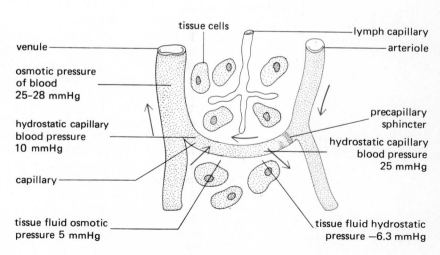

tissue cells

lymph capillary

venule

arteriole

osmotic pressure of blood 25-28 mmHg

hydrostatic capillary blood pressure 10 mmHg

precapillary sphincter

hydrostatic capillary blood pressure 25 mmHg

capillary

tissue fluid osmotic pressure 5 mmHg

tissue fluid hydrostatic pressure −6.3 mmHg

Under abnormal conditions, the balance of flow between the capillaries and the interstitial fluid may be altered. For example, if a person consumes too much salt in the diet, a greater amount of salt will be absorbed into the blood. Sodium ions pass readily through the capillary wall and into the interstitial fluid, but the tissue cell membrane is nearly impermeable to salt. Therefore the salt tends to remain in the interstitial fluid and increases its osmotic pressure. Water then moves from the cell into the interstitial fluid, causing edema.

In another situation, high arteriole pressure may increase capillary hydrostatic pressure and cause more fluid to move out of the capillary and into the interstitial fluid spaces, thus causing swelling.

When there is considerable inflammation, capillaries may become more permeable to proteins, permitting them to pass through the capillary wall and enter the interstitial fluid. In this case, the osmotic pressure of the blood plasma falls and the interstitial osmotic pressure rises, drawing water from the cell into the interstitial spaces and resulting in swelling. Also, liver disease may disrupt protein metabolism and cause edema.

THE MEASUREMENT OF ARTERIAL BLOOD PRESSURE   The first measurement of blood pressure is commonly credited to an English clergyman named Stephen Hales. Between the years 1731 and 1733 he is said to have measured the blood pressure of a horse by permitting its blood to flow into a vertical glass tube several feet long. In this experiment, he measured the pressure of the blood against the weight of a column of blood in the vertical glass tube.

Using mercury in the vertical glass tube makes it possible to read blood pressures in millimeters instead of in feet and inches. The weight of mercury, of course, is many times the weight of blood or water. In the *mercury manometer*, blood pressure is measured against the weight of mercury in a U-shaped glass tube, causing the column of mercury to rise in one side of the tube.

The modern device for measuring blood pressure, called a *sphygmomanometer*, comes in several forms. In the mercury sphygmomanometer, mercury rises in a vertical glass tube as a result of the force of air pressure transmitted from a rubber sac fitted like a tourniquet around the arm. This is obviously an indirect method, but it is the method most commonly used for obtaining blood pressure readings on human subjects (Fig. 16.7).

If the rubber cuff around the arm is inflated until the air pressure causes the walls of the artery underneath to collapse, the air pressure in the cuff is greater than the blood pressure. Holding the fingers on the radial artery at the wrist will indicate when the artery is collapsed, because the pulse can no longer be felt at this point. If the cuff is now deflated very gradually, a faint pulse can be felt at the wrist. This means that the systolic pressure of the heart is just great enough to pry the walls of the artery apart. At this time the air pressure in the rubber cuff is approximately equal to the systolic blood pressure, and the blood pressure reading is taken from the height of the mercury on the millimeter scale. A blood pressure reading of 118 taken at this time means that the pressure of the blood is great enough to raise a column of mercury 118 mm. This is the systolic blood pressure, the highest pressure developed by the heart during the ventricular systole.

The lower diastolic pressure is the pressure maintained even though the heart is at rest in the diastolic phase. In taking the diastolic pressure, a stethoscope is placed over the artery just below the rubber cuff. The sound of blood coming through the constricted portion of the artery and making contact with the more or less stationary blood below can be heard through the stethoscope. As the cuff is gradually deflated, more blood gets through and a very evident sound is heard. As the diastolic level is reached, around 80 mmHg in an average adult, the constriction on the artery has been released to such an extent that the sound disappears. The diastolic reading is now taken. The diastolic pressure is important because it indicates the degree of arterial tension during diastole.

Blood pressure values, like heart rates, vary with physiological conditions. There is no standard blood pressure. A child after the first month should have a systolic pressure of around 70 to 80 mmHg. By the time the child is twelve years old, the blood pressure average is around 105 mmHg. Blood pressure in both sexes is low until puberty, when there is usually a sharp rise. Young adult males, in a resting position, have blood pressures of 120/80 mmHg on an average. Blood pressure readings for young women are commonly 8 to 10 mmHg less than for men of equivalent age. There is a gradual increase in blood pressure with advancing age after twenty years. The difference between the systolic and diastolic pressures is termed the *pulse pressure*. It is usually around 40 mmHg but varies with changing physiological conditions.

FIGURE 16·7

The mercury sphygmomanometer used to measure arterial blood pressure. (*Courtesy of the Taylor Instrument Company.*)

EFFECTS OF EXERCISE ON BLOOD PRESSURE   The blood pressure rises considerably during physical exertion. There is some increase in diastolic pressure, but the greatest rise is seen in the systolic pressure. Exercises such as weight lifting, which require a maximum effort of strength, raise the blood pressure rapidly. Systolic pressures of 180 or even 200 mmHg are not unusual with this type of effort. The pulse does not accelerate greatly, and the blood pressure returns to normal quickly after exertion has ceased.

In exercises requiring speed, such as running a 100-yd dash, the heart rate might be expected to increase from 70 beats/min to around 120 beats/min and the systolic blood pressure to rise to around 150 mmHg. Physical training should be taken into consideration, but in general a greater effort produces an increased heart rate and higher blood pressure. The blood pressures of runners quickly return to normal and frequently drop below normal after a race.

Endurance races produce a more moderate rise in blood pressure, but the recovery phase requires a longer time. It is difficult to obtain blood pressure readings during exercise, but such readings are highly desirable. Readings taken after exercise has ceased may not mean very much, since blood pressure returns to normal quickly in many individuals. The blood pressure may fall below normal after exercise because the heart slows, while vascular dilatation continues in the blood vessels supplying recently active skeletal muscles. A rise in body temperature is also accompanied by vasodilatation of blood vessels located in the skin to provide for heat loss.

EFFECT OF GRAVITY ON BLOOD PRESSURE   Probably a great many persons have had the experience of hearing the telephone ring while they are lying down and jumping to their feet, only to find that things black out momentarily. The "black out" is associated with some degree of light-headedness, or dizziness. In older persons, especially, attaining a proper circulatory adjustment adequate to supply the brain properly in the vertical position requires an interval of several seconds. Circulatory adjustments are rapid in children, who can turn somersaults without coming up dizzy. Loss of elasticity in the circulatory system and greater body size make this more difficult for older adults.

The blood is pumped to all tissues of the body under pressure so that all parts of the body receive an adequate supply regardless of their position. Thus the brain ordinarily receives an adequate blood supply even though the body is in a vertical position. If the brain does not receive an adequate amount of blood, fainting occurs. The body falls to a horizontal position, thus making it easier for the heart to force blood to the brain. Fainting is a protective device to prevent serious damage to brain cells through lack of blood supply. A person who has fainted should not be lifted to a sitting or standing position.

The blood pressure in leg arteries is 50 to 60 mmHg higher than that in the brachial artery when the individual is standing. Aside from convenience of taking blood pressure measurements from the brachial artery, such measurements avoid the complications of gravity because this artery is at heart level. Blood pressures taken below heart level show the effect of gravity on a column of blood from the heart level to the point where the reading is taken.

Venous pressure also shows the effect of gravity. If the arm is allowed to hang idly beside the body, the surface veins can be seen to enlarge. Holding the hand above the head for some time causes the veins to collapse partially. The veins of the leg often enlarge for a variety of causes, but such enlargement is influenced by the force of gravity. The pressure on the walls of veins of the lower extremities is greater than in the upper part of the body.

VENOUS PRESSURE   The so-called *venous pump* consists of the compressing action of muscles. Muscular movement compresses the veins, and since valves in the veins of the lower extremities prevent backflow, the blood moves forward or upward. Venous pressure varies greatly in different veins, depending upon their location. When an individual has been standing still for some time, pressure in the neck veins approaches zero. The subclavian and brachial veins have a low pressure of 6 to 8 mmHg. Venous pressures are quite relative, but with the hand hanging at the side, the wrist veins may have a pressure of 35 mmHg. If the hand is held over the head for some time, the venous pressure drops to a low level. When the vena cava is in a recumbent position, close to the sucking action of the heart, pressure has been estimated to be only 0 to −2 mmHg. The venous pressure at the ankle in a standing position is 90 to 100 mmHg because of the effect of gravity. If a person is seated, the pressure at the ankle is only about half that measured at the ankle when the person stands. The venous muscular pumping action is normally so efficient that venous pressure in the leg veins while walking is only 15 to 30 mmHg (Fig. 16.8).

The respiratory pump   The movement of blood in veins returning blood to the heart is aided by the contractions of skeletal muscles that massage and squeeze the veins lightly. As we shall see in Chap. 17, when air is inhaled, the diaphragm is pushed downward and abdominal pressure is increased. This pressure tends to compress abdominal veins slightly, causing the blood to move forward. As the thoracic cage expands, a lowering of thoracic pressure aids the venous flow. All in all, these actions constitute what is known as the *respiratory pump,* which assists the return flow of venous blood to the heart.

CLINICAL ASPECTS

HYPERTENSION   *Hypertension,* or high blood pressure, affects a great number of persons. It may have many causes, but constriction of arterioles is the major one. Vasoconstriction is a sympathetic nervous system effect, but other factors may contribute to this effect. Renal hypertension involves the renin-angiotensin system, for example. Many older persons have what is called *essential hypertension,* or high blood pressure of unknown origin. A *resting blood pressure of* $\frac{160}{90}$ mmHg or more indicates the presence of hypertension. The borderline for determining hypertension seems to be $\frac{140}{90}$ mmHg. The kidneys are thought to be involved in essential hypertension, but the exact mechanism is uncertain. Perhaps aging kidneys require a higher pressure to excrete urine at a normal rate. Excessive salt intake increases the retention of water in the tissues and seems in some way to be concerned with vasoconstriction. Reduction of salt intake frequently lowers the blood pressure in essential hypertension.

The heart may gradually weaken under the strain of continually pumping against high arterial resistance. Since blood pressure rises in emotional

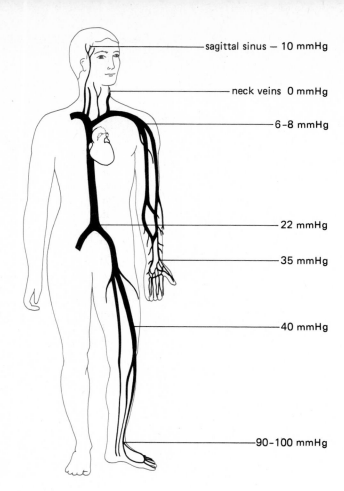

sagittal sinus — 10 mmHg

neck veins 0 mmHg

6–8 mmHg

22 mmHg

35 mmHg

40 mmHg

90–100 mmHg

FIGURE 16·8

Variation in venous pressures
in the standing position.

situations and with physical activity, older persons with high blood pressure
are often advised to live moderately.

Arteriosclerosis Many older persons are subject to hardening of the arteries, or
arteriosclerosis, and this condition often causes hypertension. In arteriosclerosis, the arterial walls harden and thicken, losing their elasticity. Their
diameter is reduced, and they cannot properly provide for the nutrition of
the tissues they supply. Since arteriosclerosis is ordinarily associated with
high blood pressure, there is always the danger that one of the affected,
fragile blood vessels may break. A consequence of such a break may be a
cerebral hemorrhage, or "stroke," caused by the rupture of a blood vessel in
the brain. The released blood can seriously damage delicate brain tissue.
Such brain injuries often result in paralysis.

   Atherosclerosis is a common form of arteriosclerosis. Atherosclerosis is
characterized by abnormal cholesterol-containing deposits called plaques on

the inner layer (intima) of the blood vessels (Fig. 16.9). In older persons, considerable calcification and hardening of the arterial walls may also be present. The deposits of cholesterol and other lipids tend to occlude the blood vessels. If this occurs in the arteries that supply the heart muscle, it is conducive to the coronary conditions discussed in Chap. 15. It is thought that a diet low in saturated fats is advisable for those persons who have atherosclerosis, but there is a difference of opinion concerning the value of low-cholesterol diets.

Other factors affecting blood pressure Rest, restricted diet, and relaxation may reduce hypertension in many cases. Systolic blood pressure during sleep may be around 25 mmHg less than during daytime activity. Worry, fear, excitement, and the tenseness with which many persons conduct the day's business are emotional conditions that tend to raise blood pressure. The effect is mediated largely through the sympathetic nervous system. Diseases of the kidneys or renal occlusion often are associated with high blood pressure.

Antihypertensive drugs Drugs used in the control of hypertension are ordinarily directed against peripheral resistance to blood flow in arterioles. Diuretics and derivatives of the plant genus *Rauwolfia* are commonly used. Diuretics increase the flow of urine and help the kidneys to excrete excess sodium chloride. The excretion of excess sodium chloride tends to lower blood pressure. Sodium retention increases water absorption by the tissues and, if uncontrolled, leads to edema. Swelling of the feet and ankles is a commonly recognized sign of edema caused by hypertension. (Aldosterone, a hormone secreted by the adrenal cortex, increases sodium reabsorption by kidney tubules.)

FIGURE 16·9

a Plaque developing in the lumen of an artery. *b* A cross section of an artery showing plaque nearly filling the lumen.

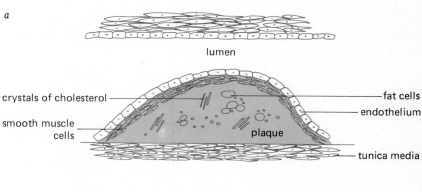

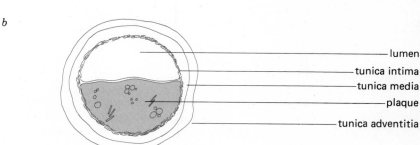

Rauwolfian derivatives such as reserpine act directly to relax arterioles; such drugs also have a tranquilizing effect.

Hypertension may also be treated with drugs that act on the sympathetic nervous system to reduce vasoconstriction or with others that affect postganglionic nerve endings. Unfortunately, all such drugs have some side effects. Common side effects of such agents are nasal congestion, fatigue, depression, headache, and depression of sexual activity. Side effects vary for different drugs and with the drug dosage, but since hypertension is dangerous and must be alleviated, some discomfort is perhaps permissible in its treatment.

HEMORRHAGE AND CIRCULATORY SHOCK *Hemorrhage* is the loss of blood from external or internal bleeding. Severe loss of blood may lead to *circulatory shock*. If there is considerable loss of blood volume or a marked decrease in cardiac output, the tissues do not receive enough oxygen and the individual becomes pale, cold, and weak. In shock the pulse is weak but rapid, the breathing is shallow and rapid.

Shock may be compensated or nonprogressive, if the bleeding is not too severe and body systems are able to reestablish the blood volume.

Progressive shock occurs in cases of severe, unchecked bleeding. The heart weakens, blood pressure drops to extremely low levels, and eventually the heart fails. The coronary arteries receive decreasing blood supplies as the heart weakens and blood pressure drops, which further weakens the heart and decreases blood pressure, setting up a vicious cycle.

Progressive shock is an example of a positive feedback mechanism. Like most such mechanisms in physiology, it is an abnormal condition that has potentially fatal effects.

SUMMARY

1 The blood makes a complete cycle around the body in about 60 s. The mean velocity of blood in the larger arteries is about 20 cm/s. Since the total diameter of all the capillaries is greater than the total diameter of all the arteries, the blood slows to 0.5 mm/s as it passes through capillaries. The velocity increases to around 130 mm/s as it returns through the veins.

2 The pulse is a pressure wave, which travels along the arteries at a rate of 5 to 8 m/s, much faster than the velocity of the blood itself. The pulse is lost as the blood filters through the capillaries.

3 Blood pressure is maintained by the force of the heart beating against the frictional resistance of blood vessels. Several factors affect blood pressure: (1) cardiac output, which in itself is dependent upon heart rate and stroke volume; (2) viscosity of the blood; (3) quantity of the blood to be moved; and (4) elasticity of the arterial walls.

4 Vasoconstriction is a decrease in the diameter of blood vessels caused by contraction of smooth muscles in the vessel walls. The greatest effect of vasoconstriction is in the arterioles. Relaxing of these muscles causes vasodilatation. Vasoconstriction control centers are located in the medulla.

5 Baroreceptors are found in the walls of the carotid sinus, around the base of the aortic arch, and in the great caval veins near the heart. As the blood pressure rises, afferent stimuli travel to vasomotor centers in the medulla. These centers send impulses that depress the heart rate and produce peripheral dilatation, thus lowering blood pressure.

6 Increased venous pressure stimulates baroreceptors located in the caval veins close to the heart. Afferent impulses transmitted via the vagus nerves depress the parasympathetic innervation of the heart, causing an increase in the heart rate. The heart responds to increased venous pressure by an increase in stroke volume and heart rate.

7 Chemoreceptors located in the carotid and aortic bodies react to changes in the chemical composition of the blood. They are especially sensitive to any lowering of oxygen pressure in the blood or to a change in the acidity of the blood. As the oxygen content of the blood falls, the chemoreceptors are stimulated and send impulses to vasoconstrictor centers in the medulla. Vasoconstriction is stimulated, and the blood pressure rises.

8 The heart still responds to an increased amount of blood by increasing its workload even when its nerves have been severed. Within limits, the ventricles enlarge, stretching the muscle fibers, and the fibers respond by moving a greater amount of blood. This has been called the law of the heart.

9 The arterial hydrostatic (blood) pressure forces fluids out of capillaries, whereas plasma-protein osmotic pressure attracts fluids toward the bloodstream. Thus two opposing forces regulate the exchange of fluids between the blood and the tissues. The capillary wall is the membrane between the blood and the tissue fluid. The hydrostatic (blood) pressure is much lower at the venous end of the capillary, about 10 mmHg, thus favoring absorption of fluid.

10 The instrument for measuring blood pressure is called a sphygmomanometer. Young men, seated and resting, commonly have systolic blood pressures around 118 to 120 mmHg. Young women, under the same conditions, commonly have systolic readings 8 to 10 mmHg less than those recorded for young men. The diastolic reading is about 40 mmHg less than the systolic reading. The difference between the two readings is called the pulse pressure.

QUESTIONS

1 Why does the blood flow more slowly through the capillaries than through the veins?
2 What is the pulse?
3 Indicate some of the factors influencing blood pressure.
4 Describe the function of the vasoconstrictor center in the medulla.
5 Discuss the baroreceptor mechanism.
6 Where are the chemoreceptors located? What is their function?
7 Explain the meaning of filtration pressure in regard to the movement of fluids through the capillary wall.
8 What pathological conditions may be caused by hypertension?

SUGGESTED READING

Baez, Silvio: Microcirculation, *Annu. Rev. Physiol.*, **39**:391–415 (1977).

Benditt, E. P.: The Origin of Atherosclerosis, *Sci. Am.*, **236**(2): 74–85 (1977).

Fein, J. M.: Microvascular Surgery for Stroke, *Sci. Am.*, **238**(4): 58–67 (1978).

Lundgren, Ove, and Mats Jodal: Regional Blood Flow, *Annu. Rev. Physiol.*, **37**:395–414 (1975).

Oberg, Bengt: Overall Cardiovascular Regulation, *Annu. Rev. Physiol.*, **38**:537–570 (1976).

UNIT FIVE

HOMEOSTATIC GROWTH AND THE REPRODUCTIVE SYSTEMS

17

RESPIRATION

CONSIDERATIONS FOR STUDY

The human body is such a large multicellular organization that oxygen cannot reach the cells by diffusion. We need specialized membranes, therefore, such as those in the lungs, for the exchange of respiratory gases. The respiratory system performs this function. In this chapter, we take up the study of respiration, the exchange of gases, with a discussion of the following topics:

1 The flow of air through the structures of the respiratory system
2 The control of respiration by two centers in the medulla and the pneumotaxic center of the pons
3 The effects of air pressure in the alveoli, the regulation of pulmonary circulation, and the factors that directly affect the respiratory center
4 The transport of respiratory gases and a study of acid-base balance in body tissues and chemical buffers
5 Recommendations on artificial respiration

A continual and abundant supply of oxygen is one of the most urgent needs of body chemistry. To supply this need and to remove the waste product carbon dioxide, a respiratory system has been developed. The meaning of the term *respiration* is often misunderstood. Respiration involves the provision of oxygen to the tissues and the removal of carbon dioxide from the tissues. The exchange of these two gases in the lungs is accomplished by breathing; this exchange is called pulmonary ventilation or external respiration. Respiratory functions may be divided into three parts: (1) the movement of air in and out of the lungs, *ventilation;* (2) the exchange of oxygen and carbon dioxide in the lungs, *pulmonary respiration;* and (3) the relationship of carbon dioxide to acid-base balance.

The structures of the respiratory system include the nasal passageway, pharynx, larynx, and trachea and the bronchus, bronchial tubes, bronchioles, air sacs, and alveoli of the lungs.

Embryonic lungs arise as a ventral evagination from the pharynx region of the primitive gut. The bilobed structure is called the *lung bud* (Fig. 17.1).

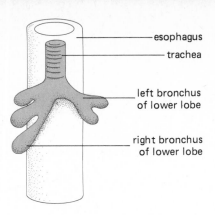

esophagus

trachea

left bronchus
of lower lobe

right bronchus
of lower lobe

FIGURE 17·1

Lung bud. Early embryonic
development of the lung.

RESPIRATORY STRUCTURES: HEAD AND THROAT

The common pathway for air entering or leaving the lungs is through the nose. There the air is filtered, moistened, and warmed as it passes over warm, moist membranes lining the nasal passageway. The nasal cavity opens posteriorly into the *nasopharynx;* the air passes through the *oropharynx,* or posterior part of the mouth (Fig. 17.2). (The pharynx is described in Chap. 19.)

LARYNX, EPIGLOTTIS, AND GLOTTIS At the entrance to the *trachea,* the tube leading to the lungs, is a cartilaginous structure, the *larynx.* It contains the vocal folds and is protected from above by a movable cartilaginous lid called the *epiglottis,* which opens as one breathes and closes when one swallows. Below the epiglottis and revealed by a narrow fissure in the larynx is the *glottis,* across which are the *vocal folds.* The opening between them is V-shaped when the individual is resting. When one sings a high note, the vocal folds are stretched and their inner edges lie close together. Singing a low note, or taking a deep breath, causes the vocal folds to be pulled apart, and the aperture between them, which opens into the glottis, becomes oval (Fig. 17.3).

The vocal folds are responsible for the voice. The sounds of the voice are caused by air being forced over the folds and causing them to vibrate; their vibrations, in turn, affect a column of air above. When the vocal cords become inflamed, as in *laryngitis,* they cannot vibrate freely, which often causes a temporary loss of voice. Choking on food or liquid in the larynx also prevents speech while the vocal folds are so irritated. Raucous singing or shouting, if prolonged, may damage the vocal folds.

The difference between the male and female voice is due to the influence of sex hormones. In childhood there is little difference between a boy's voice and a girl's voice, but after puberty the male larynx enlarges, and the vocal folds become longer and thicker, giving rise to the deeper voice of the adult male.

Pressure can be built up in the thorax by strong expiratory action of muscles concerned with breathing. With the glottis closed and the diaphragm held firmly, pressure can be exerted downward to promote defecation or to exert pressure upon the urinary bladder. Pressure may be released upward through the throat as a result of irritation to produce a cough. If the air is directed forcefully through the nose, the action is called *sneezing.*

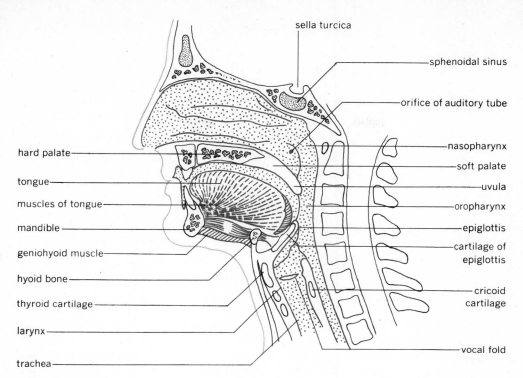

sella turcica

sphenoidal sinus

orifice of auditory tube

hard palate

nasopharynx

soft palate

tongue

uvula

muscles of tongue

oropharynx

mandible

epiglottis

cartilage of
epiglottis

geniohyoid muscle

hyoid bone

cricoid
cartilage

thyroid cartilage

larynx

vocal fold

trachea

FIGURE 17·2

A midsagittal section through
the anterior portion of the
head and neck.

THE TRACHEA The glottis leads into the *trachea*, a long membranous tube supported by rings of cartilage and fibrous connective tissue (Fig. 17.4). The rings of cartilage are not quite complete posteriorly but resemble the letter C. The cartilage rings and connective tissue permit great flexibility in the trachea and yet are strong enough to resist compression. The tracheal tube is lined with a mucous membrane and ciliated epithelium. Particles of foreign matter inhaled in the air become engulfed in mucus. Constant beating of cilia tend to wave the mucus upward toward the pharynx.

ANATOMY OF THE LUNGS

EXTERNAL APPEARANCE OF LUNGS The *lungs* occupy the thoracic cavity. The heart and other structures within the mediastinum lie between the lungs. There is, therefore, a concave cardiac impression on the medial surface of each lung. The upper surfaces, or apices, are more or less conical

FIGURE 17·3

The interior of the larynx as
viewed from above: *a* Glottis
dilated as during inspiration.
The glottis and vocal folds are
also more relaxed in the production of sounds of low frequency. *b* During vocalization,
to produce sounds of higher
frequency. The aperture of the
glottis is greatly reduced, and
the vocal folds are stretched.

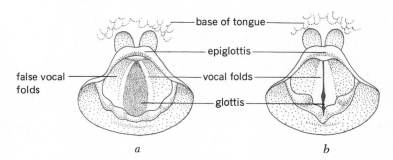

base of tongue

epiglottis

false vocal
folds

vocal folds

glottis

a

b

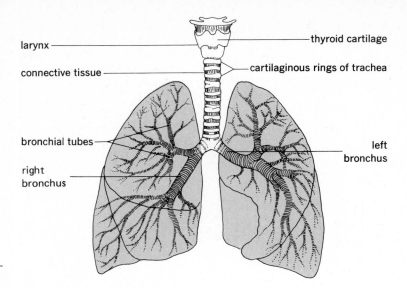

FIGURE 17·4

The larynx, trachea, and bronchial tubes (anterior view).

External structure of lungs
   Right: three lobes
   Left: two lobes

as they rise a little above the level of the clavicles; the lower surfaces are flattened and concave where they rest on the diaphragm.

The two lungs vary in shape and size. The right lung is the larger, and consists of three lobes. Its width is greater than that of the left lung, but it is about an inch shorter. The left lung is divided into two lobes and is somewhat narrower and longer (see Fig. 17.5 and Plate 12).

FIGURE 17·5

Anterior view of the lungs in the thoracic cavity.

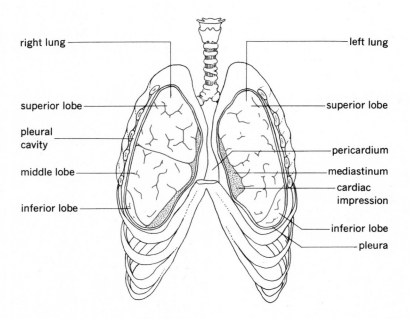

The pleura A thin serous membrane covers each lung and continues over the thoracic wall, the diaphragm, and the lateral aspects of the mediastinum. This membrane is called the *pleura* and consists of two layers. The *pulmonary pleura* covers the lung; the *parietal pleura* lines the wall of the chest and covers the diaphragm. Since the lungs fill the thoracic cavity, the two layers are in contact and the pleural cavity between the two layers is only a potential one. The serous membrane normally secretes only enough serum to moisten the surfaces that move upon each other with each respiratory movement.

INTERNAL STRUCTURE OF THE LUNGS The trachea divides into two bronchi, which enter the right and left lung, respectively. Within the lungs the bronchi subdivide into smaller branches, called bronchial tubes, and finally into bronchioles (Figs. 17.4 and 17.5). Bronchitis is an inflammation located primarily in the bronchial tubes.

The acinus The terminal bronchiole is called an acinus. It is composed of respiratory bronchioles, alveolar ducts, and alveolar sacs. This is the respiratory part of the lung; the passageways leading to the acinus are non-respiratory portions.

Alveoli Each bronchiole leads into a thin-walled sac with numerous pouches. The pouches are called alveoli, and they resemble a bunch of grapes clustered around the stem. The wall of each alveolus is very thin and highly vascular.

The alveolar capillaries Alveoli of the acinus are interconnected by minute openings called the *pores of Kohn*. These pores probably help distribute gases in breathing. The alveolus is lined with squamous epithelium, which is covered with tissue fluid (Figs. 17.6 and 17.7). The *alveolar capillaries* are where the exchange of gases between the blood and the air takes place. In addition to the elaborate branching of the bronchial tubes, the lungs contain

FIGURE 17·6

Bronchioles and air spaces. A capillary network is shown around the alveolar sacs. The atria and alveoli have an epithelial lining which is very extensive. The total surface area for gas exchange is approximately 40 ft². During normal breathing only about 20 percent of this area is utilized.

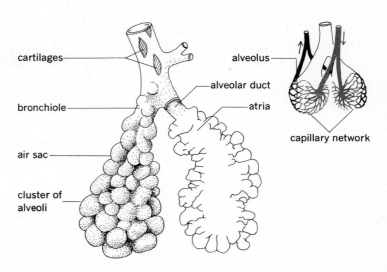

cartilages

bronchiole

air sac

cluster of alveoli

alveolus

alveolar duct

atria

capillary network

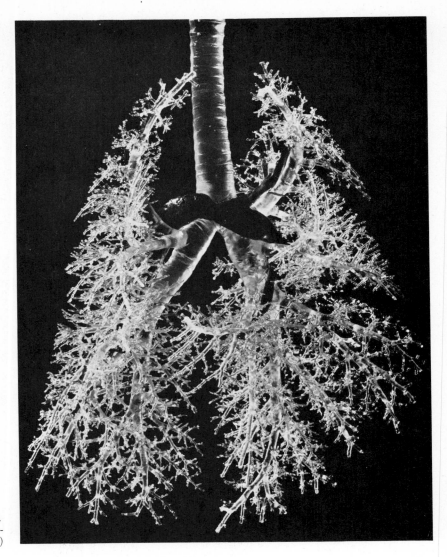

FIGURE 17·7
The bronchiotracheal tree of a dog. (*Courtesy of Ward's Natural Science Establishment, Inc.*)

a very abundant circulatory tree. The pulmonary artery divides into smaller and smaller branches until finally it resolves itself in capillary networks around bronchioles and alveoli. Pulmonary capillaries anastomose to form tiny venules as the blood leaves the alveoli and give rise to the pulmonary veins, which carry freshly oxygenated blood back to the left atrium of the heart.

The alveolar capillary membrane, although minute, consists of three parts: the endothelium (0.2 to 0.6 μm in width), an interstitial space (0.2 μm), and the alveolar epithelium (0.5 μm). Recall that a red blood cell is comparatively large at 7.5 μm.

Lung tissue is supplied by bronchial arteries directly from the aorta. Most of this blood is returned to the azygos veins by bronchial veins.

RESPIRATORY MOVEMENTS

There are two respiratory movements in breathing. Drawing air into the lungs is called *inspiration*. Forcing air out of the lungs is called *expiration*. Breathing occurs rhythmically at an average rate of about 16 times/min. The lungs do not take any active part in respiratory movements. Since the thorax is a closed cavity filled by the lungs, any change in the volume of the cavity will affect the volume of air in the lungs. During inspiration the volume of the chest cavity is increased by the descent of the diaphragm and the elevation of the ribs (Fig. 17.8). The diaphragm is always dome-shaped but is depressed during inspiration. The ribs normally slope downward, but during inspiration they are brought upward to a more horizontal position by muscular action. Throwing out the chest is accomplished by elevating the ribs. Since the movements resulting in inspiration require that the ribs and diaphragm move from their normal position, inspiration is said to be active, whereas expiration, in which these structures return to their normal position, is passive. Forced expiration, however, is active.

THE EXCHANGE OF AIR IN THE LUNGS   The lungs fill with air as the newborn child takes its first breath; they are never completely devoid of air thereafter. Even after a forced expiration, the lungs of an adult still contain a *residual volume* of air equal to 1000 to 1200 ml.

When a person is breathing quietly, there is a constant flow of *tidal air* in and out of the lungs. The volume of this tidal air for the average person is about 500 ml (1 pint).

After a normal inspiration of tidal air, an additional 3000 ml can be inhaled by taking as deep a breath as possible. The additional amount is called the *inspiratory capacity*.

The additional amount of air that can be exhaled forcibly after a quiet expiration of tidal air is the *expiratory reserve volume*. It amounts to about 1100 ml in the adult.

If a person inhales the maximum amount of air and then exhales as much as possible, the total exchange is called the *vital capacity*, which measures

FIGURE 17·8

Changes in the thoracic cavity during inspiration and expiration: lateral view of thoracic cage. During inspiration the diaphragm descends and the ribs are elevated, thus increasing the size of the thoracic cavity. In expiration the ribs fall back into place and the diaphragm becomes more dome-shaped.

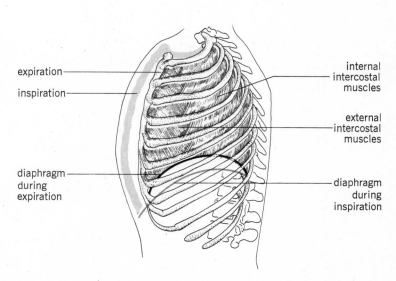

approximately 4500 ml in the average adult. Vital capacity varies with the size of the individual, among other factors. The average young woman may have a vital capacity of 4000 ml, whereas the average young man may have a vital capacity of 5000 ml. Vital-capacity measurements are clinically important, since vital capacity diminishes with certain diseases that affect the chest muscles and the lungs (Fig. 17.9).

MINUTE VOLUME   The total amount of air exchanged during one minute is termed the *minute ventilatory volume*. This quantity is obtained by multiplying the tidal volume (500 ml) by the average breath rate (16/min). The average adult has a minute volume of 8000 ml/min. The determination of minute volume and vital capacity are useful in diagnosing such lung diseases as emphysema, in which the deterioration of alveoli reduces the normal ventilation of the lungs.

DEAD AIR SPACE   A small portion of the air inhaled does not reach the alveoli since some remains in the air passageways (the trachea, the bronchi, and the bronchioles) and since some alveoli may not be fully expanded or some bronchioles may be blocked, as in asthma. Such air is known as *dead air space*. Dead air space amounts to some 150 ml in normal breathing.

Surfactant   The inner surface of the alveolar wall presents an interface between the air and a liquid. Each alveolus is lined by a thin layer of fluid. Liquids display surface tension, which would tend to constrict or collapse

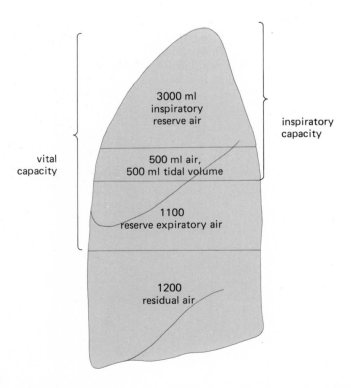

FIGURE 17·9
Volumes of air in the lungs.

the alveoli. If the liquid were simply pure water, the surface tension would make breathing very difficult and would require considerable effort. However, the alveolar air-surface liquid is covered by a phospholipid called *surfactant* (surface active agent), produced by cells lining the alveoli, which lowers the surface tension and makes breathing easier. The *respiratory distress syndrome of the newborn,* or hyaline membrane disease, especially likely to occur in premature infants, is caused largely by a lack of adequate amounts of surfactant. In these cases, there is a lack of elasticity in the lungs and the alveoli may collapse, causing breathing to become extremely difficult and exhausting. Surfactant enables the alveoli to avoid collapsing. *Compliance* (distensibility) is a term used to indicate the amount of expansion and elasticity of the lungs and chest. A considerable part of the elasticity is due to a lowering of surface tension of the fluid lining the alveoli.

Even though the movement of oxygen and carbon dioxide takes place through the alveolar and capillary walls, the respiratory membranes are very thin, averaging less than 1 μm in thickness. The capillaries are so minute that red cells touch the capillary wall as they pass through, thus aiding the diffusion of respiratory gases.

PHYSIOLOGY OF ALVEOLAR GAS EXCHANGE

PARTIAL PRESSURES OF RESPIRATORY GASES In dealing with respiratory gases it is necessary to consider the partial pressure of gases. Dalton's law of partial pressures indicates that the total pressure exerted by a mixture of gases is equal to the sum of partial pressures of the gases in the mixture. In the case of respiratory gases, the partial pressure represents the force of a particular gas against a membrane such as the alveolar membrane. The greater the partial pressure of a gas, the greater the force of movement of that gas through a membrane.

The body must function under an atmospheric pressure of 760 mmHg at sea level. The atmosphere is composed largely of nitrogen (78 percent) and oxygen (21 percent), but small amounts of water vapor, carbon dioxide, and several inert gases are also found. If we take the more exact figure for oxygen of 20.84 percent and multiply it by the total atmospheric air pressure, we find that the *partial pressure of oxygen* in inspired air is 158 mmHg; that is, the amount of pressure exerted by oxygen in inspired air is 158 mmHg. The partial pressure of oxygen at the alveoli, however, is only 104 mmHg (13.6 percent). The alveolar oxygen pressure is lower because there is considerable dead air space in the air passageways and some of the inspired air (and its oxygen) never reaches the alveoli (Table 17.1).

TABLE 17·1
Average composition of dry inspired and expired air with the human subject at rest

	O_2, vol. %	CO_2, vol. %	N_2, vol. %
Inspired air	20.9	0.04	79
Expired air	16.3	4.5	79

TABLE 17·3
Average amounts of gases in arterial and venous blood

	O_2, vol. %	CO_2, vol. %	N_2, vol. %
Arterial blood	20	48	1.7
Venous blood	13	55	1.7

Carbon dioxide partial pressure in the atmosphere is low; it is only 0.15 mmHg (0.04 percent). Carbon dioxide partial pressure in alveolar air is much higher. It is 40 mmHg (5.3 percent) because of the diffusion of carbon dioxide from venous capillaries into the alveoli.

Diffusion of gases through the alveolar membrane is a constant, two-way mechanism. Oxygen diffuses out of the alveoli, combines with hemoglobin in the arterial blood, and is carried to the body tissues; carbon dioxide diffuses into the alveoli from venous blood and is exhaled. The gases diffuse independently according to their partial pressures (Table 17.2).

The average amounts of oxygen, carbon dioxide, and nitrogen present in arterial and venous blood are shown in Table 17.3. Actual values vary with the activity of individuals or various organs or with the rate of metabolism of tissues. Even with an individual in a recumbent position, considerable variation is still found.

TABLE 17·2
Differences in air pressures shown in millimeters of mercury (atmospheric air pressure 760 mmHg at sea level)

Inspired air P_{O_2}	158+
Inspired air P_{CO_2}	0.3
Alveolar air P_{O_2}	105
Alveolar air P_{CO_2}	40
Expired air P_{O_2}	120
Expired air P_{CO_2}	32

TRANSPORT OF RESPIRATORY GASES

Oxygen and carbon dioxide are transported to and from the tissues and the lungs in the blood. Very little gas is transported in the blood as dissolved gas. Most is transported by carrier molecules in chemical combination with ions.

All the reactions that oxygen and carbon dioxide undergo in the blood are readily reversible. Because oxygen is continuously consumed by the tissues, a diffusion gradient from blood to tissue always exists for this gas. Likewise, carbon dioxide is continuously produced by the tissues so that a diffusion gradient from tissues to blood always exists. The exchange of respiratory gases at both tissues and alveoli is always a passive diffusional process. The arterial blood remains very nearly 100 percent saturated with oxygen because of the regulation of lung ventilation rate and cardiac output. Adjustment of the oxygen demand of each tissue is controlled through vasomotor regulation for blood flow to the tissue.

TRANSPORT BY HEMOGLOBIN   Very little oxygen and carbon dioxide could be transported if these gases were simply dissolved in the blood plasma. For example, oxygen dissolves in plasma according to oxygen partial pressures. In the alveoli, oxygen pressure is about 105 mmHg. At this pressure about 0.3 ml oxygen can be carried in 100 ml plasma. However, because of the association of oxygen with *hemoglobin*, saturated whole blood can carry 20 ml oxygen/100 ml plasma. In other words 98.5 percent of the oxygen transported by blood is carried by hemoglobin:

$$\frac{20 \text{ ml} - 0.3 \text{ ml}}{20 \text{ ml}} \times 100\% = 98.5\%$$

Hemoglobin is also necessary for the transport of carbon dioxide, which can combine directly with hemoglobin. But more importantly, by taking up excess hydrogen ions, hemoglobin allows the enzyme *carbonic anhydrase* to rapidly convert carbon dioxide to bicarbonate ion, which, unlike carbon dioxide, is highly soluble in physiological solutions. About 90 percent of carbon dioxide transport requires the presence of hemoglobin.

Inside the erythrocyte, all major reactions that involve the transport of oxygen and carbon dioxide are interrelated. The structure of the hemoglobin molecule was discussed in Chap. 14. Here we discuss some reactions hemoglobin undergoes in the process of gas transport.

Each hemoglobin molecule contains three major functional groups: *heme groups, amino groups,* and *imidazole groups* (Fig. 17.10). The heme portions contain iron with which oxygen can loosely associate. The association of oxygen with the *heme* is highly reversible and thus greatly influenced by oxygen partial pressures:

$$\text{Hb-heme} + O_2 \rightleftharpoons \text{Hb-heme-}O_2$$
$$\text{deoxyhemoglobin} \qquad\qquad \text{oxyhemoglobin}$$

In the tissues, oxygen tends to diffuse *out of* the blood, causing the equilibrium of this reaction to shift to the left and forming a greater amount of *deoxyhemoglobin.* In the lungs, oxygen tends to diffuse *into* the blood, thus shifting the reaction equilibrium to the right and forming *oxyhemoglobin.*

Oxygen binding by hemoglobin depends upon the partial pressure of oxygen. This relationship is best illustrated by an *oxygen dissociation curve,* which shows the relationship of the percent saturation of oxygen sites on hemoglobin versus the partial pressure of oxygen. Note the S shape (sigmoid shape) of this curve (Fig. 17.11). As oxygen partial pressure rises, binding by hemoglobin initially increases slowly. In the middle range of oxygen partial pressures, however, oxygen binding to hemoglobin per unit rise in partial pressure greatly increases. At high partial pressures large changes in oxygen level are required to produce small changes in the amount of oxygen bound to hemoglobin.

The oxygen dissociation curve for *myoglobin* (the form of hemoglobin found in muscle tissue) has a simple hyperbolic shape. As you recall from

FIGURE 17·10
Functions of the various groups within the hemoglobin molecule and the reactions they undergo in the tissues and lungs.

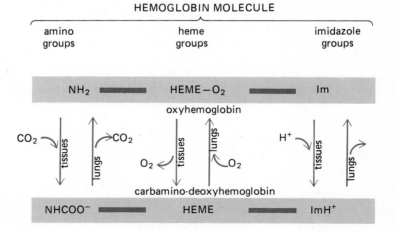

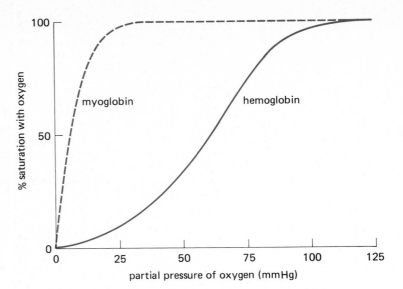

Chap. 14, the hemoglobin molecule comprises four subunits, each of which
binds oxygen, whereas myoglobin has a single heme unit. The difference in
the shape of the two oxygen dissociation curves is attributed to *subunit
interactions;* that is, the binding of an oxygen molecule by one of the heme
units in hemogobin influences the affinity for oxygen of the other heme units.
The significance of such subunit interactions rests with the fact that the
steepest portion of the hemoglobin oxygen dissociation curve, the portion of
the curve where oxygen binding is most sensitive to oxygen partial pressure,
is located in the range of oxygen partial pressures normally encountered in
body tissues. For example, tissue oxygen partial pressure must be very low
before myoglobin begins to release its oxygen, whereas, because of subunit
interactions, hemoglobin releases most of its oxygen at the much higher
plasma oxygen partial pressures.

Factors that influence oxygen binding to hemoglobin can be recognized
by their effect on the oxygen dissociation curve. Elevated temperature,
increased acidity, or a rise in the partial pressure of carbon dioxide pushes
the oxygen dissociation curve to the right; that is, these factors *decrease* the
affinity of hemoglobin for oxygen (Fig. 17.12). These characteristics of
hemoglobin facilitate the release of oxygen to very active tissues, which, of
course, have a higher oxygen demand. Note, however, that these factors do
not influence the oxygen loading of hemoglobin because their effect is not
significant at the oxygen partial pressure encountered by the blood in the
alveoli (105 mmHg).

Human erythrocytes contain a phosphorylated compound, 2,3-diphospho-
glycerate (DPG), which will shift the oxygen dissociation curve to the right
when present at elevated levels. Oxygen release to the tissues is facilitated at
high altitudes, where oxygen levels are lower, by a significant elevation of
this compound.

Similarly, the red blood cells of pregnant females contain elevated levels
of DPG, which apparently facilitates release of oxygen to the placenta for

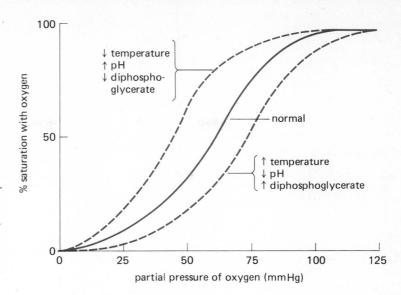

FIGURE 17·12

Oxygen dissociation curves for hemoglobin. Note that increases in certain factors shift the curve to the right and cause hemoglobin to more easily release its oxygen. Similarly, if these factors decrease, the curve will shift to the left. These factors do not prevent complete saturation of hemoglobin at the oxygen partial pressure of the alveoli (104 mmHg).

the developing fetus. Furthermore, fetal hemoglobin has a higher affinity for oxygen than that of the adult partially because of *decreased* DPG levels in the fetal red blood cells.

CARBAMINO HEMOGLOBIN   Certain amino groups in the globin portion of hemoglobin react reversibly with carbon dioxide. In the tissues, carbon dioxide tends to diffuse along its diffusion gradient into the red blood cells, where about 25 percent of it reacts with an end amino nitrogen of hemoglobin, forming *carbamino hemoglobin:*

$$\text{Hb—NH}_2 + \text{CO}_2 \rightleftharpoons \text{Hb—NHCOO}^- + \text{H}^+$$
carbamino
hemoglobin

The formation of carbamino hemoglobin releases hydrogen ions. To prevent a drastic change in intracellular pH, this hydrogen ion must be buffered. Most buffering inside the red blood cell is accomplished by the *imidazole* groups on hemoglobin (Fig. 17.13):

$$\text{Hb—Im} + \text{H}^+ \rightleftharpoons \text{Hb—ImH}$$

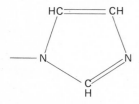

FIGURE 17·13

An imidazole ring from the histidine residue of hemoglobin.

In the lungs, carbon dioxide diffuses out of the blood and the two reactions shown above are shifted to the left. The reaction of carbon dioxide with amino groups and the subsequent release of hydrogen ions not only assures transport of part of the tissue carbon dioxide to the lungs, but it also facilitates the release of oxygen by hemoglobin at the tissue site (Fig. 17.14).

BICARBONATE IONS   Only about 25 percent of tissue carbon dioxide is transported in the blood as carbamino hemoglobin. Another small fraction, about 8 percent, is carried as dissolved carbon dioxide in plasma. The

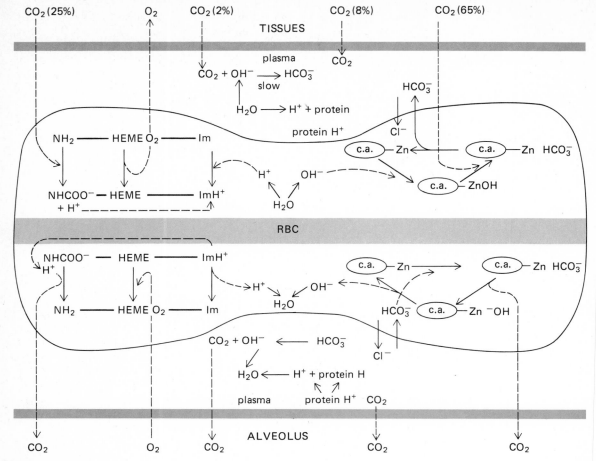

FIGURE 17·14

Summary of the means of gas transport and exchange at the site of tissues and lungs. ca, carbonic anhydrase, Im, imidazole.

remaining 67 percent is carried to the lungs as bicarbonate ions. Only 2 percent of this carbon dioxide is converted to bicarbonate ions in the plasma; the remaining 65 percent is converted inside the red blood cells.

Carbon dioxide reacts with water to form carbonic acid (the acid in soda pop), which dissociates to hydrogen and bicarbonate ions. This reaction is generally written as follows:

$$CO_2 + H_2O \rightleftharpoons H_2CO_3 \rightleftharpoons H^+ + HCO_3^-$$

Expressing the reaction between carbon dioxide and water this way, however, conceals the part of the reaction in which carbon dioxide reacts with the hydroxyl ion of water to produce the bicarbonate ion:

$$CO_2 + OH^- \rightleftharpoons HCO_3^-$$

In an aqueous solution, a very small amount of water dissociates into hydrogen ions and hydroxyl ions. The presence of carbon dioxide in such a solution thus increases the amount of free hydrogen ions by tying up hydroxyl ions and pushing the equilibrium of the reaction to the right.

The reaction of carbon dioxide with water normally occurs spontaneously but very slowly. Remember, only about 2 percent of the carbon dioxide that diffuses into the blood reacts with water in the plasma. The same reaction occurs very rapidly within the red blood cell, however, because of the catalytic activity of the enzyme *carbonic anhydrase* (Fig. 17.14). Carbonic anhydrase behaves as a hydroxyl activator inside the cell. Because the enzyme has zinc (Zn) at its active site, it is labeled E—Zn in the following equations:

(1) $$H_2O \rightleftharpoons H^+ + OH^-$$

(2) $$E—Zn + OH^- + H^+ \rightleftharpoons E—Zn—OH + H^+$$

(3)
$$CO_2 + E—Zn—OH + H^+ \rightleftharpoons E—Zn—O—\overset{\overset{\displaystyle O}{\|}}{C}—OH + H^+ \rightleftharpoons$$
bicarbonate

$$E—Zn + HCO_3^- + H^+$$

The enzyme is unchanged by the reaction, of course, and all steps are reversible. The concentration of the products determines the direction of the reaction. If hydrogen ions are allowed to accumulate, therefore, the formation of bicarbonate cannot continue in this way, but hydrogen ions do not accumulate to a significant extent because the hemoglobin imidazole groups buffer them. Thus, hemoglobin plays a dual role as it transports carbon dioxide.

Bicarbonate ions, formed by the action of carbonic anhydrase, rise in concentration inside the red blood cell until a concentration gradient is formed between the red blood cell and the plasma. The bicarbonate ions then diffuse along the concentration gradient into the plasma. Bicarbonate movement through the red-cell membrane is an example of facilitated diffusion. A transport protein in the membrane exchanges intracellular bicarbonate ions for plasma chloride ions on a one-for-one basis so that the electrochemical potential across the cell membrane is not disturbed. This exchange is called the *chloride shift;* it is reversed when the blood reaches the lungs.

CONTROL OF RESPIRATION

NEURAL CONTROL *The inspiratory and expiratory centers* Several centers influence the respiratory rate and depth of breathing. Two respiratory control centers are located in the medulla: an inspiratory center and an expiratory center. Both centers lie beneath the posterior part of the floor of the fourth ventricle; the expiratory center lies immediately below this area, and the inspiratory area somewhat deeper and more posterior. Although these respiratory areas have been identified in experimental animals, there is also evidence that the bulbar type of poliomyelitis affects similar areas in human beings.

Motor tracts from the inspiratory center descend through the spinal cord to the third, fourth, and fifth cervical levels, where they synapse with motor neurons to form the two phrenic nerves, which innervate the diaphragm. Stimulation of the phrenic nerves causes muscles of the diaphragm to contract, thus lowering the diaphragm and increasing the volume of the thoracic cavity during inspiration. Cutting the phrenic nerves stops the movement of the diaphragm (Fig. 17.15).

Another set of neurons descending from the medulla synapses with thoracic motor neurons supplying the external intercostal muscles. The external intercostals raise the ribs and enlarge the thoracic cavity during inspiration.

The internal intercostal muscles are innervated by a similar set of neurons, but their function is expiratory. They stimulate the lowering of the ribs during a deep or forced expiration.

The inspiratory and expiratory centers in the medulla are believed to reciprocally inhibit each other. In normal respiration the inspiratory center is dominant, because it has the lower threshold and is therefore more readily stimulated.

The vagus nerves are an important pathway by which afferent impulses arising from sensory receptors within the lungs regulate rhythmical expiration. As inspiration proceeds, elastic lung tissues are stretched and sensory receptors in these tissues are stimulated, producing afferent impulses. This type of stimulation, which apparently resembles the stretch reflex in muscles and tendons, sends an ever-increasing flow of nerve impulses through the

FIGURE 17·15

Diagram illustrating the centers and nerves that exercise control over breathing. The inspiratory and expiratory centers of the medulla are shown in the inset. The arrows indicate the close interrelation of nerve impulses between these centers.

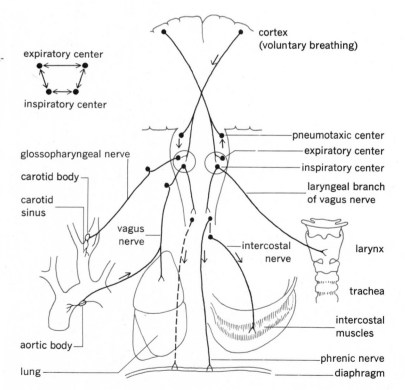

vagus nerves to the expiratory control center. Since the expiratory center has a high excitatory threshold, the lungs continue to inflate until the control center is finally stimulated. The inspiratory center is then reciprocally inhibited, and expiration follows. This mechanism appears to protect the lungs from overinflation. The lung reflex is an important mechanism and is called the *Hering-Breuer vagal reflex,* after the two physiologists who studied it and published an account of their findings in 1868.

When both vagus nerves are cut, breathing becomes slower and deeper but still retains its rhythm. The change in breathing is caused by the loss of vagal afferent impulses, which normally stimulate the expiratory center and inhibit the inspiratory center.

Delay in stimulating the expiratory center permits the lungs to fill deeply, so the rhythm of breathing is slow. The fact that a rhythm still exists seems to indicate that some other neural center must influence respiration.

The pneumotaxic center Located in the upper part of the pons is a third control area that substitutes as an inspiratory inhibitory mechanism when the vagal reflexes are eliminated. The existence and function of the *pneumotaxic center* are deduced from the observation that in an animal whose vagal nerves have been severed, deep breathing continues, as though other controls had taken over for the Hering-Breuer lung reflex. If a section is made across the brainstem between the pons and the medulla in an animal whose vagal nerves have previously been cut, thus severing connections between the pneumotaxic center and the inspiratory-expiratory centers, there is an immediate cessation of rhythmic breathing. The animal may maintain a deep inspiration until death or deep inspirations may be interspersed with short expirations, but breathing rhythm is destroyed. Such experiments indicate that all inspiratory-inhibiting centers have been destroyed.

It is probable that in ordinary breathing, afferent impulses from the lungs are the most important stimuli in maintaining the rhythm of breathing. However, with increased respiratory activity, stimulated by a rise in body temperature, the pneumotaxic center may become more dominant.

Factors directly affecting respiratory centers Among the factors that affect the neurons of the respiratory centers directly are changes in carbon dioxide and hydrogen concentration, temperature of the blood at the medulla, and rate of blood flow through medullary tissue.

Only in carefully controlled experiments can one factor be demonstrated to the exclusion of other factors. Ordinarily all chemical and nervous mechanisms are coordinated in the control of respiration. If any chemical or physical regulators are to be singled out for special consideration, changes in carbon dioxide concentration and changes in blood pH level are probably the most important in the control of breathing.

Simple experiments illustrate the effects of varying oxygen or carbon dioxide concentration in the air. If an animal breathes air directly from a small chamber, but exhaled CO_2 is removed, the O_2 in the chamber will gradually diminish. Even when O_2 concentration diminishes markedly, however, respiratory rate does not increase to any great extent.

If the experiment is varied so that the exhaled CO_2 is returned to the

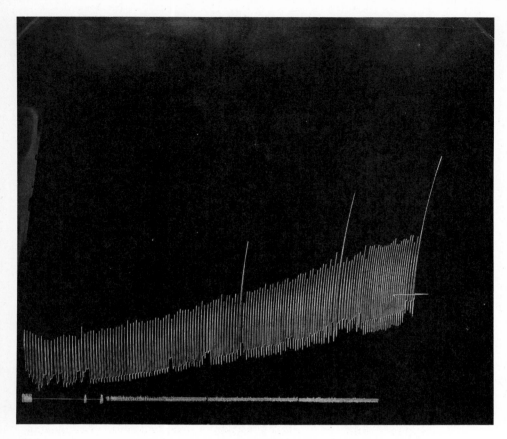

FIGURE 17·16
Effect of carbon dioxide on rat respiration. The rat breathes in a closed circuit. As the CO_2 content of the air in the breathing device rises, the respiratory rate increases and the breathing becomes stronger and deeper. (*Courtesy of E.G. Boettiger.*)

breathing chamber, the CO_2 concentration in the air in the chamber rises as the O_2 concentration diminishes. Under these circumstances, the respiratory rate increases markedly, and breathing becomes deep and vigorous (Fig. 17.16).

Adding even a little CO_2 to the inhaled air accelerates respiration. This effect is not caused entirely by the direct action of CO_2 on the respiratory centers, since increasing inspired CO_2 decreases blood pH, a condition which also elevates respiratory rate. A decrease in blood pH (increase in acidity) directly affects the respiratory center, but it also acts reflexly on chemoreceptors in the carotid and aortic bodies (see Chap. 16). Impulses from the chemoreceptors reach the respiratory centers in the medulla through afferent fibers of the glossopharyngeal and vagus nerves.

The CO_2 concentration in arterial blood can be reduced by breathing deeply and rapidly for a minute or two, a procedure called *voluntary hyperventilation*. After such forced breathing, there follows a period up to 2 min when breathing is inhibited. When breathing is resumed, it remains irregular or periodic until oxygen and carbon dioxide concentrations again return to normal in the arterial blood.

Though it is possible to hold the breath for 30 s or so by voluntary nervous control through the cerebral cortex, involuntary control centers soon overpower all voluntary inhibition and force a breath. Since breathing is controlled from centers in the medulla, one is ordinarily not conscious of its rate or depth. Under resting conditions, breath rate varies considerably in different individuals; the average rate is about 16 times/min in an adult. Like heart rate, respiratory rate varies with age, physical activity, and many other factors. Respiratory rate is much faster in little children than in adults, and exercise increases it.

The lungs of the human fetus are not functional until birth. The fetal blood is well supplied with oxygen through the placenta, and carbon dioxide is efficiently removed. At birth the placental circulation ceases to function, and it is assumed that the increase in CO_2 concentration in the blood stimulates the newborn child to gasp for air and take the first breath.

CHEMICAL REGULATION OF RESPIRATION   In addition to the neural mechanisms that exert control over breathing, there are chemical control mechanisms that adapt respiration to changes in metabolism. Chemicals affect the respiratory centers in the medulla, and they affect receptors located in the carotid and aortic bodies. A rise in carbon dioxide concentration, increased acidity of arterial blood, or a decrease in oxygen content stimulates the chemoreceptors and increases respiration reflexly.

Chemoreceptors   Ingenious experiments to isolate the carotid and aortic chemoreceptors have been performed. It has been found that respiration can be increased reflexly in experimental animals by increasing blood CO_2 concentration or by increasing the acidity of a fluid perfusing isolated carotid and aortic chemoreceptors. The variations in the perfusion fluids used in these experiments, however, were greater than the variations normally seen in the chemical composition of the blood.

In one experiment, a normal dog was permitted to inhale pure nitrogen for 30 s. Pulmonary ventilation was approximately doubled. The chemoreceptors in the dog's carotid areas were then denervated. After inhaling nitrogen for 60 s, the dog showed no increase in ventilation. Ventilation was actually reduced, since a lack of oxygen depresses the respiratory center. In a supplementary experiment, it was shown that the respiratory response to inhalation of 5 percent carbon dioxide in both the normal and the carotid-denervated animal was essentially the same. Both showed an increase in respiration rate, but the increase was caused by the direct effect of CO_2 on the respiratory centers of the medulla.

In ordinary breathing, the chemoreceptors exert only a small regulatory effect upon breathing. When the oxygen content of the arterial blood is lowered and a consequent rise in CO_2 and H^+ concentration occurs, these receptors become a more important source of respiratory control. They also control the deep, powerful respiratory movements observed when a person or animal is under the influence of ether anaesthesia.

Baroreceptors and respiration   It is well known that a sudden, substantial rise in arterial blood pressure depresses the respiratory rate. A rise in blood pressure stretches the walls of the aortic arch and the carotid sinus areas where the baroreceptors are located (Fig. 17.17). The baroreceptors are stimulated and send nerve impulses that inhibit the respiratory center reflexly (Fig. 17.18). Conversely, a considerable drop in arterial blood

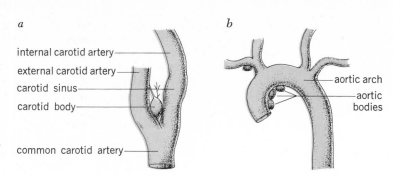

FIGURE 17·17

Diagrammatic representation indicating the location of *a* carotid and *b* aortic chemoreceptors and pressoreceptors.

pressure causes an increase in breathing rate above the normal (hyperpnea). This may be merely the result of loss of inhibitory impulses. Though baroreceptors play their part in control of respiration, they do not appear to be as essential as chemoreceptors.

REGULATION OF PULMONARY CIRCULATION   The distribution of blood to alveolar capillaries is largely controlled by the action of smooth muscle in

FIGURE 17·18

Effect of a rise in blood pressure on the rate and depth of breathing in a rabbit. The rise in blood pressure (upper tracing) was stimulated by an injection of epinephrine (Adrenalin). Note that breathing (lower tracing) is depressed. The point at which epinephrine was injected is indicated by the triangle at the bottom. (*Courtesy of E. G. Boettiger.*)

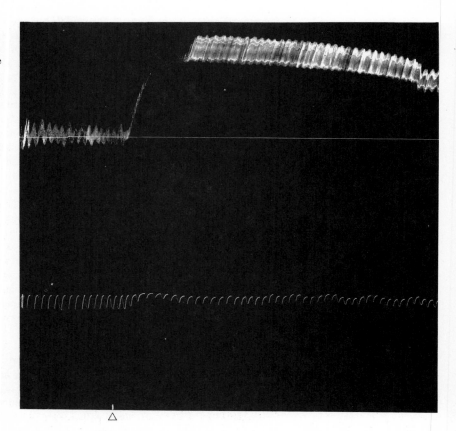

the lung arterioles. This tissue is quite sensitive to oxygen and hydrogen ion concentrations. As the oxygen concentration decreases in the alveoli, the arterioles constrict. Conversely, as oxygen concentration increases, the arterioles dilate. As hydrogen ion concentration increases, the arterioles constrict, and as the hydrogen ion concentration decreases, the arterioles dilate. It is important to note that these effects are the *opposite* of those exerted on arterioles throughout the rest of the body.

The behavior of pulmonary arterioles may be summarized as follows:

1 A decrease in oxygen concentration and an increase in hydrogen ion concentration cause arteriolar constriction.
2 An increase in oxygen concentration and a decrease in hydrogen ion concentration cause arteriolar dilatation.

In review, we also note that pulmonary arterioles are dilated by impulses from the sympathetic nervous system and constricted by impulses from the parasympathetic nervous system.

EFFECT OF EXERCISE ON RESPIRATION Vigorous exercise increases respiration markedly. All factors that stimulate respiration are brought into action. These include increased acidity of the blood (decreased pH), a rise in body temperature, and increased secretion of epinephrine into the blood. Nerve impulses from the cerebral motor area, from the heart and lungs, and from muscles and joints stimulate the respiratory centers.

During exercise, the removal of CO_2 from the body is so efficient that very little of this substance accumulates in the blood, even though the need for oxygen may increase the alveolar ventilation 10 to 20 times over the resting level. There is, in addition, no marked increase in hydrogen ion concentration from CO_2 exchange. However, muscular activity does increase the lactic acid and hydrogen ion concentration of the blood. At the present time, it is not clear which factor or factors are responsible for inducing the deep and rapid breathing associated with exercise.

REGULATION OF pH

ACID-BASE BALANCE The term *acid-base balance* refers to those homeostatic mechanisms that maintain a relatively stable hydrogen ion concentration in the body fluids. Hydrogen ion concentrations rapidly equilibrate across most cell membranes so that alterations in extracellular pH result in immediate alterations in the intracellular H^+ concentration, a characteristic that affects intracellular enzymes. Many of these life-sustaining enzymes are very sensitive to pH and function properly only within a very narrow pH range. If plasma pH drops below 7.0 or exceeds 7.7, death may result. Although this may at first seem to be a small change, it actually represents a threefold change in H^+ concentration. Intracellular pH averages about 7.0; however, it may vary between 4.5 and 7.4 in different cell types.

Many factors influence the pH of body fluids. Acids and bases (primarily acids) are continuously produced by the body itself during normal metabolism. Acids include carbon dioxide—which combines with water to form carbonic acid—lactic acid, and acetoacetic acid, as well as sulfates and phosphates, which contribute to the acid load of the blood. In addition to being produced by metabolic processes, acids and bases are ingested. Ingested acids include acetic, citric, tartaric, and fatty acids. Salts, such as potassium, influence indirectly the pH of extracellular fluids by their effect on the excretion of H^+ by the kidneys. The primary bases ingested are bicarbonate salts.

Three primary mechanisms are responsible for maintenance of stable plasma pH: (1) chemical buffering systems; (2) the respiratory system; and (3) the kidneys.

CHEMICAL BUFFERS Buffering systems, or *buffers*, are composed of weak acids and salts of weak acids. Whether a substance is considered to be a weak or strong acid depends upon its ability to dissociate to form hydrogen ions and corresponding negative ions. For example, a strong acid such as hydrochloric acid (HCl) completely dissociates to form H^+ and Cl^-, and virtually none remains in the form of the nondissociated HCl:

$$HCl \rightleftharpoons H^+ + Cl^-$$

However, a weak acid such as carbonic acid (H_2CO_3) undergoes only incomplete dissociation:

$$H_2CO_3 \rightleftharpoons H^+ + HCO_3^-$$

A weak acid obviously has a much smaller effect on the pH of a solution than a strong acid. Therefore, if hydrogen ions can be incorporated into weak acids rather than into strong acids, their effect on pH will be minimized. For example, if a strong acid such as HCl is added to sodium bicarbonate, the salt of a weak acid, the following reaction occurs:

$$HCl + NaHCO_3 \longrightarrow NaCl + H_2CO_3$$

The hydrogen ions are switched from a strong acid to a weak acid, and, thus, far fewer hydrogen ions are dissociated or free in solution.

If a strong base is added to the buffering system, the increased concentration of hydroxyl ions will elevate the pH. The effect is minimized, however, as follows:

$$NaOH + H_2CO_3 \longrightarrow NaHCO_3 + H_2O$$

The strong base NaOH is converted to the weak base $NaHCO_3$, and the change in pH is thus minimized.

Such examples show that the efficiency of a buffering system in minimizing changes in pH depends upon the relative amounts of the weak acids and their salts. An ideal buffer for human blood plasma would be one with a 1:1 ratio; that is, in the ideal system 50 percent of the buffering system would be present as the weak acid, while the other 50 percent would be maintained as the salt of that acid at normal plasma pH (7.4).

Acid-base buffering systems are found in all body fluids. The three most important chemical buffering systems are: (1) the phosphate system; (2) the proteinate system; and (3) the bicarbonate system. These chemical buffers are not functionally distinct but work together as a unit because all such systems share hydrogen ions. Thus, it is impossible to alter one buffering system without influencing all the others.

The phosphate buffering system Phosphate is present in plasma in several forms. The forms most active in the *phosphate buffering system* are the acid monosodium dihydrogen phosphate (NaH_2PO_4) and the salt *disodium monohydrogen phosphate* (Na_2HPO_4). Although the plasma concentration of the components of this buffering system is six times less than that of the elements of the bicarbonate system, the phosphate buffering system is still important. At pH 7.4, the $NaH_2PO_4:Na_2HPO_4$ ratio is 1:4, a ratio much nearer the ideal than that of the bicarbonate system. This ratio makes phosphate a much more efficient plasma buffer than other buffering systems.

Phosphates are found in intracellular fluids at much higher concentrations than in extracellular fluids. Inside cells, where its components exist as potassium salts (K_2HPO_4), the phosphate buffer system is of utmost importance.

The phosphate and bicarbonate systems work together, as well as with other buffering systems, as shown here:

(1) $$Lactic\ acid + NaHCO_3 \rightleftharpoons Na\text{-}lactate + H_2CO_3$$

(2) $$H_2CO_3 \rightleftharpoons H^+ + HCO_3^-$$

(3) $$Na_2HPO_4 + H^+ + HCO_3^- \rightleftharpoons NaH_2PO_4 + NaHCO_3$$

This example illustrates the interactions of the buffering systems.

The proteinate buffering system The *proteinate buffering system* is a critical one because it has the greatest capacity of all such systems. Three-fourths of the buffering capacity of the body fluids is found inside cells, and most intracellular buffering is done by intracellular proteins. Hemoglobin accounts for most of the blood buffering capacity. Imidazole, a weak base that is part of the histidine residue of the globin portion of hemoglobin, exists in a 1:3 ratio of hydrogen imidazole. The imidazole buffering system is present in large quantity, and its proportions make it very nearly an ideal buffer at the pH of plasma. All other blood proteins also exist in weak acid and salt forms so that they also behave as buffers, with varying degrees of effectiveness.

The bicarbonate buffering system All buffering systems of the body are important. However, the *bicarbonate buffering system* is especially important because the concentrations of both *sodium bicarbonate* ($NaHCO_3$) and *carbonic acid* (H_2CO_3), the components of this system, can be rapidly regulated. The bicarbonate system is not an ideal plasma buffer, since at pH 7.4, the ratio of H_2CO_3 to $NaHCO_3$ is 1:20. However, carbon dioxide is continuously produced by normal metabolic processes so that the amount of H_2CO_3 is constantly renewed. In addition, the enzyme *carbonic anhydrase*, which is present in red blood cells, greatly accelerates the formation of bicarbonate from carbon dioxide and hydroxyl ions.

The respiratory system strongly influences this buffer system by increasing or decreasing the rate of carbon dioxide removal from the blood. The kidneys influence the system by adjusting the reabsorption rate of filtered HCO_3^- and Na^+ or by adjusting the rate of H^+ secretion.

THE RESPIRATORY SYSTEM AND ACID-BASE BALANCE The respiratory system adjusts acid-base balance in body fluids by working through the bicarbonate buffering system. Alterations of ventilatory rate and tidal volume change the carbon dioxide level of extracellular body fluids.

An increase in hydrogen ion concentration greatly stimulates rate of ventilation and increases the tidal volume. Increasing gas exchange has the primary effect of lowering the carbon dioxide level in the blood. Because all the buffering reactions are readily reversible, a change in the concentration of any given reactant changes the concentrations of all other reactants. Thus, an increase in carbon dioxide increases formation of carbonic acid; a decrease in carbon dioxide depresses the formation of carbonic acid. By increasing the elimination of carbon dioxide, the lungs, in effect, lower hydrogen ion concentration in the extracellular body fluids.

Hydrogen ions from sources other than carbonic acid stimulate the formation of carbon dioxide from carbonic acid, producing a higher concentration of carbon dioxide in the body fluids. At the same time, raising hydrogen ion concentration stimulates respiration rate and increases the rate at which carbon dioxide is eliminated from the body fluids. The free hydrogen ions are then incorporated into water.

The ability of the respiratory system to restore acid-base balance is thus limited by the relative concentrations of the components of the bicarbonate buffering system. In addition, the respiratory system cannot completely restore balance when hydrogen ion perturbations are caused by fluctuations in concentrations of substances other than carbon dioxide. As hydrogen ions are eliminated by the bicarbonate buffering system, the stimulus to ventilate the lungs decreases, and restoration of acid-base balance can never be completed by this mechanism alone. The kidneys are necessary for complete restoration of balance.

THE KIDNEYS AND ACID-BASE BALANCE   Although the chemical buffering systems and the respiratory system quickly counteract alterations in body fluid pH, they can never completely restore acid-base balance. The kidneys respond more slowly to an acid-base imbalance, but they can *completely* restore this balance. The kidneys act primarily through the bicarbonate buffering system. The phosphate buffers and ammonia are also important, however, in renal acid-base regulation.

As we will explain in Chap. 18, urine formation involves the production of an ultrafiltrate of plasma and its subsequent modification by reabsorption of substances from it or by secretion of substances into it. The adjustment of plasma hydrogen ion concentration by the kidneys involves all these processes, i.e., filtration, reabsorption, and secretion.

The kidneys adjust the pH of body fluids primarily by varying the rate at which hydrogen ions are secreted into the urine. The secretion of hydrogen ions serves two purposes: (1) direct elimination of hydrogen ions, and (2) facilitation of bicarbonate reabsorption from the filtrate formed by the kidneys from plasma.

The rate of hydrogen ion secretion is determined largely by the relative acidity of the extracellular fluid. Any increase in hydrogen ion concentration in the extracellular fluid produces an increase in carbon dioxide level. An increase in carbon dioxide in the extracellular fluid produces an increase in carbon dioxide inside the kidney tubule cells. Some tubule cells contain a large amount of carbonic anhydrase, which, as we have pointed out, greatly accelerates the formation of bicarbonate ion from carbon dioxide and hydroxyl ions. As bicarbonate ions are synthesized, many free hydrogen ions are produced. These are transported into the urine in exchange for filtered sodium ions.

The bicarbonate ions, along with the sodium ions, are then transported from the renal tubule cells into the extracellular fluid. In the fluid, or filtrate, of the renal tubules, the secreted hydrogen ions combine with the filtered bicarbonate ions, producing carbon dioxide and water. The carbon dioxide freely diffuses back into the extracellular fluid; the secreted hydrogen ions are carried away in the urine as water. At this point, the hydrogen ion concentration of the extracellular fluid has been lowered while the bicarbonate concentration of the extracellular fluid remains unchanged, with its buffering capacity intact.

Because during normal metabolism the body usually produces excess acid, hydrogen ion secretion by the kidneys normally exceeds bicarbonate reabsorption. When HCO_3^- ions have been completely reabsorbed so that none are available to react with secreted H^+ ions, the pH of the urine declines; that is, the urine hydrogen ion concentration increases. Hydrogen ion secretion is an active transport process and thus can continue against an electrochemical gradient. The capacity of the system is limited, however, so that net hydrogen ion secretion normally ceases, regardless of the need for hydrogen ion elimination, when urine pH reaches 4.5. This limiting pH is normally never reached because phosphate buffers and dissolved ammonia in renal tubular fluid prevent the urine pH from going below 4.5. When the pH remains above 4.5, hydrogen ion secretion is continuous. The phosphate buffering system components Na_2HPO_4 and NaH_2PO_4 are present in the plasma filtrate formed by the kidneys, and these compounds buffer free hydrogen ions in the urine.

Ammonia is formed in renal tubule cells by the deamination of amino acids, mostly glutamine, which is taken up from the blood. Ammonia (NH_3) is an uncharged gas that freely diffuses through cell membranes. It is transported into the tubular fluid by passive diffusion, but does not diffuse back because it reacts with hydrogen ions to form ammonium ions (NH_4^+). Most hydrogen ions excreted by the kidneys are thus in one of four forms: free hydrogen ions, water, phosphoric acid, or ammonium chloride, an ammonium salt.

ACID-BASE IMBALANCE Hydrogen ion concentration is one of the most closely regulated characteristics of the body fluids. Acid-base imbalance occurs only when the limits of the compensatory mechanisms outlined here have been exceeded. Inadequate removal of excess hydrogen ions lowers pH, a condition termed *acidosis*. Excessive removal of hydrogen ions or an accumulation of basic substances in the body fluids elevates pH, a condition termed *alkalosis*. Both acidosis and alkalosis may be classified as either *respiratory* or *metabolic*.

Respiratory acidosis and alkalosis Any condition that hinders the respiratory removal of carbon dioxide from the blood may produce *respiratory acidosis*. The accumulation of hydrogen ions in the body fluids caused by inadequate removal of carbon dioxide may be the result of diseases that decrease the efficiency of the lungs. Emphysema and bronchial asthma both decrease the rate of carbon dioxide removal from the blood. Under resting conditions, the kidneys compensate for respiratory acidosis by excreting excess hydrogen ions in the urine. However, as we have noted, the kidneys require a long time to adjust body fluid pH and, thus, cannot prevent acidosis during active periods when carbon dioxide production greatly increases.

Respiratory alkalosis is caused by excessive elimination of carbon dioxide by the lungs. This condition may occur at high altitudes when the partial pressure of oxygen in the atmosphere is decreased. In some individuals, emotional anxiety may cause an inappropriately increased rate of ventilation (hyperventilation), thus lowering carbon dioxide concentration in the body fluids. If respiratory alkalosis is chronic, the kidneys may compensate by decreasing hydrogen ion secretion and by eliminating bicarbonate in the urine.

Metabolic acidosis and alkalosis When acidosis or alkalosis are caused by factors other than the failure of normal respiration to properly regulate carbon dioxide, they are considered to be of metabolic origin. Many different bodily disorders may lead to *metabolic acidosis*. In *diabetes mellitus,* a disease in which glucose cannot be properly taken up by many types of cells, the cells must utilize fats for energy (see Chap. 13). The metabolism of fats produces ketone bodies, mainly acetoacetic acid, which are compounds that add to the plasma hydrogen ion concentration. *Starvation* causes a similar increased production of acidic fat metabolites, as may some extreme weight-reducing regimens. Metabolic acidosis may also be caused by *diarrhea* when the sodium bicarbonate secreted by the pancreas passes through the intestine too quickly to be reabsorbed. All these conditions may be partially compensated for by increased elimination of carbon dioxide by the respiratory system and by increased hydrogen ion secretion by the kidneys.

Kidney failure, a condition referred to as *uremia* (urine in the blood), usually causes metabolic acidosis because the kidney cannot satisfactorily eliminate a normal quantity of the metabolic acids produced by the body.

Metabolic alkalosis may be caused by such conditions as excessive ingestion of alkali or vomiting. During vomiting the hydrochloric acid secreted by the stomach is lost. Because this acid is normally reabsorbed, contributing to the hydrogen ion level in body fluids, an alkaline imbalance in the body fluids is produced. Metabolic alkalosis is partially compensated for by decreased ventilation, but, of course, this decrease is limited by the need to oxygenate the blood. The kidneys respond to this type of alkalosis by excreting excess bicarbonate in the urine.

CLINICAL ASPECTS

INFECTIOUS DISEASES OF THE LUNG *Tuberculosis* Tuberculosis is an ancient disease, which attacked humans even in prehistoric times. It still poses a formidable health problem, although prevention, detection, and successful treatment have diminished its incidence considerably. Probably millions of older people throughout the world harbor active tubercle bacilli, and thousands of new active cases are reported yearly. Nearly 90 percent of the cases are pulmonary tuberculosis, an inflammation of the lungs caused by the tubercle bacillus.

Resistance to the spread of the invading tubercle bacilli is provided by T lymphocytes, which destroy great numbers by phagocytosis. Some people have a natural resistance to tuberculosis and are able to develop an immune response to the initial infection.

Among the symptoms generally observed are fatigue, a cough, congestion in the lungs, and blood in the sputum. Chest x-rays may confirm the diagnosis.

Chemotherapy is used extensively to treat tuberculosis today. Several drugs are available, and these are generally used in combination since the bacilli mutate and tend to become resistant to individual drugs. Antituberculous drugs are effective only against replicating bacilli. Since tubercle bacilli may not replicate frequently, therapy is usually extended over many months.

The tuberculin skin test The skin and other tissues of an infected person become hypersensitive to the specific protein of the tubercle bacillus. This protein, *tuberculin,* can be produced by growing the bacilli in a broth and then extracting the protein from the culture. Injecting the protein fraction into the skin causes a skin reaction in a tuberculous individual. A typical local allergic reaction indicates a positive test. The test is so accurate that a positive reaction can be obtained before the symptoms of the primary infection are recognized.

Pneumonia A lung infection usually caused by a coccal bacteria is called *pneumonia.* The pneumococcus causes inflammation of the lung tissue. Infected alveoli become flooded with fluid, making breathing difficult. The incidence of pneumonia has been sharply reduced by treatment with antibiotic drugs. Penicillin G is especially effective.

Pleurisy *Pleurisy* is a condition in which there is an inflammation of the pleural membranes. In one type considerable fluid is secreted into the pleural cavity.

DUST, POLLUTION, AND DISEASE There is a growing concern about the purity of the air that we breathe. Smoke from factories, the burning of coal with high sulfur content, and the exhaust fumes from automobiles contribute to air pollution. Some of these constituents are known to be carcinogenic (cancer causing), while others have been implicated as causative agents for other diseases.

Dust diseases The constant inhalation of dust is a hazard in many kinds of manufacturing and mining. Silica (SiO_2), a common constituent of the earth's soil and an often-used industrial agent, may cause *silicosis,* a disease in which silica fibers become embedded in lung tissue, preventing adequate ventilation. The *black lung disease* of coal miners is a similar disease caused by inhaling coal dust.

Asbestos contains fibrous minerals that are highly resistant to heat. It has numerous uses in insulation, building products, and textiles. When inhaled, the asbestos fibers lodge in the lungs, where they are irritating and highly indestructible. Not only those who work with asbestos are likely to be infected, but others who only casually come in contact with this product are shown to carry the fibers in their lungs. Recently, asbestos has been shown to be the cause of a slow-growing, but fatal, lung cancer.

Cigarette smoking Although almost everyone knows that cigarette smoking is dangerous to health, many disregard this warning. The smoke that is inhaled contains injurious hydrocarbons, carbon dioxide, carbon monoxide, and small smoke particles produced by the burning tobacco and cigarette wrapper. The "tars" and nicotine probably cause the most damage to the lungs, but smokers should not rely on claims of low tar and nicotine content to protect their health.

Research has shown that smoking reduces the movement of cilia, retards the effectiveness of alveolar macrophages, and causes obstruction of the small airways. "Smoker's cough" is a result of an increase in mucus in the bronchial tubes. The impaired cilia are unable to raise the mucus to the throat in the normal manner.

Smoking has been shown to increase the likelihood of cancer and of cardiac conditions.

Emphysema *Emphysema* is a lung condition of increasing importance in our aging population. In this condition, the alveolar walls break down progressively, thus reducing the area available for the exchange of gases. Many alveoli become confluent, leaving open spaces surrounded by fibrous tissue. As the alveoli break down, their capillaries also degenerate, tending to raise the pulmonary blood pressure. There seems to be little doubt that smoking increases the likelihood of developing emphysema.

Chronic bronchitis Excessive mucous secretion in the bronchi and in the larger cartilaginous passageways is seen in *chronic bronchitis*. It is characterized by a cough, abundant sputum production, and mucous obstruction of the air passageways. The small airways of the acini are also involved; at these sites, bronchitis is commonly associated with emphysema. Cigarette smoking, exposure to noxious gases or dusts, and air pollution are contributing factors.

Asthma Asthma is a nonspecific disease characterized by difficulty in breathing, a cough, and wheezing as air passes through partially occluded passageways. Allergic asthma is caused by exposure to pollen, various other proteins, dust, fur, feathers, and many other antigenic substances. There are various types of asthma, but basic to all are constriction of airways and edema of bronchial membranes with thickened secretions. A series of attacks is commonly followed by essentially normal breathing.

Hay fever "Hay fever" is an allergy to pollen. Technically it is termed as an allergic rhinitis. People with hay fever have watery eyes and a runny nose. The allergy is usually seasonal if it is caused by pollen from weeds and grasses. In this case it disappears after a frost or freezing weather. The name is hardly appropriate since the symptoms are not produced by hay and fever is not a common occurrence.

Carbon monoxide Inhalation of carbon monoxide may require quick application of artificial respiration. Hemoglobin has a much greater attraction for carbon monoxide than it does for oxygen. The blood is depleted of oxyhemoglobin by carboxyhemoglobin, which takes its place. Carbon monoxide is an extremely dangerous gas, and since it has no odor, it is difficult to detect.

Carbon monoxide is a component of the exhaust in gasoline engines, in the combustion of charcoal and other fuels, and in the fumes from leaking burners, stoves, furnaces, or flues. Careful ventilation of rooms in which there are any exhaust fumes or leaking of fumes is very important. Garage doors should always be open if the automobile engine is running.

ABNORMAL PULMONARY CONDITIONS   *Dyspnea*   Difficulty in breathing or a feeling of being unable to obtain enough air to breathe is termed *dyspnea*. The causes are numerous, but dyspnea may be caused by cardiac failure, pulmonary edema, or emphysema.

Pulmonary edema   When fluid collects in the alveoli and interstitial spaces of the lungs, a condition called *pulmonary edema* develops. It may be associated with pneumonia or with cardiac failure, especially of the left side of the heart. If, as a result of congestion, the capillary blood pressure rises above that of the plasma osmotic pressure, fluid begins to fill the alveolar spaces. Ordinarily, the lungs remain "dry" because the capillary blood pressure is low and fluid is absorbed from the lung tissue under the influence of the higher osmotic pressure exerted by the plasma proteins. The blood pressure in the pulmonary artery is only about 25 mmHg systolic (8 mmHg diastolic), and the resistance of the lung capillaries is remarkably low (15 mmHg).

Pneumothorax   Puncture of the chest wall, which permits air to enter between the two layers of the pleura, will cause the lung to collapse. Artificial *pneumothorax* (opening the pleural cavity) is used to rest an infected lung in certain types of tuberculosis.

EFFECTS OF RAPID DECOMPRESSION   Persons who work under conditions of increased air pressure, such as divers and workers in caissons, are subject to a different set of conditions from those who live and work in areas of normal or low atmospheric pressures. It is not possible to acclimate to these conditions as it is to the low pressures of high mountain areas. Exposure to increased atmospheric pressure does not cause discomfort, but when the subject is returned to normal atmospheric pressure too quickly, dangerous conditions can develop.

The symptoms caused by rapid decompression are called *caisson disease*, or the *bends*. Symptoms include pain in muscles and joints and disturbances of the central nervous system involving both motor and sensory nerves. Under conditions of high atmospheric pressure, the nitrogen dissolved in the blood diffuses slowly through cell membranes and is absorbed into tissues, especially fatty tissues. Rapid decompression is analogous to suddenly opening a bottle of a carbonated beverage. Carbon dioxide has been dissolved in the beverage under pressure; when the bottle is uncapped, the carbon dioxide comes out of solution as a gas and causes the drink to foam. Similarly, in rapid decompression excess nitrogen dissolved in the blood during exposure to high pressure comes out of solution and causes bubbles of the gas to form in the blood and tissues. Gas bubbles collect in the lung capillaries and other areas, causing severe symptoms.

A similar situation occurs when rapid decompression occurs during steep ascent in an airplane from sea level to a high altitude. In this case, the change is from a normal pressure to a low pressure instead of from high pressure to normal. Since nitrogen is an inert gas, it causes no damage even if absorbed in increased amounts, so long as it stays in solution. The rapid decompression that affects pilots, causing nitrogen to come out of solution, is called *aeroembolism*. The symptoms are identical with those of caisson disease. Descending to a lower altitude brings about recompression and relieves distress.

ARTIFICIAL RESPIRATION   Though there are several mechanical devices, such as oxygen masks and respirators, for establishing artificial respiration, these machines are often not available in an emergency. If the victim of an accident has stopped breathing, it is unlikely that there will be time enough to obtain a respirator or to remove the victim to a hospital. It would be advantageous, therefore, if a high percentage of the public were able to give artificial respiration manually.

The Schafer method   Many persons are acquainted with the *Schafer* (prone pressure) *method.* In this technique, the subject is placed on the ground in a prone position with arms outstretched or one arm flexed to cushion the head. The head is turned to one side and the tongue pulled forward. A heavy, folded garment may be placed underneath the abdominal area. The operator kneels astride the prone form, facing forward, and places the hands on the lowest ribs. The operator then leans forward, letting his or her weight gently compress the subject's chest and abdomen, thus causing air and perhaps water to be expelled from the respiratory tract. The pressure is maintained for about 2 s, and then the operator leans back for about 2 s, leaving the hands in position. Relaxing the pressure permits the thorax to expand to its normal position and draw air into the lungs. Rhythmical movements are established at 12 to 16 times/min to simulate the rate of normal respiration. Artificial respiration may have to be continued for a half-hour or more before natural breathing is established.

A more positive intake of air may be encouraged by lifting the hips a few inches off the ground. The hips can be raised by taking hold of the belt or clothing or by putting a towel under the hips. Raising the hips, alternated with pressure on the lower ribs, as in the Schafer method, provides both "a pull and a push" to accentuate the movement of air in and out of the lungs. This method quickly fatigues the operator, however. Two persons lifting can perform the operation much more easily. If alone, the operator can grasp the prone person by the clothing at the side of the hip, roll that side upward, and let it fall back in place. The operator can then move forward to apply pressure on the chest to induce expiration. A team of operators can apply this method more efficiently. It is superior to the Schafer method.

The back-pressure–arm-lift method   A fairly adequate method of artificial respiration is the *back-pressure–arm-lift* (Holger Nielsen) *method,* because it provides a positive push-and-pull movement with the subject in a favorable position. The subject is placed face down in the prone position. The elbows are bent, and one hand is placed on top of the other. The head is turned to one side, the cheek resting upon the hands.

The operator kneels at the head of the prone person and places the hands on the subject's back above the ribs. The operator then rocks forward until the arms are nearly vertical and his or her weight exerts pressure on the subject's chest, forcing air out of the lungs.

The operator then rocks slowly backward, grasping the subject's arms just

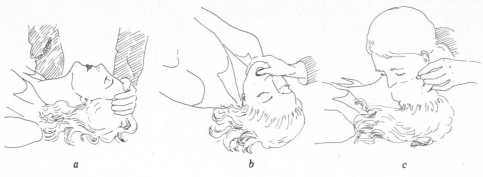

a   b   c

FIGURE 17·19

Mouth-to-mouth method of artificial respiration: *a* Push the subject's head back so that the chin points upward. *b* Pull the jaw forward into a jutting-out position (this helps to move the base of the tongue away from the back of the throat and tends to keep the tongue from blocking the air passageway). *c* Place your mouth over the subject's open mouth and pinch the subject's nostrils to close them and prevent air leakage. Blow into the subject's mouth until you see the chest rise. Breathe into the mouth regularly about 12 times/min for an adult or about 20 times/min for a child.

above the elbows. Just enough pull and lift is applied to feel the resistance and tension in the prone person's shoulders. The operator keeps his own elbows straight while drawing the subject's arms toward himself. Completing the pull or expansion phase, the subject's arms are dropped as the push-pull cycle is completed. The cycle is repeated at a steady rate of 12 times/min, allowing about equal time for the compression and expansion phases and a minimum rest period. The operator's movements should be adapted as nearly as possible to the rhythm of normal breathing.

The mouth-to-mouth respiration technique   The mouth-to-mouth respiration technique is now considered to be superior to any of the manual methods of artificial respiration. Tilting the head back, the operator blows into the subject's mouth, watching the chest rise. The subject's nose is pinched shut with the operator's free hand. The resting respiratory rate for adults is around 16/min. When working to resuscitate a child, the breath required to inflate the lungs is not as great as in an adult and the breathing can be a little more rapid, at a rate around 20/min.

Foreign matter may occlude the mouth or throat, and the mouth should be cleared manually. Children may be held briefly by the ankles, in the case of drowning, to let water run out. Sometimes a slap on the back will dislodge obstruction. If the teeth are clenched, it may still be possible to blow through the teeth or through the nose.

The volume of carbon dioxide in the operator's breath does not appear to be a serious problem, since there is still a considerable amount of oxygen in an exhaled breath. A special mouthpiece has been devised to aid the operator, but it is not necessary in an emergency.

The American Red Cross has adopted the mouth-to-mouth technique and has issued a manual for instruction in this method (Fig. 17.19).

HEIMLICH MANEUVER AND CHOKING   When a person chokes, almost everyone thinks of hitting the victim on the back, but a better method of treating choking has been recently introduced. This technique is called the *Heimlich maneuver.* In this technique, the victim's abdomen is compressed with a quick upward thrust to expel air from the lungs and force the obstructing object out of the victim's trachea or larynx. If the victim is standing or sitting, the rescuer stands behind him and puts his arms around him, with clasped fists located in the center of the abdomen below the rib cage and just above the belt line. The rescuer squeezes the abdomen quickly with an upward thrust to force the diaphragm upward and expel air from the lungs. The action may be repeated if needed. If lying down, the victim is placed in a supine position and the rescuer presses upward on the abdomen.

If alone, the victim may be able to apply the pressure with his or her own hands or to bend over the back of a chair to force the diaphragm upward and press residual air out of the lungs. Objects causing obstruction sometimes may be removed by inserting the fingers into the victim's throat and grasping the object.

INADEQUATE VENTILATION   When people are crowded into inadequate space, they may feel uncomfortable and believe there is not enough oxygen available. Normally in houses, theaters, and meeting places, enough oxygen is available, but unless there is adequate ventilation, the temperature and humidity may rise under crowded conditions. In such cases, evaporation from the skin becomes more difficult, and this causes discomfort. If fans are available to circulate the air, evaporation is enhanced and people become more comfortable. For the most part it is the same air that is being circulated, but easier evaporation makes the difference.

SUMMARY

1   The structures through which air must pass in order to reach the alveoli of the lungs are as follows: the nasal passageway, nasopharynx, pharynx, larynx, trachea, bronchi, bronchioles, and alveolar ducts. The epiglottis is a movable lid that closes above the glottis during the act of swallowing and opens during breathing. The vocal folds are located in the larynx.

2   The acinus is the terminal bronchiole in which most gas exchange occurs.

3   The alveoli are thin-walled, very vascular structures that branch off the acinus. It is there that the exchange of oxygen and carbon dioxide between the blood and the air takes place.

4   The lungs are located in the thoracic cavity. The heart and related structures lie in a cavity between the lungs, called the mediastinum. The right lung is the larger and consists of three lobes; the left lung has only two lobes. Each lung is covered with a thin serous membrane, the pleura. The pleura consists of two layers; the pulmonary pleura covers the lungs, whereas the parietal pleura lines the thoracic wall and covers the diaphragm. There is a potential pleural cavity between these two layers, but normally it contains only a little serous fluid.

5   During inspiration, the chest cavity is enlarged by muscular elevation of the ribs and lowering of the diaphragm, permitting air to flow into the lungs. Since it requires muscular effort, inspiration is said to be active. Expiration, in which the ribs fall back into place and the diaphragm returns to its normal position, is said to be passive.

6   Dead air space refers to the air that remains in the upper air passageways. It amounts to about 150 ml.

7   Two respiratory control centers are located in the medulla. There is an inspiratory center and an expiratory center. In addition, a pneumotaxic center in the pons appears to exert some regulatory control over rhythmic breathing.

8   Nervous control of respiration is supplemented by the activity of

chemoreceptors, which are stimulated by a rise in carbon dioxide concentration, increased acidity of arterial blood, or a decrease in oxygen pressure. Such stimulation causes an increase in breathing rate.

9 Baroreceptors react to a rise in blood pressure and cause the respiratory rate to be depressed.

10 Oxygen partial pressure affects the affinity of hemoglobin for oxygen. As the oxygen pressure rises, hemoglobin combines more readily with oxygen.

11 The distribution of blood to alveolar capillaries is largely controlled by the sensitivity of the stomach muscle in lung arterioles to oxygen and hydrogen ion concentrations.

12 An increase in the carbonic acid content of the blood increases the respiratory rate.

13 Inspired air contains about 20.9 percent oxygen, 0.04 percent carbon dioxide, and 79.2 percent nitrogen.

14 Oxygen is transported in the blood as oxyhemoglobin. Carbon dioxide also enters into the loose chemical combination with hemoglobin. Approximately 25 percent of the carbon dioxide in the blood is carried as carbamino hemoglobin.

15 A small amount of carbon dioxide, about 10 percent, is carried dissolved in the blood plasma or as carbonic acid, H_2CO_3. Carbon dioxide reacts slowly with the water of the blood plasma to form carbonic acid, which dissociates into bicarbonate ions, HCO_3^-, and hydrogen ions, H^+. However, this reaction within the red cell occurs very rapidly, catalyzed by the enzyme carbonic anhydrase. About 65 percent of the CO_2 of the blood is transported as bicarbonate ions, HCO_3^-.

16 The bicarbonate ion concentration within the red cell soon becomes much higher than in the plasma outside the cell. Bicarbonate ions therefore diffuse out through the red-cell membrane. Negative chloride ions move inward to take the place of the bicarbonate ions to restore the electric balance within the red cell in a process called the chloride shift.

17 During exercise, the removal of CO_2 is very efficient and little of this substance accumulates in the blood. There is no marked increase in hydrogen ion concentration from the exchange of carbon dioxide.

18 The flow of tidal air in and out of the lungs measures about 500 ml. Taking a deep breath indicates the inspiratory reserve capacity. It averages around 3000 ml. The expiratory reserve volume amounts to approximately 1100 ml. Vital capacity averages 4500 ml. The residual volume is about 1000 to 1200 ml. The total amount of air exchanged during one minute is called the minute ventilatory volume.

19 The pH of body fluids is regulated by chemical buffers, the respiratory system, and the kidneys. The respiratory system functions in maintaining an acid-base balance in body fluids through the bicarbonate buffering system.

20 The major chemical buffers are the proteinate, phosphate, and bicarbonate systems. The proteinate system is predominantly intracellular and has the largest capacity. The phosphate system is both extracellular and intracellular. The bicarbonate system is important because its components can be readily adjusted by the respiratory system and the kidneys.

1 Describe some ways in which the respiratory system is closely related to the circulatory system.
2 Trace the passage of air from the nose to the alveoli of the lungs.
3 How do the vocal folds produce the voice?
4 Explain how respiratory movements draw air in and out of the lungs.
5 Discuss the nerve centers and their control of respiratory movement.
6 Discuss the chemical regulation of breathing.
7 How do chemoreceptors and baroreceptors function?
8 Explain the effect of exercise on breathing.
9 Indicate the differences in the composition of inspired and expired air. Is there a corresponding difference between arterial and venous blood?
10 What are the effects of rapid decompression?
11 What is the relative importance of each of the chemical buffer systems of the body? How are they all related?
12 What prevents the inhibition of hydrogen ion transport by the kidneys when bicarbonate is no longer present in the renal tubular filtrate?

SUGGESTED READING

Berger, A. J., R. A. Mitchel, and J. W. Severinghaus: Regulation of Respiration, *N. Engl. J. Med.*, **297**:92–97, 138–143, 194–201 (1977).

Bonhuys, A.: "The Physiology of Breathing," Grune and Stratton, Inc., New York, 1977.

Brain, J. D., D. F. Proctor, and L. M. Reid (eds.): "Respiratory Defense Mechanisms," Marcel Dekker, Inc., New York, 1977.

Comroe, J. H.: "Retrospectroscope," Von Gehr, Menlo Park, Ca., 1977.

Forester, R. E., and E. D. Crandall: Pulmonary Gas Exchange, *Annu. Rev. Physiol.*, **38**:69–93 (1976).

Guz, A.: Regulation of Respiration in Man, *Annu. Rev. Physiol.*, **37**:303–323 (1975).

Morgan, T. E.: Pulmonary Surfactant, *N. Engl. J. Med.*, **284**:1185–1192 (1971).

Perutz, M. F.: Hemoglobin Structure and Respiratory Transport, *Sci. Am.*, **239**:92–195 (1978).

Strang, L. B.: Growth and Development of the Lung: Fetal and Postnatal, *Annu. Rev. Physiol.*, **39**:253–276 (1977).

Wagner, P. D.: Diffusion and Chemical Reaction in Pulmonary Gas Exchange, *Physiol. Rev.*, **57**:257–312 (1977).

West, J. B. (ed.): "Bioengineering Aspects of the Lung," Marcel Dekker, Inc., New York, 1977.

18

THE KIDNEYS

**CONSIDERATIONS
FOR STUDY**

In this chapter we review the structure and function of the kidneys. Rapid progress has been made recently in our understanding of these ingeniously designed organs. For this reason, as well as to correct some popular misconceptions, we have included considerable detail in our discussion. The following topics are included:

1 The basic external and internal anatomy of the kidneys and urinary tract
2 The anatomy and ultrastructure of the nephron
3 The anatomical basis and dynamics of glomerular filtration
4 The clearance concept and the processes of reabsorption and secretion
5 The functional aspects of each nephron segment
6 The major hormones influencing fluid and electrolyte balance
7 The role of the countercurrent multiplier in concentrating the urine

The kidneys, more than any other organs, are responsible for maintaining the stability of the body's internal environment. The composition of the extracellular fluid is primarily controlled not by what we ingest but by what the kidneys retain. The lungs regulate oxygen and carbon dioxide concentrations, but the kidneys are the major regulators of inorganic elements and of many organic constituents of plasma. The kidneys must be versatile in function to exert this control over practically every plasma component.

The general mechanism by which the kidneys perform their functions can be characterized as *selective elimination*. Each major substance in the extracellular fluid is handled separately by the kidneys, so that one substance may be completely eliminated, or cleared, from the blood, while another may pass through the kidneys seemingly untouched. Selective elimination involves the filtration of the blood to form a protein-free fluid (Fig. 18.1). From this filtrate different substances are returned to the blood in varying amounts. *Reabsorption* of these filtered substances and alterations of the chemical composition of the filtered fluid depend upon both energy-consuming active transport processes and the passive permeability characteristics of the renal tubular epithelium.

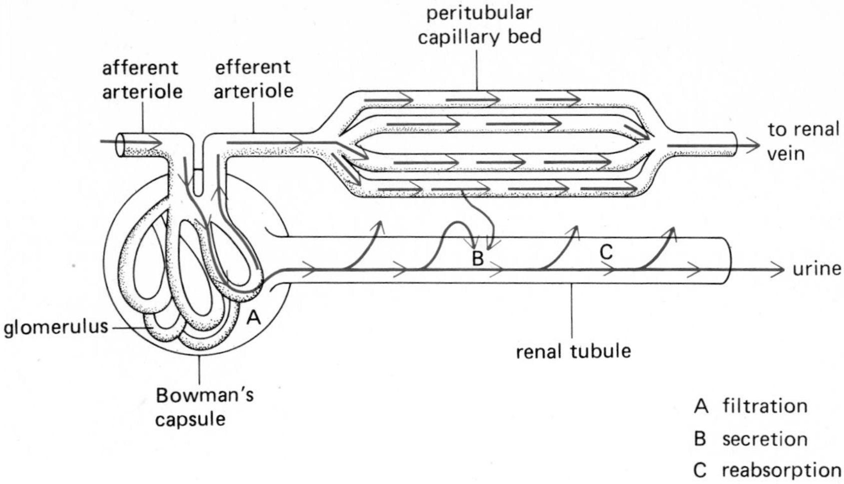

FIGURE 18·1

The three basic renal processes: *a* filtration; *b* secretion;
and *c* reabsorption. The primary urine is formed by
filtration of plasma at the
glomerulus. This filtrate is
subsequently modified by reabsorption of most of the filtered
materials. Some substances
enter the tubular urine directly by tubular secretion.

A filtration
B secretion
C reabsorption

Several substances whose blood concentrations are regulated by the kidneys do not lend themselves to the filtration-reabsorption mode of control. In many cases, such substances are not effectively eliminated by simple filtration, so they are cleared from the blood by active transport, which takes them directly from the extracellular fluid to the urine, a process usually referred to as *tubular secretion.*

The kidneys must be regarded primarily as regulatory organs. However, they also participate directly and indirectly in other important bodily processes such as control of red blood cell production, activation of vita-min D, and blood pressure regulation. The kidneys are truly vital organs, and as with many necessary systems in the human body, they are more than adequate to their tasks. Effective regulation can be maintained even with two-thirds of the kidney tissue removed.

ANATOMY OF THE KIDNEY AND THE URINARY TRACT

LOCATION AND EXTERNAL APPEARANCE The kidneys are paired structures located behind the peritoneal cavity very near each side of the vertebral column (Fig. 18.2). They lie embedded in fat on the posterior vertebral diaphragm under the last two ribs and on the abdominal muscle wall below. The right kidney is partially covered by the liver at its upper end and the colon on its lower end. The left kidney is slightly higher than the right and is in close contact with the spleen.

The kidneys are normally bean-shaped organs, about 2.5 × 5 × 10 cm, with rounded, full lateral borders and deeply concave medial borders (Fig. 18.3). The medial cavity is referred to as the *hilus.* This area is occupied by the expanded, funnel-shaped end of the *ureter* that forms the *renal pelvis,* a urine collection space within the *hilus.* The renal artery and renal vein are also present in the hilar area. Smaller spaces, continuous with the renal pelvis, extend deeply into areas of the medial kidney. These are the *calyces.*

URETERS AND URINARY BLADDER Each kidney is drained by a tube that extends from the kidney to the urinary bladder. These tubes are the *ureters.*

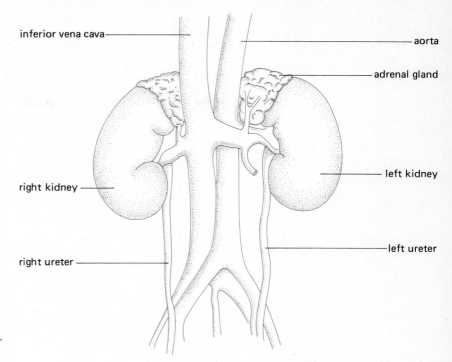

inferior vena cava

aorta

adrenal gland

right kidney

left kidney

right ureter

left ureter

FIGURE 18·2

Location of the kidneys. They rest embedded in fat on the posterior abdominal wall at the level of the last thoracic and the first few lumbar vertebrae.

FIGURE 18·3

External appearance of the right kidney showing the structures in the hilar region.

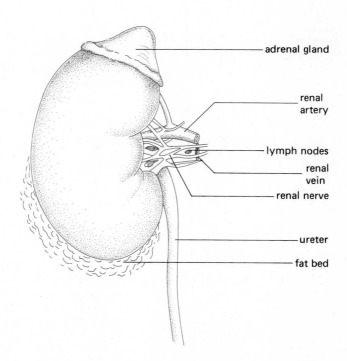

adrenal gland

renal artery

lymph nodes

renal vein

renal nerve

ureter

fat bed

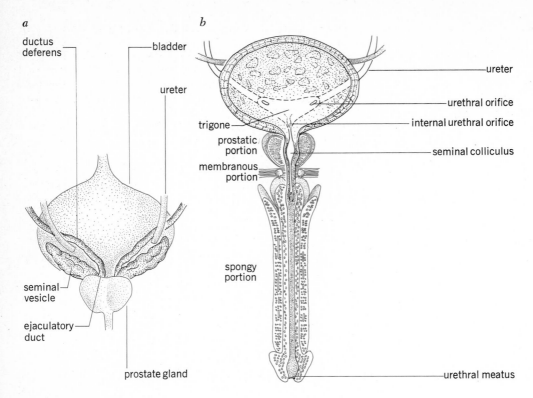

a

ductus
deferens

bladder

ureter

seminal
vesicle

ejaculatory
duct

prostate gland

b

ureter

urethral orifice

internal urethral orifice

trigone

seminal colliculus

prostatic
portion

membranous
portion

spongy
portion

urethral meatus

FIGURE 18·4

Urinary excretory pathways: *a*
bladder and ureters (male, posterior view); *b* bladder and
urethra (male, frontal section).

They pass anteriorly near the psoas major muscles and, descending into the
pelvis, enter the bladder obliquely on the posterior side. The wall of the
ureter is composed of three coats: an outer connective tissue layer, a
muscular coat, and an inner mucous membrane. The smooth muscular layer
propels the urine into the bladder by peristaltic movements (Fig. 18.4).

The urinary bladder is a muscular structure that varies in size according
to the amount of urine it contains. The adult bladder can hold about 500 ml
of urine without overdistension. When full, the bladder is nearly spherical in
shape and rises into the abdominal cavity. It is located posteriorly to the
pubic symphysis when empty but can rise well above the level of the pubic
bones.

The muscular coat, which makes up the greater portion of the wall of the
urinary bladder, is composed of three layers. An inner and an outer layer of
longitudinal muscles are separated by a medial layer of circular muscles.
The internal mucous membrane is a soft reddish lining, which is thrown into
folds when the bladder is empty.

A transitional type of stratified epithelium lines most of the urinary tract.
It is especially evident in the urinary bladder, which undergoes considerable
expansion and contraction. The tissue itself then undergoes a change in
appearance. When the tissue is contracted, the cells consist of several layers
and are more or less rounded or cuboidal in shape. In the extended tissue,
the cells of the deeper layers appear compressed, but the cells of the surface
layer become elongated and flattened, resembling stratified squamous

epithelium. Some histologists do not regard transitional tissue as a true type of epithelium.

The ureters enter the urinary bladder through narrow slits that remain closed except when a peristaltic muscular movement forces urine through the openings. The internal urethral orifice lies in an anterior and medial position to the openings of the ureters, forming a triangular area known as the *trigone*.

The *urethra* is the duct that leads from the urinary bladder to the exterior. It is considerably longer in the male since it must traverse the penis; it can reach a length of 20 cm. As it leaves the bladder, the urethra extends through the prostate gland for about 3 cm. The internal orifice at the urinary bladder is surrounded by circular smooth-muscle fibers that form the *internal urethral sphincter*. The *external urethral orifice* is a vertical slit at the distal end of the penis.

The female urethra extends from the internal orifice in the urinary bladder along the anterior wall of the vagina to the external orifice located between the clitoris and the vaginal opening. It is about 4 cm long.

GENERAL INTERNAL STRUCTURE   The kidney is composed of several anatomically distinct regions (Fig. 18.5). A tough layer of fibrous connective

FIGURE 18·5

Longitudinal section of the kidney.

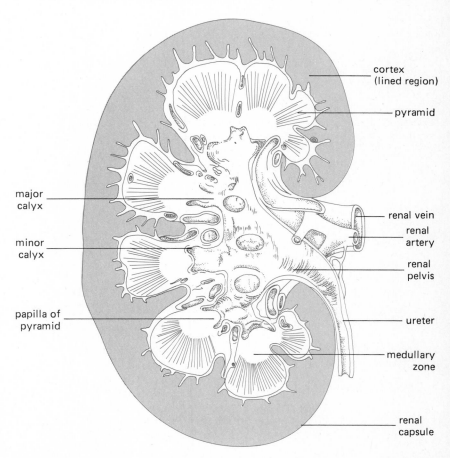

cortex (lined region)

pyramid

major calyx

minor calyx

renal vein

renal artery

renal pelvis

papilla of pyramid

ureter

medullary zone

renal capsule

tissue surrounds and protects the kidney. This *renal capsule* confines the kidney tissue and is effective in preventing leakage of fluid into the body cavity. Within the renal capsule, the kidney tissue is divided into an outer region about 2 cm thick called the *cortex* and a more medial region called the *medulla,* or medullary zone. The medulla is divided into the *outer medullary zone* adjacent to the cortex, and the *inner medullary zone,* medially. If the kidney is viewed in cross section, five or six cone-shaped regions are seen as triangular areas with a striped appearance. These are the part of the medullary tissue called *renal pyramids.* The *renal artery* runs along the border between cortex and medulla, giving rise to smaller arteries that radiate into the surrounding tissue. The *arcuate arteries* supply the cortex and give rise to the *glomerular arterioles.* Most of the kidney is supplied with a lymphatic system that drains into the *hilar nodes.*

THE NEPHRON Each kidney consists of about 1 million individual tubules called *nephrons* (Fig. 18.6). Each nephron comprises several functionally and anatomically distinct parts. All tubules originate in the renal cortex. The *renal corpuscles,* or *malpighian bodies,* located at the beginning of each nephron are responsible for the granular appearance of the renal cortex. These structures consist of compact masses of capillaries called *glomeruli* and *Bowman's capsules,* which are the expanded ends of the tubules. Each Bowman's capsule is continuous with a *proximal convoluted tubule,* which is

FIGURE 18·6
A single nephron with its blood supply.

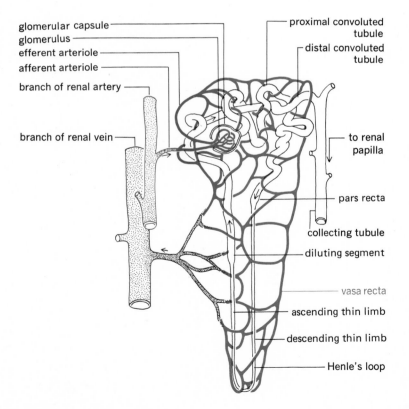

glomerular capsule
glomerulus
efferent arteriole
afferent arteriole
branch of renal artery

branch of renal vein

proximal convoluted tubule
distal convoluted tubule

to renal papilla

pars recta

collecting tubule
diluting segment

vasa recta

ascending thin limb

descending thin limb

Henle's loop

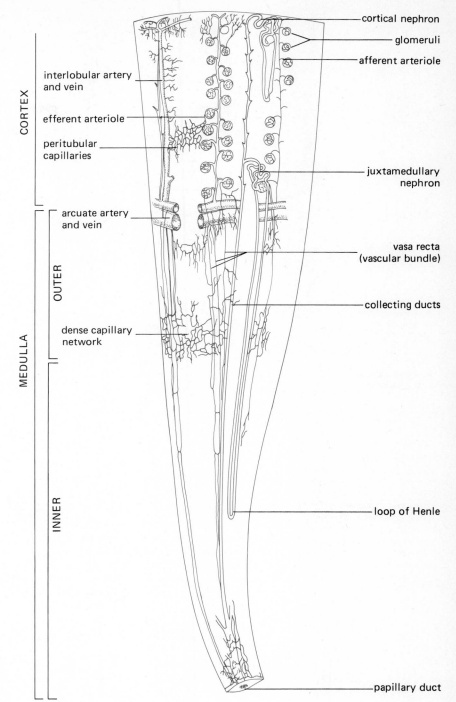

CORTEX

interlobular artery
and vein

efferent arteriole

peritubular
capillaries

arcuate artery
and vein

MEDULLA

OUTER

dense capillary
network

INNER

cortical nephron

glomeruli

afferent arteriole

juxtamedullary
nephron

vasa recta
(vascular bundle)

collecting ducts

loop of Henle

papillary duct

FIGURE 18·7

The arrangement of the nephrons with their vascular systems as seen in cross section
through a renal lobe.

a

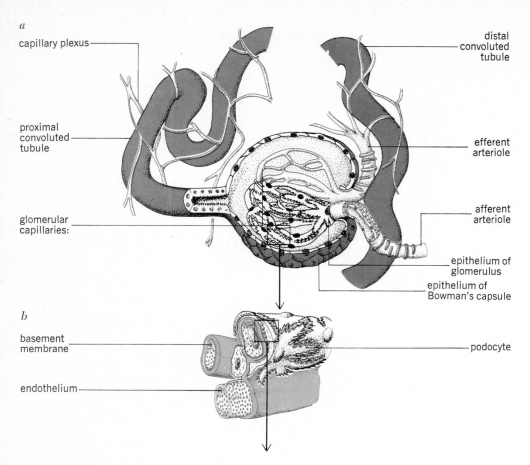

capillary plexus

distal convoluted tubule

proximal convoluted tubule

efferent arteriole

afferent arteriole

glomerular capillaries:

epithelium of glomerulus

epithelium of Bowman's capsule

b

basement membrane

podocyte

endothelium

FIGURE 18·8

Renal corpuscle. *a* Section through renal corpuscle. *b* Glomerular capillaries enlarged, showing podocytes covering the vessels. *c* Cross section through the filtering surface of the capillary. *d* Ultrastructure of the filtering surface showing multiple layers and pathway of filtered material.

followed by the *pars recta*, or proximal straight tubule, and the *loop of Henle*. The pars recta descends into the kidney medulla, and the loop of Henle is located entirely in the medulla. The loop ascends toward the cortex, where it is continuous with the *distal convoluted tubule*. The striped appearance of the renal pyramids is due to the presence of *collecting ducts*, which connect with each distal tubule in the cortex and coalesce as they descend toward the apex of the pyramid. The apical tip of each pyramidal portion of the medulla is referred to as a *renal papilla*, and about 12 to 20 collecting ducts empty at each papilla. The papillae are surrounded by the calyces.

Structure and function are intimately related in all parts of the nephron; thus, to understand nephron function one must be familiar with its anatomical detail.

Microanatomy Each *glomerulus* consists of a tuft of capillaries extending from the *afferent arteriole*. These capillaries reconverge and leave the *renal corpuscle* as the *efferent arteriole*. The efferent arteriole again breaks into a *peritubular capillary* bed, which is intimately associated with the renal

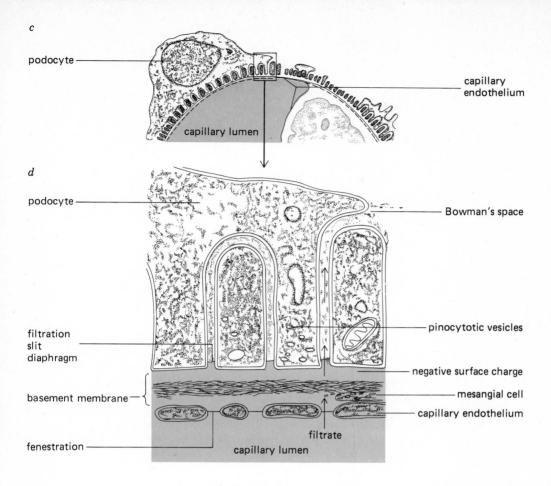

c

podocyte

capillary
endothelium

capillary lumen

d

podocyte

Bowman's space

pinocytotic vesicles

filtration
slit
diaphragm

negative surface charge

basement membrane

mesangial cell

capillary endothelium

fenestration

filtrate

capillary lumen

tubule. A few efferent arterioles do not immediately break up into capillaries, but turn and descend into a renal pyramid. In the pyramid, they become a unique capillary bed in which the capillaries form loops parallel to the tubule in this region (Fig. 18.7). This unique vascular bed is called the *vasa recta*, and will be discussed further when we take up the countercurrent multiplier system.

Bowman's capsule is formed from the blind end of the tubule, much as though a fist were pushed (glomerulus) into a balloon (end of renal tubule) (Fig. 18.8). Two epithelial layers make up Bowman's capsule: the outer, or *parietal layer*, and the inner, or *visceral layer*. The two layers are separated by *Bowman's space*.

The proximal convoluted tubule lies entirely within the kidney's cortex and represents the site at which most of the glomerular filtrate is reabsorbed. This segment consists of a single layer of cuboidal epithelial cells, which are characterized by a very highly developed microvillus *brush border* on their lumenal faces (Fig. 18.9). These cytoplasmic extensions greatly amplify the surface area over which reabsorption can occur. These cells contain abundant mitochondria, and their basal and lateral surface areas are increased by

tubular lumen

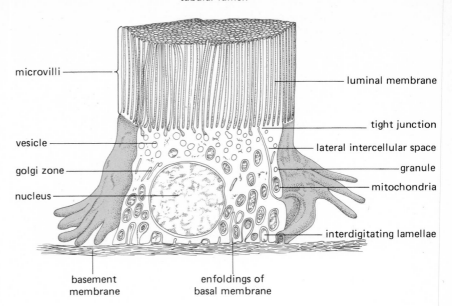

microvilli ——— luminal membrane

tight junction

vesicle ——— lateral intercellular space

golgi zone ——— granule

nucleus ——— mitochondria

interdigitating lamellae

basement   enfoldings of
membrane   basal membrane

FIGURE 18·9

Cells of the proximal tubule.
Note the tremendous amplification of the surface area
caused by microvilli.

membrane enfoldings. The many pinocytotic vesicles present in such cells
are believed to be primarily involved with reabsorption of the minute
amounts of protein that escape the glomerulus.

The *pars recta* descends from the cortical region into the outer medullary
zone. Although the pars recta cells function differently from those in the
proximal convoluted tubule, there is very little structural difference between them.

The *loop of Henle* is that portion of the nephron that extends from the
pars recta to the *distal tubule*. Two general types of renal nephrons may be
distinguished by the development of their loop of Henle segments. *Cortical
nephrons*, which are the predominant type, are characterized by very short
loops. *Juxtamedullary nephrons* have very long loops that extend into the
inner medulla; these make up about 12 percent of all nephrons. The loop of
Henle actually comprises three distinctly different segments. The *descending thin limb* of the loop of Henle represents an abrupt structural transition
from the cuboidal, mitochondrial-rich cells of the pars recta to a flattened,
squamous-type epithelium (Fig. 18.10). Almost no mitochondria are found in
this segment. The descending limb passes to varying depths in the renal
pyramid before it turns and becomes the functionally distinct *ascending thin
limb* of the loop of Henle. No epithelial transition is apparent at this point.
In the outer medulla, the epithelium of the ascending limb abruptly changes
to cuboidal, mitochondria-rich cells as the tubule expands in diameter. This
thick portion of the ascending limb is termed the *diluting segment*.

The *distal convoluted tubule* begins where the diluting segment ends and
winds through the cortex before interconnecting with the collecting duct
system. At its very beginning, the distal tubule touches the afferent arteriole
near the renal corpuscle, where a group of specialized cells called the

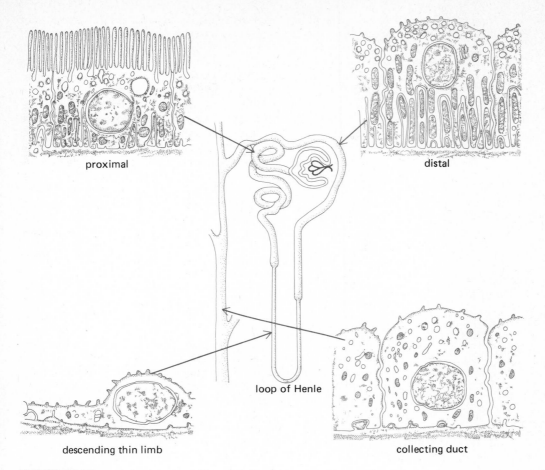

proximal

distal

loop of Henle

descending thin limb

collecting duct

FIGURE 18·10
Various cell types and their location within the nephron.

juxtaglomerular apparatus is located. The distal tubule is distinguished histologically from the proximal tubule primarily by a comparatively sparse number of microvilli on the lumenal surface of the cells.

The *collecting ducts* are not properly considered part of the nephron unit because they are of different embryological origin. The collecting ducts are more than conduits for the urine, however; they actively transport several ions and are intimately involved in determining the final urine concentration. The ducts have a branched appearance because they coalesce as they descend to the papillae. The cuboidal cells of the collecting ducts contain mitochondria and have very little surface-area amplification.

MECHANISMS OF RENAL FUNCTION

As noted earlier, the three basic processes by which the kidneys regulate the composition of the extracellular fluid involve an initial selective filtration process followed by selective reabsorption of filtered material. In some cases materials are not only filtered but are also secreted into the renal tubules. Our discussion of kidney function begins with the formation of the primary urine at the glomeruli. Subsequently, we will examine the contributions of the reabsorptive and secretory processes to the formation of the final urine.

ANATOMICAL BASIS OF GLOMERULAR FILTRATION   The renal corpuscle is the site of ultrafiltration of the blood. An understanding of the detailed ultrastructure of the renal capsule is essential to the understanding of urine formation. The visceral layer of Bowman's capsule and the capillary endothelium of the glomerulus form a multilayered barrier across which selective filtration of plasma occurs (see Fig. 18.8). Starting from the lumen of the glomerular capillary, the first cellular layer encountered is made up of *capillary endothelial cells.* These cells form the glomerular capillary wall, which contains *fenestrations* (windows) between and through cells. The fenestrations determine the permeability of the wall to the plasma contents. The endothelial cells act as a crude filter by preventing the cellular elements of the whole blood from escaping and by retaining some plasma proteins such as albumin. Plasma, including most of the plasma proteins, may penetrate the capillary wall.

The next layer is an *inner cement layer,* which lies between the capillary endothelial cells and the *basement membrane.* The cement layer consists mainly of a stiff, gel-like mucopolysaccharide called hyaluronic acid. The so-called basement membrane, which is actually not a membrane at all, is the primary *size-selective* filtration barrier in the renal corpuscle. It is made up of a dense network of reticular fibers that bar the passage of many high-molecular-weight molecules (greater than 20 Å in radius) while allowing free passage to everything else.

An outer cement layer lies between the basement membrane and the visceral epithelial cell layer of Bowman's capsule. The cells of this layer, also called podocytes, have numerous cytoplasmic extensions called *podia* (or foot processes) that form a sievelike pattern over the outer cement layer. Such cells possess numerous pinocytotic vesicles. Thin, membranous structures called *filtration slit diaphragms* are found between the podia. Proteins trapped by these structures are engulfed by the podia.

A third cell type, the *mesangial cell,* also plays a part in ultrafiltration. These cells seem to have ameboid movement (as do some leukocytes). They apparently migrate continuously between the capillary endothelium and the surface of the basement membrane. Although their function has not been clearly demonstrated, some investigators have suggested that the mesangial cells phagocytize the filtration residue entrapped in the basement membrane meshwork and thus clean the filter.

Experiments have shown that size-selective filtration is primarily performed by the basement membrane, which prevents passage of neutral molecules with a molecular radius greater than 20 Å (Fig. 18.11). The renal corpuscle also has a *charge-selective filter.* The capillary endothelial cells, the cement layers, the basement membrane, and the cells of the visceral layer of Bowman's capsule all normally have a *negative surface charge.* Thus, molecules with a molecular radius greater than 20 Å and a net negative charge tend to be repelled by the filtration apparatus. Most plasma proteins, which must be conserved, are negatively charged. Large, positively charged molecules are actually drawn into the filter by electrical forces, but such particles tend to stick to the podia of the visceral layer, where the negative surface charge seems to be greatest. The molecules are

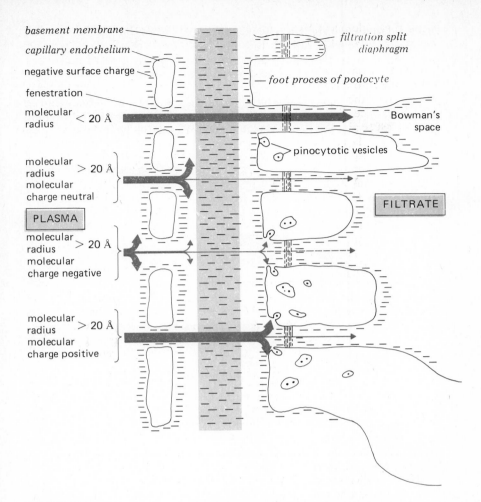

basement membrane

capillary endothelium

negative surface charge

fenestration

molecular radius < 20 Å

molecular radius > 20 Å
molecular charge neutral

PLASMA

molecular radius > 20 Å
molecular charge negative

molecular radius > 20 Å
molecular charge positive

filtration split diaphragm

foot process of podocyte

Bowman's space

pinocytotic vesicles

FILTRATE

FIGURE 18·11
Roles played by the layers of the glomerular filtration apparatus during filtration. The widths of the arrows indicate the degree to which various plasma components may be filtered.

subsequently engulfed by the podocytes in the pinocytotic process and digested by lysosomes, and the digested products are finally returned to the blood.

GLOMERULAR FILTRATION DYNAMICS In 1843, Karl Ludwig proposed that the initial step in urine formation was the production of a plasma ultrafiltrate as plasma was drawn across the walls of the glomerular capillaries. Ludwig also suggested that the driving force for such ultrafiltration was the capillary hydrostatic pressure, a pressure generated of course by the pressure waves set up by the beating heart. A. N. Richards and associates provided an important proof of this hypothesis by inserting tiny micropipettes into Bowman's space and collecting the ultrafiltrate. These investigators found that the fluid in Bowman's space was almost identical in composition to plasma except that it was virtually protein-free.

For filtration to occur, the *capillary hydrostatic pressure* must exceed the *colloid osmotic pressure* of the plasma. Because large molecules, such as the colloids albumin and gamma globulin, do not pass easily out of the capillary, they tend to draw water from Bowman's space toward the capillary lumen. The presence of such colloids thus opposes the net capillary hydrostatic pressure, which tends to force fluid out of the capillary toward Bowman's space. The balance between these two forces determines the rate of glomerular filtration.

The *net driving pressure* for ultrafiltration is actually the algebraic sum of several opposing forces (Fig. 18.12). The hydrostatic pressure in the afferent arteriole at a point just before the glomerular capillaries has been measured

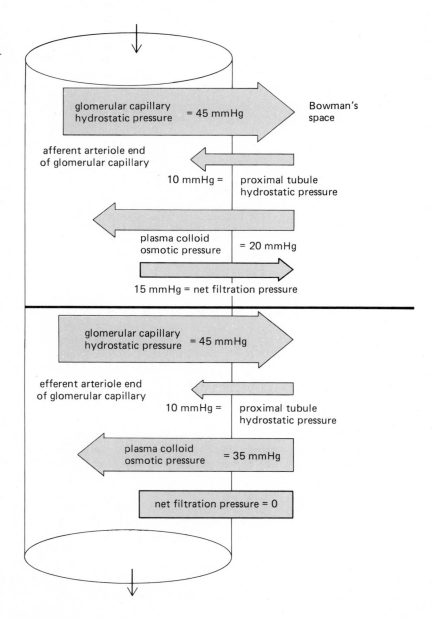

FIGURE 18·12
Pressures involved in formation of the ultrafiltrate. Average net filtration pressure throughout the capillary is about 5 mmHg.

directly in several mammals. This pressure is 40 percent of the mean systemic arterial pressure, or approximately 45 mmHg. At the efferent arteriole end of the glomerular capillaries, the hydrostatic pressure is about 43 mmHg, or slightly lower. There is very little change in capillary hydrostatic pressure, therefore, as the blood flows through the glomerulus, and this pressure can, in effect, be assumed to be constant. Capillary hydrostatic pressure is opposed by the hydrostatic pressure of fluid in Bowman's space, a quantity termed *tubular pressure*. Tubular pressure has been found to be about 10 mmHg. The hydrostatic pressure gradient thus operates to move plasma *out* of the glomerular capillaries, with a net hydrostatic pressure of 35 mmHg.

Another factor at work in ultrafiltration is the plasma colloid osmotic pressure, which is approximately 20 mmHg at the afferent end of the glomerulus. As the blood passes through the glomerulus, however, filtered fluid volume is lost to the renal tubule so that the protein concentration in the capillary *increases*. At the efferent end of the glomerulus, the colloid osmotic pressure is thus about 35 mmHg. The colloid osmotic pressure of the protein-free fluid in Bowman's space is negligibly small. Thus, by the time the blood has traversed the glomerular capillaries, the net hydrostatic pressure forcing fluid out of the capillaries equals the capillary colloid osmotic pressure, which tends to draw fluid into the capillary.

The net glomerular filtration pressure, therefore, is 15 (35 − 20) mmHg at the afferent end and is 0 (35 − 35) mmHg at the efferent end of the glomerulus. The *mean (average) net filtration pressure* in the glomerulus has been calculated to be about 5 mmHg. Glomerular filtration proceeds at a relatively rapid rate in spite of this low pressure because the glomerular capillaries are 10 to 100 times more permeable than capillaries in many other body tissues, such as muscle. In peripheral capillaries, such as those in muscle, the pressure profile from artery to vein favors filtration at the arteriole end and reabsorption at the venous end. Reabsorption apparently never occurs in the glomerular capillaries, however, because capillary hydrostatic pressure does not drop in these capillaries. The fine control over resistance to flow exerted by the afferent arterioles in the glomerulus apparently prevents a decline in pressure.

The point within the glomerular capillary at which the plasma colloid osmotic pressure exactly equals the net capillary hydrostatic pressure is called *filtration pressure equilibrium* (Fig. 18.13). If the *flow rate* of plasma through the glomerulus is increased, the point along the length of the glomerular capillary at which equilibrium is reached is pushed toward the efferent end, and more filtration occurs prior to equilibrium.

Three main factors are thus important in determining *glomerular filtration rate (GFR)*: *mean arterial blood pressure, glomerular plasma flow,* and *plasma colloid osmotic pressure*. Colloid osmotic pressure of plasma normally varies very little, so the primary influences on GFR are arteriole control of hydrostatic pressure and glomerular plasma flow rate. When mean arterial blood pressure increases, glomerular capillary hydrostatic pressure also increases, although not as much as might be expected because glomerular arterioles regulate flow in response to pressure changes. An increase in arterial blood pressure nevertheless increases net filtration pressure and glomerular filtration rate. Because glomerular capillary hydrostatic pressure is normally effectively autoregulated by the glomerular

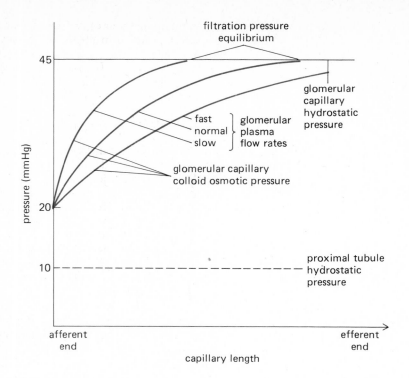

FIGURE 18·13
Glomerular filtration rate is
dependent upon the rate of
plasma flow in the glomerular
capillaries.

arterioles, glomerular plasma flow rate usually plays the most important role
in determining GFR. An increase in glomerular plasma flow rate is, therefore, usually the most significant factor contributing to an increase in GFR.

Measurement of glomerular filtration rate (*GFR*) Renal physiology is primarily
concerned with the quantitative study of filtration, reabsorption, and
secretion. Quantitation of these processes requires a reliable method of
measuring glomerular filtration rate. To measure GFR, a marker substance is
introduced into the plasma, usually by intravenous infusion. The substance
used must be: (1) physiologically inert, so that it does not interfere with
normal body metabolism and is not destroyed or used; (2) freely filtered at
the glomerulus; and (3) neither reabsorbed nor secreted by the renal tubule.
The substance most often used to measure GFR is the fructose polysaccharide *inulin*. (*Note:* Inulin is *not* the hormone insulin.)

To calculate the GFR, the urine (U_{inulin}) and plasma (P_{inulin}) inulin
concentrations (mg/l) and the rate of urine flow (*V*, l/day) must be known.
The total amount of inulin excreted per day is then determined by this
relationship:

$$
\begin{array}{ccc}
\text{Total inulin excreted} = & U_{\text{inulin}} \times & V \\
\text{(mg/day)} & \text{(mg/l)} & \text{(l/day)}
\end{array}
$$

The question to be asked is: How much plasma must be "cleared" of inulin
to account for the total amount of inulin in the urine? Knowing the number

of milligrams of inulin in a liter of plasma, the number of liters necessary to supply the amount of inulin in the urine may then be calculated:

$$\underset{\substack{(\text{inulin clear-} \\ \text{ance rate,} \\ \text{l/day})}}{C_{\text{inulin}}} = \frac{U_{\text{inulin}} \times V}{P_{\text{inulin}}} = \frac{\text{mg/l} \times \text{l/day}}{\text{mg/l}}$$

Thus the volume of plasma obtained represents the amount that was cleared of inulin. Because all the filtered inulin appears in the urine, the clearance of inulin *is* the glomerular filtration rate.

THE PROCESS OF SECRETION   Heavily secreted substances are usually not reabsorbed to any great extent. Such substances are commonly metabolic by-products that cannot be used by the body and would, in many cases, be toxic if they were allowed to accumulate. Some substances are so heavily secreted that their concentrations in plasma leaving the kidney via the renal veins are reduced practically to zero. Substances of this type are often used to determine *renal plasma flow (RPF)*. One such substance is *para-aminohippuric acid (PAH)*. This substance is freely filtered and heavily secreted. All PAH that enters the kidney is therefore removed, and the clearance of PAH is equal to the total renal plasma flow as calculated by the following formula:

$$C_{\text{PAH}} = \frac{U_{\text{PAH}} \times V}{P_{\text{PAH}}} = \text{RPF (l/day)}$$

In humans, this measurement shows that about 900 l of plasma pass through the kidney in 1 day. Thus, when the skeletal muscles are at rest, the kidneys receive 20 percent of the total cardiac output, and about 20 percent of the RPF is filtered. GFR thus equals 180 l/day.

THE REABSORPTIVE PROCESS   GFR determinations show that a volume equivalent to the entire extracellular fluid volume filters through the kidneys about 12 times per day. Urine production during this same time, however, amounts to less than 2 l. Thus, about 99 percent of the filtered volume is reabsorbed. The kidney, then, expends most of its energy in reabsorbing materials from the filtrate to prevent their excretion. At first thought this mechanism may seem to be an energy-wasting way of regulating extracellular fluid volume and composition. However, the filtration-reabsorption process allows very fine control over both the solute composition and the *volume* of extracellular fluid.

Some important nutrient materials, such as glucose, are completely reabsorbed so that their clearance value is zero. The plasma levels of these nutrients are not regulated by the kidneys although, by reabsorbing these materials, the kidneys help maintain the given plasma concentration. The renal tubules contain very powerful active transport systems for the reabsorption of substances like glucose. However, the transport process depends upon a transport protein located within the tubule cell membranes, and there are a finite number of these transport molecules. Because of this

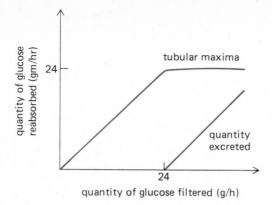

FIGURE 18·14
The concept of tubular maxima. Virtually all filtered glucose is reabsorbed up to a rate of about 24 g/h at which point no more reabsorption can occur and glucose appears in the urine.

limitation, it is possible that the amount of a substance filtered may occasionally exceed the capacity of the reabsorption system; that is, all transport sites may be occupied. The rate of transport at which the tubular reabsorptive process is saturated is called the *tubular maxima*. As an example, the normal amount of glucose filtered from plasma can easily be reabsorbed by the renal tubules. However, soon after the ingestion of a high carbohydrate meal, blood glucose may rise sharply for a short time, and the amount of glucose filtered may exceed the tubular maxima, in which case glucose will be found in the urine (Fig. 18.14).

NEPHRON FUNCTIONS

Before we begin our discussion of the function of each nephron segment in filtration, secretion, and reabsorption, it is helpful to examine the concept of fluid transport. Although fluid transport across epithelial tissue has been extensively studied, we cannot yet explain the mechanism. It is generally agreed among physiologists that active transport of water, as such, costs too much in energy to account for fluid transport by epithelia. Fluid transport might be explained if the epithelial cell layer separated two osmotically different solutions. However, most fluid transport in the body appears to occur between isosmotic compartments. Fluid-transporting epithelia have several characteristics in common, and these have been used to develop a model to account for fluid movement between isosmotic compartments (Fig. 18.15).

SOLUTE-LINKED FLUID TRANSPORT   Transepithelial fluid movement is always linked to solute movement. The solute, usually Na, is actively transported by the tissues. The transporting epithelium typically consists of a single layer of columnar cells, separated by lateral intercellular spaces that extend from the basal region to a "tight junction," or zone of adherence, near the apical or lumenal end of the cells. The tight junction forms a blind-ended channel between every cell and its neighbor. Fluid transport from the lumen of an organ to the surrounding interstitium is believed to occur as follows:

1 Na^+ is actively transported from the lumen to the lateral intercellular spaces, followed passively by Cl^-.
2 The accumulation of NaCl in each space creates a hyperosmotic region

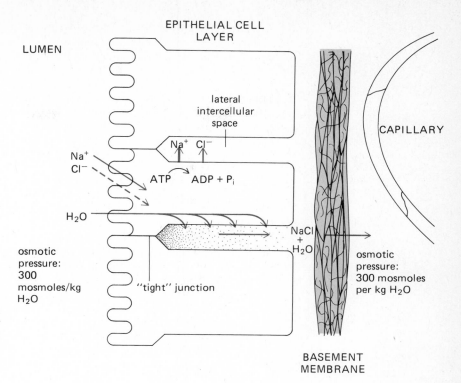

EPITHELIAL CELL
LAYER

LUMEN

lateral
intercellular
space

CAPILLARY

Na^+ Cl^-

Na^+
Cl^-

ATP ⟶ ADP + P_i

H_2O

osmotic
pressure:
300
mosmoles/kg
H_2O

"tight" junction

NaCl
+
H_2O

osmotic
pressure:
300 mosmoles
per kg H_2O

BASEMENT
MEMBRANE

FIGURE 18·15

Schematic representation of
the standing local osmotic gradient hypothesis of solute-
linked fluid transport. Water
follows an osmotic gradient
created by transport of NaCl.

that tends to draw water from the surrounding cells, which, in turn, take
up water from the tubular lumen.

3 The increased volume of the intercellular spaces creates a small hydrostatic pressure that forces fluid through the basement membrane and into
the interstitium.

4 As fluid moves through the lateral space, it constantly gains water until, at
the end of the channel, it is isosmotic to the cell and extracellular fluid.

This sequence of events is called the *standing osmotic gradient theory* and is
believed to be the mode of fluid transport in the ciliary body of the eye, the
intestine, the gallbladder, and the proximal tubule of the kidney nephron, as
well as in other tissues.

FUNCTIONAL CHARACTERISTICS OF THE NEPHRON SEGMENTS *Proximal convoluted
tubule* Approximately 80 percent of the ultrafiltrate is reabsorbed by the
nephron's proximal convoluted tubule. This is the longest segment of the
nephron, and it does most of the work. Almost all filtered nutrient materials
are reabsorbed in the proximal tubule. Active transport systems for glucose,
amino acids, and protein are found here, but most substances are reabsorbed
passively down a concentration gradient that extends from the tubule lumen
to the peritubular extracellular space. The concentration gradients are
produced by two factors:

1 The plasma colloid osmotic pressure in the peritubular capillaries is

higher than that of the protein-free ultrafiltrate; this pressure gradient draws water across the proximal tubule epithelium by osmosis.

2 Na^+ is actively transported out of the segment, and through solute-linked fluid transport, more water leaves the lumen.

The removal of water leaves a higher concentration of solutes in the ultrafiltrate inside the tubule than outside, so the solutes tend to diffuse back into the plasma. Very little control of the fluid reabsorption occurs in the proximal tubule. Reabsorption in this segment is *obligatory*. The tubular fluid that remains after the proximal tubule has been traversed is sufficient raw material for the kidney to regulate plasma composition.

The proximal convoluted tubule also actively secretes H^+ and reabsorbs HCO_3^-. As was pointed out in Chap. 17, the kidney is intimately involved with regulation of the H^+ concentration of the plasma.

Pars recta Organic acid and base secretion occur in the pars recta of the proximal segment. About 38 different organic acids and bases are usually secreted into the urine, and as many as 52 substances may be found in disease conditions. The secretory system of the pars recta can secrete almost any organic substance, whether it is endogenous to the body or introduced from outside. The kidneys are normally so efficient in removing these organic substances from the plasma that their concentrations never approach a toxic level. Some of these compounds, such as the hippuric acids, are highly toxic if allowed to accumulate, and in fact, when the kidneys fail, the accumulation of organic acids may cause death before the resulting acid-base and electrolyte imbalances can do so.

Urea, the major nitrogenous waste product produced by metabolic processes, is handled passively by the nephron in most segments with the possible exception of the pars recta, where recent experiments indicate that urea may be actively secreted into the tubular fluid. About 60 percent of the urea filtered is excreted when urine flow rate is average (about 1 to 2 l/day). Urea is produced by the body mainly to detoxify ammonia, which is an end product of catabolism of proteins and amino acids. Without urea the concentrating ability of the kidneys is greatly hindered. Thus, although urea is a "waste" product, it is also an essential material when normal water intake is restricted. Urea will be discussed again when we take up the countercurrent multiplier system.

Thin descending and ascending limbs of the loop of Henle The important function of these segments is their participation in the countercurrent multiplier system, which is discussed below. The most recent evidence indicates that no active transport occurs in these segments; all solute and water movement is believed to be passive. The *descending limb* has a relatively high permeability to water and a somewhat lower permeability to sodium, chloride, and urea. About 6 percent of the ultrafiltrate volume is reabsorbed here. The *ascending limb* has an extremely low permeability to water and a relatively high permeability to sodium and chloride. The significance of these characteristics will be discussed below.

Diluting segment The thick portion of the ascending limb of the loop of Henle is important mainly for its role in the countercurrent multiplier system.

Recent investigations, in which this segment was removed from the kidney and perfused, have shown that the urine is diluted here by active chloride reabsorption. Sodium follows the chloride passively out of the tubule, but almost no water or urea is reabsorbed here.

Distal convoluted tubule Active sodium reabsorption in this segment and the osmotic gradient created by the hypoosmotic tubular fluid that enters from the diluting segment cause further water reabsorption. About 9 percent of the ultrafiltrate is reabsorbed here, but the segment is impermeable to urea. By the time the tubular fluid traverses this segment, it is isosmotic or slightly hypoosmotic to the plasma depending upon the level of antidiuretic hormone, which controls the water permeability in the distal convoluted tubule. In this segment, H^+ is also actively secreted, as is K^+ when this ion is present in excessive levels in the plasma. The distal convoluted tubule is the major site for regulation of plasma K^+ concentration. Na^+ and K^+ movements in this segment are controlled by the hormone aldosterone, which will be discussed below.

Collecting ducts Reabsorption of Na^+, Cl^-, and water continues in the collecting ducts, a process driven by active sodium transport. Some H^+ secretion occurs, but K^+ is not secreted. Antidiuretic hormone acts here as well as in the distal convoluted tubule, and this is the site at which final urine concentration is determined. The amount of water reabsorbed in the collecting ducts is thus variable but averages about 4 percent of the filtered volume. About 1 percent of the filtered volume appears in the final urine.

Diuresis and antidiuresis: control of final urine concentration The final urine concentration is determined by how much water is extracted from the tubular fluid as it progresses through the distal tubules and the collecting ducts. A *countercurrent multiplier system* is created by the anatomical arrangement of the tubules and their blood supply in the kidney's inner medulla. This countercurrent system creates an osmotic gradient within the renal tissue that favors the extraction of water from the collecting ducts.

Tissue fluid is isosmotic to plasma in the kidney cortex, but as the fluid descends through the medulla toward the papilla tip, it steadily increases in osmolality. The collecting ducts carry the tubular fluid through this region of high osmolality just before the urine enters the renal pelvis.

The antidiuretic hormone (ADH), formerly called vasopressin, increases the permeability to water of the distal convoluted tubules and collecting ducts. When normal water intake is restricted, plasma osmolality rises slightly and blood pressure decreases slightly, two factors that stimulate the release of ADH from the neurohypophysis. If water intake increases above normal levels, plasma osmolality decreases slightly and blood pressure rises slightly, so that the release of ADH is repressed. In the absence of ADH, the distal tubules and collecting ducts are much less permeable to water. Dilute urine is produced when the hypoosmotic fluid from the diluting segment passes through the distal convoluted tubule and collecting duct while they are relatively impermeable to water. Additional removal of Na^+ and Cl^- still takes place in these segments, but virtually no water is removed. Under these conditions, the final urine is dilute, and urine volume is increased. The individual is said to be in a state of *diuresis*.

Conversely, concentrated urine is produced when ADH is present to stimulate the distal tubules and collecting ducts to reabsorb the maximal amount of water. Under these conditions, the final urine is isosmotic to the interstitial fluid of the medullary tissue at the tip of the papilla and urine volume is low. The individual is said to be in a state of *antidiuresis*.

COUNTERCURRENT MULTIPLIER SYSTEM   Although individual nephrons modify the composition and volume of the ultrafiltrate to a significant degree, the *countercurrent multiplier system,* produced by anatomical cooperation of the various components of the kidney, greatly increases the effectiveness of the kidney's reabsorptive function. The system is called "countercurrent" because the tubular fluid in the loops of Henle flows in opposite directions as it passes through the descending and ascending limbs. There are actually *two* countercurrent loops in this system. The first is formed by the loop of Henle. The second is composed of the *vasa recta* capillaries, which form loops within the renal medulla that essentially parallel the loops of Henle (Fig. 18.16). This blood supply system is located within the *center* of the hairpin loops of the tubules.

Blood enters the pyramid from the descending vasa recta, and as it dips into the medulla, the plasma rapidly equilibrates its osmotic concentration with the medullary interstitial fluid so that it loses water and gains salts and urea (Fig. 18.17). As it flows up towards the outer zone of the medulla, the plasma gains water and loses NaCl and urea, thus maintaining osmotic equilibrium with the surrounding interstitial fluid. By the time the blood has flowed down and up the vasa recta, it is only slightly more concentrated than when it entered. The anatomical arrangement of the blood supply to the medulla assures the stability of the osmotic gradient between the nephrons and the surrounding tissue. If these blood vessels simply passed through this region of the kidney in one direction, the unidirectional flow would wash out the solutes that produce the osmotic gradient.

The vasa recta capillaries have three special characteristics that facilitate maintenance of osmotic equilibrium between the blood and the medullary interstitial fluid.

1 The walls of these capillaries are highly fenestrated so that the plasma

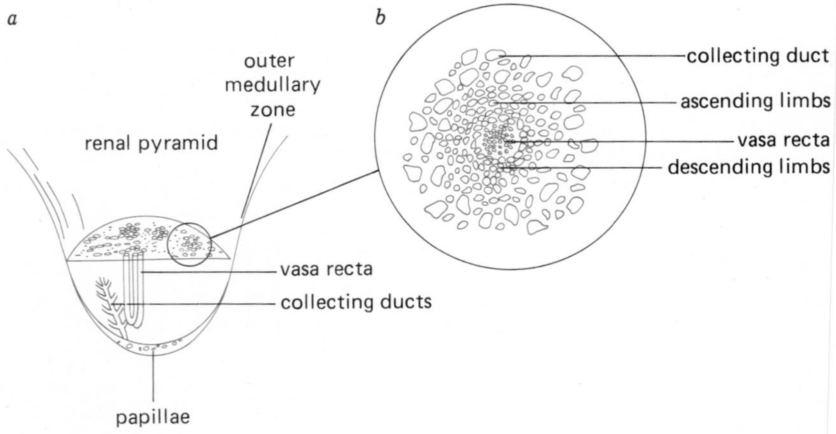

FIGURE 18·16

Organization of the tubule segments and blood vessels in the outer medullary zone. *a* A longitudinal section. *b* A cross section.

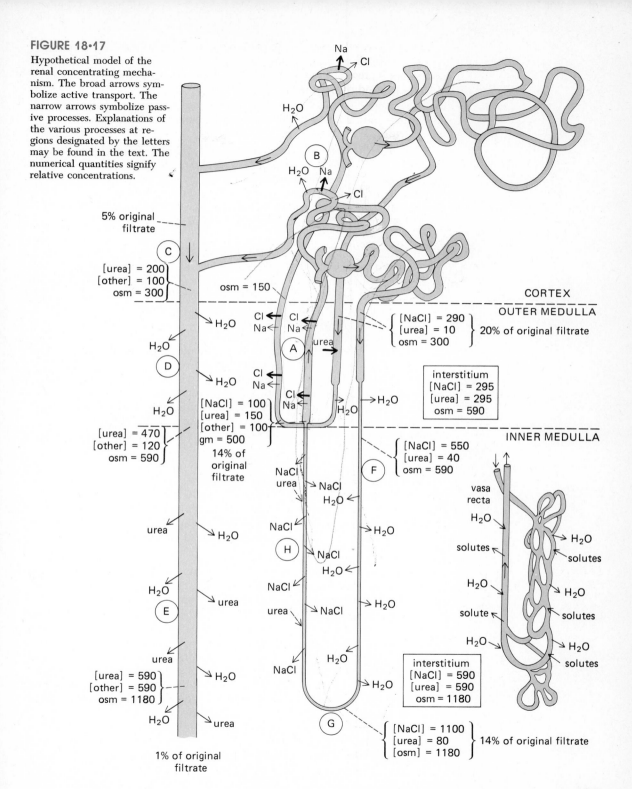

FIGURE 18·17

Hypothetical model of the renal concentrating mechanism. The broad arrows symbolize active transport. The narrow arrows symbolize passive processes. Explanations of the various processes at regions designated by the letters may be found in the text. The numerical quantities signify relative concentrations.

Na
Cl

H₂O

B

H₂O Na

Cl

5% original filtrate

C

[urea] = 200
[other] = 100
osm = 300

osm = 150

CORTEX
OUTER MEDULLA

Cl Cl
Na Na

A

urea

[NaCl] = 290
[urea] = 10
osm = 300

20% of original filtrate

H₂O

H₂O

D

H₂O

Cl
Na

Cl
Na

interstitium
[NaCl] = 295
[urea] = 295
osm = 590

H₂O

[NaCl] = 100
[urea] = 150
[other] = 100
gm = 500
14% of original filtrate

Cl
Na

→H₂O

H₂O

INNER MEDULLA

[urea] = 470
[other] = 120
osm = 590

[NaCl] = 550
[urea] = 40
osm = 590

F

vasa recta

H₂O

NaCl
urea

NaCl
H₂O

H₂O

solutes

solutes

urea

H₂O

NaCl

H₂O

H

NaCl

H₂O

H₂O

solutes

H₂O

H₂O

urea

NaCl
H₂O

solute

solutes

E

urea

NaCl

H₂O

H₂O

solute

H₂O

urea

[urea] = 590
[other] = 590
osm = 1180

H₂O

NaCl

H₂O

interstitium
[NaCl] = 590
[urea] = 590
osm = 1180

H₂O

solutes

H₂O

urea

G

[NaCl] = 1100
[urea] = 80
[osm] = 1180

14% of original filtrate

1% of original filtrate

components pass freely back and forth across the capillary walls. In effect, the capillaries become continuous with the medullary interstitial fluid.

2 The flow rate of blood in the vasa recta is very slow, which allows more time for equilibrium to occur.

3 The amount of blood entering the vasa recta is comparatively small because only a small number of vessels penetrate the area.

The other countercurrent system in the kidney is produced by the arrangement of the descending and ascending loops of Henle and the collecting ducts. To understand how this system creates an osmotic gradient within the inner medulla, we must understand the special permeability and active transport characteristics of each segment of the nephron, beginning with the thin descending limb of the loop of Henle.

1 The thin descending limb of the loop of Henle is highly permeable to water but apparently less permeable to NaCl and urea.

2 The thin ascending limb of the loop of Henle is almost impermeable to water, but it is permeable to NaCl and, to a lesser extent, to urea. Na^+ may be actively transported out of the tubule here.

3 The diluting segment actively transports Cl^-, which is passively followed by Na^+, out of the tubular fluid. Because the diluting segment is relatively impermeable to water, very little water accompanies the solute. This segment has a very low permeability to urea.

4 The distal tubule actively reabsorbs NaCl, which is passively followed by water, from the tubular fluid, but it is impermeable to urea.

5 The collecting duct in the cortex and outer medulla continues to reabsorb NaCl and water from the tubular fluid while retaining urea.

6 The collecting duct in the inner medullary zone near the papilla tip is permeable to urea.

The countercurrent multiplier system of the mammalian kidney is presently under very active investigation, and it must be emphasized that the model presented below is hypothetical; the numbers given in Fig. 18.17 are intended only to illustrate how this system may work. We begin our description with the active chloride transport in the diluting segment.

1 In the diluting segment, active reabsorption of Cl^-, followed by passive sodium reabsorption, makes the tubular fluid hypoosmotic to plasma (Fig. 18.17a). Urea cannot pass out of the lumen in this segment.

2 The fluid next enters the distal convoluted tubule where water reabsorption occurs. We will assume that ADH is present so that the distal convoluted tubule and collecting ducts are permeable to water. In the distal tubule water follows its osmotic gradient out of the tubule. Water reabsorption is helped also by sodium chloride removal (Fig. 18.17b). Thus by the time it reaches the end of the distal tubule, the tubular fluid is isosmotic to plasma, while the urea concentration is greatly elevated because urea cannot pass out of the distal tubule (Fig. 18.17c).

3 The concentration of urea in the tubular fluid continues to increase in the cortical collecting duct because salt and water are removed. In the outer medullary zone, urea concentration in the collecting duct increases still

more because here the duct passes through an area of high osmolality produced by the active chloride reabsorption in the diluting segment (Fig. 18.17d). In the inner zone of the medulla, the collecting duct becomes permeable to urea, which diffuses down a concentration gradient out of the collecting duct and into the medullary interstitium. The movement of urea helps to create a high osmotic pressure in the interstitium of the inner zone of the medulla (Fig. 18.17e).

4 The thin descending limb of the loop of Henle passes down into this hyperosmotic region (Fig. 18.17f). The tubular fluid entering the thin descending limb is isosmotic to plasma so that, as it passes into the hyperosmotic medulla, water is osmotically extracted from it and a small amount of sodium chloride and urea diffuse into it. Osmotic equilibrium is achieved at all points along this segment. However, because the dominant solute in the tubular fluid entering the thin descending limb is sodium and the dominant solute in the interstitial fluid is urea, the removal of water from the tubular fluid produces a much higher concentration of sodium inside the tubule than outside (Fig. 18.17g).

5 The tubular fluid then passes into the thin ascending limb, which is permeable to Na^+ and somewhat permeable to urea but virtually impermeable to water. Na^+ diffuses down its concentration gradient into the interstitium, followed by Cl^-, but the salts are not followed by water so that there is a net increase in the interstitial fluid osmotic pressure (Fig. 18.17h).

6 The concentrations of NaCl and urea in the interstitial fluid then increase Na^+ concentration in the tubular fluid in the thin descending limb even more by extracting more water from fluid in this tubule segment. Thus, a higher concentration of Na^+ moves into the ascending thin limb so that an even higher osmotic pressure is produced in the interstitial fluid and so on.

This cycle *multiplies* the osmotic gradient. The maximum osmotic gradient the countercurrent multiplier system can generate is limited by the length of the loops of Henle. The small amount of urea that enters the thin ascending limb is recycled into the interstitial fluid at the inner zone of the medullary collecting duct. As the tubular fluid moves up the thin ascending limb, the diffusion gradient along which Na^+ diffuses out of the tubule diminishes, so less and less Na^+ enters the interstitial fluid. The interstitial fluid osmotic pressure diminishes, therefore, in the region of the renal cortex. The high concentrations of NaCl and urea in the interstitial fluid act as the main osmotic driving force for extraction of water from the collecting ducts just before the urine exits the kidney. Without this mechanism for water conservation our survival would be much more dependent upon a continuously available source of drinking water.

In brief, then, the permeability characteristics of the nephron segments establish the osmotic pressure gradient in the medulla. The energy for this process is supplied by active chloride transport in the diluting segment, which indirectly increases urea concentration in the terminal portion of the collecting duct. The urea leaves the duct and enters the interstitium, where it becomes the osmotic driving force for water extraction from the descending limb of the loop of Henle. This in turn allows NaCl to diffuse out of the thin ascending limb of the loop, where its accumulation in the interstitial fluid extracts water from the collecting ducts.

WATER BALANCE   As we discussed in Chap. 2, water is the medium of life, and regulation of its concentration in body fluids is essential. If the body is to remain in water balance, intake must equal output. With a normal diet, the average person loses water at the rate of about 2.4 l/day. About 40 percent of this quantity is excreted in the urine; however, the major path of water loss is evaporation through the skin and exhalation in expired air. Expired air is 100 percent saturated with water, and water lost in this way accounts for about 21 percent of the total output. The unnoticed loss of water from evaporation through the body surface is called *insensible evaporative water loss,* and it accounts for about 25 percent of the daily water loss. Smaller amounts of water are lost through sweating (4.2 percent) and in the feces (8.3 percent). These values are only averages, of course, and the percentages vary greatly with activity, environment, and diet.

Because of these continual losses, we must ingest a certain amount of water each day. The kidneys, which are predominantly responsible for regulating body water content, cannot completely stop urine formation when water intake is restricted. Some water must be used to excrete substances removed from the blood by the kidneys. For example, normal metabolic processes produce urea and creatinine, as well as inorganic phosphates and sulfates. The maximum concentration of human urine is the same as that at the tip of the inner medulla, or about 1200 mosmol/kg water. The minimum urine volume is then about 500 ml/day since approximately 600 mosmol of metabolic wastes must be excreted.

WATER REGULATION   Control of body water content is primarily mediated through the antidiuretic hormone (ADH). The circulating level of this hormone is reflexly regulated by both *osmoreceptors* and *baroreceptors.*

For example, excess ingestion of pure water slightly lowers the osmolality of extracellular fluid (Fig. 18.18). A change in plasma osmolality of as little

FIGURE 18·18

Effect of ingestion of pure water on secretion of antidiuretic hormone.

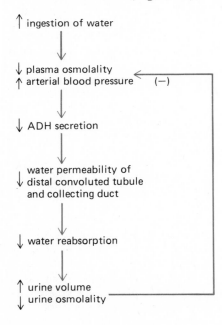

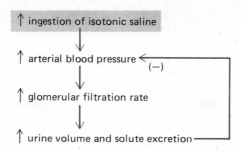

FIGURE 18·19

Effect of extracellular volume
increase on urine formation.

as 0.1 percent can be detected by special osmoreceptors located in the supraoptic nuclei of the anterior hypothalamus outside the blood-brain barrier. An osmolality that is lower than normal causes the osmoreceptors to swell and to decrease the frequency of impulses they send to the neurohypophysis, where ADH, which is produced by cells in the hypothalamus, is stored. The rate at which ADH is released from the neurohypophysis is partially determined by the frequency of impulses from the osmoreceptors. When the frequency of these impulses declines, the rate of ADH release decreases. When the amount of ADH present in the kidney decreases, the permeability to water of the distal convoluted tubule and the collecting duct declines, less water is reabsorbed from the tubular fluid, and urine output increases. ADH is currently thought to have no effect on tubular sodium reabsorption. Salts, then, are reabsorbed normally, and the increased volume of urine is very dilute. In these circumstances, decreased water reabsorption and constant salt reabsorption increase the osmolality of extracellular fluid.

ADH is also controlled by changes in blood pressure. For example, when an excess fluid volume isosmotic to the plasma is ingested, the volume of extracellular fluid is increased. An increase in the extracellular fluid volume slightly increases systemic blood pressure. The increased pressure is detected by baroreceptors located in the large arteries near the heart, and these receptors respond by increasing the frequency of their signals to the hypothalamus. An increased signal rate from these receptors decreases the rate of ADH release by the neurohypophysis, thus causing production of an increased volume of urine. The excretion of additional fluid volume in the urine decreases extracellular fluid volume and thus blood pressure, causing ADH level to rise to normal again.

Another factor to be considered in the example is the effect that elevated blood pressure has on the glomerular filtration rate. As already noted, increased blood pressure increases GFR, which increases urine volume as well as solute excretion (Fig. 18.19).

SODIUM CHLORIDE: KEY TO EXTRACELLULAR VOLUME AND OSMOLALITY When we discussed the functional characteristics of each nephron segment, we noted the approximate percentages of the ultrafiltrate reabsorbed in each segment and pointed out the importance of solute-linked fluid transport in the reabsorptive process. The major driving force for fluid reabsorption along the nephron is active sodium reabsorption. Sodium and chloride together constitute about 90 percent of the solutes determining extracellular fluid osmolality; thus, fluid reabsorption is controlled largely through sodium reabsorption. Ninety-nine percent of the filtered sodium is normally reabsorbed and, with it, about 99 percent of the filtered water. Small adjustments in the rate of sodium reabsorption thus greatly affect the amount of water excreted by the kidneys. Many factors influence sodium reabsorption, but one of the major systems involved is the *renin-angiotensin system* (Fig. 18.20).

Renin (pronounced ree-nin) is produced by myoepithelioid cells located in the afferent arterioles just before they enter the renal corpuscles. This area of an afferent arteriole lies close to the distal tubule. The cells of the distal tubule at this point are slightly modified and are called *macula densa*.

FIGURE 18·20
The renin-angiotensin system and its effects.

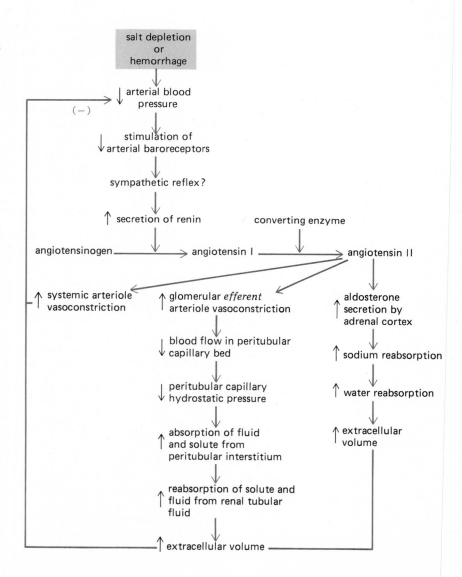

This region of the afferent arteriole, together with the macula densa, is called the *juxtaglomerular apparatus.*

The precise stimuli for renin release are not fully understood. It is believed, however, that a sympathetic reflex is involved. Adrenergic nerve fibers innervate the kidney, especially in association with the blood vessels. Sympathetic stimulation of the kidney produces marked constriction of the glomerular afferent arterioles, and such constriction greatly decreases glomerular filtration and blood flow in the peritubular capillaries.

In addition, when body sodium concentration or blood volume drops acutely, the arterial baroreceptors, acting through sympathetic fibers, may stimulate the release of renin by the juxtaglomerular cells. Renin's effect, through a sequence of reactions, is to convert *angiotensinogen* to *angiotensin II.* Angiotensinogen is produced by the liver (and perhaps by the kidney) and is always present in the blood, but it remains inactive until converted to angiotensin II by renin.

Angiotensin II has several important effects on kidney function. It is a vasoconstrictor that causes a general increase in systemic blood pressure. Within the kidney, angiotensin II seems to cause local vasoconstriction primarily of the efferent arteriole of the glomerulus. Both of these actions tend to increase the GFR by increasing capillary hydrostatic pressure; however, constriction of the efferent arteriole restricts glomerular plasma flow, which tends to reduce GFR so that it remains virtually unchanged. Angiotensin's net effect is to reduce the blood volume flowing into the peritubular capillary bed. The decreased flow reduces capillary hydrostatic pressure, thus stimulating reabsorption of filtered fluid and lowering urine output. All these activities help to restore extracellular volume.

Working through the mechanisms described above, angiotensin II also stimulates the release of the hormone *aldosterone* from the adrenal cortex. Aldosterone acts mainly on the distal tubule to stimulate active sodium reabsorption and potassium secretion. Aldosterone release is also stimulated directly by a rise in plasma potassium concentration (Fig. 18.21). Aldosterone is, therefore, extremely important in the maintenance of extracellular fluid mineral balance.

Another function of the juxtaglomerular apparatus, which may or may not be involved with the renin-angiotensin system, is the autoregulation of glomerular filtration rate in individual nephrons. This regulatory system seems to operate through the macula densa. When the rate at which fluid travels through the diluting segment decreases, the availability and thus the quantity of chloride ions transported by the macula densa also decrease. Decreased chloride concentration, through an as yet unknown feedback mechanism, increases the GFR (Fig. 18.22). Similarly, if an increased amount of chloride ion is available for transport, the GFR decreases, which, in turn, decreases the rate at which chloride is delivered to the macula densa. The sensitivity of the macula densa to chloride varies depending upon the conditions prior to stimulation. When extracellular volume is too high, the sensitivity of the system decreases so that GFR is not autoregulated. Thus, excess extracellular fluid can be eliminated through an increase in GFR. However, when extracellular fluid is decreased, autoregulation of GFR is near perfect. This mechanism produces an ample filtrate for the kidney to act upon, while preventing potentially damaging fluctuations in GFR.

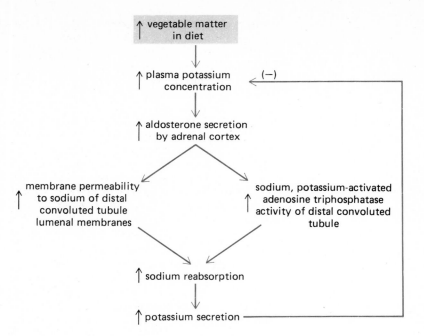

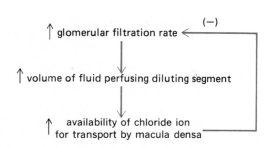

FIGURE 18·21
Interrelationship of potassium ingestion and aldosterone activity.

FIGURE 18·22
a Autoregulation of glomerular filtration rate. *b* Preconditioning factors affecting autoregulation.

THIRST Finally, we must consider one other aspect of water balance: the control of water intake by thirst. We normally drink an adequate water intake simply through a learned anticipatory response. However, if water intake is restricted to the point that extracellular volume declines or osmolality increases, reflex pathways, through hypothalamic control centers, induce drinking. The precise control of this reflex has not been determined.

KIDNEY DISEASES Various diseases of the kidney are responsible for tens of thousands of deaths in the United States each year. Kidney disease can take many forms and give rise to numerous symptoms. Many factors ranging from throat infection to obstruction of the ureters by kidney stones may cause bacterial infection of the kidneys, a condition termed *pyelonephritis.* Deterioration of the glomeruli because of infection is termed *glomerulone-phritis.* This type of condition impairs the ability of the kidney to excrete toxic metabolic wastes and to regulate normal plasma constituents. The bodily symptoms produced by this lack of regulation are termed *uremia,* or urine in the blood.

Uremia, regardless of its cause, may be treated by use of the artificial kidney or *dialysis machine.* In dialysis, the patient's blood is passed through dialysis tubing immersed in an electrolyte bathing medium similar in composition to the blood. Substances that have built up in the blood diffuse through the dialysis tubing into the bath fluid. High-molecular-weight molecules, such as the plasma proteins, cannot pass through the dialysis tubing and are thus retained in the blood. This cleansing process can greatly extend the life of an individual suffering from kidney failure.

Diabetes insipidus is a rare disease caused by the failure of adequate secretion of antidiuretic hormone (ADH) by the neurohypophysis. Lack of ADH produces a high volume of low-concentration urine and severe thirst.

Kidney stones consist primarily of calcium phosphate and are caused by abnormal conditions in the tubule fluid that precipitate this salt. These stones, or *renal calculi,* cause extreme pain when they travel through the urinary tract, and they may even obstruct major ducts, causing severe kidney disorders. They can be removed by surgical procedures.

Urinalysis A routine urinalysis usually consists of a test for specific gravity and chemical tests for the presence of glucose and protein. Just as analysis of ashes tells a great deal about what was burned in a furnace, so a chemical and microscopic examination of urine gives considerable information concerning the state of metabolism in both normal and pathological conditions.

The acidity of the urine can be readily tested. Microscopic examination may reveal blood cells, epithelial cells, pus cells, casts, calculi, bacteria, or parasites. Chemical analysis indicates the presence of bile pigments, bile salts, hemoglobin, or albumin. Traces of indican (a potassium salt) and ketone bodies may be present in normal urine, but if they are found in greater than trace amounts, their presence indicates an abnormal state of metabolism.

Micturition The act of voiding urine, called *micturition,* is an autonomic reflex act upon which voluntary control is superimposed. The baby voids urine reflexly whenever the pressure of accumulating urine in the bladder becomes great enough to provide an adequate nerve stimulus. Gradually the young child learns to control these impulses, at least when not asleep. Control over micturition even during sleep is usually achieved by the three-year-old child. Inability to establish such control at night is called *nocturnal enuresis*, or *bed-wetting*.

SUMMARY

1 The general mechanism of kidney action is selective elimination. This process involves the formation of a protein-free filtrate from plasma that is then processed to form urine. The kidneys regulate the plasma concentration of many substances by varying the amount of those substances reabsorbed from the filtrate. Several substances, such as organic acids, are secreted into the urine by the renal tubules.

2 The kidneys are bean-shaped organs located in the posterior part of the abdominal cavity. They are embedded in fat, covered by peritoneum, and supplied by the renal arteries. The kidney is divided internally into cortical and medullary portions. The cortex contains renal corpuscles, convoluted tubules, and blood vessels. The medullary portion forms wedge-shaped sections termed pyramids in which are located the long loops of Henle, the vasa recta, and the collecting ducts, which empty at the tip of the pyramid, the papilla. Urine is collected in the renal pelvis and conducted by the ureters to the urinary bladder.

3 The functional unit of the kidney is the nephron. It consists of Bowman's capsule, in which is located the glomerulus. Attached to the capsule is the proximal convoluted tubule, which reabsorbs 80 percent of the ultrafiltrate. The straight portion of the proximal tubule is the pars recta, from which arises the loop of Henle. The thick ascending portion of the loop of Henle is the diluting segment. The diluting segment ends and the distal convoluted tubule begins at the macula densa, which forms the juxtaglomerular apparatus together with the juxtaglomerular cells of the afferent glomerular arteriole. The distal tubule empties into the collecting duct, which merges with other collecting ducts toward the papilla.

4 The glomerular filtration apparatus is both size- and charge-selective. The major determinants of glomerular filtration rate are glomerular capillary hydrostatic pressure, plasma colloid osmotic pressure, and glomerular plasma flow. The latter is usually the most significant factor.

5 Measurements of functions of the intact kidney involve the determination of the volume of plasma from which a substance is completely removed or cleared. The clearance of inulin equals the glomerular filtration rate because inulin is freely filtered and neither reabsorbed nor secreted by the tubules. Nutrients such as glucose usually have clearances of zero, meaning they are completely reabsorbed from the filtrate. Substances with clearance values higher than that of inulin have been filtered as well as secreted into the urine.

6 Water and salt balances are dependent upon the interaction of glomerular filtration rate, the renin-angiotensin system, antidiuretic hormone, and aldosterone. The volume and concentration of the final urine is determined mainly by the countercurrent multiplier system, which generates an osmotic gradient in the renal medulla. This gradient helps to extract water from the tubular fluid in the collecting ducts.

7 Lack of proper regulation of the composition of body fluids by the kidney produces symptoms called uremia.

QUESTIONS

1 What is the basic principle of renal function? What three basic processes are involved?
2 Describe the location and appearance of the kidneys. What is the basic unit of function? How many of these units are there in each kidney?
3 What are the two basic divisions of the internal kidney structure? What types of nephrons form the long loops in the renal pyramids?
4 Why is the clearance of inulin equal to the glomerular filtration rate?
5 Describe the components of the glomerular filtration apparatus and the function of each component.
6 What are the major determinants of glomerular filtration rate?
7 What are the effects of ADH, aldosterone, and angiotensin II on the renal tubule?
8 List the various segments of the nephron and state their main functions.
9 What are the permeability, active transport, and anatomical characteristics of the tubules which make possible the countercurrent multiplier system?
10 During water restriction, what are the major avenues of water loss from the body?

SUGGESTED READING

Andreoli, T. E., et al.: Questions and Replies: Renal Mechanisms for Urinary Concentrating and Diluting Processes, *Am. J. Physiol.*, 4:Fl–Fll (1978).

Arruda, J. A. L., and N. A. Kurtzman: Relationship of Renal Sodium and Water Transport to Hydrogen Ion Secretion, *Annu. Rev. Physiol.*, **40:**43–66 (1978).

Davis, J. O. (ed.): Advances in Our Knowledge of the Renin-Angiotensin System, *Fed. Proc.*, **36:**1753–1787 (1977).

Deetjen, P., et al.: "Physiology of the Kidney and of Water Balance," Springer-Verlag New York Inc., New York, 1975.

Jacobsen, H. R., and D. W. Seldin: Proximal Tubular Reabsorption and Its Regulation, *Annu. Rev. Pharmacol. Toxicol.*, **17:**623–646 (1977).

Thames, M. D.: Neural Control of Renal Function: Contribution of the Cardiopulmonary Baroreceptors to the Control of the Kidney, *Fed. Proc.*, **37:**1209–1213 (1977).

Thurau, K.: Kidney and Urinary Tract Physiology, vol. 6, Physiology Series One, MTP International Review of Science, University Park Press, Baltimore, 1974.

Thurau, K.: Kidney and Urinary Tract Physiology II, vol. 11, International Review of Physiology, University Park Press, Baltimore, 1976.

Wright, F. S., and J. P. Briggs: Feedback Regulation of Glomerular Filtration Rate, *Am. J. Physiol.*, **233**(1):F1–F7 (1977).

19

THE DIGESTIVE SYSTEM

In our study of the digestive system, we consider the structure of the gastrointestinal tract (GI tract) and the function of the organs that compose it. In this chapter we take up the following topics:

1 The structure and function of the mouth, the teeth, and the salivary glands
2 The conduction of food to the stomach by way of the pharynx and esophagus
3 The functions of the stomach, including those of the specialized secretory cells that line the stomach
4 The structures of the small intestine: the duodenum, the jejunum, and the ileum
5 The digestive functions of the pancreas
6 The intricate structure of the liver and its complex glycogenic and glycogenolytic functions
7 The absorption of food materials by the small intestine
8 The structure of the large intestine as it extends from the ileocecal valve to the anus and its principal function, the absorption of water

To live, we must take in food, provide for its digestion and absorption, and utilize the energy derived therefrom. The principal part of the digestive system consists of a long tube extending from mouth to anus, which is greatly coiled in the abdominal region. Within this tube and its derivatives, food materials are prepared for digestion and absorption. The tube, in different regions, is known as the *mouth, pharynx, esophagus, stomach, small*

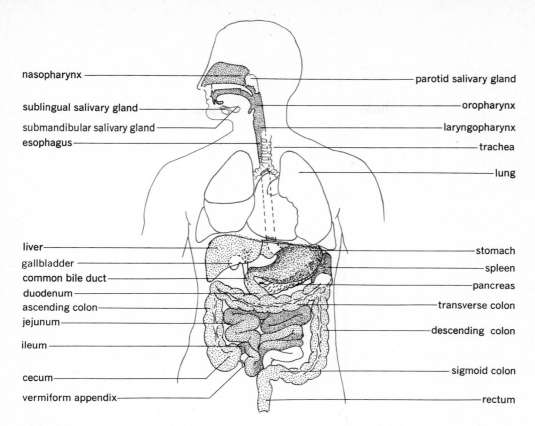

nasopharynx

sublingual salivary gland

submandibular salivary gland

esophagus

parotid salivary gland

oropharynx

laryngopharynx

trachea

lung

liver

gallbladder

common bile duct

duodenum

ascending colon

jejunum

ileum

cecum

vermiform appendix

stomach

spleen

pancreas

transverse colon

descending colon

sigmoid colon

rectum

FIGURE 19·1
General topography of the digestive system. The alimentary canal, anterior view.

intestine, and *large intestine.* Large glandular organs such as the stomach, liver, and pancreas contribute to the process of digestion (see Fig. 19.1 and Plate 13).

The musculature of the digestive tract is adapted for the special functions it performs. Striated muscles of the mouth and pharynx provide the voluntary motive force for chewing and swallowing. Involuntary smooth muscles propel the food materials along the tube and churn the food in the stomach and intestine.

Digestive enzymes, produced by glands of the digestive system, are protein in nature. They act as organic catalysts to help break down the complex chemical stucture of food materials into simpler compounds. When digestion has been completed, these simpler compounds are absorbed through the wall of the intestine and carried away by the blood. Considerable water is reabsorbed by the body as it passes through the large intestine, and finally the waste products are eliminated as fecal matter.

The embryonic digestive system is shown in Fig. 19.2. It is essentially an uncoiled tube closed at both ends. An oral cavity (stomodeum), which will later become the mouth, invaginates from the anterior end of the embryo to meet the anterior end of the tube. The surface ectoderm is pulled in with this invagination and lines the anterior part of the oral cavity. The partition between the oral cavity and the digestive tube soon breaks down. In the

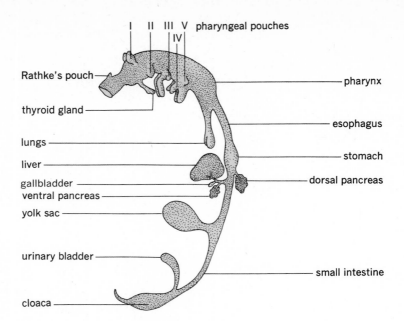

I II III V pharyngeal pouches
IV

Rathke's pouch

thyroid gland

lungs

liver

gallbladder

ventral pancreas

yolk sac

urinary bladder

cloaca

pharynx

esophagus

stomach

dorsal pancreas

small intestine

FIGURE 19·2

Derivatives of the embryonic digestive tract, shown diagrammatically. (*After C. D. Turner, "General Endocrinology," W. B. Saunders Company, Philadelphia, 1971.*)

same way, a posterior ectodermal invagination pushes in to meet the posterior end of the digestive tube and later lines the anal canal. Derivatives of the embryonic digestive tract are shown clearly in the figure.

MOUTH

Mouth
  Tongue
  Teeth
  Hard palate
  Soft palate
  Uvula

The digestive tract begins at the oral opening, or mouth. The oral cavity is lined with mucous and submucous layers and contains, among other structures, the *tongue* and *teeth*. The roof of the mouth is the *hard palate,* which is formed by the maxillary and palatine bones and covered by mucous membrane. The *soft palate* is a posterior continuation of membrane covering muscle and nerve fibers. A tip of membrane hangs down into the throat cavity at the back of the mouth. This is the *uvula*. It is drawn back out of the way during swallowing or when we open the mouth and say "ah."

TONGUE

The tongue is a muscular structure of many functions. It is known primarily as a special organ for the sense of taste, a function we discussed in Chap. 12, but the sense of touch and temperature sense are also well developed on the tongue. The tongue starts the digestion process by mashing the softer particles of food and guides other particles between the upper and lower teeth to be crushed. In the act of swallowing, the tongue propels a ball (or bolus) of solid food and liquids back into the pharynx.

TEETH

Teeth
  Incisors
  Cuspids (canines)
  Bicuspids
  Molars

Hard, enameled teeth located in sockets of the jawbones cut, tear, and crush food. Human beings have two sets of teeth: a deciduous, or milk, set of 20 teeth that fit the jaw of a child and an adult set of 32 teeth for the jaw of an adult. The deciduous teeth appear gradually and are shed gradually as adult teeth begin to emerge. This means that a child of eight or ten years

FIGURE 19·3

The teeth of a five-year-old child. The jaws have been partially dissected to show the roots of the deciduous teeth and the developing adult teeth. The deciduous teeth are shown in white, the permanent teeth in color.

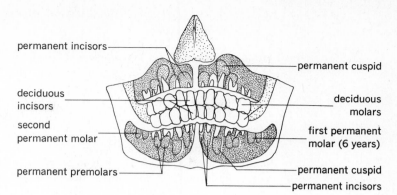

permanent incisors

permanent cuspid

deciduous incisors

deciduous molars

second permanent molar

first permanent molar (6 years)

permanent premolars

permanent cuspid

permanent incisors

can have some deciduous teeth and some adult teeth in use at the same time (Fig. 19.3).

Mammalian teeth are differentiated into incisors; canines, or cuspids; premolars; and molars. *Incisors* are the front teeth. They are flat, chisel-like teeth for cutting. There are eight of them, four in each jaw. The *cuspids,* or *canines,* are long and rounded for tearing food. They are located on either side of the lateral incisors, two in each jaw. Cuspids form the fangs of carnivores. Premolars, or bicuspids, are found only in the adult, or permanent, set. There are four in each jaw. There are commonly two cusps to a tooth; since they are located just anterior to the molar teeth, they are called *bicuspids* or *premolars.* The *molar teeth* are the large, flat teeth at the back of the jaw. The crushing surface is provided with four or five tubercles, which aid in mastication. There are eight molar teeth plus four *third molars,* or *wisdom teeth,* in the adult (Fig. 19.4).

DECIDUOUS TEETH   A newborn baby does not ordinarily have any erupted teeth, but at birth, teeth are forming in the jawbones. Before the end of the first year the child probably will have cut its lower central incisors and upper incisors. At the age of two to two and one-half years the child may be expected to have a complete set of deciduous teeth. The second molars are

FIGURE 19·4

The teeth: permanent teeth of the upper and lower jaw (lateral view).

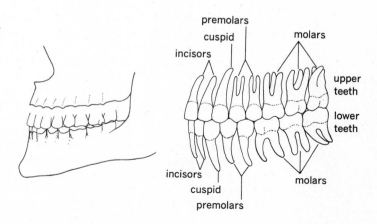

premolars

cuspid

molars

incisors

upper teeth

lower teeth

incisors

cuspid

molars

premolars

usually the last of the deciduous teeth. There are no premolars in the deciduous set.

Deciduous dental formula The number of deciduous teeth may be calculated by the following formula for one-half of each jaw:

$$\text{Incisors}\frac{2}{2} \quad \text{cuspids}\frac{1}{1} \quad \text{molars}\frac{2}{2} = \frac{5 \times 2}{5 \times 2} = 20$$

The deciduous teeth are smaller than adult teeth, and their roots are smaller. The incisors are the first to be shed, around the seventh and eighth years. Children of this age present a peculiar toothless appearance before the adult incisors take their place. The second molars are the last of the baby teeth to be lost, usually when a child is around eleven or twelve years of age.

PERMANENT TEETH   Children of six or seven years actually possess two sets of teeth. The adult, or permanent, set is forming in sockets in the jawbones: the enamel caps of these teeth are still below the gums. By the time a child is six or seven years old, the jaw has lengthened considerably, and there is room for the first adult molar to appear immediately behind the second baby molar. The first molar of the adult set is the so-called sixth-year molar.

Adult dental formula The dental formula for the human adult with an entire set of permanent teeth is as follows:

$$\text{Incisors}\frac{2}{2} \quad \text{cuspids}\frac{1}{1} \quad \text{premolars}\frac{2}{2} \quad \text{molars}\frac{3}{3} = \frac{8 \times 2}{8 \times 2} = 32$$

The deciduous teeth are gradually replaced by the permanent teeth, with the exception of the third molar, during the period between six and thirteen years of age. The second molars usually appear around twelve or thirteen years of age. The third molars, or wisdom teeth, are very late in making their appearance. If they appear at all, it is usually between seventeen and twenty-five years, often at around eighteen or nineteen years. The jaws are usually long enough to accommodate them by this time, but sometimes they arise at an angle, grow into the second molar, and turn sideways. When this happens the tooth is said to be *impacted,* and the tooth usually must be removed by a dentist. Wisdom teeth are sometimes formed with a poor coating of enamel, in which case the dentist commonly advises their removal.

Proper occlusion of the teeth is important for their chewing function and in the configuration of the face. A receding chin or outjutting lower jaw may be due to faulty occlusion of the teeth. In extreme old age, after the loss of all the teeth, the alveolar processes of the jawbones wear down, and the chin is drawn up close to the nose when the jaws are closed.

Tooth structure
Crown
Neck
Root

TOOTH STRUCTURE   A tooth may be divided into three regions. The visible part above the gums is the *crown;* the *neck* is a constricted portion between the crown and the root; the *root* is the part embedded in a socket in the jawbone.

The crown of a tooth is composed of enamel, the hardest substance in the body. The developing tooth is covered with an ectodermal epithelium (the enamel organ) that forms enamel, beginning with the top of the crown. The enamel is thickest on the wearing surface. The histological structure of enamel consists of hexagonal rods called *enamel prisms*. Their arrangement accounts for the radial lines and striations characteristic of this substance. The enamel covering becomes thin as it terminates in the neck region of the tooth. The chewing surfaces of premolar and molar teeth, especially, wear down with use, and the cusp patterns change with age.

Teeth develop in sockets, or alveoli. The *gums,* or *gingiva,* are composed of dense connective tissue covered by mucous membrane of stratified squamous epithelium. The developing tooth and socket are at first covered by this connective tissue. When the tooth is "cut," it emerges through the gums. The alveolus is lined with periosteum, or periodontal membrane, which has several functions. It helps attach the tooth to its socket, acts as a nourishing membrane, and forms the bonelike substance called *cementum* over the root of the tooth (Fig. 19.5).

The cement is a thinner and softer layer than enamel. It covers the tooth from the termination of the enamel to the tip of the root. Cement is described as bonelike because it contains some of the lacunae, canaliculi, and haversian canals characteristic of bone.

Dentine   The greater bulk of the tooth is formed of a bonelike structure called *dentine*. When sectioned, dentine appears striated because of the fine parallel tubes called *dental canaliculi*. They radiate outward from dentine-forming cells (odontoblasts) located along the border of the pulp cavity. Unlike bone, dentine contains no cells. It is formed instead from long protoplasmic fibrils from odontoblast cells, which extend through the tiny canals. Dentine is a yellowish color. It is sometimes called *sensitive dentine,* because the dentist's drill causes pain here. Since there are apparently no nerve fibers in the dentine, one explanation is that the protoplasmic strands within the canaliculi act as nerve fibers. Dentine forms slowly throughout the life of the tooth. Injury accelerates the formation of new, or secondary, dentine. Normally in older persons the pulp cavity is reduced by the continued formation of dentine.

Pulp cavity and pulp   The *pulp cavity* lies in the central portion of the tooth

FIGURE 19·5

Section of a tooth *in situ,* showing its structure.

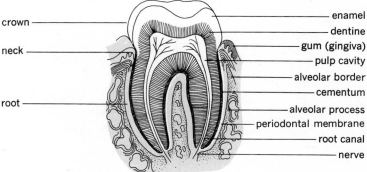

crown

neck

root

enamel
dentine
gum (gingiva)
pulp cavity
alveolar border
cementum
alveolar process
periodontal membrane
root canal
nerve

and extends down into the root. The *pulp* is composed of loose connective tissues. It is highly vascular and contains a great many nerve fibers. Odontoblast cells are located around the periphery. Blood vessels and sensory nerves enter through a tiny orifice at the tip of the root and pass through the root canal to the pulp cavity proper.

SALIVARY GLANDS

As mastication proceeds, the *salivary glands* pour their secretion, called *saliva,* over the food.

There are three pairs of large salivary glands: the *sublingual, submandibular,* and *parotid glands.* Their secretion is supplemented by the activity of numerous minute glands located in the mucous membrane lining the mouth. These are the *buccal glands,* the secretions of which are considered to be essentially the same as those produced by the principal salivary glands.

The *sublingual gland* is the smallest of the salivary glands. It is located below the anterior portion of the tongue and beneath the mucous membrane in the floor of the mouth (Fig. 19.6). A varying number of small ducts open separately into the mouth below the tongue. The larger sublingual duct, the duct of Bartholin, may join the submandibular duct or may open separately into the mouth below the anterior part of the tongue.

The *submandibular (submaxillary) gland* is located posterior to the sublingual gland and deeper in the floor of the mouth. Its position, roughly speaking, is inside and a little below the mandible, close to the insertion of the masseter muscle. A long submandibular duct, Wharton's duct, opens through the floor of the mouth beneath the anterior portion of the tongue. The duct is thin-walled and about 5 cm long. The papilla, through which it opens, can be seen if the tongue is raised.

The *parotid gland,* the largest of the salivary glands, lies at the angle of the jaw and in front of the ear. The external carotid artery and the posterior facial vein pass through it. The duct of the parotid gland (Stensen's duct) passes across the outer surface of the masseter muscle and turns inward to

FIGURE 19·6
The location of the salivary glands and their ducts, lateral view.

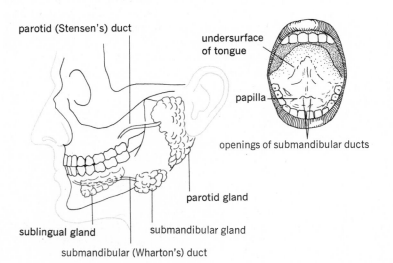

parotid (Stensen's) duct

undersurface of tongue

papilla

openings of submandibular ducts

parotid gland

sublingual gland

submandibular gland

submandibular (Wharton's) duct

open through the mucous lining of the cheek at a small orifice opposite the crown of the second upper molar tooth.

The parotid gland is involved in the viral disease known as *mumps,* in which the gland becomes greatly swollen and inflamed. All the salivary glands may become affected as well as other glandular tissues such as the testes. Serious inflammation and swelling of the testes resulting in sterility, usually unilateral, may be a serious complication of mumps in adolescent and young adult males.

SALIVARY SECRETION   Saliva moistens and softens the food, holds particles together, and aids in the digestion of starch. Many dry food substances cannot be tasted unless they are mixed with or dissolved in saliva. The bacterial enzyme *lysozyme* protects the mucous membrane of the mouth against bacterial infection.

Saliva is a watery fluid with a specific gravity only a little greater than that of water. Under normal conditions various samples range from slightly acidic to slightly alkaline. The average amount secreted per day is between 1000 and 1500 ml. The chemical composition of saliva includes several inorganic salts such as sodium and potassium chloride, sodium bicarbonate, and phosphates of sodium and calcium. Some oxygen and carbon dioxide are included as gases. Organic substances include some of the blood proteins, urea, mucin, and the enzyme *salivary amylase,* or *ptyalin.* Mucin is secreted largely by the sublingual and submandibular glands. The composition of saliva varies considerably, depending upon the type and strength of the stimulus that elicits its secretion. Blood cells in the saliva are usually lymphocytes, the so-called salivary corpuscles. The crust that commonly collects on the teeth is called *tartar,* or *plaque.* It is a deposit of calcium carbonate and phosphate plus some organic materials from the saliva.

CONTROL OF SALIVARY SECRETION   The salivary glands are innervated by two sets of nerves from the autonomic nervous system. The facial (VIIth cranial) nerve carries parasympathetic fibers by way of the chorda tympani nerve to the sublingual and submandibular glands. The parotid gland is innervated by a branch of the glossopharyngeal (IXth cranial) nerve. The sympathetic innervation to the three pairs of salivary glands is through the superior cervical ganglia at the cephalic end of the chain of sympathetic ganglia (see Fig. 10.8).

Thinking of the taste of sour food, such as lemons or vinegar, will usually cause a flow of saliva. Salivation at the thought or sight of food is a good example of a conditioned reflex. Placing substances in the mouth stimulates the taste buds and elicits salivation as a direct reflex act. Pain impulses, such as those invoked by the dentist's drill, cause a salivation of the ropy type because mucus is secreted. Salivation can be inhibited by fear and by embarrassment.

Experimental stimulation of salivary glands compared with observations of normal behavior leads to some similar conclusions. The parasympathetic system promotes a normal flow of saliva during digestion. There is no question that stimulation of the parasympathetic nerves induces vasodilatation in the arterioles supplying the salivary glands and that stimulation of sympathetic nerves causes vasoconstriction. Conditions that would be unfavorable to digestion, such as pain, fear, or unpleasant associations, also activate the sympathetic system.

THE DIGESTIVE FUNCTION OF SALIVA   The salivary enzyme that acts upon starches was originally called *ptyalin.* Chemical nomenclature now requires that enzymes have the ending *-ase,* and so this enzyme is now called *salivary amylase. Amyl* is derived from the Greek word for starch. Since food does not remain long in the mouth before being swallowed, much starch digestion is carried on in the stomach until the salivary amylase swallowed with the food is inhibited by the acid of the stomach. The period in which such digestion takes place is probably 15 to 30 min. Cooked starch is much more easily digested, since uncooked starch granules have a cellulose covering. Salivary digestion of starch is not fully understood, but the larger starch molecules break down through dextrins to maltose and glucose.

It appears that human salivary amylase, acting in vitro, can convert starch to glucose if the relative amount of the enzyme is great enough. Salivary amylase does not normally convert maltose to glucose, however, and there appears to be no maltose in human saliva. The saliva of many animals, including the dog, seems to have no digestive action at all.

EXCRETORY FUNCTION OF SALIVA   Many substances can be excreted in the saliva. Mercury and lead can be excreted, and in lead poisoning a deposition of lead salts, known as the *blue line,* appears on the gums close to the teeth. Cyanides, which are found as trace elements in our food, are very toxic but are readily converted into thiocyanates, which are nontoxic. Thiocyanates are excreted by the salivary glands as well as by the kidneys. Urea is excreted by the salivary glands, especially in certain kidney diseases. Diabetics sometimes notice a sweet taste in the mouth caused by the excretion of sugar into the mouth. Viruses are sometimes excreted in the saliva; a well-known example is the rabies virus.

PHARYNX AND TONSILS

The *pharynx* is the aperture back of the mouth. It is lined by a continuation of the mucous membrane that lines the mouth and nasal passageways. The mouth opens into the pharynx through an isthmus called the *fauces.* The opening is bounded by the palatine arches on either side; the *uvula* hangs down from the velum of the soft palate above; and the base of the tongue lies below. The nasopharynx communicates with the nasal passageways and continues posteriorly. The internal openings of the auditory tubes are nearby. The pharynx communicates below with the larynx and esophagus.

At the back of the mouth on either side of the uvula are the palatine arches. Tucked in between the arches on either side are reddish masses of lymphoid tissue commonly known as the *tonsils.* These are the *palatine tonsils.* Small mounds of lymphoid tissue below the tongue are called *lingual tonsils.* There are also *pharyngeal tonsils,* which are often referred to as *adenoids.* The tonsils are specialized masses of lymphoid tissue whose surface is marked with pits and depressions. Lymphocytes produced in the tonsils pass to the surface through the epithelium. They can be observed in saliva as salivary corpuscles. The tonsil pits, however, occasionally become inflamed by the action of bacteria harbored there. Infected tonsils are more common in children than in adults. Adenoids can become so much enlarged, especially in children, that they interfere with nasal breathing; they can also compress the opening of the auditory tubes.

ACT OF SWALLOWING Swallowing is the muscular propulsion of food from the mouth through the pharynx and into the esophagus, the tube leading to the stomach. The bolus, or mass of food, has usually been chewed and mixed with saliva prior to swallowing, and mucin from the saliva helps to hold the particles together. The muscular action of the tongue propels the food backward through the fauces into the pharynx. The constrictor muscles of the pharynx contract and force the food into the esophagus. During the act of swallowing, the soft palate is raised to prevent food from entering the nasal passageways. At the same time, the larynx is raised to a position under the base of the tongue and is covered by the epiglottis. The elevation of the larynx during swallowing can be readily observed.

There can be no intake of air through the trachea during swallowing, and the respiratory center is momentarily inhibited. A brief inspiration just precedes the act, and the glottis is then closed during the passage of food materials. Sometimes when persons are talking, laughing, and eating as well, a mistake in timing occurs and solids or liquids are directed into the larynx. A violent coughing reflex is set off by such events.

ESOPHAGUS

The muscular tube extending from the pharynx to the stomach is the *esophagus*. It is about 25 cm long and about 2.5 cm wide. The wall of the tube is composed of four coats.

1 The outer coat is thin, fibrous connective tissue.
2 The muscular portion of the esophagus contains an outer layer of longitudinal fibers and an inner layer of circular muscles. The musculature of the upper part of the tube is thick, striated muscle like that of the pharynx, while that in the lower part of the tube is smooth, visceral muscle.
3 An areolar submucous layer supporting blood vessels lines the muscular layer.
4 A final mucous layer lines the tube. (Fig. 19.7).

Liquids and solids are propelled through the esophagus by wavelike (peristaltic) contractions of the muscular wall. Liquids are moved very rapidly, with gravity playing some part in their descent to the stomach in humans. In animals that drink with the head down, peristalsis evidently overcomes the force of gravity. The food passes through the upper striated portion of the tube more rapidly than through the lower smooth-muscle portion. Some animals, such as the dog, have striated muscle throughout the length of the esophagus; such animals can swallow very quickly. As the bolus of food approaches the gastroesophageal sphincter muscles, which guard the entrance to the stomach, these muscles relax and permit the food to enter the stomach. The gastroesophageal sphincter may not be a true sphincter muscle in the strict sense, but muscle fibers in this region act as sphincters in opening and closing the upper orifice of the stomach.

The first stage of swallowing is voluntary, but when the food reaches the pharynx, the second and third stages are purely reflex and involuntary.

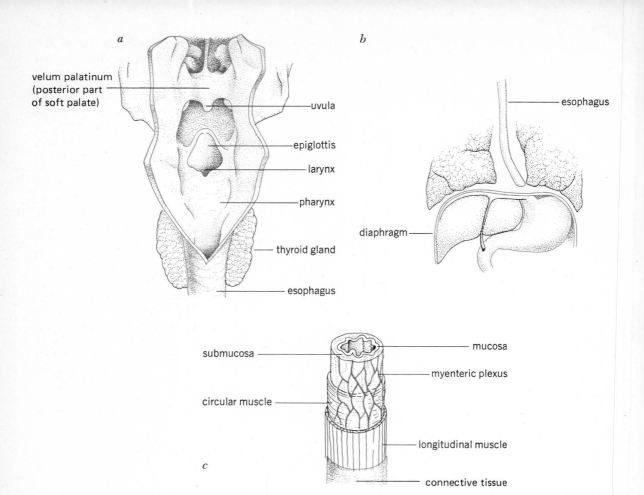

velum palatinum (posterior part of soft palate)

uvula

epiglottis

larynx

pharynx

thyroid gland

esophagus

a

b

esophagus

diaphragm

submucosa

circular muscle

mucosa

myenteric plexus

longitudinal muscle

connective tissue

c

FIGURE 19·7

The esophagus, *a* superior opening; *b* inferior portion; *c* structure.

Motor nerves concerned with swallowing are branches of the hypoglossal, trigeminal, glossopharyngeal, and vagus.

ABDOMEN

The abdominal cavity is the largest body cavity. It extends from the dome-shaped diaphragm above to the lesser pelvis below. Posteriorly are the vertebrae and the deep muscles of the back. Abdominal muscles form the sides and the front walls. Contained within the abdomen are the stomach, liver, pancreas, intestine, spleen, and kidneys. The abdominal wall has several openings that permit the passage of various structures. The umbilical opening transmits the umbilical blood vessels in the fetus. The esophagus penetrates the diaphragm to enter the stomach. The vena cava and the aorta also pass through the diaphragm. The thoracic duct and the azygos vein accompany the aorta through the diaphragm. The floor of the abdominal cavity has two openings on either side; one transmits the femoral artery and vein, and the other transmits the spermatic cord in the male and the round ligament in the female.

The abdominal cavity is lined with a serous membrane called the *peritoneum*. It is folded and reflected over abdominal organs. The intestine is held in place by folds of peritoneum called the *mesentery*. The mesentery is a double layer; between the two layers it supports mesenteric blood vessels, lymphatic vessels, lymph nodes, and nerves. A large fold, called the *greater omentum*, is connected to the greater curvature of the stomach and to the transverse colon. This fold covers the intestine anteriorly like an apron. It is usually well supplied with fat, and it functions as a protective and insulating layer. The *lesser omentum* is a double layer of peritoneum attached between the stomach and liver.

STOMACH

The stomach is an enlarged portion of the digestive tract located in the abdominal cavity directly below the diaphragm on the left side. It is continuous with the esophagus and the gastroesophageal sphincter, which surrounds its upper orifice. The shape of the stomach varies. When empty, it is more like a tube; when filled, it is more saclike. Certain areas of the stomach can be readily identified. The *fundus* is the rounded, upper portion that extends above the gastroesophageal sphincter (Fig. 19.8). At the stomach's lower end is the *pyloric antrum,* a slight enlargement above the pyloric orifice, that opens into the *duodenum,* or upper part of the intestine. The pyloric opening is guarded by the pyloric sphincter muscle. The body of the stomach lies between the pylorus and the fundus. The lesser curvature of the stomach is concave toward the right and faces the liver. The greater curvature is convex on the opposite side and is several times longer.

Stomach wall
   Serous layer
   Muscular layer
   Submucous layer
   Mucous layer

STOMACH WALL   The *stomach wall* is composed of four layers typical of the digestive tract.

FIGURE 19·8
The stomach (anterior view).

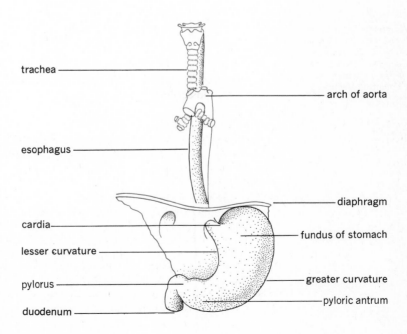

trachea

arch of aorta

esophagus

diaphragm

cardia

fundus of stomach

lesser curvature

greater curvature

pylorus

pyloric antrum

duodenum

a

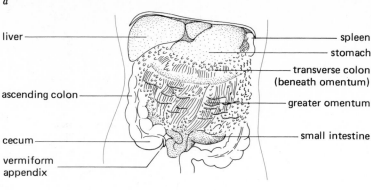

liver —————————————————— spleen

————————————— stomach

————————————— transverse colon
(beneath omentum)

ascending colon —————————— greater omentum

cecum ————————————————— small intestine

vermiform
appendix

b

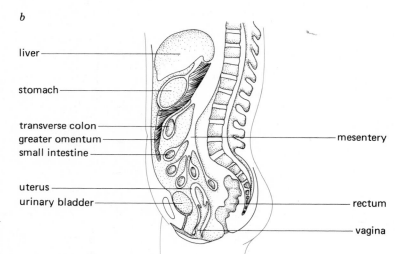

liver ——————————————

stomach ——————————————

transverse colon —————————— mesentery
greater omentum ——————————
small intestine ——————————

uterus ——————————————
urinary bladder —————————— rectum

————————————— vagina

FIGURE 19·9

a The greater omentum covering the abdominal viscera (anterior view). *b* The trunk in sagittal section showing the location of the mesentery and greater omentum (female).

1 The outer *serous layer* is formed from the peritoneum. It is reflected over the organ in such a way that the two layers come together at the lesser curvature and extend to the liver as the *lesser omentum*. The two layers join along the greater curvature and extend downward as the *greater omentum* (Fig. 19.9).

2 The thick *muscular layer* of the stomach wall consists of three layers of visceral muscle: longitudinal, circular, and oblique. The position of these muscle layers makes possible the churning and peristaltic motions that prepare the food for further digestion. The longitudinal layer is the outermost layer. At the cardiac opening it is continuous with the longitudinal muscles of the esophagus. The circular layer is the intermediate layer and is continuous with the circular layer of the esophagus. At the lower end of the stomach, the fibers form a strong ring of muscular tissue, the pyloric valve. The internal oblique layer is composed of fibers radiating from the cardiac area.

3 The *submucous layer* consists of loose, areolar tissue. It connects the muscular and mucous layer.

4 The *mucous layer* is thick and glandular. If the stomach is not distended, the mucous lining typically lies in mostly longitudinal folds. The surface of this layer is pitted with the openings of glands, which secrete the gastric fluid. The inner surface of the mucous layer is covered by columnar epithelium containing goblet cells (Fig. 19.10).

STOMACH GLAND CELLS   Three kinds of cells are found in the small, tubular principal glands of the stomach. First, the *mucous cells,* which are located in the neck region of the gland or in the lower part of the pit, secrete mucin. Second, the zymogenic, or *chief cells,* line the tubule and secrete *pepsinogen.* Zymogen granules, thought to be the source of the pepsinogen, are found at the bases of these cells. Microanalysis indicates that the zymogenic cells also contain dipeptidase, usually considered to be an intestinal enzyme acting on dipeptides. The third type is the *parietal* cell.

Parietal, or *oxyntic, cells* are large spheroid cells distributed at intervals between the zymogenic cells. These cells produce hydrochloric acid. The cells secrete both H^+ and Cl^- ions from secretory canaliculi. Parietal cells are not present in the glands located at the pyloric end of the stomach.

STOMACH FUNCTION   The saclike stomach enables us to eat a considerable quantity of food at one time. When it reaches the stomach, the food is moved back and forth by muscular contractions and is mixed with gastric juice. Through the action of digestive enzymes and hydrochloric acid, the food becomes partially liquefied; in this state it is called *chyme.* When chyme reaches the proper consistency, it is forced out through the pylorus and into the duodenum a little at a time.

Pepsin is the principal digestive enzyme found in the stomach. It is a protease, which breaks down complex protein molecules into cleavage stages known as *proteoses* and *peptones.* Pepsin, however, does not complete the work of breaking down proteins to their end products, amino acids. Protease enzymes of the pancreas and intestine accomplish this final breakdown before the protein is absorbed into the blood. Pepsin is produced by the zymogenic cells in an inactive form called *pepsinogen,* which is activated upon contact with hydrochloric acid.

Hydrochloric acid activates pepsinogen and provides the proper acid medium for digestion by pepsin. The concentration of HCl in the gastric juice secreted by the parietal cells is around 0.5 percent. The acidity of the stomach contents is seldom as great as this, however, since the hydrochloric acid is diluted by saliva, food, and mucin. In addition, the secretions of the pyloric glands are slightly alkaline and somewhat neutralizing. Even so, the acidity of the gastric juice is usually found to be about pH 1.6 to 2.6. When the gastric contents are regurgitated into the throat, the hydrochloric acid causes the burning sensation.

In addition to activating pepsinogen and providing an acid medium for pepsin, hydrochloric acid swells and softens proteins such as those in meat fibers. It also renders gastric juice highly antiseptic, and it initiates the breakdown of cane sugar and the curdling of milk.

The enzyme *rennin* acts upon milk. It is probably produced in an inactive form called prorennin. Rennin, or a similar synthetic product, has been used in cheese making for many years. It coagulates milk, producing a white mass

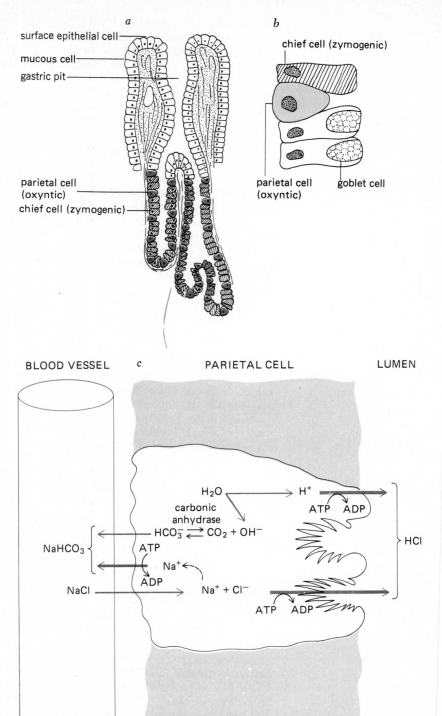

a
surface epithelial cell
mucous cell
gastric pit
parietal cell (oxyntic)
chief cell (zymogenic)

b
chief cell (zymogenic)
parietal cell (oxyntic)
goblet cell

BLOOD VESSEL   c   PARIETAL CELL   LUMEN

$H_2O \longrightarrow H^+$

carbonic anhydrase

$HCO_3^- \rightleftharpoons CO_2 + OH^-$

$ATP \quad ADP$

$NaHCO_3$
ATP
Na^+
ADP
NaCl
$Na^+ + Cl^-$

HCl

$ATP \quad ADP$

FIGURE 19·10
a Section of stomach wall showing glands. *b* The cells shown enlarged are surface epithelial cells, mucous cells, parietal cells, and zymogen or chief cells. *c* Secretion of hydrochloric acid by the parietal cells of the stomach.

called *curd*, which separates from the fluid material called *whey*. Rennin acts upon the soluble protein *casein*, converting it into *paracasein*, which combines with calcium ions to form the curd. Rennin is found in the gastric juice of infants, but it does not seem to be effective in the stomach of the adult, in which the acid content of the gastric juice is high. In the adult, milk is curdled by the direct action of HCl. In the infant, however, the acid concentration in the stomach is much lower, thus favoring the action of rennin. Rennin is thought to be produced by the zymogenic cells.

The functions of digestive enzymes of the stomach and other portions of the digestive system are summarized in Table 19.1.

REGULATION OF GASTRIC DIGESTION   The taste of food, or even the sight of food, causes gastric-fluid secretion. The latter is a good example of a conditioned reflex. This response is thought to be primarily mediated by the vagus nerve. If the vagus is cut, the sight of food no longer stimulates much gastric secretion. It is possible, however, that a hormonal mechanism may be partially responsible for this reflex. This response has physiological value because the stomach is thereby better prepared to receive and digest food. This process is commonly referred to as the *psychic phase* of gastric digestion.

The flow of digestive juices can be reduced or inhibited by unpleasant sights, tastes, or smells. This effect is mediated through the sympathetic nervous system. A strong emotional reaction is also unfavorable to good digestion.

Pleasant surroundings, good food, and peace of mind favor good digestion. The phrase "laugh and grow fat" has some basis in fact. The rush of present-day living is often not conducive to good digestion.

When food reaches the stomach, distension of the stomach wall, plus the influence of *secretagogues* in the food itself, initiates the release of the hormone *gastrin*. Secretagogues are chemical substances in certain foods that are released as the food is digested. Meat and some other proteins have a high secretagogue content. Gastrin is absorbed by the blood, and when it returns to the gastric mucosa, it stimulates the parietal cells to secrete HCl. Histamine has a similar effect, but gastrin normally is much more effective. The secretion of gastric juice after food has entered the stomach is known as the *gastric phase* of digestion.

A third phase of gastric digestion is commonly called the *intestinal phase*. When food that has never been in the stomach is experimentally introduced into the upper part of the small intestine of an animal, the presence of food causes the gastric glands to secrete. The effect appears to be hormonal, but no hormone has been isolated.

When unusual concentrations of food substances, notably fats and carbohydrates, are passed into the duodenum (the first part of the intestine) too rapidly, a hormonal factor that inhibits gastric digestion is released. The hormonal factor is *enterogastrone,* and it is secreted by glands in the mucous lining of the duodenum. Since all hormones are absorbed by the blood, enterogastrone is absorbed and carried to the stomach. Inhibition of gastric digestion appears to be a protective device to prevent overloading of the duodenum, but it also functions in reducing gastric acidity.

The question is often asked: Why doesn't the stomach digest itself? The question is a difficult one, and the complete answer is not known. Pepsin is

TABLE 19·1

Digestive fluids and principal digestive enzymes

Source	Fluids	Enzymes	Substrate	End products
Mouth	Saliva	Amylase (ptyalin)	Starch	Dextrins and maltose
Stomach	Gastric juice (acidified by HCl)	Pepsin	Proteins	Proteoses Peptones Polypeptides
		Lipase	Emulsified fats (as in cream, egg yolks)	Glycerol Fatty acids
		Rennin (in infants)	Caseinogen	Paracasein (precipitated by calcium to form curd)
Pancreas	Pancreatic fluid	Trypsin Chymotrypsin	Undigested proteins and proteoses	Peptones Polypeptides Amino acids
		Amylase	Starch, dextrins	Disaccharides maltose
		Lipase	Lipids	Glycerol Fatty acids
Small intestine		Carboxypeptidases	Polypeptides Dipeptides	Amino acids
		Ribonuclease, deoxyribonuclease	RNA, DNA	Nucleotides
		Disaccharidases		
Brush border of intestinal villi and breaking down of epithelial cells		Lactase°	Lactose	Glucose (dextrose) Galactose
		Maltase°	Maltose	Glucose (dextrose)
		Sucrase°	Maltose	Glucose (dextrose)
		Sucrase° (invertase) (several of lesser importance)	Sucrose (cane sugar)	Glucose and fructose
Small intestine	Intestinal fluid	Enterokinase	Activates pancreatic trypsinogen	
Liver	Bile	(Contains no enzymes)	Fats are not absorbed unless emulsified by bile salts	

° Lactase, maltase, and sucrase are inverting enzymes.

considered to be in an inactive form when it is produced by the zymogenic cells. After it is activated by hydrochloric acid, copious amounts of mucin help to protect the stomach wall. It is possible that the gastric mucosa produces a protective antienzyme to inactivate pepsin, but no such substance has been isolated. It is well known that after death the digestive enzymes do attack the stomach.

The act of vomiting forces the stomach contents upward through the esophagus and mouth. Vomiting is a reflex act controlled from a vomiting center in the medulla. Stimuli of many different kinds may cause vomiting. Among the better-known stimuli are disturbances of the stomach and intestine, stimulation or irritation of the back of the mouth or pharynx, and disturbances of the semicircular canals, as in motion sickness. The act is preceded by nausea; the gastroesophageal sphincter and esophagus are relaxed, while the diaphragm and muscles of the abdomen supply the propelling force. Physical signs of nausea are pallor, perspiration, a feeling of weakness, and excessive salivation.

THE DUODENUM AND ASSIMILATION

Partially digested food leaving the stomach enters the upper part of the small intestine, or *duodenum*. The small intestine extends from the pylorus to the ileocecal valve, where it enters the large intestine. The duodenum is the widest portion, extending to the *jejunum*, or middle portion. The *ileum* is the terminal part. There is no strict morphological demarcation among these different parts; a gradual change in size and other morphological characteristics are the only indications of difference. The length of the small intestine in the human adult is commonly given as 7 m, but there are indications that the living intestine is held in a state of contraction and is therefore less than half this length. The intestine is sometimes divided into two parts: the duodenum and the mesenteric intestine. The duodenum is not supported by the mesentery and only partially covered by peritoneum. The mesenteric intestine, which is covered by mesentery, comprises the jejunum and ileum.

The duodenum is said to have derived its name from the practice of estimating its length by the width of 12 fingers, which is about 25 cm. This shortest and widest part of the small intestine makes a loop downward a little beyond the pylorus; the distal end of the loop rises almost as high as the proximal end. The greater part of the pancreas lies in the loop of the duodenum. The pancreatic and bile ducts enter the duodenum about 10 cm beyond the pylorus. The fluid within the duodenum, unlike that in the stomach, is alkaline. Just beyond the pylorus are numerous mucus-secreting glands called *Brunner's glands*. The mucus secreted by these glands protects the duodenum from the strongly acidic gastric juice emerging from the stomach.

PANCREAS

The pancreas is a slender gland extending from the loop of the duodenum upward behind the stomach to a length of 12 to 15 cm. It is a compound gland, somewhat pink in color, and similar in many respects to the salivary glands. We have already discussed its hormonal function in Chap. 13. Digestive enzymes secreted by the pancreas enter the duodenum through the pancreatic duct. The main pancreatic duct extends the length of the gland, receiving many tributary branches. Near its base it is commonly joined by a small accessory pancreatic duct. Close to the duodenum, the pancreatic duct lies beside the common bile duct, the two joining and opening into the duodenum by a common orifice, the ampulla (of Vater) (Figs. 19.11 and 19.12).

The pancreas secretes three major enzymes, one for each of the three kinds of food. *Trypsin* (or *tryptase*) digests proteins; *amylase* acts upon starches; and *lipase* acts upon fats. Other proteolytic enzymes such as *chymotrypsin,* which is similar to trypsin, are also secreted. The *carboxy-peptidases,* which act upon reduced proteins called peptides, are still another group. All enzymes exhibit a high degree of specificity for a certain type of substrate upon which they act; that is, protein-digesting enzymes do not digest carbohydrates or fats. Some enzymes are limited to acting only on certain stages in the breakdown of a molecule.

Control over the stimulation of pancreatic fluid is of two types: nervous and hormonal. Stimulation of the vagus nerve causes the pancreas to secrete a fluid rich in digestive enzymes. Experimental stimulation of the sympathetic splanchnic nerves, has produced varying results. Some investigators have obtained a fluid similar to that seen upon stimulation of the vagus (parasympathetic) nerve. Stimulation of the sympathetic nerves brings about vasomotor changes that complicate interpretation of the results. Hormonal control of pancreatic secretion is exercised largely by the hormones secretin and cholecystokinin.

PANCREATIC ENZYMES *Trypsin* This protein-digesting enzyme is secreted by pancreatic glandular cells in an inactive form called *trypsinogen.* When collected from the pancreatic duct, trypsinogen has little ability to digest proteins. If trypsinogen is mixed with intestinal juice, however, it becomes active trypsin. It is activated by the coenzyme *enterokinase,* which is produced by cells of the intestinal mucosa. Trypsinogen also may be activated by calcium. Trypsin digests proteins in an alkaline medium, and its optimum pH is near 8.

Trypsin acts upon the proteoses and peptones produced by gastric digestion, breaking them down to peptides. Proteins that have not been

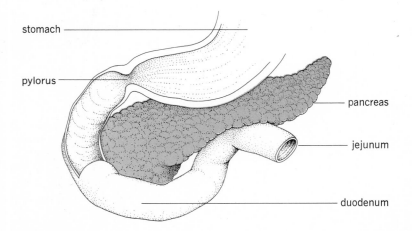

FIGURE 19·11

A portion of the small intestine: the duodenum and the proximal part of the jejunum.

stomach

pylorus

pancreas

jejunum

duodenum

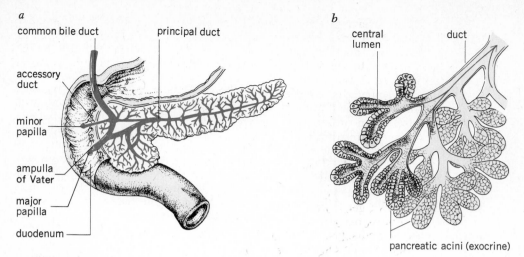

a

common bile duct

principal duct

accessory duct

minor papilla

ampulla of Vater

major papilla

duodenum

b

central lumen

duct

pancreatic acini (exocrine)

FIGURE 19·12

The pancreas: *a* system of ducts; *b* the exocrine portion (these are the acinar cells that produce the pancreatic digestive enzymes with an output of about 1200 ml/day).

acted upon by pepsin in the stomach are also broken down. The degree of digestion produced by trypsin is determined by optimum conditions and the length of time the material spends in the intestine. Apparently trypsin does not ordinarily complete the breakdown of proteins to amino acids. The final digestion is accomplished by the peptidases secreted by the small intestine.

Amylase The starch-digesting enzyme of the pancreas, amylase, is similar to salivary amylase but somewhat more active. In an alkaline-to-neutral medium, amylase hydrolyzes starch to various dextrins and to maltose. In the light of recent work on salivary amylase, it is possible that glucose may also be produced by the action of amylase on starch.

Lipase Only one enzyme is involved in the digestion of fats. Lipase, the lipolytic enzyme, splits fat molecules into fatty acids and glycerol. Its action is aided by bile salts from the liver. Bile salts coat fat particles with a detergent action that reduces their surface tension; this process is called *emulsification*. Emulsified fats present a greater surface area for the action of the digestive enzyme. Bile salts also help in the absorption of fatty acids. The fatty acids involved are organic acids that are highly insoluble in water, although they are soluble in bile-salt solutions. Emulsification of fats with high melting points is difficult, and such fats are not easily digested. Fats with low melting points appear to be more easily digested. Glycerol, a by-product of fat breakdown, does not present any special digestive problems, since it is readily soluble in body fluids.

Lactase The enzyme that hydrolyzes lactose, the sugar in milk, is lactase. Some individuals are unable to digest milk; they may have abdominal cramps and diarrhea after meals containing milk. This condition is especially difficult to manage in infants whose diet largely consists of milk. Cow's milk contains about 4.8 percent lactose, while human milk contains about 6 percent. The inability to digest lactose is usually an inherited condition in which the individual fails to synthesize the enzyme lactase.

Most children have lactase, but many lose it as adults. Lactase deficiency is more common in certain racial groups. Whereas it has been estimated that about 5 percent of adult white Americans are affected, in American blacks and Orientals the incidence of lactase deficiency may reach 60 to 90 percent.

HORMONAL CONTROL OF PANCREATIC SECRETION   Hormonal control over pancreatic secretion involves the hormone *secretin*. When the acid contents of the stomach enter the normally alkaline contents of the duodenum, secretin is released from the duodenal mucous lining. This hormone is absorbed by the blood and carried to the pancreas, where it causes this organ to secrete a fluid that is largely water and bicarbonate. The secretion is thin and low in enzyme content, but it is more alkaline than that obtained by stimulation of the vagus nerve. The hormonal mechanism stimulates the pancreas to secrete at the time food is present to be acted upon. The discovery of secretin goes back to the work of Bayliss and Starling in 1902. When bile is introduced into the duodenum, secretin is absorbed and the pancreas is stimulated to secrete. There is now evidence that another hormone is involved in regulating pancreatic enzyme secretion. This hormone, called *cholecystokinin,* stimulates the flow of digestive enzymes into the pancreatic fluid.

LIVER

The liver is the largest glandular structure in the body (Fig. 19.13). It is located just under the diaphragm in the upper right portion of the abdominal cavity. Somewhat wedge-shaped, the thicker portion is on the right, with the thin part of the wedge to the left, lying over the stomach. It is also in contact with the right kidney, the duodenum, and the right colon at its flexure. The surface is largely covered by peritoneum. The color is dark reddish-brown. The gland is a soft tissue, easily torn. It has an abundant blood supply, receiving around 1500 ml blood/min. The liver is proportionately very large in the embryo and fetus.

FIGURE 19·13
Liver and gallbladder (anterior view).

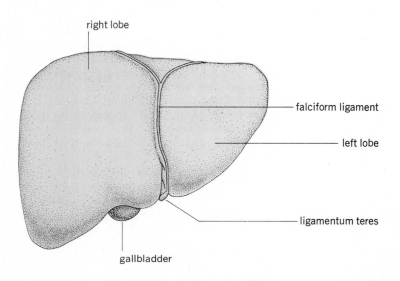

right lobe

falciform ligament

left lobe

ligamentum teres

gallbladder

The liver is an organ of many complex functions.

1 It secretes bile.
2 It plays a role in glycogenesis.
3 It plays an important part in protein and fat metabolism.
4 It detoxifies many harmful metabolites.
5 It produces serum albumin, serum globulin, fibrinogen, and heparin.

This list is by no means a complete summary of its functions, however.

HISTOLOGICAL STRUCTURE The liver is arranged in *lobules* 1 to 2 mm in diameter. The lobule is considered to be the unit of gross structure. It consists of cords of cells that radiate from a center. The columns of cells anastomose freely and are quite irregular. Between the rows of cells is a more or less open capillary network, the *liver sinusoids*. The lobules are held together by interlobular connective tissue, which also supports the blood vessels, lymphatics, and nerves. A serous membrane, derived from the peritoneum, and a thin fibrous coat lie over the entire structure.

The description given here applies to a typical section of a liver lobule (Fig. 19.14). Actually, this description is too simplified and stylized. The lobules, especially in the human being, are not well defined, and the cords of cells do not necessarily lie in a flat plane. A more realistic description would show the microscopic anatomy as somewhat resembling the three-dimensional architecture of a sponge. However, the simplified drawing is useful for teaching purposes.

The liver is supplied from the portal vein and the hepatic artery. Within the liver the portal vein and the much smaller hepatic artery give off numerous branches. Their terminal branches are the interlobular veins and

FIGURE 19·14

A portion of a liver lobule: detailed structure.

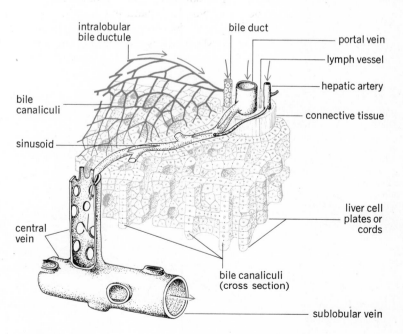

intralobular bile ductule

bile duct

portal vein

lymph vessel

hepatic artery

connective tissue

bile canaliculi

sinusoid

liver cell plates or cords

central vein

bile canaliculi (cross section)

sublobular vein

interlobular arteries, respectively, which form a network of blood vessels around the lobules. From these peripheral blood vessels, the blood filters through the liver sinusoids, which are capillarylike vessels with incomplete walls. The sinusoids permit the blood to flow toward the center of the lobule and into the central veins. From the central veins, the blood flows into the sublobular veins and so into the hepatic veins. The hepatic veins empty into the inferior vena cava.

Located in the sinusoids are the large, stellate Kupffer cells. These cells are phagocytic macrophages belonging to the reticuloendothelial system. They ingest bacteria or other foreign material in the blood. Kupffer cells play an important part in filtering the blood that flows through the hepatic sinusoids. It has been demonstrated that the macrophages can store India-ink carbon particles until these cells are practically black in appearance.

GALLBLADDER AND BILE DUCTS The gallbladder is a large, conspicuous green sac located on the undersurface of the liver, with its distal extremity close to the anterior border. It commonly is 7 to 10 cm long and 2 to 2.5 cm wide and holds about 30 ml of bile. The gallbladder may be regarded as an enlargement of the *cystic duct,* which drains it. The cystic duct and the neck of the gallbladder have a peculiar twisted appearance because of the attachment of a spiral valve within (Fig. 19.15).

The *hepatic duct* arises in the liver. Interlobular bile ducts drain into larger ducts and finally give rise to two large tributaries, one from the right portion of the liver and one from the left. These large ducts unite to form the hepatic duct. The bile descends from the liver and through the hepatic duct. It may either pass up the cystic duct to the gallbladder or pass down the common bile duct.

The *common bile duct* arises from the union of the hepatic and cystic ducts. The duct is about 7.5 cm long; it passes downward and enters the duodenum from the left, along with the pancreatic duct, 7 to 10 cm below the pylorus. The lower portion of the duct is embedded in pancreatic tissue, and in a majority of individuals it unites with the pancreatic duct at its base. The short, dilated, common passageway is known as the *ampulla (of Vater).*

FIGURE 19·15

The gallbladder and hepatic ducts. The common bile duct and the principal pancreatic duct form the ampulla (of Vater). The combined products of the liver and pancreas enter the duodenum through the major papilla, located just below the ampulla of Vater.

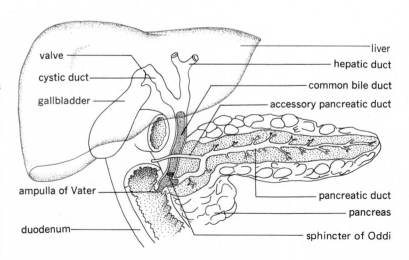

valve

cystic duct

gallbladder

ampulla of Vater

duodenum

liver

hepatic duct

common bile duct

accessory pancreatic duct

pancreatic duct

pancreas

sphincter of Oddi

A sphincter muscle at the base of the bile duct and in the ampulla closes the duct. Closure of the sphincter causes the bile to back up into the gallbladder.

Control of gallbladder emptying The smooth muscle of the gallbladder is innervated by a branch of the vagus nerve as well as by axons of the sympathetic system. Though some nervous control exists, hormonal control appears to be of greater importance. The gallbladder can be caused to empty experimentally by introducing fat or fat-digestion products into the intestine, even though all nervous connections have been severed. *Cholecystokinin,* the hormone involved in this reaction, is secreted by the intestine. Injections of cholecystokinin cause the gallbladder to contract. Normally, as the contents of the stomach enter the duodenum, fats and fat-digestion products stimulate the formation of cholecystokinin. The hormone is absorbed by the blood and carried to the gallbladder, where it causes that structure to contract.

BILE The liver secretes a yellowish fluid called *bile.* It is an alkaline, watery fluid containing bile pigments, bile salts, cholesterol, mucin, and other constituents, both inorganic and organic. Bile is not as concentrated as it leaves the liver as it is after it has been held in the gallbladder, indicating that the gallbladder absorbs a considerable amount of water and some salts while storing the bile. The inner lining of the gallbladder also secretes mucus into the bile.

Bile pigments *Bile pigments* are derived from the breakdown of hemoglobin in old red cells. The disintegration of red blood cells, which is a continuous process in the blood, releases hemoglobin, which is broken down in the spleen, liver, and connective tissues, where it is converted into *bilirubin,* the principal pigment in human bile. *Biliverdin,* another pigment, is formed as an oxidation product of bilirubin. These pigments are carried by the blood to the liver, where they become bile pigments. Bilirubin is an orange color, and biliverdin is green. A reduction product of bilirubin in the intestine is urobilinogen, which causes the brown color of fecal material. Most bile pigment is excreted with the feces, but a small amount is absorbed through the capillary network in the lining of the intestine. This pigment is returned to the liver by way of the portal system to be used again.

Bile salts The *bile salts* sodium glycocholate and sodium taurocholate play an important part in the digestion of fats. These salts are formed in the liver by the chemical combination of cholic acid with the amino acid glycine in one case and with the amino acid taurine in the other. Bile contains no enzymes, but fat digestion by pancreatic lipase improves when bile is present. Although fat digestion and absorption are poorly understood, it appears that bile salts act as detergents to emulsify fats. The fat is broken down into small globules, each with decreased surface tension. Such globules provide a much greater surface area for the digestive action of lipase. Bile salts also play an important role in the absorption of glycerol and the insoluble fatty acids. The bile salts, too, are absorbed and used over again by the liver.

Cholesterol *Cholesterol,* a substance present in tissues and fluids of the body, is a sterol, or a type of alcohol. It is related chemically to vitamin D, to the sex hormones, and to hormones of the adrenal cortex. Cholesterol is fairly soluble in bile, though it is highly insoluble in water. It is usually considered to be an excretory product in bile. The concretions that sometimes form in the gallbladder (gallstones) are largely composed of cholesterol. Small stones passing down the bile ducts often cause severe pain. Larger stones can block the flow of bile through the ducts and cause obstructive jaundice. The function of the gallbladder is such that it can be removed, if necessary, without serious consequences. Obviously, bile can no longer be stored if the gallbladder is removed, but bile continues to flow down the hepatic and common bile ducts. The digestion of fats is not as efficient after removal of the gallbladder because the same quantity of bile cannot be released at just the time that fats enter the duodenum. By reducing the amount of fat in the diet, however, many persons are able to lead normal lives after removal of the gallbladder.

GLYCOGENIC AND GLYCOGENOLYTIC FUNCTIONS OF THE LIVER Absorbed glucose is carried by the portal system to the liver, where it is stored in the form of glycogen (animal starch). The process of converting glucose to glycogen is called glycogenesis, a chemical pathway we described in Chap. 3. As the tissues use glucose in their metabolism, the liver converts some of its stored glycogen back to glucose, thereby helping to maintain a constant blood sugar level. The process of breaking down glycogen to glucose is called *glycogenolysis,* a pathway also described in Chap. 3.

Glycogenesis The blood carries a small amount of glucose, and the liver contains some, but there is no great amount of glucose stored in the body. Glucose is converted to glycogen. In this form, it is stored largely in the liver, but a small amount is stored in skeletal muscle.

Glycogenolysis As the body requires more glucose, glycogen is broken down step by step by action of the enzyme *phosphorylase,* found predominantly in the liver and in muscle. Phosphorylase must be activated by the hormones epinephrine and glucagon before it is effective. When the blood glucose level falls below normal, these hormones activate the enzyme *adenyl cyclase,* which is present in liver cells. This enzyme catalyzes the conversion of ATP to cyclic AMP, and cyclic AMP activates liver phosphorylase.

The liver can also form glucose from certain amino acids derived from food proteins. During starvation, glucose is derived from protein in the tissues themselves. It can also be obtained from glycerol, which is split off from the fat molecules during digestion. The ability of the liver to form glucose from noncarbohydrate sources is called *gluconeogenesis.*

The liver plays an important role in maintaining a normal blood glucose level. This level is around 0.1 percent, or 80 to 120 mg/100 ml, but it varies considerably with various bodily conditions. Any great reduction in blood glucose level (hypoglycemia), as in the case of insulin shock, causes serious disturbances that may lead to convulsions and coma. A rise in blood glucose level (hyperglycemia), as is seen in uncontrolled diabetes mellitus, causes no discomfort. If blood glucose level rises above a certain level (the renal threshold for glucose), the kidneys excrete the excess glucose in the urine (see Chap. 13).

OTHER FUNCTIONS OF THE LIVER *Deamination of amino acids* Proteins are not stored in the body as such, but amino acids are found in the blood, liver, and in tissues generally. In the liver, amino acids undergo deamination, in which the amino group, —NH_2, is split off from the amino acid molecule. The enzyme *deaminase* catalyzes the reaction. The amino group is used to form urea, $CO(NH_2)_2$, which is mainly produced in the liver and excreted by the kidneys. Some deaminated amino acids are converted into glucose. Glucose, as we have seen, can be utilized or stored in the liver as glycogen. Deamination of glycogenic amino acids can also be a step in the formation of fat. The liver breaks down the waste products of protein metabolism into substances that are nontoxic and that may be excreted by the kidneys.

Fat metabolism The liver plays an important role in the metabolism of fats (see Chap. 3). Among the enzymes concerned is *acetyl coenzyme A,* which functions not only in the breakdown of fatty acid chains but also in the synthesis of lipids. The necessary hepatic enzymes and coenzymes are located in the mitochondria of the liver.

Fats, of course, are a good source of energy. When they are completely oxidized, they produce approximately 9 kcal/g as compared with 4 kcal/g for carbohydrates or proteins.

After eating a meal containing a large amount of fat, the amount of fat in the liver increases as a result of the steady flow of absorbed fat to this organ. In the process of breaking down fats, some intermediate products of fatty acid metabolism are formed. These substances are called *ketone bodies;* they are composed of acetoacetic acid and betahydroxybutyric acid. Acetone is a breakdown product of either of these acids. Small amounts of ketone bodies are normal breakdown products of fatty acids, and they are oxidized in cellular metabolism by way of the tricarboxylic acid cycle.

Ketosis occurs when ketone bodies are produced in large quantities by the liver. This state occurs especially in conditions of diabetes and starvation.

Ordinarily in the process of metabolism, energy is derived from the three classes of foodstuffs—carbohydrates, proteins, and fats—and fats are not metabolized at the expense of carbohydrates and proteins. In diabetes, carbohydrate metabolism cannot proceed normally because glucose is not utilized. The body therefore utilizes more fat and proteins to meet its energy requirements. Excess ketone bodies are then formed from fats and some of the amino acids. Body tissues cannot oxidize all the excess ketone bodies formed by the liver, and as they accumulate in the blood, they are excreted by the kidneys. Since acetone is a breakdown product, this volatile substance may give the urine, or even the breath, a characteristic odor, which is often noticed in uncontrolled diabetes.

During starvation, stores of carbohydrates are soon depleted, and the body's metabolic needs are met by utilizing fats and proteins for energy. As in diabetes, ketosis results from inability of the tissues to utilize all the ketone bodies in excess of their normal metabolic needs. Their accumulation causes a rise in the acidity of the blood by depleting the alkaline reserve.

The protective function of the liver is associated with its ability to detoxify products of catabolism, which might accumulate in dangerous proportions. These products are changed chemically into substances that can be excreted by the kidneys or through the intestinal tract. We have noted that the macrophages found in liver sinusoids help to filter foreign matter from the blood.

The blood proteins serum albumin and serum globulin are formed in the liver. Some of the substances involved in blood clotting, such as fibrinogen and heparin, are formed there too. Liver tissue also contains an antianemia factor. The discovery of this factor gave rise to the practice of using liver and liver extracts to treat anemia. The antianemia factor in liver has been determined to be the same as the extrinsic factor found in food, or vitamin B_{12} (see section on vitamins in Chap. 20 and see Table 20.1). The functions of the liver are summarized in Table 19.2.

TABLE 19·2
A brief summary of liver functions

1 *Intermediary metabolism*

Glycogenesis, conversion of glucose to glycogen
Glycogenolysis, the process of breaking down glycogen to glucose
Gluconeogenesis, formation of glucose from noncarbohydrate sources
Deamination of amino and nucleic acids, removal of an amino group by a deaminase (produces a portion of body heat)
Lipogenesis, formation of fats from glucose, glycogen, or amino acids (enzymes present are in liver mitochondria)
Lipolysis, breakdown of triglycerides into fatty acids and glycerol (acetyl coenzyme A), formation of ketone bodies
Cholesterol metabolism, synthesis, and breakdown, from ingested fats

2 *Storage*

Glucose, stored as glycogen (release of blood sugar involves cyclic AMP)
Bile, stored in gallbladder
Vitamins: fat-soluble, A, D, E, K; water-soluble, thiamine, riboflavin, folic acid, and B_{12}
Fats and fatty acids, lipids such as phospholipids and cholesterol
Amino acids, in proteins such as albumin, beta globulin, and in liver protoplasm
Water, contributes to blood volume
Minerals: iron and copper; most of the iron is stored as ferritin (iron combined with protein)
Transfer of blood from hepatic portal system to systemic circulation

3 *Secretion and synthesis*

Secretion of bile salts into bile
Formation of blood-clotting factors, prothrombin, fibrinogen
Blood proteins, serum albumin, serum globulin
Hemopoietin, antianemic substance
Elaborates heparin, an agent which acts in the prevention of blood clotting
Erythrocytes, produced only in the embryo or before birth

4 *Excretory functions*

Breakdown of hemoglobin in old erythrocytes to form bilirubin and biliverdin
Ingestion of bacteria, old red cells, or other particles by reticuloendothelial cells (Kupffer cells)
Detoxification of harmful acids and drugs and excretion of their breakdown products
Breakdown of sterols such as sex hormones
Formation of urea from amino acids, and uric acid from nucleic acids

JEJUNUM AND ILEUM

The small intestine is the area where most of the absorption of food materials occurs. The duodenum, which we have already discussed, is the anterior portion of the small intestine. The *jejunum* forms the intermediate portion of the small intestine, and the *ileum*, the distal part. There is little actual difference in structure between these parts, and there is no strict dividing line between the jejunum and the ileum, which terminates at the *colic*, or *ileocecal*, *valve*. The ileum enters the side of the colon (or large intestine); the aperture by which it enters is guarded by a sphincter muscle, which forms the ileocecal valve.

Like the stomach, the small intestine is covered by a serous membrane and lined by a mucous membrane. The muscular wall is composed of a thin outer layer of longitudinal fibers and a thicker inner layer of circular muscle. Rhythmic muscular waves pass over the stomach and intestine, helping to propel the food along. This action is called *peristalsis* (Fig. 19.16).

CONTROL OF PERISTALSIS   Peristaltic movement of the intestine is largely independent of extrinsic nerve connections. Although stimulation of the vagus increases peristalsis, and stimulation of the sympathetic innervation tends to inhibit peristaltic waves, with all extrinsic innervation severed, peristaltic waves still continue to pass along the intestine, though in somewhat altered form. A network of neurons within the intestinal wall, known as the *enteric system*, is probably responsible for the independent muscular activity of the intestine; this autonomic innervation has a largely regulatory function.

A layer of areolar connective tissue called the submucous layer lines the muscular layers. This layer is covered with a mucous layer, which lines the intestine. The submucous and mucous layers exhibit circular folds, which project into the cavity of the intestine. These folds increase the absorptive area, and they also retain the food and prevent it from passing through too rapidly.

INTESTINAL VILLI   The lining of the small intestine is filled with fine processes projecting into the lumen. These processes of the mucosa, called *villi*, increase the absorptive area of the intestine very greatly. They function in the absorption of food. Villi are freely movable. The fact that they wave back and forth probably makes them more efficient in absorption (Fig. 19.17).

Within a villus lies a capillary network of the terminal branches of the mesenteric arteries and veins. A central lymph vessel called a *lacteal* is also present. Carbohydrates, proteins, and some fats are absorbed through the capillary network of the villus. The greater part of fat absorption is through the lacteal, or lymph channel (Fig. 19.17*b*).

FIGURE 19·16
Diagram illustrating peristalsis in the intestine.

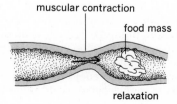

muscular contraction

food mass

relaxation

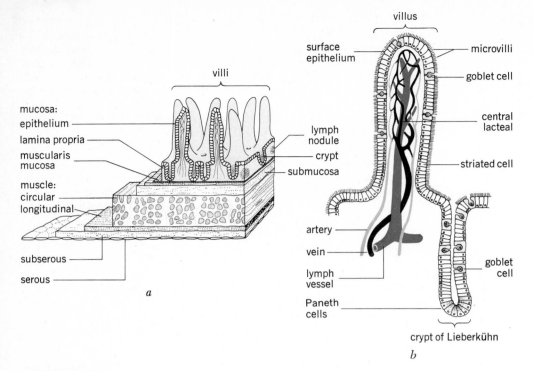

villi

mucosa:
epithelium
lamina propria
muscularis
mucosa

muscle:
circular
longitudinal

subserous

serous

a

lymph
nodule

crypt

submucosa

villus

surface
epithelium

artery

vein

lymph
vessel

Paneth
cells

microvilli

goblet cell

central
lacteal

striated cell

goblet
cell

crypt of Lieberkühn

b

FIGURE 19·17

Detailed structure of the small intestine: *a* layers of the intestinal wall and villi; *b* longitudinal section of a villus and a crypt of Lieberkühn. Paneth cells contain large eosinophilic granules in their cytoplasm. Their function is not well understood, but they may be concerned in the manufacture of digestive enzymes and their secretion.

Columnar epithelial cells, including goblet cells, cover the surface of the villi. The absorptive surface is further augmented by microvilli that constitute the brush borders (see Figs. 4.4 and 4.5). The cells at the bases of the villi have a high rate of mitosis. They appear to divide and start moving up the sides of the villi, while the older cells are sloughed off at the top continuously. At the bases of villi are numerous tubular glands called the *crypts of Lieberkühn,* which secrete a portion of the intestinal fluid. It was thought formerly that the disaccharidases and proteases found in the intestinal fluid were secreted as the digestive part of the fluid. It appears now that the only enzymes found in pure intestinal fluid (succus entericus), uncontaminated by cell fragments, are *enterokinase* and a small amount of *amylase.* The other enzymes in intestinal fluid come from broken-down epithelial cells cast off from the surface of the villi and from brush-border cells. Functional disaccharidases have been shown to be present in the brush borders of the columnar epithelium of the villi. Mucus-secreting goblet cells appear to be merocrine rather than apocrine in function. *Paneth cells* are thought to function in the synthesis of digestive enzymes.

The intestinal fluid produced in quantity by the crypts of Lieberkühn is thought to circulate constantly until it is absorbed by the villi. This fluid holds food substances in suspension in order to enhance enzymatic digestion. The lining of the small intestine should not be considered as an undifferentiated membrane, since differences in secretory and absorptive abilities of the crypts and villi in the jejunum and ileum are now recognized.

The hormone *enterocrinin,* isolated from extracts of the intestine, stimu-

TABLE 19·3
Hormones of the digestive tract

Hormone	Origin	Action
Gastrin	Mucosa of pyloric antrum	Causes an increase in the secretion and acidity of gastric juice
Enterogastrone	Mucosa of duodenum and jejunum	Inhibits gastric motility and secretion
Secretin	Duodenal mucosa	Stimulates pancreatic secretion which is weak and watery and low in enzyme content. High concentration of bicarbonate. Stimulates the volume of bile secretion by the liver
Cholecystokinin	Duodenal mucosa	Causes the pancreas to secrete pancreatic juice containing enzymes. Stimulates contraction of gallbladder
Enterocrinin	Mucosa of small intestine	Controls the secretion of the intestinal juice
Villikinin	Mucosa of small intestine	Stimulates movement of villi

lates the small intestine to produce succus entericus in greater volume (Table 19.3).

DIGESTION

Food as taken into the mouth is not immediately available for use by the tissues. For the most part, the molecules of undigested foods are too large and complex to be absorbed from the intestine. From a functional point of view, the food materials in the digestive tube are not considered really to enter the body until they have been absorbed. The process of *digestion* reduces food materials to a condition suitable for absorption. Large complex molecules are reduced to smaller and simpler molecules by the action of digestive enzymes at various places in the digestive tract. Finally these end products of digestion are absorbed through the mucosa of the intestine and into the bloodstream to make them available to the tissues.

The process of absorption is more complicated than can be explained by a simple diffusion of molecules through the mucosa. The absorption rates of various simple monosaccharide sugars have been shown to vary considerably. Glucose and galactose are absorbed readily, but fructose and mannose are absorbed slowly. If the process of absorption through a membrane were the only problem involved, all monosaccharide sugars might be expected to be absorbed at equal rates. Inverting enzymes break down sugars to monosaccharides before they can be absorbed in the intestine. Specific transport proteins are found for all hexose sugars, amino acids, and some polypeptides. This method of absorption is called *sodium-dependent secondary active transport,* a process we discussed in Chap. 18.

The large protein molecules must be converted to amino acids by digestion before they can be absorbed. Proteins, as they are found in food substances, are not water-soluble. Even the breakdown products such as proteoses and peptones are not absorbed. Amino acids, however, are readily absorbed, after which they become the building blocks by which new protein materials can be constructed.

Fats are broken down into glycerol and fatty acids by the action of lipase and bile salts before being absorbed. Glycerol is water-soluble, but fatty acids are highly insoluble. Bile salts are largely responsible for splitting free fatty acids into minute water-soluble droplets known as *micelles*. Micelles enter the membrane of the villi. In passing through the epithelial cells of the villi, fatty acids are resynthesized to triacylglycerols; they enter the bloodstream in this form. Triacylglycerols also enter the lacteals of the villi as minute fat globules called *chylomicrons*. Absorbed fat in the form of chylomicrons gives the lymph a milky appearance.

LARGE INTESTINE

The ileum enters the *large intestine* laterally. The distal end of the ileum is guarded by circular muscles, and its entrance is protected by the ileocecal valve. Two horizontal folds of the valve project into the large intestine. The valve prevents fecal material from backing up into the ileum; probably it also controls the rate of flow from the ileum into the large intestine, preventing the ileum from discharging its liquid contents too rapidly (Fig. 19.18).

The large intestine extends from the ileum to the anus. In succeeding areas it is called the *cecum, ascending colon, transverse colon, descending colon, sigmoid colon, rectum,* and *anal canal.* The various parts of the colon form a sort of frame around the abdominal cavity—the ascending colon on the right, the transverse colon along the top, and the descending colon on the left.

The large intestine has a much greater diameter than the small intestine,

FIGURE 19·18

The large intestine (anterior view); cut away to show interior of cecum.

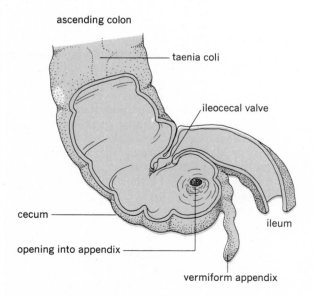

but it is only about 1.5 m long. It is not greatly convoluted, and in some respects its structure is quite different from that of the small intestine. The longitudinal muscles of the large intestine are gathered, for example, into three bands of fibers, the *taenia coli*. Contractions of the longitudinal muscles cause the colon to pucker into a series of pouches called *haustra*. The large intestine has no villi.

The *cecum* is a blind saclike portion below the entrance of the ileum. It is located in the lower right area of the abdomen just above the right iliopsoas muscle. Occasionally it becomes distended with accumulated waste material. A small prolongation of the cecum is called the *vermiform appendix*.

FUNCTIONS OF THE LARGE INTESTINE The material that enters the large intestine rapidly assumes the characteristics of waste matter. One of the principal functions of the colon is to absorb water. The removal of water reduces the liquid intestinal contents that enter the large intestine to the more solid consistency of the *fecal matter* that exits from the distal part of this organ. The large intestine secretes mucus in considerable quantity, which tends to hold the fecal particles together in a solid mass. Mucus also protects the lining of the intestine and acts as a lubricant.

The contents of the colon are made up not only of food residues but also of materials excreted from the blood and the digestive glands. The waste materials are subject to bacterial decomposition, and the bacteria themselves make up a small percentage of the waste product.

The large intestine is not involved in digestion. Some digestive enzymes are found in the contents of the large intestine but they probably enter with food materials from the ileum. The digestion and absorption of food in the small intestine are remarkably efficient, since most of the available food products are removed before they enter the large intestine. Plant foods contain cellulose, which is not digested in human beings and therefore contributes to the bulk of the fecal matter.

Bacterial decomposition produces a number of gases, acids, and amines characteristic of fecal matter. Gases include carbon dioxide, ammonia, hydrogen sulfide, hydrogen, and methane. Acetic, lactic, and butyric acids are commonly present; however, the secretion of bicarbonate into the large intestine prevents the feces from becoming acid. Amines such as *indole* and *skatole* are breakdown products of amino acids, which contribute to the odor of the feces.

Some food materials in the diet affect the color of the fecal matter. Plant pigments, such as those derived from beets and spinach, are examples of such foods.

While bacterial decomposition produces toxic substances, it is thought that such substances, under normal conditions, are detoxified in the liver and excreted by the kidneys. It has also been shown that intestinal organisms can synthesize some vitamins of the B complex and vitamin K. This function is regarded as a significant contribution to the nutrition of the individual.

DEFECATION REFLEX The rectum is usually empty until just before *defecation,* when the feces enter this region of the intestine. Unlike the segmental churning of the small intestine, the fecal material in the *ascending* and *transverse colon* is propelled by *mass movements.* Large segments of these colon regions contract simultaneously, moving a considerable quantity of the contents along at one time. The defecation reflex is initiated when mass movements force fecal material into the lower sigmoid colon and the rectum, causing distension of these regions. This distension causes afferent signals to travel to the sacral spinal cord, where several effector responses are produced. Parasympathetic impulses traveling through sacral spinal nerves stimulate the sigmoid colon and rectum to undergo peristalsis. Simultaneously, the individual reflexly takes a deep breath, closes the glottis, and contracts the abdominal muscles. Distension of the sigmoid colon and rectum also stimulates the myenteric plexus in this region, producing peristaltic waves.

The anal canal is closed by an internal sphincter of involuntary smooth muscle and an external sphincter of voluntary skeletal muscle. During the defecation reflex, peristalsis causes *receptive relaxation* of the internal anal sphincter. If the individual voluntarily relaxes the external anal sphincter, elimination occurs. If the external sphincter is not opened, the peristaltic waves gradually dissipate, but they will begin again a short time later.

It is usually considered that regular bowel movements are conducive to good health. It is a good hygienic practice to have a regular time for elimination. Adequate amounts of fruits and vegetables in the diet help prevent constipation, as does drinking a reasonable amount of water. Daily exercise, adapted to the needs of the individual, also promotes regularity. There is no reason to believe that it is good practice to take laxative medicines regularly. Every effort should be made to secure proper elimination habits by training, proper diet, and good health habits.

CLINICAL ASPECTS

DENTAL CARIES AND PERIODONTAL DISEASES   Tooth decay is a common condition associated with eating sugar-laden soft foods that stick to the teeth and form *dental plaque*. The species of bacteria most involved in tooth decay is *Lactobacillus acidophilus*. Acids produced by the bacteria tend to dissolve gradually the protective covering of enamel, and a *cavity*, or *caries*, is formed in the outer layer of the tooth. If the cavity invades the sensitive dentine, the person will probably develop a toothache. Dentists drill out the decayed portion of the tooth and pack the cavity with a porcelain composition or metal filling.

Good care of the teeth involves brushing them after meals and using dental floss to clean between the teeth down to the gum line. Fluoride, which is added to drinking water in some areas or which may be applied directly to the tooth surface by a dentist, appears to make tooth enamel more resistant to caries. Fluoride treatment is most beneficial when applied to very young children in whom the tooth enamel is forming.

Periodontal diseases involve inflammation of the gums and a breakdown of the alveolar bone that surrounds the base of the teeth and holds them firmly in place. There are many causes of periodontal disorders. Among them are the destructive action of bacteria, the presence of impacted food particles, poor nutrition, lack of vitamins, and certain systemic diseases. Periodontal disease is the most serious adult dental problem. *Gingivitis* is an inflammation of the gums.

HEARTBURN   Many people experience a burning sensation in the throat or esophagus from time to time. *Heartburn* is usually a temporary condition caused by regurgitation of acid from the stomach into the esophagus; it is not a serious condition.

Hyperacidity of the stomach is usually a temporary condition produced by overeating or perhaps by eating under unfavorable conditions. Constant hyperacidity may indicate a more serious condition. *Gastric ulcers,* sore spots in the stomach mucosa, which resemble canker sores in the mouth, are commonly associated with long-standing hyperacidity. Normally, the mucous lining of the stomach protects the deeper layers of the stomach wall from the action of pepsin and HCl. The open ulcers lack this protection and are therefore irritated by these secretions.

PEPTIC ULCER Ulcers are unprotected, depressed areas in membrane tissue. If ulcers occur in the stomach, they are called *gastric ulcers;* if they are located in the duodenum, they are termed *duodenal ulcers.* Duodenal ulcers are usually located close to the pylorus, where the acid food from the stomach enters the normally alkaline contents of the intestine. The exact cause of peptic ulceration is not known, but certain conditions are present in most cases. These are: (1) hypersecretion of gastric juice, pepsin, and HCl; (2) emotional tension or stress; and (3) possible faulty secretion of protective mucus. Highly seasoned foods, alcohol, caffeine, and aspirin have been implicated in producing hypersecretion of acid. Drugs that tend to stimulate the vagus nerves may also be a factor.

The ulcer, if uncontrolled, may deepen and attack the muscles and blood vessels of the digestive tract, causing bleeding. The ulcer may even perforate the wall of the stomach or intestine. *Peritonitis* may follow perforation if the contents of the perforated organ are permitted to enter the body cavity.

LIVER DISEASES *Hepatitis* Hepatitis is a viral inflammation of the liver. There are at least two kinds of acute viral hepatitis. Hepatitis A is called *infectious hepatitis,* and hepatitis B is known as *serum hepatitis.* The two viruses can be distinguished by their antigenic characteristics, but clinically their effects are much the same. Infectious hepatitis is caused by ingestion of food or water containing the hepatitis A virus. It is spread when unsanitary conditions exist. Shellfish taken from water polluted by untreated sewage is a likely source of infection. Serum hepatitis is usually caused by blood transfusions of infected blood. Drug addicts who use contaminated needles to inject themselves may also be infected. A typical hepatitis infection causes enlargement of the liver and jaundice.

Cirrhosis of the liver Cirrhosis is another type of inflammation of the liver. *Laennec's cirrhosis* is the most common type of cirrhosis. In this condition, functional liver cells are gradually replaced with fatty or fibrous (scar tissue) infiltration. The exact cause is not known, but it seems to be related to ingestion of large amounts of alcohol and to poor nutrition.

Biliary cirrhosis is usually caused by inflammation of the bile ducts or by gallstones that obstruct the bile ducts. Removal of the gallbladder and bile duct is commonly recommended. *Cholecystitis* is a term applied to inflammation of the gallbladder.

Jaundice The normal outlet for the excretion of bile is through the intestine. If the liver does not function properly, bile pigments may be absorbed by the blood in excessive amounts. In some cases even though the liver functions properly, the bile may be prevented from passing readily into the intestine by a stoppage of the bile ducts by gallstones.

The absorption of abnormal amounts of *bilirubin* by the blood causes the tissues to take up unusual amounts of this pigment. A yellow coloration is especially noticeable in the skin and white of the eye. *Jaundice* is the name given to such a condition. There are various types of jaundice, such as obstructive, infectious, and hemolytic jaundice. In *obstructive jaundice* the fecal matter is light-colored because the bile is blocked from its normal outlet through the intestine. Unusual amounts of bilirubin are usually excreted by the kidneys, and the urine is dark-colored.

Infectious jaundice usually indicates the presence of a toxic substance, which reduces the ability of the liver to excrete bile pigment. Bilirubin is therefore absorbed by the blood plasma in excessive amounts.

Hemolytic jaundice may occur when there is an abnormal breakdown of red blood cells because of the effects of such diseases as malaria or the presence of toxic substances. The release of unusual amounts of hemoglobin causes the bilirubin content of the blood to rise considerably above normal limits.

COLON AND RECTUM DISEASE *Diverticulosis* Outpocketings occasionally occur in the wall of the colon. Usually consisting of small bulges located in a weakened area, these *diverticula* are seldom single, and they usually appear like the imprint of a small bunch of grapes. Probably they are more common in older people who suffer from constipation, although this is not necessarily so. They are usually not treated unless inflammation is present. When diverticuli become inflamed, the condition is termed *diverticulitis*.

Polyps Benign tumors of the colon and rectum are called polyps. They are growths of various shapes and sizes that project into the lumen of the large intestine. They cause no symptoms when small but are usually removed surgically because they may become cancerous.

Diarrhea Diarrhea is a common disorder of the lower bowel that may have several causes. Aside from being quite unpleasant, it may be quite dangerous, especially in young infants. Diarrhea occurs when the mucosal lining of the ileum and colon secretes massive amounts of fluid, electrolytes, and mucus. This fluid causes distension of the colon, which stimulates mass movements and peristalsis. Diarrhea has two basic causes: (1) irritation of the mucosal lining, usually because of bacterial infection; and (2) emotional distress, which causes excessive parasympathetic stimulation of the bowel. Diarrhea may be dangerous if severe dehydration results from excessive loss of fluid and electrolytes.

Appendicitis The appendix, or *vermiform process,* is a blind sac, a process off the lower part of the cecum. The appendix is therefore located in the lower right portion of the abdomen. It is an extension of the intestinal wall and is essentially nonfunctional in human beings. The appendix may become infected and inflamed, giving rise to the condition known as *appendicitis.* Inflammation and swelling weaken the wall of the appendix, and it may rupture, spewing its bacteria-laden contents into the abdominal cavity. The mesentery and peritoneum may be infected, which produces a much more severe condition called *peritonitis*. So long as harmful intestinal bacteria are confined to the closed sac of the appendix, infection is limited, so surgical removal of the appendix before rupture is the obvious treatment for appendicitis. Since the appendix is a part of the intestine, anyone who suspects appendicitis should refrain from using cathartics (laxatives), which can act unfavorably on an inflamed appendix.

Cathartics *Cathartics* are agents used to treat constipation. Some cathartics stimulate peristaltic movements of the intestine, some stimulate the mucous glands to secrete, and others prevent the absorption of water or attract water into the intestine. Some salts, such as the chlorides, are readily absorbed if their concentration in the intestine is higher than that of the blood. The tartrates, citrates, and sulfates are commonly used as cathartic salts because they are not absorbed in water-soluble form. Furthermore, in hypertonic solutions these salts cause more water to enter the intestine, thus increasing the bulk and fluidity of the contents.

Hemorrhoids *Hemorrhoids,* commonly called piles, are dilated veins in the anal region. They may be internal (located in the rectum) or external (located outside the external anal sphincter muscles). They can usually be treated medically, but severe cases may require surgical excision.

SUMMARY

1 The mouth is the anterior opening of the digestive tract. The tongue is a muscular organ that propels the food back into the pharynx. It is also an organ that detects taste, touch, and temperature. There are two sets of teeth in mammals, a deciduous set and an adult set. The human deciduous teeth number 20, and the set is composed of 4 incisors, 2 cuspids, and 4 molars in each jaw. The human adult set consists of 4 incisors, 2 cuspids, 4 premolars, and 6 molars in each jaw, 32 teeth in all.

2 The structure of the tooth includes the enamel-covered crown, a neck region, and a root covered with cementum. Under the enamel and cementum is a sensitive dentine. The pulp cavity contains blood capillaries and the nerve. The teeth lie in alveolar sockets attached by the periodontal membrane.

3 There are three pairs of salivary glands: the sublingual, submandibular, and parotid. The sublingual glands lie below the tongue in the floor of the oral cavity. The parotid gland is located at the angle of the jaw. Saliva contains mucin and the enzyme amylase, as well as other constituents. Salivary amylase initiates the digestion of starch.

4 The esophagus is the muscular tube that extends from the pharynx to the stomach. The smooth muscles of the esophagus propel solids and liquids to the stomach by peristalsis.

5 The abdominal cavity is lined by a serous membrane, the peritoneum. Folds of the peritoneum hold the intestine like an arm in a sling. These folds are the mesentery. A large fold called the greater omentum covers the intestine anteriorly.

6 The stomach is a muscular saclike organ located below the diaphragm and to the left of the liver. The upper, or esophageal, opening is the cardiac orifice; the lower opening is the pylorus, which is guarded by the pyloric sphincter. The stomach is lined by a mucous gland-containing

membrane. The principal glands of the stomach are composed of chief cells, which secrete pepsinogen, and parietal cells, which initiate the formation of hydrochloric acid.

7 The duodenum is the anterior part of the small intestine beyond the pylorus. It is only about 25 cm long. The pancreas lies in the loop of the duodenum and extends upward behind the stomach. It secretes the enzymes trypsin, amylase, and lipase. Trypsin is a protease, breaking down proteins to proteases, peptones, and amino acids. Amylase is a starch-digesting enzyme, hydrolyzing starch to the sugar maltose. Lipase acts upon fats, breaking them down to fatty acids and glycerol.

8 Secretin, produced by the duodenum, stimulates the pancreas to secrete a thin watery secretion low in enzyme content. The actual substance secreted by the duodenum is the inactive prosecretin, which is changed to secretin by the action of the acid food entering the duodenum.

9 The liver is an organ of many functions. It secretes bile, has glycogenic and glycogenolytic functions, and can perform gluconeogenesis. It is also concerned with the chemical breakdown of proteins and fats. The liver is one of the principal blood filters; it also detoxifies products of catabolism, which might otherwise accumulate in injurious quantities. The liver produces serum albumin and serum globulin, as well as fibrinogen and heparin.

10 The small intestine is divided into three parts: the duodenum, jejunum, and ileum. Most absorption of food materials takes place in the small intestine. The fluid of the small intestine is called the succus entericus.

11 The villi are minute processes projecting into the cavity of the intestine. They are concerned with the absorption of food materials and greatly increase the absorptive surface. Each villus contains a capillary network and a lymph vessel called a lacteal.

12 The ileum enters the large intestine laterally. Below the point of entrance are the cecum and the vermiform appendix. The entrance to the large intestine is guarded by the ileocecal valve. The ascending colon is on the right side of the abdomen; the transverse colon lies just below the stomach; the descending colon is on the left side of the abdomen. Other parts of the large intestine are the sigmoid colon, the rectum, and the anal canal. One of the principal functions of the large intestine is the absorption of water. As water is removed, the food residues are acted upon by bacteria, and the material in the large intestine assumes the characteristic of fecal matter.

QUESTIONS

1 Trace the digestion of a piece of bread and butter as it passes through the digestive tract. What will the end products be?
2 Name and locate the larger glands of the digestive tract. Give the functions of each.
3 What is the relation between the portal system and the digestive system?
4 Show how the deciduous teeth are gradually replaced by the permanent set.
5 Why is it possible for a horse to drink with the head held lower than the stomach?
6 Explain the functions of the stomach.
7 What part does the pancreas play in digestion?

8 List some of the numerous functions of the liver.
9 When food is present in the duodenum, the pancreas is stimulated to secrete and the gallbladder is stimulated to contract. Explain the mechanism by which this is accomplished.
10 Indicate the arrangement of bile ducts that permits the removal of the gallbladder without stopping the flow of bile from the liver to the intestine.
11 What is the source of bile pigments?
12 The bile salts have what function?
13 Discuss the causes of jaundice.
14 Explain the glycogenic and glycogenolytic functions of the liver.
15 What is peristalsis?
16 The villi have what function?
17 Describe the function of the small intestine in the absorption of food materials.
18 List the functions of the large intestine.

SUGGESTED READING

Banting, F. G., and C. H. Best: The Internal Secretion of the Pancreas, *J. Lab. Clin. Med.*, **7:**251–266 (1922).

Carlson, A. J.: Chemistry of Normal Gastric Juice, *Am. J. Physiol.*, **38:**248–268 (1915).

Cori, C. F.: The Fate of Sugar in the Animal Body, I, The Rate of Absorption of Hexoses and Pentoses from the Intestinal Tract, *J. Biol. Chem.*, **66:**691–715 (1925).

Davenport, J. W.: Why the Stomach Does Not Digest Itself, *Sci. Am.*, **226:**87–93 (1972).

Dickerson, R. E.: The Structure and History of an Ancient Protein, *Sci. Am.*, **226:**58–72 (1972).

Hendrix, T. R., and T. M. Bayless: Digestion: Intestinal Secretion, *Annu. Rev. Physiol.*, **32:**139–164 (1970).

Leloir, L. F.: Two Decades of Research on the Biosynthesis of Saccharides, *Science*, **172:**1299–1303 (1971).

Lipmann, F.: Evolution of Peptide Biosynthesis, *Science*, **173:**875–884 (1971).

Rothman, S. S.: The Digestive Enzymes of the Pancreas: A Mixture of Inconstant Proportions, *Annu. Rev. Physiol.*, **39:**373–389 (1977).

20

FOODS, NUTRITION, AND METABOLISM

Foodstuffs such as bread, butter, and eggs are ordinarily broken down into simpler chemical components by enzymatic action in the process of digestion. It is difficult for the average person to regard common foods as chemical substances that may release stored energy. As we have seen in the preceding chapter, however, the end products of digestion are absorbed and carried to the cells, where a final breakdown of their molecular structure releases energy, the principal driving force of metabolism. The release of energy within the cell is also catalyzed by enzymes.

The sun is the ultimate source of energy for life on earth. Irradiation of plant life by the sun creates the energy resources of coal, oil, and wood. These rich energy sources are not available directly as foods, largely because human beings have no enzyme systems that can break down their molecular structure and release the stored energy. Cellulose, a polysaccharide found in the cell walls of plants, is perhaps a little closer in chemical structure to being a food material for us, but we are still unable to digest it and use its energy. Small insects called *termites* can eat and digest woody substances because their digestive tract contains certain kinds of protozoa that produce the enzymes which can digest wood tissue.

The chemical elements contained in foods are building blocks for the growth of new tissue and the maintenance and repair of older tissues. In metabolic processes chemical elements are continually used up and continually replaced. Food provides the materials for such replacement.

Substances used as food fall into three major groups: carbohydrates, proteins, and fats. Other food substances such as vitamins and certain minerals are also essential to life. Water is also an important constituent of foods. Fruits and vegetables contain a high percentage of water, and the body tissues of higher animals are around 75 to 90 percent water. Metabolic processes utilize water, which is also produced as a by-product of metabolism.

Though only plants have the ability to synthesize glucose, both plants and animals use glucose in their metabolism and derive energy from the breakdown of its molecular structure. Both plants and animals can also convert glucose into a more complex substance for storage. Thus plants convert glucose into starch, and animals convert it into glycogen, which is often called "animal starch." Glucose can also be converted into fats by both plants and animals, or it may form the basic substance from which proteins are derived.

CARBOHYDRATES

Sugars and starches are examples of *carbohydrates,* an important group of compounds whose chemical structure we described in Chap. 1. Carbohydrates represent one of the principal fuels for body metabolism. Complex carbohydrates, or polysaccharides, are broken down finally into simpler compounds, or monosaccharides, such as glucose, fructose, and galactose. Fructose is converted to glucose in the cellular wall of the intestine. After it is absorbed from the intestine, glucose is carried to most body tissues by the blood, and it passes through cell membranes in the tissues by facilitated diffusion. Glucose constitutes a prime source of energy. We have mentioned in a previous chapter that insulin helps move glucose through the cell membrane. Within the cell, sugars are phosphorylated with the help of ATP and certain specific enzymes.

Galactose cannot be absorbed directly. It is first converted to glucose in the liver. Several enzymes plus ATP and uridine diphosphate (UDPG) are required for this conversion.

Table sugar (sucrose), milk sugar (lactose), and malt sugar (maltose) are *disaccharides*. Sucrose is formed in plant cells by the combination of two monosaccharide molecules and the removal of one molecule of water. Sucrose is found widely in fruits and vegetables.

Lactose in milk from cows represents about 4.9 percent concentration, in human milk it constitutes nearly 6 percent, while rabbit's milk contains only 1.95 percent.

Maltose is obtained by the action of the enzyme maltase on starch. Maltose can be split into two glucose molecules by hydrolysis. Much of the glucose we use is derived from starch obtained from plant foods. Breakfast cereals contain a large amount of starch.

The more complex carbohydrates such as starch, glycogen, and cellulose are termed *polysaccharides*. Glycogen is stored in the liver and to a much smaller extent in muscles. If the blood glucose level rises too high, glucose is converted to glycogen and stored in the liver. If the blood glucose level falls after several hours of fasting, liver glycogen is then converted back to glucose. In this way, the liver is able to regulate blood glucose level.

Sugars provide a quick source of energy, but the American diet usually contains too much of it. Sugar is injurious to the teeth, and it does not contribute to the development of muscles and bones.

Cellulose is a plant product and a starch. It is found with other substances in woody tissues and in the cell wall of plants. It is not available to us as a food because, as we have pointed out, we have no enzyme to break down its molecular structure. The woody parts of plants may have some value, however, by increasing the bulk of the foods we consume.

FATS AND RELATED COMPOUNDS

LIPIDS Fats are composed of the same basic elements as carbohydrates, but the proportion of the elements is different, as we showed in Chap. 1. The amount of oxygen is less in fats than in carbohydrates. When completely oxidized, both fats and carbohydrates yield energy and break down to carbon dioxide and water. More energy can be derived from the combustion of a given amount of fat than from combustion of a similar amount of carbohydrate.

In Chap. 19, we pointed out that fats are largely digested in the intestine after they are mixed with pancreatic lipase and bile salts from the liver. Bile acts as a detergent in this process. A small amount of fat may be digested in the stomach. Fats break down by enzymatic action into glycerol, fatty acids, and glycerides. These products of fat digestion are absorbed through the cells of the villi and pass into the lacteals. Fat particles can pass through the cells of the intestinal wall, but as soon as they reach the villi they combine into neutral fat. As we have seen, fats are gathered up by the lymphatic system but are distributed to the tissues by the blood.

Ingested fats are largely absorbed by the lymph as small fat droplets, or *chylomicrons*. Fats are stored in adipose tissue almost entirely in the form of triacylglycerides. The greatest amount of fat is stored as "fat pads" in the subcutaneous adipose tissue. Considerable amounts are stored around the kidneys, on the mesentery and omentum. The stored fat can then be dispersed for energy as needed. Adrenocortical hormones increase the permeability of fat cell membranes and permit fat droplets to be released to the blood.

Some fats are essential for growth and nutrition, but others are not. Rats on a fat-free diet fail to grow normally and show the effects of this dietary deficiency. The symptoms of fat deficiencies do not improve upon administration of fat-soluble vitamins, but rats recover when given any one of several unsaturated fatty acids, such as linoleic acid or linolenic acid. These conditions would not be likely to develop in most people because adequate amounts of fat are usually present in the human diet.

STEROLS *Sterols* are not fats but complex secondary alcohols. They often combine with fatty acids to form esters. (An ester is a compound formed by the condensing of an acid and an alcohol, with elimination of H_2O.) The sterol *cholesterol* is found in tissues throughout the body. It can be present in its free state or as an ester. Bile cholesterol is in the free state, as is the cholesterol found in erythrocytes. However, much cholesterol in blood plasma is present as the ester. Cholesterol is related chemically to other substances of great biological significance such as the bile acids, the sex hormones, the adrenocortical hormones, and vitamin D.

PROTEINS

Protein substances occur in all living matter; they are essential to the living protoplasm of every cell. The protein molecule is often large and extremely complex. It commonly contains a great number of carbon atoms to which are attached atoms of oxygen, hydrogen, and nitrogen. Other elements associated with proteins in small amounts include sulfur, phosphorus, iron, copper, and iodine. For example, iron is included in hemoglobin, and iodine forms a part of the thyroid hormones. Compared with the formulas of common carbohydrates, the protein molecule appears to be larger and more intricate. It commonly contains more carbon and relatively less oxygen. Protein molecules often represent an unusually great molecular weight, but there are also relatively simple proteins. The structural formulas of proteins vary widely because of the number and position of carbon atoms and the elements attached to them. For this reason, proteins differing only slightly in their molecular formulas exist in almost endless variety.

Protein foods are necessary for growth and repair of tissues, but they vary in the number and kind of amino acids of which they are composed. Not all protein foods have the same nutritional value. Some amino acids are absolutely essential for growth; others can be manufactured by the body. Eggs (ovalbumin), milk and cheese (lactalbumin and casein), meat, and glutenin of wheat contain all the essential amino acids, and these foods are called *complete proteins*.

The proteins of corn and gelatin and many others are called *incomplete*, because they do not contain all the essential amino acids. You do not need to be greatly concerned as to whether the various proteins in a normal diet are complete or incomplete. A wide range of amino acids is found in most common food proteins, but some contain more of the essential amino acids than others.

AMINO ACIDS   Amino acids have been called the building blocks of proteins. Though many amino acids have been described, food proteins are usually considered to be derived from 23 amino acids in various combinations.

Essential amino acids are those the body is unable to synthesize. These amino acids must therefore be consumed in foods. If the 10 essential amino acids are provided in adequate amounts in the diet, the body will be able to synthesize the others. Experimental evidence along these lines has been derived mostly from nutritional studies on rats. There are indications that various animals and human beings have somewhat different requirements. Following is a list of essential amino acids:

Threonine	Arginine
Valine	Lysine
Leucine	Phenylalanine
Isoleucine	Tryptophan
Methionine	Histidine

Unlike most carbohydrates and fats, the proteins of an animal or plant species are likely to be specific for that species. Organs or pieces of tissue cannot ordinarily survive transplantation from one species to another. The structural formulas of proteins indicate the possibility of almost unlimited combinations of molecules. This may partially explain the great variety of tissue proteins and how it is possible for proteins to be specific for a given species.

Amino acids are absorbed from the intestine and carried by the blood to various organs and tissues. They may combine with other substances within the cell. The colloid nature of proteins permits them to be retained within the cell membrane. Those amino acids that are not utilized directly may undergo deamination, a process in which NH_2 groups are removed. The nitrogen is combined with carbon dioxide to form urea, which is excreted in the urine. Though deamination can occur in the kidneys and other tissues, the liver is the principal organ concerned in this process.

After deamination, an amino acid molecule can either be oxidized to carbon dioxide and water with a consequent release of energy or its chain of carbon, hydrogen, and oxygen atoms can be used in the conversion of the molecule into a fat or a carbohydrate. Conversion of protein to fat enables the organism to store the energy from excess amino acids not utilized directly by the tissues. Conversion of protein to glucose is especially interesting in diabetic patients, who continue to excrete glucose although carbohydrates are excluded from their diet. However, not all amino acids form glucose or glycogen after deamination; over half are glycogenic, but some are only so to a limited degree.

INORGANIC REQUIREMENTS IN THE DIET

The inorganic composition of a tissue can be determined by burning it and analyzing its ash constituents. Carbon, hydrogen, oxygen, and nitrogen are volatilized and driven off by this process, but the remaining ash includes calcium, sodium, potassium, magnesium, phosphorus, iron, sulfur, and traces of several other elements. Apparently the organism needs these elements to build tissue.

MINERALS   Most minerals are present in common foods, and adequate amounts are obtained from a normal diet. Sodium, potassium, magnesium, and phosphorus fall into this category. We include considerable quantities of *sodium chloride* in our diet, perhaps more than is healthful. Low-salt diets are sometimes recommended for those who have high blood pressure. Sodium chloride is very important, however, in maintenance of blood osmotic pressure and for proper kidney function, as we saw in Chap. 18. Salt is lost by excretion in urine and perspiration. People who work under conditions of unusual exposure to high temperature and who lose a considerable amount of salt by sweating are often given salt tablets to help make up this loss.

Sodium is found largely in interstitial fluids. Sodium ions, along with potassium ions, play a principal part in the excitability of the nervous system. Many foods contain sodium, and adequate amounts are readily available.

Potassium, as you may recall, takes part in many physiological reactions, especially in neuron excitability and muscle contraction. *Magnesium* acts as a counterbalance against calcium in muscle contraction and in neuron excitability. It is also present in the chemical composition of many coenzymes. Adequate amounts of potassium and magnesium are usually obtained from plant foods.

Phosphorus and sulfur are largely obtained from protein foods. *Sulfur* is a necessary element for synthesis of the amino acids methionine and cystine. If adequate amounts of protein are included in the diet, enough phosphorus and sulfur for physiological needs will be automatically ingested.

Phosphorus is found in all cells but especially in nerve and muscle cells. It also plays an important part in development of bones and teeth. Many biological compounds such as phosphocreatine, the phospholipids, and phosphoglucomutase contain phosphorus. Phosphorylations are a common occurrence in enzyme-catalyzed reactions and in the absorption of carbohydrates from the intestine.

Calcium is essential to development of bones and teeth, clotting of the blood, contraction of muscle, and nerve excitability. Calcium is usually found in the body as carbonate or phosphate salts. Calcium, potassium, and sodium are maintained at a nearly constant level in the bloodstream and tissues. Though calcium is a common element in food, many foods contain only minute amounts, and it is apparently not readily absorbed from some foods. Milk and milk products are good sources of calcium. Growing children need an abundant supply of calcium for the development of good skeletal structures. Our diets are more apt to be deficient in calcium than in the other elements. Calcium cannot be used efficiently without vitamin D, which is essential to proper absorption and utilization of mineral salts deposited in bone.

Iron is necessary for the formation of hemoglobin and myoglobin and for the operation of the cytochrome system. A small amount of iron combines with protein and is stored in liver, bone marrow, and spleen. When bound to protein, iron is known as *ferritin*. Dietary iron is ionized by the acid of the gastric juice in order to be absorbed.

Iron is not stored in the body in any great amount, so we are dependent upon ingesting an adequate amount in our daily diet. Eggs, meat, and some cereals are good sources of iron. Many foods contain iron in small amounts. When old cells are broken down in the liver, the iron of the hemoglobin is largely conserved to be used over again. A considerable loss of blood may produce an iron deficiency. Iron deficiency is more common in women than in men. Iron is commonly added to the diet in iron-deficiency anemias to ensure an adequate amount for hemoglobin formation.

Iodine is needed only in minute amounts, but it is essential to the normal functioning of the thyroid gland. The hormone of this gland, thyroxine, is an amino acid containing iodine. In iodine deficiency the thyroid gland may enlarge, causing a condition called simple goiter (see Chap. 13). Traces of iodine occur in drinking water and in vegetables grown where iodine is available in the soil. Sea foods are good sources of iodine. The use of iodized salt should supply adequate amounts of this element.

Minute amounts of other minerals are also found in cells; these have been called *trace elements*. Some, such as copper, manganese, and zinc appear to be essential for certain enzymatic reactions. Trace elements are not to be confused with radioactive substances commonly called *tracers*. Magnesium is an essential element in muscle metabolism, and fluorides produce hard, cavity-resisting enamel in teeth.

VITAMINS   At the beginning of the twentieth century it was generally assumed that proper proportions of carbohydrates, proteins, fats, and minerals in the diet would provide all the essentials of a good diet. It became more evident around 1911 that something else was needed to provide for normal growth and nutrition. The science of food chemistry existed before 1911, but the period since that date has been especially noteworthy in the discovery of many essential chemical substances known as *vitamins*. The literature on the subject of vitamins has become exceedingly extensive, and many good books and special articles are available.

The word *vitamin*, which may be translated as an "amine essential for life," is not strictly a correct term, since not all vitamins are amines and not all are vital for the maintenance of life. Vitamins are not necessarily closely related chemically, and their physiological effects are quite diverse. Before the chemical nature of vitamins was determined, these substances were designated by letters of the alphabet as vitamins A, B, C, and so forth. This practice is still followed, but now that the chemical structure of vitamins is better understood, it is considered good usage to refer to them by descriptive names such as niacin, thiamine, or ascorbic acid. It has not been possible to break away from the convenient alphabetical classification completely, however. Vitamin A is a good example of this confusion of nomenclature. The structural formula is known, and the chemistry of vitamin A substances has been studied extensively, but no chemical name for this vitamin has been generally accepted.

Fat-soluble vitamins   VITAMIN A   This important vitamin has the formula $C_{20}H_{30}O$. *Vitamin A* is formed from precursors, or provitamins, of which four are well known. The vitamin A provitamins are alpha-, beta-, and gamma-carotene and cryptoxanthin. These substances are hydrolyzed by the animal organism to produce vitamin A. Beta carotene is the best source of vitamin A among carotenes. Beta carotene is hydrolyzed to two molecules of vitamin A, whereas the other precursors yield only one molecule.

$$C_{40}H_{56} + 2H_2O \longrightarrow 2C_{20}H_{30}O$$

Carotene   Vitamin A

Strictly speaking, plants do not contain vitamin A but only its precursors. Certain plant foods may, therefore, have vitamin A value, even though they do not actually contain any vitamin A. In general, the carotenes are found in yellow vegetables and fruits and in the green leaves of vegetables in which the yellow color is covered over by the green of chlorophyll. Cryptoxanthin is found in orange peel. The very common yellow pigment xanthophyll, which shows so brilliantly in yellow leaves of trees in the fall, is not a precursor of vitamin A. The best sources of vitamin A are green, leafy vegetables, yellow fruits and vegetables, and animal products such as egg yolks, butter, cream, or whole milk. Fish-liver oils are excellent sources.

Animals take the yellow precursor substances from plants and transform them into colorless vitamin A. We consider that chickens, cows, and fish are important converters of plant provitamins into vitamin A. Animal products that are yellow-colored do not necessarily contain a greater amount of vitamin A than do some colorless substances. White whole milk does not necessarily contain less vitamin A than milk with more yellow coloring matter, for example. Animal products may contain both carotenoid substances and vitamin A. Furthermore, the yellow color of some foods is caused by carotenoid substances that are not precursors of vitamin A.

Vitamin A and the carotenes are fat-soluble, and their absorption and metabolism are similar to those of fats. Bile is essential for their absorption. Not all fats are good sources of vitamin A, however. Most vegetable oils used for cooking as well as the animal fats in beef and lard are poor sources. Since animals can store vitamin A in the liver, this organ may contain relatively large amounts. Fish which consume quantities of plant food, or larger fish which feed upon smaller plankton-feeding fish, can store considerable concentrations of vitamin A in their livers. Cod-liver oil is an important commercial source of the two fat-soluble vitamins A and D. Oil from halibut livers is a somewhat better source.

VITAMIN D This vitamin promotes growth and development of bones. It helps mineralize bones by increasing the absorption of calcium from the intestine and by influencing the deposition of calcium. *Vitamin D* cooperates with the hormone of the parathyroid gland to regulate calcium metabolism. A vitamin D deficiency condition that produces a malformation of growing bones is called *rickets* (Fig. 20.1).

Provitamin D substances are activated by certain wavelengths of light within the ultraviolet range. Such light waves have a weak penetrating power so that they are not effective through clothing or even through ordinary window glass. Such light waves apparently penetrate the skin to a depth of only about 0.5 mm. The wavelengths of light that have antirachitic value are also the wavelengths that cause sunburn and tanning of the skin.

The most abundant sources of vitamin D are liver oils of bony fish, such as halibut, cod, tuna, and others of the percomorph group. The average diet includes such vitamin D sources as fresh and canned fish, milk, eggs, and butter. Green vegetables also contain small amounts. Though mammalian liver stores adequate amounts of vitamins A and D, the liver is not a good source of vitamin D, although it is an excellent source of vitamin A. These two fat-soluble vitamins are commonly found in the same food sources but not necessarily in equal amounts. Since vitamin D is concerned with the utilization of calcium and phosphorus salts, it is essential that the diet contain these salts in adequate amounts.

Coupled with sunlight and adequate amounts of cholesterol in the skin, the average diet probably contains enough vitamin D for an adult. It is doubtful whether or not this amount is adequate for growing children. Infants receiving a quart of whole milk per day should be protected from rickets, a vitamin D deficiency disease (Fig. 20.1), but the diet during infancy is commonly supplemented with additional vitamin D from a commercial source. If children are growing rapidly, their need for calcium and vitamin D is increased. Perhaps not all growing children need to have vitamin D added to their diet, but for many it would be helpful. During pregnancy and lactation vitamin D intake should also be increased.

VITAMIN E (*the tocopherols*) A search for another fat-soluble vitamin evolved around the observation that rats reared upon certain restricted diets were often sterile. Adding green vegetables, wheat germ, or vegetable oils to the diet corrected this defect. Investigation finally showed that *vitamin E,*

FIGURE 20·1
Rickets in a young child. The bone curvatures and bowlegs are caused by a deficiency of vitamin D. (*From "The Vitamin Manual," published by the Upjohn Company: courtesy of Rosa Lee Nemir, M.D.*)

included under the general name *tocopherols,* was the antisterility vitamin in rats.

Vitamin E deficiency in the male rat is manifested by degeneration of the germinal epithelium of the testes, so that the animal cannot produce living, motile spermatozoa. A female on the same diet can mate and develop fertile eggs, but the young embryos die and disintegrate. Their substance is then reabsorbed into the tissues of the mother. The addition of tocopherols to the diet enables the female to produce normal living young. The progress of damage to the testes in the male is not reversible upon adding tocopherols to the diet.

Rats and other animals also show muscular weakness and paralysis as a result of a vitamin E deficiency that extends over many months. It has been observed also that vitamin E exerts a "sparing" effect on metabolism of vitamin A and that more vitamin A is stored in the liver of animals given vitamin E.

Little is known about the effect of the tocopherols in the metabolism of human beings, but it appears to be an antioxidant that prevents or inhibits the oxidation of unsaturated fats. Vitamin E is found in many plant food materials, and there is little chance of a deficiency with an average diet. It is found in green leafy vegetables and also in peas and beans. Wheat-germ oil and vegetable oils are regarded as excellent sources. Animal tissues contain only small amounts.

Attempts to alleviate muscular weakness or to improve reproductive functions in human beings by administering vitamin E substances have not been successful. It is not an aphrodisiac.

VITAMIN K Chicks kept upon a restricted fat-free diet were studied by Dr. Henrick Dam, of the University of Copenhagen. These animals hemorrhaged and were anemic, and the clotting ability of their blood was very greatly reduced. Since these conditions were not relieved by any of the known vitamins, Dam postulated that an unknown vitamin was involved. He called it the *Koagulations vitamin*, and it has become known as *vitamin K*.

Two naturally occurring compounds of vitamin K exist. The first was isolated from alfalfa, and it is known as K_1. Vitamin K_2 was isolated from fish meal subjected to bacterial putrefaction. It had been known for some time that bacteria can synthesize this vitamin, not only in bacterial preparations of various sorts but also in the living intestine. The empirical formula for vitamin K_1 is $C_{31}H_{46}O_2$; that for K_2 is $C_{41}H_{56}O_2$. There are, in addition to K_1 and K_2, several compounds that have vitamin K properties. One of the best known of these synthetic compounds is 2-methyl-1,4-naphthoquinone. It has had considerable use as a substitute for the more expensive K vitamins. The K vitamins are chemically related to ubiquinone (coenzyme Q).

Vitamin K_1 is found chiefly in green plants. Most laboratory animals synthesize and absorb the vitamin readily and do not, therefore, show vitamin K deficiency, even though they are fed a diet lacking this vitamin. Chicks, however, do not absorb K_2 readily; when K_1 is withheld from their diet, they become hemorrhagic, a manifestation of vitamin K deficiency.

Vitamin K acts as a coenzyme necessary for the synthesis of prothrombin, a substance involved in blood clotting. It raises the prothrombin level in the blood, however, only if the liver is functioning, since prothrombin is synthesized in the liver. In the absence of bile, very little vitamin K is absorbed into the bloodstream. The administration of bile salts plus vitamin K has proved helpful in cases of vitamin K deficiency caused by bile deficiency.

The prothrombin level of the blood is likely to be low in the newborn infant, and the clotting ability of the blood, therefore, may be impaired. Administration of vitamin K to the mother before birth or to the infant at birth tends to increase the clotting ability of the blood (Table 20.1).

Water-soluble vitamins THE VITAMIN B COMPLEX The search for the vitamin that would come to be known as vitamin B_1 centered around a cure for a polyneuritis that had been known for centuries to affect populations whose chief dietary staple was polished rice. This vitamin-deficiency condition is called *beriberi*. The peripheral nerves become inflamed in this condition, resulting eventually in paralysis of the appendages and circulatory and

TABLE 20·1
The principal vitamins

Vitamins	Major sources	Function	Deficiency
Fat-soluble			
A (beta-carotene) (retinol)	Yellow vegetables, green leafy vegetables, egg yolk, butter, fish-liver oils	Maintenance of epithelial tissues; essential for formation of rhodopsin	Xerophthalmia, night blindness
D_2 (calciferol) D_3 (cholecalciferol)	Fish-liver oils / Egg yolk, milk, butter	Increases absorption of calcium and phosphorus from intestine and their utilization in bones and teeth	Rickets in children
E (tocopherol)	Green leafy vegetables, wheat germ	Increases the stability of membranes (?); acts as an antioxidant in fat metabolism	Degeneration of the germinal epithelium of testes in white rats
K (naphthoquinone)	Leafy vegetables	Promotes synthesis of prothrombin and blood factor VII in liver	Impairment of blood clotting
Water-soluble			
B complex: B_1 (thiamine)	Whole grain, cereals, nuts, eggs, pork	Acts as cocarboxylase in oxidation of pyruvic acid	Beriberi, some forms of neuritis
B_2 (riboflavin)	Liver, meat, milk, eggs, whole grains	Acts as coenzyme in oxidative phosphorylation (electron transport)	Some forms of dermatitis
Niacin	Liver, wheat germ, soy beans, peanuts	Acts as a coenzyme in cellular respiration (NAD, NADP)	Pellagra
B_6 (pyridoxine)	Seeds, liver, meat, fish	Forms pyridoxal phosphate, coenzyme in amino acid and fatty acid metabolism	Dermatitis around eyes in animals
B_{12} (cyanocobalamin)	Liver, meat, milk, eggs	Essential to production and maturation of erythrocytes	Pernicious anemia
Folic acid (pteroylglutamic acid) (folacin)	Green leafy vegetables, yeast, liver	Essential to synthesis of RNA and DNA, also for normal erythrocyte maturation	Degenerative changes in bone marrow
Pantothenic acid	Vegetables, liver, eggs	Used in synthesis of coenzyme A	Metabolic functions, especially in carbohydrates and fats
Biotin	Yeast, milk, liver, vegetables (synthesized by bacteria in intestine)	Takes part in fatty acid synthesis and in carboxylation reactions	No definite deficiency in humans
Choline	Plant and animal tissues as a constituent of phospholipids	Used in metabolism of fats, oxidation; forms acetylcholine, a neurotransmitter	Prevents fatty livers in animals
C (ascorbic acid)	Citrus fruits, tomatoes, cabbage, most fresh fruits and vegetables	Regulates formation of collagen and other intercellular substances	Scurvy, in severe cases

sensory effects. The condition also occurs in animals fed restricted diets. Birds are especially susceptible.

When the nature of the vitamin was determined, it was called *vitamin B* because it was a water-soluble substance that would cure or prevent beriberi. Later experimentation showed that this water extract of food substances contained not just one vitamin but a series of vitamins. The original vitamin, which effected the cure of beriberi, was then called B_1, and the other vitamins in this series were called B_2, B_6, and B_{12}. All these vitamins are now termed the *vitamin B complex*, which includes a number of chemical substances. Now that the chemical nature of these substances is known, the chemical names are often used instead of the alphabetical symbols. Vitamin B_1 now is known as *thiamine*.

THIAMINE   *Thiamine* plays an important part in the proper utilization of carbohydrates. The breakdown of pyruvic acid requires the enzyme carboxylase. Thiamine unites with phosphoric acid to form a pyrophosphate. Thiamine pyrophosphate then acts as a coenzyme with carboxylase in the oxidation of pyruvic acid and is thus called a *cocarboxylase*.

Vitamins of the B complex and vitamin K are produced synthetically by bacterial action in the large intestine. In most instances the amount synthesized is not adequate to meet the nutritional needs of the individual, but it does tend to offset dietary vitamin deficiencies.

The amount of thiamine required by an individual varies slightly with the person's sex, age, physical condition, and the amount of work done. In general, the requirement is 1 or 2 mg/day. This amount can readily be obtained from food sources. Thiamine is synthesized by plants but is found in most fruits and vegetables only in minute amounts. A greater concentration is usually found in seeds, and so whole grains, beans, peas, and nuts are good sources. The best animal sources include eggs and lean meat, especially pork muscle. In general, the more highly refined the flours and cereals, the less thiamine is retained in them. Therefore whole grain bread and breakfast cereals contain more thiamine unless white bread and cereals have been enriched by adding a commercially prepared product. Although a good diet should provide enough thiamine, there is a tendency to use too much highly refined flour, breakfast cereals, sugar, and fats, foods that are low in content for this essential vitamin.

Thiamine deficiency results in a loss of appetite, which, in turn, reduces the intake of food. This sequence complicates studies of thiamine deficiency, since some effects may be caused by malnutrition rather than by a thiamine deficiency. Vitamin B_1 has a favorable influence on growth, but, again, an increased intake of food including other vitamins makes it difficult to assess the effect of this particular vitamin. Deficiencies of thiamine, as well as other vitamins of the B group, may develop during pregnancy and lactation. Thiamine is stored by the liver to a limited extent.

RIBOFLAVIN   The discovery of this vitamin is somewhat unique, for it first became known as a respiratory coenzyme in an oxidation-reduction reaction. This was the "yellow enzyme" described by Warburg and Christian, in 1932. Riboflavin was at first called *lactoflavin*, because it was isolated from milk. Other yellow fluorescent pigments called *flavins* were discovered in various food sources. When it was determined that these flavins were identical and that all contained the sugar ribose, or a derivative, the substance was called *riboflavin*. Riboflavin is an orange-yellow color. In aqueous solution it shows a greenish-yellow fluorescence, and elongate orange-yellow crystals can be obtained upon drying the solution.

When riboflavin is phosphorylated, it forms two coenzymes, flavin mononucleotide (FMN) and flavin adenine dinucleotide (FAD). Both are active as hydrogen carriers in oxidation-reduction reactions. FAD acts as an electron acceptor from coenzymes NAD and NADP in the electron-transport system. As the flavoprotein is oxidized or dehydrogenated, it passes electrons along to the cytochrome system, with the release of hydrogen atoms.

Since riboflavin is present in such a wide variety of foods, it is quite likely that noticeable effects do not commonly occur. Intestinal bacteria can synthesize the vitamin, thus adding to the amount available, but persons who live on inadequate diets may become deficient in this vitamin. Good food sources are leafy vegetables, fruits, yeasts, milk, liver, muscle, and egg white.

NIACIN, OR NICOTINIC ACID   *Nicotinic acid* had been known for at least 70 years before the discovery of its value as a vitamin. It was obtained from nicotine by chemists in the latter half of the nineteenth century. In the search for a cure for beriberi, nicotinic acid was isolated from rice polishings as early as 1911, but it was discarded when it failed to cure this condition. Nicotinic acid was thus the first vitamin to be isolated in pure chemical form, even though it was not known as a vitamin until many years later. Nicotinic acid should not be confused with nicotine. The body does not derive the vitamin from nicotine absorbed by smoking tobacco. Although prepared chemically through oxidation of nicotine, nicotinic acid has very low toxicity compared with that of nicotine.

In tissues, nicotinic acid occurs in the form of its amide. The carboxyl group, COOH, of the acid is changed to include an amino group, NH_2, and the molecule then becomes nicotinic acid amide, or nicotinamide. Since tissues or foods contain only the amide, strictly speaking, this substance is the vitamin, which is also called *niacin*. The chemistry of the body, of course, has the ability to transform nicotinic acid into nicotinamide. Nicotinamide adenine dinucleotide (NAD) and nicotinamide adenine dinucleotide phosphate (NADP) act as coenzymes in the electron-transport system involved in cellular respiration.

Niacin is relatively heat-stable, although there is always the chance that some of the water-soluble vitamins will be lost in water used for cooking. The vitamin is found in a wide variety of vegetables, whole wheat, milk, eggs, and lean meat. Among the best sources are liver, kidney, salmon, brewer's yeast, wheat germ, soy beans, and peanuts.

The amino acid tryptophan is a source of niacin, and the amount and kind of protein in the diet affect the amount of niacin available. The amount of tryptophan in animal protein is about 1.4 percent; in vegetable protein, it is about 1 percent.

PYRIDOXINE (*Vitamin B_6*)   *Pyridoxine*, or pyridoxal phosphate, is essential as a coenzyme in many phases of amino acid and fatty acid metabolism. It has been used to prevent a form of dermatitis in experimental animals. Human deficiency effects have not been demonstrated with certainty. This may be largely because of the wide distribution of pyridoxine in common foods. It is water-soluble and unstable in light. In animal and plant tissues it is mostly found combined with protein or starch rather than in the free form. Seeds, legumes, wheat germ, liver, kidney, meat, and fish are food sources of pyridoxine. Pyridoxine is essential in the metabolism of certain amino acids and in their transport across cell membranes. It also is involved in the conversion of proteins to fats and in the metabolism of some of the fatty acids.

PANTOTHENIC ACID *Pantothenic acid* is a part of coenzyme A. Coenzyme A takes part in several known acetylation reactions; however, to be effective, it must be in the form of acetylcoenzyme A. In the presence of acetate and ATP and catalyzed by enzymatic reaction, coenzyme A can be converted to acetylcoenzyme A. As acetylcoenzyme A, the compound acetylates choline to form acetylcholine. In carbohydrate metabolism, pyruvic acid is acted upon by acetylcoenzyme A as it enters the tricarboxylic acid cycle; it then combines with oxaloacetic acid to form citric acid.

BIOTIN This vitamin substance functions as a coenzyme in several systems. *Biotin* takes part in deamination processes, the synthesis of fats, and in the formation of glycogen. It is found in small amounts in many foods. It is present in fairly large amounts in yeast, milk, liver, kidney, and raw potatoes. Other sources are egg yolk and common garden vegetables, such as carrots and tomatoes.

FOLIC ACID (*Pteroylglutamic acid*) This vitamin was originally isolated from the green leaves of plants such as spinach, peas, and clover. Since its source was the foliage of plants, it was called *folic acid*. It is found in yeast concentrations and in liver and is essential for the growth of certain kinds of bacteria and protozoa. In vertebrate animals and human beings, it functions in the normal development of red blood cells. Its also takes part in the synthesis of purines and pyrimidines, the building blocks of nucleic acids and nucleoproteins. For example, it is involved in the formation of thymidine, which, in turn, is a component of DNA.

VITAMIN B_{12} (*Cyanocobalamin*) A red crystalline substance, isolated from the liver, exerts a powerful effect in promoting the maturation of red cells and also relieves nervous and digestive disturbances associated with pernicious anemia. This factor is called *vitamin B_{12}*. It is the erythrocyte maturation factor that has been sought for many years. Vitamin B_{12} is a cobalt compound. Cobalt itself, when added to the food of rats in small amounts, has been shown to increase the number of erythrocytes and the amount of hemoglobin, but cobalt alone has not proved of value in treating pernicious anemia. Vitamin B_{12}, acting as a coenzyme, is effective in minute amounts and is many times more effective than folic acid in promoting the production of red blood cells. It is used also, in a general way, to promote appetite and growth.

The effectiveness of vitamin B_{12} depends upon the presence of another enzyme, the *intrinsic factor*, secreted by the stomach. This enzyme, a mucoprotein secreted, and interacting with HCl, by the parietal cells is necessary to facilitate the absorption of B_{12} through the mucosal membrane of the intestine. If B_{12} cannot be absorbed in adequate amounts, pernicious anemia develops. Subcutaneous injections of B_{12} produce an effective response in cases of pernicious anemia. The vitamin taken by mouth is ineffective if the amount of intrinsic factor is inadequate.

In the blood, vitamin B_{12} is bound to proteins of the alpha globulin type;

the vitamin is stored mainly in the liver. Vitamin B_{12} probably functions in the metabolism of all cells, but it is especially effective in stimulating the production of red cells in the bone marrow. It also is essential in the normal metabolism of nerve cells and the cells of the digestive tract.

Vitamin B_{12} has a high potency, so that 1 to 1.5 μg/day is adequate to satisfy the needs of normal persons. This vitamin is found in small amounts in meat, milk, and eggs, but plant foods are deficient in it.

CHOLINE   As a chemical compound, *choline* has been known for many years. It was first isolated from bile in 1894. Only recently has investigation shown that it plays an important part in nutrition. There is no known condition of human deficiency, but deficiencies of choline can be demonstrated in experimental animals. Choline occurs in both plant and animal tissues largely as a constituent of phospholipids. It prevents excessive accumulation of fat in the liver of experimental animals fed diets of high fat content. Choline is not stored in the body, but is absorbed from the intestine. Choline is not considered by some to be a true vitamin, but it is essential to nutrition.

Choline plays a part in various phases of metabolism. Acetylcholine, a derivative, is the chemical substance involved in the transmission of the nerve impulse across cholinergic synapses and at neuromuscular junctions. Choline is the basic substance in the phospholipid lecithin. As a lipotropic substance, it reduces the fat content of the liver. It also accelerates the formation of phospholipids from the fats present in the liver.

ASCORBIC ACID, OR VITAMIN C   A substance found in fresh fruits and vegetables prevents development of a deficiency condition called *scurvy*. This substance has been named, appropriately, *ascorbic acid,* or *vitamin C.* The chemical name of vitamin C is long and difficult, and so the coined name "ascorbic acid" is probably a fortunate choice. The term "vitamin C" is so well established that it is still in common use, and we shall use both names interchangeably.

Early experimental work proved that limes and lemons contained a substance that prevented scurvy. The list of foods with antiscorbutic properties was finally enlarged to include all fresh fruits and vegetables, but citrus fruits remain one of the best sources. Sprouting grain and bean sprouts contain vitamin C, but dried grains and legumes do not. The ascorbic acid content of foods is rapidly reduced by oxidation. Prolonged cooking or even prolonged standing and drying of fruits and vegetables reduces the potency of their vitamin C concentration. In general, commercial canning retains the vitamin better than home-canning methods, although home-canned tomatoes retain the vitamin fairly well. Milk is low in vitamin C; hence we have the recommended practice of feeding fruit juice or vegetable juice to infants to supplement their diet.

Ascorbic acid plays a part in numerous biological functions, including tissue respiration, the healing of wounds (especially in cartilage and bone), promotion of capillary stability, resistance to infection, and formation of blood cells in bone marrow.

Most animals can synthesize vitamin C, but primates and guinea pigs are known to be exceptions. The recommended daily intake for adults is 70 mg. For infants and growing children it is 30 to 80 mg. One-third to one-sixth of this amount should provide protection against the more evident signs of scurvy. Although considerable amounts of vitamin C are found in the adrenal cortex, it is not stored in the body.

NEED FOR VITAMINS It is possible to consume an excessive amount of fat-soluble vitamins; however it is unlikely that one could consume an excessive amount of the water-soluble ones since these are readily excreted by the kidneys. In all probability it will eventually be determined that nearly all vitamins act as enzymes or coenzymes in body metabolism. The sources, functions, and effects of deficiencies of vitamins are summarized in Table 20.1.

METABOLISM

Life processes, or those processes performed by all living forms in order to maintain life, are grouped under the general heading of *metabolism.* In a very broad sense, the energy reactions involved in the assimilation and utilization of food and in the promotion of growth are considered to be building-up processes and are often referred to as *anabolism. Catabolism,* on the other hand, is taken to mean the breakdown of stored energy reserves, which leaves the organism with a reduced store of energy. Very often it is difficult to classify some reactions accurately under these headings. For example, how do we classify the breakdown of a food substance into simpler substances, so that these, in turn, can be converted into fat and stored? Is this an example of anabolism, catabolism, or both?

Metabolism is essentially an expression of energy relationships. Green plants use the radiant energy of the sun in the process of photosynthesis in which they incorporate sun energy in the synthesis of food material. Animals cannot manufacture their food directly and so are dependent upon energy stored in the molecules of plants that can be used for food. If energy is utilized in building up the molecules of a food substance, then that energy can be released, provided the animal organism has the ability to break down molecular structure. Enzyme systems catalyze a series of chemical reactions that enable the animal to release this energy in the case of food substances. Wood and coal, of course, are sources of abundant energy, but the higher organisms have no enzymes capable of releasing this energy. Wood and coal, therefore, are not available directly as foodstuffs.

The attempt to show how much energy is derived from foods presents several well-known problems. In the first place, the law of the conservation of energy indicates that while the total energy of the universe remains constant, one form of energy may be converted into another. Energy in the body can be used in various ways. It can be used, for example, to synthesize chemical compounds; some of these compounds, such as glycogen and fat, are a source of stored energy. Energy can also be used to do work, such as that performed by the contraction of muscles. In this case only a small amount of energy is used in work; a far greater amount is given off as heat. Other cells utilize energy and perform work in various ways. The production of a secretion by secretory cells represents work and requires an outlay of energy. The best way to demonstrate and measure the energy derived from foods is to measure it in the form of heat. These concepts are discussed in greater detail in Chap. 3.

DIRECT CALORIMETRY One approach to measuring energy contained in foods is *direct calorimetry*. Measured amounts of food substances can be burned in an apparatus called a *calorimeter*, and the energy given off as heat can be determined with a high degree of accuracy. Calorimeters have various designs. The bomb calorimeter consists of a strong steel cylinder immersed in a weighed amount of water. The cylinder contains the food sample and is filled with oxygen to obtain complete and immediate oxidation. A platinum wire completes an electric circuit within the cylinder. When an electric current is introduced, foods are oxidized to carbon dioxide and water, except proteins, in which case oxides of nitrogen, sulfur, and phosphorus also remain. The heat liberated causes the temperature of the water to rise. The temperature change of a known amount of water can then be calculated in terms of calories.

A *calorie* is an energy unit defined in terms of heat. The *gram calorie* (cal) is the amount of heat required to raise the temperature of one gram of water one degree Celsius. More specifically it must raise the temperature of water from 14 to 15°C. The unit measurement commonly used in physiology is the *kilogram calorie*, which is written in abbreviated form kcal, or Cal. This is the large calorie. It is 1000 times larger than the small calorie, and it represents the amount of heat required to raise the temperature of one kilogram of water one degree Celsius.

The kilogram calories produced by complete oxidation of different food substances as measured in a bomb-type calorimeter are averages based on a knowledge of the relative amounts of these substances in foods. It is known that glucose, when completely oxidized, produces only 3.75 kcal/g, whereas starch and glycogen produce 4.22 kcal/g. The average for the oxidation of carbohydrates in the bomb calorimeter is 4.1 kcal/g. The average for a similar oxidation of fats is 9.5 kcal/g and that for proteins is 5.6 kcal/g.

Experimentation shows that the oxidation of carbohydrates and fats within the body is as complete as the oxidation of these foods by actual burning. These two kinds of foods are oxidized to carbon dioxide and water in both the calorimeter and the body. Proteins are not completely oxidized in the body. The nitrogenous by-products are discarded mainly as urea and ammonia. These end products contain a certain amount of energy, which is released by further oxidation in the calorimeter but not in the body. The body in its metabolism, therefore, derives only 4.3 kcal/g from protein.

Not all food materials are completely digestible. Plant foods, for example, contain cellulose, which is indigestible. Carbohydrate foods are estimated to be about 98 percent digestible, fats 95 percent, and proteins 92 percent. The figures often quoted for the energy value of foods per gram are carbohydrates, 4.1 kcal/g; fats, 9.3 kcal/g; and proteins, 4.1 kcal/g. These figures were originally determined by experimentation with dogs who were fed carefully controlled diets. The figures, therefore, may not be entirely accurate for human beings eating an average mixed diet. Figures for practical use in human nutrition can be simplified as shown in the table.

Food	kcal/g
Carbohydrates	4
Fats	9
Proteins	4

INDIRECT CALORIMETRY One of the problems of direct calorimetry is that it is difficult to determine the exact value in calories for each of the three kinds of food. Even if a person is fed only carbohydrate food while reclining in a large calorimeter, body fats and proteins are utilized also. We have already observed that most nitrogen obtained from the breakdown of protein foods is eliminated in the urine. Nitrogen composes about 16 percent of the protein molecule, or 1 part in 6.25. In the person who is fasting or resting in a postabsorptive state, all nitrogen recovered from the urine will be a by-product of the breakdown of protein. The weight of protein utilized may be calculated by multiplying the weight of the urinary nitrogen in grams by 6.25. Since the oxidation of protein in a calorimeter produces approximately 4.3 kcal/g, the weight of the protein multiplied by 4.3 gives the caloric value of the protein utilized. The caloric value of carbohydrates and fats during the experimental period also can be determined for use in direct calorimetry. A study of the respiratory quotient explains how this can be accomplished.

Respiratory quotient A definite ratio, called the *respiratory quotient*, exists between the volume of oxygen used and the volume of carbon dioxide given off during the oxidation of carbohydrates and fats. This invariant ratio provides an indirect method of calculating the caloric value of these foods. The equation for the production of glucose by photosynthesis is reversible. Repeating an equation used in Chap. 3, we find the following:

$$C_6H_{12}O_6 + 6O_2 + \text{catalysts} \rightleftharpoons 6H_2O + 6CO_2 + \text{energy}$$

In this balanced equation, it is evident that just as much oxygen is used as carbon dioxide is given off. Therefore

$$\frac{6 \text{ (volumes of } CO_2)}{6 \text{ (volumes of } O_2)}$$

gives a respiratory quotient of 1 for carbohydrate.

The respiratory quotient for fats is less than that for carbohydrates. The proportion of oxygen to hydrogen and carbon is much smaller in a molecule of fat. This necessitates the utilization of a greater proportion of hydrogen and carbon to produce a greater amount of energy (9 kcal/g). When fats are oxidized, oxygen combines with the hydrogen to form water, and more oxygen is needed to complete the oxidation. Since the volume of oxygen used is greater than the volume of carbon dioxide eliminated, the respiratory quotient for fats is less than 1, or about 0.7.

The calculation of the respiratory quotient for proteins is more difficult because of the varied and complex nature of these food substances. The protein respiratory quotient lies between the figure given for carbohydrate and that given for fats, and is usually considered to be 0.8. When, in addition to protein, equal quantities of carbohydrate and fats are oxidized, as in a mixed diet, the respiratory quotient is approximately 0.85. The respiratory quotient in the postabsorptive state after fasting for 12 h is 0.82.

The nonprotein respiratory quotient may be expressed as the amount of oxygen consumed in kilogram calories per liter. The burning of a given amount of nonprotein food requires the consumption of a certain amount of oxygen. If this ratio is constant, the heat value for any respiratory quotient can be determined by measuring the volume of oxygen consumed. Taking the respiratory quotient at 0.82, as in fasting, we find that 4.825 kcal of heat are produced per liter of oxygen consumed. Protein metabolism can be determined by calculation from the amount of nitrogen in the urine, as we have stated. This is done in the most exact calculations of metabolism, but since the amount of error is slight, ordinary calculations do not include this phase of protein metabolism. The measurement of oxygen consumption based on the respiratory quotient is a form of indirect calorimetry rather than a direct measurement of the heat given off by the body.

BASAL METABOLISM   The study of metabolism has led to attempts to establish a standard test of metabolism that would apply to all individuals. Upon first consideration, it seems that a number of factors may vary considerably from one individual to another. For example, one person might have eaten a heavy meal just before the test, while another might have eaten little or nothing. This variable is accounted for by requiring all persons to fast for 12 to 16 h before taking a basal metabolism test. Everyone is tested, then, while in a postabsorptive state. Ordinarily metabolism tests are conducted in the morning before the subject has eaten breakfast. Fasting permits the food from the last meal to be digested and absorbed and eliminates food energy as a variable factor. Secondly, some persons exercise more than others; therefore some individuals produce more heat as a result of muscular activity. This source of variability can be reduced to a minimum by having the subject lie quietly on a cot for 15 to 30 min before submitting to the test. Of course, there is always some muscular activity in the heart, breathing muscles, and other viscera, but this factor does not change from the beginning to the end of the test period. A basal metabolism test measures the energy output of the resting body by recording the amount of heat given off. Such testing does not measure the lowest energy output, because that occurs during sleep. Suggestions that basal metabolism be called *standard metabolism* have not met with general acceptance.

The subject inhales oxygen from the breathing chamber for a recorded length of time, usually 6 to 15 min (Fig. 20.2). At the end of the test period, the subject's basal metabolism is calculated either from the volume of oxygen used or from the carbon dioxide exhaled and absorbed by the soda lime in the basal metabolism apparatus. Ordinarily the computation is based on the volume of oxygen used. Basal metabolism, or the *basal metabolic rate* (BMR), is usually calculated as calories per square meter of body surface per hour, or $cal/(m^2 \cdot h)$.

FIGURE 20·2

Basal metabolism apparatus. The diagram indicates the circulation through the apparatus. (*Courtesy of the Sanborn Company.*)

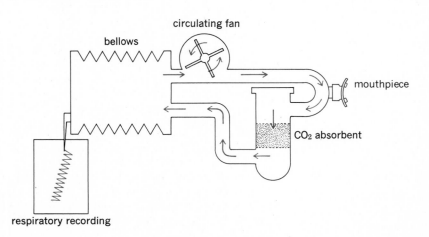

The person who consumes the greater volume of oxygen should be expected to have the greater rate of metabolism, but a large person should also be expected to utilize more oxygen than a small person. A large individual having a greater body surface should also give off more heat than a small individual. Body size obviously has to be taken into consideration in determining the basal metabolism rate.

An increase in weight means an increase in cubic measurement, as well as in the surface measurement, but these two do not increase proportionately. It has been found that increases in basal metabolism are proportional to increases in body surface area rather than to increases in weight. The surface of the body can be accurately measured, but for purposes of measuring metabolism, it is estimated from a chart that shows the body surface in square meters in relation to height and weight. A person weighing 137 pounds and standing 5 ft 6 in tall would have a body surface of 1.7 m^2. Let us suppose that this person took a basal metabolism test and consumed 2400 l of oxygen over a test period of 10 min. The nonprotein respiratory quotient during fasting is 0.82. The number of kcal/l O$_2$ for this respiratory quotient is 4.825. The total heat per hour can then be calculated as follows:

$$2400 \times 4.825 \times \frac{60}{10} = 69.48 \text{ kcal}$$

To find the number of kilogram calories per square meter of body surface per hour, we divide 69.48 by 1.7, which gives 40.87 kcal/(m^2·h). The BMR is usually close to 40 kcal/(m^2·h) in young men. It is often expressed on the basis of a 24-h day, although this is hardly accurate, since one does not remain at the basal level for that length of time. However, expressing it this way provides a useful comparison with the calories needed for a 24-h day that includes the usual daily activities. The calculation on the basis of a 24-h day for the individual mentioned above who had a heat production of 40.87 kcal/(m^2·h) and 1.7 m^2 of body surface is as follows:

$$24 \text{ h} \times 40.87 \text{ kcal/(m}^2\text{·h)} \times 1.7 \text{ m}^2 = 1667.5 \text{ kcal/day BMR}$$

Women of eighteen or nineteen years can be expected to have an average BMR of around 37 kcal/(m^2·h). A woman 5 ft 2 in tall and weighing 115 lb would have 1.5 m^2 of body surface. Women of this age commonly have basal rates of 1200 to 1400 kcal/day.

$$24 \text{ h} \times 37 \text{ kcal/(m}^2\text{·h)} \times 1.5 \text{ m}^2 = 1332 \text{ kcal/day}$$

Basal metabolism rates are considerably higher in children. The BMR decreases sharply from childhood until around the age of twenty, when it tends to level off. It then decreases gradually with age.

Physicians commonly express BMR as the percentage above or below normal standards worked out for different sex and age groups. According to this system, a BMR of +5 means that the total heat production is 5 percent above the average for individuals of that sex and age group. Since ±15 percent is considered to be a normal limit of deviation, a +5 rate is not significantly different from normal. This system is open to criticism in that it may not indicate an increase in heat production for the person who is overweight. The BMR for the overweight person determined by this method may be considered normal in relation to body surface, but the actual heat production in kilogram calories per hour may be greatly above normal.

The secretions of some of the endocrine glands, notably the thyroid, have a marked effect upon metabolism. When the thyroid is overactive, the metabolic rate is commonly high.

As we have seen, the basal metabolism test measures energy expenditure by measuring the amount of oxygen used during a certain time period in the resting state. The BMR, however, has limited diagnostic value since many factors affect it and since it is a time-consuming test.

There are several tests that are more accurate from a clinical standpoint and that can be given in less time. We know that the secretions of the thyroid gland have a marked effect on metabolism. We have already mentioned the *protein-bound iodine* serum test (PBI) (Chap. 13). This blood test measures the uptake of iodine by the thyroid gland and indicates the amount of hormone secretion by the thyroid gland. Another test of metabolic rate involves measuring the uptake of *radioactive* iodine by the thyroid gland. A tracer ^{131}I is introduced and its uptake is measured. Iodine uptake in the normal person is between 15 to 40 percent in 24 h. A *triiodothyronine* T_3 test measures the activity of red blood cells in binding T_3. A small amount of blood is incubated with radioactively labeled T_3 for several hours. The blood cells are then washed several times, and the remaining radioactive activity is calculated. This type of test has the advantage of not requiring the introduction of a radioactive substance into the bloodstream of the patient.

SPECIFIC DYNAMIC ACTION   Foods, as we have seen, are a source of energy; not only do they provide for immediate needs, but they enable the body to conserve its energy resources and store an energy reserve in the form of stored fat. One interesting effect of food utilization is a rise in total body-heat production, which is not accounted for by the energy required in the digestion or storage of food. This effect raises the metabolism above the basal level and is called the *specific dynamic action* (SDA) of food. Carbohydrate and fats raise the metabolism a small percentage above the basal level, but when protein is fed to a *fasting individual,* a very considerable rise in total heat production occurs. The rise is equal to 25 to 30 percent of the energy value of the protein, and the effect lasts for several hours. Carbohydrate causes a rise in heat production of about 6 percent and fats of about 4 percent when fed separately.

If the total heat production rises above the energy value of the protein fed, the body must use its own energy reserve in the utilization of this food substance. Furthermore, if loss of weight is to be prevented, this energy loss has to be taken into consideration. For practical purposes in calculating the total energy requirement, the specific dynamic effect accounts for only 6 to 10 percent of the calories used by a nonfasting individual on a mixed diet.

The source of heat produced by SDA appears to be the energy released by the metabolism of degraded food products in the liver. This extra heat production is not seen in an animal whose liver has been removed.

ENERGY REQUIREMENTS FOR WORK   *Mental work,* strangely enough, requires only a very small outlay of energy. F. G. Benedict and C. G. Benedict found that the extra calories required for 1 h of intense mental effort could be obtained by eating one oyster cracker or by consuming one-half of a salted peanut.

Muscular work has a pronounced effect, however, in raising the metabolism and therefore the need for more food energy. A man sitting quietly has a total metabolism of about 100 kcal/h. Moderate exercise can increase the rate of metabolism to around 300 kcal/h. Heat is produced as a result of muscular activity; some of this heat is used to keep the body warm, and some is lost from the body surface. Muscular exercise on a hot day produces more heat than is needed, and the body's heat-regulating mechanism attempts to provide greater heat loss through evaporation of sweat, for example. In a very cold environment we may exercise just to keep warm. Muscular activity is the most important single factor affecting the expenditure of energy and thus in determining the amount of food required to maintain something of a balance in metabolism (Table 20.2).

Total calories for a 24-h day We could not be active at the basal rate of metabolism; neither could a person work continuously for a 24-h day. Obviously the average day is divided into periods for sleeping, working, and resting. For many, the working day may be subdivided into periods of light exercise and periods of moderately heavy muscular work.

Some men in sedentary occupations may need less than 3000 kcal/day; many who perform heavy muscular work need more. Women engaged in moderately active work need about 2000 kcal/day. Growing girls from 15 to 18 years require about 2100 kcal/day. Boys in the same age group require about 3000 kcal/day (see Table 20.3.)

Reducing diets Consuming more food calories than are expended in energy during the day should result in an increase in weight, while working harder and eating less should cause one to lose weight. Even though this is basically true, losing weight by working off fat is usually a difficult way to reduce. Studies show that a brisk walk for an hour burns only 250 kcal. It is interesting that walking upstairs requires 2.5 times more energy than is expended in walking downstairs (Table 20.4). After exercise a person usually develops an increased appetite and is likely to eat more than his or her normal amount of food. The best way to reduce is to limit the intake of food

TABLE 20·2
Total calories needed for a 24-h day°

Activity	kcal
8 h sleep, 65 kcal/h	520
2 h preparation for work, 125 kcal/h	250
8 h moderate work, 200 kcal/h	1600
2 h evening chores, 125 kcal/h	250
4 h sitting at rest, 100 kcal/h	400
	3020

° These are estimates of calories needed for a 24-h day based on the probable energy requirements of a man of average size (70 kg, or 154 lb).

TABLE 20·3
Calories needed for energy at various ages °

	Age (years)	Weight (kg)	Weight (lb)	Height (cm)	Height (in)	Energy (kcal)
Infants	0.0–0.5	6	14	60	24	kg × 117
	0.5–1.0	9	20	71	28	kg × 108
Children	1–3	13	28	86	34	1300
	4–6	20	44	110	44	1800
	7–10	30	66	135	54	2400
Males	11–14	44	97	158	63	2800
	15–18	61	134	172	69	3000
	19–22	67	147	172	69	3000
	23–50	70	154	172	69	2700
	51+	70	154	172	69	2400
Females	11–14	44	97	155	62	2400
	15–18	54	119	162	65	2100
	19–22	58	128	162	65	2100
	23–50	58	128	162	65	2000
	51+	58	128	162	65	1800
Pregnant						+300
Lactating						+500

° Food and Nutrition Board, National Academy of Sciences, National Research Council, revised 1974.

calories to a level somewhat below the normal energy requirement. Doing this enables the dieter to lose weight slowly, at the rate of a pound or so per week. Severe dieting can be harmful. Even moderate dieting probably should be undertaken only upon the advice of a physician, for a person cannot simply stop eating. The body can supply some reserve fats for fuel, but it cannot supply enough protein, vitamins, and minerals essential to health (Table 20.5).

The proper balance between the three kinds of food eaten is desirable whether one is dieting or not. Carbohydrates represent a readily available source of energy and usually constitute 40 to 50 percent of the diet. It is common for Americans to include a greater than optimum percentage of starch and sugar in their diets. Fats are the best energy foods for conversion into heat. In addition, they supply certain essential acids (linoleic, linolenic). Various sources recommend that fats compose 35 to 45 percent of the daily diet. Protein is used for the growth and repair of tissues. The normal diet should provide an assortment of protein foods that will contain all the essential amino acids. The recommended allowance for protein is 10 to 15 percent of the total calories consumed. Protein foods supply nitrogen, which is essential to the growth and maintenance of protoplasm.

An average diet usually contains the three kinds of food in proper proportion and an adequate supply of vitamins and minerals. A reducing diet should contain the same essentials, but with a reduced amount of carbohydrates and fats. Minerals such as sodium, potassium, and phosphorus are usually present in adequate amounts in both normal and reducing diets. A reducing diet should not be deficient in calcium, iron, and iodine. Other considerations in dieting include spacing of meals to avoid a feeling of

TABLE 20·4
Metabolic cost of various activities

Activity	Body wt, kg	kcal/min	kcal/(kg/10 min)
Sleeping	68.1	1.17	0.172
Resting in bed	73.2	1.26	0.174
Sitting, normally	73.2	1.29	0.176
Sitting, reading	73.2	1.29	0.176
Lying, quietly	68.1	1.33	0.195
Sitting, eating	73.2	1.49	0.204
Sitting, playing cards	73.2	1.53	0.210
Standing, normally	73.2	1.50	0.206
Classwork, lecture	68.1	1.67	0.245
Conversing	68.1	1.83	0.269
Personal toilet	73.2	2.02	0.278
Sitting, writing	82.0	2.20	0.268
Standing, light activity	73.2	2.60	0.356
Washing and dressing	68.0	2.60	0.382
Washing and shaving	62.0	2.60	0.419
Driving a car	64.0	2.80	0.438
Washing clothes		3.13	
Walking indoors		3.11	
Shining shoes	73.2	3.20	0.437
Making bed	59.1	3.38	0.572
Dressing	73.2	3.40	0.466
Showering	73.2	3.40	0.466
Driving motorcycle	64.0	3.40	0.531
Cleaning windows	61.0	3.70	0.607
Sweeping floors	73.2	3.91	0.535
Ironing clothes	67.0	4.20	0.627
Mopping floors	73.2	4.86	0.665
Walking downstairs	73.2	7.14	0.976
Walking upstairs	73.2	18.58	2.540

SOURCE: Courtesy of Consolazio, Johnson, and Pecora, "Physiological Measurements of Metabolic Functions in Man," McGraw-Hill Book Company, New York, 1963.

weakness or fatigue and eating enough plant or vegetable material to keep up the bulk of food material along the digestive tract.

FASTING AND STARVATION *Fasting* and *starvation* are both postabsorptive states, but fasting is usually considered to be of shorter duration than starvation. In either case the gastrointestinal tract empties and the fasting individual must depend upon stored food products. A healthy individual can fast for many weeks if supplied with water to drink. In a prolonged fast, vitamins are also helpful.

To maintain an adequate blood glucose concentration for proper functioning of the nervous system during fasting, liver glycogen is used, but this source is rapidly depleted. Glycogen in muscle can be used, but it must first be broken down be glycolysis to pyruvate and lactate. These products are absorbed by the blood and carried to the liver, where they are synthesized into glucose.

Even though glucose represents a ready source of energy, the quantity of

TABLE 20·5
Metabolic cost of manual labor activities

Activity	Body wt, kg	kcal/min	kcal/(kg/10 min)
Metal working	68.1	3.50	0.514
House painting	68.1	3.50	0.514
Carpentry	68.1	3.84	0.564
Farming chores	68.1	3.84	0.564
Plastering walls		4.10	
Truck and automobile repair	68.1	4.17	0.612
Farming, planting, hoeing, raking	68.1	4.67	0.686
Mixing cement		4.70	
Repaving roads	68.1	5.00	0.734
Gardening, weeding	65.0	5.60	0.862
Stacking lumber	68.1	5.83	0.856
Stone, masonry	68.1	6.33	0.930
Pick-and-shovel work	68.1	6.67	0.979
Farming, haying, plowing with horse	68.1	6.67	0.979
Shoveling (miners)		6.80	
Hewing with a pick (miners)		7.00	
Chopping wood	68.1	7.50	1.101
Gardening, digging	63.0	8.60	1.365

SOURCE: Courtesy of Consolazio, Johnson, and Pecora, "Physiological Measurements of Metabolic Functions in Man," McGraw-Hill Book Company, New York, 1964.

glucose produced by the liver is not nearly adequate for the energy needs of the body, which would normally be around 3000 kcal/day. To meet energy needs during fasting, most tissues of the body, with the exception of the nervous system, utilize fats. This arrangement has a sparing effect on glucose produced by the liver and permits the nervous system to function normally.

Adipose tissue is a source of glycerol. Triacylglycerides break down into glycerol and fatty acids. The former can be converted into glucose by the liver.

Proteins can provide amino acids to be converted to glucose by the liver. In prolonged fasting or starvation, the nervous system uses ketone bodies as well as glucose for energy, thus sparing body proteins. In prolonged fasting and starvation depletion of body protein brings about serious problems.

WATER BALANCE   Water is not a food, but it plays an important part in the chemical reactions involved in metabolism. Water balance must be considered in relation to any study of a proper diet. The average person has an intake of about 2500 ml water/day, and water loss is about the same. Only 100 to 1200 ml of this amount is acquired by drinking. Food accounts for a similar amount. A potato, lean meat, or a tomato may appear as more or less solid food, but all contain water in considerable quantity. If the diet contains soup and fruits, for example, the necessary intake by drinking may be somewhat less. Finally there is the water of oxidation, which is produced as a result of metabolic processes through the oxidation of hydrogen in the food or tissues. The water of oxidation amounts to approximately 300 ml.

HEAT PRODUCTION AND BODY-TEMPERATURE REGULATION We have touched briefly in earlier chapters on temperature regulation by the hypothalamus. The skin, muscle activity, and blood flow are accessories to this regulation. However, we should consider the source of body heat. Nearly all chemical reactions in the body produce some heat, but most of the heat comes from the breakdown of absorbed food materials. We know that the hypothalamus regulates body heat and keeps the body temperature constant at 37°C or 98.6°F.

If the temperature outside the body rises, sensors in the skin are evidently able to stimulate thermal receptors in the hypothalamus to make adjustments for cooling the body. If the ambient temperature falls, shivering may occur and the rapid muscular contractions produce heat. At the same time, circulatory adjustments tend to conserve heat.

Human beings, as well as all other warm-blooded animals, prefer that the surrounding air temperature should be several degrees below their body temperature so that their bodies can lose heat into the environment. For example, we prefer a room temperature of around 72°F. Loose clothing provides an air space for insulation against changes in ambient temperature.

Heat stroke, or *sunstroke,* occurs when the body's cooling mechanism is unable to cope with hot surrounding air temperature and high humidity. The victim's skin is dry, and perspiration ceases. The body temperature rises very high, reaching 104°F or higher. Rapid breathing, a weak pulse, and dizziness are characteristic of heat stroke. This condition should be treated as an emergency. Rapid cooling of the victim is imperative.

Heat exhaustion presents a different picture. The body temperature is not inordinately high, profuse perspiration occurs, and the skin feels moist. Nausea and a feeling of weakness may be the chief complaints. Heat exhaustion is more a state of collapse with depletion of blood plasma volume, dilatation of capillaries in the skin, and general exhaustion. The victim should lie down and rest. Sips of water and stimulants such as coffee or tea may be given. A physician may be required if the patient does not respond, but heat exhaustion is ordinarily not as serious as heat stroke.

When a *fever* occurs, the hypothalamus heat center seems to be "set" at a higher point. The body continues to conserve heat, the skin is dry, and the body temperature rises. The mechanism for resetting the body thermostat in fever is not entirely clear. It is known that the toxins produced by disease and inflammation release protein substances called *endogenous pyrogens* into the extracellular fluid. The pyrogens cause the thermoreceptors in the hypothalamus to raise the set point and produce a fever.

It is interesting that fever is often preceded by a "chill." Since the heat center is set at a higher level than the normal blood temperature, the patient may shiver and feel cold until the body temperature rises to a fever level. Often when the fever "breaks," bed covers are thrown off and the patient perspires as the body temperature returns to normal.

It has been suggested that fever, within limits, is actually beneficial because it enhances the individual's resistance to bacterial disease. The body's defense mechanism may thus be aided indirectly by the increase in temperature. The mechanism of this protective action, however, has not been determined.

CLINICAL ASPECTS   Numerous disorders of metabolism are connected with inability to use certain nutrients. Many of these are disorders of amino acid metabolism.

GENETIC DISORDERS *Phenylketonuria* *Phenylketonuria* is an inherited condition resulting from inability of an individual to oxidize phenylalanine to tyrosine. Phenylalanine is an essential amino acid normally used in the synthesis of proteins. It must be obtained from foods, since the body is unable to synthesize it. Most phenylalanine that is not used for building protein is normally converted to tyrosine, but in individuals suffering from phenylketonuria, a metabolic block prevents this conversion. These individuals are homozygous recessives for a gene that cannot direct the production of the enzyme phenylalanine hydroxylase, which is essential for the conversion of phenylalanine to tyrosine.

Under these conditions an abnormal amount of phenylpyruvic acid is produced and excreted in the urine. The condition must be detected as soon as possible in infancy since serious injury to the brain results from this defect. A simple diaper test may detect this condition. A drop or two of 10 percent ferric chloride solution is placed on a diaper wet with the infant's urine. If the treated spot retains an orange color this indicates the urine is normal, but if phenylpyruvic acid is present, a blue-green color appears, indicating that the ferric chloride is reduced by phenylpyruvic acid.

This test cannot be used with newborn infants since phenylpyruvic acid would not be excreted until phenylalanine in the blood reaches a high level. The time required for phenylalanine to rise to this point varies from a few days to a few weeks in infants with phenylketonuria.

A clinical fluorimetric test for the determination of phenylalanine in serum can be performed in a laboratory equipped with a photofluorometer.

As we have seen, phenylalanine must be obtained from food, but tyrosine may either be obtained from food or from the oxidation of phenylalanine of the diet. In phenylketonuria, phenylalanine cannot be converted to tyrosine, and phenylpyruvic acid accumulates in abnormal amounts. To alleviate this undesirable effect, the infant is given a special diet containing a level of phenylalanine low enough to prevent a buildup of phenylpyruvic acid but high enough to enable the body to manufacture proteins for growth. On this delicately balanced diet, the child is able to grow and develop normally. The relatively new diet, if used early enough, greatly lessens the danger of brain damage, whereas a normal diet in these cases may very well lead to severe mental retardation.

Cystic fibrosis   A disease that affects infants and children primarily, *cystic fibrosis* is inherited as an autosomal recessive trait. The condition is characterized by thick mucous secretions in the lungs, the bile ducts, and the pancreatic ducts. Pancreatic enzymes are prevented from reaching the intestine, where they are needed for the digestion of food. Little or no lipase is therefore available for the digestion of fats. The child who suffers from cystic fibrosis is often emaciated because of the inability to absorb fats or the fat-soluble vitamins A, D, and K.

Another aspect of the disease, which is of diagnostic value, is an unusually high loss of sodium and chloride in the perspiration.

Treatment consists of supplying pancreatic extracts; water-soluble forms of vitamins A, D, and K; additional table salt; and antibiotics to control infections. Pulmonary congestion is treated by vigorous physical therapy. A strictly controlled diet low in fats and rather high in carbohydrates and proteins is recommended.

VITAMIN-DEFICIENCY CONDITIONS *Vitamin A deficiency* As we have seen, vitamin A deficiency may lead to night blindness (Chap. 11). Vitamin A deficiency also affects the epithelial tissues in various parts of the body. Perhaps the most striking changes are those that occur in the cornea and conjunctiva of the eye, causing "dry-eye disease," or xerophthalmia. Inflammatory conditions of the eye are produced in a high percentage of rats fed on a diet lacking vitamin A. Epithelial tissues of the skin and those of the respiratory, digestive, and urinogenital systems are also affected. There is a tendency toward squamous hyperplasia, a condition in which stratified squamous epithelium proliferates. The sweat glands and sebaceous glands of the skin become inflamed; such inflammation is seen as eruptions or sores. Early experimentation showed that rats fed on a vitamin A–free diet failed to gain weight, and were subject to malformation of bone.

Vitamin D deficiencies *Rickets* is a vitamin D–deficiency condition that produces a malformation of growing bones. Rickets affects children and young animals who have not received adequate amounts of fat-soluble vitamin D or sunlight. Calcium and phosphorus metabolism are disturbed in this condition, and poor calcification of bone results. The poorly calcified leg bones do not support the weight of the body well, and they bend, producing knock-knees or bowlegs. Other skeletal malformations ascribed to rickets are narrow chest, scoliosis, and malformations of the skull and pelvis. In young growing bones, a very evident defect in calcification at the junction of the epiphysis and the shaft of the bone is seen. In adults such deficiencies cause more calcium and phosphorus salts to be removed than are deposited. Loss of strength in the bones can lead to deformity. The condition is essentially adult rickets, or *osteomalacia*.

Several substances have antirachitic properties. The two of greatest importance are *ergosterol* and *7-dehydrocholesterol*. These substances are activated by irradiation with ultraviolet light. Ergosterol is a sterol originating in plants. Activated ergosterol is known as vitamin D_2, or ergocalciferol. Commercial forms are called *viosterol* and *calciferol*. Yeast is a good source for commercial preparations of ergosterol. Activated 7-dehydrocholesterol produces cholecalciferol or vitamin D_3, which is found in animal fats and is the natural vitamin of fish-liver oils. This is also the substance in the mammalian skin that is activated by ultraviolet light. There is no vitamin D_1.

Riboflavin deficiency The effects of riboflavin deficiency are not as definite as those of some other vitamin deficiencies. Animals do not grow properly, and a loss of hair is seen in rats (Fig. 20.3). Vascular changes in the cornea, conjunctivitis, and cataract have been described for various animals. Human deficiency conditions include cracks and sores at the corners of the mouth and ocular changes involving the cornea and the conjunctiva. Several investigators have reported dry and scaly skin as a riboflavin deficiency.

Thiamine deficiency It is thought that the accumulation of pyruvic acid caused by the failure of an intermediary metabolic process is one of the factors

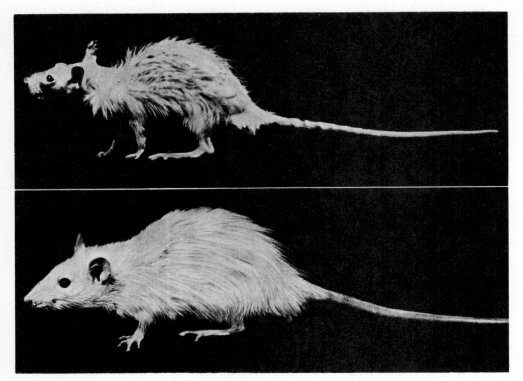

FIGURE 20·3

Riboflavin deficiency. The rat in the upper photograph is 28 weeks old and has had no riboflavin. Note the loss of hair, especially around the head. Weight is 63 g. The lower photo shows the same rat 6 weeks later, after receiving food rich in riboflavin. It has recovered its fine fur and now weighs 169 g. (*Courtesy of U. S. Department of Agriculture, Bureau of Human Nutrition and Home Economics, Washington, D.C.*)

producing symptoms of *polyneuritis*. This condition is relieved by adding thiamine to the diet.

Niacin (nicotinic acid) and nicotinic acid amide deficiency   Pellagra is a condition characterized by red lesions of the skin, especially on the backs of the hands and on the forearms, legs, and feet. The tongue assumes a bright red color, and eventually symptoms of depression or even dementia appear, indicating involvement of the central nervous system. Pellagra is found predominantly among persons of poor economic status whose diet is confined largely to cornmeal, white flour, polished rice, and sugar. Niacin is considered to be specific for the treatment of pellagra.

Pantothenic acid deficiency   A deficiency of pantothenic acid in animals is indicated by a type of dermatitis and by failure to grow. Deficiencies seldom occur in human beings, probably because adequate amounts of the vitamin are found in the foods eaten.

Vitamin C deficiency   Scurvy, a vitamin C–deficiency disease, is characterized by weakness and lassitude. Marked tenderness and swelling of the joints is seen, and the gums become red and swollen. The teeth become infected around their bases, causing them to loosen. Change in the integrity of the capillaries occurs, permitting hemorrhagic conditions beneath the skin, in mucous membranes, and under the periosteum. Eating fresh fruits and vegetables should provide enough vitamin C to ward off scurvy.

SUMMARY

1 There are three kinds of food: carbohydrates, fats, and proteins. Carbohydrates may be classified as monosaccharides, disaccharides, and polysaccharides. Monosaccharides are simple sugars, such as glucose; table sugar, or sucrose, is an example of a disaccharide; polysaccharides are complex carbohydrates, such as starch or cellulose.

Neutral fats are produced by the chemical combination of glycerol with fatty acid groups. Examples of compound lipids or lipoid substances are the phospholipids, lecithin, and cephalin. Sterols are secondary alcohols. However, they often are combined with fatty acids as esters; cholesterol is an example.

Proteins are used for building new tissue and for tissue repair. Their building blocks are the amino acids. Protein molecules are usually large and complex, exhibiting almost endless variety.

2 Vitamins are food substances, present in minute amounts but essential to health. Many have been demonstrated to be coenzymes essential to vital metabolic processes. Vitamins A, D, and E are fat-soluble. Vitamin A is necessary for the formation of retinal and rhodopsin. Vitamin D is concerned with calcium and phosphorus metabolism. A deficiency condition may give rise to rickets.

Vitamins of the B complex are water-soluble. Thiamine is one of the B vitamins. It plays a part in cellular metabolism. The deficiency condition is beriberi.

Riboflavin is a respiratory coenzyme. It is included in a group of flavoproteins that act as electron acceptors from the coenzymes NAD and NADP. Flavoproteins pass electrons along to the cytochrome system.

Nicotinamide acts as a coenzyme with NAD and NADP in cellular respiration.

Other vitamins of the B group are pyridoxine (B_6), pantothenic acid, inositol, biotin, folic acid, B_{12}, and choline. Pantothenic acid is the precursor of coenzyme A. Vitamin B_{12} is an erythrocyte maturation factor, which is effective in the treatment of pernicious anemia.

Ascorbic acid, or vitamin C, prevents the development of scurvy. It also promotes the healing of wounds, the stability of capillaries, and resistance to infections. It is found in fresh fruits and vegetables.

The tocopherols, or vitamin E substances, have been shown to act as antisterility substances for white rats.

Vitamin K stimulates the production of prothrombin by the liver and so increases the clotting ability of the blood.

3 The energy from the breakdown of foodstuffs is stored as high-energy bonds of the ADP-ATP system. This is largely accomplished by way of the tricarboxylic acid cycle, which is a common pathway in the metabolism of all three kinds of foods.

4 The amount of heat given off in the utilization of food can be measured by either direct or indirect calorimetry. By either method the approximate energy values are as follows: carbohydrates, 4 kcal/g; fats, 9 kcal/g; proteins, 4 kcal/g.

5 Basal metabolism tests attempt to determine the energy given off by the resting body in a postabsorptive state by measuring the amount of heat given off. Basal metabolic rates for young women are commonly in the range of 1200 to 1400 kcal/day. The BMR for young men is more likely to be around 1600 kcal/day. Calorie needs vary with a number of factors

such as age, sex, and weight. In general, the active young man probably needs 3000 or more kcal/day, whereas young women, moderately active, probably need only about 2100 kcal/day. Boys in the 13- to 18-year age group need a higher caloric intake than the average person; older persons need less.

6   The average American diet consists of carbohydrates, 40 to 50 percent; fats, 35 to 45 percent; and proteins, 10 to 15 percent.

7   The average intake of water is around 2500 ml/day. About 1200 ml of this amount comes from drinking; the rest is acquired from food and from the water of oxidation.

QUESTIONS

1   In what forms are the different kinds of food absorbed? Where are they absorbed?

2   Discuss the functioning of the liver with respect to the end products of carbohydrate, protein, and fat digestion.

3   Is glucose or a related substance stored by animal organisms? If so, where and in what form? Give reasons for avoiding obesity.

4   What is the energy source for green plants? What is the energy source for animal organisms?

5   Discuss the chemical relationship of various food carbohydrates with which you are acquainted.

6   What are fats? Name some of the different kinds of lipids. Of what use are fats in metabolism?

7   Discuss the kinds and chemical composition of proteins.

8   Why are minerals necessary in an adequate diet? Which vitamin contains cobalt?

9   Describe the essential facts regarding vitamin sources and deficiencies.

10 Describe the varied functions of vitamins of the B group.

11 What is meant by direct calorimetry?

12 When a person is placed inside a calorimeter in order to measure the heat given off from the body, is this measurement technique called direct or indirect calorimetry?

13 How does indirect calorimetry measure heat loss? Describe the technique.

14 How is the respiratory quotient used in indirect calorimetry?

15 Discuss the concept of "basal metabolism." What is the difference between metabolism and basal metabolism?

16 Under what conditions can the basal metabolic rate be accurately determined?

17 What does the term "specific dynamic action" mean?

18 Considering your daily activities, estimate your food requirement in kilogram calories per day.

SUGGESTED READING

Corradino, R. A., and R. H. Wasserman: Vitamin D_3: Induction of Calcium-binding Protein in Embryonic Chick Intestine in Vitro, *Science,* **172:**731–733 (1971).

Forker, E. L.: Mechanisms of Hepatic Bile Formation, *Annu. Rev. Physiol.,* **39:**329–347 (1977).

Heller, H. C., L. I. Crawshaw, and H. T. Hammel: The Thermostat of Vertebrate Animals, *Sci. Am.*, **239:**102–113 (1978).

Hendrix, T. R., and T. M. Bayless: Digestion: Intestinal Secretion, *Annu. Rev. Physiol.*, **32:**139–164 (1970).

Johnson, L. R.: Gastrointestinal Hormones and their Functions, *Annu. Rev. Physiol.*, **39:**135–158 (1977).

Kluger, M. J.: The Evolution and Adaptive Value of Fever, *Am. Sci.*, **66:**38–43 (1978).

Lieber, C. S.: The Metabolism of Alcohol, *Sci. Am.*, **234**(3):25–33 (1976).

Masoro, E. J.: Lipids and Lipid Metabolism, *Annu. Rev. Physiol.*, **39:**301–321 (1977).

Nyquist, S. E., F. L. Crane, and D. J. Morré: Vitamin A: Concentration in the Rat Liver Golgi Apparatus, *Science*, **173:**939–940 (1971).

Rothman, S. S.: The Digestive Enzymes of the Pancreas: A Mixture of Inconstant Proportions, *Annu. Rev. Physiol.*, **39:**373–389 (1977).

Scrimshaw, N. S., and V. R. Young: The Requirements of Human Nutrition, *Sci. Am.*, **235:**50–64 (1976).

Woolfson, A. M. J., R. V. Heatly, and S. P. Allison: Insulin to Inhibit Protein Catabolism After Injury, *N. Engl. J. Med.*, **300**(1):14–17 (1979).

Young, V. R., and N. S. Scrimshaw: The Physiology of Starvation, *Sci. Am.*, **225:**14–21 (1971).

21

THE REPRODUCTIVE SYSTEM

The reproductive system is unique among the body's organ systems, because the organs vary greatly between the sexes. Differentiation of the external genitalia is interesting from the standpoint of homologous or

686

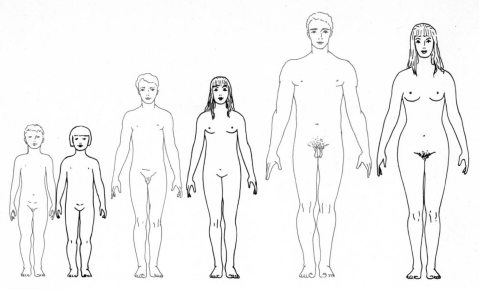

FIGURE 21·1

The male and female figures, in children and after puberty. Note that male and female children before puberty do not differ markedly in body form.

structurally similar forms. Male and female children do not differ remarkably in body form until they reach the age of puberty. At this time, under the influence of hormones, striking changes occur in several systems. In the male, the voice gradually changes to a deeper, masculine tone; the beard becomes a little stronger; pubic, axillary, and body hair develop; and the boy gradually assumes the characteristics of the adult male. The body form of the adult male shows increased musculature, with broader shoulders and narrow hips. In the female, hormonal secretions at puberty promote development of a feminine contour, caused largely by deposition of subepidermal fat, enlargement of the mammary glands, and a slight deepening of the voice. The internal and external genitalia of both sexes approach maturity at this time, and the gonads begin to produce mature sex cells (Fig. 21.1).

The *gonads* are the male and female primary sex organs, which produce the germ cells, or gametes. The male gonads, or testes, produce spermatozoa, and the female gonads, or ovaries, produce ova. The gametes are derived from undifferentiated germ cells, which develop by a process called *gametogenesis*. The development of male gametes, or spermatozoa, is called *spermatogenesis;* in the female the development of ova is referred to as *oogenesis.*

GAMETOGENESIS

The gonads form in the early embryo from two parallel ridges of peritoneal epithelium called the *genital ridges.* Primordial germ cells differentiate and appear in the area in which the reproductive organs will form. The exact origin of the primordial germ cells remains a biological problem, and the further development, or fate, of the primordial germ cells is also not completely understood. Germinal epithelium containing primordial germ cells is seen in 10-mm human embryos, but it is questionable whether these cells produce mature sperm or ova. The problem is of especial interest in the female, where the ovary contains thousands of potential ova at birth. Are these the cells that form mature ova many years later, or are new germ cells formed constantly from the germinal epithelium? Many primordial germ cells degenerate, while others divide and associate closely with the germinal epithelium of the gonads. At present there is no definitive answer to this question.

The development of mature spermatozoa and ova are complicated processes involving nuclear and cytoplasmic changes of great significance. There are marked differences between the development of spermatozoa and that of ova, but here we shall emphasize the similarities in the behavior of the sex cell chromosomes. Gametogenesis, the production of gametes, begins with the multiplication of cells derived from undifferentiated germ cells. The cells increase in number by mitotic division. These cells contain the complete, or diploid, number of chromosomes, which in a human being is 46. The number of human chromosomes was once thought to be 48, but more recent studies indicate that 46 is the correct number.

The developing germ cells pass through a period of growth and multiplication, after which certain cells may be designated in the male as *primary spermatocytes* and in the female as *primary oocytes* (Fig. 21.2).

MEIOSIS

In Chap. 3, we considered the behavior of the chromosomes during the process of mitosis. In mitotic division, each daughter cell has a chromosome number equal to that of the original cell. The primary objective of meiotic division, however, is the reduction of the chromosome number to half that existing in the original germ cells. Cells with the complete number of chromosomes are said to be in the *diploid* condition, whereas those with the reduced number are *haploid*. During meiosis the chromosome number is reduced from the diploid condition found in the undifferentiated germ cells to the haploid condition seen in the gametes (Fig. 21.3).

STAGES OF MEIOSIS *Metaphase I* The bivalent chromosomes (tetrads) move to the equatorial plate of the spindle.

Anaphase I and the first meiotic division The two tetrads belonging to each dyad move away from their homologues, often presenting a V-shaped appearance as they move toward opposite poles. The separation of the chromosome pairs reduces the number of the chromosomes to half in each cell produced in the subsequent cell division. The first meiotic division is thus termed a *reductional division*. The secondary spermatocyte or oocyte produced by this division contains the haploid number of double-stranded chromosomes. The centromeres of the individual chromosomes do not divide at this time.

Telophase I The telophase of the first meiotic division is usually a short period preceding the initiation of the second meiotic division.

The second meiotic division The secondary spermatocyte or oocyte passes through another prophase, metaphase, and anaphase. During the second metaphase, a spindle forms, and the chromosomes arrange themselves near the equator. The second anaphase is characterized by the division of the centromeres and the separation of the chromatids. This is commonly called an *equational division,* since no further reduction of chromosome number

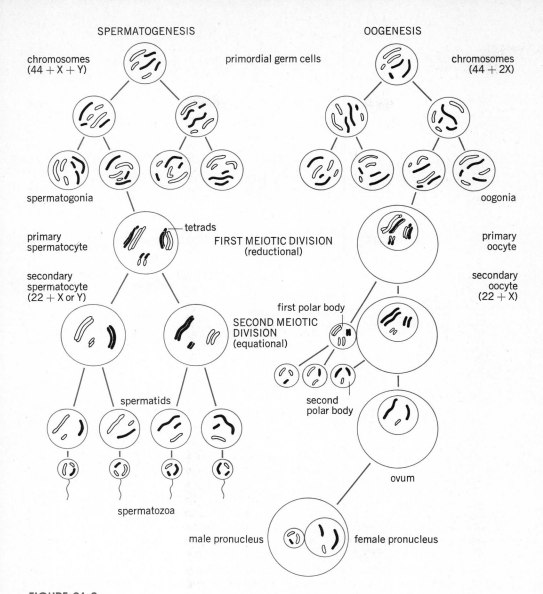

SPERMATOGENESIS

OOGENESIS

chromosomes
(44 + X + Y)

primordial germ cells

chromosomes
(44 + 2X)

spermatogonia

oogonia

primary
spermatocyte

tetrads

FIRST MEIOTIC DIVISION
(reductional)

primary
oocyte

secondary
spermatocyte
(22 + X or Y)

secondary
oocyte
(22 + X)

first polar body

SECOND MEIOTIC
DIVISION
(equational)

second
polar body

spermatids

ovum

spermatozoa

male pronucleus

female pronucleus

FIGURE 21·2

The behavior of chromosomes
during gametogenesis.

occurs. Each chromatid in the second meiotic division becomes a separate
chromosome. The spermatid or ovum contains the haploid number of
individual chromosomes.

ESSENTIAL DIFFERENCES BETWEEN MITOSIS AND MEIOSIS   During mitosis
homologous chromosomes do not pair; the individual chromosomes replicate themselves, and the duplicates then separate during cell division. Each
cell produced by a mitotic division receives the same number of chromosomes, which have the same genetic composition as those of the parent cell
(Fig. 21.4).

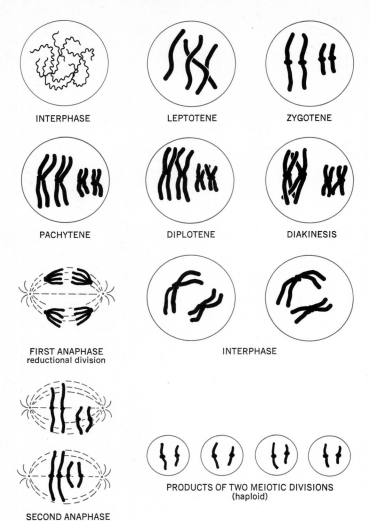

INTERPHASE   LEPTOTENE   ZYGOTENE

PACHYTENE   DIPLOTENE   DIAKINESIS

FIRST ANAPHASE
reductional division

INTERPHASE

SECOND ANAPHASE
equational division

PRODUCTS OF TWO MEIOTIC DIVISIONS
(haploid)

FIGURE 21·3

Various stages of meiosis in the development of spermatocytes or oocytes.

Meiotic division is characterized by the pairing of homologous chromosomes during the first prophase and by the subsequent separation of chromatids at anaphase II. This anaphase separation is the reductional division, which results in the production of secondary spermatocytes or oocytes that have the reduced, or haploid, chromosome number. The gametes, therefore, are haploid. The diploid condition is restored by fertilization of the ovum, when the male and female pronuclei unite. The union of the sex cells at fertilization thus brings about new combinations of genetic material.

DETERMINATION OF SEX

Although the sex chromosomes are of special interest in studies of reproduction, it should be remembered that all cells have chromosomes and their units are called *genes*. All chromosomes other than the sex chromosomes are

MITOSIS

PROPHASE
(chromosomes
become visible)

METAPHASE
(no pairing of
chromosomes)

ANAPHASE
(daughter
chromosomes
move toward
opposite poles)

LATE ANAPHASE
(complete number
of chromosomes
go to each
resulting cell)

MEIOSIS: first meiotic division

PROPHASE
(chromosome pairs
appear together)

METAPHASE
(bivalent chromosomes
on equatorial plate as
tetrads)

ANAPHASE
(paired chromosomes
move away from
their homologues)

LATE ANAPHASE
(the secondary
spermatocyte
or oocyte contains the
haploid number
of chromosomes)

MEIOSIS: second meiotic division

PROPHASE

METAPHASE

ANAPHASE
(separation of
chromatids)

LATE ANAPHASE
(each spermatid
or ovum receives
the haploid
number of individual
chromosomes)

FIGURE 21·4
The differences between mitosis and the first and second meiotic divisions.

called *autosomes*. Human beings have 44 autosomes plus two sex chromosomes. Each male has one X and one Y sex chromosome; each female has two X chromosomes. The total number of chromosomes may be indicated as 44 + X + Y or 44 + X + X.

The two X chromosomes of the female follow exactly the same divisional pattern in meiosis that the autosomes do. The net result provides the developing ovum with one X chromosome. The XY chromosomes of the male follow essentially the same meiotic procedure. Reduction division (when the chromosome pairs separate) produces secondary spermatocytes that contain either an X or a Y chromosome. The spermatids, and ultimately the spermatozoa, therefore carry either X or Y chromosomes. At fertilization, when male and female gametes unite to form the fertilized egg, or zygote, the chromosome pairs are restored. If the ovum is fertilized by a Y sperm, the sex chromosome pair consists of one X and one Y chromosome, and the offspring will be male; fertilization of the ovum by an X sperm restores the XX condition, and the offspring will be female. Many animals, such as reptiles, birds, moths, and butterflies, have a different sex chromosome pattern. In these animals the male is of the XX type, while the female is either XY or XO.

Though the sex pattern is established by certain combinations of sex chromosomes, sexual differentiation is influenced by hormones and possibly other factors in the internal environment.

The behavior of the sex chromosomes is important in understanding sex-linked characteristics. Two such common characteristics are red-green color blindness and hemophilia. The genes that determine these conditions are located on an X chromosome. Sons do not inherit the condition from their father, since they do not receive an X chromosome from the male parent. Daughters can receive the X chromosome, which carries the gene responsible for the defect, from either parent, but since they will probably have only one defective gene, they will not manifest the trait. Such females are called "carriers" because they can pass their defective gene to their children. Males will be color blind or have hemophilia if they receive the defective gene from their mother (see Chaps. 11 and 14). The method of inheritance of sex-linked characters provides additional proof that genes are located on chromosomes. The Y chromosome is thought to contain only a few genes, and in some animals the Y chromosome is described as "empty," or containing no genes.

MALE REPRODUCTIVE SYSTEM

Male reproductive system
Testes
Spermatozoa
Epididymis
Vas deferens
Seminal vesicles
Prostate gland
Bulbourethral glands
Penis

The organs of the male reproductive system are the testes, the epididymis, the vas deferens, the seminal vesicles, the prostate gland, and the penis.

TESTES   The *testes,* the male gonads, have two primary functions: spermatogenesis and the production of the male sex hormones. These processes take place at different sites in these organs. The testes descend from an abdominal position before birth and come to lie in a sac called the *scrotum* (Fig. 21.5). The scrotum is composed of two compartments; a median seam (or raphe) indicates where the embryonic scrotal folds have grown together. The skin is wrinkled into transverse ridges by the contraction of dermal and subcutaneous muscle fibers, and it bears a sparse coat of hair. Occasionally the testes fail to descend into the scrotum, a condition known as *cryptorchism.* Undescended testes are almost invariably sterile, although they produce the male sex hormone. Spermatozoa are very sensitive to heat. It is generally believed that the temperature within the body cavity is unfavorable for spermatogenesis. When the body is exposed to a cold environment, the testes are drawn up close to the body; in a warm environment, muscular relaxation permits them to lie deep in the scrotum away from the body.

The two testes are essentially ovoid in shape, being a little flattened laterally. They are about 4 cm in the longer axis and 3 cm in width. The internal structure consists of compartments or lobules filled with seminifer-

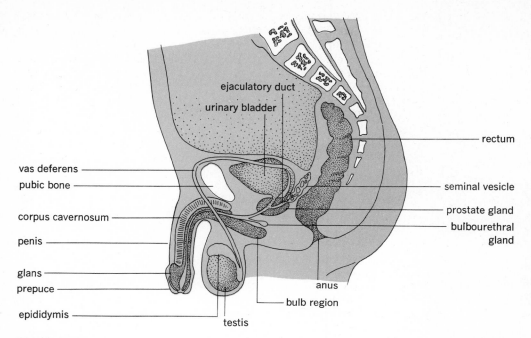

ejaculatory duct
urinary bladder
rectum
vas deferens
pubic bone
seminal vesicle
corpus cavernosum
prostate gland
bulbourethral gland
penis
glans
prepuce
anus
bulb region
epididymis
testis

FIGURE 21·5

Male reproductive system in longitudinal section.

ous (convoluted) tubules. The tubules are lined with germinal epithelium, and it is there that spermatogenesis takes place (Fig. 21.6*a*). The germinal epithelium also contains *Sertoli cells,* which are modified columnar cells with large oval nuclei and prominent nucleoli. They are thought to secrete a nourishing fluid for the spermatids, which attach themselves to these cells (Fig. 21.6*b*).

Interstitial cells (Leydig cells) lie between the seminiferous tubules and produce the hormone testosterone. The adenohypophysis produces the gonadotrophic hormone ICSH (also called LH), which stimulates the interstitial cells to secrete testosterone.

The male sex hormones are responsible for the development of the secondary sexual characteristics of the male. These include the beard, the deep voice, and body-hair patterns. The pubic-hair pattern of the male commonly resembles a triangle with the apex extending upward toward the umbilicus, whereas in the female pubic hair extends upward only to a horizontal line along the superior border of the mons veneris.

Spermatozoa Human spermatozoa are microscopic in size, with a length of only 0.06 mm. The male gamete is not a somatic cell; it is a motile specialized germ cell that carries its genetic material in a condensed condition. The spermatozoon consists of a head, a middle piece, and a tail. The head is oval in its widest aspect, but narrow and oblong in side view. The head is nearly filled with a large nucleus containing DNA and is covered with a cap called the *acrosome.* The cap is thought to contain several enzymes that enable the spermatozoon to penetrate the gelatinous layers surrounding the ovum. The acrosome is produced by the golgi apparatus of the cell (Fig. 21.6*c*).

The middle piece of the spermatozoon consists of a coiled mitochondrial

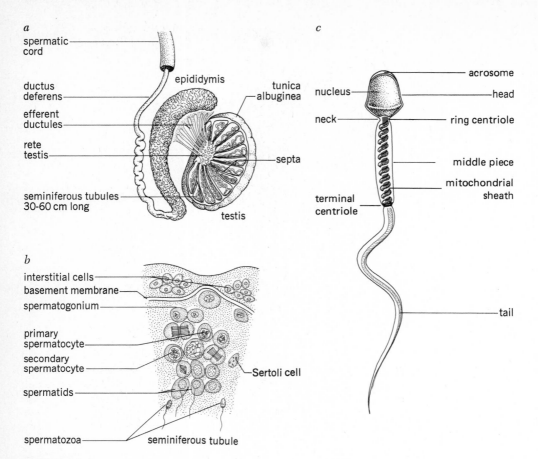

a

spermatic cord

ductus deferens

efferent ductules

rete testis

seminiferous tubules 30-60 cm long

epididymis

tunica albuginea

septa

testis

b

interstitial cells

basement membrane

spermatogonium

primary spermatocyte

secondary spermatocyte

spermatids

Sertoli cell

spermatozoa

seminiferous tubule

c

nucleus

neck

terminal centriole

acrosome

head

ring centriole

middle piece

mitochondrial sheath

tail

FIGURE 21·6

Testes and spermatozoa: *a* section of the testis, lateral view; *b* section of seminiferous tubule illustrating spermatogenesis; *c* enlargement of spermatozoon showing structure.

sheath packed with mitochondria that supply energy, but it also contains an anterior ring centriole and a posterior centriole.

The tail is composed of contractile filaments that propel the spermatozoon with whiplike movements (Fig. 21.7).

There are 200 million to 500 million spermatozoa in a single seminal discharge. The spermatozoa are not motile in the testes and are found clustered around tall Sertoli cells (or sustentacular cells), which support them and perhaps nourish them.

The length of time required to produce mature spermatozoa from spermatogonia has been studied in various ways. Human spermatogenesis occurs in four cycles and has been estimated to require 64 to 74 days.

EPIDIDYMIS   Sperm are not motile when they are released by the testes. After production, the sperm are propelled up through the convoluted seminiferous tubules into a network of fine tubules (the rete testis) and on into the efferent ducts of the *epididymis*. The epididymis is a body containing a tightly convoluted tubule. It is located along the posterior surface of the testis. The tube is only 0.4 mm in diameter, but it is 18 to 20 ft in length.

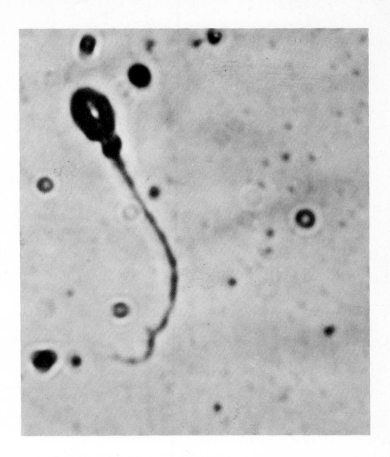

FIGURE 21·7

Human spermatozoon. (*Courtesy of Dr. L. B. Shettles.*)

VAS DEFERENS   The duct of the epididymis is continuous with a larger duct, the *vas deferens,* or *ductus deferens* (see Plate 44), which leads the sperm away from the testis. The vas deferens extends upward from the testis through the spermatic cord; it passes through the inguinal canal, over the pubic arch, and posteriorly over the urinary bladder to terminate in the ejaculatory duct. The roundabout course of the vas deferens is explained by the fact that the testes have changed their location. When the testes descend through the inguinal canal into the scrotum, they carry with them the vas deferens, blood vessels, lymph vessels, and nerves. From the abdominal inguinal ring to the testis these structures form the spermatic cord. The enclosing fasciae contain muscle fibers of the cremaster muscles, which help draw the testes close to the body.

The vas deferens joins the duct from the seminal vesicle and enters the prostate gland as the ejaculatory duct. The right and left ejaculatory ducts open into the urethra within the prostate gland. These ducts are 2 cm long, much smaller and shorter than the vas deferens.

SEMINAL VESICLES   The *seminal vesicles* are lobulated sacs located at the posterior surface of the bladder. They secrete a fluid that forms a part of the semen. The fluid, which passes down a small duct and enters the ejaculatory duct, is thought to contribute to the viability of the spermatozoa.

PROSTATE GLAND   The *prostate gland*, a muscular and glandular organ, is located below the bladder and anterior to the rectum. The base of the urethra passes through it. The prostatic secretion is alkaline, somewhat milky, and it contributes to the odor of semen. The gland measures about 4 cm in its transverse, or horizontal, plane and about 3 cm in its vertical plane. The base of the urethra runs almost vertically through the anterior portion of the gland when the body is in a standing position. The lobules of the gland discharge their secretion through 20 to 30 small ducts, which open into the urethra through minute pores.

The prostate gland is somewhat unfortunately located, since it surrounds the base of the urethra. It tends to enlarge in older men, often constricting the urethra and making it difficult to empty the bladder.

BULBOURETHRAL (COWPER'S) GLANDS   The *bulbourethral glands* are two small, yellow glands about the size of peas. They are located in the bulb region at the base of the penis, and they empty into the urethra from below. Their secretion is a clear, mucoid fluid discharged during sexual stimulation. The secretion precedes ejaculation, and it has been suggested that its function is to lubricate the urethra and glans penis as well as to neutralize the uric acid in the urethra before the spermatozoa pass through. A small amount of this secretion is contributed to the seminal fluid (see Fig. 21.5).

PENIS   The *penis* is the copulatory organ of the male. It is attached to the pubic arch and covered with skin that is continuous with the integument covering the scrotum.

The body of the penis is composed of three longitudinal columns of *erectile tissue* (see Plate 44). Two of these bodies are in the dorsolateral part of the penis and are called the *corpora cavernosa;* the third, the *corpus spongiosum urethrae*, is midventral and contains the urethra. Erectile tissue is composed of blood spaces. These are ordinarily not distended with blood, and the penis is then soft and flaccid. Sexual excitement causes blood to pour into these spaces faster than it can be drained away by the veins. As the spongy tissue fills with blood, it expands and, in doing so, compresses the veins, permitting only a small amount of blood to leave the area. The vascular expansion of the erectile tissue is controlled by the parasympathetic nervous system, and during parasympathetic stimulation, the sympathetic nerves to the arterioles are inhibited. As the walls of the erectile tissue are distended with blood, the penis becomes hard. This process is called *erection*.

The erect penis is inserted into the vagina in the act of sexual intercourse. After sexual excitement has passed, blood drains out of the erectile tissue, and the penis becomes soft again. Erectile tissue is also present in the clitoris, the female homologue of the penis.

The corpus cavernosum urethrae is reflected back over the end of the penis like a cap and contains a vertical slit, which is the external orifice of the urethra. The smooth tip of the penis is the glans portion and is covered by loose skin called the *foreskin*, or *prepuce*. Sometimes the foreskin covers the glans too tightly or becomes adherent. *Circumcision* is an operation to remove the foreskin. The operation may be complete or partial. The area posterior to the glans contains modified sebaceous glands that secrete a soft, whitish substance, which soon deteriorates. Circumcision exposes the surface of the glans and makes it easier to cleanse the area where the secretion (*smegma*) has collected.

Ejaculation The amount of fluid in a single ejaculation averages 3 ml. The average number of spermatozoa is around 120 million/ml. The 300 million spermatozoa found in an average seminal discharge seems to be a remarkably large number, especially since only one is permitted to enter the ovum at fertilization. When the number of spermatozoa drops below 60 million/ml, however, the fertility of the individual is considered to be far below average.

Ejaculation occurs when the vas deferens, the seminal vesicles, and the prostate gland are stimulated to pour their accumulated contents into the base of the urethra by way of the ejaculatory ducts. Skeletal muscles of the bulb region then contract, and with the relaxation of the urethral sphincter, the semen is ejected. The ejection is intermittent, coming in several waves. The ejection of semen is a reflex act involving both the parasympathetic and sympathetic divisions of the autonomic nervous system, but the basic muscular reactions are largely parasympathetic. At the time of ejaculation the sphincter of the bladder is closed, excluding sperm from the bladder and preventing urine from contaminating the semen. A variety of sensations resulting from the ejaculation of semen constitutes the *orgasm* in the male. The act is accompanied by a rise in blood pressure and heart rate. It is followed by relaxation of the reproductive organs and general lassitude. Orgasm in the female is essentially the same reflex phenomenon following a high degree of sexual stimulation, but there is no similar ejection of fluid.

The spermatozoa become highly motile upon ejection. If they are ejected into the vagina of the female, the sperm move up through the uterus and the fallopian tubes at a surprising rate of speed. Recent estimates indicate that the sperm reach the upper part of the uterine tubes within 30 minutes after being deposited near the cervix of the uterus. The sperm may maintain their motility in various parts of the female genital tract for at least 50 h. The fact that sperm are still motile, however, does not mean necessarily that they would be able to effect fertilization. The fertilization life of the ovum has been estimated at 6 to 12 h. It may be less than 6 h. It would seem that the period of high fertility in the female is very closely linked to the time of ovulation.

Sexual intercourse is not necessary to maintain the reproductive organs in a good state of health; neither is masturbation. Healthy young men may experience occasional nocturnal emissions, commonly called *wet dreams*, which eliminate excess accumulations of semen.

FEMALE REPRODUCTIVE SYSTEM

The reproductive organs of the female are the ovaries, the uterine tubes (or oviducts), the uterus, the vagina, the external genitalia, and the mammary glands (Fig. 21.8).

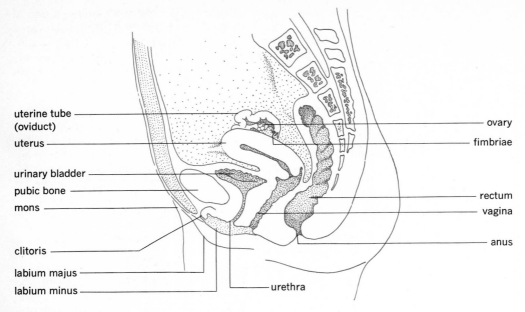

uterine tube
(oviduct)

uterus

urinary bladder

pubic bone

mons

clitoris

labium majus

labium minus

ovary

fimbriae

rectum

vagina

anus

urethra

FIGURE 21·8

The female reproductive organs in sagittal section.

Female reproductive system
  Ovaries
  Fallopian tubes
  Uterus
  Vagina

OVARIES The paired *ovaries* lie on either side of the uterus and below the uterine tubes (Fig. 21.8). They are oblong bodies 2.5 to 4 cm in length and about 1.5 to 2 cm in their anterior-posterior measurement. Their thickness is a few millimeters less than their depth. They lie posterior to a supporting fold of the peritoneum called the *broad ligament* and are attached to the uterus by the ovarian ligaments.

The internal structure of the ovary consists of a connective tissue framework, which supports the developing germ cells, muscle cells, blood vessels, and nerves. The cortex of germinal epithelium contains numerous germ cells and follicles in various stages of development. The ovary is covered with a delicate membrane of columnar epithelium.

Oogenesis The ova develop within the ovarian follicle. The various stages of oogenesis occur there, and the developing ovum in one of the more mature follicles is, in reality, a primary oocyte. Follicles develop under the influence of the follicle-stimulating hormone (FSH) and the luteinizing hormone (LH). From puberty to the menopause, mature follicles approach the surface of the ovary and rupture mature ova through the surface at fairly regular monthly intervals in the process known as *ovulation*. It is assumed that the ovaries alternate in producing mature ova, but little is known about the regularity of the process. It is certain that occasionally more than one mature ovum is produced at the time of ovulation, as in the case of fraternal twins, but these ova could have been produced by one ovary. Of the thousands of potential ova found in the ovaries before puberty, only a few ever reach maturity. The mature follicle is 10 to 12 mm in diameter and bulges from the surface of the ovary. Ovulation occurs about the middle of the 28-day menstrual cycle, but the follicle cells persist, undergoing a transformation into the corpus luteum.

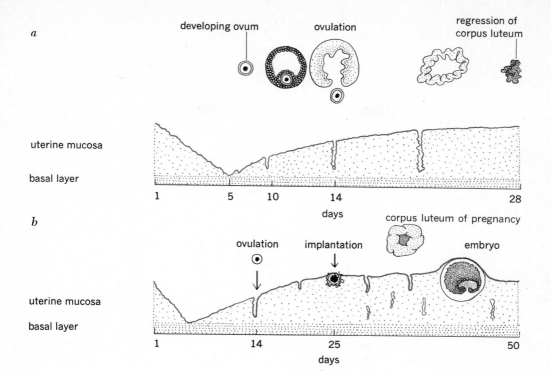

a

developing ovum   ovulation   regression of
corpus luteum

uterine mucosa

basal layer

1   5   10   14   28

days

b

ovulation   implantation   corpus luteum of pregnancy   embryo

uterine mucosa

basal layer

1   14   25   50

days

FIGURE 21·9

The sequence of events during
an ordinary menstrual cycle
and during pregnancy: *a* ordinary menstrual cycle, showing
the development of uterine
mucosa and its deterioration at
menstruation; *b* development
of the uterine mucosa in the
case of fertilization, implantation, and development of the
embryo.

Corpus luteum The follicular cells, after ovulation, enlarge and increase in
numbers so that the number of cell layers increases. The cavity of the old
follicle fills with blood, but the blood is gradually resorbed as new cell layers
fill in the cavity. Connective tissue and blood vessels grow in from a
connective tissue layer surrounding the old follicle.

A yellowish thick-walled body called the *corpus luteum* replaces the old
follicle (Fig. 21.9). The cell cytoplasm of the corpus luteum contains a
lipoidal substance known as *lutein*. It gives the cell mass a slightly yellowish
color, especially after the corpus luteum is fully formed. In the period
between ovulation and menstruation the corpus luteum secretes the hormones progesterone and estrogen, which apparently exert a sustaining
influence on the lining of the uterus. If the ovum is not fertilized, the corpus
luteum begins to degenerate toward the end of the menstrual cycle and
menstruation follows. If the ovum is fertilized, the corpus luteum reaches
the height of its development about the third month of pregnancy, after
which it begins to degenerate. Probably the production of gonadotrophic
hormones following implantation of the blastocyst is responsible for maintenance of the corpus luteum. After the placenta is formed, it produces
progesterone and estrogens; this may relieve the corpus luteum of its
function. The corpus luteum degenerates slowly and is still present in the
ovary at the time of childbirth.

Whether degeneration of the corpus luteum takes place in a monthly
cycle or following a pregnancy, the cellular substance is replaced by fibrous
connective tissue. The location of an old corpus luteum is marked by an area
of white scar tissue in the ovary and is called a *corpus albicans*.

UTERINE TUBES The tubes that conduct the ova from the ovaries to the uterus are usually called oviducts in animals; in humans they are more commonly referred to as *fallopian tubes,* or *uterine tubes.* They lie in a horizontal plane above the ovaries (Fig. 21.10). The distal ends near the ovaries flare out in a funnel-like fashion. The funnels bear fringed processes (fimbriae) that guide the ovum into the tube. The tube is not passive at the time of ovulation. The fimbriae are erectile, and at ovulation the fimbriae and the funnel move closer to the ovary to receive the ovum. A ciliated epithelium lines the tube, and the beating of these cilia moves the ovum along toward the uterus. Smooth muscles in the wall of the tube also propel the ovum.

UTERUS The *uterus* is a thick-walled organ located in the upper part of the pelvic region (see Plate 43). Its function is to receive the blastocyst and to provide protection and nourishment to the developing embryo and fetus after implantation. It is a small organ during childhood, but after puberty it is usually about 3 in long, nearly 2 in wide, and about 1 in thick. It is somewhat larger after the first pregnancy. The cavity of the nonpregnant uterus is always small.

The uterus greatly enlarges during pregnancy, and it extends high into the abdominal cavity. An increase in the number of muscle fibers and the lengthening of fibers permit the uterus to expand. The soft mucosal lining is called the endometrium.

The position of the uterus varies, but it is usually tipped forward over the urinary bladder. It is supported by the broad ligament and the round ligament. The lower part of the uterus is more cylindrical in shape and is called the *cervix.* Its external orifice opens into the vagina (Fig. 21.10).

VAGINA The canal leading from the vestibule of the external genitalia to the cervix of the uterus is called the *vagina.* It is a muscular canal lined with mucous membrane and capable of considerable distension. It lies almost at a right angle to the plane of the uterus, extending inward about 7 to 9 cm (Fig. 21.8). The anterior wall is shorter than the posterior wall, which extends behind the cervix. The projection of the cervix into the vagina creates a pocket or fissure called the *fornix,* the posterior fornix being the deeper recess.

The vagina receives the penis of the male during sexual intercourse; an

FIGURE 21·10

The uterus and appendages.

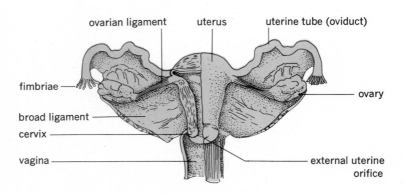

ejaculation releases sperm near the external orifice of the uterus. At childbirth the vagina becomes greatly distended to form the birth canal, which extends from the cervix to the exterior.

The external orifice of the vagina is partially occluded in young girls by a fold of membrane known as the *hymen*. The hymen varies considerably in its shape and degree of extensibility. It may be distended or torn slightly at the first sexual intercourse, but its presence is not regarded as a very reliable sign of virginity, since it may also be stretched by heavy exercise or the use of tampons during menstruation.

EXTERNAL GENITALIA   The female external genitalia comprise the labia majora, the labia minora, the clitoris, and the vestibule (Fig. 21.11). The *labia majora* are two outer fleshy folds covered with pubic hair. They are continuous with the *mons pubis* above. The mons is an eminence of fat over the symphysis pubis and is covered with pubic hair. The *labia minora* are two membranous folds underneath and medial to the labia majora. They are red or pink in color, devoid of fat, and without pubic hair. At their upper extremity the labia minora extend around the *clitoris*, a structure homologous to the penis of the male. The clitoris contains erectile tissue, blood vessels, and nerves. It is a small stucture, 2 to 2.5 cm long, and is largely embedded in tissue. Only the tip, or glans portion, protrudes, and it is ordinarily covered by the upper portion of the labia minora. The glans contains sensory receptors and is an erogenous zone. The *vestibule* is the space bounded by the labia minor and the clitoris. It is evolved from the urogenital sinus of the embryo and contains the external orifices of the urethra and the vagina. The major vestibular glands, or Bartholin's glands, also empty into the vestibule. Their secretion is believed to function as a lubricant.

FIGURE 21·11

Female external genitalia, anterior view.

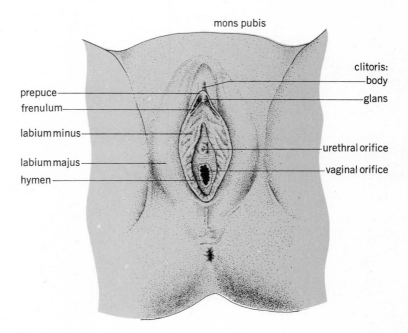

MAMMARY GLANDS The *mammary glands* are modified skin glands that develop from two rows of differentiated ectodermal epithelium in the embryo. The so-called milk line extends from the axillae to the inguinal region in the embryo; in some animals a row of mammary glands develops along each milk line. The human breast glands develop in the pectoral region at the level of the fourth and fifth ribs. They are present in both sexes but develop fully under the influence of female sex hormones (Fig. 21.12*a*).

The female breasts remain relatively underdeveloped until puberty, when the accumulation of fat adds materially to their size. The glandular

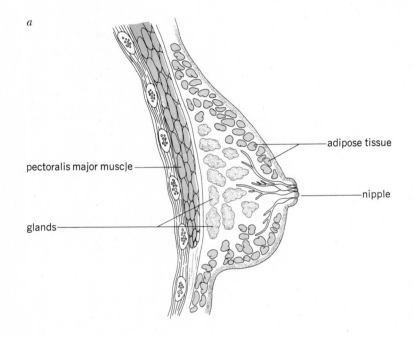

a

pectoralis major muscle

glands

adipose tissue

nipple

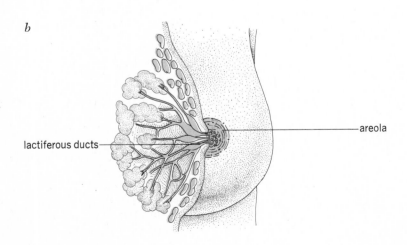

b

lactiferous ducts

areola

portion does not mature and become secretory until the termination of pregnancy. Lactation is stimulated and maintained under the influence of the adenohypophyseal hormone prolactin. The secretion of the mammary glands is at first a thin, yellowish substance called *colostrum*. It contains nutrient materials, but the composition is different from milk. Within a few days milk is secreted, and lactation may continue for several months. Human milk is presumably best adapted for feeding infants. Though cow's milk is frequently substituted, it should be realized that the composition of cow's milk is somewhat different from that of human milk.

The breast, or mammary gland, is covered with thin, soft skin. At the apex is a nipple, which contains 15 to 20 depressions representing the individual openings of ducts. Surrounding the nipple is a pigmented circular area called the *areola*. The pigmentation varies with the complexion of the individual and deepens during pregnancy (see Fig. 21.12*b* and Plate 43).

The glandular portion of the breast is composed of 15 to 20 lobes, each with an individual lactiferous duct opening through the nipple. Each lobe is subdivided into lobules with their ducts emptying into the larger lactiferous duct. The lobules, in turn, are compound glands composed of small glandular sacs (or alveoli). The glands regress during the later years of life, the alveoli being largely resorbed. The tubular structure remains, supported by connective and adipose tissue.

MENSTRUAL CYCLE

The female reproductive functions are greatly influenced by the female sex hormones, which we discussed in Chap. 13. In the female, secretion of these hormones is cyclical, regulating the regular *menstrual cycle* in which ova mature and are released from the ovaries. The "average" menstrual cycle may be summarized as follows:

1st–4th day	Menstruation
5th–12th day	Preovulatory period
	Development of ovarian follicle and growth of endometrium
	Rise of estrogen level
13th–15th day	Ovulation
15th–20th day	Migration and breakdown of unfertilized ovum
	Development of corpus luteum
	High progesterone level
21st–28th day	Premenstrual period
	Regression of corpus luteum
	Progesterone level falls
	Deterioration of endometrium

If menstruation represents the beginning of the monthly cycle, then the cycle's first to the fifth day encompasses the final deterioration of the soft mucous lining of the uterus and its removal in the menstrual flow of blood. During the preovulatory period the junctional layer of the mucosal lining gradually thickens and presents a soft, highly vascular bed for the implantation of the ovum, should it be fertilized. In the ovary a graafian follicle grows, as the ovum reaches maturity. The production of estrogen is high at this time. It should be stressed that although ovulation ordinarily occurs about the middle of the menstrual cycle, it cannot be said to occur precisely on the fourteenth day or necessarily within the 13- to 15-day period. A

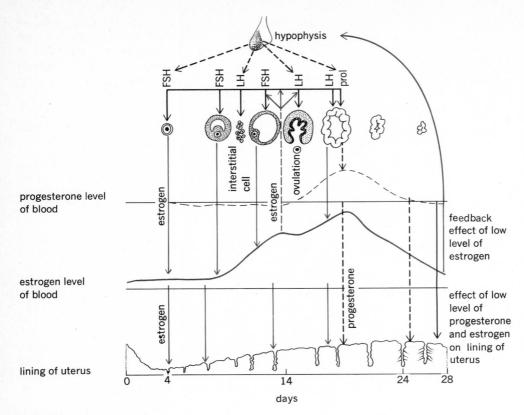

feedback
effect of low
level of
estrogen

effect of low
level of
progesterone
and estrogen
on lining of
uterus

progesterone level
of blood

estrogen level
of blood

lining of uterus

days

FIGURE 21·13

The influence of hormones on
female reproductive cycle. A
rise in blood estrogen level
tends to inhibit production of
FSH by the anterior lobe of
the hypophysis. The rise in
blood estrogen level also stimulates secretion of LH and
prolactin. At the end of the
monthly cycle the low estrogen level stimulates renewed
secretion of FSH by a feedback mechanism. The low levels of estrogen and progesterone are associated with
deterioration of the lining of
the uterus, and menstruation
follows.

statistical average seems to indicate that ovulation most frequently occurs
during this period. Ordinarily there are no outward signs of ovulation, and it
is difficult to determine just when it does occur. There is, however, a slight
rise in rectal temperature of about 0.5°F following a slight drop below
normal at ovulation.

The period following ovulation is a secretory phase in which the endometrium is in a state of heightened activity and continues to grow. The
coiled arteries of the endometrium are distended, and its mucus-secreting
glands are enlarged. The corpus luteum is active, and blood progesterone
and estrogen levels are high (Figs. 21.13 and 21.14). Toward the end of the
period, if the ovum is not fertilized, a regression in the endometrium takes
place. The circulation in the capillaries greatly diminishes during the latter
part of the premenstrual period. Leukocytes begin to migrate into the area,
and a breaking up of the superficial tissue takes place; bleeding from
degenerating capillaries occurs, and the menstrual period follows.

HUMAN OVUM   The ova of placental mammals, unlike the eggs of reptiles
and birds, are minute and contain little yolk. The human ovum is so small
that it is barely visible to the unaided eye. It is estimated to weigh about
0.000001 g. When the oocyte ruptures from the ovarian follicle at ovulation,
it is surrounded by numerous cells that constitute the *corona radiata* (Fig.
21.15). The limiting noncellular layer around the ovum is the transparent
zona pellucida (Fig. 21.16).

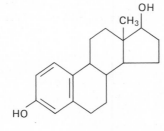

cholesterol

progesterone

testosterone

estradiol

steroids

FIGURE 21·14

Structural formulas of cholesterol and sex hormones.

FIGURE 21·15

A living human oocyte, surrounded by from 3000 to 4000 corona radiata cells, photographed in phase contrast. The oocyte was removed from an ovarian follicle by aspiration with a syringe and needle as it neared ovulation. Original diameter of oocyte proper, 100 µm. (*Courtesy of Dr. L. B. Shettles.*)

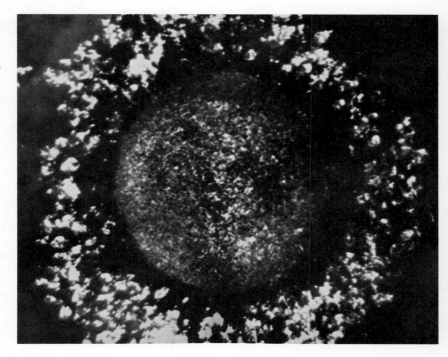

FIGURE 21·16

a Living human ovum, minus corona radiata cells, surrounded by spermatozoa. The outer noncellular layer which acts as a limiting membrane is the zona pellucida. Here, under phase-contrast microscopy, it appears as the outer dark zone. Original magnification, x400. *b* Living human ovum with two polar bodies. The corona radiata is absent. *c* Living two-cell stage, surrounded by the zona pellucida. One polar body can be seen. The difference in the size of the two cells appears exaggerated because they are at different focal levels. Phase contrast; original magnification, x100. *d* Dividing human ovum. Phase-contrast microscopy; original magnification, x200. (*Courtesy of Dr. L. B. Shettles.*)

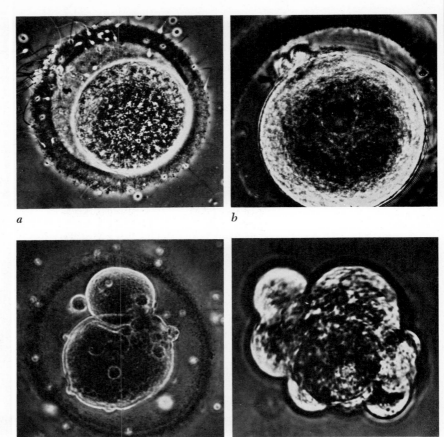

a   b

c   d

HORMONAL REGULATION OF THE CYCLE   The functions of the ovary are directly influenced by gonadotrophic hormones of the adenohypophysis (anterior pituitary gland). FSH influences the development of ovarian follicles, while LH helps stimulate ovulation and the development of the corpus luteum. It appears that several hormones take part in the luteotropic process, rather than LH alone. The hormones thought to be involved, including LH, are prolactin, estrogens, FSH, and their releasing factors. These hormones help maintain the corpus luteum and enable it to secrete progesterone.

The fertilization of the ovum, the early cleavage stages, and the implantation of the blastocyst are discussed later in this chapter. If fertilization occurs and is followed by implantation, the corpus luteum grows larger and persists as the corpus luteum of pregnancy. Ovulation does not ordinarily occur during pregnancy, and menstruation is also suppressed. The blood estrogen level is high, suppressing FSH, which is ordinarily secreted by the adenohypophysis. Progesterone is secreted first by the corpus luteum and later by the placenta. Progesterone stimulates the growth of the endometrium, producing a favorable environment for implantation and for the maintenance of pregnancy.

The discomfort that some women experience at the onset of the menstrual period is believed to be caused by hormonal imbalance. The corpus luteum regresses and stops secreting progesterone at this time, so the blood level of progesterone falls to a low level. The blood level of estrogen is also low, since graafian follicles are stimulated to develop actively only after menstruation.

MENOPAUSE The female reproductive cycle comes to an end when menstruation gradually ceases, during the *menopause*. This condition is reached ordinarily at an age of approximately forty-seven years. Some women experience "hot flashes," which are caused by the sudden dilatation of arterioles in the skin and are accompanied by a feeling of undue warmth and by excessive perspiration. Hormone imbalance may also lead to temporary emotional disturbances until the physiological adjustment is accomplished.

BIRTH CONTROL METHODS

Birth control may be accomplished by the use of several methods. Oral contraceptives that suppress ovulation are widely used. Spermicidal chemicals, placed in the vagina shortly before intercourse, are another type of contraceptive. These include foams, jellies, creams, and suppositories, but they are not considered to be very reliable. Washing out the vagina with a douche after intercourse is also an unreliable form of contraception since sperm may travel quickly beyond the reach of the douche. Contraception by physical means includes intrauterine devices (IUDs), condoms, or diaphragms. Several methods are considered below.

ORAL CONTRACEPTION *Oral contraceptives* today are commonly referred to as "the pill." In the original experimentation the natural hormones progestin and estrogen were used, but these were not satisfactory because they produced undesirable side effects. Oral contraceptives now include synthetic estrogen and progestin components in lower doses than formerly, thus diminishing side effects in most women. These hormones inhibit FSH production, which normally causes ovulation to occur at monthly intervals. Use of oral contraceptives increases the likelihood that a blood clot may form, causing the condition known as thrombophlebitis, which is associated with inflammation of the veins, usually leg veins. Oral contraceptive use is also associated with some threat of hypertension. This method remains an effective one, in spite of the dangers, but it should only be used under close supervision.

A long-acting hormonal contraceptive is now available, but it must be injected. It is said to provide protection for about three months.

INTRAUTERINE DEVICES (IUDS) An IUD is a contraceptive appliance that is inserted into the uterus by a physician. IUDs come in a great variety of sizes and shapes. They are generally made of plastic and shaped into a loop, coil, or shield. Some contain copper, which appears to add extra contraceptive protection. The IUD is thought to act as a foreign body, causing changes in the lining of the uterus that prevent implantation of the fertilized ovum. There are some disadvantages to the use of an IUD. The IUD must be inserted (and removed) through the cervix, a minor surgical procedure. It may cause discomfort if it is too large, or it may be expelled if it is too small. Some women experience severe cramping and excessive menstrual bleeding when wearing an IUD. The advantage of the IUD is that once inserted, it provides continuous contraceptive protection without further action.

CONDOM The *condom* is a rubber sheath that is placed over the penis prior to intercourse. When properly applied and if it does not break, the condom is very reliable. It also provides some protection against venereal disease.

DIAPHRAGM The contraceptive *diaphragm* is fitted over the cervix of the uterus. It is filled with contraceptive jelly and put in place before each act of intercourse. The diaphragm, when inserted correctly and used during each intercourse, is very reliable.

STERILIZATION The effects of castration were discussed in Chap. 13, but sterilization by other means is often confused with castration. Sterility in the male may be caused by accidental injury to the conducting ducts or by sectioning and tying off the vas deferens in the scrotum (*vasectomy*), a procedure used as a birth control technique. In the female, sectioning and tying off the fallopian tubes near the uterus requires an abdominal operation. It is important to understand that sterilization by these operative procedures does not affect the gonads, which continue to produce germ cells and sex hormones. Since the hormones are absorbed by the blood, there is no change in their availability and therefore no change in the sexual characteristics they support. Sectioning of the conducting tubes merely prevents the germ cells from being available for fertilization.

ABORTION The removal of the fetus from the uterus constitutes *abortion*. If abortion is spontaneous, it is commonly called a *miscarriage*. In most spontaneous abortions, the embryo has some abnormality that makes it nonviable. Abortions are sometimes performed to preserve the health of the mother if her life would be endangered by pregnancy. An abortion may also be advisable if there are indications that the fetus may be malformed or defective.

Abortions are sometimes used as a means of birth control by women who do not wish to have a child at the time they find themselves pregnant. The use of abortions for birth control is controversial for both moral and medical reasons. With the widespread availability of contraceptives, abortion, which is a surgical procedure involving some risk, should be a birth control technique of last resort.

GROWTH AND DEVELOPMENT

FERTILIZATION The intricate stages of gametogenesis reach a logical culmination at the union of male and female gametes in the process known as *fertilization*. Fertilization is preceded by the first meiotic division of oogenesis, with the extrusion of the polar body; by ovulation; and by the beginning of the ovum's journey down the uterine tube. Fertilization takes place high in the uterine tube, as the spermatozoa travel rapidly through the neck and body of the uterus and into the uterine tube to meet the descending ovum. The ovum is waved down the tube by the beating of cilia, which line the tube, and by the muscular contractions of the walls.

Only one sperm enters the egg, although there is evidence that numerous spermatozoa must be present to ensure fertilization. The cells of the corona radiata are held together by an organic acid called *hyaluronic acid*. The individual sperm contains an enzyme called *hyaluronidase*, which breaks down hyaluronic acid, but the amount of enzyme contained in a single sperm is so minute that it is ineffective. If several thousand sperm surround the ovum, however, the cells of the corona can be loosened so that one sperm can successfully enter the egg and fertilize it.

When the sperm enters the egg, the tail of the sperm is absorbed and its nucleus, called the *male pronucleus*, begins to move through the cytoplasm of the ovum. The entrance of one sperm into the ovum causes a change in the surface membrane, and no additional sperm can enter. The entrance of the sperm provides the stimulus for the second maturation division to take place in the ovum. The second polar spindle is already formed by then in the mammalian ovum, and after division, the second polar body is promptly extruded (Fig. 21.17).

The male and female pronuclei then approach each other, a mitotic spindle forming between them. The nuclear membrane disappears, and the chromosomes are observed on the spindle where the chromosome pairs are restored. The process of fertilization is then completed, and the fertilized ovum, or *zygote*, is restored to the diploid condition. It may be well to emphasize that, in restoring the chromosome pairs, the male parent contributes 23 chromosomes representing his own characteristics, and the female parent provides 23 chromosomes representing her line of descent. The zygote therefore contains a new arrangement of chromosomes never before duplicated in any other individual. The offspring destined to develop from the fertilized ovum will have a genetic constitution different from that of either parent and of anyone else in the world.

FIGURE 21·17

Fertilization of the ovum and the formation of the first cleavage spindle, shown diagrammatically. (*Redrawn from William Patten in Bradley M. Patten, "Human Embryology," McGraw-Hill Book Company, New York.*)

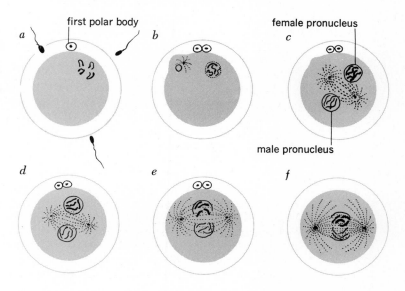

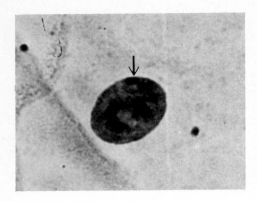

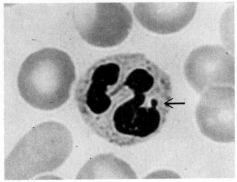

a   *b*

FIGURE 21·18

a Nucleus of normal female squamous epithelium showing Barr body (see arrow); *b* white blood cell of normal human female showing drumstick on nucleus (see arrow). x1,400. *(Courtesy of Carolina Biological Supply Company.)*

The union of the paired chromosomes also brings together the sex chromosome pairs. As we have seen, if the ovum is fertilized by a sperm bearing a Y chromosome, the paired chromosome will be XY and the offspring will be male; if the sex chromosome pairs are reconstituted as XX, the offspring will be female. Thus sex determination is effected at fertilization. The union of the male and female pronuclei also provide the stimulus for further mitotic divisions of the zygote as it passes through a series of divisions known as *cleavage stages.*

SEX CHROMATIN DIFFERENCES   Certain cells of the female show a dark-staining body near the edge of the interphase nucleus. This *sex chromatin,* or *Barr body,* is related to the female X chromosome and is not found in male cells. Though present in various tissues, the Barr body is commonly demonstrated in cells scraped from the mucosal membrane lining the cheek. A similar structure is found in polymorphonuclear leukocytes of the female. This is a stalked body, resembling a drumstick, attached to one of the lobes of the nucleus. It is found in only a small number of these leukocytes and only in females. These examples of sex chromatin can be demonstrated in cells that are not undergoing mitosis (Fig. 21.18).

ABNORMALITIES IN CHROMOSOME NUMBER   Occasionally the chromosome number varies from the normal 46. To count chromosomes, of course, they must be observed when the cell is undergoing division. Dividing cells are usually prepared, stained, and mounted on a microscope slide. One chromosome abnormality is known as *Klinefelter's syndrome* (Fig. 21.19). The individual is morphologically male, but his sex chromosomes are XXY, so his diploid chromosome number is 47. The testes are underdeveloped, there is some enlargement of the breasts, and there is often some mental retardation.

In *Down's syndrome (mongolism),* there are 47 chromosomes but the extra one is an autosome, not a sex chromosome. The extra chromosome is at number 21 where there is *trisomy* (three chromosomes in a group). These individuals are of short stature, and their fingers and toes are short and thick. The face is rounded, and the degree of intelligence is very low (mongolian idiocy). The chromosomes of a normal female are shown for comparison (Figs. 21.20 and 21.21).

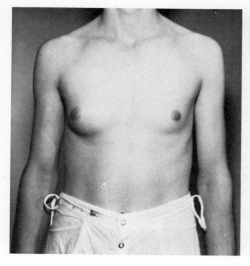

a   b

FIGURE 21·19

Klinefelter's syndrome. (*Courtesy of Armed Forces Institute, Washington, D.C.*)

Another abnormality is called *Turner's syndrome*. These individuals are female but have only one X chromosome. They are sexually infantile, with immature ovaries. They are usually quite short and have peculiar "webbed" necks. They are XO individuals, and their 2n chromosome number is 45.

Nondisjunction in the paired X chromosomes of the mother can cause Klinefelter's syndrome if the ovum is fertilized by a Y sperm. Turner's syndrome probably arises when the ovum does not contribute an X chromosome and the sperm provides the X chromosome. If there is only a Y chromosome the ovum fails to develop.

FIGURE 21·20

Chromosomes of a normal female. (*Courtesy of Earl H. Newcomer.*)

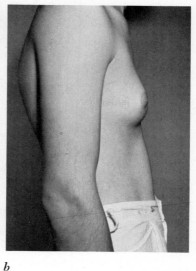

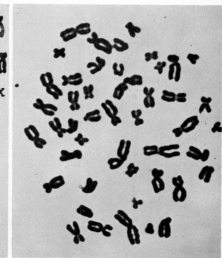

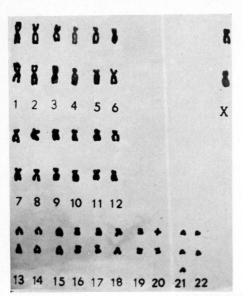

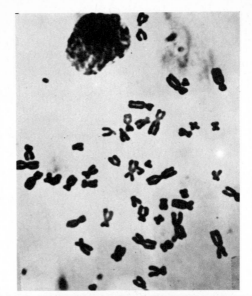

FIGURE 21·21

Chromosomes of a female with Down's syndrome (top). Note trisomy of number 21. (*Courtesy of Earl H. Newcomer.*) Boy with Down's syndrome (bottom). (*Courtesy of Maurice Whittinghill, "Human Genetics and Its Foundations," Reinhold Publishing Corporation, New York, 1965.*)

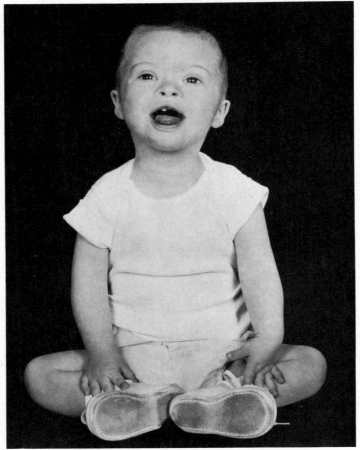

The chromosome abnormality XYY is found in some males characterized as being unusually tall. There are some indications that this abnormal karyotype may be accompanied by mental retardation.

Amniocentesis is a relatively new technique that allows the physician to use a long hypodermic needle to withdraw amniotic fluid from the pregnant uterus. This fluid may then be analyzed to determine if the fetus possesses some biochemical defects or chromosomal abnormalities. An abnormal number of chromosomes can be detected by this method. This procedure makes possible an early diagnosis of Down's syndrome and other chromosomal abnormalities in the early months of pregnancy. The tests are usually conducted in the fourth or fifth months of pregnancy.

CLEAVAGE   The zygote enters a period of rapid mitotic divisions in which it passes through stages described as two-, four-, eight-cell stages, and so forth. The mammalian ovum has been described as a very small germinal cell, but it is large compared with the average size of mammalian somatic cells. Successive cleavages produce cells of smaller size so that the developing mass of cells is only a little larger than the original zygote. The pattern of cleavage varies with ova of various kinds, but cleavage planes tend to be more complete in ova containing a small amount of yolk.

Most of our knowledge concerning cleavage in the human zygote has been derived from observing cleavage in other mammals. Detailed study of cleavage in the monkey ovum has presented an excellent picture of the probable course of events in human cleavage. The cells derived by cleavage are called *blastomeres*.

The cells continue to divide within the zona pellucida, which acts as a limiting membrane, and finally a solid ball of cells is formed. This ball is called a *morula* because of a fancied resemblance to a mulberry. The cells of the morula continue to multiply, and they eventually form a hollow ball called a *blastodermic vesicle*, a *blastula*, or a *blastocyst*. The blastodermic vesicle consists of a thin layer of cells, the *trophoblast*, that forms a cavity filled with fluid and containing an inner cell mass (Fig. 21.22).

The blastodermic vesicle   The cells of the trophoblast surround an *inner cell mass*, which is destined to form the embryo itself. The inner cell mass forms at the *animal pole* of the blastodermic vesicle, where it soon flattens and becomes the blastoderm from which the germ layers of the embryo are derived. The trophoblast plays no direct part in the formation of the

FIGURE 21·22
Early development of mammalian blastocyst (blastodermic vesicle). (*Redrawn from Bradley M. Patten, "Human Embryology," McGraw-Hill Book Company, New York, 1968.*)

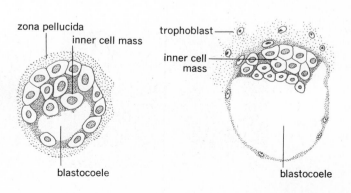

zona pellucida

inner cell mass

blastocoele

trophoblast

inner cell mass

blastocoele

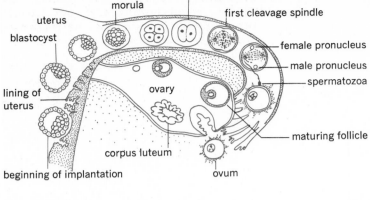

cleavage (two-cell stage)

morula

uterus

first cleavage spindle

blastocyst

female pronucleus

male pronucleus

spermatozoa

lining of
uterus

ovary

maturing follicle

corpus luteum

beginning of implantation

ovum

FIGURE 21·23

Ovulation and the passage of
the fertilized ovum through
the oviduct. Various cleavage
stages are shown diagrammatically. Finally the blastocyst is
shown at the beginning of implantation. (*Redrawn from
W. J. Hamilton, J. D. Boyd
and H. W. Mossman, "Human
Embryology," 2d ed., W.
Heffer and Sons, Ltd., Cambridge, England.*)

embryo; it is involved in the development of the fetal membranes that form
the fetal portion of the placenta. The ovum passes through its cleavage
stages as it moves down the uterine tube, but the blastodermic-vesicle stage
is reached in the uterus just before implantation (Fig. 21.23). The zona
pellucida disappears at the time of formation of the blastodermic vesicle,
and the trophoblastic cells form the attachment with the uterine epithelium.

IMPLANTATION   Little is known about the implantation of the human
blastocyst, but there is every reason to believe that it follows the same
general plan observed in other mammals. During implantation the blastodermic vesicle sinks into the soft lining of the uterus, where it later becomes
firmly attached (Fig. 21.24). The minute blastodermic vesicle of the monkey,
only 0.5 mm in diameter, is known to reach its implantation stage on the
ninth day. It usually implants in the upper part of the uterus on either dorsal
or ventral surface, as does the human blastocyst. It is known that the human
blastocyst sinks much more deeply into the lining of the uterus than does
that of the monkey. The human blastocyst becomes completely covered by
the uterine lining, but the monkey blastocyst remains exposed. The lining of
the human uterus thickens below the implanted blastocyst, and cells of the
trophoblast grow down into it. Later these cells become long, slender
processes, which are called *villi*. Villi exchange products of metabolism
between the blood of the mother and that of the embryo.

EXTRAEMBRYONIC, OR FETAL, MEMBRANES   The early cellular layers develop
far beyond the region of the embryonic disk and are shown as *extraembry-*

FIGURE 21·24

a Surface view of the lining of
the uterus, showing the implantation site of a normal
7-day ovum (*Carnegie No.
8225*). The implantation site is
pale gray and is a little less
than 1 mm in diameter. Observe the prominent openings
of the glands in the uterine
lining and the wrinkling of the
surface which is consistent and
typical of the aging lining at
the twenty-second day of the
menstrual cycle. *b* A profile
view of a 12-day ovum showing the elevation of the blastocyst from the surrounding endometrium (*Carnegie No.
8330*). [*From Hertig, Rock, and
Adams, Am. J. Anat., 98:435–
494 (1956); courtesy of Carnegie Institution of Washington.*]

a

b

714   THE REPRODUCTIVE SYSTEM

onic layers. They do not contribute directly to the formation of the embryo itself but are concerned with the development of the extraembryonic, or fetal, membranes. These membranes are the amnion, yolk sac, allantois, and chorion. The placenta comprises both fetal and maternal tissues.

Amnion The *amnion* is a thin protective membrane containing amniotic fluid. The embryo develops in this clear, watery fluid, which protects against mechanical injury. The amnion expands as the fetus grows until it fills the extraembryonic cavity. It commonly ruptures just before childbirth, releasing the fluid. Individuals said to be "born with a veil" are those who are born with the amnion, or a portion of it, covering the face (Fig. 21.25).

Yolk sac and allantois The ova of many animals contain large amounts of yolk, and the yolk sac therefore is large and of great importance. The human ovum contains very little yolk, and the yolk sac is evident only at an early embryonic stage. As the fetus develops, the yolk sac is incorporated into the umbilical cord, where it persists as a degenerate structure.

The *allantois,* like the yolk sac, is a membrane that plays an important part in the embryonic development of many animals, especially birds and reptiles. In these animals it has an excretory function, serving as a temporary storage place for urine. Combining with the chorion, it serves as a respiratory organ within the egg. The placenta takes over the excretory and respiratory functions of the human embryo, and the allantois remains minute and rudimentary. The base of the allantois, however, gives rise to the urinary bladder.

FIGURE 21·25
Development of the extraembryonic membranes in the human embryo; *a* an embryo of about 12 days; *b* embryo of about 16 days; *c* embryo of about 28 days; *d* embryo of about 12 weeks. (*After Bradley M. Patten, "Human Embryology," McGraw-Hill Book Company, New York, 1968.*)

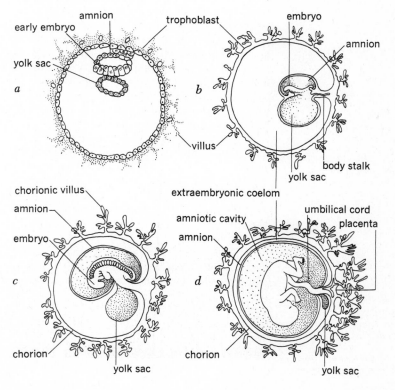

Chorion The *chorion* is a much thicker membrane than the amnion. It limits the extraembryonic cavity and forms the embryonic portion of the placenta. At an early stage in the development of the human embryo, a layer of mesodermal cells is added to the ectodermal trophoblast (*trophectoderm*), and this outer layer of the blastocyst is then called the *trophoderm*. The trophoderm combines with the allantois to form the chorion. The chorion as an outer membrane develops into a highly vascular membrane that functions in the exchange of the products of metabolism and nutrients.

Placenta Small branching processes develop over the outer surface of the chorion. These processes are vascular and are called *chorionic villi*. They cover the chorion at first, but later they are best developed at the place where the chorion makes contact with the lining of the uterus. The mucosal lining of the uterus (endometrium) becomes greatly altered during pregnancy. It is eventually cast off after the child is born and is therefore called the *decidua*, as trees that shed their leaves in the fall are called deciduous trees. The portion of the decidua to which the chorion is attached is called the *decidua basalis*. It is there that an exchange of oxygen and food from the mother and excretory products from the fetus takes place.

The *placenta* is composed of two distinct portions, even though the two parts seem to be closely associated. The chorionic villi become elaborately branched (chorion frondosum) in the area opposite the decidua basalis, and this portion of the chorion is the fetal placenta. The decidua basalis becomes filled with blood sinuses into which the chorionic villi of that region project. The decidua basalis is therefore the maternal portion of the placenta. There is no exchange of blood between the two portions. The fetus soon develops its own blood and circulatory system, and the fetal heart pumps blood out through the villi of the fetal placenta. The mother's blood fills the blood spaces and bathes the chorionic villi with blood. The maternal placenta derives its blood supply from the uterine arteries, and this blood is returned by way of the uterine veins (Fig. 21.26).

The mature human placenta is a circular disk about 20 cm in diameter and nearly 2.5 cm thick. It detaches from the uterus at parturition and forms a large part of the afterbirth.

The mother provides nourishment for the developing fetus from her food. The food—digested, and absorbed by the blood—diffuses through the membrane of the chorionic villi. Once absorbed into the blood of the fetus, simple food molecules are rebuilt into complex molecules. Oxygen from the mother's blood diffuses into the blood of the embryo, while carbon dioxide is exchanged. The waste products of metabolism, such as urea, diffuse out of the embryonic bloodstream and are disposed of by the excretory organs of the mother.

The placental membranes present an effective barrier against bacteria, and the fetus is usually well protected against diseases of bacterial origin. The viruses and some active blood-borne diseases, such as syphilis, can affect the fetus.

In review, it should be mentioned that the placenta produces estrogenic hormones, progesterone, and chorionic gonadotrophin.

Umbilical cord The connection between the fetus and the placenta is established by the umbilical cord. Its fetal attachment is at the umbilicus (or

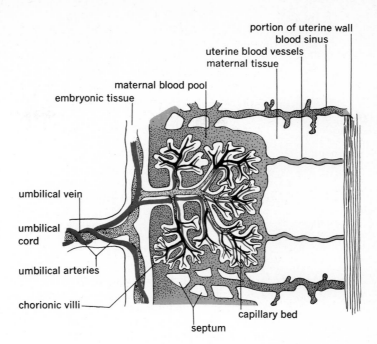

portion of uterine wall
blood sinus
uterine blood vessels
maternal tissue
maternal blood pool
embryonic tissue
umbilical vein
umbilical cord
umbilical arteries
chorionic villi
septum
capillary bed

FIGURE 21·26

A portion of the placenta, illustrating the relationship between fetal and maternal tissues.

navel). The cord is not an outgrowth of the body wall of the fetus; it is formed by the amnion. The body stalk, the first attachment of the embryo, and the yolk sac are pulled together and incorporated within the cord. The umbilical blood vessels consist of two arteries and a single large vein. The cord's substance is a jellylike mucous tissue (Wharton's jelly) peculiar to the umbilical cord. The cord at the full term of pregnancy is coiled and twisted; it measures about 1.3 cm in diameter and is nearly 0.6 m long. The cord attaches near the center of the placenta. It is shed with the placenta as part of the afterbirth.

TWINNING   In various primitive species, twins may be produced as early as the two-cell stage. If two organizing centers are established, two embryos can develop separately into identical twins. In the mammalian species, including the human species, early separation of the blastomeres is unlikely because cleavage takes place within the zona pellucida. Twinning therefore is more likely to occur at the blastocyst stage as a result of divisions in the inner cell mass. Separate organization centers can develop to provide for the development of two embryos from a single ovum.

Human *identical twins* are single-ovum twins, who are always of the same sex and who look very much alike. They develop with a single chorion and placenta (Fig. 21.27). Scientists study such twins because they represent two individuals with apparently identical chromosomes. This poses a question for scientists: How much variation is seen in two individuals who have the same chromosomal constitution? It is well known that identical twins usually resemble each other closely in appearance. Twins do develop different capabilities, however. Occasionally identical twins do not undergo completely separate development. Such twins may be physically joined at

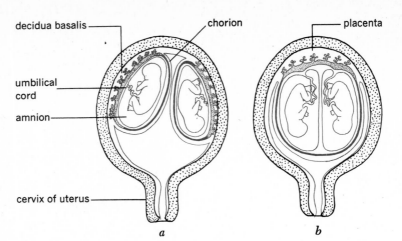

FIGURE 21·27
The differences in the fetal membranes of fraternal and identical twins; *a* fraternal twins with entirely separate membranes; *b* identical twins with common chorion and placenta. (*Redrawn from J. Kollmann, "Handatlas der Entwicklungsgeschichte des Menschen," Fischer, Jena.*)

decidua basalis

chorion

placenta

umbilical cord

amnion

cervix of uterus

a

b

various places and may have organs in common. Such twins do not ordinarily live very long, but some, such as those joined at the base of the spine, have lived to old age.

Fraternal twins are produced when two ova are fertilized and both develop. They develop in individual chorions, and each has its own placenta. Such twins may be of opposite sexes and show only a familial resemblance to each other.

Triplets, quadruplets, and quintuplets may be identical or combinations of identical and fraternal twins with single-ovum individuals.

EARLY DEVELOPMENT OF EXTERNAL GENITALIA   The external genitalia begin to develop during the sixth and seventh weeks of gestation, but it is not possible to tell the sex of the embryo at this time because the external genitalia are in an undifferentiated stage. As early as the fifth week, before the cloacal opening divides into a urogenital sinus and rectum, a slight elevation called the *urogenital tubercle* (papilla) forms above the cloacal opening. The tubercle develops a distal glans portion, which will become the penis of a male embryo or the clitoris of a female. A median urethral groove below the glans is bordered by two ridges called the *genital folds*. Lateral to the genital folds, as development of the genitalia progresses, two additional elevations, known as *genital swellings*, appear. These elevations form the scrotal fold of the male or the labia majora of the female (Fig. 21.28).

Male genitalia   In the male embryo the penis lengthens and the urethral (genital) folds unite to incorporate the urethra inside the penis. The closing of the urethral groove leaves a seam (or raphe) along the ventral side of the penis. The genital swellings enlarge and fuse to form the scrotum, the median scrotal raphe being continuous with that of the penis. These changes occur in the embryo around the tenth to the twelfth week, but the testes do not descend into the more mature scrotum until the eighth month.

Female genitalia   It is evident that the clitoris of the female is the homologue of the penis of the male, since both are derived from the genital tubercle.

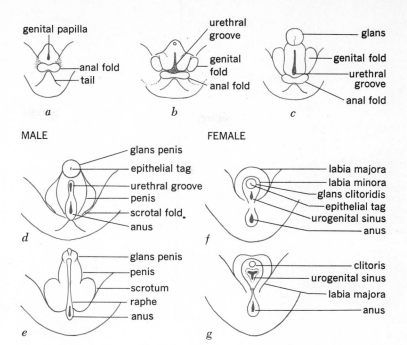

genital papilla
anal fold
tail

a

urethral groove
genital fold
anal fold

b

glans
genital fold
urethral groove
anal fold

c

MALE

glans penis
epithelial tag
urethral groove
penis
scrotal fold
anus

d

glans penis
penis
scrotum
raphe
anus

e

FEMALE

labia majora
labia minora
glans clitoridis
epithelial tag
urogenital sinus
anus

f

clitoris
urogenital sinus
labia majora
anus

g

FIGURE 21·28
Development of the external genitalia in the human embryo; *a, b, c* undifferentiated stages; *d, e* differentiation of the male external genitalia; *f, g* differentiation of the female external genitalia.

The urethral groove does not extend into the clitoris; it opens into the urogenital sinus (vestibule). The urethra of the female is shorter than that of the male, and in the mature reproductive system it opens directly to the exterior through the urethral orifice.

The female external genitalia do not differentiate greatly during development. The clitoris develops slowly and does not become large. The urethral folds become the inner membranous folds, or *labia minora*. The genital folds develop into the outer folds, or *labia majora*. The vagina is at first a slender tube opening into the urogenital sinus or vestibule, but toward the end of the fetal period it enlarges and opens separately to the exterior. The entrance to the vagina is partly occluded by a thin membrane called the *hymen*.

CHILDBIRTH

The full-term human fetus grows rapidly in the last 3 months of gestation, and the uterus expands high into the abdominal cavity. The fetus changes its position many times, but just before birth it commonly assumes a position with the head down at the opening of the uterine cervix. The date of childbirth, or *parturition*, is commonly calculated as 280 days from the beginning of the last menstrual period rather than as approximately 9 months from conception. The date of the last menstrual period usually can be determined more accurately, but by any method of calculation the baby may be born a little earlier or a little later than the estimated date.

As childbirth begins, the cervix gradually dilates, and the fetal membranes rupture. Rhythmic contractions of uterine musculature propel the fetus through the dilated cervix (Fig. 21.29). The uterine contractions are

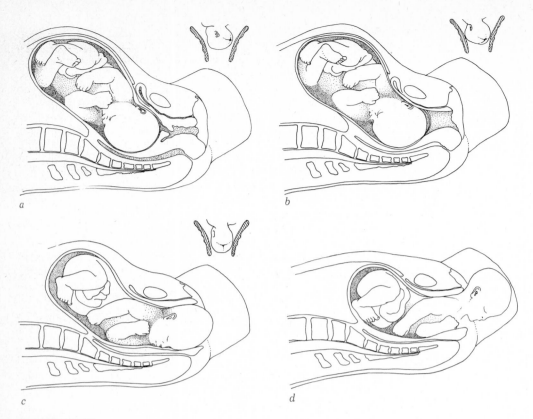

a

b

c

d

FIGURE 21·29

a The beginning of childbirth. The cervical canal still is not dilated. The inset shows the position of the head in relation to the pelvis, frontal view. *b* Early stages of childbirth. The cervix has dilated, and the membranes have broken. Note flexion and descent of head in inset. *c* The progress of childbirth. The head has reached the pelvic floor, and the occiput has rotated so that it faces the pubic symphysis. Note the position of the head in inset. *d* The progress of childbirth. The head is presented: first the occiput, then the brow, and finally the face. (*After Bryant and Overland, "Woodward and Gardner's Obstetric Management and Nursing," F. A. Davis Company.*)

involuntary but are supplemented by voluntary contractions of the abdominal muscles of the mother. The dilatation of the cervix is usually a slow process; the ultimate dilatation marks the end of first stage of labor. During the second stage, the fetus moves into the birth canal; if the baby is turned so that the head appears first, the crown of the head is "presented" at the bulging perineum.

The emergence of the infant into a new world brings about rapid changes in its physiological activities. The newborn child is now freed from its dependence on the placenta and must begin to breathe for itself. Pulsations cease in the umbilical cord, which is cut and tied off. The heart adjusts itself to assume the added work of active pulmonary circulation. Later, food must be taken by mouth for the first time. A newborn infant is well adapted to survive, but many growth processes are still incomplete at birth.

After the child is born, the uterus contracts to expel the loosened placenta and decidual membranes, which form the *afterbirth*. The uterine mucosa then undergoes a period of repair and regeneration similar to that following menstruation. The length of this period varies, but commonly menstruation is not resumed until about 3 months after parturition. Ovulation can occur within this period, however, and a new pregnancy can be started.

AGING

An individual passes through the stages of adolescence, puberty, early adulthood and eventually middle age and old age. Some authorities believe that aging starts very early in life, as indeed it does in some people. A reduction in energy is one of the effects of aging. Less obvious, perhaps, but surely more vital will be the effects of aging and disease on internal organ systems. The heart and blood vessels may show early signs of aging. The heart becomes less vigorous, the coronary arteries more or less clogged as a result of atherosclerosis and arteriosclerosis, the brain probably will show some loss of memory for recent events and perhaps be a bit slower in reacting to simple everyday problems. All organs tend to become somewhat less efficient, especially the lungs, gastrointestinal tract, and the kidneys. It has been said, however, that barring disease or injury, the vital organs are capable of functioning with reasonable efficiency for the duration of a long life of eighty years or more.

DEATH It makes us uncomfortable to even think about death, yet certainly we must encounter it in others as we go through life and eventually we all shall die. We should not view death with horror; for some it is a welcome relief from pain and suffering. Other elderly people are just too weak and tired to go on living.

Death is often compared to sleep, but it is not the same. One awakens from sleep refreshed and able to meet the coming day. Yet for many, dying is a matter of giving up life and drifting away as in sleep.

What is death? It is the irreversible loss of all vital functions, but in particular it is the cessation of brain waves. If the EEG shows no brain waves for a number of hours, the brain is dead and so are the other vital functions.

CLINICAL ASPECTS

VENEREAL DISEASES Venereal diseases are almost always acquired by sexual contact. There are several well-known diseases; *gonorrhea* and *syphilis* are the most important.

Gonorrhea Gonorrhea, commonly called "clap," is caused by pus-forming, round, diplococcal forms of bacteria. The organism *Neisseria gonorrhoeae* (Fig. 21.30) invades the urogenital tract, causing inflammation of membranes lining the passageways. The inflammation often causes closure of small ducts through which the germ cells pass and results in sterility. If untreated, the disease may spread through blood or lymph to various parts of the body. It often localizes in the joints, causing an arthritis of gonorrheal origin.

Gonorrheal ophthalmia is an infection of the conjunctiva, which may be acquired from recently contaminated towels or from the hands. It is a serious disease that may result in blindness unless promptly treated. Infants can acquire a similar infection of the eyes, called *ophthalmia neonatorum*, as the head of the newborn infant passes through an infected birth canal. Many states require that the eyes of newborn infants be treated for this disease whether or not there is any indication of gonorrhea in the mother. Such treatments have helped to reduce blindness in newborn infants. Some strains of the disease have become resistant to penicillin, but other antibiotics may be used to treat such strains.

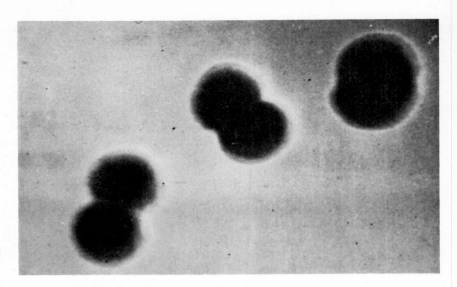

Syphilis Another serious venereal disease responsible for a great deal of human misery is *syphilis*. It causes many persons to become invalids for life and is responsible for many deaths. The causative organism, method of attack, and course of the disease are quite different from those of gonorrhea. The organism *Treponema pallidum* is a tightly coiled spirochete 6 to 14 μm in diameter (Fig. 21.31). It enters the body through a soft, moist surface; the usual points of entry are on the surface of the penis in the male or on the external genitalia and in the vagina of the female. A lesion, called a *chancre*, develops at the point of entry after an incubation period of 2 to 4 weeks. If the lesion occurs within the vagina or on the cervix of the uterus, it may not be noticed at this stage. This is the primary stage of the disease. The ulcers swarm with syphilitic spirochetes, and the host is highly infectious. The ulcer disappears after about 6 weeks, and an uninformed person may think the disease has been cured or at least arrested.

The spirochetes then enter the bloodstream and spread throughout the body in the second stage of the disease. During the next 6 to 12 weeks other symptoms of disease manifest themselves. Often a syphilitic skin rash appears, and lesions may appear in various parts of the body. These symptoms also subside after a time, and the disease may become latent for several months or several years.

The organism invades other organs during the third stage of the disease. Affected organs include the heart, the aorta, the brain, the spinal cord, the skeletal system, and others. Degenerative lesions of vital organs often cause death. Syphilitic lesions in the brain can cause insanity or paresis. Lesions in the spinal cord can cause a kind of paralysis called *locomotor ataxia*. The heart and the arch of the aorta are commonly attacked.

Several good diagnostic tests may be used to test for the presence of the syphilitic organism. Many states require a blood test before marriage, for if one of the partners in a marriage has syphilis, the other partner will very likely become infected. An infected pregnant mother can transmit the disease to the fetus. Such infection is called *congenital syphilis* and is

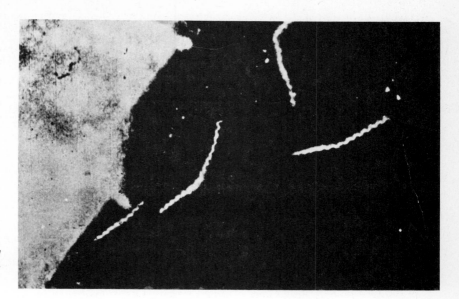

FIGURE 21·31
The spirochete of syphilis, *Treponema pallidum.* (*Courtesy of U. S. Department of Health, Education, and Welfare, Public Health Service.*)

responsible for many stillbirths. If the effects of the disease are not evident in the newborn infant, symptoms may appear later.

Gonorrhea and syphilis can now be brought under control, especially if treatment is begun during the early stages. Penicillin and other antibiotics are of great value in the treatment, but no one should try self-treatment.

Prophylaxis is largely limited to the male but is apt to be ineffective. At any rate, the gonorrhea organism in the male invades by way of the urethra, whereas the syphilis spirochete is more likely to invade through the surface of the penis. Prophylaxis, to be effective, must include both these areas.

Herpes simplex type 2 Herpes simplex is a virus that attacks many parts of the human body, but the type 2 strain usually affects the reproductive organs. This strain has been isolated from the penis, the cervix, the vagina, and the vulva. It has also been associated with cancer of the cervix.

In the male, herpes appears as painful blisters or ulcers on the penis. The eruptions usually heal within a short time, but they may return.

Chancroid A bacterial disease that produces a soft *chancre,* or ulceration, on the penis in the male and on the labia in the female is called chancroid. This disease is transmitted by sexual contact. The large ulcer in the male is usually located on the prepuce or foreskin and is very painful. Fortunately the disease is not a common one in the United States. It may be treated with sulfonamid therapy.

DISORDERS OF THE MALE REPRODUCTIVE ORGANS *Prostatitis and prostatic tumors* The prostate may enlarge in older men, a condition we have already mentioned. The prostate gland is also subject to infection and tumors. *Prostatitis,* or infection of the prostate, is more likely to be caused by a specific organism such as gonorrhea in young men, whereas in older men, prostatitis is usually nonspecific and chronic. In the nonspecific and chronic types small numbers of bacteria are present but the prostate seems to be of normal size when palpated. A chronic type is one of the most common conditions found in older men.

Benign and malignant tumors of the prostate are, unfortunately, common in elderly men. Both types tend to enlarge the gland and constrict the base of the urethra, making urination difficult. A backup of urine may involve the kidneys and cause infection. Surgery is usually required to remove the prostate, or a part of it, in order to open the urethral passageway. Malignant tumors are, of course, the most dangerous. Small malignant nodules, if detected early, may be removed by surgery. Malignant prostatic tumors are "hormone-dependent." Male sex hormones accelerate the growth of these cancers, while female sex hormones have an antagonistic action. Hormone therapy does not cure prostatic cancer, but the growth of the tumor may be slowed by this procedure.

Impotence A small number of adult males cannot keep an erection long enough to ejaculate during intercourse and are termed *impotent.* Impotence should not be confused with infertility or sterility. Usually impotent individuals produce spermatozoa and sufficient amounts of male hormones for normal sexual activity. Occasionally neurological or venereal diseases cause impotence, but the usual cause is psychological. Various fears may cause impotence: fear of women, fear of causing the partner to become pregnant, fear of religious condemnation, fear of inability to perform sexually, or perhaps fear of poor health.

Infertility Some men either produce insufficient numbers of sperm or a large number of abnormal, nonviable sperm, causing infertility. If the testes are injured by x-rays or by venereal disease, infertility may result. Sometimes excessive sexual activity can temporarily deplete the sperm count.

DISORDERS OF THE FEMALE REPRODUCTIVE ORGANS *Infertility* Female *infertility* may arise from many causes. If ovulation occurs regularly, infertility may be caused by anatomical obstruction that prevents the sperm from reaching the ovum. Occlusion of the uterine tubes is a fairly common condition. Inflammation or infection of the uterine tubes is called *salpingitis.*

Ovarian cysts There are several types of *ovarian cysts.* Basically, such cysts are thin-walled and contain serous albuminous fluid. They are initially small but may become as large as an apple. Some cysts have a slender pedicle or stem. If the cyst is twisted tightly around the stem, it may be very painful. Enlarged or painful cysts should be removed surgically.

Cervical cancer The cervix of the uterus is vulnerable to cancer. Abnormal cells in the cervical tissue often multiply, causing cervical dysplasia. These cells are not malignant, but they may develop into a malignancy. A procedure called the Papanicolaou test or "Pap smear" has been developed for early identification of abnormal cells. The Pap smear is about 90 percent reliable. In this test, the surface of the cervix is lightly scraped and cells, along with vaginal secretions, are spread onto glass microscope slides. After staining, the cells are studied to discern any abnormalities. Cancer cells stain deeply and have irregular-shaped nuclei.

The early stages of cervical cancer can be treated successfully. If the cancer is allowed to develop and metastasize, however, a hysterectomy may be necessary. Hysterectomy means removal of the uterus. The ovaries are not removed unless they are infected, and they continue to secrete the female sex hormones. Even though it may be necessary to remove the ovaries, hormone therapy should enable patients to lead essentially normal postoperative lives.

Cancer of the breast One of the most common sites for cancer in women is in the mammary glands. The breast glands seem to be highly vulnerable to cysts and tumors. Quite often a woman will feel a lump in her breast and ignore it or delay going to a physician for further examination. This, unfortunately, is the wrong course of action. Women should examine their own breasts periodically, and if any nodules or lumps are found, the discovery should be reported promptly to a physician. Most breast lumps are not cancerous. The lump may be a cyst or a benign fibrous tumor. However, breast cancer is not ordinarily painful in the early stages, and the nature of any lump should be identified.

The breasts contain numerous lymph vessels and lymph nodes. If cancer is present, it may spread (metastasize) throughout the tissue of the breast and to lymph nodes in the armpit as well as to chest muscles.

Mammography is very useful in revealing even small nodules of cancerous tissue. It is a radiographic procedure in which the breast is injected with an opaque, contrast medium and then exposed to x-rays. If a breast tumor is present, a biopsy is performed to determine whether it is malignant or benign.

Surgery is usually required to treat breast cancer. A modified radical or radical mastectomy are the most common procedures. In a radical mastectomy, the entire breast, axillary lymph nodes, and some of the pectoral musculature are removed. In a modified radical procedure, the pectoral musculature is left intact. Radiation therapy is used to help destroy any remaining cancer cells.

Some types of breast cancer, like cancer of the prostate, are "hormone-dependent." In the female, the female sex hormones stimulate the growth of the cancer whereas male sex hormones (androgens) are antagonistic. There are also anticancer drugs that do not cure the cancer but may relieve pain, delay the spread of the disease, and so have a palliative effect in prolonging comfortable life.

SUMMARY

1 Gametogenesis means the development of the gametes, or germ cells. Undifferentiated germ cells multiply by mitotic division. After a period of growth and multiplication, certain male germ cells can be identified as primary spermatocytes and similar cells in the female can be labeled primary oocytes.

2 During meiosis, unlike mitosis, the chromosome number is reduced from the diploid condition in undifferentiated germ cells to the haploid condition of the gametes. The first meiotic division is a reduction division; the secondary spermatocytes or oocytes contain the haploid number of chromosomes.

3 Telophase I is a short period, and the spermatocytes or oocytes start the

second meiotic division. The prophase and metaphase are typical of a mitotic division. Anaphase II is characterized by the division of the centromeres and the separation of the chromatids. This is the equational division; there is no further reduction in number, and each chromatid becomes a separate chromosome.

4 The spermatozoa of the male arise in the testes. The testes are ovoid organs, which lie in a sac called the scrotum.

5 The epididymis leads sperm away from the testis into the vas deferens. The vas deferens passes through the inguinal canal and over the urinary bladder. Posterior to the urinary bladder, the vas deferens receives a duct from the seminal vesicle. It then enters the tissue of the prostate gland as the ejaculatory duct. The short right and left ejaculatory ducts open into the urethra. The urethra traverses the penis to the external orifice.

6 The seminal vesicles and the prostate gland secrete a portion of the seminal fluid. The prostate gland surrounds the base of the urethra. If it enlarges, it may constrict the urethra.

7 The penis is the copulatory organ of the male.

8 The ovaries lie on either side of the uterus and below the uterine tubes. Ova develop within follicles. As the follicles and ova mature, the larger follicles appear at the surface of the ovary. About the middle of the monthly cycle the ova rupture through the surface in the process known as ovulation. After ovulation, the follicle is transformed into a yellowish body called the corpus luteum, which secretes progesterone and estrogens. When the ovum ruptures from the ovary, it passes down the fallopian tube. If fertilized, the ovum passes through its early cleavage stages and is then implanted in the soft lining of the uterus.

9 The uterus is a thick-walled muscular organ, somewhat pear-shaped, with the neck or cervix projecting into the vagina. The vagina is the canal leading inward from the vestibule of the external genitalia. The external orifice of the virginal vagina is partially occluded by a membrane called the hymen. The labia majora, the labia minora, the clitoris, and the vestibule constitute the female external genitalia. The mammary glands are the breast glands.

10 If the ovum is fertilized and implanted, the corpus luteum persists and is very active throughout the first 3 to 4 months of pregnancy. A high progesterone level is maintained, producing a favorable environment for pregnancy. After the first few months, the placenta produces enough estrogen and progesterone to supply the needs of the body; the corpus luteum then begins to regress. The placenta also secretes chorionic gonadotrophin.

11 There are 46 chromosomes in man; 44 plus X and Y chromosomes in the male, 44 plus two X chromosomes in the female. The development of sperm cells is called spermatogenesis; the development of ova is called oogenesis. Fertilization of the ovum occurs upon the union of male and female pronuclei. The fertilized ovum is a zygote.

12 The zygote passes through cleavage stages, first forming a solid ball of cells, which later becomes a hollow ball, or blastula. The mammalian blastula is specialized and is called a blastodermic vesicle, or blastocyst. The cleavage stages progress as the zygote moves down the oviduct. By

the time it reaches the uterus, it is a blastodermic vesicle. It is implanted in the uterine wall at this stage.

13 The principal fetal membranes are the amnion, chorion, and the fetal portion of the placenta. The amnion is a thin, protective membrane, which immediately surrounds the embryo. It contains the amniotic fluid. The chorion limits the extraembryonic cavity and forms the embryonic portion of the placenta. The placenta comprises a maternal portion and an embryonic portion. In the placenta the exchange of oxygen and carbon dioxide occurs. Nutritive materials are transferred across lymph spaces from the mother's blood to the fetal circulation. The placenta also functions as an excretory organ for the fetus, removing waste products from the fetal circulatory system.

14 The external genitalia develop from an undifferentiated stage consisting of a genital tubercle and two ridges called genital folds. The tubercle forms the penis of the male or the clitoris of the female. Genital swellings of the original genital folds become the scrotum of the male or the labia majora of the female.

15 The fetus grows rapidly during the last 3 months of gestation. The date of childbirth is calculated as 280 days from the beginning of the last menstrual period.

16 At the beginning of parturition, the cervix gradually dilates and the amnion commonly ruptures. Rhythmic uterine contractions propel the fetus through the dilated cervix and into the birth canal. After childbirth, further uterine contractions expel the afterbirth, which consists of the placenta and the decidual membranes.

QUESTIONS

1 Trace the development of spermatozoa or ova through various stages of gametogenesis.
2 Discuss the behavior of the sex chromosomes in gametogenesis.
3 Just when is fertilization of the ovum completed?
4 Where does fertilization occur?
5 Outline the various stages of cleavage.
6 Describe the blastodermic vesicle, or blastocyst.
7 What is meant by implantation?
8 Explain the function of each of the fetal membranes.
9 Describe the structure, function, and content of the umbilical cord.
10 Distinguish between identical and fraternal twins.
11 Outline the events of the menstrual cycle.
12 Discuss the changes that occur in the ovary and in the lining of the uterus following ovulation.
13 Trace the pathway of the sperm from their origin in the testes to the place where the ovum is fertilized.
14 Explain the mechanism that enables erectile tissue to function.
15 Discuss the changes that occur at puberty.
16 Approximately how many sperm are produced in a single seminal discharge? How many ova are produced during the monthly cycle?
17 How long do spermatozoa retain their motility? What is the estimated fertilization life of the ovum? Estimate the time of highest fertility in the female.

18 By what means does gonorrhea invade the body? How does syphilis enter? Describe the causative organism in each case.

19 How are newborn infants infected in the case of gonorrhea? In the case of syphilis?

20 Discuss the three stages in the development of syphilis.

SUGGESTED READING

Adler, Wm. H.: Aging and Immune Function, *BioScience*, **25**(10):652–657, (1975).

Biggers, J. D., and R. M. Borland: Physiological Aspects of Growth and Development of the Preimplantation Mammalian Embryo, *Annu. Rev. Physiol.*, **38**:95–119 (1976).

Diamond, M. C.: The Aging Brain: Some Enlightening and Optimistic Results, *Am. Sci.*, **66**:66–71 (1978).

Greep, R. O., M. A. Koblinsky, and F. S. Jaffe: Reproduction and Human Welfare: A Challenge to Research, *BioScience*, **26**(1):677–684 (1976).

Sacher, G. A.: Longevity and Aging in Vertebrate Evolution, *BioScience* **28**(8):497–501 (1978).

Segal, S. J.: The Physiology of Human Reproduction, *Sci. Am.*, **231**(3):52–62 (1974).

Tanner, J. M.: Growing Up, *Sci. Am.*, **229**(3):34–43 (1973).

REFERENCE BOOKS

GROSS ANATOMY TEXTBOOKS AND ATLASES

Anson, B. J.: "Atlas of Human Anatomy," 3d ed., W. B. Saunders Company, Philadelphia, 1963.

———— (ed.): "Morris' Human Anatomy, " 12th ed., McGraw-Hill Book Company, New York, 1966.

Basmajian, J. V.: "Primary Anatomy," 7th ed., The Williams & Wilkins Company, Baltimore, 1976.

Crouch, James E.: "Functional Human Anatomy," 2d ed., Lea & Febiger, Philadelphia, 1972.

Cunningham, D. J.: in G. J. Romanes (ed.), "Textbook of Anatomy," 10th ed., Oxford University Press, Fair Lawn, N.J., 1964.

Edwards, L. F., and G. R. L. Gaughran: "Concise Anatomy," 3d ed., McGraw-Hill Book Company, New York, 1971.

Francis, C. C., and A. H. Martin: "Introduction to Human Anatomy," The C. V. Mosby Company, Saint Louis, 1975.

Gardner, W. D., and W. A. Osburn: "Anatomy of the Human Body," 3d ed., W. B. Saunders Company, Philadelphia, 1978.

Gray, H.: in C. M. Goss (ed.), "Anatomy of the Human Body," 29th ed., Lea & Febiger, Philadelphia, 1973.

Netter, F. H.: "CIBA Collection of Medical Illustrations," CIBA Corporation, Summit, N.J., 1969.

Schlossberg, L., and G. D. Zuidema: "The Johns Hopkins Atlas of Human Functional Anatomy," The Johns Hopkins University Press, Baltimore, 1977.

Tortora, G. J.: "Principles of Human Anatomy," Canfield Press, San Francisco, 1977.

Wilson, D. B., and W. J. Wilson, "Human Anatomy," Oxford University Press, London, 1978.

CELLULAR BIOLOGY

Baker, J. J. W., and G. E. Allen: "Matter, Energy and Life," 3d ed., Addison-Wesley Publishing Company, Inc., Reading, Mass., 1975.

Bourne, G. H.: "Division of Labor in Cells," 2d ed., Academic Press, Inc., New York, 1970.

De Robertis, E. D. P., F. A. Saez, and E. M. F. De Robertis; "Cell Biology," 6th ed., W. B. Saunders Company, Philadelphia, 1975.

Giese, A. C.: "Cell Physiology," 5th ed., W. B. Saunders Company, Philadelphia, 1979.

Lehninger, A. D.: "Biochemistry, The Molecular Basis of Cell Structure, " 3d ed., Worth Publishers, Inc., New York, 1975.

McElroy, Wm. D., and Carl Swanson: "Modern Cell Biology," 2d ed., Prentice-Hall, Inc., Englewood Cliffs, N.J., 1976.

Nass, Gisela: "The Molecules of Life," McGraw-Hill Book Company, New York, 1970.

Porter, Keith R., and Mary A. Bonneville: "An Introduction to the Fine Structure of Cells and Tissues," 4th ed., Lea & Febiger, Philadelphia, 1973.

Price, W. E.: "Basic Molecular Biology," John Wiley & Sons, Inc., New York, 1979.

HISTOLOGY

Arey, L. B.: "Human Histology," 3d ed., W. B. Saunders Company, Philadelphia, 1968.

Bloom, W., and D. W. Fawcett: "A Textbook of Histology," 10th ed., W. B. Saunders Company, Philadelphia, 1975.

DiFiore, M. S. H.: "An Atlas of Human Histology," 3d ed., Lea & Febiger, Philadelphia, 1967.

Greep, R. O. (ed.): "Histology," 3d ed., McGraw-Hill Book Company, New York, 1973.

Windle, W. F.: "Textbook of Histology," 5th ed., McGraw-Hill Book Company, New York, 1976.

PHYSIOLOGY

Beck, W. S.: "Human Design," Harcourt, Brace, Jovanovich, Inc., New York, 1971.

Crouch, J. E., and J. R. McClintic: "Human Anatomy and Physiology," John Wiley & Sons, Inc., New York, 1976.

Eccles, J. C.: "The Understanding of the Brain," 2d ed., McGraw-Hill Book Company, New York, 1977.

Grollman, Sigmund: "The Human Body," 4th ed., The Macmillan Company, New York, 1978.

Guyton, A. C.: "Physiology of the Human Body," 5th ed., W. B. Saunders Company, Philadelphia, 1979.

———: "Textbook of Medical Physiology," 5th ed., W. B. Saunders Company, Philadelphia, 1978.

———: "Basic Human Physiology," W. B. Saunders Company, Philadelphia, 1978.

Jacob, S. W., and C. A. Francone: "Structure and Function in Man," 4th ed., W. B. Saunders Company, Philadelphia, 1978.

Langley, L. L.: "Physiology of Man," 4th ed., Van Nostrand Reinhold Company, New York, 1971.

———: "Review of Physiology," 3d ed., McGraw-Hill Book Company, New York, 1971.

Luciano, D. S., A. J. Vander, and J. H. Sherman: "Human Function and Structure," McGraw-Hill Book Company, New York, 1978.

Schotelius, B. A., and D. D. Schotelius: "Textbook of Physiology," 18th ed., The C. V. Mosby Company, St. Louis, 1978.

Spence, A. P., and E. B. Mason: "Human Anatomy and Physiology," Benjamin Cummings, Menlo Park, Calif., 1979.

Sullivan, L. P.: "Physiology of the Kidney," Lea & Febiger, Philadelphia, 1974.

Vander, A. J., J. H. Sherman, and D. S. Luciano: "Human Physiology," 2d ed., McGraw-Hill Book Company, New York, 1975.

Wooldridge, Dean E.: "The Machinery of Life," McGraw-Hill Book Company, New York, 1966.

EMBRYOLOGY

Arey, L. B.: "Developmental Anatomy," 7th ed., W. B. Saunders Company, Philadelphia, 1965.

Balinsky, B. I.: "An Introduction to Embryology," 3d ed., W. B. Saunders Company, Philadelphia, 1970.

Berrill, N. J.: "Developmental Biology," McGraw-Hill Book Company, New York, 1971.

Patten, Bradley M., and B. M. Carlson: "Foundations of Embryology," 3d ed., McGraw-Hill Book Company, New York, 1974.

Spemann, Hans: "Embryonic Development and Induction," Stechert-Hafner, Inc., New York, 1962.

ENDOCRINOLOGY

Frieden, E., and H. Lipner: "Biochemical Endocrinology of the Vertebrates," Prentice-Hall, Inc., Englewood Cliffs, N.J., 1971.

Greene, R.: "Human Hormones," McGraw-Hill Book Company, New York, 1970.

Martin, C. R.: "Textbook of Endocrine Physiology," The Williams & Wilkins Company, Baltimore, 1976.

Turner, C. D., and J. T. Bagnara: "General Endocrinology," 5th ed., W. B. Saunders Company, Philadelphia, 1971.

Villee, D. B.: "Human Endocrinology: A Developmental Approach," W. B. Saunders Company, Philadelphia, 1975.

Williams, Robert H.: "Textbook of Endocrinology," 5th ed., W. B. Saunders Company, Philadelphia, 1974.

BIOCHEMISTRY

Baker, J. J. W., and Garland E. Allen: "Matter, Energy and Life," 2d ed., Addison-Wesley Publishing Company, Inc., Reading, Mass., 1975.

Davies, J., and B. S. Littlewood: "Elementary Biochemistry: An Introduction to the Chemistry of Living Cells," Prentice-Hall, Inc., Englewood Cliffs, N.J., 1979.

Mazur, A., and B. Harrow: "Textbook of Biochemistry," 10th ed., W. B. Saunders Company, Philadelphia, 1971.

Stephenson, W. K.: "Concepts in Biochemistry," 2d ed., John Wiley & Sons, Inc., New York, 1978.

BIOLOGY AND ZOOLOGY

Boolootian, R. A.: "Zoology, An Introduction to the Study of Animals," The Macmillan Company, New York, 1979.

Eckert, R., and D. Randall: "Animal Physiology," W. H. Freeman, San Francisco, 1978.

Harrison, R. J., and Wm. Montagna: "Man," Appleton-Century-Crofts, New York, 1969.

Sherman, J. W., and V. G. Sherman: "Biology, A Human Approach," Oxford University Press, New York, 1979.

Storer, T., Robert Usinger, R. D. Stebbins, and J. W. Nybakken: "General Zoology," 6th ed., McGraw-Hill Book Company, New York, 1979.

Villee, C. A., W. F. Walker, Jr., and R. D. Barnes, "Human and Animal Biology," W. B. Saunders Company, Philadelphia, 1979.

GLOSSARY

abduction The movement of a part away from the midline or axis of the body.

accommodation The adjustment or focusing of the eye for vision at different distances.

acetabulum The round depression, or cavity, of the hip bone (os coxae) which receives the head of the femur.

acetylcholine The acetyl ester of choline. A substance considered to be secreted at cholinergic nerve endings. The transmitter substance at synapses and myoneural junctions.

acid A compound capable of dissociating in aqueous solution to form hydrogen ions.

acromegaly An endocrine condition characterized by overgrowth of bones of the face and extremities.

acromion A process of the scapula.

ACTH Adrenocorticotropic hormone from the anterior hypophysis.

actin A muscle protein which takes part in contraction.

active transport The movement of materials through a membrane against a gradient and requiring expenditure of energy.

actomyosin A muscle protein complex composed of actin and myosin.

adduction The movement of a part toward another part or toward the midline of the body.

adenine A purine base.

adenohypophysis The glandular anterior lobe of the pituitary gland.

adenoid An enlarged lymphoid growth in the nasopharynx.

adenyl cyclase An enzyme converting ATP to 3'5' cyclic AMP.

adipose Fatty tissue; fat.

adrenal (suprarenal) gland An endocrine gland located on the superior border of the kidney in man.

adrenaline See epinephrine.

adrenergic Certain nerve fibers releasing an adrenaline-like, i.e. epinephrine-like, substance as a transmitter. Especially the terminal filaments of most sympathetic postganglionic neurons.

afferent Leading toward, as sensory neurons. Vessels that progress toward or enter an organ.

albinism A congenital condition characterized by lack of pigment in skin, hair, and iris.

aldosterone A mineralocorticoid of the adrenal cortex.

alimentary Pertaining to food or aliment, as the alimentary tract of the digestive system.

alkaline A substance or solution containing more hydroxyl than hydrogen ions.

allantois An extraembryonic membrane arising as an outgrowth of the embryonic hindgut.

allele One of a pair of genes. Each has the same locus on homologous chromosomes.

allergy A condition of hypersensitivity to substances known as allergens.

alveolus The bony socket of a tooth or the alveolar border of the jawbone. A pouch in the air sac of the lung.

amacrine cells Retinal neurons with long lateral processes.

ameba A colorless, unicellular, protozoan organism which constantly changes its form, progressing by means of pseudopodia. Ameboid, resembling an ameba.

amino acid An organic compound with the basic formula NH_2—R—COOH.

amniocentesis Removal of the amniotic fluid by inserting a long needle into the amniotic cavity.

amnion An embryonic membrane filled with amniotic fluid, which immediately encloses the embryo.

amphoteric Applied to fluids that possess qualities of both alkalies and acids.

ampulla A dilatation of a canal or duct.

amygdala The almond-shaped end of the caudate nucleus.

amylase (Greek, *amyl*, starch.) Any starch-digesting enzyme that hydrolyzes starch to sugar.

amyloid Starchy or starchlike.

amylopsin The starch-digesting ferment of pancreatic fluid.

anabolism The synthesis of more complex structures from simpler ones.

anastomosis The opening of one vessel into another.

ancon The elbow.

anconeus A muscle of the elbow joint.

androgen A male sex hormone that influences the development of secondary sex characteristics in the male.

anemia A severe reduction in the number of red cells or reduced hemoglobin concentration in the blood.

angiotensin (angiotonin) A polypeptide formed by the action of the proteinase renin on renin substrate. It causes a rise in blood pressure.

angstrom A unit of length; 1 angstrom (Å) = 0.0001 micron (μm) = 1×10^{-7} millimeter (mm).

anlage The embryonic primordium from which a body part or an organ develops. A blastema.

annulus A ring-shaped opening. Annulus ovalis, the margin of the foramen ovalis of the fetal heart.

anode The positive pole of a battery.

antebrachium The forearm.

antecubital Applied to the space in front of the elbow.

antibody A protective substance formed to react with foreign substances or antigens that may be introduced into the body.

antidiuresis The state in which a low volume of concentrated urine is formed.

antigen A substance that stimulates the production of antibodies or reacts with them.

antitoxin A substance in the blood capable of neutralizing a specific toxin.

antrum A cave. The cavity or sinus in the maxilla is called the antrum of Highmore.

aorta The largest of all the arteries in the body.

aortic bodies Small epithelioid structures containing chemoreceptors and located in the arch of the aorta.

apnea Suspension of breathing.

apoenzyme The protein part of an enzyme.

aponeurosis A layer of strong white fibrous tissue.

aqueous Watery.

arachnoid Like a spider's web, for fineness. One of the membranes of the brain and spinal cord. The primitive or embryonic digestive cavity.

areolar Having little spaces.

arrectores pilorum Cutaneous muscles attached to the bases of hairs. Arrectores pilorum cause the hairs to "stand on end."

arteriosclerosis Hardening of the arteries.

artery A vessel carrying blood away from the heart.

arthrosis A joint or articulation.

ascorbic acid Vitamin C.

asphyxia A condition in which the blood is deprived of oxygen.

assimilation The taking up of nutriment by the body tissues in such a manner that it becomes a part of them.

astrocyte A star-shaped neuroglial cell.

atherosclerosis A form of arteriosclerosis characterized by the formation of plaques in arteries.

atlas The first cervical vertebra, upon which the skull rests. Named after the fabled giant who bore the globe upon his shoulders.

ATP Adenosine triphosphate.

ATPase Adenosine triphosphatase.

atrium A hall; a chamber of the heart where blood enters.

atrophy Wasting. Reduction in size, often with degeneration.

Auerbach's plexus A sympathetic nerve network in the wall of the intestine.

auricular Shaped like, or belonging to, an ear.

axilla The armpit.

axis The second cervical vertebra. Named because of the pivot around which the atlas revolves (like a wheel around an axis).

axolemma A membrane covering an axon.

axon The efferent fiber of a nerve cell.

azygos Without a yoke. The name of certain vessels or nerves that are not in pairs.

baroreceptor A neuroreceptor reacting to a change in blood pressure.

Bartholin, glands of Glands located on either side of the vaginal orifice. Vestibular glands.

basal ganglia A group of nuclei located below the third ventricle of the brain; concerned with intricate muscle movements.

basal metabolic rate (BMR) The amount of energy required to maintain the body under resting conditions.

basket cells Branching cells of the cerebellum.

basophil A granular leukocyte. The granules and nucleus stain blue with basic or alkaline dyes.

biceps Having two heads, as the biceps femoris, biceps brachii.

bicuspid Having two points or cusps. A bicuspid tooth.

blastocoele The cavity within the blastula.

blastocyst A type of blastula characteristic of Mammalia.

blastoderm The primitive germ layer, which gives rise to the primary germ layers.

blastodermic vesicle The blastocyst.

bolus A rounded mass, as a bolus of food in the intestine.

Bowman's capsule The invaginated distal portion of a uriniferous tubule, which contains the glomerulus.

brachialis Belonging to the arm, or brachium.

bradycardia Slow heart action.

bradykinin A vasoconstrictor in the blood plasma.

Broca's area The motor speech center in the brain.

bronchus (Plural, bronchi.) An air tube. The smallest air tubes are called bronchioles.

Brunner's glands Mucus-secreting glands of the duodenum.

brush border Closely packed microvilli, usually on the lumenal or apical cell surfaces, which serve to amplify surface area.

buccinator From a word meaning trumpet. The blowing or trumpeting muscle. It tightens the cheeks and lips.

buffer A substance that, when added to a solution, tends to maintain the hydrogen ion concentration against the action of either excess acid or alkali.

bulbourethral glands Cowper's glands, the ducts of which open into the urethra.

bulk flow Mass movement of all the molecules in a medium in the same direction, usually in response to a hydrostatic pressure gradient.

bundle of His The atrioventricular, neuromuscular bundle of the heart.

bursa Literally, a purse. The bursae are small sacs containing fluid and are found in the fascia under skin, in muscles, or in tendons.

calcaneus The heel bone. The tendo calcaneus, or tendo Achillis, is attached to the calcaneus.

calcitonin A thyroid hormone.

calculus A stonelike body formed in some fluid of the body. Renal calculus, in the kidney; biliary c., in the gallbladder, etc.

callus A thickened portion of the skin. The material thrown out (provisional callus) for the repair of fractured bone, to become the permanent callus when the bone is completely ossified.

calorie, large A term referring to the amount of heat required to raise one kilogram of water from 15 to 16°C. A kilogram calorie.

cancellous Resembling lattice work. A cancellous or spongy bone.

canine Resembling a dog. Canine teeth, like a dog's long, pointed teeth.

canthus (Plural, canthi.) The angle at the meeting of upper and lower eyelid.

capillary Resembling a hair in size.

capitellum or capitulum A little head, an eminence on the lower extremity of the humerus.

capsule A structure that encloses an organ or part. (The capsule of a joint.)

carbohydrate An organic substance composed of carbon, hydrogen, and oxygen, as in sugars and starches.

carboxyl The COOH group characteristic of organic acids.

carcinogen A chemical substance capable of producing cancerous growths.

cardiac Referring to the heart or cardia.

caries Decay of bone or teeth. Dental caries.

carotene A yellowish carotenoid, precursor of vitamin A.

carotid The name of the large arteries of the neck.

casein A milk protein, precipitated by the action of rennin or acids.

cast An albuminous structure molded in tubular form. Renal casts.

castration Removal of the gonads.

catabolism The breakdown of large, complex molecules to simpler, smaller molecules.

cataract A condition in which the lens of the eye becomes clouded or opaque.

cathode The negative pole of a battery.

cauda equina A horsetail. The name given to the bundle of spinal nerves in the lower portion of the spinal canal.

caudate nucleus A comma-shaped portion of the basal ganglia at the base of the brain.

cecum Blind. The blind pouch at the beginning of the large intestine.

celiac (coeliac) Pertaining to the celia or belly.

center In the nerve system, a center is a collection of gray cells. The central nervous system comprises the brain and spinal cord, which contain the large nerve centers, e.g., respiratory center.

centrifugal Referring to a force that is exerted from the center outward.

centriole A minute body in the centrosome. During mitotic cell division, a centriole is found at either end of the spindle.

centripetal Applied to a force that seeks a center.

cephalic (Greek, kephalē, head.) Toward the head. Referring to the head.

cerebellum Little brain. Lower, posterior, coordinating portion of the brain.

cerebrum Two hemispheres arising from the forebrain and containing higher sensory and motor centers.

cerumen The wax of the ear.

cervix Neck. Cervical, belonging to or resembling a neck. The cervix of the uterus.

chemoreceptor A receptor reacting to chemical stimuli.

chiasma A crossing or decussation of nerve fibers within the central nervous system. The optic chiasma.

chief cells Cells of the gastric mucosa that secrete pepsinogen.

chlorophyll The green coloring matter in plants that is concerned with photosynthesis.

choana A funnel. The choanae are the posterior openings from the nose into the pharynx.

cholesterol A sterol commonly present in animal fats but also present in the protoplasm of cells. An important constituent of nervous tissue, blood, and bile.

choline A base, commonly a constituent of phospholipids. Considered to be a vitamin of the B complex. See also acetylcholine.

cholinergic Referring to nerve fibers that, when stimulated, release acetylcholine.

cholinesterase An enzyme that catalyzes the hydrolysis of acetylcholine.

chorda tympani A branch of the facial (VIIth cranial) nerve that passes through the tympanic cavity to unite with the lingual branch of the trigeminal (Vth cranial) nerve.

chordae tendineae Tendinous cords attaching the heart valves to the papillary muscles.

choroid The intermediate, vascular coat of the eyeball.

chromaffin Cells having an affinity for, and staining deeply with, chromion salts. Usually refers to cells of the adrenal medulla or other cells of sympathetic origin.

chromatin Material within the nucleus that stains deeply with basic dyes.

chyle Lymph containing absorbed fat.

chylomicron A minute emulsified fat droplet absorbed from the intestine.

chyme Acid, partially digested food material as it leaves the stomach.

cicatrix A scar. It is formed of fibrous connective tissue.

cilia Eyelashes. Ciliated, having tiny hairlike projections, as ciliated epithelium.

ciliary The ciliary region of the eye presents radiating lines, caused by folds of tissues composing it.

circle of Willis An arterial circle at the base of the brain.

circumduction Leading around. This is the motion made when a part is moved around in a circle, one end being stationary. The extremities, the digits, and the head can be circumducted.

circumflex To bend around. Circumflex arteries wind around the arm or thigh.

circumvallate Walled around. The circumvallate papillae at the base of the tongue are encircled by a ridge.

cisterna chyli Chyle cistern. An enlargement at the base of the thoracic duct.

clavicle The clavicula, which resembles a Roman key. Bone of the shoulder girdle.

climacteric A time of life when the system is believed to undergo marked and permanent changes; usually applied to the time of cessation of menstruation, the menopause.

clitoris A small erectile organ in the upper part of the vulva, the homologue of the male penis.

clone A group of cells descended from a single cell.

coagulation The clotting of blood.

coccyx A cuckoo's beak. The fused bones at the end of the spinal column, named for their shape.

cochlea A conch shell. A cavity of the internal ear resembling a snail shell in form.

coelom The embryonic body cavity formed between somatic and splanchnic mesoderm.

coenzyme The nonprotein or prosthetic group of an enzyme.

collagen A protein found especially in white fibrous tissues, cartilage, and bone. On being boiled, it produces gelatin.

collateral A side branch of an axon. Collateral circulation is secured by the union of branches of two vessels, whereby the main current or fluid may be carried by this side route if necessary.

collecting ducts Structures which receive renal tubular fluid from many nephrons and deliver the fluid to the papilla tip.

colloids Large molecules such as proteins which do not readily penetrate cell membranes and tend to remain suspended indefinitely.

commissure A joining. A commissure connects right and left parts, as the commissures of the brain and spinal cord.

communis Common.

concha A shell.

condyle A rounded eminence of bone.

congenital Existing at or before birth.

conjunctiva Connecting. The mucous membrane that lines the undersurfaces of the eyelids and covers the anterior surface of the eyeball.

convoluted Twisted.

coracoid Like a crow's beak. The coracoid process of the scapula.

corium Leather. The deep portion of the skin from which leather is made; the dermis.

cornea The transparent anterior portion of the scleroid coat of the eye.

cornua Plural of cornu, a horn.

coronal, coronoid Pertaining to, or resembling, a crown.

coronary The coronary arteries encircle the crown of the heart.

corpus callosum The transverse commissure of the cerebral hemispheres.

corpus luteum Yellow body. The tissue formed in a ruptured graafian follicle of the ovary.

cortex Bark. The superficial layer, as the cortex of the brain.

cortical nephron A type of renal tubule which has a short loop of Henle extending only as far as the outer medulla of the kidney.

cortisone An adrenal cortical extract, 17-hydroxy-11-dehydrocorticosterone.

costal Relating to a rib or costa. Costal cartilage.

coxae Plural of coxa, the hip; also the genitive form, as os coxae, the bone of the hip.

cranium The part of the skull that contains the brain.

cretinism The condition of a cretin or undeveloped person, both mentally and physically, resulting from deficient activity of the thyroid gland.

cribriform Resembling a sieve. Cribriform plate of the ethmoid bone.

cricoid Like a ring. The cricoid cartilage of the larynx is shaped like a seal ring.

crista A crest. The crista galli of the ethmoid bone. The crista acustica, a ridge of sensory hair cells in the ampulla of a semicircular canal. The cristae of mitochondria.

crucial Like a cross. The crucial ligaments cross each other.

crural Belonging to or like the lower extremity, from crus, a leg; the crura (or legs) of the diaphragm. The crura cerebri, or cerebral peduncles, descending nerve tracts from the cerebral hemispheres.

cryptorchism A condition referring to the failure of the testes to descend.

crypts of Lieberkühn Tubular glands at the bases of villi in the intestine.

cumulus oophorus A mound of cells supporting the maturing ovum within the follicle.

cyanosis A bluish coloration of the skin and mucous membranes caused by lack of oxygen in the blood.

cyclic AMP Adenosine 3′5′-monophosphate.

cystic Relating to a cyst, or a sac containing fluid (cystic duct). A cystic ovary has cysts developed from its substance.

cytochromes Heme proteins involved in cellular respiration.

cytoplasm The protoplasm of the cell exclusive of the nucleoplasm.

cytosine A pyrimidine base present in nucleic acid.

deamination The removal of an amino group (NH_2) from an amino acid or from another organic compound.

decidua (Latin, *deciduus,* that which falls off.) The mucous membrane lining the uterus, especially the part that is shed at menstruation or following parturition as a part of the afterbirth.

decussation A crossing, as of nerve tracts. Decussation of the pyramidal tracts.

deglutition The act of swallowing.

dehydrogenase An enzyme that catalyzes the oxidation of a substrate by the removal of hydrogen.

deltoid Shaped like the Greek letter delta, Δ.

dendrite A process of a neuron carrying impulses toward the cell body.

dental From dens, a tooth, belonging to a tooth. Dentated, toothed.

dentine The sensitive substance of the tooth between the enamel and the pulp.

dentition The eruption or cutting of the teeth. The kind, number, or arrangement of the teeth.

dermis The inner layer of the skin, the corium.

desmosome An electron-dense zone of attachment between apposing plasma membranes.

diabetes insipidus A condition due to malfunction of the hypothalamus and characterized by excessive thirst and excessive excretion of urine of low specific gravity.

diabetes mellitus A condition in which carbohydrates are not oxidized properly by the tissues. Usually insulin is lacking. There is an excess of sugar in the blood, and sugar is present in the urine.

diapedesis The passing of blood cells through the walls of capillaries.

diaphoretic An agent that increases the amount of perspiration.

diaphragm A wall across a space. The muscle that separates the cavity of the thorax from that of the abdomen.

diaphysis The greater part of the shaft of a bone.

diarthrosis A movable joint.

diastole A Greek word meaning a drawing apart. The dilatation of the chambers of the heart.

digastric Double-bellied, as the digastric muscle.

digit A finger or toe.

diopter The refractory power of a lens having a focal distance of one meter.

distal Farthest from the head or trunk.

distal convoluted tubule The segment of the renal tubule located between the macula densa and the collecting duct.

diuresis The state in which a high volume of dilute urine is formed.

diuretic An agent that increases the quantity of urine.

diverticulum An outpocketing or sac; the cecum.

DNA Deoxyribonucleic acid.

dopa Considered to be a possible neurotransmitter.

dopamine 3,4-Dihydroxyphenylalanine.

dorsal Belonging to the dorsum, or back.

ductus arteriosus A blood vessel of the fetal circulatory system that diverts blood from the left pulmonary artery to the aorta.

duodenum Meaning twelve. The duodenum is 12 fingerwidths long.

dura mater The fibrous outer membrane of the brain and spinal cord.

dyspnea Difficult breathing.

ECG A record of heart action made by an electrocardiograph.

ectoderm The outermost germ layer of the embryo.

ectoproteins Those proteins present in cell membranes which protrude from the phospholipid matrix on both the inside and outside.

edema Swelling caused by effusion of serous fluid into areolar tissues.

EEG Electroencephalogram. A record of rhythmic brain waves, largely from the cerebral cortex, indicating the electrical activity of the brain.

efferent Meaning away from. Efferent vessels leave organs. Efferent or motor neurons leave the brain or spinal cord.

Einthoven's triangle Refers to the placing of electrodes in electrocardiography.

electrolyte A substance that ionizes in solution and is capable of conducting an electric current.

electron The smallest unit of negative electricity.

eliminate To excrete substances that are useless.

embolism The blocking of a blood vessel by the formation of a clot.

embryo An organism in the early stages of development.

emesis Vomiting.

emetic An agent that induces vomiting.

emission A seminal discharge.

endo- Within. Endocardium, within the heart. Endothelium, the epithelial lining of circulatory organs.

endocrine gland A gland of internal secretion. A ductless gland.

endoderm The innermost germ layer of the embryo (entoderm).

endolymph The fluid contained in the membranous labyrinth of the inner ear.

endometrium The mucous membrane lining the uterus.

endomysium The sheath of a muscle fiber.

endoplasmic reticulum A series of tubules and vesicles forming a network, granular or agranular, throughout the cytoplasm of the cell.

endoproteins Those proteins present in cell membranes which protrude from only one surface of the phospholipid bilayer.

endosteum The lining of medullary canals in long bones.

enema A fluid introduced into the rectum.

ensiform Sword-shaped. The process of the sternum.

enteric Pertaining to the enteron or intestine.

enterokinase An enzyme of the small intestine that changes trypsinogen to active trypsin.

enzyme An organic compound produced by living cells, which acts as a catalyst in chemical reactions. Digestive enzymes.

eosinophil A granular leukocyte whose granules stain eosin red with acid dyes.

epi- Upon, as epicondyle, epidermis, epiglottis.

epidermis The outer or ectodermal layer of the skin.

epididymis A group of coiled tubules on the testis, continuous with the tubules of the testis and with the vas deferens.

epimysium The connective tissue muscle sheath.

epinephrine The hormone of the adrenal medulla.

epiphysis A part of a bone that is formed independently and joined later to complete the whole bone.

epithelial Pertaining to epithelium.

epithelium A tissue that forms the epidermis and the lining of ducts and hollow organs.

EPSP Excitatory postsynaptic potential.

erythroblast A nucleated red bone marrow cell, a precursor of the mature erythrocyte.

erythrocyte A red blood cell.

erythrocytopoiesis The production and development of erythrocytes. Synonym: erythropoiesis.

erythropoietin A hormone that stimulates the formation and development of erythrocytes.

esophagus From a Greek word meaning to carry food. The esophagus transmits food from pharynx to stomach.

ester A compound formed from an alcohol and an acid by the elimination of water.

esterase An enzyme that catalyzes the hydrolysis of an ester into an alcohol and an acid.

estrogen An estrus-producing hormone. A female sex hormone.

ethmoid Sievelike. The ethmoid bone has many openings on its surface.

Eucaryota Organisms with distinct nuclei. Protista.

eunuch A male castrate.

eversion Turning outward. To evert an eyelid is to fold it back so as to expose the interior surface.

excretion A waste substance to be removed from the body. The process of removing waste products from the tissues.

extension Stretching out or extending. (Bending backward is overextension.)

extirpation The complete removal of a part from the body.

extrinsic Originating outside a structure. Extrinsic muscles of the eyeball.

exudate A collection of material that has filtered through the walls of vessels into surrounding tissues.

FAD Flavin adenine dinucleotide.

falciform Sickle-shaped.

fallopian tube Uterine tube or oviduct.

falx A sickle. Falx cerebri, a membrane covering the brain.

fascia (Plural, fasciae.) A band. The tissue that binds organs or parts of organs together.

fasciculus In the nervous system, a nerve tract.

fasciculus cuneatus A laterodorsal afferent (ascending) tract of the spinal cord.

fasciculus gracilis A mediodorsal afferent tract of the

spinal cord. Proprioceptive from the lower part of the trunk and from the legs.

fauces (Latin, *faux,* throat.) Isthmus of, the space bounded by the soft palate, tonsils, and tongue. Pillars of, the folds connecting the soft palate with the tongue and pharynx. (The tonsil is between the pillars of either side.)

feedback mechanism A regulatory arrangement whereby a portion of the output of a system is recycled back into the system to control any further output. Negative feedback.

femoral Belonging to the femur or thighbone.

fetus The mammalian organism in the later stages of development, especially after the beginning of the third month.

fibrin Protein threads that form the framework of a blood clot.

filiform Threadlike in shape, slender; as filiform papillae of the tongue.

fimbria A fringe; fimbriated, having a fringelike appearance.

fissure A cleft or groove, as a fissure of the brain surface.

fistula A tubelike passage formed by incomplete closure.

flavus Yellow.

flexion Bending. Flexure, a bend.

follicle A very small sac (or bag) containing a secretion.

fontanel A little spring. A membranous spot in the infant's skull; the name suggested by the rising and falling caused by the child's respirations.

foramen (Plural, foramina.) An opening.

foreskin The prepuce.

fornix A vaultlike space. An arch of nerve fibers below the corpus callosum.

fossa A depression or concavity.

fourchette A little fork. The fold of mucous membrane at the posterior ends of the labia majora.

fovea A small pit. The fovea centralis is a tiny depression in the macula lutea of the retina.

frenum A curb or bridle. The frenum linguae is the fold of mucous membrane attaching the tongue to the floor of the mouth.

FSH The follicle-stimulating hormone secreted by the anterior lobe of the hypophysis.

fulcrum The place on which a lever rests or moves.

fundus The base.

fungiform Shaped like a fungus or mushroom.

fusiform Spindle-shaped.

GABA Gamma-amniobutyric acid. Considered to be a possible synaptic transmitter.

gamete The mature ovum or spermatozoon.

gametogenesis The origin and development of the gametes.

ganglion (Plural, ganglia.) A group of nerve cell bodies usually outside the brain and spinal cord.

gaster The stomach. Gastric, belonging to the stomach, or gaster.

gastrocnemius The belly of the leg. The prominent muscle of the calf of the leg.

gene A hereditary unit having a definite location on a chromosome.

genotype Having the same genetic constitution.

genu A knee.

gigantism A condition of abnormal growth involving excessive secretion of the anterior lobe of the hypophysis.

gingiva The gum. The tissue that encloses the neck of the tooth and covers the jawbone.

glabella A little smooth space. The smooth space between the eyebrows.

gladiolus A little sword. The body of the sternum.

gland A collection of cells that can form a secretion or an excretion.

glans The head of the clitoris or penis.

glaucoma A condition of the eye characterized by a marked increase in intraocular pressure.

glenoid Having the form of a shallow cavity. Belonging to a cavity.

globulin A group of animal and plant proteins. Alpha, beta, and gamma globulins are fractions of blood serum globulin. Gamma globulin contains antibodies.

glomerulus The minute coiled mass of capillaries within a Bowman's capsule of the kidney. Also a network of synaptic branching within the olfactory bulb.

glossopharyngeal Belonging to the tongue and pharynx.

glottis The upper opening of the larynx. Epiglottis, the leaf-shaped cartilage upon the upper border of the larynx.

glucagon A hormone produced by alpha cells of the pancreatic islet group. The hormone causes a rise in blood glucose levels.

glucokinase An enzyme of the liver that catalyzes the phosphorylation of glucose.

gluconeogenesis The formation of glucose from noncarbohydrate materials such as protein.

glucose A monosaccharide sugar, $C_6H_{12}O_6$.

gluteal Referring to the gluteus, or buttock.

glycogen A polysaccharide, $(C_6H_{10}O_5)_n$, formed from glucose and found in various tissues but stored especially in the liver and muscles.

glycogenesis The formation of glycogen from glucose in the liver.

glycogenolysis The breakdown of glycogen to glucose.

glycolysis The breaking down of glucose in the tissues, to pyruvic or lactic acid, by enzymatic action.

glyconeogenesis The formation of glycogen from non-carbohydrate sources such as from amino acids.

glycosuria The presence of an abnormally high proportion of sugar in the urine.

goblet cells Mucus-secreting cells in the lining of the intestine.

goiter The enlargement of the thyroid gland.

golgi apparatus A smooth membranous structure in the cytoplasm of animal cells, usually located near the nucleus.

gonad A reproductive organ. The testis or the ovary.

gonadotrophic A hormone secreted by the anterior hypophysis that influences the development and maintenance of the gonads.

graafian follicle An ovarian follicle.

gram molecular weight The weight of a substance in grams equivalent to its molecular weight.

groin The depressed area of the abdomen adjacent to the thigh.

guanine A nitrogenous base, a purine.

gustatory Associated with the sense of taste.

gyrus (Plural, gyri.) A circle. Convolutions of the brain cortex.

haploid The reduced number of chromosomes after meiosis in the maturing germ cells.

haversian Name applied to the central canals in bone tissue, from the English anatomist Havers.

helicotrema The confluence at the apex of the cochlea between the scala tympani and the scala vestibuli.

helix A spiral form.

hematocrit The percentage by volume of blood cells in a unit volume of blood.

heme An iron-containing red pigment that has an affinity for oxygen.

hemoglobin The oxygen-carrying substance of red blood cells, to which their color is due.

hemolysis Destruction of red blood cells.

hemophilia A sex-linked hereditary condition characterized by reduced clotting ability of the blood, resulting in prolonged bleeding.

hemopoiesis The formation of blood.

hemorrhoidal From a word meaning flowing with blood. Pertaining to a hemorrhoid, or pile.

Henle's loop A specialized turn in the uriniferous tubule.

heparin A substance found in the liver that inhibits the coagulation of blood.

hepatic Referring to the liver, or hepar.

hernia The protrusion of a part of an organ through an opening.

heterozygous Having different alleles representing a given character.

hilum Literally, a little thing. Applied to the depression where vessels enter and leave an organ.

homeostasis The maintenance of a steady state in the internal environment.

homologous Having the same origin, development, and structure.

homozygous Having common alleles representing a given character or referring to the whole individual.

hormones Chemical substances formed in endocrine glands and conveyed by the blood to other organs, to influence their activity.

hyaline Resembling glass, clear. Hyaloid has a similar meaning.

hydration Saturating with water.

hydrocephalus A collection of fluid either within the ventricles or outside the brain.

hymen A membrane partially covering the orifice of the vagina.

hyoid U-shaped, as the hyoid bone.

hyperemia An increase of blood in a local area of the body.

hyperglycemia An excess of sugar in the blood.

hypermetropia Farsightedness.

hyperplasia Excessive formation of tissue.

hypertrophy Overgrowth. Derived from two Greek words meaning too much nourishment.

hypochondrium Under the cartilage. The hypochondriac region is under the cartilages of ribs.

hypogastric Under the stomach.

hypoglossal Under the tongue.

hypophysis (Greek, growing under.) An endocrine gland located under the hypothalamus. The pituitary gland.

hypothalamus An area that is located in the floor of the diencephalon.

ileum A roll or twist; the portion of small intestine that appears rolled or convoluted.

ilium The upper portion of the hipbone, or os coxae.

immunity A state of being protected from the invasion of foreign substances.

immunoglobulin A blood protein capable of acting as an antibody.

incisor A cutting instrument. The front teeth are incisors.

index Indicator. The first finger, named from its common use.

infra- Beneath.

infundibulum A funnel-shaped space or part. The stalk of the pituitary gland.

inguinal Belonging to or near the thigh, or inguen. Inguinal canal.

inhibition The restraining or stopping of normal action.

insertion The attachment of a muscle to a bone at the more freely movable end.

in situ In position.

instep The arch of the foot, dorsal aspect.

insulin   The hormone of the pancreas secreted by the islet cells (of Langerhans).

integument   The skin.

inter-   Between, as intercostal, between the ribs; intercellular, between the cells, etc.

intercalated   Placed between, as the electron-dense areas at opposing plasma membranes in cardiac tissue.

intermediary metabolism   The metabolism of cells or body fluids after absorption and before excretion.

interstitial   Tissue or fluid located in the interspaces of a structure. The endocrine tissue of the testis.

intima   The lining of blood vessels.

intrinisic   Located within a part or organ.

inversion   A turning in, as inversion of the eyelashes; inversion of the foot.

in vitro   In glass; referring to a process or reaction carried on in a test tube, that is, outside the body.

in vivo   Referring to a process or reaction carried on within a living organism.

involution   The changing back to a former condition, of an organ that has fulfilled a function, as the involution of the uterus after parturition.

IPSP   Inhibitory postsynaptic potential.

iris   A circle or halo of colors. The colored circle behind the cornea of the eye.

ischemia   A decrease in blood supply to a part of the body.

ischium   The lowest part of the hipbone, or os coxae.

islands of Langerhans   Small islet cells of the pancreas, which secrete the hormone insulin.

isometric   The same measure. A type of contraction wherein the tension increases but the length of the muscle remains essentially the same.

isotonic   Having the same tonicity as another fluid taken as a standard. Contraction of muscle that permits the muscle to shorten.

isotopes   Elements that have the same number of atoms but different atomic masses.

jaundice   A condition characterized by yellowing of tissues caused by the absorption of bile pigments into the blood.

jejunum   Empty. The second portion of the small intestine, usually found empty.

jugular   Belonging to the neck, or jugulum.

juxtaglomerular apparatus   Smooth muscle cells in the walls of glomerular afferent arterioles which synthesize, store, and release renin.

juxtamedullary nephron   A renal tubule which has a long loop of Henle extending deeply into the inner medulla.

karyokinesis   Mitosis.

keratitis   Inflammation of the cornea.

keto-   A prefix indicating that an organic compound contains a carbonyl (CO) group.

ketone   A compound containing a carbonyl (keto) group.

kidney   An important organ of regulation and excretion, in which the urine is formed.

kinase   An activator of a zymogen to form an enzyme.

kinesiology   The science of muscular movement.

kinetic   Referring to motion. Kinetic energy.

kinetochore   The constriction in the chromosome indicating the attachment of the spindle fiber. A centromere.

Kupffer cells   Fixed macrophages lining sinusoids of the liver.

kymograph   An instrument consisting of a revolving drum covered with paper in which a record of some physiological activity can be traced.

kyphosis   Hunchback. An exaggerated dorsal spinal curvature in the thoracic region.

labia majora   The large outer folds of the female external genitalia.

labia minora   The small inner membranous folds of the female external genitalia.

labium   (Plural, labia.) A lip.

lacrimal   Having to do with tears, or lacrimae, as the lacrimal gland.

lactase   An enzyme that acts upon lactose.

lactation   The secretion of milk by the mammary glands.

lacteal   Like milk (from *lac*, milk). The lacteals are lymph vessels that carry milky-looking chyle.

lactic acid, $C_3H_6O_3$ An acid formed from carbohydrates, as in sour milk.

lactose   Milk sugar.

lacuna   A minute cavity, as in cartilage and bony tissue.

lambdoid   Resembling the Greek letter lambda, λ.

lamella   A little plate or thin layer.

lamina   A plate or layer.

lanugo   The fine downy hair that covers the fetus at about 5 months.

larynx   The part of the air passage extending from the base of the tongue to the trachea. The cartilaginous organ of the voice.

latissimus   Broadest. Latissimus dorsi, broadest muscle of the back.

lens   A glass or crystal, curved and shaped to change the direction of (or refract) rays of light.

lentiform   Shaped like a lens.

lesion   An injury to a tissue that changes its structure or function.

leukemia   A pathological condition of the blood-forming organs characterized by an uncontrolled production of leukocytes.

leukocyte   A white cell of the blood or lymph. Leukocytosis, an increase in the number of leukocytes.

leukopenia   A decrease in the number of leukocytes.

levator   A lifter. Levator palpebrae, lifter of the eyelid.

limbic system   A series of structures in the diencephalon and cerebrum concerned with emotion and influencing behavior.

linea   A line.

linea alba   A median ventral line on the abdomen indicating the line of junction of the tendons of the external oblique muscles.

linea aspera   A rough line along the posterior surface of the femur.

lingual   Belonging to the tongue.

lipase   A digestive enzyme that acts upon fats.

lobule   A little lobe.

lumbar   Belonging to the loin. The lower part of the back.

lumen   The cavity of a hollow tube or organ.

luteal hormone   The hormone secreted by the corpus luteum of the ovary.

lymph   The clear fluid of the lymphatic system and tissue spaces.

lymph node   A nodule of lymphoid tissue occurring along lymphatic vessels.

lymphocyte   A nongranular leukocyte arising in lymphoid tissues.

lysosomes   Minute cell particles containing hydrolytic enzymes.

lysozyme   An enzyme found especially in tears. It is mildly antiseptic.

macrophage   A large phagocytic cell of the reticuloendothelial system.

macula   A spot. Macula lutea, yellow spot. Macula of inner ear.

major   Greater or larger.

malar   Belonging to the cheek.

malleolus   A little hammer. The two malleoli are processes located at the lower extremities of the tibia and fibula.

malleus   Mallet-shaped. The ear bone that is attached to the tympanic membrane.

maltase   A digestive enzyme acting upon maltose.

mammary   Pertaining to the breast.

mandible   (Latin, *mandere,* to chew.) The lower jawbone.

manubrium   A handle. The first part of the sternum.

masseter   A chewer. One of the muscles of mastication, or chewing.

mast cell   A large granular cell found in connective tissues.

mastitis   Inflammation of the breast, or mammary glands.

mastoid   Shaped like a breast.

maxilla   The jawbone. Applied to the upper jawbone.

meatus   A passageway.

medial   Toward the middle line.

median   Middle, as the median line of the body.

mediastinum   The space in the middle of the thorax.

medulla   Marrow. The central part of a gland or organ. Medulla oblongata, the posterior part of the brain.

medullary   Pertaining to, or like, marrow. The medullary canals contain marrow.

megakaryocytes   Giant cells found in bone marrow. They are thought to produce blood platelets by fragmentation.

melanin   A dark pigment.

melanocyte   A pigment cell containing melanin.

menarche   The time when menstruation begins.

meninges   Membranes. Membranes of the brain and spinal cord.

meningitis   Inflammation of the meninges.

menopause   The cessation of menstruation at the close of the reproductive period.

menstruation   A periodic discharge of blood from the genital canal of a woman, associated with changes in the lining of the uterus.

mesenchyme   Embryonic connective tissue.

mesentery   From two Greek words, meaning middle and bowel. (The mesentery supports the intestine from the posterior abdominal wall.)

mesoderm   The middle germ layer of the embryo.

metabolism   The chemical changes associated with the assimilation of energy materials into cell protoplasm and the elimination of the waste products of cellular activity. Life processes.

metastasis   (Greek, to transpose). The translocation of a disease from its primary focus to some other part of the body through the medium of blood or lymph.

micron 1μ. It is 0.001 millimeter.

microvilli   Minute fingerlike projections on the lumenal surface of certain epithelial cells, formed as modifications of the plasma membrane.

micturition   The act of emptying the urinary bladder.

minimus   Least or smallest.

minor   Lesser.

mitochondria   Minute bodies in the cytoplasm of the cell.

mitral   Resembling a miter in outline. Mitral valve, the left atrioventricular valve of the heart. Also mitral cells of the olfactory bulb.

molar   Like a millstone, or mola. The molar teeth grind the food.

molar solution   A solution containing in one liter as many grams of the substance (solute) as the molecular weight of the solute.

mucous   Containing or resembling mucus. Mucosa, a mucous membrane.

mucus   A thick clear fluid secreted by the cells of mucous membranes.

myelin sheath   The white, fatlike inner covering of a myelinated nerve fiber.

myenteric plexus   A network of sympathetic fibers in the wall of the intestine.

myocardium   The muscular wall of the heart.

myoglobin   A muscle protein capable of short-term storage of oxygen.

myopia   Nearsightedness.

myosin   One of the principal proteins of muscle.

myotatic reflex   A stretch reflex of muscle.

myotic   A drug causing constriction of the pupil.

myxedema   A condition resulting from hypothyroidism.

nanometer   One-millionth of a millimeter.

naris   The nostril.

navicular   Boat-shaped, as the navicular, or scaphoid, bone.

necrosis   The death of a portion of tissue, while still surrounded by living structures.

nephron   The basic functional unit of the kidneys. A tubule extending from Bowman's capsule to the collecting ducts.

neural   Pertaining to nerves. The neural axis is the spinal cord. The neural canal is the spinal canal. The neural cavity contains the brain and spinal cord.

neurilemma   The outermost membrane covering a nerve fiber.

neuroglia   The ectodermal connective tissue of the brain and spinal cord.

neuron   A unit of the nerve tissues. It consists of a cell body, dendrites, axon, and terminal filaments.

neutrophil   A granular leukocyte staining with neutral dyes.

Nissl substance   Dark-staining material in the cytoplasm of nerve cells.

nodes of Ranvier   Constrictions in the myelin sheath of nerve fibers.

nucha   The nape of the neck.

nucleolus   A smaller nucleus within the nucleus of a cell.

nucleus   A small round body near the center of a cell enclosed in the nuclear membrane. The vital part of a nucleated cell. A group of cell bodies within the central nervous system.

nutrient   Nourishing.

nutrition   The process of nourishing the cells of living tissues.

nystagmus   Abnormal, involuntary, rhythmic oscillation of the eyeballs.

occipital   Belonging to the back of the head, or the occiput.

odontoid   Resembling a tooth in shape.

olecranon   The large process at the upper end of the ulna. The head of the elbow.

oligodendrocytes   A type of neuroglial cell found along neurons and capillaries of the central nervous system. Synonym: oligodendroglia.

omentum   A fold of peritoneum connected with the stomach.

omos   The shoulder. Omohyoid, belonging to shoulder and hyoid bone, as the omohyoid muscle.

oocyte   An immature ovum.

ophthalmic   Belonging to the eye or ophthalmos.

orbicular   Ring-shaped. A ligament or muscle that resembles a little circle.

organ   A structure designated for a particular function or use.

organ of Corti   An organ of hearing located in the cochlear duct of the cochlea. The spiral organ.

orifice   An aperture or opening.

os   A bone. Ossicle, a little bone.

os   A mouth.

osmoreceptors   Receptors in the hypothalamus which are sensitive to changes in plasma osmolality.

osmosis   The movement of water from an area of low particle concentration to an area of higher particle concentration through a semipermeable membrane.

osseous   Bony.

ossification   The formation of bone.

osteoblasts   Cells concerned with bone formation. The precursors of osteocytes.

osteoclasts   Large multinucleate cells thought to be involved in the breakdown of bony tissue.

osteocyte   Bone cell.

osteon   Refers to a Haversian system of bone.

otoconia   Minute, calcareous concretions of the vestibular apparatus.

osteogenic   Referring to the development of bony tissue.

osteology   The science that treats of bones.

outlet   The inferior opening, or strait, of the pelvis.

oviduct   The fallopian or uterine tube.

ovulation   The rupture of the ovum from the ovary.

ovum   (Plural, ova.) Female germ cell.

oxytocin   A hormone of the pituitary gland stimulating the contraction of uterine muscle.

pacemaker   The sinoatrial node of the heart. Controlling the heart rate.

pachytene   The stage in meiosis that follows synapsis. The homologous chromosome threads appear shorter and thicker.

pacinian corpuscles   Tactile or pressure receptors in the skin, mesentery, and other tissues.

palpebra   An eyelid. Palpebral fissure, the fissure between the eyelids.

pancreas A digestive gland located below the stomach in the loop of the duodenum.

Paneth cells Granular cells in the crypt of Lieberkühn.

papilla (Latin, nipple). A soft conic eminence.

papilla of Vater An enlargement at the common opening of the pancreatic and bile ducts into the duodenum.

parenteral Other than by way of the intestine or alimentary tract.

parietal Referring to the wall of a body cavity or an outer membrane lining the wall, as opposed to an inner or visceral membrane. The parietal bones of the skull.

parietal cells Found along the border of fundic glands of the stomach. They produce a substance that is later converted to hydrochloric acid.

parotid Near the ear. The parotid salivary gland is below the external ear.

pars recta A segment of the nephron. The portion of the proximal tubule which is straight.

parturition The act of bringing forth, or giving birth to, young.

patella A little pan. The sesamoid bone in front of the knee joint; the kneecap.

pectoral Connected with the breast, as pectoral muscles.

pedicle A little foot. Peduncle has a similar meaning.

pelvis A basin. The cavity in the lowest part of the trunk.

penis External genital organ of the male.

pepsin The protease of the stomach.

peptic cells Also called chief cells of the stomach. They secrete pepsinogen.

pericardium The membrane that encloses the heart.

perichondrium The nourishing membrane that covers cartilage.

perilymph The lymph of the perilymphatic spaces between the membranes and bony labyrinths of the inner ear.

perimysium A connective tissue around small bundles of muscle fibers.

perineal Pertaining to the perineum, that region of the body in front of the anus.

periosteum The nourishing membrane around bone.

peristalsis (From two Greek words, meaning around and constriction.) The intestinal movements that propel the food.

peritoneum (From two Greek words, meaning around and to stretch.) The serous membrane around abdominal organs. The lining of the abdominal cavity.

peritubular capillary A capillary distributed around a renal tubule.

peroneal Relating to the fibula. Peroneal nerves supply peroneal muscles on the fibula.

petrous Hard, like a rock.

Peyer's patches Aggregations of lymph nodules in the mucous lining of the ileum.

phagocyte White blood cells having the power to take microorganisms into their substance and to digest them.

phalanges Plural of phalanx, a body of troops drawn up closely together. Bones of the fingers and toes.

phallus The penis.

pharynx That part of the food passage that connects the mouth and esophagus. The upper part is the nasopharynx, an air passage.

phenotype A group of individual organisms that look alike but differ in their genetic constitution.

phenylketonuria (PKU) A condition resulting from inability of an individual to oxidize phenylalanine to tyrosine.

phlebitis Inflammation of a vein.

phrenic Pertaining to the phren, or diaphragm, as the phrenic nerves.

pia mater Tender mother. The delicate membrane that bears the blood vessels of brain and spinal cord. The innermost membrane of the meninges.

pigment Coloring matter.

pineal body A cone-shaped body arising from the roof of the diencephalon.

pinocytosis The process by which extracellular fluid and certain substances such as protein molecules are taken into the cell by invagination of the plasma membrane.

pituitary gland An endocrine gland that lies beneath the brain in the sella turcica of the sphenoid bone. The hypophysis.

placenta A membranous structure that provides the exchange of food materials and oxygen between the blood of the mother and that of the embryo or fetus. Waste products of fetal metabolism are removed through the placenta.

plantar Belonging to the sole of the foot.

plasma The name given to the fluid portion of circulating blood.

plasma cell A lymphoid cell that secretes immunoglobulin or antibody.

platysma Broad. Platysma muscle.

pleura A side. The name of the serous membrane that lines the thorax and covers the lungs.

plexus A network. An arrangement in which vessels or nerves appear to be woven together.

pneumogastric Referring to the lungs and stomach.

podia Cytoplasmic extensions, or "foot processes," of podocytes.

podocytes The cells of the visceral layer of Bowman's capsule.

polycythemia A condition in which there is a marked increase in the number of erythrocytes.

polymorphonuclear Having nuclei of various shapes, such as granular leukocytes.

poples A space behind the knee (popliteal space).

popliteal Belonging to the back of the knee.

porta A gate. The portal vein enters the porta or gate of the liver.

prehension Taking hold of.

premolar Applied to the teeth that stand immediately in front of the molars.

prepuce The foreskin of the penis or clitoris.

presbyopia The farsightedness of aging, due to diminished elasticity of the lens.

pressor Producing a rise in blood pressure.

process In anatomy, a projection, a prominence, or an outgrowth.

proctodeum The invaginated portion of the hindgut, which is lined with ectoderm.

progesterone The hormone of the corpus luteum.

prolactin A lactogenic hormone of the anterior pituitary gland that stimulates the production of milk in nursing mothers.

pronation Literally, bending forward. The position of the hand when the thumb is toward the body. The act of turning the hand palm downward, or in the prone position.

prone A position of the body. Face downward.

prostaglandins Hormonelike substances present in various tissues and having a wide range of physiological functions.

prostate From Greek words meaning to stand before. The prostate gland is in front of the neck of the bladder and surrounds the base of the urethra in the male.

prothrombin The inactive form of thrombin.

protoplasm The essential living matter of all cells.

protuberance A knoblike projection.

proximal Nearest to the trunk, median line, or center.

proximal convoluted tubule The first and largest segment of the nephron.

psychic Pertaining to the mind.

pterygoid Wing-shaped.

ptosis (Greek, a falling.) Prolapse or lowering of an organ or part.

ptyalin A salivary amylase.

puberty The age at which the reproductive organs become functional.

pubes The anterior portion of the os coxae, the pubic bones. The hairy region above the pubic bones.

pudendum The external genitalia, especially of the female.

pulmonary Pertaining to the lung, or pulmo.

purkinje cells Neurons of the cortex of the cerebellum with great arborization of the dendrites.

putamen The outer, darker layer of the lentiform nucleus of the brain.

pylorus The lower opening of the stomach into the duodenum.

pyramidal Shaped like a pyramid. Pyramidal cells of the motor area of the cerebral cortex.

pyramidal tracts Corticospinal nerve fibers from the cerebral motor area and extending downward into the spinal cord.

pyridoxine Vitamin B_6.

pyrogen A substance that produces a fever.

quadrate Four-sided; square or rectangular.

quadri- A combining form meaning four.

quadriceps Four-headed.

quadrigeminal Consisting of four parts.

quinti- Referring to the fifth.

racemose Resembling a cluster of grapes.

rachitis Rachis, the vertebral column. A vitamin D deficiency, rickets.

radius A rod or spoke. The lateral bone of the forearm.

ramus A branch, as the ramus of the mandible.

raphe A seam. Seamlike union of two parts in a line.

reaction Response to a stimulus or test. The iris reacts to the stimulus of light. Urine reacts to the litmus test.

receptor The specialized end organ of an afferent neuron.

recession Withdrawal, as the margin of the gums from the teeth.

rectus Straight, as rectus muscles.

red nuclei Two masses of gray matter in the midbrain.

refractory period The period of reduced irritability, as in muscle and nerve.

relaxin A hormone secreted by the corpus luteum of pregnancy which relaxes the cervix of the uterus.

renal Pertaining to the kidney.

renal capsule The tough fibrous connective tissue outer lining of the kidney.

renal corpuscle A bulbous structure consisting of a glomerulus inside a Bowman's capsule.

renal papilla The medial apex of the inner medulla of the kidney.

renal pelvis The enlarged, funnel-shaped end of the ureter into which urine is collected.

renin A protease from the kidney that releases angiotensin from angiotensinogen.

rennin A milk-coagulating enzyme of the gastric juice.

rete A net.

reticular Resembling a network. A diffuse mixture of gray matter throughout the white matter of the brainstem, the reticular formation.

retina A network. The complex inner nervous tunic of the eye.

retinaculum A fibrous band of fascia that holds down the tendons in various parts of the body, especially at the wrists, ankles, and feet.

retinal A rod cell pigment. (Retinene.)

retro- Prefix meaning behind. Retroperitoneal, behind the peritoneum.

Rh factor (rhesus factor) An antigen known to occur on the red cells of approximately 85 percent of the white population of the United States and Great Britain.

rhinencephalon The olfactory part of the brain.

rhino- A combining form referring to the nose.

rhodopsin The visual purple pigment of rod cells.

riboflavin One of the vitamins of the vitamin B complex.

ribose A five-carbon sugar, a pentose.

ribosomes Particles consisting of ribonucleic acid, commonly found on the endoplasmic reticulum.

rickets A vitamin D deficiency causing malformation of bone and cartilage.

rigor mortis Rigidity of death. The muscular stiffness that occurs after death.

RNA Ribonucleic acid.

rod cells Visual cells of the retina adapted for night vision.

roentgenogram An x-ray photograph.

rugae (Plural of ruga.) Folds. Wrinkles.

saccade Rapid eye movements.

saccule A little sac.

sacral Relating to the sacrum, or the bone that protects the pelvic organs.

sagittal Like an arrow, straight. The straight suture of the skull. A plane that divides the body into right and left portions. Midsagittal, a median longitudinal plane.

saline Salty.

saliva The mixed secretions of glands of the mouth and salivary glands.

saphenous Manifest or plainly seen. The large superficial vein on the medial side of the lower extremity and the longest vein in the body.

sarcolemma The delicate sheath around a muscle fiber.

sarcomere In muscle, a segment limited by a Z membrane at either end.

sarcoplasmic reticulum A network of minute channels within the striated muscle fiber.

sartorius (Latin, *sartor*, a tailor.) An anterior muscle of the thigh.

satellite cells Neuroglial cells clustered around cell bodies of neurons.

Schwann cell A type of cell that covers myelinated and unmyelinated fibers of vertebrate peripheral nerves.

sciatic Pertaining to the ischium. The sciatic nerve.

sclerotic Hard. The sclerotic layer is the tough fibrous coat of the eye; the sclera.

scoliosis Lateral curvature of the spine.

scrotum The pouch containing the testes.

scurvy A vitamin-deficiency condition due to lack of vitamin C.

sebaceous Applied to the glands that produce the oil, or sebum, of the skin.

secretin An intestinal hormone that stimulates the pancreas.

secretion A substance formed by glandular cells.

sella turcica Turk's saddle. A saddle-shaped depression in the sphenoid bone.

semen The secretion of male reproductive glands containing spermatozoa.

semilunar Shaped like a half-moon. The semilunar valves of the heart.

seminal vesicle A secretory gland of the male reproductive system.

septum A partition.

serous Of the nature of serum, a thin watery fluid derived from the blood.

serrated Having teeth like a saw.

Sertoli cells Modified cells present in male germinal epithelium.

serum The fluid portion of the blood after clotting has taken place.

sesamoid Resembling a grain in form. Applied to small nodules of bone sometimes found in tendons.

shaft The main portion of a long bone.

sigmoid Curved like the letter S. As the sigmoid (or transverse) sinus; the sigmoid colon.

sinoatrial node A group of nerve cells and fibers in the wall of the right atrium near the opening of the superior vena cava, which functions as a pacemaker for the heartbeat.

sinus A hollow space, or cavity. A cavity within a bone. A cavity containing blood, as a venous sinus.

sinusoids Capillarylike blood vessels.

soluble That which can be dissolved or made into a solution.

somatic Pertaining to the body. Somatic cells, body cells exclusive of germ cells.

somatomedin Peptides that promote growth.

specific gravity The weight of a substance, judged in comparison with an accepted standard. In the case of urine, the standard is an equal volume of distilled water.

spermatozoa Male germ cells.

sphenoid Wedge-shaped. The sphenoid bone.

sphincter A muscle that closes an orifice.

spiral organ (of Corti) An inner ear membrane that contains the receptors for the sense of hearing.

sphygmomanometer An instrument for measuring blood pressure.

spirometer An instrument for determining the amount of air respired.

splanchnic Pertaining to the viscera or internal organs.

squamous   Shaped like a scale.

stenosis   The narrowing or contraction of a passageway.

stereognosis   The faculty of recognition of objects by the sense of touch.

sternum   Breastbone.

stimulus   That which excites activity or function.

stomodeum   The embryonic ectodermal invagination that leads to the formation of the mouth.

stratum   A layer.

striated   Striped. The alternate light and dark bands of striated muscle.

sty   An inflammation of a sebaceous gland of the eyelid.

styloid   Pointed, like the stylus that was used for writing in ancient times.

sub-   Prefix meaning below or under.

subarachnoid space   The space beneath the arachnoid membrane filled with cerebrospinal fluid.

subcutaneous   Under the skin.

submucosa   A layer of fibrous connective tissue beneath or adjacent to a mucous membrane.

subserous   Under a serous membrane.

substantia nigra   A dark-pigmented area located in the cerebral peduncles.

substrate   The substance acted upon by an enzyme.

succus entericus   The digestive juice of the small intestine.

sucrase   An enzyme that acts upon sucrose.

sudoriferous   Bearing sweat, as sudoriferous glands. (Sudoriparous has the same meaning.)

sulcus   (Plural, sulci.) A furrow or groove. The depressions between the convolutions of the brain.

super-   Prefix meaning above.

supercilium   The eyebrow, or prominence above the eyelashes. (Adj., superciliary.)

supination   The attitude of one lying on the back. The position of the hand when the little finger is next to the body, or turning the palm upward.

suppuration   The formation of pus.

supra-   A prefix meaning above.

suprarenal   Above the kidney. Adrenal. An endocrine gland located on the superior surface of the kidney.

surfactant   A lipoprotein which lowers the surface tension in lung alveoli.

sustentaculum tali   A process of the calcaneus (heelbone) that supports the talus.

suture   (Latin, sutura, seam) A seam. The joints of the cranium are sutures.

sympathomimetic   To mimic the action of the sympathetic nervous system. To cause physiologic actions similar to those produced by the sympathetic nervous system.

symphysis   A joining of two bones, especially in a sagittal plane, as the symphysis of the mandible and the pubic symphysis.

synapse   The region where the end knobs of the terminal filaments of one neuron come in close physiologic relationship with the dendrites and cell body of a succeeding neuron.

synapsis   The conjugation of homologous chromosomes during meiosis.

synaptic cleft   A space of about 100 to 200 Å between the presynaptic and postsynaptic membranes.

synarthrosis   An immovable joint.

syncytium   A multinucleate mass of protoplasm.

syndrome   A typical set of conditions that characterize a deficiency or disease.

synergic, synergetic   Two or more agents acting as one. The cooperative action of certain muscles.

synovia   A fluid resembling the raw white of an egg, found in joint cavities, bursae, and tendon sheaths.

synovial membrane   A membrane lining a joint cavity, a bursa, or tendon sheath and concerned with the secretion of synovial fluid.

systole   (Greek, contraction.) The contraction of the chambers of the heart.

tachycardia   Increased heart rate.

tactile   Referring to the sense of touch.

taeniae coli   Three tapelike longitudinal muscle bands of the colon.

talus   The ankle bone upon which the tibia rests, also called astragalus.

tectorial membrane   A covering over the spiral organ of the inner ear.

telodendria   Terminal filaments of a neuron.

tendo Achillis   The tendon of Achilles, or calcaneus tendon. The tendon of leg muscles attached to the calcaneus or heelbone by which Achilles was held when his mother submerged him in the river Styx to render him invulnerable. Only the heel remained vulnerable.

tensor fasciae latae   A muscle that tenses the fascia of the thigh.

tentorium   A tent. The tentorium cerebelli (of the cerebellum) covers the cerebellum.

teres   Round. (Ligamentum teres, round ligament.)

testes   The male reproductive glands that produce spermatozoa.

testosterone   The male sex hormone secreted by the interstitial cells of the testis.

tetanus   (Physiol.) A sustained contraction of muscle, especially when produced experimentally.

thalamus   (Greek, a bed.) The optic thalamus is in the base or bed of the brain.

thenar   Relating to the palm or sole.

thiamine hydrochloride   An essential vitamin of the vitamin B complex.

threshold stimulus   The least strength of stimulus that will cause a reaction. A minimal stimulus.

thorax   The chest. The portion of the trunk that contains the heart and lungs.

thrombin   An enzyme acting upon fibrinogen to produce fibrin.

thrombocyte   A blood platelet.

thromboplastin   A thromboplastic substance which, along with calcium ions and other factors, converts prothrombin to active thrombin.

thrombus   A blood clot formed within the heart or blood vessels, as in coronary thrombosis.

thymus   A lymphoid structure located beneath the sternum in the mediastinum.

thyrocalcitonin   A hypocalcemic factor; a regulator of calcium in blood plasma.

thyroid   Shield-shaped. The thyroid gland, an endocrine gland.

thyrotropin   An anterior pituitary hormone that regulates the secretion of the thyroid gland.

thyroxine   A hormone of the thyroid gland.

tidal air   The amount of air inspired or expired during quiet breathing.

tissue   A group of cells of similar origin, structure, and function.

tocopherol   Vitamin E.

tonus   A state of mild contraction exhibited by muscle tissue.

torticollis   Twisted neck, wryneck.

trabeculae   Little beams. The cross bands of connective tissue that support soft structures, as in the spleen.

transducer   An organism or mechanism capable of changing one form of energy to another. Example, receptors.

trapezius   A muscle of the back.

trauma   A wound or injury.

triceps   Three-headed.

trigone   A space or surface having three angles or corners.

trochanter   From a word signifying a wheel. (The muscles that are attached to the trochanters roll the femurs.) A bony process of the femur.

trochlea   A pulley. A trochlear surface is a grooved convexity, as the trochlea of the humerus. A ring of connective tissue in the upper margin of the orbit.

trochlear   Pertaining to a pulley. The trochlear cranial nerve innervates the superior oblique (pulley) muscle of the eye.

tropomyosin   A muscle protein.

trypsin   The enzyme of the pancreas that digests proteins. A protease.

tuber   A swelling or bump.

tubercle   A small projection like a swelling.

tuberosity   A large projection on a bone.

turbinated   Rolled, like a scroll. Turbinate bones, the nasal conchae.

tympanum   Pertaining to the middle ear or to the eardrum. Tympanic cavity. Tympanic membrane.

tyrosine   An amino acid.

ubiquinone   Coenzyme Q, a factor in electron transport.

ulna   A cubit; the elbow. The longer bone in the medial side of the forearm.

umbilicus   (Latin, *umbo*, the elevated or depressed point in the middle of an oval shield.) The navel.

uncinate   Hooked. A process shaped like a hook.

ungual   Belonging to the nail.

uracil   A constituent of nucleic acids; a pyrimidine base.

urea, $CO(NH_2)_2$ A substance representing the chief nitrogenous product of tissue waste.

ureter   The duct of the kidney, which conveys urine to the bladder.

urethra   The passage through which urine is expelled from the bladder.

uterus   A pear-shaped, muscular organ of the female reproductive system, in which the fetus develops.

utriculus   A membranous sac in the vestibule of the inner ear, connected with the semicircular canals.

uvula   The median, posterior tip of the soft palate.

vagina   A sheath. The passageway from the uterus to the external orifice.

vagus   The Xth cranial nerve.

vallate   Situated in a cavity surrounded by a ridge. The vallate papillae of the tongue.

valvula   A little valve, a flaplike fold.

valvulae conniventes   Little valvelike folds. Seen on the mucous coat of the small intestine.

varicose veins   Abnormally swollen and tortuous veins.

vas   A tube or duct that conveys a fluid.

vasa efferentia   Tubules that lead from the testis into the vas deferens.

vasa recta   The capillary network which parallels the loops of Henle of the kidney tubules.

vascular   Having many blood vessels.

vas deferens   The efferent duct of the testis.

vasectomy   Resection of the vas deferens.

vasomotor   Literally, vessel mover. Applied to the nerves that dilate blood vessels or contract them, or vasodilators and vasoconstrictors.

vasopressin   A posterior pituitary antidiuretic hormone.

velum   The veil, or soft hanging portion of the palate or roof of the mouth.

vena cava   A large, hollow vein. Carries blood to the right atrium of the heart.

venesection   Cutting a vein.

ventral   Toward the front of the body, as the ventral cavity.

ventricle   A cavity in the brain or in the heart.

venule   A very small vein.

vermiform   Worm-shaped. The vermiform appendix.

vermis   An area between the two cerebellar hemispheres.

vertebrae   From a Latin word meaning to turn. Certain movements of the vertebrae turn the body from
side to side. The bones of the spinal column.

vertex   The crown of the head.

vesicle   A liquid-filled sac or cavity.

vestibule   A cavity of the internal ear through which
impulses are transmitted to auditory and vestibular
nerves.

villus   The villi of the intestine are hairlike in shape
and belong to the mucous coat. Vascular, fingerlike
processes of the chorion.

viosterol   Activated ergosterol. Vitamin D$_2$. An antirachitic compound.

viscus   (Plural, viscera.) An internal organ of the head
or trunk.

vital capacity   The greatest volume of air that can be
moved in or out of the lungs.

vitamin   An organic compound usually present in
minute amounts of foods and essential for growth
and nutrition.

vitreous   Glassy. The vitreous humor resembles glass in
appearance. The vitreous layers of the skull are
brittle like glass.

volar   Belonging to the palm.

Volkmann's canals   Minute openings permitting small
blood vessels to enter bone.

vulva   The external genitalia of the female.

Wernicke's area   An area located in the temporal lobe
of the brain close to the auditory area. Injuries to
this area result in failure to understand the meaning
of spoken words.

Wharton's jelly   Mucoid connective tissue of the
umbilical cord.

wheal   Elevated lesions of the skin.

white matter   Nerve tissue composed chiefly of nerve
fibers.

Wormian bones   Small supernumerary bones in the
sutures of the skull.

xantho-   A combining form meaning yellow.

xanthophyll   A yellow pigment found in plants.

X chromosome   One of the sex chromosomes.

xero-   A combining form meaning dry.

xerophthalmia   A vitamin A–deficiency condition
that is characterized by a dry and thickened conjunctiva.

xiphoid   Sword-shaped. The third piece of the sternum
is the xiphoid, or ensiform, process.

Y chromosome   A male sex chromosome in man.

yellow spot   The macula lutea of the retina.

yolk   Nutritive material of the ovum.

yolk sac   An extraembryonic membrane containing the
yolk if yolk is present.

Z line   A narrow zone of dense material present in
skeletal muscle tissue.

zein   A prolamine of maize.

zona pellucida   A transparent, noncellular, secreted
layer surrounding the ovum.

zoology   The science of animal life.

zygapophysis   An articular process of a vertebra.

zygoma   A yoke. The arch of bone at the side of the
face formed between zygomatic and temporal
bones.

zygote   The fertilized ovum before cleavage.

zymase   An enzyme present in yeast concerned with
alcoholic fermentation.

zymogen   The inactive precursor of an enzyme.

zymogenic cells   The chief cells of the stomach.

INDEX

INDEX

Page numbers in **boldface** indicate illustrations.

Axis, **150**
  odontoid process, **150**

B cells, 461, 462, 472
Babinski reflex, 311
Baroreceptor mechanism,
  535, 608
Baroreceptors, 567
Barr body, 710
Bartholin's gland, 701
Basal body, 98
Basal ganglia, **292**
Basement membrane, 93,
  594
Bases, basic solutions, 11–13
  strong base, 13
  weak base, 13
Basilar membrane, **385**
Basophils, 461
Beta cells of pancreas, 427,
  429
Beta-oxidation, 68
Beta receptors, 334
Bicarbonate ions, 561
Biceps brachii muscle, 221,
  222, 223
Biceps femoris, 239, **240**
Bile, 639–640
  canaliculi, **637**
  pigments, 639
  salts, 639
Bilirubin, 639
Biliverdin, 639
Binocular vision, 365
Biotin, 667
Birthmarks, 119
Blastocyst, 713, **714**
Blastodermic vesicle, 713
Blood, 454, 484
  circulation of: cardiac cy-
    cle, 487, **495, 496**
  flow: in arteries, 528,
    529
    in capillaries, 530,
      531
    in veins, **529**
  hepatic portal system,
    487
  pulmonary, 568, 569
  systemic, 528–531

Blood:
  clotting of: extrinsic sys-
    tem, 465
    factors, 465
    formation of, 467
    intrinsic system, 465
    phases, 465
  constituents, 454–475
    cells, 454–459
    plasma, 454, 455
    platelets, **460**, 463
    salt and water balance,
      454
  grouping, the ABO sys-
    tem, 468–470
Blood pressure, 531–541
  effect of exercise on,
    540
  effect of gravity, 541
Blood transfusion, 467
Blood volume, 453
Body cavities, 132, 133
  abdominopelvic, 132
  cranial, 133
  spinal, 133
  thoracic, 133
Body temperature regula-
  tion, 679
Boils, 114
Bone(s), 124–178
  arm, 156–160
  chemical structure, 112
  cranial, 134–148
  foot, 165–167
  hand, 160
  leg, 163–165
  pelvic girdle, 160–163
  shoulder girdle, 153
  skeleton, **154–155**
  skull, 134–147
  thorax, 152–153
  vertebral column, 148–
    152
  wrist, 160
Bone cancer, 175
Bowman's capsule, 588
  parietal layer, **590**, 591
  visceral layer, **590**, 591
Bowman's space, **591**
Brachialis muscle, 221, **222,
  223**

Brachioradialis muscle, 221,
  225
Bradycardia, 523
Bradykinin, 116, **535**
Brain, 278–310
  association areas of, **290**
  forebrain, 278
    cerebrum, 284–299
  hindbrain, 300–310
    cerebellum, 300–303
    medulla, 303
  midbrain, 299
    cerebral aqueduct, 296,
      299
    cerebral peduncles,
      299
    colliculi, 296
Brain tumors, 315
Brain waves, 298
Bronchi, **552**, 553
Bronchial tubes, **552, 553**
Bronchioles, **553**
Bronchiotracheal tree, **554**
Brunner's glands, 633
Brush border, 591
Buccinator muscle, 211, 212
Buffers, 14, 22, 570, 571
  chemical, 570, 571
    bicarbonate, 571
    phosphate, 571
    proteinate, 571
Bulbourethral glands, 696
Bulk flow, 35, **36**, 38
Bunion, 167
Burns, 118
Bursae, 112, 168
Bursitis, 164

Calcaneus, **165**
Calciferol, 681
Calcitonin, 128
Calcium, 659
Calories, 20, 54
  requirements, **675**, 676
Calorimetry:
  direct, 670
  indirect, 670
Calyces, 584, **587**
Canaliculi, 125
Cancellous bone, 125

Glycine, 18
Glycogen, 14–16, 26, 193, 194
Glycogenesis, 57, 58
Glycogenolysis, 58
Glycolysis, 53, 58–60, **59**, 62, 67
Glycoprotein, 30–31, 33, 39
Goblet cells, **95**, 96, **97**
Goiter, 420
Golgi apparatus, 24, 29–**30**, 45–46, 51
Gonads, 438
Gouty arthritis, 175
Gracilis muscle, 238, **242**
Gram molecular weight, 10
Granular leukocytes, 459, **460**
Growth, cell, 26–27, 50, 53
Growth hormone (GH), 411, 416
Growth hormone abnormalities, 415
Guanine, 70, **71**
Gustatory receptors, **394**, 395

Hair development, 115
Hair follicle, **100**
Haploid, 70
Hard palate, 618
Haversian canals, 125
Haversian system, **126**, 127
Hay fever, 576
Headache, 315
  migraine, 316
Hearing, theories of, 384
    place, 385
    resonance, 384
    traveling wave, 385
Heart, 48, 49, 488–500
  arteries and veins, 488
  atria, 488
  autonomic innervation, 498
  blood pathway, 488, 489
  chambers and valves, 489, **490**
  contractions, 493, 495
  law of, 536

Heart:
  veins of: cardiac, 508
    coronary sinus, 508
  ventricles, 488
Heart rate, 497
Heart sounds, 496
Heartburn, 648–649
Heat production, 679
Heat stroke, 679
Heimlich maneuver, 579, 580
Helicotrema, **382, 385**
Hemoglobin, 457, 561, **558–563**
  gas transport, 558–563
  carbamino, 561
Hemophilia, 481
Hemorrhage, 544
Hemorrhoids, 651
Hemostasis, 464
Heparin, 103, 466
Hepatic duct, 638
Hepatic portal system, **512**, 513
  inferior mesenteric vein, **512**
  portal vein, **512**, 637
  splenic vein, **512**
  superior mesenteric vein, **512**
Hering-Breuer vagal reflex, 565
Hernia, 247–**250**
Hexose, 14
Hilar nodes, 588
Hilus of kidney, 584
Hindbrain, 300–310
Hippocampus, **291**
Histamine, 103, 535
Histones, 70
Homeostasis, 24, 46, 50–51
Homeostatic mechanism, 46–48, 50, 53
Horizontal section, 132, **133**
Hormones, digestive tract, 645
Humeral tuberosities, 157, **158**
Humerus, 156, **157, 158**
Hyaline cartilage, **110**
Hyaluronic acid, 709

Hyaluronidase, 709
Hydrides, 10
Hydrochloric acid, 629, **630**
Hydrogen, 4–11, 14, 16–17, 19–21
Hydrogen bonding, 10
Hydrogen ion, 11–14, 22
  hydrated, 12
Hydrolase, 21
Hydrolysis, 21
Hydrophilic molecule, 17, 30, 31
Hydrophobic molecule, 17, 30, 31
Hydrostatic pressure, 35, 37–38, 46
Hydrostatic pressure gradient, 35–36
Hymen, **701**
Hyoid bone, 147, 148
Hyperemia, active, 534
Hyperglycemia, 429
Hyperinsulinism, 431
Hyperopia, 372
Hyperosmotic solution, 38
Hyperparathyroidism, 425
Hypertension, 542
Hyperthyroidism, **421**
Hypertonic solution, 38–39
Hyperventilation, 566
Hypoglossal, **301**, 307
Hypoosmotic solution, 38
Hypoparathyroidism, 425
Hypophysectomy, 415
Hypophysis, **408**–416
Hypophysis hormones, 410
Hypothalamus releasing hormones, 409
Hypothermia, 523
Hypotonic solution, 38–39

I band of muscles, 184, **185**
ICSH (interstitial cell-stimulating hormone), 441
Identical twins, 717
Ileocecal valve, 643, **646**
Ileum, 643
Iliopsoas muscle:
  iliacus, **234, 235**

Iliopsoas muscle:
  psoas major, **234**, 235
  psoas minor, **234**, 235
Imidazole, **561**
Immunity, 471–475
  active, 473
  cellular, 472
  humoral, 472
  passive, 475
Immunoglobin E (IgE), 483
Implantation, 714
Impotence, 724
Inferior mesenteric artery,
  506
Inferior nasal conchae, 147
Infertility, male and female,
  724
Inflammation, 478
Inguinal ligament, 231
Inorganic compounds, 8
Insensible evaporative
  water loss, 608
Inspiration, 555
Inspiratory center, 563
Insula, 285, **286**
Insulin, 427, 428, 430
Insulin shock, 431
Integument, 113
  dermis, **113**, 115
  epidermis, **113**
  hypodermis, 113
Intercalated disks, 203, **204**
Intercellular fluid, 46–47
Intercostal muscles, 216,
  555
  external, 555
  internal, 555
Interferon, 475
Internal oblique muscle,
  232, 233
Interoceptors, 340, 341
Interossei muscles:
  dorsal, **231**
  palmar, **231**
  testis, **443**
Interstitial cell-stimulating
  hormone (ICSH), 413
Interstitial (Leydig) cells,
  693, 694
Interstitial fluid, 46
Interstitial lamellae, 127

Interventricular foramen,
  281
Intervertebral disk, rupture,
  169
Intestine, 617
Intrauterine devices (IUDs),
  707
Intrinsic factor, 667
Inulin, 598
Inversion of image, 359, **364,
  365**
Iodine, 659
Ion, 8, 10, 13–14, 42
Ionic bond, 8–9, 22
Ionization, 8, 13–14, 22
Iris, 346, **347**, 348
Iron, 659
Irritability, cell, 26, 50
Ischemia, 522
Ischial tuberosities, **161**
Islet cells of Langerhans,
  427
Isometric contraction,
  199
Isosmotic solution, 38
Isotonic contraction, 199
Isotonic solution, 38–39
Isotopes, 7–8, 22

Jaundice, 649, 650
  hemolytic, 650
  obstructive, 650
Jejunum, 633, 643–645
Joints, 167
  freely movable, 168
  immovable, 167
  slightly movable, 167
Jugular veins, 508
  external, 508
  internal, 508
Juxtaglomerular apparatus,
  593, 611
Juxtamedullary nephrons,
  592

Keratitis, 371
Ketosis, 641–642

Kidney, 583–614
  anatomy of, 583–594
    external, 584, **585**, **586**
      location of adrenal
        glands, **585**
      ureters and urinary
        bladder, 584, **585,
        586**
    internal renal struc-
      tures, 584–593
      arteries, arcuate and
        renal, 588
      capsule, **588**
      corpuscles, **587**, 588
      cortex, **588**
      medulla, **588**
      papilla, 590
      pelvis, 584, **587**
      pyramids, 588, 604
  renal function mecha-
    nisms, 593–614
Kidney diseases, 613
  glomerulonephritis, 613
  pyelonephritis, 613
  renal calculi (kidney
    stones), 613
  uremia, 613
  use of artificial kidney or
    dialysis machine, 613
Kinins, 635
  bradykinin, 116, **535**
Klinefelter's syndrome, 710,
  711
Kupffer cells, 105, 638
Kyphosis, 152

Labia major, **701**
Labia minora, **701**
Lacrimal bones, **138**, 147
Lacrimal structures, 342
Lactase, 635–636
Lacteal, 518
Lactose, 15
Lacunae, 109, **111, 112**, 125
Lambdoidal suture, 135, 136
Lamellae, 125
Lanugo, 114
Large intestine, **646–649**
  functions of, 647–648
Laryngitis, 550

Larynx, 550
Latent period in muscle, 196, **197**
Lateral, 131, 132
Lateral intercellular space, **592**
Lateral sacs of muscle, 186
Latissimus dorsi, **219**, 220
Law of the heart, 536
Length-tension concept, 200
Lens, **345, 347,** 351, **352**
Lens function, 358
  accommodation, 359
  for near and distant vision, 363
Leucopenia, 481
Leucotaxine, 461
Leukemia, 481
Leukocytes, 457, 459, **460**
  in inflammation and disease, 463
  nongranular, **460,** 461, 462
Leukocytosis, 481
Levator ani muscle, 235, **236**
Levator palpebrae superioris muscle, 211
Levator scapulae muscle, 218, **219**
Levers of body, 209
Ligaments, 106
Light, physical characteristics, 365–366
Light adaptation, 356
Light reflex, 360
Limbic system, 291
Linea alba, **232,** 233
Lipase, pancreatic, **632,** 634, 635
Lipids, 14, 16–17, 26, 50, 656
Lipoprotein, 30
Liver, **636,** 642
  cirrhosis, 649
  diseases, 649
  functions, **642**
  glycogenic functions, 640
  lobules, **637**
  sinusoids, **637**
Locomotor ataxia, 313
Loop of Henle, 590, 592, **589**
  thin ascending limb, 602
  thin descending limb, 602

Lordosis, 152
Lumbar artery, 506
Lumbar puncture, 313
Lumbar vertebrae, 150, **155**
Lumbricales muscle, **229,** 230, 247, **248,** 249
Lungs, **551, 552**
  buds, 549, **550**
  structure, internal, **553**
Luteinizing hormone (LH), 413
Lymph, 515
Lymph nodes, 518, **519, 520,** 521
Lymph sinuses, 520
Lymphagogues, 524
Lymphatic capillaries, **516, 517**
Lymphatic ducts, 518, **519**
  left, 518, **519**
  right, 518, **519**
Lymphatic system, 515–521
Lymphatics, 516, **517,** 518
Lymphoblasts, 461
Lymphocytosis, 481
Lymphoid tissues, 520
Lysosomes, 24, 29–30, 44, 45

M, N, and MN blood factors, 471
Macrophages, **105,** 463
Macula densa, 610
Macula lutea, 357
Magnesium, 658
Male reproductive system, 692, **693**–697
  testes, 692
Male sex hormones, 441
Malpighian bodies, 588
Maltose, 15, 16
Mammary glands, **702**
Mammography, 725
Mandible, **136,** 142
  condyloid process, 142
  coronoid process, 142
  mental foramen, 142
  rami, 142
Manubrium, 153

Marrow, 125
  red, 125
  yellow, 125
Mass, 4
Masseter muscle, 212
Mast cells, 103, **105**
Mastoiditis, 136
Matrix, 61–62
  of mitochrondria, 61, 62
Matter, 4
Maxillae, **136,** 143
  alveolar process, 143
  frontal process, 143
  maxillary sinuses, 143
  palatine process, 143
  zygomatic process, 143
Medial, 131, 132
Mediastinum, 552
Medulla, 303
Medullary zone, 588
  inner, 588
  outer, 588
Megakaryocytes, **462,** 463
Meiosis, 686, **690, 691**
Meissner's corpuscles, **391,** 392
Melanin, 116, 346
Melanocyte-stimulating hormone (MSH), 414
Melanocytes, **100,** 116
Melanosomes, 116
Melatonin, 426
Membrane bones, 128
Meniere's disease, 399
Meninges, 279, **280**
  arachnoid, 279, 280
  dura mater, 279, 280
  pia mater, 279, 280
Menopause, 707
Menstrual cycle, 703, **704**
  sex hormones in, **439**
Mesangial cell, **591,** 594
Mesenchyme, 102
Mesentery, 102
Mesothelium, 93, 103
Messenger RNA (mRNA), 53, 76
Metabolic cost, activity, **677, 678**
Metabolic pathways, 57, 58

Nonhistone component proteins (NHC), 70, 79
Norepinephrine, 433, 534
Nuclear membrane, 29–30, 45
Nucleic acids, 14, 18–19
Nucleolus, 29–30, 69
Nucleoplasm, 25, 30, 60
Nucleoprotein, 26
Nucleoside, 70
Nucleotides, 18–19, 70, **71**
Nucleus, cell, 29–30, 53, 69

Obturator foramen, **161**, 162
Occipital bone, 135, 136
　foramen magnum, 135
　occipital condyles, 135
Oculomotor nerve, **301**, 305
Olecranon process, **159**
Olfaction, 396
Olfactory bulbs, 290
Olfactory disorders, 400
Olfactory nerve, 305
Olfactory receptors, **397**
Olfactory tract, 397, **398**
Omentum, 627
　greater, **627**
　lesser, **627**
On and off retinal response, **364**
Oogenesis, 698
Open heart surgery, 523
Opponens digiti minimi muscle, 230
Opponens pollicis muscle, 227, **229**
Optic chiasma, 362
Optic nerve, **301**, 305
Oral contraception, 707, 708
Orbicularis oculi muscle, 211, **212**
Orbicularis oris, 211, 212
Orbitals, atomic, 5, 6, 22
Organelles, cellular, 24–25, 29–30, 50, 53
Organic compounds, 14
Oropharynx, 550
Osmoles, 38
Osmoreceptors, 608
Osmosis, 24–25, 35–38, 51

Osmotic pressure, 37–38, 46, 51, 537
Ossein, 112
Osteoarthritis, 175
Osteoblasts, 128
Osteoclasts, 127
Osteocytes, 125
Osteogenesis, 128
Osteogenic layer, 128
Osteomyelitis, 174
Osteon, **126,** 127
Osteoporosis, 174
Otoconia, 387
Ovarian artery, 503, 506
Ovarian cysts, 724
Ovulation, 698, 706, **714**
Ovum, human, 704
Oxidase, 21
Oxidation, 21, 55–57
Oxidation-reduction, 53, 55–56
Oxidoreductase, 21
Oxygen:
　as an element, 4–11, 14–21
　in respiration: dissociation curves, 559–561
　partial pressure, 557–559
　transport by hemoglobin, 558–560
Oxygen debt, 195
Oxyhemoglobin, 557–559
Oxytocin, 414

Pacemaker cells, 202
Pacemaker potential, 202
Pacinian corpuscles, **391,** 392
Pain, 393
Pancreas, 427, 633, 634, **635**
Paneth cells, **644**
Pantothenic acid, 667
　deficiency, 682
Paranasal sinuses, 147
Parasympathetic cranial nerves, 328
Parathyroid glands, **423**
Parathyroid hormone (PTH), 423

Parenteral hyperalimentation, 39
Parietal bones, **136**
Parietal cells, 629, **630**
Parkinson's disease, 315
Pars recta, 590, 592, 602
Parturition, 719, **720**
Past pointing, 390
Patella, 163, **164**
Patellar reflex, 311
Patellar tendon reflex, 273
Pectoralis major muscle, 214, **215**
Pectoralis minor muscle, 214, **215**
Pediculosis, 120
Pelvic girdle, 160
　ilium, 160
　ischium, **161**
　pubis, 162
Penis, 696
Pentose, 14
Pepsin, 629
Pepsinogen, 629
Peptic ulcer, 649
Peptide, 18
Peptide linkage, **18**
Pericardium, 102, 490
Perichondral bone, 129
Perichondrium, 109
Perilymph, 381, **386, 387**
Perineum, 236
Periodontal disease, 648
Periosteum, 127
Peristalsis, **643**
　control of, 643
Peritoneum, 101, **627**
Peritonitis, 649
Peritubular capillary, **590**
Pernicious anemia, 479
Peroneus brevis muscle, 243, **244**
Peroneus longus muscle, 242, **243**
Peyer's patches, 521
pH, 11, **13,** 22, 46
　of blood, 14
　regulation of, 569, 574
Phagocytes, 459
Phagocytosis, **44,** 51
Phalanges, foot, **166**

Respiratory pump, 592
Respiratory quotient, 641
Reticular formation, 302, 303
Reticular tissue, 106
Reticuloendothelial cells, 105
   Kupffer cells, 105
Retina, 345, **347**, 348, 349, **350**
Reversible reaction, 13
Rh factor, 470, 471, 482
Rheumatoid arthritis, 175
Rhodopsin, 355, **356**
Rhomboideus muscles, major and minor, 218, **219**
Riboflavin, 665
   deficiency of, 681, **682**
Ribonucleic acid (*see* RNA)
Ribosomes, 29–30, 51, 53
Rickets, 170
Rigor mortis, 191
Ringworm, 119
RNA, 14, 19, 69, 70, 71
   messenger (mRNA), 53, 76
   polymerase, 73
   ribosomal, 74, 75
   transfer, 75, 77
Rods, **353**, 354, **355, 358**

Saccule, **381**, 387
Sacral autonomics, 330
Sacroiliac joint, **161**, 163
Sacrum, 150, **155, 161**
Sagittal plane, 132
Saliva, 624
   digestive function, 624
   excretory function, 624
Salivary glands, 622–624
   classification of: parotid, **622**
      sublingual, **622**
      submandibular, 622
   innervation of, 328
   secretion, 623
Salts, 12–14
Sarcolemma, 181, **183**
Sarcomere, **183**, 184, **185**
Sarcoplasm, 181, **183**

Sarcoplasmic reticulum, 185, **186**
Sartorius muscle, 238
Scala media (cochlear duct), **382, 383**
Scala tympani, **381, 382**
Scala vestibuli, 381, 382
Scapula, 153, **156, 157**
Scarring, 120
Schwann cell, **260**
Scoliosis, 152
Scurvy, 682
Sebaceous glands, **100,** 101
Secretin, 634, 636
Selective elimination, 583
Selective permeability, 30
Semen, 697
Semicircular canals, **380, 381**
Semimembranosus muscle, 239, **240**
Seminal vesicles, 695
Semitendinosus muscle, 239, **240**
Sensors, 47–48
Sensory receptors, 340
Sensory unit, 340
Serous membranes, 101
Serratus anterior muscle, **215**, 216, **219**
Sertoli cells, **442**, 694
Serum, **467**
Sex chromatin, 710
Sex determination, 690
   sex chromosomes, 690, 691, 692
Sex hormones in menstrual cycle, **439**
Shingles, 314
Sickle-cell anemia, **479, 480**
Sinoatrial node, 487, 493
Skin cancer, 120
Sliding filament theory, 184, **185**
Small intestine, 633
Sodium chloride, 658
Sodium-dependent secondary active transport, 645

Sodium-potassium-activated adenosine triphosphatase (Na, K-ATPase), 42
Sodium-potassium exchange pump, 42–44, **43**
Soft palate, 618
Soleus muscle, **244**, 245, **246**
Solute, 10
   pressure, 37–38
Solute-linked fluid transport, 600
   standing osmotic gradient theory, **601**
Solution, 10, 12, 14
Solvent, 10, 17
Spasm, 205
Specific dynamic action, 674
Specificity, 39
Spermatic artery, **503**, 506
Spermatozoa, 693
Sphenoid bone, 142, **144**
   greater wings, 142
   lesser wings, 142
   pterygoid process, 142
   sella turcica, 142
Sphincter pupillae, 348
Sphygmomanometer, 539
Spinal cord, 269, 271, 307–309
   conduction pathways, 307
      ascending tracts, 307, **309**
      descending tracts, **309**
Spinal nerve structure, 270, **271**
Spinal nerves, 310, **312**
Spinal plexuses, 311
Spinous process, 134
Spiral ganglion, **382**
Spiral organ, **383**
Spleen, 476, **477**
   functions, 476
Splenius capitus muscle, 214
Sprain, 169, 205
Squamous epithelium, 93, **95**
Stapedius muscle, 380
Starches, 14, 15
Starvation, 677

Transmission deafness, 399
Transport:
  active, 24, 39–41, 44, 51, 64
  passive, 39–40, 51
Transverse plane, 132, **133**
Transverse tubules of muscle, 186
Transversus abdominis muscle, 233, **250**
Trapezius muscle, 218, **219**
Treppe, **198**
Triacylglycerols, 16–**17**, 26
Tricarboxylic acid (TCA) cycle, 53, 57–58, 60, 62, **63**, 64, 69
Triceps brachii muscle, 221, **223**
Triceps surae muscle, **243**, **244**
Trigone, 587
Trisomy, 710
Trochanter, 134
Trochlear nerve, **301**, 305
Tropocollagen, 103
Trypsin, **632**, 634
Tubercle, 134
Tuberculosis, 574–575
Tubular maxima, **600**
  reabsorption, 583, **584**
  secretion, **584**
Twinning, 717, **718**
Tympanic membrane, 379, 380
Tyrosinase, 116
Tyrosine, 116

Ubiquinone, 64–67
Ulna, **159**
Umbilical cord, 513, **514**, 716
Umbilicus, 513
Uracil, **73**, 74
Urea, 602
Ureter, 584, **585**
Urethra, 587
  urethral orifice, 587
  urethral sphincter, 587
Uridine diphosphate, 655
Uridine triphosphate (UTP), 58

Urinalysis, 613
Urinary tract anatomy, 586
Uterine tubes, **700**
Utricle, 381, 387
Uvula, 618

Vaccines, 474
Vagina, **700**
Vagus nerve, **301**, 306
Valence, 9, 14
Vallate papillae, **394**
Varicose veins, 524
Vasa recta, **604**, **605**
Vascular spasm, 464
Vas deferens, 695
Vasoconstrictor center, 533
Vasomotor control of blood vessel diameter, 533
Vastus intermedius muscle, 237, 238
Vastus lateralis muscle, 237, 238
Vastus medialis, 237, **238**
Veins:
  of arm, 510
    basilic, 510
    cephalic, 510
    radial, 510
    ulnar, 510
  of head and neck 508–510
    azygos, 510
    brachiocephalic, **509**
    external jugular, 508
    hemiazygos, 510
    internal jugular, 508, **509**
    superior vena cava, **509**, 510
  of heart, 508
    cardiac, 508
    coronary sinus, 508
  of pelvis and abdomen, 511, 512
    common iliac, 511
    external iliac, 511
    hepatic, 512
    inferior vena cava, 511
    internal iliac, 511
    renal, 511
    uterine, 511

Veins:
  of thigh and leg, 510, 511
    deep, 511
      anterior tibial, 511
      femoral, 511
      popliteal, 511
      posterior tibial, 511
    superficial, saphenous, 510, 511
Venereal diseases, 721–723
  chancroid, 723
  gonorrhea, 721
  herpes simplex, 723
  syphilis, 722, **723**
Venous pressure variation, **539**, 541
Venous pump, 541
Venous system, pulmonary veins, 508
Ventilation, 549
Ventricles, 280
  of brain, **285**
Ventricular fibrillation, 523
Vermiform appendix, **646**, 647
Vertebral column, 148, **149**, 151, 155
Vertebral foramen, **149**
Vertebrochondral ribs, 152
Vesicle, 29–30, 44–46, 51
Vestibular apparatus, 386
Vestibule, 380, **381**, 701
Vestibulocochlear nerve, **301**, 306
Villi, 518
  intestinal, 543, **644**
Visceral afferent neurons, 335
Visceral pain, 336
Vision physiology, 361
Visual cortex, 362, **363**
Visual neural pathway, **362**
Vital capacity, 555
Vitamin A, 662
  deficiency of, 681
Vitamin B complex, 63
Vitamin B_6, 666
Vitamin B_{12}, 667
Vitamin C, 668
  deficiency of, 682